Intelligent Energy Field Manufacturing

Interdisciplinary Process Innovations

Edited by WENWU ZHANG

CRC Press
Taylor & Francis Group
Boca Raton London New York

CRC Press is an imprint of the
Taylor & Francis Group, an **informa** business

CRC Press
Taylor & Francis Group
6000 Broken Sound Parkway NW, Suite 300
Boca Raton, FL 33487-2742

First issued in paperback 2019

© 2011 by Taylor & Francis Group, LLC
CRC Press is an imprint of Taylor & Francis Group, an Informa business

No claim to original U.S. Government works

ISBN-13: 978-1-4200-7101-6 (hbk)
ISBN-13: 978-0-367-38383-1 (pbk)

Library of Congress Cataloging-in-Publication Data

Intelligent energy field manufacturing : interdisciplinary process innovations / edited by Wenwu Zhang.
    p. cm.
  Includes bibliographical references and index.
  ISBN 978-1-4200-7101-6 (hardcover)
  1. Micromachining. 2. Electromagnetic fields--Industrial applications. 3. Ultrasonic waves--Industrial applications. 4. Lasers--Industrial applications. 5. Water jets--Industrial applications. 6. Technological innovations. 7. Intelligent control systems. I. Zhang, Wenwu, 1969- II. Title.

TJ1191.5.I58 2010
670--dc22
    2009051649

Visit the Taylor & Francis Web site at
http://www.taylorandfrancis.com

and the CRC Press Web site at
http://www.crcpress.com

To my wife, Alice, and my children, Samson and Melody,
Who give me overflowing love and happiness despite my busy work.

To my parents,
Who raised me in central China and showed me
why we should respect the Universe.

And to all of my teachers and mentors,
Who helped me and inspired me on the way to realize my dreams.

With love and respect.

Contents

PART I   Fundamentals of Intelligent Energy Field Manufacturing

PART II   Classic Nonmechanical Manufacturing Processes

PART III   Interdisciplinary Process Innovations

PART IV   Selected Innovative Processes

PART V   Toward Intelligent EFM

Preface

THE FUNDAMENTAL PURPOSE OF ENGINEERING

Sun Tzu said: "The art of war is of vital importance to the State. It is a matter of life or death, a road either to safety or to ruin; hence it is an inquiry which on no account can be neglected."

In this book, we emphasize, "Engineering is of vital importance to human beings. It is a matter of being in harmony or in conflict with nature, a road either to long-term sustainability or to short-term disastrous consequences for our civilization; hence the art and philosophy of engineering is an inquiry which on no account can be neglected."

In a certain sense, engineering is all about decision making. Each of our decisions affects the final outcome of an engineering activity. Unfortunately, the current foundation of engineering decision making, the criteria of engineering optimization, is biased. We are taught to or forced to optimize our engineering solutions to best meet customer requirements so that we or our organization or state can gain the best advantage in market competition. We live in a market-driven economy, or, more accurately, a fossil fuel–based market-driven economy. Under these criteria of engineering optimization, those "qualified" engineers can lead the world to disaster. Mass production can reduce manufacturing cost and position an organization at the forefront of world market competition, but mass production can turn into mass destruction if it results in uncontainable pollution to the environment.

Engineers directly interact with nature. Before wielding our power, we must understand the fundamental purpose of engineering. Since the Industrial Revolution, human beings have used increasingly powerful tools and harnessed energy resources. The impact of human beings has been increasingly global. Today, the earth has become a village, and human beings are exploring the deep space and deep sea. In the meantime, human beings are also facing many urgent challenges, including climate change, energy shortage, pure water shortage, pollution, food shortage, extinction of species, eruption of new diseases, etc.

Engineering has brought human civilization to an unprecedented level of prosperity and crisis. Many people have realized that the time for change has come. We must reshape our engineering philosophy to maintain the beautiful cycles of Mother Earth. Market advantage is not the most important thing for human beings; sustainability is.

In reality, however, the power of the market may be so strong that it could easily stifle many of our valuable efforts. In the case of international initiatives to control greenhouse gas emission, many world powers tried to avoid, reduce, or delay their duties to gain an advantage in global competition. Environment-friendly technologies will find it difficult to be widely adopted unless they show market advantages over competing technologies.

Over the years, our education system has been following the command of market-driven economy. Engineers who receive such an education usually become defenders of this system. Finding a way out of this cycle is one of the motivations of this book.

In this book, we try to establish the new philosophy of engineering, with the belief that we should adopt the strategy of "market-driven sustainable economy." Purely sustainable solutions might be too weak to survive in a market-driven economy. All successful engineering solutions have to pass the test of customers and society. In this book, we argue that we should adopt the "from nature to nature" philosophy, extending our engineering optimization criteria to cover the whole value chain of engineering activities, instead of only a very narrow segment of this chain (from factory to customer). Furthermore, we argue that the proper use of this philosophy will lead to huge market competition advantages, while neglecting this philosophy will sooner or later lead to risks or disasters to organizations. The fundamental philosophy ensures that we will act for both the short- and long-term benefits of human beings.

REGARDING TECHNOLOGICAL INNOVATION

Engineers are at the front line fighting the challenges of human beings. Technological innovations are needed to solve new challenges. While many people related to engineering talk about innovation, surprisingly, only a very small fraction of these people actually understand innovation. What is the difference between invention and innovation? Is the capability of innovation inspiration-driven or can it be systematically improved through training? What are the categories of innovations? What innovation strategy should an organization adopt?

I wish I had learned technological innovations earlier. I only received systematic training in technological innovation after my PhD study. This was achieved through corporate training and self-learning.

As a Chinese proverb says, 授人以鱼，三餐之需；授人以渔，终生之用 (Give a man a fish and you feed him for a day. Teach a man to fish and you feed him for a lifetime). Knowledge is about proven things; only when we know how to use knowledge to solve practical issues can we say we have grasped it, and only when we can skillfully innovate can we say we have grasped the art of engineering. Classical education normally does not include a systematic introduction to technological innovation; knowledge is conveyed as facts, while the thinking and methodology of these great achievements are not well explained to interested learners.

Technological innovation is a broad topic, involving both engineering and management aspects. Introducing innovation itself requires a dedicated book. However, we believe it is valuable to at least give engineers an overview of technological innovations before they dive into technical details. This can help them avoid unnecessary barriers in their careers.

One barrier of technological innovation is the interdisciplinary nature of engineering activities. Over the years, I have had discussions with many scholars in both academic and industrial circles. A common topic is that despite the lengthy engineering education, fresh graduates are overwhelmed by interdisciplinary engineering challenges in the real world. Even veteran engineers are frequently overwhelmed by certain technological innovations.

Many may be well trained in their specific areas; however, they may feel uncomfortable in interdisciplinary discussions. We call such people experts. Modern education is good at training experts but systematically falls short on preparing engineers with a big picture of engineering. Many of us are used to sitting in a cell in our work and research; seldom do we cross the borders to seek opportunities, although people who do cross the borders achieve many "surprising" breakthroughs. The challenges of the twenty-first century require us to be systematically interdisciplinary. This necessitates major changes in our education system, since a new culture has to be fostered.

From 1999 to 2001, I had the chance to work on the National Science Foundation–funded project, "Combined research and curriculum development—Nontraditional manufacturing and process innovations." This project, along with my real-world R&D experience at GE Global Research Center (GE GRC) and my long-term research in energy field manufacturing (EFM) led me to the idea of writing a book that would lower the threshold of process innovations and better prepare engineers and engineering school students for modern engineering challenges.

This book tries to shed light on the philosophy of modern engineering, with a focus on interdisciplinary manufacturing process innovations. Targeted readers are engineering school students, both senior undergraduates and graduates, engineers, and the leaders of technological innovations.

THE METHODOLOGY OF MANUFACTURING

To this point, manufacturing processes in mechanical engineering are generally divided into traditional manufacturing and nontraditional manufacturing (processes mainly involving mechanical force and mechanical contact vs. processes involving nonmechanical force and nonmechanical contact). Textbooks on manufacturing are heavily focused on traditional manufacturing processes, with nontraditional manufacturing processes overly simplified. On the other hand, the majority of books on nontraditional manufacturing discuss processes on a case-by-case basis, lacking a systematic approach or a general methodology. Books dedicated to certain processes may be too lengthy, too advanced, or too detailed for engineering education.

The consequences of this situation are multifold. First, the large amount of information on nontraditional manufacturing is not delivered to the readers efficiently. In certain cases, this deprives them of the precious opportunity of expanding technical background during our formal education period.

Second, the biased training in manufacturing processes unnecessarily blocks innovations. All manufacturing processes are inherently EFM processes; that is, all processes use various energy fields to convert material into objective configurations. All energy fields should be treated equally for process optimization.

Third, due to the lack of appropriate methodology, graduates entering real world R&D may be shocked by the interdisciplinary nature of engineering tasks. The imperfect education system dictates that many individuals require additional time to adapt to reality.

Finally, we are in the age of knowledge explosion. We are immersed in a sea of information. The urgent task is to find ways to filter the valuable part of information, instead of being passively soaked. Each year there are many international conferences on many specific topics, but no one has the chance to attend all of them. There are so many papers and talks even within one conference, and the real challenge is to transfer the value of fresh information into real value in work and study. To make things worse, only a small fraction of people can attend these frontier discussions; therefore, there is a big delay before the general public in engineering can access such knowledge.

We need a book that cuts across the engineering disciplines and allows readers to gain a larger view of manufacturing before diving into the details of individual processes. We need a book that would inform readers of the frontiers of engineering in limited time. We also need a book to explain how innovations are carried out. The majority of books report the facts of proven knowledge, while the behind-the-scene thinking is usually neglected. In this book, we hope we can show you how many of the innovations are achieved.

Through the history of engineering, there have been many valuable methodologies, such as standardization and interchangeability, mass production, lean manufacturing, Six Sigma quality control, Lean Six Sigma, concurrent engineering, CAD/CAM/CIMS, automation, design for assembly and design for manufacturing, MEMS, rapid prototype manufacturing, hybrid processes, intelligent manufacturing, systematic innovation, globalization, green engineering and sustainability, from nature to nature philosophy, etc. These are legacies that should be properly inherited. The trouble is that there are too many of them, and few people really have the time to understand all of them.

Can we have a simplified frame of manufacturing that integrates all of the important methodologies of engineering?

This question arose in 1988 when I was tired of the lengthy processes required to make a precision gage block. Finally, I realized that all manufacturing processes in all engineering disciplines are actually the same. They consist of three flows: the material flow, the energy flow, and the intelligence flow. Any manufacturing process is actually a process of injecting human intelligence into the interaction between the material and the various energy fields in order to transfer the material into desired configurations. In this sense, every energy field, be it mechanical force, gravity, chemical solutions, laser light, or ultrasound, is equally important for the optimization of engineering solutions. All manufacturing employs EFM processes, including natural processes such as crystallization and the growth of apples. The purpose of engineering is to inject human intelligence into these EFM processes. We call this intelligent EFM.

This primitive thinking survived the test of time and has evolved over time. Initially, EFM was proposed to solve the challenges of 3D manufacturing. Later on, it was applied to the general analysis of manufacturing processes. The concepts of general energy field, general logic functional materials, general intelligence, and the dynamic M-PIE model were proposed. Intelligent EFM is an open system and can naturally unite with other methodologies if we introduce the new criteria of engineering optimization (the new CEO). Finally, we have established an engineering system that could give us a big picture of engineering and get into details without getting lost.

This methodology is powerful. It immediately removes the barriers between disciplines, since engineering disciplines are basically the art of EFM processes with some featured energy field and material applications. It naturally connects with the frontiers of engineering, since engineering is evolving toward increased levels of intelligence and integration. The essence of engineering is simple now. It is the art of utilizing the dynamic M-PIE flows. With this methodology, one can better appreciate the behind-the-scene thinking of technological innovations. When we combine this methodology with the legacy of engineering, we see the hope of a simplified frame of engineering that could meet the challenge of our new era.

Acknowledgments

Giving a decent introduction to nontraditional processes is challenging, and combining this with interdisciplinary process innovations is even more difficult. We also wanted to include an overview of technological innovations, and combine it with other engineering methodologies. Writing such a book was a daunting task, far beyond the capability of a single person.

To be honest, I felt I was not ready to write this book yet, although it has been my dream since 1988. A friend encouraged me and said: "You can never be fully ready for something that is challenging in nature." So I got started. Dr. Shuting Lei and I successfully organized EFM symposiums in ASME/MSEC. These symposiums won the BOSS Awards in ASME/MSEC (International Conferences on Manufacturing Science and Engineering organized by American Society of Mechanical Engineers) 2006/2008. With the book contract from CRC Press, I started the journey of organizing the first book dedicated to intelligent EFM and interdisciplinary process innovations.

Knowing my limitations, I decided to share the load with the true experts in many areas. I have written Part I of this book and have organized and coordinated the remaining chapters, while many of the technical directions have been contributed by scholars worldwide.

Here, I would like to express my heartfelt gratitude to all the contributing authors. I am honored to have worked with you toward a common dream. Many of the authors were extremely busy, but all of them treated their chapters with a sacred belief. We believe that we should act together to make a change in engineering philosophy and engineering education. With this belief, scholars from the United States, Canada, China, Britain, Germany, and Japan formed a world-class team and gave birth to this book. Writing for a technical book is basically a volunteer work. This dedicated group sacrificed much of their spare time. I want to take this opportunity to express my sincere thanks to the family members of the contributing authors. Without your support and understanding, this book might have been delayed and may never have reached such a high quality.

I would like to thank some of my friends who wanted to help but could not do so due to personal preoccupations. Thank you for your interest and support to this effort. I hope you can continue to support this research direction and help advocate this book.

I would also like to thank four professors who helped in the evolution of EFM. My special thanks go to Professor Jiqing Gao of the University of Science and Technology of China (USTC), my undergraduate advisor, who inspired me to the road of "innovation" and planted the seeding idea of EFM; to Professor Yongnian Yan of Tsinghua University, China, who encouraged me to adhere to the ideal, helped coin the name of this direction, and shared his great methodological thinking; to Professor Y. Larry Yao of Columbia University, my PhD advisor, who gave me rigorous training and broad exposure to nontraditional manufacturing processes, which greatly promoted my career and guided me to a practical way of realizing my dream; to Professor Shuting Lei of Kansas State University, a great friend who helped organize the earlier EFM symposiums in ASME and contributed in popularizing this field to a worldwide audience.

My special thanks go to GE GRC and many of my colleagues. GE GRC was a unique place where I accumulated abundant engineering experience, received advanced technical training, and enjoyed attending international academic activities. My immediate manager, Magdi Azer, was very supportive of my spare-time mission of writing a book. He was both a friend and a mentor. My colleague, Dr. Marshall Jones, was always there to help whenever I needed him. He was my mentor in many ways. We spent many enjoyable moments discussing various interesting topics covered in this book.

I must thank ASME for fostering the growth of this research direction. I met many talented individuals in ASME/MSEC conferences. Without the EFM symposiums in ASME, this book might have been postponed.

Words cannot express my gratefulness to my family. Organizing a book like this requires a huge amount of dedicated time. My wife took over the majority of the large and small tasks at home. As long as I mowed the lawn every two weeks, I was allowed to enjoy writing the book and playing with the kids. Thank you, Alice, Samson, and Melody. With your love, writing this book became a happy journey.

The idea for this book was mine, but the contributing authors are the real heroes who brought to the readers the rich technical content and in-depth analysis of engineering frontiers. Different from other books, they purposely shared their methodological thinking of technical directions. As promised, we wanted to offer a book that taught how to fish rather than giving you one or two fishes. We hope that readers will appreciate this.

I am happy to see that with great teamwork and with help from CRC Press, especially with the guidance of our editor, Michael Slaughter, and project coordinator, Amber Donley, this book is finally complete. It has all the ingredients we initially hoped for.

Overview

This book has been written to establish a new philosophy and methodology of engineering, lower the barriers of technological innovations, to meet new engineering challenges, and to quickly yet systematically introduce interdisciplinary process innovations. The book is organized into five parts.

Part I (Chapters 1 through 4) describes the methodology and has been written by me. Chapter 1 gives a short yet systematic introduction to technological innovations. The role of process innovations is highlighted. With a comprehensive picture of innovation, engineers may better equip themselves for various roles in the future.

Chapter 2 introduces the fundamentals of intelligent EFM. The evolution of EFM is first reviewed; the core concepts in intelligent EFM, such as general energy field, general logic functional materials, general intelligence, the dynamic M-PIE model, etc., are then established. Finally, sustainability and the new CEO are discussed.

Chapter 3 tries to merge the legacy of engineering with intelligent EFM. It reviews the major engineering methodologies and philosophies in history. We justify that the mission of engineering should be to help establish a market-driven sustainable economy through technological innovations. Chapter 4 discusses the representative principles and techniques in intelligent EFM, illustrated with practical examples. The remaining chapters of the book introduce specific technical directions and have been written by contributing authors, whose bios have been provided at the end of their respective chapters.

Part II (Chapters 5 through 11) covers the classic nonmechanical manufacturing processes. These processes include waterjet-based machining, electrical- and electrochemical-based machining, ultrasonic machining, and laser-based machining. Each topic contains sufficient detail so that readers can not only have a comprehensive picture of these processes, but can also reach the level for further studies if interested. Pay attention to the state-of-the-art and methodological parts of these processes.

Part III (Chapters 12 through 17) introduces multiple interdisciplinary process innovations. Topics include the methodology and process innovations in materials science and engineering, nanotechnology, near field optics, coating processes, additive manufacturing, and bioengineering. Many engineering frontiers are covered. We have been careful not to confuse readers with too many technical details. Our focus is to introduce the area, give an overall picture, and expose you to the latest innovations.

Part IV (Chapters 18 through 22) covers the so-called innovative processes. These processes may be unfamiliar to some people, but they reflect the methodology of intelligent EFM and can inspire readers to further technological innovations. Therefore, they should be widely studied. Topics include EM dynamic forming, electric-assisted forming, laser-assisted machining, advanced polishing, and progress in tribology.

Part V (Chapters 23 through 26) covers the intelligence aspect of manufacturing processes. Metrology and quality control, MEMS-based sensor and process control, and progress in CAD/CAM and design are introduced, with the final chapter discussing the open system nature and the future trend of intelligent EFM.

Again, the focus of this book is not on specific details of individual studies. We are trying to give readers a high-level historical and methodological view of technical directions and lead them to the analysis and further innovations of manufacturing processes. The book will be a valuable reference book for the study of interdisciplinary manufacturing processes and for people who want to know the latest process innovations in active engineering directions.

This book can also be used as a textbook for graduate and advanced undergraduate education in engineering schools, especially for nontraditional manufacturing process education. For this reason, each chapter contains a Questions section. These questions are not time-consuming theoretical and analytical questions. They are tips to lead readers to further explorations of process innovations, or suggestions for small projects. Teachers can handle the classroom material flexibly. Interactive discussions are encouraged. The final exam could be the reader's version of the methodology for engineering innovations, or the details of a process invention illustrating the principles of intelligent EFM.

TO THE READERS

We would like to point out that intelligent EFM is still evolving. You can and you should be part of it. It is aligned with the evolution of modern engineering, which is increasingly interdisciplinary and integrated. It is meant to meet the challenges of our world, with sustainability being the highest priority. It tries to inherit the legacy of engineering and lower the threshold of technological innovations.

This book is the first trial to meet the lofty objectives stated above. I have tried my best with the invaluable help of experts worldwide. There were many things we could have done better. Anyway, this is the first step into a new area. Your feedback is sincerely welcomed.

I would like to thank again all the people who helped in this effort and who showed interest in this book. I hope you enjoy reading it and find it useful.

Dr. Wenwu Zhang
GE Global Research Center
Schenectady, New York

Editor

Wenwu Zhang was born in central China. He received his bachelor of engineering degree from the University of Science and Technology, Hefei, China, in 1992; his master of engineering degree from Beijing Institute of Control Devices in 1995; and his PhD from the Department of Mechanical Engineering at Columbia University, New York, in 2002. He is currently a lead engineer in Laser and Metrology Systems Lab., Materials Systems Technologies, General Electric Global Research Center (GE GRC), Schenectady, New York.

Dr. Zhang is a pioneer in the research of microscale laser shock peening and energy field manufacturing, and the inventor of liquid core fiber laser material processing. He is currently leading the laser micro/nano R&D work in GE GRC. His research interest includes laser material processing, intelligent EFM, and sustainability of engineering and technological innovation methodologies. Dr. Zhang won the 2005 SME Robert A. Dougherty Outstanding Young Manufacturing Engineer Award, the 2006 ASME Blackall Machine Tool and Gage Award, and the 2006/2008 ASME/MSEC BOSS Award. He is a member of the Optical Society of America, ASME, Sigma Xi, the Society of Manufacturing Engineers (SME), and the Laser Institute of America. Dr. Zhang is a successful inventor as well as a science fiction writer.

Contributors

Jian Cao
Department of Mechanical Engineering
Northwestern University
Evanston, Illinois

Gary J. Cheng
School of Industrial Engineering
Purdue University
West Lafayette, Indiana

Sabine Claußen
Materials & Processes Department
Laser Zentrum Hannover e.V.
Hanover, Germany

Glenn Daehn
Department of Materials Science and
    Engineering
Ohio State University
Columbus, Ohio

Tao Deng
Chemical Nanotechnology Laboratory
GE Global Research Center
Schenectady, New York

Randy Gilmore
The Ex One Company, LLC
Irwin, Pennsylvania

Kevin Harding
GE Global Research Center
Schenectady, New York

Mohamed Hashish
Flow International Corporation
Kent, Washington

Marshall G. Jones
Laser and Metrology Systems Laboratory
GE Global Research Center
Schenectady, New York

Chen-Chun Kao
Cummins Fuel Systems
Columbus, Indiana

Shuting Lei
Department of Industrial and Manufacturing
    Systems Engineering
Kansas State University
Manhattan, Kansas

Lin Li
School of Mechanical, Aerospace and Civil
    Engineering
University of Manchester
Manchester, United Kingdom

Xiaochun Li
Department of Mechanical Engineering
University of Wisconsin–Madison
Madison, Wisconsin

Yongfeng Lu
Department of Electrical Engineering
University of Nebraska–Lincoln
Lincoln, Nebraska

Judson S. Marte
Ceramic & Metallurgy Technologies
GE Global Research Center
Schenectady, New York

Andreas Ostendorf
Department of Mechanical Engineering
Ruhr-University Bochum
Bochum, Germany

Frank E. Pfefferkorn
Department of Mechanical Engineering
University of Wisconsin–Madison
Madison, Wisconsin

Kamlakar P. Rajurkar
Department of Industrial and Management
    Systems Engineering
University of Nebraska–Lincoln
Lincoln, Nebraska

John T. Roth
Department of Mechanical Engineering
Penn State Erie, The Behrend College
Erie, Pennsylvania

Wesley A. Salandro
International Centre for Automotive Research
Clemson University
Clemson, South Carolina

Takashi Sato
Department of Machine Intelligence and
    Systems Engineering
Faculty of Systems Science and Technology
Akita Prefectural University
Yurihonjo, Japan

Albert Shih
Department of Mechanical Engineering
University of Michigan
Ann Arbor, Michigan

Yung C. Shin
School of Mechanical Engineering
Purdue University
West Lafayette, Indiana

Murali Meenakshi Sundaram
School of Dynamic Systems
University of Cincinnati
Cincinnati, Ohio

Q. Jane Wang
Department of Mechanical Engineering
Northwestern University
Evanston, Illinois

Ronald Xu
Biomedical Engineering Department
Ohio State University
Columbus, Ohio

Lijue Xue
Industrial Materials Institute
National Research Council Canada
London, Ontario, Canada

Hitomi Yamaguchi
Department of Mechanical & Aerospace
    Engineering
University of Florida
Gainesville, Florida

Y. Lawrence Yao
Department of Mechanical Engineering
Columbia University
New York, New York

Kaijun Yi
Deparment of Electrical Engineering
University of Nebraska–Lincoln
Lincoln, Nebraska

Wenwu Zhang
Laser and Metrology Systems Laboratory
Materials Systems Technologies
GE Global Research Center
Schenectady, New York

Zuozhi Zhao
Corporate Technology of Siemens Ltd.
Beijing, China

Dalong Zhong
Coatings and Surface Technologies
    Laboratory
GE Global Research Center
Schenectady, New York

Dong Zhu
State Key Laboratory of Tribology
Tsinghua University
Beijing, China

Part I

Fundamentals of Intelligent Energy Field Manufacturing

1 Technology Innovations and Manufacturing Processes

Wenwu Zhang

CONTENTS

1.1 THE PURPOSE OF ENGINEERING

Before discussing specific manufacturing processes, let's ask the following questions:

1. What is engineering and can people in the engineering field communicate efficiently?
2. What should be the purpose of engineering?

The philosophy behind asking these questions is the starting point of engineering. Answers to the first question have gained relative consensus, while answers to the second question vary strikingly depending on whom you ask.

1.1.1 COMMUNICATION ISSUES IN ENGINEERING

In Wikipedia, "engineering" is defined as the discipline of acquiring and applying scientific and technical knowledge to the design, analysis, and/or construction of works for practical purposes. The American Engineers' Council for Professional Development, also known as ECPD, defines engineering as "The creative application of scientific principles to design or develop structures, machines, apparatus, or manufacturing processes, or works utilizing them singly or in combination; or to construct or operate the same with full cognizance of their design; or to forecast their behavior

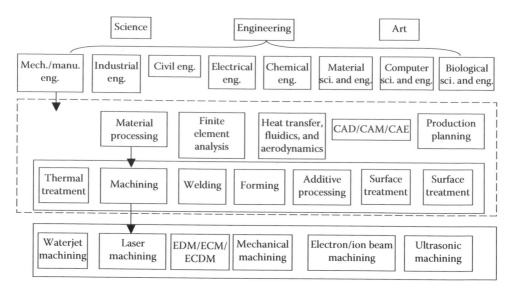

FIGURE 1.1 The landscape of engineering.

under specific operating conditions; all as respects an intended function, economics of operation and safety to life and property."

In my own words, engineering is the social activity that converts natural resources into human-desired configurations under the guidance of science and technology; and engineers are the initiators and enablers of such conversions.

As illustrated in Figure 1.1, there are many disciplines in engineering, such as mechanical engineering, industrial engineering, civil engineering, chemical engineering, material science and engineering, biological engineering, electrical engineering, computer science and engineering, earth engineering, etc.

Within each branch of engineering, there are many directions and frontiers. For example, in mechanical engineering, people may be working on computer-aided design (CAD), computer-aided manufacturing (CAM), finite element analysis of aerodynamics and mechanics, metal forming, composite processing, strategies of quality control, laser material processing, electrochemical machining (ECM), and electro-discharge machining (EDM), to name a few. The richness of knowledge in individual disciplines has reached such high levels that people in each discipline will go to distinguished journals and conferences to acquire information, and people use professional languages or terminologies to communicate. These terminologies and many of the abbreviations are formidable barriers to outsiders. Thus, the many different academic circles dedicated to a narrow branch of engineering, such as composite processing or laser material processing has been formed.

Unfortunately, even within a small branch of engineering, there are still many sub-directions and enough important frontiers that may excite some people for a long time, and they become experts, people who have expertise in certain areas of engineering.

Take laser material processing for example—people may be working on laser machining, laser welding, laser additive processes, and laser surface treatment. For laser machining, people may be working on thick section high-speed cutting or micro/nano machining with further focus on process development or modeling and simulation. People will use heat affected zone (HAZ), depth of focus (DOF), and line energy (E_{line}) to communicate.

If you fully understand these words and you are not someone who is working on or who has worked on lasers, you are abnormal!

The normal case is that each of us in engineering is like someone sitting in a cell of a company located on a certain floor of one of the many buildings in Manhattan, New York. We may see each other at the elevators, but seldom do we have the chance to really cross the borders and understand each other in depth.

This is the current communication situation in engineering. In short, we delve so much into the "valleys" of engineering, focusing on some near-term things that we frequently miss the chance to see the whole picture and weaken the ability to cross over the borders between different engineering branches and disciplines.

The reason that we can still communicate between different disciplines is because we have the common language of science, logic, and mathematics. To improve the efficiency of engineering communication, we should find the common methodology in different engineering disciplines.

One common ground between the different disciplines in engineering is energy field engineering, which views engineering as the art and social activities of controlling various energy fields to convert materials into desirable configurations. One objective of this book is to improve the communication efficacy in engineering through the methodology of intelligent energy field manufacturing (EFM).

1.1.2 THE PURPOSE OF ENGINEERING

What should be the purpose of engineering?

The answer from an automobile manufacturer may be maximizing business profit through cost cutting based on technological innovations. For a plastic supplier, the answer may be occupying the largest share of the world market through diversifying applications of their plastics products. A lawn chemical company may wish that all lawns were taken care of with their lawn treatment chemicals. A customer may hope for getting higher quality products with many functions and a lower cost to better enjoy his or her personal life. A nation may consider engineering as the pillar of state economy and may encourage certain strategic technologies.

It looks like a general answer to the question is to maximize the market competition advantages for individual groups or organizations. This answer is no surprise in a market-driven economy.

A more lofty answer may be to expand the capability limits of human beings and to improve their standard of living through technological innovations and engineering activities. Understanding nature entices the scientists, while conquering nature entices engineers and great inventors.

In the twenty-first century, we have to double-check these answers for any possible *misguidance*, because this is the starting point and is a chain reaction.

All human activities are connected with nature. Thus, the first important thing in choosing a career in engineering is to understand that *we may have larger power and more opportunities for changing nature, but we also have the responsibility to maintain the healthy sustainability of nature*. Without this philosophy, our engineering education system may produce more guilty engineers than qualified ones judged by history and nature.

In June 2008, the middle-west states of the United States and the southern provinces of China were struggling with severe flooding disasters. The ice-cap of the arctic may completely melt for the first time in human history! The glaciers on Greenland were retreating and shrinking and the glaciers on the Antarctic continent were breaking into the sea. While there are still many people arguing about whether human activities really caused global warming, the above facts are undeniable.

The economy of human beings has been riding on fossil fuels, such as coal, oil, natural gas, etc., since the industrial revolution 200 years ago. As this book was written in 2008, the world oil price climbed to greater than $140 per barrel, the gas price in New York state changed from ~$1.0 per gallon in 2002 to over $4.3 per gallon in 2008. The world population has reached 6.5 billion and food crises have appeared in many countries.

Industrialization based on fossil fuel was based on the glorious inventions of power plants, steam engines, automobiles, railways, airplanes, warships, highways, etc. But the negative aftermath of a fossil-fuel-based economy was only strongly felt in the beginning of the twenty-first century. Industrialization continues in the twenty-first century in the form of globalization. While the economies of China and India are achieving high increasing rate, the environmental quality in many areas in these countries is degrading at an alarming speed. To ensure the air quality of the 2008 Olympics in Beijing, the Chinese government had to temporarily shut down many high pollution factories. Cars offer the convenience of family transportation, but they also create issues of traffic jams in big cities and international tensions of oil-supply.

Looking back historically, could we have done better? How can we avoid making the same mistakes in the future?

Let's check some examples. Technological innovations are powerful if strong market support exists. The IT industry shows that performance can be steadily progressing, while the cost of manufacturing can be extremely low. A 4 GB memory stick was only $40 in 2008, which was unbelievable 5 years earlier. The technology challenges behind these products are huge, but these challenges were conquered. On the other hand, gas-based cars are bringing in more trouble than they should.

There was good competition between electric cars and gas-based cars when the automobile business took shape in the beginning. Electric cars lost the competition in the 1950s, and the development of relevant technologies was halted for decades. The recent high oil price ignited the competition for hybrid cars, which combine gas energy and fuel cell energy.

Technology innovation is not the only important factor deciding the future direction of technology. Politics, major interest groups, the public, and the engineering realm all play their roles. But engineers, as the initiators and enablers of technology, should take the responsibility for influencing society to develop technology in the most favorable ways for a healthy and sustainable nature.

We are living in a market-driven and fossil-fuel-based economy. We say our economy is market driven because currently the success of a technology or organization is judged by its performance in the market and is measured by money or profit. The criteria of engineering optimization is thus to maintain a strong position for a country, an organization, or a company in market competition.

Normally, the optimization and the choice of technology in engineering considered the interest of customers and the company and stopped right there, ignoring the rest of the cycle of nature. Establishing the new criteria of engineering optimization (New CEO) is one of the reasons why this book was written.

Nature has many beautiful cycles; human activities are part of these cycles. The beauty of nature lies in its self-sustainability. This should be the goal of long-term engineering development.

In my opinion, *the ultimate purpose of engineering is to innovate technology and to carry out production to improve the living standards of human beings while maintaining the healthy self-sustainability of nature.*

Following this purpose, the objectives of engineering are

1. To transform scientific knowledge into practical applications (technology innovation)
2. To optimize various processes and win competition advantages in a market-driven economy (optimization and win competition)
3. To expand the capability limits of human beings (exploration)

Technology innovation is essential in achieving the objectives of engineering. Although the word *innovation* appears frequently in daily life, a good understanding of innovation is generally missing, even for many people whose work is relevant to engineering. There are many misconceptions about innovation, such as

- Invention is the same as innovation (very wrong—invention is only one of the early steps in the innovation process).
- A great inventor is naturally a great innovator (wrong—a great inventor may become bankrupt and never harvest the value of his or her inventions. A great innovator has to be great in the complete cycle of innovation).

- Technological innovations serve the market and follow the trend of economy (not exactly—technological innovations drive the trend of economy and decide the long-term cycles of economy).
- A company must be inventive to be innovative (not exactly—frequently we see big companies innovate from the great inventions of small creative companies).
- Technological innovation is the work of scientists and engineers (not exactly, anyone involved in modern economy can be involved and can contribute to technological innovations).

These misconceptions may come back hurting both the economy and the technological innovations. For this reason, before the detailed discussion of intelligent EFM, we will give a short yet systematic introduction of technological innovations.

1.2 INTRODUCTION TO TECHNOLOGICAL INNOVATIONS

1.2.1 INVENTION AND INNOVATION

Who invented the incandescent lamp?

Many people would answer Thomas Edison. In fact, Humphry Davy created the first incandescent light by passing battery generated electric current through a thin strip of platinum in 1802. It was not bright enough and it didn't last long enough to be practical. Many people worked to improve the incandescent lamp in the next 75 years until Thomas Edison's creation of the first practical incandescent lamp in 1879. Edison should be remembered for his innovation of the entire system of electric lighting, in which the lamp was only one component.

Invention is the creation of a feasible way of doing or making something, be it new material, a new device, a new product, a new process, or a new strategy of service. Invention is different from innovation—invention is only one of the early stages of innovation.

The differentiation between invention and innovation is very important. Technological innovations are science, nature and society based. Invention is the first bridge that connects scientific knowledge with social needs. There can be many inventions, good or bad, trying to meet the challenge of social needs. Only a small fraction of these inventions may get into production, even fewer of the inventions reach the commercial stage, and very few of the inventions turn out to be commercially successful.

Before an invention enters commercial production, it has to march through multiple stages, which is called *the cycle of innovation.* The cycle of product innovation is illustrated with Figure 1.2. Technological innovations originate from market needs, technology provided opportunities, or new

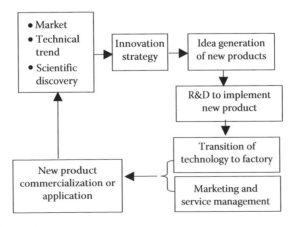

FIGURE 1.2 Cycle of product innovation.

thoughts from scientific discoveries. Innovation strategies are established to exploit the potential commercial opportunities. Technological intelligence is accumulated around these strategies; research and development teams are set up to invent, to integrate, and to overcome the technical barriers of production or implementation; finally, the technology is transferred to commercial production. Marketing is used to deliver the product (or service) to customers, reaching the final stage of new product commercialization. Commercialization is judged by the social system. This cycle goes on and on, with some technologies thriving while other technologies are filtered out.

Real-life examples show that the later the innovation stages, the higher the necessary level of investment. The investment actually may increase more than 10 times from stage to stage! Imagine filing a patent for $5000, the prototype development may cost $50K, the pilot development of the patented idea may cost $500K–$1MM, the final factory production may cost $5MM, while the final marketing sale may cost $50MM. Only at this point can some revenue be seen. This is why many great ideas may not make their way to commercialization. Technological innovations are very expensive and should be treated very carefully.

Critical contributors to great technological innovations are remembered as technology heroes in history, while many names were forgotten if they simply contributed the idea but did not make the idea an innovation success. It is time to give the definition of innovation and discuss the fundamentals of innovations.

1.2.2 THE DEFINITION OF INNOVATION

Innovation is the systematic introduction into the social system of new products, processes, services, or strategies.

Some books define innovation as *the introduction in the marketplace of new products, processes, or services* (Frederick Betz, 1987/1995). Limiting the definition to *marketplace* may be too narrow. Such a definition will exclude innovations that are nonprofit based, such as a new method of public health enhancement, a new strategy of education, a new technology for environment control, a new technology introduction in international high energy physics research, etc. There are many national and world level innovation efforts that are not market driven. There are social systems that are not market driven, but there are plenty of innovations, such as the new ways of cooking, living, etc. Thus, social system is more applicable in covering all kinds of innovations.

This definition also emphasizes the system side of innovation. Innovation is a social event—it is a systematic introduction of something new into the social system in order to impact society in certain ways. Invention can be individual based, but innovation has to be social system based as well as nature and science based. For example, one may personally think of new ways of harnessing solar energy and file patents, but to apply the invention in the real world, one has to transfer the invention into practical products and circulate the products in the social system. Understanding the interactions among social systems, engineering systems, and nature systems are critical to the success of technological innovations.

1.2.3 TYPES OF TECHNOLOGICAL INNOVATIONS

Technology is the technical knowledge of manipulating nature for human purposes. Technological innovation is a subset of innovation. It is the systematic introduction into the social system of new products, processes, services, or strategies based on new technology.

Technological innovations can be classified by applications and by the degree of social impacts. The degrees of social impacts can be radical, system, or incremental:

- Radical innovations are breakthrough innovations that can change or create whole new industries, such as the invention of printing, guns, steam engines, electric lighting, lasers, the Internet, etc.

- System innovations are a series of innovations that forms a system of technologies in support of an industry or a business, such as the power delivery system, e-marketing system, etc.
- Incremental innovations are small but important improvements to certain technologies.

According to applications, technology innovations can be classified into product innovations, service innovations, process innovations, and strategy innovations:

- A product innovation is the introduction into the social system of a new technology-based product. Examples include the first commercialization of personal computers, bicycles, cars, cell phones, etc.
- A service innovation is the introduction into the social system a new type of technology-based service. A typical example is the e-ticket booking service, which relies heavily on Internet and computer technology. Another example is global cell phone service, which is not possible without the use of communication satellites.
- A process innovation is the introduction of a new process into the manufacturing of a product or the implementation of a kind of service. Process innovation is the fundamental element of product and service innovations. We will discuss this point later on
- A strategy innovation is the systematic implementation of a new technology strategy of an organization in order to win a competition advantage of the organization. For example, the General Electric Company carried out a company-wide Design for Six Sigma (DFSS) strategy to improve quality control and to cut production costs. Recently, the company advocated Ecomagination, which means technological innovation with a strong focus on long-term ecological system sustainability.

Table 1.1 gives examples of various technological innovations. Readers can use these examples to further understand the classification of technological innovations. For example, laser material processing is a radical process innovation, which provided new possibilities for high-speed, high-resolution, and high-quality machining of difficult to machine materials. A new business has been formed around this technology. System innovations in this area include galvanometer scanner systems that make use of the high repetition rate of laser systems and mirrors' ability to reflect laser energy at high speeds. There are many incremental innovations in the laser material processing, such as the annual power increase of laser systems.

Why is the understanding of technological innovations so important? The following sections will answer this question.

TABLE 1.1
Types of Technological Innovations

Degree of Impact	Radical	System	Incremental
Product innovation	Electricity generator	Power delivery	High efficiency lighting
Service innovation	e-Marketing	Computer network support for e-marketing	Next generation high-speed Internet
Process innovation	Laser material processing	Support of motion system and beam delivery for laser material processing	Speed increase due to high-speed scanning and higher laser power
Strategy or organic innovation	Growth through globalization and green technology innovation	Strategy of renewable energy	A commercial plan to improve the module efficiency of thin film solar cells above 15%

1.2.4   THE TECHNOLOGY IMPERATIVE

As Frederick Betz pointed out in his book *Managing Technological Innovation*, technology changes drove the historical transformation of human society.

> The grand theme of managing technological innovation is the whole story of technological change and its impact on society. Historically, this story is both dramatic and ruthless. ... *The ruthless in technological change has been its force, which no society was able to resist and which has been called "technology imperative"*. For the last five hundred years, technological change has been irresistible in military conflict, in business competition, and in societal transformation. ...*The imperative in technology is that the superior technology of a competitor cannot be ignored by other competitors, except at their peril.*

Technology change drives long economic cycles, while the central concept of managing technological change is how to implement technological innovations.

One can easily find out many examples manifesting the imperativeness of technological changes. The innovation of the gun ended the era of feudal warriors, because new soldiers could be quickly trained to reach certain level of military power and could defeat another troop with inferior weapons. The steam engine powered human society into the industrial age. Large-scale production was possible with the innovation of steam engines, while home-based handcraft businesses had to yield the central stage to teamwork and modern power-based factories. The paper-based news business was one of the major sources of public information 20 years ago, but the Internet is now cornering this business. The Internet is faster, flexible, has multimedia content, and has lower cost. Thus, it is bound to take over the traditional market of newspapers. The traditional news business can either choose to adapt or perish.

Understanding the imperative in technology change is the first step toward a proactive response to the force of technology change.

The trouble is, the imperativeness of technology change is not well appreciated. No matter what the case, when the time comes, one has to face the consequences. Mind inertia may naturally allure people to maintain or improve current systems, but the ruthless power of technology change is bound to destruct the old system and move the wheel of history. One might optimize the technological system of animal-based transportation, but how could it compete against the gas-based transportation technology in the twentieth century?

Despite the importance of technological innovation and the imperativeness of technology, there is a big culture gap between the business world and the technical world, which puts up unnecessary barriers to technological innovations. Most engineering schools focus on the scientific training of students, ignoring the management and the system aspects of engineering. Business schools can be just the opposite, which focus on business management while ignoring the uniqueness of managing technological research and development. Accordingly, the education of engineers, managers, and scientists are incomplete, which leads to unnecessary frictions and misunderstandings between the groups.

One purpose of this book is to bridge this gap from the engineering side. This is why, in this book, before the discussion of manufacturing processes, we spend time in this chapter trying to understand the importance of innovations, the types of innovations, the relation between technology and economy, the system and dynamic nature of innovations, the sources and the bases of innovations, and some curves of technological innovations.

1.2.5   SOURCES OF TECHNOLOGICAL INNOVATIONS

Technological innovation is complex, risky, and requires huge investment. Among all the complexities, it is important to understand the social, natural, and scientific bases of innovation. They are the origin of technological innovations; they also put constraints on technological innovations.

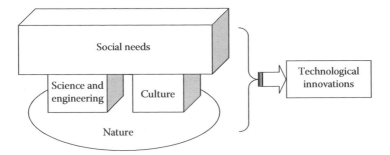

FIGURE 1.3 Origin and constraints of technological innovations.

Market volume is limited for any technology, because it is decided by limited social needs and the limited availability of natural resources; there are theoretical limits to any physical phenomena as well. As Figure 1.3 shows, nature is the foundation of all the activities of human beings. Nature has energy, mass, and intelligence (some readers may suspect the validness of the intelligence of nature. We will introduce the concept of general intelligence in Chapter 2). Human beings are part of nature. We share nature with other existences on earth, and we coexist with other existences on earth. We try to understand natural laws through scientific study. Science, engineering, and culture (such as tradition, belief, and religion) are the pillars of our civilization. Human beings currently dominate the earth. Let's hope we do not go too far out of our share of natural resources.

All these factors are both sources and constraints for technological innovations. Human beings are still in the infancy of imitating nature. Many of the inventions and technological innovations are *nature inspired*. For example, renewable energy is the hot spot of technological innovation, but this is exactly how nature has been functioning for billions of years. Could we say we invented composite material? Well, check any plant and we should yield the honor to nature. There is limited usable space and material from nature for any technology, and nature is a beautifully balanced system. Thus, any technological innovation should consider these limits and try to maintain the sustainability of nature.

Breakthroughs in science normally lead to waves of technological innovations. New knowledge triggers new ways of thinking on how to meet social needs. Inventions with good market potential will be launched through technological innovations. Technology innovation is theoretically limited by the physics it is relying on. For example, metal-wire-based communication couldn't compete with optical-fiber-based signal transmission due to the fundamental difference in physics—the first is electron-based signal transmission, while the latter is photon-based signal transmission.

Technology based on certain scientific knowledge will mature over time and may approach the theoretical limit, as shown in Figure 1.4, the S-curve of technology evolution. In the wake of scientific discoveries (stage I, discovery stage), early movers match science with social needs, invent for targeted applications, and carry out early research and development to overcome technical barriers (stage II, new invention stage). Once some feasibility is proven, the promise of new technology would attract more social resources. More people enter the competition, application grows quickly, and technology maturity is improved at a much faster speed (stage III, fast improvement stage). Finally, technology matures and the market is divided among major competitors. Technology may or may not approach the theoretical limit, which is decided by the constraints from society, nature, and science. It all depends on whether there are new replacing technologies entering the competition.

The success of technological innovation is normally judged by its market performance. This market performance judgment in a market-driven economy is theoretically flawed. Later on, we will discuss what remedy should be considered.

Technological innovations can create more values in the industrial value chain (Figure 1.5) than the existing technologies.

The industrial value chains start from the resource acquisition sector, which includes raw material mining and labor hiring, etc. Value is added by extracting the raw materials from nature. Raw

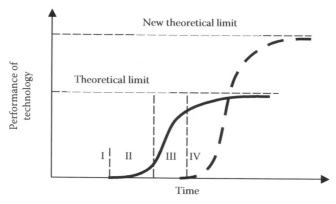

FIGURE 1.4 Evolution of technology—the S-curve.

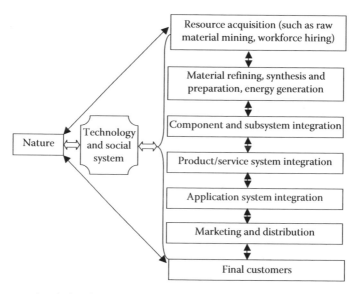

FIGURE 1.5 The complete industrial value chains—from nature to nature.

materials are refined; industrial materials are synthesized and prepared. Value is added by adding more desirable properties of the processed materials. Industrial materials are used to make components and subsystems of products; these subsystems are integrated into the product or service systems, such as entertainment products and service systems. To be directly usable, the complete application system and industrial structure have to be formed. And finally, the products and/or services are sold to the final customers through marketing and distribution efforts.

It is important to understand that the resource acquisition sector and the final customers interact with nature and all the other sectors interact with nature as well—either directly or through the technological system. *This complete value chain is a cycle from nature to nature, with many couplings and interactions. Engineering activities add value at each loop, while technology innovations add more values than existing technologies.*

1.2.6 Understanding Technological Innovations to Be Proactive

Technological innovations are high risk, with very few inventions turning into successful innovations. Due to the imperativeness of technology, an enterprise has to be proactive to maintain competitive advantages.

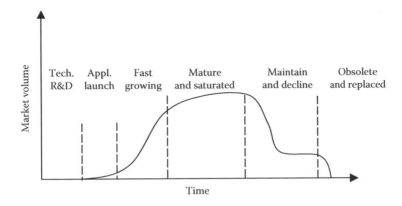

FIGURE 1.6 Curve of market volume over time.

The ideal condition of innovation is the commercialization of new technology with no resistance, no competition, no lead time, immediate profit and unlimited market volume, unlimited product/service lifetime, and at no cost.

A more practical version of the ideal innovation process could be the commercialization of new technology with no fatal resistance, dominance over competition, minimal lead time to profit, continuous growing of market volume during the lifetime of the product/service line, the ability to connect smoothly with the next wave of innovation, and at minimal cost.

The real world is far from ideal. There may be strong resistance both internally and externally, there may be multiple competing technologies, a huge investment is required and a long period of time should be expected before the break-even point of investment, there is normally a limited lifetime of a product or service, and when the time comes, one has to forget past successes and embrace new technologies.

Figure 1.6 illustrates the market volume of a product over time. It is normally an asymmetric bell shape. Some lead-time is used for the invention, research, and development of a new technology before the application is launched in the market. The growth of market volume is slow in the beginning, which connects to a fast growing period. This is due to the wide acceptance of the technology through the exploratory work of First Movers. The quick growth may be also due to breakthroughs in key processes and scale up and/or the entering of competition. Technology matures with time and becomes commodity technology, with market saturates gradually. After this, the market volume declines due to competing new technologies and may maintain a reduced market volume due to technology service and used part replacement. Finally, new competing technologies dominate the market, while the once innovative technology become obsolete and retires.

We should analyze this curve along with other technology curves.

Figure 1.7 is the typical financial chart of a successful technological project. The vertical axis is the accumulated profit and the sales volume. Accumulated profit is negative and increases in quantity until the start of sales. If the sales can gain a positive margin over cost, accumulated profit starts going upward until the point when all the investment is recovered. This point is called the "break-even point" of investment. The sales volume increases with time and so does profit. The first period is the innovation period, featured by quick product and process innovations and many competitors. In the second period, the market is fully developed, featuring continued market growth and a reduced number of major players. Accumulated profit continues to climb up. The remaining players are usually those who launched technological innovation with the right technology, sufficient resource, the right strategy, and at the right time. Finally, sales decline due to market saturation or technology being obsolete. Accumulated profit may drop if the product sales continue, thus the product or service will be withdrawn eventually. This is a sketch of the successful technical project. The real world can be more brutal. Unsuccessful technological

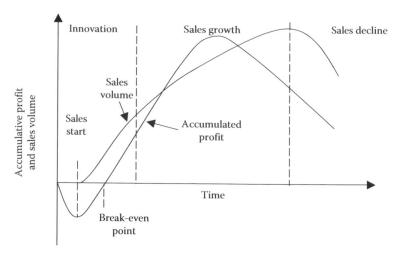

FIGURE 1.7 Financial chart of technical project. (After Betz, F., *Managing Technological Innovation—Competitive Advantage from Change*, John Wiley & Sons, Inc., New York, 1997, Figure 12.1.)

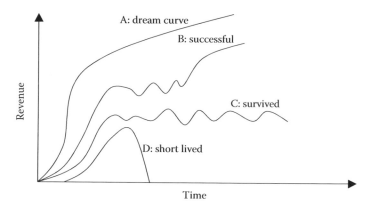

FIGURE 1.8 Typical growth trend of new ventures.

innovations may not reach the break-even point or may end much earlier before the peak point of the profit curve.

Figure 1.8 sketches the typical growth trend of new technology ventures. Pattern A is a "dream curve," which has a fast and unlimited potential of revenue growth. In reality, an exceptionally successful new venture is like curve B, which has one increase period after another. It has the capability to break away from a sluggish state to another quick growing state through technological innovations. Unfortunately, pattern C and D are more common, with C being the barely survived and D being the short-lived technological innovations.

One critical thing deciding the fate of new ventures is the sustainability of their profit margin. Profit margin equals revenue minus cost. A bigger margin is beneficial, while a negative margin is disastrous.

Figure 1.9 plots out the relation between profit margin and market volume. Zone 1 is a small-volume market with a high profit margin; Zone 2 is a medium-volume market with a low profit margin; and Zone 3 is a large volume market with either a high or a low profit margin. Zone 1 includes products or services covering a niche market, such as the market of a special instrument and the market of international satellite launching. The profit margin can be high if it has very little competition. Zone 2 is where intense competition exists. The technology has spread out. With so many competitors, they have to lower the price to take a share of the market, thus, a low

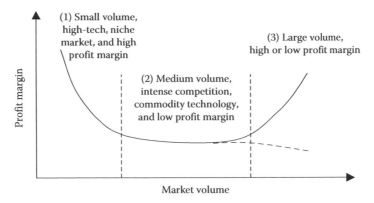

FIGURE 1.9 The U-shape of profit margin.

profit margin. This normally is due to the maturity of core technology. Ideally, one wants to have high volume and a high profit margin. This is possible only when one owns the dominating patented technology or the crucial business secret. The early days of Xerox is an example of strong patent protection, while Coca-Cola is an example of a well-maintained business secret. These types of technology dominance or know-how create monopoly competition advantages, allowing high volume and high profit margin. Otherwise, the profit margin usually shrinks when the technology matures, although the volume may still be big. The automobile industry after the 1950s is a good example.

Figure 1.10 further explains the affecting factors of profit margin. As time goes on, technology matures following an S-curve. A sale price may or may not cover the cost in the beginning. When early market success is important, special pricing strategies may be used to acquire market share. There are many uncertainties, many of which are out of the control of the individual organizations, especially in the time of globalization. However, profit margins can be increased through technology innovations, especially through cost-cuts enabled by product or process innovations. For example, conventional labeling uses print and glue. With the adoption of laser marking, the process steps are reduced, labor is saved, and the processing cost is greatly reduced. Process innovations can lower the cost and win competition advantages.

Profit margin is what matters in the end. Management, design, manufacturing, pricing strategy, external environment, etc. all play important roles in profit margin. This book will focus on how we can improve profit margins through process innovations.

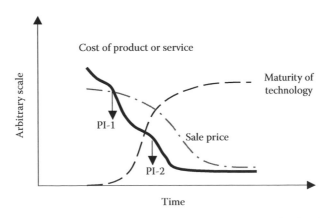

FIGURE 1.10 Profit margin and influencing factors—cost, price, maturity of technology, and process innovations (PI).

1.3   THE ROLE OF PROCESS INNOVATIONS—*ALL INNOVATIONS ARE CONSISTED OF PROCESS INNOVATIONS*

As mentioned earlier, there are four kinds of technological innovations: product innovation, service innovation, process innovation, and strategy innovation. It is interesting to take a look at the rate of innovation as shown in Figure 1.11. For convenience, the trend of market volume is shown along with the curves of product and process innovations.

Before a new technology is commercialized and introduced into the market, the enterprise has to invest in product/service innovations and process innovations. There is a lead-time between the start of a new project and the announcement to the market. Actually, big companies have the ability to shape the market through the control of technology reserves. For example, Microsoft released the new Windows operating platforms following a business strategy. When one model was on the market, there were lots of already achieved features reserved for future products.

In the early days of fundamentally new technology, such as in the early days of the automobile business, there is strong competition in product or service innovations. Early movers of a new technology compete to win territories in the intellectual property (IP) space. But commercialization of new technology won't be feasible until critical process barriers are cleared. Once the market is revealed and the critical technology is widely known, more social resources are attracted into the competition. The market will ruthlessly down-select the technology routes, transforming many of the competitors into the final several major players and dependent supporting businesses. At this point, the product or service is relatively standardized. For example, there were 69 auto firms in the United States in 1909, but in 1918, Ford's new Model T automobile started putting many of the companies out of business. The Ford technology set a new industry standard. In 1960, the number of the remaining U.S. domestic auto firms was four! After the Japanese cars entered the American market in the 1970s, only the big three remained: General Motors, Ford, and Chrysler.

Naturally, enterprise put a lot of effort into product or service innovation in the beginning. There is a big lag in the support on process innovations. However, market competition will force all the players to ramp up process innovation to remain in the game.

Ideas can easily be learned, but the secrets of processes are difficult to duplicate. Personally, I think this is one of the major reasons why many companies went out of business in the competition. Without a good manufacturing process, product quality couldn't be ensured, production cost may be too high to keep the business, and production volume couldn't scale up to occupy the market. Such a company will fail, even though it may hold critical product innovation patents. It will disappear or get acquired by more powerful companies.

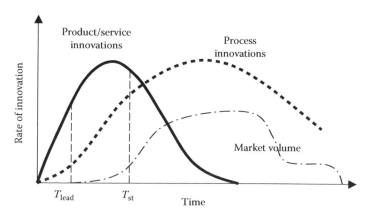

FIGURE 1.11   Rate of innovation vs. market volume. T_{lead} is the lead time for new product/service introduction, T_{st} is the time when certain standard is formed for a given market.

Process innovations continue to be important after several major players dominate the market. Process innovations and product innovations are used to increase profit margins and to extend product lifetime. These may be incremental innovations; however, these innovations will decide the market position of the players. When the market matures, the role of process innovation is more important because there is very little room in the IP space for new models of products or services—the competition leans toward those who can provide better products at a cheaper rate when the functionalities are similar. The profit margin shrinks when the market matures, as shown in Figure 1.10. Some new technologies will move the competition to a new ground. At this point, strategic innovation is critical to the fate of the enterprise.

Here, let's think about how an enterprise can do better in technological innovations. With globalization, modern communication technologies, and myriad platforms of technology reports, new ideas of technology took a much shorter time to spread out. As Thomas J. Peters said in his popular book, *The Circle of Innovation*, "Distance is dead," each of us has six billion next-door neighbors! The immature introduction of a product is simply attracting competitors to defeat you earlier. Thus, a wise IP strategy for all technology innovations is important. But IP can only protect to a certain level and in some countries. An ideal situation is that no defense is needed because no competitors could keep up with your pace and you are always one of the players to decide the standard. This is only possible if one makes correct decisions on all four kinds of technological innovations.

It is important to point out that the importance of process innovation is commonly underestimated. With the above analysis, we understand that process innovation may become *the deciding factor* in market shake out and competition. It may also become the bottleneck of brilliant ideas. For example, the surface of a lotus is super-hydrophobic, which shows functions of self-cleaning. Study of such structures showed that micro features are needed along with the modulation of surface tension. If such features can be used on metal or other materials, tremendous value can be added to many of the existing products. For example, how about self-cleaning cars? How about low friction ships in water? The idea has been known for decades, but this is still a high-tech research project. People are striving to develop cost-effective processes to create these structures.

Our major question is: when should process engineers be involved? How many resources should be allocated to develop new processes?

If the purpose of the project is fundamental research, scholars and scientists can take the lead, focusing on simulations, modeling, and certain feasibility studies. But if the target is commercialization, process engineers should be involved in the very beginning and more than enough resources should be allocated to develop the critical processes. This emphasize on the intense development of critical processes can offer solid competition advantages for the organization.

People frequently argue that small companies are on average more creative than big companies, thus they are also more innovative than big companies. But size matters in technology innovations. Big companies may be slow in reaction to new technologies, but they can be quicker in overcoming the process barriers and they can normally acquire key product ideas. With their size, they can tolerate failures. A start-up can beat big companies only if they also act quickly on process innovations.

The focus of this book is on interdisciplinary manufacturing process innovations. Manufacturing processes directly influence quality, performance, cost, and scale. Let's go one step further. It is meaningful to think that

All activities can be divided into small process steps, and all innovations are consisted of and are supported by process innovations.

In this way, one can decompose big innovation tasks into manageable process innovations, treating each process as a unit to be optimized under the New CEO.

As shown in Figure 1.12, technological innovations are motivated by market pull or technology push. Although product or service innovations directly generate revenue, they ride on process innovations. Of course, all human activities are supported and restrained by nature. Thinking of dividing innovations into process steps can help manage technological innovations. It can also help cut

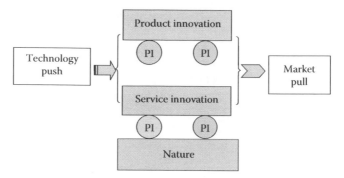

FIGURE 1.12   The importance of process innovations (PI).

the cost of innovation projects following similar rules of lean design and lean manufacturing. Lean design focuses on preventing waste from happening in the early stage of innovations and engineering activities, while lean manufacturing is waste cutting when waste has already been generated. Bart Huthwaite wrote a very good book on lean design. This book will try to integrate the philosophy of lean engineering with those of intelligent EFM.

1.4   TECHNOLOGY CHANGE AND THE LONG WAVES OF ECONOMY

Many factors contribute to the dynamics of economy, such as money, politics, international market, resource availability and cost, labor, education, war, natural disaster, etc. There are short- and long-term cycles in the modern economy, which show the cycles of expansion and contraction. It is generally agreed upon that the major factor in the long-term cycle of economy is technological innovation, with the cycle time between 45 and 60 years. This was first described by Leontieff Kondratieff, a Russian economist, in the 1930s. His work was rediscovered by an American economist, Joseph Schumpeter, in the 1940s, and was further developed by many scholars in the 1980s. The far reaching impact of technology on economic systems is that new basic technologies can create new functionality through new industries, or pervade existing industries, and such changes are imperative.

Robert Ayres did the up-to-date empirical correlation between European industrial expansion and contraction and the occurrences of new technology-based industries (Ayres, 1990). The invention of the steam engine required support from both science and engineering. The new disciplines of physics and chemistry formed the necessary base for new technological innovations. The period between 1770 and 1800 was the first economy expansion period in Europe based on the new technologies of steam power, coal-fired steel, and textile machinery. The 1830–1850 economic expansion accelerated the European industrial revolution, which was based on the technologies of railroads, telegraphs, steamships, and gas lighting. New physics of electricity and magnetism enabled the inventions of electrical power, light, and telephone, and the advances in chemistry gave birth to the industry of chemical dyes and petroleum. This induced the 1870–1895 economic expansion. The invention of automobiles fueled the fourth long economic expansion along with the invention of radio, plastics, and airplanes, which was from 1895 to 1930.

How the economic long wave evolves is as follows. Scientific discoveries provide the new thinking of manipulating nature, triggering important technological inventions. New basic inventions develop into new businesses and form new industry structures. Such a new high-tech industry provides rapid market expansion and economic growth. The market continues growing with technology improvement, with more competitors entering. As technology matures, production capacity exceeds market needs, triggering price wars. Excess production capacity and reduced profit margin increases business bankruptcy, causing chaos in financial markets; the economy evolves from a recession to a depression until new science and new technology induce the next economy expansion.

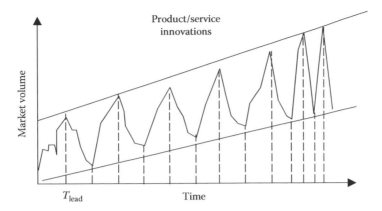

FIGURE 1.13   The long waves of economy.

This trend is illustrated in Figure 1.13. There are two takeaways in this illustration. First is that the general trend of market volume is increasing, which means human beings are using more and more natural resources. In reality, the trend of increase might follow an exponential curve rather than a linear slope. Second is that the period of economy long waves may have shortened with the explosion of technology. In simple daily words, human beings are trying to use more and waste more. This is basically a cyclic process with positive feedback. *Such development modes won't be sustainable in the long term unless some self-correcting mechanism is introduced into the social system.*

The biggest challenges of our time include climate change, energy, population, water, and security. Engineering must meet these challenges. Technological innovations may have caused many issues in the past; they might and should be the solutions to the above challenges in the future.

In my opinion, *the most urgent and important task in engineering is to establish and implement the new CEO to guide engineering activities and technological innovations onto the track of long-term sustainable development.* This will be further discussed in Chapters 2 through 4.

1.5   STRATEGIC INNOVATIONS—THE SUMMIT OF INNOVATION

For completeness, let's briefly talk about strategic innovations. Strategic innovations are normally the decision of senior leaders of organizations, but innovation is everybody's task and will affect everybody. Strategic innovation is needed to break through the growth limit of existing technologies.

Strategic innovations are innovations affecting organizational strategy and involving new, unproven, and significantly different tests to the following: Who will be the customer? What is the value to the customer? And, how will the value be delivered?

Compared with other kinds of innovations, strategic innovations are the most challenging. It is basically a one trial test—you have possibly just one or two chance(s) in your lifetime to prove its success or not. It involves a radical departure from existing businesses, it creates values discontinuously rather than incrementally, it lasts much longer than other innovations, and there is no clear model you can follow. It will compete internally against existing businesses, creating tension. It may be unprofitable for many years and etch away the strength of existing businesses. Yet, it has high growth potential, and it is the only way to break the growth bottleneck of matured technology.

For example, a company is currently a major player in energy generation; the major products are steam engines. These steam engines convert fossil fuel energy into electricity. It is still quite profitable, but the profit margin has been constantly decreasing. The company leaders are facing a crucial decision: shall the company improve the technology and cut costs to maintain growth or shall the company go for new technologies? One option is renewable energy. It is clear that renewable energy is the future of energy generation, while fossil-fuel-based energy generation is facing more and more restrictions. After lots of market research and feasibility study, the company decides to launch a solar energy business. This is the start of a strategic innovation for the company.

The company now faces many unknown challenges. Its customer changes, the customer value varies, and its past sales chains are not valid anymore. It has to set up a new department dedicated to the R&D of solar energy. People have to forget past success. The production line is totally different from that of steam engines; a lot of new technologies have to be introduced or developed. Instead of a major player in the world market, the company is a newcomer in the world of solar energy. The IP landscape is not friendly at all.

So, shall the new solar business follow the same culture and report structure as the rest of the company? What resources can the new business borrow from the current businesses? Who should the new business report to? How should the progress of the new business be measured and rewarded? To what degree should the new business be supported? Should new technology be developed or should the company acquire some strategic IP and work with partners? When can the new business be profitable? Imagine a company with $500MM profit, and each year, at least $100MM needs to be spent on the nonprofiting new business. Huge tension may arise between the profiting current business and the ambitious new business. How to handle this tension in a positive way? The homerun is to make the new business profitable and grow quickly. How do you establish quick learning cycles to speed up this process?

In the book *10 Rules for Strategic Innovators—From Idea to Execution*, the authors studied the challenges of strategic innovations and proposed an iterative procedure, namely, *theory-focused planning*, to increase the opportunity of success in this kind of innovation. Any organization has her unique organizational structure, culture and spirit, or the organizational DNA as the authors put it. The challenges are how to forget, burrow, and learn to improve.

Strategic innovation is not the focus of this book and it is impossible to cover the details in several pages. Hopefully, this short introduction gives you a peek onto the "summit" of innovation. If you are interested in becoming a brilliant business leader, find more about this topic and prepare yourself earlier in your career.

1.6 SUMMARY—THE FUNDAMENTAL PHILOSOPHY OF ENGINEERING

What should be the purpose of engineering? The thinking behind this question is the starting point of engineering. Engineering consists of the social activities that convert nature resources into human-desired configurations under the guidance of science and technology; and engineers are the initiators and enablers of such conversions. The beauty of nature lies in its balanced self-sustainability. It is important to realize that the ultimate purpose of engineering is innovating technology and carrying out production to improve the living standards of human beings while maintaining the healthy self-sustainability of nature. Communication in engineering faces many barriers that should not be there. A general philosophy is needed to improve the communication in engineering.

Innovation is the systematic introduction into the social system of new products, processes, services, or strategies. Although many people talk about innovation, there are many misunderstandings. Thus, before dedicating the book to the discussion of manufacturing process methodologies and innovations, this chapter gives a short yet systematic introduction to technological innovations. There are four classes of technological innovations. Product and service innovations directly create revenue, strategic innovations drive the whole process of innovation, and process innovations are the fundamental elements of all innovations.

Technological innovations are complex and risky. Among all the complexities, it is important to understand the social, natural, and scientific base of innovation. They are the origin of technological innovations; they also put constraints on technological innovations.

Technology change is imperative. *The imperativeness of technology means that the superior technology of a competitor cannot be ignored by other competitors.* Furthermore, technology changes drive the long economic cycles, while the central concept of managing technological change is how to implement technological innovations. To be proactive, one should understand the important curves in technological innovations.

We are living in a market-driven and fossil-fuel-based economy. We say our economy is market driven because the success of a technology or organization is judged by its performance in the market, which is measured in money or profit. The criteria of engineering optimization is thus to maintain a strong position for a country, an organization, or a company in market competition. Normally, the optimization and choice of technology in engineering considered the interest of customers and the company and stopped right there, ignoring the rest of the nature cycle. Establishing the New CEO is one of the reasons why this book was written.

Despite the importance of technological innovation and the imperativeness of technology, there is a big culture gap between the business world and the technical world, which put up unnecessary barriers to technological innovations. Most engineering schools focus on the scientific training of students, ignoring the management and system aspects of engineering. Business schools can be just the opposite, which focus on business management while ignoring the uniqueness of managing technological research and development. Accordingly, the education of engineers, managers, and scientists are incomplete, which leads to unnecessary frictions and misunderstandings between the groups. One purpose of this book is to bridge this gap from the engineering side. This is why in this book, before the discussion of manufacturing processes, we spend time trying to understand the importance of innovations, the types of innovations, the relation between technology and economy, the system and dynamic nature of innovations, the sources and the bases of innovations, and some curves of technological innovations.

Technological innovations created glories and issues in the past; it will continue to do so and shape the future. The bottom line is that all human activities are part of the nature cycle, from nature to nature and sustainable development should be the fundamental philosophy of engineering. This is a topic beyond one book; this also requires an effort from all levels of society.

With this, we are ready to discuss *intelligent EFM* in Chapter 2.

QUESTIONS

Q.1.1 Why do we run into difficulties when communicating between different engineering disciplines and different branches of a discipline? How shall we improve?

Q.1.2 What are the differences between invention and innovation?

Q.1.3 What should be the purpose of engineering? Why is this question important?

Q.1.4 What are the sources of technological innovations?

Q.1.5 Analyze two innovations, explaining why technology change is imperative.

Q.1.6 What are the categories of technological innovations?

Q.1.7 Please explain the complete industrial value chain.

Q.1.8 What are the factors affecting profit margin? How?

Q.1.9 What is the role of process innovation?

Q.1.10 Why are strategic innovations challenging?

Q.1.11 How can you be proactive to technology changes?

Q.1.12 Establish your database of technological innovation.

Q.1.13 Discussion: what are the changes needed in education regarding technological innovations?

REFERENCES

1. Geng, H., *Manufacturing Engineering Handbook*, McGraw-Hill, New York, 2004.
2. National Academy of Engineering, Greatest engineering achievements of the 20th century, 2010, http://www.greatachievements.org/

3. National Academy of Engineering, Grand challenges for engineering, 2010, http://www.engineering challenges.org/cms/challenges.aspx
4. Jacks, M., The history of the light bulb—An electric dawn. TheHistoryOf.net, 2008, http://www.the historyof.net/the-history-of-the-light-bulb.html
5. Nicholson-Lord, D., The biggest challenges of our time, *The Independent*, Mar. 21, 2005, http://www. independent.co.uk/environment/the-biggest-challenge-of-our-time-529294.html
6. Dorf, R. (ed.), *The Engineering Handbook*, 2nd edn., CRC Press, Boca Raton, FL, 2005.
7. Watson, P., *Ideas: A History of Thought and Invention, from Fire to Freud*, Harper Perennial, HarperCollins, New York, 2006.
8. Peters, T. J., *The Circle of Innovation*, Vintage Books, New York, June 1999.
9. von Baranov, E., Kondratyev theory letters, 2007, http://www.kwaves.com/kond_overview.htm
10. Betz, F., *Managing Technology—Competing through New Ventures, Innovation, and Corporate Research*, Prentice-Hall Inc., London, U.K., 1987.
11. Betz, F., *Managing Technological Innovation—Competitive Advantage from Change*, John Wiley & Sons, Inc., New York, 1997.
12. Govindarajan, V. and Trimble, C., *10 Rules for Strategic Innovators—From Idea to Execution*, Harvard Business School Press, Boston, MA, 2005.
13. Huthwaite, B., *The Lean Design Solution—A Practical Guide to Streamlining Product Design and Development*, Institute for Lean Design, Southfield, MI, 2004.
14. Ayres, R. U., Technological transformations and long waves: Part I and II, *Technology Forecasting and Social Change*, 37, 1–37, 111–137, 1990.
15. Zhang, W. and Mika, D. P., Manufacturing and energy field method, *Transactions of NAMRI/SME*, 33, 73–80, 2005.
16. Fey, V. R. and Rivin, E. I., *The Science of Innovation*, TRIZ Group, Southfield, MI, 1997.
17. Rantanen, K. and Domb, E., *Simplified TRIZ, New Problem Solving Applications for Engineers & Manufacturing Professionals*, Times Mirror, London, U.K., 2002.

2 Introduction to Intelligent Energy Field Manufacturing

Wenwu Zhang

CONTENTS

In Chapter 1, we argued that the ultimate purpose of engineering is to innovate technology and carry out production to improve the living standards of human beings while maintaining the healthy self-sustainability of nature. There are four classes of technological innovations, with process innovations being the fundamental elements of all innovations. Technological innovations are complex and risky, but technology changes are imperative. All human activities are part of the nature cycle; the fundamental philosophy of engineering should be from nature to nature and sustainable development.

On the other hand, we are living in a market-driven and fossil-fuel-based economy. Any technological innovation has to survive the test of the market to be successful. A new philosophy of engineering is needed to transfer our economy from purely market driven to market driven and sustainable. Intelligent energy field manufacturing (EFM) is one of such efforts.

In this chapter, we will introduce the origination, evolution, and fundamentals of intelligent EFM.

2.1 THE EVOLUTION OF ENERGY FIELD MANUFACTURING (1988–2008)

2.1.1 THE STORY OF A SCARED UNDERGRADUATE

It was the September of 1988. The campus of USTC (University of Science and Technology of China, Hefei, AnHui Province) was quiet and beautiful. Professor Jiqing Gao was giving a lecture to the sophomore students of mechanical engineering. Professor Gao was a well-known expert in tolerance and interchangeability in China. The class was on measurement and tolerance control. He showed off a precision measurement gage block like a treasure; his eloquent voice resonated in the class of 36 students. The gage blocks were used as the standard of length measurement. Thus, they had to be very precise and very stable during usage. The students were shocked to see that the segments of the gage blocks could connect together like one piece of solid metal by just gently pushing the shiny surfaces together! The metal was not magnetic. The surfaces were simply so flat that the molecular force could bring the different pieces together like magic.

Wearing gloves and dangling the connected gage blocks, Professor Gao asked: "Class, what processes are needed to make these pieces?"

We already had our first manufacturing process class in the previous semester. We proposed several processing steps. "Just these?" Professor Gao stared at us in disbelief. He then explained the over 20 major process steps necessary to make the measurement gage!

You have to pick the right material; the material must have excellent temperature stability, otherwise one degree of temperature change would make it meaningless. The block has to be highly flat and parallel in all three directions; so, precision positioning of the workpiece is needed. Several steps from coarse machining to precision machining to polishing are used. Some stress-relieving steps are needed. The pieces have to be calibrated with a laser interferometer and have to be reworked several times. Finally, they need to go through the thermal stabilizing process and be stored with good protection against any potential chemical changes. The manufacturing cost increases strikingly with the increased size and accuracy.

I was one of the students in his lecture. I like Professor Gao. He is knowledgeable and very enlightening. I dreamt of becoming a great inventor one day and he was already an inventor. I couldn't forget the delicious goose meat he shared with me when I worked in his lab. However, he possibly didn't realize how much his lecture had influenced me.

While some of my classmates dozed in the class, I was scared! Would this be my future career, spending my whole life on perfecting these lengthy steps to make just such a simple component?

2.1.2 THE RIDDLE AND THE BIG DREAM

For several days, I felt very sad. Are there any simpler ways of making things?

The surface flatness requirement of the gage is 10 nm (nanometer) over 20 mm, or 2,000,000:1 relative flatness. What happens when both larger area and higher accuracy are needed? Let's ask, is

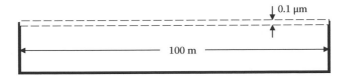

FIGURE 2.1 The riddle of achieving high flatness over large area.

it possible to create a flat surface as big as a football field with a relative flatness of 1,000,000,000:1 (Figure 2.1)? How can we make such a surface cost effectively?

The manufacturing cost with normal processing methods may climb like a rocket and still not make it. There must be a way. Several days later, I was eager to discuss this challenge with my teachers.

When we are facing challenges that are impossible to solve using conventional methods, we tend to be more creative. Afterward, we may find that the answer is actually pretty simple. For the above riddle, the answer is just like this:

> Fill a 100 m wide pond with pure water and shield it from any wind or vibration disturbance. Under the action of surface tension and gravity, the water surface is smooth to molecular scale. To further improve the surface smoothness, a thin layer of oil can be applied on the water surface (floating glass is made in similar ways). Then, gradually lower the temperature until it freezes. In this way, we should have a solid surface with one billion to one relative flatness!

At that moment, it occurred to me that manufacturing could be very interesting. I was no longer a "mechanical" engineering student!

This was an important turning point in my life. To do manufacturing, we should integrate various energy fields to find out the optimal solution of an engineering task. On the other hand, if we are limited to one kind of energy such as mechanical energy, we may find a best solution based on this energy form, but this solution may be a very poor one when all energy fields are considered.

There was no Internet on the campus in 1988, and I didn't know that someone had proposed layered manufacturing to realize complex three-dimensional (3D) prototype manufacturing (stereolithography was invented in 1986) [1]. This did not turn out to be a bad thing. Otherwise, I would have lost my independent thinking. My biggest dream when I was an undergraduate was to realize truly 3D manufacturing, like growing an apple out of water, air, and dirt. The processes I was aware of are processes taught in classic manufacturing textbooks. I felt that direct mechanical machining might be way too difficult for truly 3D manufacturing, because the programming to control the tool is too complex, and I hate complex solutions. Honestly, I was not a good programmer. I was in favor of the mold-casting process. Casting is a parallel and truly 3D process. The only trouble is that the mold has to be made first.

We need a magic mold and a magic forming process!

My imaginative mind coined the concepts of "virtual mold" and "dissociation forming." A virtual mold is a computer-controlled energy field that functions like a real mold in casting. It has the information of 3D geometry and all the means of energy field control. In other words, it is an intelligent energy field generator. The suitable energy fields are generated at the location of forming. A special material receives the energy from several directions, the intersecting point of the material dissociates from the rest of the bulk material due to a higher dose of energy or other "logic" effects. In this way, one can continuously vary the 3D energy fields and carry out truly 3D manufacturing. This was an idea before I watched any relevant science fiction or read anything about rapid prototype manufacturing (RPM). For good reason, I was very excited. I proudly included it as an appendix in my undergraduate thesis, titled "Virtual Mold 3D Manufacturing." I felt that the essence of manufacturing was very simple, it was controlling energy fields with some kind of "virtual mold" to convert materials into desired objects.

TABLE 2.1
The Evolution of Intelligent EFM

1988–1992	Origination of the seed concept of EFM: Virtual Mold 3D Manufacturing, University of Science and Technology of China, Hefei, China
1993–1997	Evolution into "Energy Field 3D manufacturing"; proposition of Logic Functional Materials, energy field generators, M-PIE flows and energy field method, China
1999–2001	NSF project on "Combined Research and Curriculum Development on Nontraditional Manufacturing (NTM)," led by Columbia University, United States; Study of hybrid process innovations, University of Nebraska–Lincoln and Columbia University, United States
2002–2005	Concept of general logic functional materials, general energy fields, and dynamic M-PIE model, GE Global Research Center, United States
2006	The concept of general intelligence; incorporation of TRIZ and lean six sigma into intelligent EFM; the success of the first international symposium on EFM in ASME/MSEC2006 conference, United States
2007	Won the contract from CRC Press to organize *Intelligent Energy Field Manufacturing: Interdisciplinary Process Innovations*
2007–2008	Concept of a market-driven sustainable economy and new criteria of engineering optimization (new CEO); second international symposium on EFM in ASME/MSEC2008, United States

This is the seed of EFM. This seed sprouted and evolved over the years, as shown in Table 2.1. I benefited a lot from this kind of thinking. It is still not fully developed yet, but it might be meaningful to share it with more people.

2.1.3 Virtual Mold 3D Manufacturing

The year 1988 marked the beginning of EFM. A science-fiction-like process—virtual mold 3D manufacturing—was proposed, independent of the work of RPM. Due to many limitations, this idea didn't develop into any real products, but the fundamental concepts matured over time. Let's briefly go over how it evolved into today's concept of intelligent EFM.

There are several key points in virtual mold 3D manufacturing.

1. *The hypothesis of material and energy field interaction*: Material will react to applied energy fields. As long as the energy fields are different, the response will be different. Such difference might be used in manufacturing.

This is a simple yet important belief. If you couldn't find out the difference, you possibly haven't found the right quantities to measure yet or possibly your tool is not sensitive enough. As shown in Figure 2.2, steel bar C is constrained between two solid walls. When the temperature changes slightly from T to $T+dT$, would anything different happen? Yes. Due to the thermal expansion of the material, any temperature change will change the dimension of the bar, thus inducing additional

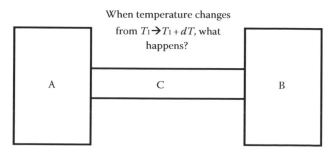

FIGURE 2.2   Illustration of the material and energy interaction hypothesis.

stress among components A, B, and C. This is straightforward for many people with knowledge of thermal expansion.

Let's ask another question: when light radiates on steel bar C, would anything different happen from the condition without light radiation? Without hesitation, the answer is YES. Many things can be different. If the light is a focused laser beam, it can actually drill a hole in the bar in less than 1 ns.

The key takeaway here is that we should design energy fields to magnify the difference to a certain level, keeping in mind that any difference an energy field induces may be useful for certain purposes, either for manufacturing or sensing. For example, oxygen gas sensors are based on the chemical reactions between oxygen and a sensing medium; the different response to different levels of gas content is used for safety purposes. Although many phenomena have been used for various detection purposes, their potential in manufacturing is far from well explored.

2. *Dissociation forming*: manufacturing technology is evolving from two-dimensional (2D) to 3D and beyond 3D. Truly 3D manufacturing requires new methodology relative to 2D manufacturing. Dissociation forming is proposed to address this challenge.

Truly 3D manufacturing processes are processes that can directly produce 3D objects with any transition of curvature. In this regard, a majority of the current processes are 2.5D, which can build up 3D objects based on 2D processing. For example, RPM produces 3D objects layer by layer, but each layer is basically 2D. Numerical control machining centers can mill out 3D objects directly, but this relies on complex tool path planning and the results are not very satisfactory.

What methodology and hardware developments are needed to realize and simplify truly 3D manufacturing with low cost, high speed, and high quality?

A basic belief is that 3D manufacturing requires newer and more complex methodologies than 2D and 2.5D manufacturing. A simple analogy between these methodologies can be understood by contrasting airplane transportation, which follows the contour of the earth, to automobile transportation, which follows the local contour of the land. The breakthrough in aviation technology gave human beings the truly 3D freedom of motion.

Virtual mold 3D manufacturing was proposed to address this challenge. Important concepts in virtual mold 3D manufacturing include: *energy field generator, logic functional material,* and *dissociation forming.*

As shown in Figure 2.3, energy field generators produce programmable energy fields in different directions and act on logic functional material. At the intersection of energy fields, the material is physically or chemically modified and dissociates from the bulk material, thus 3D geometry can be continuously grown up or carved out from the rest of the material. This way of forming is termed *dissociation forming.* The energy field generators are the virtual molds, which have been replaced with energy fields in recent years. Here, energy fields can be parallel fields, not necessarily a point

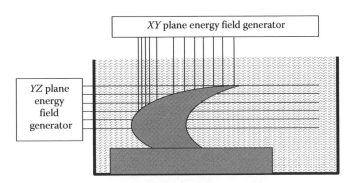

FIGURE 2.3 Illustration of virtual mold 3D manufacturing.

energy source. The use of programmable energy fields aims to simplify the path-planning work and speed up the process through parallel processing.

3. Virtual mold and energy field generator

Mold casting is a process with a long history. The earliest casting dates back to 3200 BC. In casting, a liquid material is poured into a mold, which contains a hollow cavity of the desired shape, and then the liquid solidifies. The solid is then released to complete the process. Casting is often used to make complex shapes that would be otherwise difficult or uneconomical to make by other methods. Casting is actually a parallel 3D process. The limitation is that the mold has to be made first.

If we can digitally control energy field generators to form a virtual mold, we can then have a truly flexible 3D manufacturing process. Here we talk about energy fields instead of energy sources, trying to establish the field philosophy in the very beginning. An energy field is the temporal and spatial distribution of energy. People are used to using "point" energy sources for manufacturing, such as a laser beam, a cutting tool, etc. But an energy field is a more accurate description of the real world.

An *energy field generator* is the integrated system of energy source and the relevant control unit. With proper control, specific energy fields can be generated in space with a certain temporal duration.

Energy field generator is just a term to summarize the various energy devices. A TV set, a light bulb, a computer system, a magnet, etc., are all examples of energy field generators. The energy field generators used in 3D manufacturing are required to form the necessary fields in space to induce dissociation forming. To make the forming process feasible, a special property material is necessary. Such materials are called logic functional materials.

4. Logic functional materials

Logic functional material is defined as a material with three states: normal state, excited state, and qualitatively changed state, as shown in Figure 2.4. Under normal state, the material demonstrates certain properties, such as its phase, its mechanical properties, etc. The normal state is the physical and chemical environment in which the material exists in a normal natural condition. With suitable

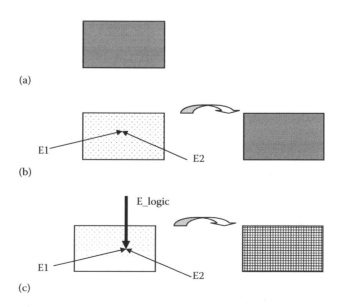

FIGURE 2.4 Three states of logic functional materials: (a) normal state, (b) excited state, and (c) qualitatively changed state.

levels of applied energy, such as energy field E1 and E2, the material is excited close to the permanent modification state, but after the applied energy fields are removed, it recovers to normal state. Elastic deformation would provide a relevant example. Material may transit into a qualitatively changed state when a certain additional energy field, energy field E-logic, is applied or when a further increase in the intensity of the energy field is used, preferably a selective energy field. In this case, the material changes permanently. Certain effects can be used to dissociate the affected zone from the bulk material. This bulk material can be gas, liquid, or solid. This additional energy field acts like a logic key that can trigger permanent property changes. Such materials are thus called logic functional materials. Ideally, the property change of the material is limited to the zone of the intersection of energy fields. The affected zone defines the resolution of the fields. It is also desirable to make the material sensitive to the linear propagation of the energy fields.

When these concepts were initially proposed, they were regarded as too ideal to be practical. There are certain materials that qualify this definition, such as the materials used in selective laser sintering (SLS). In the SLS of thermoplastics, laser irradiation raises the temperature locally and cures the polymer locally, thus building up the 3D objective layer by layer. Such polymers are in a liquid state under a normal room environment. But once the temperature goes beyond the thermosetting point, it cures and changes into a solid state. In SLS, the polymer can be heated to a temperature beyond room temperature to lower the energy requirement of the laser unit. Thus, the bulk of the liquid pool is in an excited state, while only a small zone of laser irradiation is in the qualitatively changed state. The same thing can be said about SLS of metal powders.

Some materials are called *smart materials* because these materials have one or more properties that can be significantly changed in a controlled fashion by external stimuli, such as temperature, stress, EM fields, moisture, etc. Examples include piezoelectric materials, shape memory alloys, UV curable polymers, liquid crystals, electrorheological fluid, and magnetorheological fluid [2]. Look up information about these materials. All these materials demonstrate the logic functionality described above.

Smart is a relative word. Logic functionality is more objective. Later on we will argue that all materials have a certain level of logic functionality. Logic functional material at this point was mainly used for the convenience of 3D manufacturing.

2.1.4 FROM VIRTUAL MOLD 3D MANUFACTURING TO ENERGY FIELD MANUFACTURING

After attending a conference in Beijing in 1995, I visited Prof. Yongnian Yan's group at Tsinghua University. Prof. Yan is a pioneer in the research of RPM technology in China. We discussed the concept of virtual mold 3D manufacturing. Prof. Yan agreed with many aspects of my thought, but he felt that "virtual mold" could be a misleading terminology; what I really meant was a digitally controllable energy field. Together we decided that we should name this direction "energy field 3D manufacturing."

Energy field 3D manufacturing actually provided a new strategy for 3D manufacturing, quite different from the existing method of RPM. It is inherently a field-based parallel method. It requires computers to extract and generate computer-aided design (CAD)/computer-aided manufacturing (CAM) information, energy field generators to implement the forming field, and the right materials to form the final 3D objects [3,4]. We jointly filed the Chinese patent application in 1997 to protect the key ideas [5]. Unfortunately, we did not get enough support to implement this promising patent. Maybe the concept was too ahead of its time in 1997, and I was not ready to take on the task at all. I was an active science fiction writer at that time, but my technical expertise was too shallow and too narrow. Strange or not, I simply felt that it should be my life-long objective to fully implement energy field 3D manufacturing. This has not changed over the years, and this book is a step toward this lofty goal. I decided to improve myself through PhD study first.

It was the Thanksgiving of 1998 when Prof. Y. Lawrence Yao, my future PhD advisor at Columbia University, interviewed me. Prof. Yao gave me very valuable advice: to realize the lofty goals of energy field 3D manufacturing, I should have a good understanding of nontraditional processes; to understand the essence of manufacturing, I should study in-depth one of the representative nontraditional processes first. I happily joined Prof. Yao's group in 1999 to study laser micromachining for my PhD thesis, and in the mean time, worked on a National Science Foundation (NSF)-funded project that deeply affected the later development of energy field manufacturing.

From 1999 to 2001, Prof. Y. Lawrence Yao's group at Columbia University, Prof. K.P. Rajurkar's group at the University of Nebraska-Lincoln, and Prof. Radovan Kovacevic's group at Southern Methodist University participated in an NSF-funded project (CRCD EEC-98-13028)—Combined Research and Curriculum Development on Nontraditional Manufacturing (NTM) [6]. Representative NTM processes, such as laser material processing, abrasive water-jet machining, and ECM/EDM, were studied, cross-process innovations were discussed, and a Web-based NTM curriculum was made available to the public [6,7]. This project revealed that

- NTM processes are normally regarded as the alternative to traditional manufacturing processes. They are employed when traditional methods do not meet the processing requirement. This way of thinking, in addition to the capital investment considerations required for NTM, hampers the widespread use of NTM processes.
- Current NTM teaching is process based and lacks a systematic way of introducing the technology concepts. As a result, a change in NTM teaching is needed that will better prepare the next generation of engineers for future challenges.
- Process innovations underscored the importance of breaking the barriers among various energy fields and processes. When trying to find the optimal solution to an engineering task, all energy forms should be considered. In reality, unfortunately, the old thinking of dividing people and resources into traditional and nontraditional disciplines may seriously hamper successful innovation.

These deficiencies are understandable given that the majority of managers, engineers, and workers are only trained in limited manufacturing processes, and trepidation toward the unknown world is human nature.

Manufacturing technologies typically fall under two broad categories: the so-called traditional and nontraditional manufacturing processes. Traditional manufacturing relies on direct mechanical contact between the tool and the workpiece, such as the processes of forging, turning, milling, etc. In contrast, NTM processes are (1) processes in which there are nontraditional mechanisms of interactions between the tool and the workpiece and (2) processes in which nontraditional media are used to enable the transfer of energy from the tool to the workpiece [8].

Such definitions of traditional and nontraditional manufacturing, however, change with the maturity of technology and are historically biased toward mechanical methods. For example, casting processes have been utilized for thousands of years. It is generally regarded as conventional manufacturing although it mainly involves thermally controlled phase transformations, a nonmechanical energy process. On the other hand, diamond precision machining holds the highest machining quality, yet it is unconvincing to claim it as a traditional process.

A strict distinction between traditional and nontraditional manufacturing doesn't have much engineering value, but it is important to notice that improper education and research philosophy based on the historically biased vision of manufacturing may implicitly hamper technology innovation and integration. To better equip researchers and engineers for technology follow-up, improvement, and innovation, and to better meet the challenges of modern technology development, the methodology and philosophy of manufacturing should be continually studied.

A general philosophy of manufacturing will lower the difficulties of process innovation by providing a framework for systematic and guided thinking, lowering the shock of technology progress, and providing a common vernacular for communicating engineering ideas, thus bringing engineering closer to the public.

Innovation has been regarded as the single-most important factor that decides the global technology leadership for a country or an organization in the global economy. Big-impact product innovations normally require some key process innovations; thus, the success of a product innovation plan ties closely to the manufacturing processes or process innovations. Despite tools such as brainstorming, trade-off analysis, and benchmarking that help break the barriers of process innovation, it is observed that process innovation is still mainly based on trial and error, lacking a systematic approach. TRIZ, or creative problem solving in Russian, gained certain success in the United States, Japan, and Europe. TRIZ proposed a systematic approach using patterns, principles, and knowledge base to solve system contradictions using tool–object interaction analysis [9–11]. TRIZ is a good methodology for general problem solving; however, engineering innovations need not only the operable procedures to generate creative ideas, but also the clear skills and methodology to implement and optimize the creative solutions.

Over the years, it has been felt that the methodology initially proposed for energy field 3D manufacturing applies to general manufacturing processes. When combined with other engineering thoughts, a high-level methodology for engineering can be derived to facilitate systematic process innovation and more efficient technical communication.

Why shall we label mechanical contact processes as "traditional" and other processes "nontraditional"? This historically biased terminology might have produced some negative impact in reality. People working on contact processes think that contact processes are mainstream while noncontact processes are supplementary; individuals working on noncontact processes emphasize the advantages of noncontact and might be reluctant to integrate contact processes.

In reality, mechanical contact or mechanical force is just one common form of energy, a relative to sonic, electric, magnetic, photonic, and chemical energy. None of the various energy forms should have an unconditional biased priority if a truly optimal solution is to be found for engineering tasks.

The strength of big R&D centers, such as GE Global Research, IBM Research, world class engineering schools, etc., is that talents in all areas can work together dynamically under the incentive of innovation and growth, empowered by the best practices of management and leadership. Simply put, success can be attributed to an environment that facilitates the sharing of "gems" of individuals. Technological isolation is prevalent. Such isolation arises from the administrative structures that tend to be divided along core technologies, funding, or other cultural structures that limit cross-lab pollination or from result-driven time constraints. Thus, more efficient innovation is possible by removing these common barriers.

The energy field is the spatial and temporal distribution of energy. "Point energy" is a simplified concept—"point energy" is actually a localized energy field. Energy field is a more general concept and a more strict description. A focused laser beam might be regarded as a point energy source; however, in micromachining, laser spatial and temporal distributions must be considered. Traditional machining commonly relates to stress fields and thermal fields. Gravitational fields and environmental pressure fields are ever present.

One common feature of various manufacturing processes is the extensive use of various energy forms and energy fields. In fact, engineering is the art of energy field utilization and manipulation. Rather than relying on personal inspiration, a methodology can be developed to systematically improve our skills of energy field manipulation and integration. We define engineering solutions featuring meaningful energy field manipulation and integration *energy field methods*.

2.1.5 THE DYNAMIC M-PIE MODEL OF MANUFACTURING

There are many engineering solutions that are good examples of the energy field method. This book will report representative examples, such as the production of single crystal alloys, abrasive water-jet machining, laser shock peening, electromagnetic dynamic forming, texturing for friction reduction, lithography in electrical engineering, soft lithography in nano manufacturing, etc. The

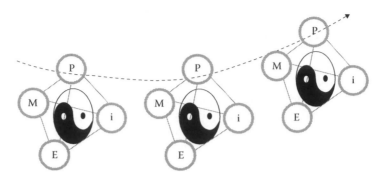

FIGURE 2.5   The dynamic M-PIE model.

challenge is how to find a systematic way to analyze and implement these methods, instead of relying on scattered inspirations.

To address this challenge, we need to have a concise model of manufacturing that is able to bridge the past of engineering to the future of engineering, and to break the barriers between traditional and nontraditional manufacturing. This is an unending challenge. This book is only one of the attempts to answer this challenge. Our starting point is the dynamic M-PIE model, as illustrated in Figure 2.5.

Figure 2.5 illustrates the flows existing in any manufacturing processes and systems: information and intelligence flows (I-flows), energy flows (E-flows), material/resource flows (M-flows), and the system level flows of products/processes/projects (P-flows). The word "flow" reflects the dynamic nature of information, energy, material, and processes.

The I-flows include (1) the knowledge database of manufacturing; (2) the design, monitoring, and control of manufacturing processes; and (3) the tracking of the complete cycle of energy and material/resource flows. Human beings and labor can be treated as part of the I-flows.

The E-flows are the flows of all energy fields affecting manufacturing processes. In mechanical drilling, for example, one should consider which energy forms are involved. Mechanical interaction is the major energy form; EM energy is used to drive the motor; the drilling system draws electrical energy from the power grid; lubrication may be used to cool down the drill bit, thus the thermal field is involved; etc. How about environmental vibration, gravity, and pressure? In E-flows, we should also consider the impact on nature. How can we optimize the integration of E-flows to lower the impact on nature? Along this thought, a lot of topics can be listed. The analysis of E-flows may immediately guide us to new processes. For instance, how about integrating vibration energy with a conventional constant rotation process? This leads to ultrasonic vibration-assisted cutting or drilling technology. Ultrasonic vibration-assisted cutting has been effectively used for foam and composite machining [12].

The M-flows are the material and resource flows. Manufacturing converts materials from one state to another, adding values to make it useful as a product. The resource flows include capital and human resource flows. Thus, money, people, hardware, and workpiece materials are all part of the M-flows. Again, we need to analyze the complete cycle of M-flows and the impact on nature. All materials and resources are from nature and will finally become a part of nature. Conventional textbooks normally address the issue of how to select materials for product and production. The M-flow analysis in EFM extends the scope to a much wider range.

The M-E-I flows can be both process-level and system-level, which are linked together by the P-flows—the system level flows of products/processes/projects. This includes the planning and management of manufacturing projects. Existing manufacturing methodologies, such as concurrent engineering, digital manufacturing, lean manufacturing, six-sigma quality control, TRIZ, etc., can be naturally integrated into the P-flows.

All factors are relative in nature and have their positive and negative sides. Is water a liquid and is water soft? Is laser machining more precise than mechanical machining? Is silicon rigid and brittle?

Think twice and do enough homework before answering them. Although silicon is usually brittle, it can bend when it is made thin enough (a couple of hundred nanometers). Intel commercialized strained silicon. Electrons move through strained silicon 80% faster than in conventional silicon, and transistors switch on and off up to about 30% faster. Researchers have set out to make flexible forms of strained silicon and are turning it into many applications. How would this change our future? Simply type in "flexible silicon" and search the Web.

To meet practical needs, we frequently run into contradictions, which are conditions that we desire something to meet certain requirements but not other requirements. For example, we need enough metal materials for cars to ensure safety and strength, but we don't want the mass at all if we want to minimize energy consumption.

The Tai-Chi symbol in Figure 2.5 highlights the pervasive existence of contradictions, the relative and the dynamic nature of these contradictions and the interactions of different flows. This is called the dynamic M-PIE model of manufacturing. Individual cells of M-PIE flows can be formed into complex structures to describe systems of products, processes, and projects, while the bigger system can have the analysis of M-PIE flows at the system level as well.

The pivot of EFM is to find a systematic way to address the contradictions and reach a higher level of optimization. Understanding the generality and relativity of M-PIE flows is the first step. By following the basic principles of EFM, one can have a quick understanding of a process or one can elaborate on a process to consider all aspects of manufacturing.

2.1.6 Toward Intelligent Energy Field Manufacturing

The essence of manufacturing is utilizing information to control an energy field to transfer materials into desired configurations. Energy fields carry information and convert materials into final products. Thus, energy field manipulation is central in all manufacturing processes. Rather than dividing manufacturing into traditional and nontraditional processes, we should treat them equally as processes of *EFM. EFM is defined as methodologies and activities of manufacturing featuring the systematic application of energy field methods and the optimal integration of dynamic M-PIE flows.*

In a certain sense, all manufacturing processes are EFM and all engineering processes are energy field engineering, because they all involve the M-PIE flows and energy fields are always essential. So, why bother even proposing energy field manufacturing? We are practicing EFM! The difference is whether we do it systematically and consider the optimal integration of the manufacturing flows.

Nature is full of examples of EFM. We are part of nature, and our knowledge comes from nature. I was amazed at the crystallization process when I first learned how that happened. With the proper change of temperature and composition, some phase will be formed under the interaction between fundamental particles and various energy fields.

How about the eruption of a volcano or the growth of plants? These are also processes of EFM. Then what are the differences between such natural processes and human processes?

The trouble comes from the terminology.

EFM only covered the energy flow and the material flow, but the information flow and the P-low are not explicitly covered. When asked the question "which is more fundamental, energy, material, or information," many people answered "energy." That was my answer in the beginning. But I was not fully confident in this answer. Energy and mass are mutually convertible, but information seems to be able to stand alone by itself. We need energy and mass to detect, create, process, and transmit information, but information should be there even if we are not doing anything. A deeper discussion of this question might bring us to the relation between philosophy and religion.

The point is that the role of information is not fully reflected in the terminology of energy field manufacturing. Better terminology should be used to reflect all aspects of human beings' manufacturing activities.

Human beings' engineering activity is a process of injecting intelligence into the interactions between energy fields and materials. We create value by increasing the intelligence level of the configurations of energy and materials. Thus, to better differentiate human manufacturing from natural processes, we could call it *intelligent EFM.*

Detection and control have become an inseparable part of modern manufacturing. *Spontaneous interactions between energy fields and materials can happen in nature, but only when we use our intelligence to control how they interact can we add value to the flows.* Intelligent control of processes is a big progress in manufacturing. Controlling the information flow is very important in the modern economy. When some process techniques become common know-how, with the information of innovative design controlled, one can move onto outsourcing for manufacturing.

The terminology of intelligent EFM was first published in the 2006 ASME/MSEC conference.

2.1.7 FROM ASME/MSEC SYMPOSIUMS TO THIS BOOK

From 2002 to 2006, the author realized the generality of logic functional materials, energy fields, and intelligence. The dynamic M-PIE model was published in the SME/NAMRC2005 conference [13]. The concept of general intelligence was conceived in 2006, and the frames of intelligent EFM were published in ASME/MSEC2006 conference [14].

The first international symposium on EFM was successfully held in AMSE/MSEC2006 at the University of Michigan. Dr. Wenwu Zhang and Dr. Shuting Lei co-organized the symposium and won the ASME/MSEC 2006 BOSS Award (Best Organizer of Symposiums and Seminars Award). The symposium took three sessions to finish. Scholars around the world talked about laser shot peening, thermal-assisted brittle material processing, magnetic polishing, innovative nano-manufacturing processes, etc.

Should intelligent EFM continue to be of personal interest or shall it be more extensively studied worldwide? Honestly, I felt I was not fully ready to write a book like this. Several things encouraged me to go for a book proposal. First is the feeling of obligation to future engineers. Many innovative engineers may have formed their own unique ways of technological innovation, but they learned these through success and failures in the real world, usually with the big cost of time and money. These experiences are scattered and difficult to pass onto future generations. A new graduate from a university will have a period of culture shock—he or she has to get used to many new things in the practical engineering world. To be a successful engineer, one has to establish the capability of innovation. Being imaginative and creative is not sufficient. Personally, I changed a lot after I studied books on technological innovations. But this only happened accidentally 5 years after my PhD study. Things could be very different if the knowledge of innovation was learned earlier in our career.

Second, our world has plenty of brilliant inventions and technologies. How can we quickly analyze and absorb the sea of knowledge, both the past and the state of art? I still remember the excitement and uneasiness when I first attended an international conference. Relying on personal experience is usually not efficient enough. Due to the flooding of information, we are kind of in a mode of picking up and throwing out along the way, wasting a majority of the value of technological information. How can we improve the way we communicate in engineering and how can we better inherit from the existing system of knowledge? There are some good ways of achieving this, but they are not widely known to the public.

Third, thinking of manufacturing as EFM is still totally new to many people. We see unnecessary barriers for new process implementations in daily life. These barriers might actually originate from the inertia of engineering education. How can we better cross the borders between disciplines? How can we lower the threshold or better prepare people for process innovations?

And finally, the current criteria of engineering optimization are missing a natural mechanism to protect the sustainable development of mankind. Manufacturing and engineering in general are at the very front of the nature–human interaction. Other social activities follow the flow of engineering and finally get back to nature. Engineering is the driving force of this cycle. People in engineering should take the responsibility of ensuring the natural sustainability of our society. Unfortunately, this fundamental point has not been sufficiently reflected in a majority of the engineering books.

Several friends kindly reminded me that one could never be fully ready for something new and challenging. With the success of the first international symposium on EFM in 2006 and with the 20 years anniversary of the concept of EFM approaching, I tried to submit book proposals to publishers. In May of 2007, I was offered the book contract from CRC Press to organize the first book on *Intelligent Energy Field Manufacturing and Interdisciplinary Process Innovations*. The strategy is to make this a worldwide collective effort, inviting experts in different areas to shed light on what is the state of art in their topics and what is the philosophical thinking behind these technological innovations.

In 2008, Dr. Wenwu Zhang and Dr. Shuting Lei organized the second international symposium on EFM in ASME/MSEC2008. Figure 2.6 shows a picture of authors attending the first EFM symposium. Luckily, we won the ASME/MSEC BOSS Award in both 2006 and 2008. Figure 2.7 is a picture taken when Dr. Wenwu Zhang and Dr. Shuting Lei received the 2008 BOSS Award.

It is important to point out that intelligent EFM is a research direction still in dynamic evolution [15]. You can and should be part of it. The rest of the chapter will discuss the generality of energy fields, logic functional material, and intelligence. The following chapters will discuss the principles of intelligent EFM and the suggested ways of implementation. This is followed by contributions from worldwide scholars, who will report their understanding of how a specific discipline or process evolved.

FIGURE 2.6 Authors attending the first EFM symposium at ASME/MSEC2006.

FIGURE 2.7 Dr. Wenwu Zhang and Dr. Shuting Lei receiving the BOSS Award at ASME/MSEC2008.

2.2 THE PHILOSOPHY OF GENERAL ENERGY FIELDS

Why shall we talk about energy fields in manufacturing processes? In manufacturing, we convert materials from one configuration to another to fulfill certain human purposes. In this conversion, we use our knowledge to apply forces, to control energy fields, to change the status of materials, and to imbed intelligence (or functions) in the new configurations. The more we understand energy fields and the interactions between energy forms and materials, the more powerful we are in manufacturing.

Historically, human beings interpreted the world in forces, energy, and fundamental elements. Let's first define the fundamental concepts of force, work, and energy before we introduce the concept of general energy fields [16,17].

2.2.1 THE CONCEPT OF FORCE

- In physics, force is the factor that can cause an object with mass to accelerate. Force is a vector quantity with both magnitude and direction.
- Many forms of forces had been defined, but they can all be derived from the four fundamental forces in universe.

To explain the mechanisms of motion, the concepts of force and mass were developed. This followed an intuitive descriptive approach. Some quantities can be observed and measured, such as the measurement of dimension, time, weight (mass), and speed. Force is then defined to correlate the observed results. Through experimentation, the definitions of forces were validated and this gave rise to many kinds of forces.

For example, a lever can extend our ability of lifting a heavy load. The lever is in contact with the load to transfer force. Many tools were invented to help us in a similar manner. This gave rise to the concept of mechanical forces. Archimedes was famous for formulating a treatment of buoyant forces in fluids. Elastic force was initially defined to describe the force exerted by a spring when the spring was displaced from its equivalent position. Normal force and frictional force were proposed to model the motion or resistance of one object in contact with another. Centrifugal force, impact force, tension, contraction, adhesion, capillary force, attraction, air resistance, etc., are all similar examples.

Will a heavy stone ball fall faster than a light stone ball under the same conditions? Does the motion of the stars in orbit follow the same rules as the small objects on earth? Early physicists, including Gallileo and Newton, had given us the answers. When laboratory experimental observations were fully consistent with the conceptual definition of force offered by the models, such as Newtonian mechanics, these models became physical laws of our time until their limitations or defects were discovered.

Forces are vector quantities that are very useful in the analysis of motion and energy–material interactions. With the accumulation of our knowledge, there are so many forces defined in very specific ways (special models of some phenomena), that one may get lost in the details and run short of time and energy to see the big picture. This is why we are trying to develop the concept of a general energy field in this book.

It is meaningful to point out the following facts about forces:

- There are four fundamental forces: gravity, electromagnetic force, strong force, and weak force. The electromagnetic force acts between electric charges and the gravitational force acts between masses. The strong and weak forces act only at very short distances and are responsible for holding nucleons and compound nuclei together. All forces can be derived from the four fundamental forces [17,18]. It is this clear. Mechanical force, elastic force, hardness, friction, adhesion, surface tension, attraction and expellation, and formation of crystals can all be explained from the electromagnetic structure of atoms and molecules.

- Force is defined with Newton's second law of motion, which is mass times acceleration, or the derivative of momentum. With the development of quantum field theory and general relativity, it was realized that "force" is a redundant concept arising from the conservation of momentum. The conservation of momentum is considered more fundamental than the concept of force. Thus, the currently known fundamental forces are considered more accurately to be "fundamental interactions."
- An energy field is a more powerful tool in describing energy–material interactions. Force can describe the motion of a system, but motion is just part of the whole picture. How about other quantities of a system, such as temperature, spectra, density, etc.? This is one of the reasons that we choose energy fields as the fundamental concept in the discussion of manufacturing processes.

- Work: mechanical work is the amount of energy transferred by a mechanical force. In thermodynamics, work is the amount of energy transferred from one system to another without an accompanying transfer of entropy. It is a generalization of mechanical work in mechanics.
- Energy is the ability to do work.
- Energy is conserved in nature, but energy can be converted from one form (or state) to another.
- There are many forms of energy. The more control we have in energy use, the more flexibility we have in manufacturing process optimization.

2.2.2 Work and Energy

There are many definitions of energy. People talk about energy crisis, energy safety, energy efficiency, etc. So, what is your definition of energy?

Normally, energy is defined as the ability to do work. Then what is the definition of work?

As pointed out earlier, human beings initially used force and material interaction to explain the dynamics of the world. Force can do work. In the SI system of measurement, work is measured in joules (symbol: J). The rate at which work is performed is power. Work can be in the sense of mechanical work or thermodynamic work.

Mechanical work

Mechanical work equals force times the distance of force action, as defined by

$$W = \int_c F \bullet dS \tag{2.1}$$

where
   c is the path traversed by the object
   F is the force vector
   S is the position vector

Mechanical energy is the summation of potential energy and kinetic energy. The work done by an external force is equal to the change of mechanical energy of the system when other energy transitions are negligible.

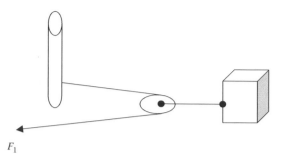

FIGURE 2.8 Force and work.

Machines can increase the magnitude of an input force, but not the total work. Since energy dissipation always accompanies any mechanical process, the total work done by the machine has to be larger than the work necessary when a direct force is used. For example, a pulley can double the pulling force on the block in Figure 2.8, but the distance of travel is doubled at F_1 relative to the displacement of the pulley. Due to frictional energy loss etc., the work needed to move the block is no less than the work needed when pulling the block directly. The engineer's task is to increase the energy efficiency of the machine and make it easier to use for useful works. Imagine how little work we do when we actually move a train on the rail or a car on the road.

No mechanical work is done without motion, although force may exist. For example, you don't do mechanical work when sitting still in your chair. Mechanical energy is not transferred between you and the chair, although your body weight and the reacting force are experienced by you and your chair.

Note that mechanical energy does not include thermal energy and rest-mass energy (which is constant as long as the rest mass remains the same). Mechanical energy is the energy of a system at macroscale, relative to the molecular and atomic scale.

Thermodynamic work

In thermodynamics, work is the quantity of energy transferred from one system to another without an accompanying transfer of entropy. Thermodynamic work is slightly more general than mechanical work because it can include other types of energy transfers. Thermodynamic work can include electrical work (the work done by an electric field when a charged particle is moved) and pressure-volume work (work done by a fluid when its volume changes) etc., in addition to mechanical work (such as the work done by a moving piston in an engine). Electrical work is defined by

$$W = q \int E \bullet dr \tag{2.2}$$

where
q is the charge of the particle
E is the strength of the electric field
r is the distance

Both E and r are vector quantities. Pressure-volume work is defined by

$$W = -\int_{V2}^{V1} P dV \tag{2.3}$$

where

W is the work done on the system
P is the fluid pressure
V is the volume

The first law of thermodynamics is a more universal physical law—the conservation of energy. It states: The increase in the internal energy of a system is equal to the amount of energy added by heating the system minus the amount lost as a result of the work done by the system on its surroundings. In mathematical form,

$$dU = \delta Q - \delta W \tag{2.4}$$

where

dU is a small increase in the internal energy of the system
δQ is a small amount of heat added to the system
δW is a small amount of work done by the system to the surroundings

Internal energy

In thermodynamics, the internal energy of a thermodynamic system with well-defined boundaries is the total of the kinetic energy due to the motion of molecules (translational, rotational, vibrational) and the potential energy associated with the vibrational and electric energy of atoms within molecules or crystals. Internal energy is a state function of a system.

Is heat equivalent to work in certain ways? This is not a trivial question. It was a very important question in the 1800s, the glorious time of heat engines.

What is the nature of heat? Ancient people could feel the heat of fire. It was so mysterious that almost all cultures have legendary stories about fire. Is heat a kind of invisible mass (the substance of heat called *caloric*)? If it is a kind of mass, it is conserved, how does it enter a body and leave a body? Friction can raise the temperature of rubbing surfaces, what is really going on? These questions were not well answered until the year of 1845 when James Joule proved the equivalence between mechanical work and heat. Figure 2.9 shows the setup Joule used to prove the mechanical equivalence of heat.

The caloric theory and the kinetic theory of heat coexisted for many years until the vision of thermodynamics matured. The modern day definitions of heat, work, temperature, and energy all

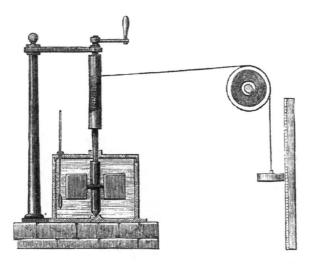

FIGURE 2.9 Engraving of Joule's apparatus for measuring the mechanical equivalent of heat. (From Abbott, J., *Harper's New Monthly Magazine*, 39, 327, August 1869.)

have connections to Joule's experiment. In thermodynamics, heat is a form of kinetic energy produced by the motion of atoms and molecules and is transferred from one body or system to another due to a difference in temperature. Heat can flow spontaneously from a higher temperature body to a lower temperature body, while the opposite can happen only with the assistance of a heat pump.

Work and heat transfer are the two means of energy transfer in thermodynamics. Work refers to energy transfer in the macroscale, for example, the energy that is used to expand the volume of a system against an external pressure. Heat transfer is through the *microscopic* thermal motions of particles.

2.2.3 A HISTORICAL VIEW OF ENERGY

To explain natural phenomena, many kinds of energy have been defined or discovered, such as thermal energy, kinetic energy, potential energy, mechanical energy, chemical energy, nuclear energy, radiation energy, wind energy, sound energy, strain energy, etc.

The engineering history of human beings is a history of the understanding, harnessing, and utilizing of various energy forms [18]. Manufacturing is part of engineering. Currently, manufacturing processes are normally categorized based on the dominating energy forms or dominating physical effects. For example, laser or ultrasonic material processing is material processing mainly relying on laser or ultrasonic energy. It should be understood that this is just a convenience of terminology, and we should not limit our thoughts merely on this energy field. In many cases, engineering optimization and process innovations require the consideration of a wider range of physical laws and effects. Let's briefly highlight the historical events in engineering before we introduce the concept of general energy field.

The Earth inherited potential energy at birth. Such potential energy is stored in the form of mass. Mass is actually a condensed form of energy. The other major energy source on Earth is the energy from the Sun. Solar energy is stored in forms of fossil fuel energy, wind energy, tide energy, biochemical potential energy (plants and animals), gravitational potential energy (the cycle of water), etc. Energy transformation occurs when a suitable triggering mechanism is used.

Of all the energy forms, biochemical energy is possibly the first energy we utilized—our body is a perfect machine for biochemical energy conversion. In the twilight of human beings' civilization, mechanical energy was explored. Tools made of bone, wood, or sharpened stone were exploited in the living competition in the wild. These tools helped to concentrate or magnify human beings' manual force (manufacturing is actually a word derived from Latin *manu factus*, meaning *made by hand*). More complex tools were made over time and we called this branch of engineering mechanical and manufacturing engineering.

Making fire is the start of a long history of the conscientious use of available energy resources on earth. With fire, mankind easily gained advantages over other animals on earth. People lived healthier and longer and new materials such as ceramics and metals were developed. Phase changes and chemical reactions were studied and utilized. This went all the way down to the invention of steam engines, which started the age of Industrial Revolution.

Energy conversion became more and more sophisticated with the progress of technology. In ancient times, water flow and wind force were used to assist transportation. In medieval Europe, the power of creeks and rivers was harnessed to do work such as flour grinding. In 1750, J. A. Segner built a mill driven by an impulse hydraulic turbine. When coal became a big need in England for heating and metallurgy, the need for pumping water out of deep mines stimulated the research of steam driven pumping systems, which led to the maturity of steam engines and the invention of railway transportation systems. To this point, human beings were able to generate a huge amount of energy, making it possible for big workshops and plants, thus changing the way manufacturing had been practiced.

The modern age did not arrive until electromagnetic energy was harnessed. Faraday discovered the conversion mechanism between mechanical energy and electromagnetic energy in 1831. Only

2 years later, an American, Thomas Davenport, built the first electric motor. The modern electric generator was invented in 1861 by Hungarian Anyos Jedlik. The first central electric generation and distribution system was built by Thomas Edison in New York City in 1881. The ability to convert various energy forms into electricity, deliver electricity to wherever is needed, and then use electricity to initiate other processes has greatly shaped modern society.

There were still many other energy forms to be explored and harnessed. Oil and gas became the strategic energy resource in the twentieth century. Gas powered internal combustion engines laid the foundation for the modern automobile industry. Petroleum extraction and refinery gave birth to a chain of chemical businesses.

Solar radiation sustains the life cycle on earth. Can human beings generate light as bright as the sun? What is the nature of light? In 1917, Albert Einstein explained stimulated emission in his paper *On the Quantum Theory of Radiation*. This laid the theoretical foundation for the future development of masers and lasers. The first working laser was made by Theodore H. Maiman in 1960. Maiman used a solid-state flashlamp-pumped synthetic ruby crystal to produce red laser light at 694 nm wavelength. The family of lasers and the applications of lasers have exploded in the past 48 years. Multi-killowatt lasers are used to cut or weld thick section metal plates, and ultrashort and ultrabright lasers are used for precision micromachining or fusion energy study. Laser material processing has become a good sized business. How about directly converting solar energy into laser or electricity? There is active research along these thoughts [19].

Despite all these achievements, many parts of the world are still facing a serious shortage of electricity or other energy resources. This is due to the increased volume of economy, the increased world population, and the drying-out of cheap energy resources. In 2005, the world energy consumption was around 15 TW ($= 1.5 \times 10^{13}$ W), with 86.5% derived from the combustion of fossil fuels. It is estimated that the world would need an extra 10 TW in 20 years. Figure 2.10 shows the historical data of electricity production from the United States, Germany, Japan, China, and India

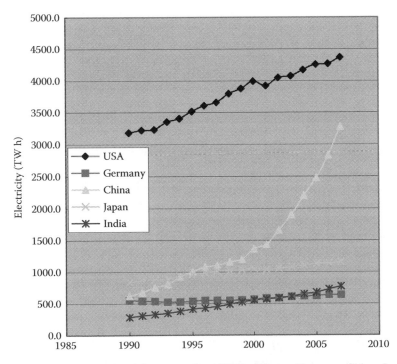

FIGURE 2.10 Historical data of electricity generation of United States, Germany, China, Japan, and India. (Based on data from British Petroleum, http://www.bp.com/sectiongenericarticle.do?categoryId=9023752& contentId=7044473.)

TABLE 2.2
Historical Data of Daily Energy Consumption

History	kcal/Day	Relative to Fundamental Needs
Primitive man, no fire, no hunting	2,000	1
Hunting man, with fire but with no agriculture	5,000	2.5
Early agriculture man	12,000	6
Advanced agriculture man with tools and domesticated animals	27,000	13.5
Early industrial man, around 1870 in Europe	70,000	35
Modern technological man, 1970 in United States	230,000	115

Source: Simon, A.L., *Energy Resources*, Pergamon Press Inc., New York, 1975.

between 1995 and 2007 [20]. All of the nations are increasing their electricity consumption, with China and India having a stronger trend of energy consumption increase.

Energy crisis and global climate change are the top challenges of our time. How can we cool down the Earth while all nations continue to consume more energy day-by-day? As engineers, what can we do to help reduce energy consumption and still improve the quality of living? While cheap fossil fuel supply may dry out in this century, energy from nuclear fission and fusion may sustain our energy needs for a long time. With the rising cost and the environmental impacts of fossil fuels and the increased energy demands, alternative energy solutions are being actively explored worldwide. Wind, solar, nuclear, geothermal, and tide wave energy are on the list.

Turning to alternative energy may not be enough. Fossil fuel economy will continue to dominate for decades and the global carbon-dioxide level has risen to very dangerous levels. We should cut back on energy consumption.

In his book *Energy Resources*, Andrew Simon made an interesting comparison about daily energy use by individuals through human history [18]. Take a look at Table 2.2. In ancient times, a primitive man needed only 2000 kcal of energy per day to be alive. Using this as the baseline, our energy consumption per day has increased quickly in the past 300 years, 115-fold that of the fundamental needs.

Discussion

How can we cool down the Earth while the world consumes more energy day-by-day? As engineers, what can we do to help reduce energy consumption and still improve the quality of living?

2.2.4 FIELD AND PHYSICAL QUANTITIES

To reduce confusion, let's define *field* in this book.

Field is the spatial and temporal distribution of physical quantities.

In physics, space and time are combined into a single construct called spacetime. By combining space and time into a single manifold, physicists have significantly simplified physical theories, and these theories have been described in a more uniform way at both the supergalactic and subatomic levels. *According to Euclidean space perceptions, the universe has three dimensions of space and one dimension of time.* The still developing string theory suggests that the spacetime has 11 dimensions (10 dimensions in space and 1 dimension in time). We will adopt the four-dimensional Euclidean space in this book.

Then what is a physical quantity?

A physical quantity is a quantifiable physical property. It can be measured and expressed with value and a specified unit. SI units are preferred in this book.

TABLE 2.3
Base Quantities in International System of Quantities (SI Units)

Name	Symbol for Quantity	SI Base Unit
Length	L	Meter (m)
Time	T	Second (s)
Mass	M	Kilogram (kg)
Electric current	I	Ampere (A)
Temperature	T	Kelvin (K)
Amount of substance	N	Mole (mol)
Luminous intensity	I_v	Candela (cd)

There are many physical quantities, however, in the SI system, there are only seven base quantities. Table 2.3 lists the base quantities and their units according to the International System of Quantities (ISQ).

Given the units of the seven base quantities, the units of the rest of the physical quantities can be derived. Those quantities are called derived physical quantities. Isn't this neat and amazing? The more we study nature, the more sophisticated it appears to us. More and more quantities are defined to describe our understanding of some part of the universe, yet, today we are using seven base quantities to describe all (or nearly all) of them, including force, energy, light intensity, color, rotation, vacuum, etc.

In this book, I would argue that we should add at least one more base quantity in Table 2.3 to include information as the additional fundamental physical quantity.

What shall we use to describe information? Not sure. This can be a future research topic. But first of all, can we treat information as a physical quantity?

A physical quantity is a quantifiable *physical property*. Then what is physical property?

Physical property is any aspect of an existence (an object or substance) that can be measured or perceived without changing its identity.

In philosophy, *identity* is whatever makes something the same or different, in other words, identity is whatever makes an entity definable and recognizable, in terms of possessing a set of qualities or characteristics that distinguish it from the entities of different types.

So to be treated as a physical quantity, it has to satisfy the following:

1. It exists
2. It has identity
3. It is quantifiable

Here is the definition of "physical" in Webster dictionary: of or pertaining to nature (as including all created existences); in accordance with the laws of nature; of or relating to natural or material things, or to the bodily structure, as opposed to things mental, moral, spiritual, or imaginary.

Based on the above criteria, in my opinion, *information is a physical quantify, and since it cannot be derived from other physical quantities, it is also a base physical quantity.* However, the laws of information and intelligence have not been fully developed; the interactions between information (or intelligence) and material are sometimes treated as supernatural. But, how could we do anything without the help of some kind of information? Information exists in nature, it is not spiritual or imaginary.

Furthermore, does information have the ability to do work? These are very fundamental questions awaiting future exploration. Experiments can be designed to prove that information can do work. This is beyond the scope of this book.

Another fundamental question is to ask whether material is a field. These questions lead to the need to propose the concept of general energy fields.

Discussion

1. What is your opinion of treating information as a physical quantity in parallel with the other familiar quantities?
2. Can material distribution be treated as energy fields? Write down two examples of the clever use of material distribution in engineering.

2.2.5   Evolution of the Field Philosophy in Physics

Energy is the ability to do work, and work is the transfer of energy from one system to another. *This conventional definition of energy and work comes in circles*—energy is defined by work and work is defined by energy.

In this book, we adopt a more general definition of energy. *In general, energy is any entity with the ability to change the state of a system. In other words, any state change of a system requires energy and involves energy.*

We have defined a field as the spatial and temporal distribution of a physical quantity. Thus, *an energy field is the spatial and temporal distribution of energy.*

Is this intuitive to you? We were and we are still used to explaining the world in terms of various forces, energies, and certain physical quantities. Actually, the field philosophy was not fully appreciated until James Clerk Maxwell (1831–1879) published the theory of electromagnetism. Note the title difference in his paper in 1861 (*On Physical Lines of Force*) and in 1864 (*A Dynamical Theory of the Electromagnetic Field*). Michael Faraday (1791–1867) was the pioneer of the field philosophy [21].

Imagine we were in the time of Michael Faraday. Faraday was an English chemist and physicist who contributed to the fields of electrochemistry and electromagnetism. In 1821, soon after Hans Christian Ørsted discovered the phenomenon of electromagnetism, many people tried to design an electric motor but failed. Faraday, just 30 years old in 1821, built two devices to produce electromagnetic rotation. These experiments and inventions form the foundation of modern electromagnetic technology.

In 1831, Faraday began his experiments that uncovered electromagnetic induction. Faraday wrapped two insulated coils of wire around an iron ring and found that upon passing a current through one coil, a momentary current was induced in the other coil. This phenomenon is known as *mutual induction*. In subsequent experiments, he found that if he moved a magnet through a loop of wire, an electric current flowed in the wire. He demonstrated that a changing magnetic field could produce an electric field. Faraday used this principle to construct his electric dynamo, the ancestor of modern power generators. Faraday went on to prove that electricity from frictional rubbing, battery, and biological bodies were of the same nature (we now take this for granted). Franklin did his famous kite experiments in 1752 to prove that lightning was due to the discharge of electricity of the clouds, similar to the discharge of electricity on the ground.

In his later years, Faraday proposed that electromagnetic forces extended into the empty space around the conductor. This mental model was crucial to the successful development of electromechanical devices for the remainder of the nineteenth century, but *this idea was rejected by his fellow scientists in his time.*

Faraday's field philosophy deeply influenced future researchers. In the 1860s, Maxwell came upon the elegant model of electromagnetism and predicted that light was also a kind of EM wave. It was the first big success of field theory in physics. Maxwell's work in electromagnetism has been called the *"second great unification in physics,"* after the first one carried out by Newton who unified the force from the stars and the Earth. In a certain sense, Newton's philosophy is force based. After Maxwell's work, field theory became an important tool in physics. Einstein pushed field theory to another height. Einstein's *special theory of relativity* reconciled mechanics with electromagnetism, and his *general theory of relativity* was intended to provide a new theory of gravitation. In his later years, Einstein attempted to generalize his theory of gravitation in order to unify and simplify the fundamental laws of physics, particularly gravitation and electromagnetism. Einstein's dream of unifying the laws of physics under a single model continues as modern scientists pursuing the grand unification theory, the string theory, etc. *Field Theory* has become a special term in physics. Readers are encouraged to dig out more on this topic if interested.

Engineering and technological innovations always progress with science. New physical laws are the important sources of technological innovations. In this book, we will try to form a general philosophy of manufacturing engineering based on the concept of a general energy field.

Discussion

1. List all the energy fields you can think of.
2. What is the role of information in doing work?

2.2.6 THE CONCEPT OF GENERAL ENERGY FIELD

We have defined *energy as any entity with the ability to change the state of a system. In other words, any state change of a system requires energy and involves energy.* This is the *generalized energy* in this book.

This is quite different from the conventional definition of energy, which claims energy as the ability to do work. Then the definition of work needs to be clarified. We reviewed mechanical energy and thermodynamic energy. Energy and work are mutually defined in this conventional definition of energy.

The concept of a general energy field is based on a fundamental assumption: some work is done, or the state of a system is changed, when a system experiences different energy conditions. In other words, when a system experiences energy condition A and energy condition A + B, the system is bound to experience some kinds of state change, no matter if it is detectable to us or not.

For example, when a piece of iron is put in air and the oxygen content is gradually increased, the iron will experience an increased rate of oxidation. Now keeping other things the same, we put stress on the metal block. Will the metal experience a different rate of oxidation? You can search the literature to find out the answer.

Living things in normal gravity and zero gravity will show many biological differences. Astronauts know it. How about a piece of metal? Will changes in gravity change the metal? Maybe your belief is not that strong when we change the object from a living thing to a piece of metal. But, the fundamental assumption of generalized energy field encourages us to have full confidence to find out the differences (the work done) whenever the state of energy changes.

The philosophy of the energy field evolved from the philosophy of force and energy. We have defined a field as the spatial and temporal distribution of a physical quantity. Thus, we define the general energy field as the following.

A general energy field is the spatial and temporal distribution of any entity (including but not limited to force, energy, mass, information and intelligence, social and financial resources, etc.) that contributes to the process of doing work or the process of changing the state of a system.

There are several important points in this definition.

First, doing work is the process of changing the state of a system. You may find that this is an extension to the familiar definition of virtual work. Virtual work on a system is the work resulting from either virtual forces acting through a real displacement or real forces acting through a virtual displacement. The process of doing work in a general energy field is not tied to the definition of virtual force or virtual displacement, although you can think in this way for convenience.

Second, the class of energy field is much broader than in the conventional definition, which only includes energy and force. Here, the distributions of material and information are also treated as an energy field. This difference will have a huge impact on the way we evaluate engineering solutions and manufacturing processes.

The distributions of forces and energy (mechanical force, EM force, ultrasonic energy, laser radiation, gravity, chemical energy, etc.) are energy fields. How about the distribution of materials? Materials and energy are closely related, with the bridge of mass-energy equivalence. Water flows from a high level to a low level under gravity, thus the ground is an entity that contributes to the state change of water—it is an energy field. Oil, coal, and dry wood are regarded as energy sources; chemical energy is stored in any material. Materials are simply stabilized or slow variation energy fields.

Material is widely used as constraints in our processes, right? Imagine your chemical reactions without a container. Think about a car running on nothing. Think about lubricants used in mechanical cutting. No doubt, materials should be thought of as stabilized or slow changing energy fields.

Treating the spatial and temporal distribution of material as an energy field is straightforward and very powerful. Yet, this is a point many people ignore in practice. People focus too much on the optimization of the familiar fields of forces.

Should the spatial and temporal distribution of information and intelligence be treated as an energy field?

We have argued that information physically exists and it should be treated as a physical quantity. Next we need to justify that information can do work and information can change the state of a system.

Think about a workshop. All the energy, materials, and hardware needed to make a product are in place. Let the workshop undergo two different conditions: (a) do the work with detailed instruction of proven production processes and (b) do the work without the knowledge of the production processes. The results will be very different. Clearly, the knowledge of production processes can bring a system (the workshop) into a different state. According to our definition of work (the ability to change the state of a system), information does contribute to the process of doing work, thus, the distribution of this entity is a kind of general energy field.

The role of information flow in engineering is becoming more and more important. We call it close loop control, intelligent manufacturing, digital manufacturing, smart products, lean manufacturing, production management, etc. However, it is rarely treated as an energy field. The information flow in engineering includes both information and intelligence. Later on in this chapter, the concept of general intelligence will be introduced.

It is interesting that very few discussions could be found regarding how information and intelligence field interact with other fields, such as the fields of energy, force, and materials. We are using such coupling everyday, but we kind of reserve information and intelligence to human beings.

Let's think about this. When heated, water will evaporate at a fixed temperature given the same ambient pressure. This is a piece of information that exists independent of human beings. All materials or energy forms possess information from nature. This can be called natural information, in contrast to the information *collected or discovered* by human beings.

Energy and mass are conserved physical quantities. How about information and intelligence? Can we *create* information and intelligence?

Information and intelligence are physical quantities. If we can create information and intelligence, then information and intelligence are very different from energy and mass. Energy and mass are conserved in a closed system.

Information and intelligence may be nonconserved physical quantities!

I wonder about the correctness of this claim. To be modest and safe, let's assume we cannot create information and intelligence, although we can collect and concentrate the field of information and intelligence (the **I–I field**), just like we cannot create energy although we can convert energy or concentrate energy.

In the definition of general energy fields, we should also include the entity of social and financial resources. When this chapter was written (in the second half of 2008), a worldwide financial crisis occurred. The automobile industry in the United States was seeking government financial aid to maintain the huge manufacturing activities in Detroit. Clearly, society and finance are important factors that affect the performance of any manufacturing activity. These are system-level factors. Although this won't be the focus in this book, it is still meaningful to include it in the definition of general energy field. This helps clear out the barriers between disciplines when we try to innovate.

In short, the definition of a general energy field relates to any factors that contribute to the change of the state of a system, be it energy, force, mass, information and intelligence, or finance and society. This is a bold extension of the conventional definition of energy field, which is limited to the distributions of various energy forms and forces.

Why shall we adopt the definition of a general energy field? Because we want to properly select the dimensions of engineering optimization.

2.2.7 THE DIMENSION OF ENGINEERING OPTIMIZATION [14,15]

Figure 2.11 illustrates the tasks to be solved in engineering. Y_0 and Y_N are the current and final state of products/processes that are functions of time t, space r, and other factors, X. The black box represents the route we need to develop to link Y_0 and Y_N in the (X, r, t) space. The task of intelligent EFM is to optimize and execute the engineering route linking Y_0 and Y_N based on the objectives, resources, and constraints from customers, manufacturing engineering, society, and environment.

Note that (X, r, t) is the optimization space. The more freedom of optimization we have, the higher chance of finding the best route. X is the general energy field. It decides the degree of freedom of our engineering solutions. There is a theoretical optimal solution given the constraints. Normally, the more fields we consider in X, the higher dimension we are in, the more solutions we can find, and the higher chance of finding the optimal solution. Integrating energy fields to find the optimal solution of engineering tasks is one of the important principles of intelligent EFM.

The reader may be familiar with some hybrid processes, such as abrasive water-jet machining, laser-assisted ceramics turning, mechanical machining with ultrasonic vibration assistance, lithography (chemical etching + light irradiation), etc.

Even if you think you are using a "pure" process, such as mechanical turning, you are still in the (X, r, t) space with X covering all the fields. Some fields are there, taking advantage of them or not. Some fields are not there, which may help if we can bring them in.

For example, gravity, atmosphere pressure, environmental magnetic field, temperature field, EM radiation field, and mechanical vibration field, etc., are effective in all engineering works. When we modulate the fields, such as applying a high voltage, focusing a laser beam, concentrating a

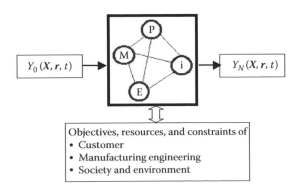

FIGURE 2.11 The black-box challenge of engineering.

chemical medium, putting on material constraints (an optical fiber, for example) etc., we change the state of the system and we guide the engineering solution to a different route.

Some engineering solutions may achieve success in a market-driven economy, but may fail badly in a sustainable economy. Y_0 and Y_N are usually the two intermediate states in a long chain of state transformations. Optimization for a local segment of the long value chain won't be sufficient for the sustainability of all chains. For example, the energy industry based on fossil fuel and the automobile industry based on internal combustion engine technology, etc., may profit in a market-driven economy for more than a century, but they may get totally retired in the next 50 years because they are not long-term sustainable solutions for human transportation.

Technological innovations drive the long-term cycles of economy. Product and service innovations generate direct market values. Both kinds of innovations ride on the wheels of process innovations, i.e., process innovations are the enabling factors for product and service innovations. The sources of innovations are either from technology push or market pull. An important point is that all technology innovations and all human activities ride in nature and are part of nature. Human beings get input from nature, change the states of the M-PIE flows, add values for human purposes, and change the state of nature. To be sustainable, we must consider the complete cycle from nature to nature.

The criteria of engineering optimization decide which solution we should use and what optimal solution we could achieve in the (X, r, t) space. We will introduce the New Criteria of Engineering Optimization (new CEO) at the end of this chapter.

2.2.8 Summary of the Philosophy of General Energy Field

Historically, we interpret the world through force, energy, and fundamental elements. With the increased sophistication of science and engineering, many branches of science and technology evolved. Sometimes we get lost in the details. For example, there are many unnecessary barriers in real life when we try to cross the borders of different disciplines. In this section, we try to pull our thinking back to the big picture, the philosophy of engineering. We used the concept of a general energy field.

A field is the spatial and temporal distribution of a physical quantity. Energy is normally defined as the ability to do work, while work is defined as the force and distance product in mechanical energy or the energy transferred from one system to another in thermodynamics. There is also the concept of virtual force, virtual work, and virtual energy. We argue that there is ambiguity in this definition of work and energy. We introduced a more generalized definition of energy. *The generalized energy is any physical entity with the ability to change the state of a system.* Physical here means to physically exist. The distribution of generalized energy is a general energy field.

Force, energy, materials, etc., are physical quantities and are entities of general energy fields. We extend the field concept to information and intelligence as well. Information and intelligence are not the privileges of human beings. They can exist without us. They are physical quantities. They can change the state of a system and we can test their effects using a different state of information and intelligence. This is the same rule we use when we study other physical quantities. Thus, they are entities of general energy fields. The entities of general energy fields should include other factors, such as social and financial resources, although these are factors at a higher system level.

Finally, we talked about the dimension of engineering optimization. In engineering, our task is to transfer a system from a starting state to the final state in the (X, r, t) space, where X is the general energy field, r is the spatial dimension, and t is the time dimension. The more freedom of optimization we have, the higher chance of finding the optimal engineering solution. All optimization is tied to certain criteria. We will discuss the new CEO after we introduce the concept of general logic functional material and general intelligence.

2.3 GENERAL LOGIC FUNCTIONAL MATERIALS [15]

Logic functional material was a concept initially proposed for virtual mold 3D manufacturing (see Section 2.1.2). Basically, it defined a category of materials with the feature of dissociation forming under the action of directional energy fields. Some directional energy fields act as the triggering mechanism to start or stop the dissociation forming process. Thus, this material is called logic functional material.

In many manufacturing textbooks, engineering materials mainly refer to nonorganic materials, including metals, metal alloys, ceramics, semiconductors, plastics, composites, and some construction materials such as stones. Organic materials, cells, and chemical and biological solutions are the focus of chemical and biomedical engineering.

There are many kinds of engineering materials and there are detailed studies on their properties and applications. We won't repeat these materials in this book, except a chapter on the general methodology of materials science and engineering. Readers are encouraged to refer to handbooks and classical manufacturing books for further information. In this section, let's first review how engineering materials evolved and then examine the concept of general logic functional materials.

Materials are an inseparable part of human history. The evolution of engineering materials shows an interesting trend.

2.3.1 THE EVOLUTION OF ENGINEERING MATERIALS [2]

The influence of materials on society is so fundamental that historians termed our history in names of materials, such as the Stone Age, the Bronze Age, the Iron Age, the Steel Age, etc. We are living in the Synthetic Materials Age and we are transitioning from synthetic functional materials to future smart materials and structures.

As shown in Figure 2.12, materials in the first generation were materials directly taken from nature and were used by early humans in the Old Stone Age. *We call these natural materials.* Major representatives include stones, bones, woods, fibers, skins, natural colors, and things to make fire. Humans learned to make various tools for hunting, clothing, and cooking. Flint stones were used as knives, spearheads, arrowheads, etc.

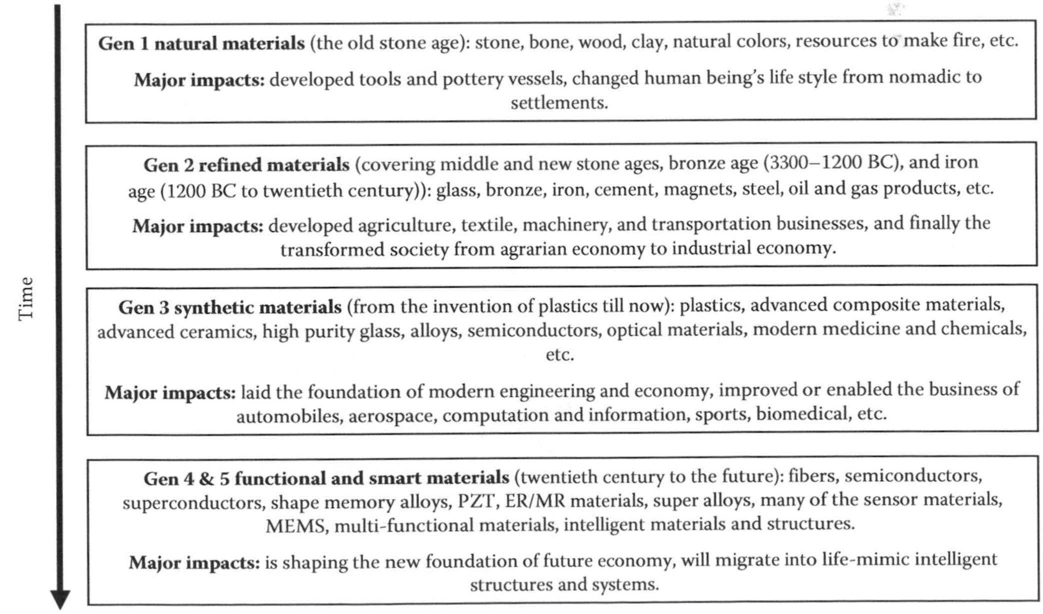

Gen 1 natural materials (the old stone age): stone, bone, wood, clay, natural colors, resources to make fire, etc.

Major impacts: developed tools and pottery vessels, changed human being's life style from nomadic to settlements.

Gen 2 refined materials (covering middle and new stone ages, bronze age (3300–1200 BC), and iron age (1200 BC to twentieth century)): glass, bronze, iron, cement, magnets, steel, oil and gas products, etc.

Major impacts: developed agriculture, textile, machinery, and transportation businesses, and finally the transformed society from agrarian economy to industrial economy.

Gen 3 synthetic materials (from the invention of plastics till now): plastics, advanced composite materials, advanced ceramics, high purity glass, alloys, semiconductors, optical materials, modern medicine and chemicals, etc.

Major impacts: laid the foundation of modern engineering and economy, improved or enabled the business of automobiles, aerospace, computation and information, sports, biomedical, etc.

Gen 4 & 5 functional and smart materials (twentieth century to the future): fibers, semiconductors, superconductors, shape memory alloys, PZT, ER/MR materials, super alloys, many of the sensor materials, MEMS, multi-functional materials, intelligent materials and structures.

Major impacts: is shaping the new foundation of future economy, will migrate into life-mimic intelligent structures and systems.

Time

FIGURE 2.12 The evolution of engineering materials.

The use of tools and fire led to a chain of social changes. People lived healthier and longer. With improved weapons, people had captured more animals than they could eat in certain times. There grew a need to exchange tools and other products. The firing of clay into hardened utensils was a big milestone in the history of engineering. The soft clay was available everywhere. When fired, it turned into porcelain. People invented the process of molding and casting to make useful tools. So after using knives to cut or carve something, the thermal energy was used to transform materials from one state to another. The curiosity of transforming materials with fire continued, thus, came the Bronze Age and the Iron Age.

The biggest impact of the Natural Material Age is that humans started to settle down in villages and cities, agriculture was developed, and merchandise exchange took shape.

The second generation of engineering material spanned the majority of human history from the Middle Stone Age to the Bronze Age (3300–1200 BC) until the Iron Age (1200 BC to the twentieth century). In this period, many new materials were made from the refining of natural resources and we call them *refined materials.*

In the exploration of using fires to smelt various minerals, glass, bronze, iron, gold, etc., were discovered. Bronze is the alloy of copper and tin. It is the oldest alloy utilized by mankind. Tools, weapons, sacred vessels, etc., were made from bronze in this period. Figure 2.13 shows the two-handled bronze *gefuding gui* from the Chinese Shang Dynasty (1600–1046 BC). The major manufacturing process is casting. It is amazing to see the level of skills humans achieved in metallurgy and in manufacturing.

Due to the higher smelting temperature of iron ore than copper ore (~1537°C vs. 1037°C), iron was refined later than bronze. But iron ore was cheaper and far more abundant than copper. Once the technology of fabricating iron was established, it spread rapidly. Iron products were wrought (worked) rather than cast. They were harder, stronger, and cheaper than their bronze counterparts. Iron-based materials dominated human history until the era of synthetic materials. Many mechanical tools, devices, and systems were developed based on iron materials. The other materials important in the Iron Age are hydraulic cement and natural magnets. Iron continued to be the dominating material until the widespread use of steel. Steel is an alloy of iron consisting mostly of iron, with a carbon content between 0.2% and 2.14% and other alloying elements such as aluminum, magnesium, cobalt, chromium, nickel, tungsten, silicon, etc. Steel could overcome many of the weaknesses of wrought iron. The economy related to iron and steel triggered the Industrial Revolution in Britain (later eighteenth century to early nineteenth century) and was repeated in other nations until the twentieth century.

The Industrial Revolution marked the quick evolution of technology in all areas, including materials science and engineering, industrial and mechanical manufacturing, chemical engineering, and biomedical engineering, etc. Scientific data accumulated and technology evolved into finer and

FIGURE 2.13   The two-handled bronze *Gefuding Gui*, from the Chinese Shang Dynasty (1600–1046 BC). (From Wikimedia Commons.)

finer disciplines. The social economy also migrated from agrarian to industrial. New machines, new energy sources, new ways of transportation (automobiles, railways, airplanes) and communication (telegraph and telephone), new modes of production (standardization and mass production in factories), and new military forces dictated the shake-up and redistribution of world wealth and power. Oil and gas became strategic energy resources at the end of the Refined Materials Age and our economy finally became an oil-based economy.

For the first time in history, human activities could make a significant impact on the environment. Human territory expanded quickly to all corners of the world. Our impact nowadays is on a global scale.

The third generation of engineering material is called the *synthetic materials.*

Until the beginning of the twentieth century, natural materials were still the dominant resources in economy. We utilize natural materials with their known weaknesses. There were some efforts to remedy the natural weaknesses, as in the case of steel. But such efforts were not systematic. This philosophy changed with the era of synthetic materials, starting from plastics. Celluloid, xylonite, cellophane, etc., were synthesized by 1910. The volume of plastics produced in the United States exceeded that of steel in 1979.

The material properties of synthetic materials can be tailored to specific applications. Such efforts are science based and resulted in the replacement of traditional engineering materials with synthetic materials, such as plastics, ceramics, and composites. There are tens of thousands of plastics in the market. The wide use of plastics in automobiles helped reduce the body weight and thus improve the oil mileage and the wide use of composites enabled the upgrade of airplanes. The use of high temperature alloys and ceramics greatly improved the safety and efficiency of engines. Synthetic plastics have become an inseparable part of the packaging and household appliances industry.

Other synthetic materials include high-purity glasses, semiconductor materials, and modern medical and chemical products. Optical fibers, computers, and the Internet were invented and brought a new era of information transfer. Optical glasses and single crystals laid the foundation for laser production. Composite materials helped in breaking many of the world records in sports. Biomedical compatible materials initiated a new business, such as artificial joints, implant organs, etc.

In short, synthetic materials laid the foundation of the modern economy. Human beings used these materials to extend our territory to deep seas and deep space. Our impact on the environment became increasingly global.

We are entering the Age of Functional and Smart Materials with the fourth generation of functional materials and the fifth generation of smart materials. It is difficult to draw a definite line between the fourth and fifth generations, thus we combine them together.

Some materials showed useful functionalities. For example, shape memory alloys can take different shapes under different temperatures, piezoelectric materials can convert external mechanical force into electric signals or vice versa, semiconductor materials show special electric performance that can be used to manipulate electric signals, nonlinear crystals can change the wavelength of light, special alloys can withstand extremely high temperature and can survive harsh environments, and superconductors can transmit electricity with negligible resistance, etc. These are called functional materials or smart materials.

With the developments in computation, micro electron mechanical systems (MEMSs), and biomedical engineering, human beings are increasingly integrating information and intelligence with energy and material flows in manufacturing. The philosophy of materials science and engineering has evolved from picking the best natural material for engineering, to refining the best natural materials for engineering, to systematically synthesizing materials for engineering, to designing and imbedding functions and intelligence in materials for manufacturing. We are learning from nature, trying to develop life-like intelligent structures and systems. The Age of Functional and Smart Materials has begun and this will shape the future economy. Imagine the future engineering materials with the capability of self-sensing, self-actuating, and self-repair. This direction is still in its infancy and it is highly interdisciplinary in nature. More details of smart materials and structures are discussed in the book of Dr. M. V. Gandhi and B. S. Thompson [2].

In this review, we noticed the integration of the material, energy, information, and intelligence flows in materials science and engineering. This is no surprise, since this integration is a general trend in engineering.

Discussion

Find out more about smart materials in literature. How many smart materials can you list?

2.3.2 All Materials Are Logic Functional Materials

All materials are logic functional materials. We call this the generality of logic functional materials.

We just reviewed how engineering materials evolved from natural materials to refined materials, synthetic materials, functional materials, and toward smart materials and structures. Now let's ask two interesting questions.

Q1: Is water a smart material?

To be regarded as a smart material, it needs to show the capability to sense the environment, process the collected information, and react accordingly.

How could we tell that a material can sense and react to external excitations? As long as the output is different from the input, a system reacts to the input.

You may argue that many materials and structures can only react passively, without a "purpose." With the same excitation, it gives the same response. Thus, they are passive, not smart. (*Passivity* is a property of engineering systems that is commonly used in electronic engineering and control systems. A *passive component* may either refer to a component that consumes but does not produce energy or to a component that is incapable of power gain.)

To keep the consistency in terminology, we won't change the definition of smart materials. In the next section, we will introduce the concept of general intelligence, pointing out that intelligence is not the privilege of human beings.

Here, it is sufficient to point out that water is a logic functional material, if it is not a qualified smart material in your mind. To be regarded as a logic functional material, it needs to have three states: the normal state, the excited state, and the qualitatively changed state. The conversion from the excited state to qualitatively changed state is under the control of some kind of energy field.

Water is one of the fundamental reasons that life exists on the Earth. We are intelligent, then, does water contribute anything to our intelligence? Under room temperature and one atmosphere pressure, water is liquid. When the temperature goes beyond 100°C, vaporization starts. Thus, one can use a selective energy field, such as a laser beam, to locally vaporize water. There are myriad applications of water. Each application is actually a logic function of water. For example, water in a container has weight and volume, so it can be used to stabilize a structure. Its weight and volume are dependent on the energy fields surrounding it. It may explode if it is heated too much in a closed container. A phase change of water can absorb or give off a lot of heat. Playing with the location of heat/work exchange, human beings invented the steam engine. Water in a liquid or gas state can flow a long distance under pressure, thus it can be conveniently used to heat a building.

Well, things get more and more complex when we talk about so many applications of water or matter in general. But it is simple in philosophy. All materials have logic functionality or physical and chemical properties as we normally call it. Metal, water, and air can be compressed, but they change differently under the same external pressure increase. Here, we treat any material as logic functional material. In a certain sense, we "highly respect" their physical and chemical properties, and link these properties to potential functionalities. We treat materials as intelligent existence, as things with logic functionality, although their intelligence may be lower than that of human beings.

Q2: Who was the earliest inventor of composites?

Composite materials are materials made of two or more constituent materials with significantly different physical or chemical properties that remain distinct on a macroscopic level within the finished structure. Advanced composite materials enabled human beings to fly back and forth from space. Space shuttles and fighter airplanes use a lot of composite materials. These materials are strong and light and some of them can survive a high heat load. In the case of a space shuttle, the temperature may go beyond 2000°C. Composite materials are used in daily life as well. Your bathtub is very possibly made of composite materials.

But who was the earliest inventor(s) of composite material?

The earliest record of man-made composite material was possibly the dried straw and mud mixture used to form bricks for building construction, as described in the Biblical book of Exodus. So the honor possibly went to the ancient Egyptians.

However, human beings are not the earliest inventor of composites at all. Look at the cross section of a small twig. Wood actually has a structure comprising cellulose fibers in a lignin and hemicellulose matrix, it is a fiber-reinforced polymer composite. Our bones are also examples of composite structures. So are our legs, arms, hands, ears, and noses!

Composite structure is the foundation of life.

Nature is the true inventor and master of composites.

Human beings use composites to meet very challenging engineering requirements, but the origination of the invention is from nature. This is a pervasive observation. We are learning from nature, mimicking nature. Even the physical laws are the logic that nature uses.

In summary, all materials are logic functional materials. We call this the generality of logic functional materials. All materials have properties that can be revealed under certain energy field conditions. Or all materials are "smart and intelligent materials" to certain degrees.

Now we are ready to discuss the third fundamental concept in intelligent EFM, the general intelligence.

2.4 GENERAL INTELLIGENCE

2.4.1 DISCUSSION WITH MY FRIEND

I was driving with my friend, Tao, during a business trip. It was evening time. Nothing to see outside, but inside the car, we had a heated debate.

Tao and I were in the same research team studying super-hydrophobicity. It is well known that lotus leaves are super-hydrophobic, which means the leaf surface repels water strongly. Water won't wet the lotus leaves—water actually rolls on the leaves. Because of this, lotus leaves are self-cleaning. After a rain shower, dusts fall off with the water droplets. The lotus leaf has microscale mounds and nanoscale hair-like structures, as shown in Figure 2.14 [22]. In our project, we want to create super-hydrophobic surfaces directly on metals so that we can make the blades of aircraft engines super-hydrophobic, reducing drag, reducing the adhesion of ice in cold weather, enhancing heat transfer, etc. It is amazing that so many applications can be found based on this nanostructure.

I asked Tao whether the structure we were studying was intelligent. Without hesitation, Tao replied that he didn't think so. Well, this triggered our heated discussion of intelligence.

"Wenwu, first thing first, what is your definition of intelligence?" Tao asked.

"Well, I am not fully satisfied with my definition yet. But I don't think intelligence is the privilege of human beings. To me, all living things have certain intelligence, and all materials and energy configurations have intelligence. I define intelligence as the ability of patterned response to surrounding energy fields."

"Patterned response?"

"Yes. A system, given the similar excitation, will show a similar response."

FIGURE 2.14   The microstructure of lotus leaves.

"Aha. To me, intelligence means the ability to think, to understand, to reason, to invent things that are nonexistent, and to react based on intuitive judgments. You make different decisions when you are given the same choices, that is intelligence, not the opposite, man!"

I know what he means. Go to a restaurant, you eat different dishes from time to time, that is smart, while eating the same dish every time is ridiculous, if not stupid.

I replied: "That's why I say it is a *patterned* response. That is your pattern of eating. That is part of our intelligence. You make different decisions under the same options, but are you sure everything is the same when you make these decisions? These differences will influence your decisions. So my patterned response definition is still valid."

"It doesn't include high-level intelligence, such as intuition and invention."

"Let's start from something simple. Suppose, you accompanied your girlfriend in a walk along the bank of a river, none of you fell into the river. It rained, but none of you got wet. We say both of you are intelligent. Now, a dog accompanied a blind girl walking along the bank of the river and didn't fall into the river, and didn't get wet because the girl used an umbrella while the dog had his self-developed super-hydrophobic fur. We say both the girl and the dog are intelligent, right?"

Tao was a little mad at my words. He retorted: "Exactly, except you accompanied me in this trip."

"Well, I don't care what you say, this is serious discussion. You know, I was amazed at the beauty of crystallization. Isn't this process smart? Atoms form various patterns under certain conditions."

"I know what you want to say, it is *patterned response*. But it is a passive process—the atoms react *passively*, without a predetermined purpose. We say something is intelligent because it knows how to achieve certain goals with thinking and reasoning."

"You want to say 'cognitive thinking', right?"

Tao nodded.

"That is the same response I got from many people when we discussed this concept. Some people agreed with me immediately, while others like you persisted that only human beings were intelligent."

"No, you and your dog." Tao wouldn't forgo any chance of getting even.

"But, we are too arrogant to assign the purpose of human beings to other existences. Of course, atoms have their own purposes; they want to stay in the most stable energy condition all the time, right, chemistry dude?"

"Mechanical dude, you can give a new definition to intelligence, but it is very different from the one people are familiar with. The bottom line is whether it is meaningful in engineering."

"I know it is different. That's why I call it general intelligence, a concept that can be used for materials, energy fields, and living things. Then it is easier to break the barriers between these systems and learn from each other."

"You are very imaginative, but I am still not fully convinced."

I paused, then said: "In your opinion, how do you become smarter than an atom, a tree, a dog, and me?"

"What do you want to say? Be straight."

"I want to say we are relying on nature and we are part of nature. We are made of small units, although we are a big, advanced system. If all elements are absolute zero in something, how can the assembled system become nonzero later on? If your body elements have no intelligence, you won't be intelligent at all, smart man."

"There are many loose ends in your argument. I wish you good luck. Watch out!"

We entered New York City. The traffic ended our debate.

As this book was written, I got more prepared for the concept of intelligence. In the dictionary, "intelligent" means having the capacity to understand, to think, and to reason, especially to a high degree. In academic research, intelligence is used to describe a property of our mind that is related to the capacities to reason, to plan, to solve problems, to think abstractly, to comprehend ideas, to use language, and to learn. In the theory of multiple intelligences, Howard Gardner defined eight categories of core intelligence: linguistic, logical-mathematical, spatial, bodily-kinesthetic, musical, interpersonal, intrapersonal, and the naturalist intelligence [23].

The other theory, called the theory of general intelligence (same terminology as ours), regards intelligence as the ability of cognitive reaction to given tasks. Cognition is related to the processing of information, applying knowledge, and changing preferences. Cognitive processes can be conscious or unconscious. Cognition is commonly considered an abstract property of advanced living organisms. It is studied as a direct property of a brain (or of an abstract mind) at the factual and symbolic levels [23].

In summary, there are basically two theories of intelligence: general intelligence and multiple intelligences, but the current research of intelligence regards intelligence as the property of advanced living organisms and measures intelligence relative to human beings.

In this book, we will introduce the concept of general intelligence in the context of intelligent EFM, extending intelligence to all systems.

Intelligence is not the privilege of advanced living organisms.

Discussion

To inject intelligence into something (this is what we do in engineering), the material and structure have to have the elements of intelligence, allowing the state of information to transit from 0 to nonzero. Do you agree with this argument? Analyze several examples.

2.4.2 General Intelligence [14,15]

In this book, *general intelligence is the ability of driving (receiving, processing, and generating) the information flows through a patterned response, or the rule- and logic-based response, to relevant energy fields.*

In short, intelligence is the ability to drive the information flows. Here, driving includes the activities of receiving, processing, and generating information flows. In this definition, we avoid the need to say learning, interpreting, reasoning, and inventing, which are too much tied to human beings or living organisms.

According to this definition, intelligence is naturally not the privilege of advanced living organisms. All entities, including but not limited to material and energy systems, have a certain level of intelligence. I am not sure whether intelligence can exist without the coexistence of materials and energy fields. Further discussion of this point is beyond the scope of this book. It is sufficient to say that any engineering entity, or any system or configuration of material and energy, has intelligence.

The intelligent level of entities can be compared or potentially quantified based on the same standard, the performance in driving the relevant information flows. So we can measure the performance using the skill of receiving, processing, or generating.

In this sense, we say piezoelectric material is a smart or functional material because it has higher performance relative to other materials in the receiving and processing of external pressure or electricity stimulates. Similarly, we quantify people's intelligence level through their performance in learning and usage of information (knowledge in human beings' words). Technically, human beings' intelligence may be more advanced than some existences, but there are no fundamental differences.

Note also that the driving of the information flow is through the patterned, or rule- and logic-based response to relevant energy fields. The energy fields are the general energy fields defined in this book. Remember? *The general energy field is the spatial and temporal distribution of any entity (including but not limited to force, energy, mass, information and intelligence, social and financial resources, etc.) that contributes to the process of doing work or the process of changing the state of a system.*

In engineering, there are three fundamental flows: the mass flow, the energy flow, and the information flow. The three flows are coupled and interact with each other. A configuration of mass and energy carries information with it and reacts to internal or external energy fields. The response to energy fields has patterns, rules, or logic. These patterns, rules, or logic are part of the intelligence of the configuration.

In this way, physical laws governing energy–material interactions are part of the intelligence of an energy-material configuration. Fundamental particles have intelligence and we are built upon them to have advanced intelligence.

A patterned response means a repeatable response given the same states and the same energy fields. Such patterned responses can be beneficial or harmful to our objectives depending on how we use them.

For example, heating a structure then cooling it down can induce thermal stress inside the material. Thermal distortion is a big issue in welding, especially in large structure metal welding such as the welding done in the shipyard industry. Thermal stress is a negative factor in the welding process. However, thermal distortion can be used in a controlled way, as in the process of laser forming, which purposely uses thermal stress–induced plastic deformations to transform a metal structure into desired geometries [24]. The metal tubes formed by laser bending are shown in Figure 2.15. Laser tube bending is a noncontact process; no hard tooling is used, no wall thinning happens, and the deformation can be precisely controlled. Thus, for the metal tube in Figure 2.15, we brazed an enhancement piece on the stainless steel tube first, then laser bent the two sides of the brazed structure. Such work is very difficult for conventional tube bending, which needs spare space to hold the tube and normally requires constant outside diameter (OD) of the tubes.

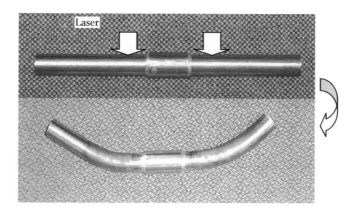

FIGURE 2.15 Laser bending of metal tubes.

Suppose we have to make the huge metal structure of a warship. After many assembling and welding processes, some critical position of the structure has certain dimensional mismatches that are unacceptable. There is no guarantee that redoing it can reduce the dimensional mismatch and it is too expensive to redo everything after the structure has been put together. Portable laser forming, or more generally, portable thermal forming, can be used to gradually bring the structure within specs. If we know how to apply the necessary thermal load, we are able to develop an intelligent thermal forming process, leading to striking manufacturing savings. Portable automated bulkheads and hull straighteners or PAS-B/H is now commercially available [25].

In the research of laser forming, people tried to find the mechanism of thermal distortion and then tried to use thermal distortion in beneficial ways. The intelligence (knowledge, rules, and logic) in laser–material interaction was gradually grasped and finally people were able to drive the process—get the initial conditions of a metal structure, process this information, decide the necessary treatment, and finally implement the treatment.

Material and energy systems are by themselves receivers, processors, and actuators of information imparted to them. For example, the material properties of some materials change quite a bit when the environmental pressure or temperature changes. These materials can sense the change of certain fields and they react according to physical laws. Thus, they make their decisions (process information) according to their logic, rules, patterns, or intelligence.

Is this definition of intelligence free of loose ends? I am not sure. It covers the intelligence of living organisms and nonorganic existences, and it covers the intelligence of both low level and high level. Physical laws, scientific discoveries, statistics, and empirical observations are patterns of energy–material interaction. Close-loop control, adaptive control, artificial intelligence, neural networks, etc., are means of improving our capability of driving the information flows in engineering.

The bottom line is that this concept of general intelligence is useful in engineering. Intelligence is no longer the privilege of human beings or advanced living organisms. All entities have a certain level of intelligence. The task of engineers is to capture the general intelligence of relevant entities, inject the intelligence of human beings into the entities, and drive the information flow in favor of human beings.

Let's be clear that the information and the intelligence of entities could exist without us. Entities drive information flows in their ways. We couple our intelligence into their processes and guide them into our ways. We can work against their intelligence (rules, logic, or pattern) or work with their intelligence. This can make a huge difference in real life. For example, making a sharp turn at a low speed might be fine, but don't do it when you are driving a car at a high speed, especially when it snows. Another example: If you want to block the heat transfer between two environments, don't use good conductors. Use porous and nonconductive materials instead.

Discussion

Compare the intelligence of an atom, a wooden door, a tree, an animal, and you. With the help of general intelligence, do you see any common features when they respond to a gust of wind?

Do you have questions regarding the definition of general intelligence? Feel free to communicate to me.

Having discussed the generality of energy fields, logic functional materials, and intelligence, it is time to introduce our new CEO.

Who is he? Keep reading.

2.5 SUSTAINABILITY AND THE NEW CRITERIA OF ENGINEERING OPTIMIZATION

2.5.1 The Power of Market-Driven Economy

Our lab just purchased a solid-state Nd:YAG laser for the study of high-speed micromachining. It costs $110K. It has an average power of 30 W at 532 nm wavelength. I was very excited about its potential applications because I finally upgraded our micromachining tool from 2 to 30 W. The International Congress on Applications of Lasers & Electro–Optics (ICALEO) is one of the most important international conferences on lasers. One hot topic in the conference has been how we could expand the business of laser applications. In 2006, the global laser business had a revenue of ~$6 billion, while Disney World in Orlando (the hosting site of ICALEO'06) had a revenue of ~$23 billion! Aren't lasers quite expensive and high tech? Yes, considering their size and power. The trouble is that a medium-sized laser vendor would be very happy if they could sell over a hundred laser models in one year. There is intense competition and there are many laser vendors selling similar products at small quantities. Many well-known laser companies are companies with less than 100 employees.

On the other hand, I purchased my first car when I graduated and purchased another car for my expanded family 2 years later. Both cars are less than $25,000. My car is 100 times larger than my new laser, uses more materials, and is actually a far more complex integration of various technologies. Think about all the fancy functions in a car. It is a matured industry, but it uses lots of advanced technologies—streamlined car body, sound system, GPS, various sensors, and state-of-the-art engine technologies. It is amazing in many ways. It is a very efficient system—given one gallon of gas, it can carry tons of load 25 miles; it works robustly in all seasons; and it is affordable for many families. It is not rare to hear of a student buying a second-hand car for less than 500 bucks. You may be surprised, at least I was surprised, to know that the automobile industry consists of over 80% of the U.S. manufacturing business. General Motors generated revenues of $193.5 billion in 2004 and Ford Motor Company generated revenues of ~$170 billion in 2005 [26]. Well, in the 2008 crisis, both companies were seeking government help and their quarterly losses were in the billions of dollars.

Let us take a look at the computer industry. Computer technology was high tech before the 1970s. At that time, only military and other big organizations could afford it. But after the marketing of personal computers exploded in the 1980s and with the age of the Internet in the 1990s, the computing power of computers kept increasing while the price kept dropping or stayed the same. With $500 in 2008, you could buy a very good computer with a 19″ LCD screen, 3 GHz microprocessor, rewritable DVD, >40 GB hard disk, and so on. It is better than the super-computers in the 1980s. In 1986, the supercomputer Cray XMP-48 had 4 CPUs, 64 Mb of memory, and a CPU frequency of 112 MHz, with a theoretical peak performance of 800 Megaflops. In 2008, the number of personal computers in use worldwide hit one billion. About 180 million PCs were expected to be replaced and 35 million were to be dumped into landfills in 2008 [27,28]. The leading giants of PCs have revenues greater than $100 billion.

The above examples indicate that technology complexity is not the major obstacle to the widespread use of a technology as long as humans can justify the product in a market-driven economy. If the products are profitable and the potential market is big, the relevant business can attract enough social resources to quickly perfect the technology. The cost of manufacturing decreases with the maturation of technology. You can easily list more examples. A Christmas light set with 1000 light emitting diodes (LED) sells for just $5? The ways that LEDs, computers, and automobiles are made are very different from that of lasers.

The world economy in 2008 was like riding a roller coaster. The gas price peaked at $147 per barrel in mid-year, but quickly dropped to below $40 per barrel by Christmas time [29]. A worldwide economy crisis occurred. We are confident in our ability of conquering the technical difficulties, but we are not sure of our future and the future of our planet.

2.5.2 THE CONCEPT OF EARTH 3.0

I went to Beijing in April 2008. The city was using any possible measures to bring back the blue skies for the sake of the 2008 Summer Olympics, including temporarily shutting down high-pollution plants, allowing only even or odd numbered cars on the road each day, etc. Isn't a blue sky a free gift from mother Earth? I went back to my hometown in central China as well. All the houses of the village had been rebuilt in the past 20 years, and the village had expanded three times into the original fertile farming fields. More farming fields were being changed into factories. Air pollution was not too bad in my hometown, but water shortage and water pollution were alarming. In many places in China, the economy has taken off in the past 30 years at the heavy price of environment degradation.

I went back to the United States and discussed this with my American friend. He said: "That is no surprise at all. Guess why many towns in many countries were built along rivers, including ours? Because the factories could get free water from the river, dump anything into the river, and transport products cheaply along the river. Guess what this area was like 40 years ago. It was worse than your hometown. River was the drain of anything. I grew up here. I am lucky I have not got cancer yet. Only recently did we protect rivers from serious pollutions. No one dares to eat the fish in this river, even now! China is repeating some of our errors."

The beautiful river in front of us changed color at that moment.

I recommend people read the October 2008 issue of *Scientific American*. It was titled *Earth 3.0*. Building Earth 3.0 should be a movement of our planet. As the editor pointed out, this planet is no longer simply the home of our species—it is also our creation [30].

Earth 1.0 was the world that persisted and evolved for billions of years, up until the Industrial Revolution. In Earth 1.0, the environment was dominated by the balanced ecological loops, and some geological and astronomical processes. Life was highly sustainable. Human beings' development of agriculture considerably enlarged our footprint on the environment, but the overall influence was fairly small and localized.

That has changed with the arrival of Earth 2.0, which starts from the Industrial Revolution until now. In this period, humans harnessed more energy forms and achieved unprecedented health and prosperity, but at the price of wanton consumption of natural resources. The world population increased from one billion in 1810 to 6.5 billion in 2006 [31]. Today humans have become the major drivers of potentially disastrous climate change.

The freezing rain in northeastern United States in December 2008 knocked out the power of millions of homes, including mine. The experience was miserable, but it was no comparison to the disasters that happened to baby Antarctic penguins. Temperatures in the Antarctic have risen by 3°C over the past 50 years to an average of −14.7°C and rain is now far more common than snow. Tens of thousands of newly born penguins were freezing to death as Antarctica experienced freaky rainstorms in July 2008. Imagine going through 6 days of freezing rain without the protection of clothes or wearing wet frozen clothes.

Is Earth 2.0 a glorious time for human beings? Maybe. But it is disastrous for other species on Earth, and it is not sustainable for human beings in the near future.

Earth 3.0 is the new Earth we need to establish, one with the prosperity of Earth 2.0 but also the sustainability of Earth 1.0. The right solutions must address both environmental and economic concerns rather than sacrificing one for the sake of the other.

In recent American Society of Mechanical Engineers/Manufacturing Science and Engineering Conference (ASME/MSEC) annual conferences, many invited speakers argued about what was the most fundamental challenge of manufacturing engineering, or engineering in general. Some people picked solving the energy crisis, since once the energy crisis is solved, many other issues disappear. Some people talked about the challenges of globalization, outer sourcing or inter-sourcing. Some talked about lean manufacturing and the agility of manufacturing. A senior NSF officer claimed it was the role change of engineers from decision-followers to decision-makers. I argued that it was none of the above. The most important challenge of engineering is to urgently migrate our criteria of engineering optimization (CEO) from the old unsustainable one to the new sustainable one [15].

To do this, we need to understand the inherent shortcomings of a market-driven economy, then introduce new genes into our economy system. In the end, engineering is a game of decision-making. Decision-making is based on certain criteria. Establishing the new CEO is the core value of intelligent EFM.

In short, we need our new CEO for Earth 3.0. Building Earth 3.0 is the urgent task for all human beings.

2.5.3 The Criteria of Sustainability

With globalization, most organizations have been forced into a cycle of continuous innovation of products, services, and production processes. Innovations can be high risk. A majority of the innovation initiatives may fail, wasting time and resources. But only organizations that innovate continuously can survive in the long term. This is due to the market impact of technology life cycles, as we discussed in Chapter 1.

The continued industrialization of the world has changed our society and environment dramatically. The economy in many nations is market driven, often referred to as a free-market economy. A market-driven economy normally judges the success of an activity on how much market value it produces. Accordingly, technology innovation in a market-driven economy would set its goals on maximizing the market value for specific groups and organizations.

Let's take a different look at the innovations of internal combustion engines. The automobile industry was the major manufacturing activity in the twentieth century and the innovations in internal combustion engines greatly reshaped the world economy map. Because the internal combustion engine helped gas-burning-based cars winning the competition over electricity-based cars in a market-driven economy, the United States consumes around 19.6 million barrels of oil per day, which is more than 25% of the world's total [32]. Once oil-based technology dominated the economy, special interest groups were formed. They set arbitrary obstacles for competing and more environment-friendly solutions of transportation, as evidenced by the sporadic activities in electric car development after the 1930s.

Because of this reality, the aftermath of technological innovations in a market-driven economy can be disastrous in many aspects if the big ecosystem of nature is not considered.

Having realized the serious impacts, sustainability has become a hot topic in many realms, such as agriculture, fishery, water, air, weather, ecosystem protection, etc.

Sustainability can be defined as the ability of a system to function usefully and indefinitely.

The most beautiful thing about nature is that it consists of many sustainable and well-balanced cycles. The sustainability in nature is featured by at least two things: (a) cycles of resources, so that there is repeatability in resource utilization and (b) the balance of such cycles—the cycles are well balanced and can last indefinitely if they are not disrupted by extreme conditions, such as large-scale human activities or exterior space disasters [33].

Engineering has picked more conservative criteria regarding sustainability.

Engineering optimization is focused on cutting resource consumption per unit, increasing production volume and speed, and improving the performance of products. *As long as an engineering activity is cost-effective in the current market environment, it is considered to be sustainable.*

Due to the influence of a market-driven economy, the pursuit of sustainability in engineering is basically focused on how an organization or business can continue the manufacturing of certain products or providing certain services while maintaining a good market margin or a better market margin than the competitors. When some engineering activities consume less resources, they are labeled *green*.

Clearly, the above engineering strategy missed both features of sustainability in nature—cycles of resources and the ability of indefinite life. Green in engineering is only a relative concept, which is way too conservative to lower the risks of technology innovations in a market-driven economy.

Discussion

Compare the life cycles of plants, animals, and engineers. Engineers are normally the worst performers among the three categories judged by how much they contributed to or negatively affected the sustainability of the ecosystem. In a certain sense, engineers and our engineering education system should feel guilty. Do you agree?

2.5.4 THE INHERENT SHORTCOMINGS OF MARKET-DRIVEN ECONOMY

So what has gone wrong with the market-driven economy regarding the challenges of sustainability?

Figure 2.16 shows the cycles in a market-driven economy. At the center are the enterprise systems, which get energy and resources from nature and use technology and innovations to serve society with either products or services. These products and services compete in the market; the market performance is used to decide whether to increase further resource investment, with the winning ones getting more resources. This is basically a cycle of positive feedback. The enterprises also produce side products and waste a certain percentage of energy and resources; these, combined with the retired products and services, will eventually go back to nature.

The big issue is that this backflow into nature in the current market-driven economy is not effectively linked to the other parts of the cycle; thus, with the cycle of positive feedback of the enterprise systems, the current style of economy is theoretically unstable and unsustainable.

Note that current engineering optimization is normally the optimization within a segment of the value chains shown in Figure 2.16. Success is judged by market performance. Despite the many efforts to achieve "*green*" or "*relative environment friendliness*" in engineering, the reality is that "green" products/services won't survive if they don't have advantages in market competition.

The root cause of the above issue is the market-driven economy. In a market-driven economy, the market is the dominant driving force. The market is very powerful in perfecting winning technologies, but unless evolved in time from the current market-driven economy to a sustainable economy in the future, damage to the sustainability of ecosystems and other natural cycles would continue to degrade until they are totally unsustainable.

Since technology innovations are the major driving forces in modern economy, such urgent conversion can only be realized through technology innovations based on the new philosophy of sustainable economy.

All the activities of human beings take resources from nature and finally return the used resources back to nature. A sustainable economy is an economy with the same definition of sustainability as other nature cycles, i.e., it recycles 100% of the used resources, and it can last forever under normal conditions.

We argue that a practical route toward the above transition is to gradually establish a market-driven sustainable economy.

Basically, we choose to maintain the market-driven nature of economy so that the basic frame of economy is not seriously disrupted. At the same time, it is not a free-market economy; it is an

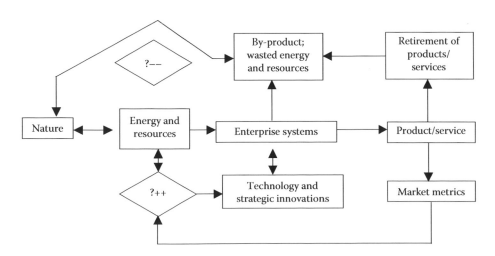

FIGURE 2.16　The cycles in market-driven economy.

economy constrained by the sustainability requirements of the general public and the whole eco-system of nature.

2.5.5 THE NEW CRITERIA OF ENGINEERING OPTIMIZATION FOR MARKET-DRIVEN SUSTAINABLE ECONOMY

New methodologies and new philosophies are needed to establish a market-driven sustainable econ-omy. Intelligent EFM is one such methodology and philosophy, with the new CEO as its core value.

One practical way is to use the market to drive the sustainability of economy by effectively con-sidering the backflows into nature and modulating the in-take from nature.

Let's introduce several concepts before defining the new CEO.

Direct market value (DMV): DMV is the direct sales revenue of a product or service, measured by money value. The higher this value, the bigger the market share. But higher direct market value may not necessarily mean higher business profit. Actually, many big enterprises run into trouble not because of direct sales, but because of net profit.

Nature economy value (NEV): NEV is the sum of the following:

1. The cost of energy and resources taken from nature; always negative in value
2. The cost of 100% recycling and environment-compatible treatment of the back flows into nature; always negative in value
3. Societal compensation or penalty to balance the impact on the big ecosystem; can be posi-tive or negative in value

NEV is a new concept. It can be quantified in market value, but how to quantify it requires the efforts from all levels of society. Note that this index is linked to the current market. Thus, the NEV of any activity can change with the economy.

One may argue that this index is not very scientific. A scientific index should not change with society. It is true that we should develop and use scientific indexes, such as the footprint of green-house gas emission, equivalent of energy and water consumption, etc. The purpose here is to develop an index that can be used to drive the sustainability in a market-driven economy. Thus, such an index must be put on the same starting line with the DMV, which is society- and time-dependent.

For example, water in a water-rich area is much cheaper than in a water-poor area. The same amount of water consumption in these two areas will have quite different social impacts. Think about building a golf course in New York State in the United States, where water is plentiful, and in the desert of the Middle East, where water is a strategic resource for national security.

Another example: In my neighborhood in the United States, you need to pay people to remove a big tree from your yard, because few people find the value of a big tree. In China, you own a big asset if you have big trees in your yard. Lumber is valuable in China, and labor is negligible com-pared with the value of the lumber. So, the seemingly scientific index has to be adjusted to market-specific values before we can objectively evaluate its impact in a market-driven society.

This is not to deny the needs of systematically developing the scientific indexes of a natural economy. In fact, the calculation of the nature economy index in this book is dependent on the scientific indexes, but they are adjusted into market- and society-specific values to evaluate their impact in a market-driven economy. Market value is a standard already established in our civiliza-tion. A smoother transition can be achieved if we follow the same rules of direct market economy and use the same rules to modulate the direct market economy. We are targeting a self-motivated mechanism for sustainability.

The index of sustainability (IS): the NEV divided by the DMV can be used as an IS,

$$IS = NEV/DMV \tag{2.5}$$

The higher this value, the higher sustainability of an entity (product/service/process). This is a number that should be decided by authorized organizations. There are already some similar indexes in our market. For example, the energy efficiency of home appliances, the noise level of electric fans, etc.

Sustainability is not only a challenge to our society, but also an opportunity for an enterprise. Major corporations are already positioning themselves for the era of a market-driven sustainable economy. That's why you see so many big companies, including GE, Sharp, Intel, Phillips, etc., invest heavily in solar energy or wind energy. Nations worldwide are implementing forced measures to protect the environment. The national policy of China has changed from "economy development is the first priority" to "innovative, scientific, and sustainable development of national economy." The index of sustainability can be used as an objective measure by the society to rate an entity in economy. Engineers can help drive up this index.

Sustainability adjusted value (SAV): The SAV is the summation of the DMV and the NEV.

$$SAV = DMV + NEV \tag{2.6}$$

The New Criteria of Engineering Optimization (New CEO):

Performance, quality, cost, cycle time, and responsiveness are the basic requirements for a process or a product. Conventional engineering optimization tasks mainly focus on how to gain the enterprise the largest market competition advantages. In intelligent EFM, we also consider the level of M-PIE flow integration and the impact of resource consumption.

The new CEO is to maximize the index of sustainability and maximize the SAV of engineering activities in a market-driven economy. This can be achieved through maximizing the DMV of products/services and minimizing the NEV.

Currently, the technology cycles in a market-driven economy are missing a correcting mechanism to be sustainable theoretically. Introducing the new CEO can help establish the self-correcting mechanism to implement a market-driven sustainable economy, as shown in Figure 2.17.

In Figure 2.17, the by-product and wasted energy and resources will impact nature eventually. Their impacts are quantified as NEVs, which is normally a negative value. The DMV of products/services/ processes are positive values. DMV tends to retire less market competitive products/services. The *DMV* is summed up with *NEV*, which equals the *SAV*. The *IS* and the *SAV* are then used to decide whether energy and resources should be adjusted for certain innovation initiatives.

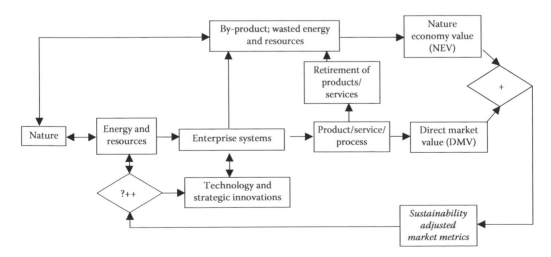

FIGURE 2.17 The self-correcting mechanism in market-driven sustainable economy.

The society should form a culture that enforces the new CEO. Enterprises can do their targeted business activities, but they must go through strict SAV certification. Government and international organizations should have corresponding economic policies to make the market competition in favor of products/processes/services with a high index of sustainability.

For example, link the tax policies with the *NEV*. The more negative the *NEV*, the higher the tax. This tax income is the asset of society and should be used to offset the debt of the *NEV*, fostering the balance of the ecosystem. Although enterprises can still sell at the DMV of their products and services, they must submit additional tax based on their NEV to the government and the public, forcing the enterprises to transfer cost to DMV.

This will form a deteriorating cycle for products and services with a low *IS*, forcing them to be eliminated through market force. At the same time, the winning of products/services with high *IS* can be accelerated through privilege national tax policies. In this way, a self-correcting mechanism for a market-driven sustainable economy can be formed, which will naturally force the current economy to evolve toward sustainability.

The new CEOs include the requirements of current engineering optimization, but extend it two levels further. First, it extends the engineering optimization domain from technical measurements to the index of market performance; secondly, it extends the optimization to the great nature system, considering all the chains of the nature cycle.

Actually, the index of sustainability defined here can be used to measure a broad range of phenomena, such as accidents, natural disasters, etc. In this book, we will use it for the optimization of manufacturing processes. These concepts are the soul of intelligent EFM.

With the new CEO and the *IS*, we can analyze our manufacturing activities. It sets penalties on low *IS* processes, while adding more value to high *IS* processes. Since the complete chain in the cycle is considered, such quantification is more objective to judge how "green" an activity is than current marketing-like treatment of green products.

For example, the index of the gas mileage of automobiles is not enough to judge their sustainability. We have to consider their intake cost from nature, as well as the final retirement and recycling of wasted energy and resources. Fuel cells and hybrid cars may improve mileage, but a complete chain analysis may indicate that its nature economy value can be highly negative, thus dragging down its index of sustainability. The comparison between a gas-based car and a hybrid car is not a simple matter.

2.5.6 Levels of Engineering Decisions

Is the new CEO far-fetched with the conventional criteria of engineering optimization? Well, they are closely tied together if we want to transfer our economy to a sustainable one. Table 2.4 shows the levels of engineering decisions. Level 1 is the new CEO and the philosophy is from nature to nature. This level should govern all the other levels of engineering decisions. Levels 2–4 are the familiar decision levels. In intelligent EFM, these levels are modified to implement the new CEO. In conventional criteria, the top-level philosophy is profit driven and is focused on winning market competition. Such criteria tend to detach fundamental engineering activities from the sustainable nature-to-nature cycles.

In intelligent EFM, we flow down the new CEO into all the levels. The main frame of conventional engineering implementation is still there, but Level 4 considers the requirement of the sustainability of nature and it is required to carry out evaluations of both market economy value and NEV. This is used to optimize the engineering solutions based on current and future national and international policies.

The philosophy of the new CEO was realized in 2008 and was first reported in the 2008 ASME/MSEC conference. She was a newborn baby. Her brothers and sisters are lean and adaptive manufacturing, six-sigma, green initiative, globalization, creative engineering solution, etc.,

TABLE 2.4
Levels of Engineering Decision

Level	Philosophy	Approach
1. New CEO	From nature to nature • Achieve a market-driven sustainable economy	• Maximize direct market value, minimize nature economy value, thus forming a self-correcting mechanism to enforce sustainability in market-driven economy
2. Market-driven strategic engineering decisions	Profit driven • Manage resources to win the largest profit and market share	• Improve efficiency of operation • Establish technology lead through continuous innovation • Maintain competitiveness using various marketing strategies
3. Top level engineering decisions	Customer and market driven • Connecting current and future market needs with engineering activities	• Convert customer and market requirements into key engineering objectives • Benchmark and idea generation • Down-select engineering solutions using various tools • R&D and commercialization with cost and performance management
4. Detailed technical decisions and engineering activities	Effective project management and execution • Team work to meet the joint requirement of performance, cost, time, and *environment*	• Recruit talents and build up team • Execute projects • Carry out science and best-practice-based engineering designs and implementations • Progress technology through invention and innovation • *Evaluate the market and nature economy value* • *Engineering optimization with new CEO* • *Build up knowledge base*

are the other widely known engineering methodologies. Her mission is to simplify our thinking on the fundamental challenges of human beings—maintaining the sustainability of our planet. All these other methodologies can be part of the philosophy of the new CEO. In Chapter 3, we will introduce the SP curve (the curve of sustainability and prosperity), review how the engineering world addressed sustainability and prosperity, and tie them with the strategy of intelligent EFM.

The study of the new CEO has just begun. We expect to drill down deeper in the future. It is good to see that there have been increased discussions on the sustainability of manufacturing in many conferences, and society has gradually accepted that we need to act quickly to avoid disastrous consequences. Maintaining sustainability is the responsibility of all human beings, with engineering in the frontline of the battle. The new CEO is helpful in establishing the self-correcting mechanism to transit our economy from market driven to market driven and sustainable. It is connected to the daily activities of engineering.

2.6 DEFINITION OF INTELLIGENT ENERGY FIELD MANUFACTURING

2.6.1 DEFINITION OF ENERGY FIELD MANUFACTURING

The essence of manufacturing is to control various energy fields using information and human intelligence to convert materials into the desired configurations. Energy fields carry information and convert materials into final products. Thus, energy field manipulation is central in all manufacturing processes. The key challenge is to find a systematic way of addressing the contradictions in manufacturing and to reach a higher level of optimization.

Rather than dividing manufacturing processes into traditional or nontraditional, we should treat various processes equally as different kinds of *EFM, which feature the optimal integration of the dynamic M-PIE flows*. The dynamic M-PIE flows were discussed in the first half of this chapter, as shown in Figure 2.5. They are the flows existing in any manufacturing processes and systems: information and I-flows, E-flows, M-flows, and the system level P-flows. The word "flow" reflects the dynamic nature of information, energy, material, and processes.

Thus, EFM is not a category of new processes, it is the methodology used to solve engineering challenges and contradictions, it is the software and hardware for manufacturing process analysis, improvement and optimization, and the knowledge base and philosophy for interdisciplinary process innovation.

2.6.2 INTELLIGENT ENERGY FIELD MANUFACTURING

With the above definition of EFM, all manufacturing processes are EFM processes, including human engineered processes and nature driven processes. The key difference between engineering and nature processes is that human beings inject human intelligence into the dynamic M-PIE flows. The trend of modern manufacturing is steadily evolving toward improving the level of integration of the M-PIE flows and increasing the intelligence level of these flows, as shown in the development of smart materials, close-loop and intelligently (neural networks, gene algorithm, fuzzy logic, expert system, etc.) controlled processes, digital-driven manufacturing, concurrent and distributed manufacturing, etc.

EFM is manufacturing based on energy fields and M-PIE flow integration. To differentiate our modern manufacturing activities from the conventional and the spontaneous nature-based EFM and to reflect the trend of engineering, we define intelligent EFM as the following.

Intelligent Energy Field Manufacturing (EFM) is *EFM featuring the effective and systematic exploitation of general energy fields, general logic functional materials, general intelligence, and the implementation of the new CEO*.

We need to differentiate intelligent EFM in this book from the intelligent manufacturing (IM) discussed in the 1990s [34]. IM focused on the introduction of human-like decision-making capabilities into the manufacturing systems to make them intelligent. IM features knowledge-based manufacturing, utilizing intelligent techniques such as sensing and feedback control, expert systems, fuzzy logic, neural networks, and genetic algorithms.

Intelligent EFM goes several steps further from IM:

1. In addition to using intelligent techniques in manufacturing, it features the methodology of EFM.
2. It emphasizes the utilization of general energy fields, general logic functionality of materials, and general intelligence of materials and systems.
3. It is the combination of methodology, intelligent techniques, and hardware innovations to implement intelligent EFM.
4. It is guided by the new CEO, which has a nature-to-nature philosophy.
5. It inherits the existing engineering methodologies and uses it in cross-discipline engineering innovations.

Intelligent EFM is proposed to meet future engineering challenges. These challenges include:

3D and intelligent structure manufacturing: Manufacturing is evolving from 2D manufacturing to 3D. Solving 3D issues using 2D technologies is usually complex and not cost effective, or is simply impossible. Beyond 3D manufacturing is intelligent structure manufacturing, which has both

3D geometry and self-sensing/self-repair functions. For example, structures with embedded sensors are being studied and used in airplanes, spacecrafts, buildings, etc., to increase system safety.

Micro/nano fabrication: Manufacturing needs to solve issues at both the macroscale and micro/nanoscale. Conventional methods may be good for macro scale applications (dimensions > 100 μm). Shrinkage to dimensions below 100 μm or submicrons requires the optimal use of various energy field methods and the general intelligence of materials and systems.

Winning in global competition through innovation: Globalization is inevitable in the information age. Any nation or organization that wants to take the lead in the competition must have constant innovation in technology and management with minimal resource waste. Intelligent EFM addresses this challenge by reducing the innovation barriers and by developing a systematic way of technology innovation and optimization.

Adapt to and exploit the explosive increase of knowledge base: The knowledge base of human beings is in explosive expansion and is becoming more accessible to a much wider population due to modern communication technologies. The effective use of information will be critical for success. Innovation is constrained by the accessible knowledge of the innovators. Many resources may be wasted in reinventing the wheel—solving the solved engineering challenges. Intelligent EFM aims to build a system that naturally adapts to and helps engineers better exploit the explosive increase in knowledge.

2.7 CONCLUDING REMARKS

In this chapter, we reviewed the evolution of intelligent EFM. The seed of EFM sprouted in 1988. Concepts such as energy field generator, logic functional materials, dissociation forming, etc., were proposed for the convenience of 3D manufacturing. This methodology was extended to a wider range of manufacturing processes over the years. The dynamic M-PIE flows, the concepts of general energy field, general logic functional materials, and general intelligence, along with the new CEO, form the foundation of intelligent EFM. We tried to give the reader a historical perspective on the fundamental concepts of work, energy, physical quantities, fields, materials, and intelligence. The mission of intelligent EFM is to lower or break the barriers of interdisciplinary technological innovations and accelerate the smooth transition of our economy from an unstable market-driven economy to a market-driven and sustainable economy.

Intelligent EFM is an engineering philosophy that many people are tacitly practicing without explicitly naming it. There are growing needs of making this philosophy more systematic and more scientific. Classic textbooks offered extensive knowledge on conventional processes, but not enough detail on nontraditional processes. Processes were normally introduced in a case-by-case style. The first two chapters of this book are designed to give the reader a high-level view of individual processes. Many abstract concepts are introduced to this point. In the next chapter, we will discuss the representative principles of intelligent EFM. The rest of the book will give more specific examples of manufacturing processes from various disciplines. The focus is not on conventional processes, which has been well covered [35–38].

We want to remind the readers that the philosophy of intelligent EFM is still in fast evolution. We presented some new concepts. It is very normal to have different opinions on these topics. We would have achieved our purpose if these new concepts could initiate the methodological thinking of manufacturing in your mind.

QUESTIONS

Q.2.1 Quickly review several manufacturing books and report on the major philosophy in these books.

Q.2.2 Do you agree with the concept of general energy fields? Can vacuum do work? Can lower temperature do work? Can useful information do work?

Q.2.3 Discuss whether it is reasonable to treat information and intelligence as physical quantities.

Q.2.4 What is general intelligence in this book? How is it different from the conventional definition of intelligence?

Q.2.5 Summarize the applications of information flow in modern engineering.

Q.2.6 What is the mission of engineering? What is your mission as an engineer or future engineer?

Q.2.7 What standard shall we use for sustainability?

Q.2.8 Find out the latest trends in economy and engineering for which sustainability means both risk and opportunities for enterprises.

Q.2.9 Explain the new CEO. Discuss why a market-driven economy is theoretically unstable and how we can develop the self-correcting mechanism in intelligent EFM.

REFERENCES

1. National Science Foundation, Manufacturing—Rapid prototyping, *America's Investment in the Future*. NSF, Arlington, VA, 2000, pp. 48–61. Available at http://www.nsf.gov/about/history/nsf0050/manufacturing/rapid.htm

2. Gandhi, M. V. and Thompson, B. S., *Smart Materials and Structures*, Chapman & Hall, London, U.K., 1992.

3. Zhang, W., 3D manufacturing technology and functional materials, *New Technology and New Process* (in Chinese), 4, 1996, 34–38.

4. Zhang, W., Study on RPM, 3DM and energy field formics, *Proceedings of International Conference of RPM* (*ICRPM98*), Beijing, China, 1998, pp. 316–320.

5. Zhang, W., Wang, Y., Wang, W., Feng, J., Chinese patent CN97104397.3, *System and Method for 3D Manufacturing Based on Energy Field Method*, 1997.

6. http://www.mrl.columbia.edu/ntm/

7. Yao, Y. L., Cheng, G. J., Rajurkar, K. P., Kovacevic, R., Feiner, R., and Zhang, W., Combined research and curriculum development of nontraditional manufacturing, *European Journal of Engineering Education*, 3 (3), September 2005, 363–376.

8. Rajurkar, K. P. and Ross, R. F., The role of nontraditional manufacturing processes in future manufacturing industries, *ASME Manufacturing International*, 1992, 23–37.

9. Fey, V. R. and Rivin, E. I., *The Science of Innovation*, TRIZ Group, Southfield, MI, 1997.

10. Rantanen, K. and Domb, E., *Simplified TRIZ, New Problem Solving Applications for Engineers & Manufacturing Professionals*, Times Mirror, London, U.K., 2002.

11. Dunphy, S., Herbig, P. A., and Palumbo, F. A., Structure and innovation, in Hussey, D. (ed.), *The Innovation Challenge*, John Wiley & Sons, New York, 1997, p. 216.

12. American GFM, 2D – Ultrasonic Cutting Machines. Available at http://www.agfm.com/Cutting/Cutting.htm

13. Zhang, W. and Mika, D. P., Manufacturing and energy field method, *Transactions of NAMRI/SME*, 33, 2005, 73–80.

14. Zhang, W. and Azer, M., Intelligent energy field manufacturing, *Proceedings of ASME/MSEC2006*, 2006, P#MSEC2006-21005.

15. Zhang, W., Review of intelligent energy field manufacturing (EFM), *Proceedings of ASME/MSEC2008*, 2008, P#MSEC_ICMP2008-72541.

16. Serway, R. A. and Jewett, J. W., *Physics for Scientists and Engineers*, 6th edn., Brooks/Cole, 2004.

17. Weinberg, S., *Dreams of a Final Theory*. Vintage Books, New York, 1994, ISBN 0-679-74408-8.

18. Simon, A. L. *Energy Resources*, Pergamon Press Inc., New York, 1975.

19. Yabe, T. and Uchida, S., Solar light pumped laser and cooling method of solar light pumped laser, U.S. patent 0,225,912, 2008.

20. British Petroleum, *BP Statistical Review of World Energy*, 2009, Available at http://www.bp.com/sectiongenericarticle.do?categoryId=9023752&contentId=7044473

21. Faraday, M., *On the Various Forces of Nature and Their Relations to Each Other*, Chatto, London, 1894. Available at http://www.archive.org/details/onvariousforceso00farauoft

22. Cheng, Y. T., Rodak, D. E., Wong, C. A., and Hayden, C. A., Effects of micro- and nano-structures on the self-cleaning behavior of lotus leaves, *Nanotechnology*, 17, 2006, 1359–1362.

23. Sternberg, R. J. and Salter, W., *Handbook of Human Intelligence*. Cambridge University Press, Cambridge, UK, 1982.
24. Zhang, J., Cheng, P., Zhang, W., Graham, M., Jones, J., Jones, M., and Yao, Y. L., Effect of scanning schemes on laser tube bending, *ASME Transactions Journal of Manufacturing Science and Engineering*, 128, 2006, 20–33.
25. National Shipbuilding Research Program, Portable automated bulkhead and hull straightener. Available at http://www.nsrp.org/Project_Information/major_projects/summaries/Portable-AutoBulkhead-Hull Straightener.pdf
26. The Auto Channel, Ford Motor Company reports 2005 net income, Jan. 23, 2006, available at: http://www.theautochannel.com/news/2006/01/23/208140.html
27. Virki, T., Computers in use pass 1 billion mark: Gartner, Reuters, Jun. 23, 2008, available at http://www.reuters.com/article/technologyNews/idUSL2324525420080623
28. European Centre for Medium-Range Weather Forecasts, ECMWF supercomputer history, Feb. 1, 2009, available at http://www.ecmwf.int/services/computing/overview/supercomputer_history.html
29. U.S. Energy Information Administration, *Crude Oil Production, 1973–2008*, data available online at http://www.eia.doe.gov/emeu/international/Crude1.xls
30. *Scientific American*, Earth 3.0 (special edition), Oct. 2008, information available online at http://www.sciam.com/special-editions/?contents=2008-10
31. Aubochon, V., World population growth history, April, 2010, available at Vaughn's Summaries, http://www.vaughns-1-pagers.com/history/world-population-growth.htm
32. Churchill, J. J., Oil consumption in North America, Oct. 2000, available online at http://maps.unomaha.edu/Peterson/funda/Sidebar/OilConsumption.html
33. Odum, E. P., *Fundamentals of Ecology*, 3rd edn., Saunders, New York, 1971.
34. For example, see back issues of *Intelligent Manufacturing*, available at http://lionhrtpub.com/IM/IM-welcome.shtml
35. Betz, F., *Managing Technology—Competing Through New Ventures, Innovation, and Corporate Research*, Prentice-Hall, Inc., Englewood Cliffs, NJ, 1987, Chapter 7.
36. Degarmo, E. P., Black, J. T., and Kohser, R. A., *Materials and Processes in Manufacturing*, 8th edn., Prentice-Hall, Inc., London, U.K., 1997.
37. Kalpakjian, S., *Manufacturing Processes for Engineering Materials*, Addison-Wesley Publishing Company, Inc., London, U.K., 1984.
38. Amstead, B. H., Ostwald, P. F., and Begeman, M. L., *Manufacturing Processes*, 8th edn., John Wiley & Sons, New York, 1987.

3 Evolution of Engineering Philosophies and the General Strategy of Intelligent EFM

Wenwu Zhang

CONTENTS

3.1 THE MISSION OF ENGINEERING AND THE PHILOSOPHY OF INTELLIGENT EFM

Technological innovations are the major forces driving long-term economic waves. Our economy is currently market driven and is becoming increasingly global. Despite many of the engineering achievements, there is increasing concern that the current mode of economy could be unsustainable.

Thus, the mission of engineering should be helping to establish a market-driven sustainable economy through technological innovations.

In Chapter 2, we discussed why a market-driven sustainable economy is necessary. Sustainable engineering solutions without a market competitive advantage won't survive the long-term test of the real world. A purely market-driven economy is theoretically unstable because it misses a stabilizing mechanism. Efforts from government, society, education, and engineering are needed to transit our economy from mainly market driven to market driven and sustainable.

After understanding the mission of engineering, it is important to choose the right philosophy to guide our activities. The fundamental philosophy that we should have in intelligent energy field manufacturing (EFM) can be boiled down to two points:

1. The from-nature-to-nature philosophy: all human activities are part of and will influence the natural cycles. Rather than considering a small segment of the value chain in the economy, we should consider the complete cycle, considering all of the impacts from nature to nature. Such a cycle should be optimized to achieve maximum sustainability, like other natural cycles.
2. The Tai-Chi Yin–Yang philosophy in engineering (Figure 3.1): the world is in dynamic change due to the interaction of opposing factors (contradictions). Factors in the contradiction pairs are called Yin (negative factors, such as cold, inside, decrease, weak, female) and Yang (positive factors, such as hot, outside, increase, strong, male). The Yin and Yang factors coexist, are relative in nature, and can convert mutually under suitable conditions. Harmony (or the ideal engineering solution) can be achieved only when the Yin and Yang factors are properly balanced.

This Yin–Yang philosophy originates from the oriental Tai-Chi Yin–Yang philosophy [1]. Nature has many beautiful cycles, such as the water cycle, the nutrient cycle, the weather cycle, the life cycle, etc. The activity of humans is part of the nature cycle. As humans grasped large-scale energy applications, our impact on nature has changed from a minor role to a more dominant role. The first step in maximizing the sustainability of our earth is to understand that we are one segment of the long chain of the natural cycles; any human activity will have many chain effects. In 2008, many people watched a great science fiction movie, *Wall-E*. In this movie, Wall-E is the robot left on earth to clean up the garbage while human beings were forced to float in space due to the unlivable environment on mother Earth. What pride can we have if all engineering does is changing the sustainable Earth into an unsustainable one? Thus, the from-nature-to-nature philosophy should be the most fundamental philosophy of modern engineering and modern society.

To maximize the sustainability of nature while continuing the prosperity of human society, we should understand the Tai-Chi Yin–Yang philosophy. The task of engineering is to solve various engineering contradictions, such as lowering the manufacturing cost while maintaining the quality of products, achieving high strength of a structure while minimizing the total weight, etc. The various situations and dynamics of the world are driven by the relevant opposing factors, which were called Yin–Yang in oriental Tai-Chi Yin–Yang philosophy. To achieve the ideal solutions of engineering, we must achieve the suitable balance of the Yin–Yang factors.

The Tai-Chi Yin–Yang symbol, as shown in Figure 3.1, has many implications. First, Yin and Yang factors (contradictions) are pervasive in any entity, big or small. They are balanced in amount. Inside Yin there is some Yang, and inside Yang there is some Yin. In other words, anything has two sides. When we build highways, we achieved a transportation convenience for human

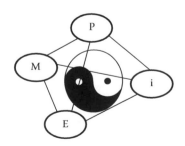

FIGURE 3.1 The Tai-Chi symbol.

beings, but we surely had isolated the important migrating route of natural animals and influenced the natural weather pattern as well. When we invented long-lasting plastics, it may be good for corrosion-resistance, but it may add additional cost to dispose after its use. We might enjoy the luxury packaging of simple merchandise, but these packages may lead to unnecessary natural resource waste.

Second, the Yin and Yang factors are relative in nature and can mutually convert under suitable conditions. A peak is neighboring its valleys, highs are followed by lows. You may assume that a rigid structure is stronger than a flexible structure. But when facing a shock impact, is a rigid structure more resistant than a flexible structure? A hammer can easily break a glass window, but it may take a while to knock out a hole in a soft mattress. Composite material is developed to have both the strength and toughness benefits. The interesting point is that extreme Yin will change into Yang and extreme Yang will change into Yin. For example, decent precision is a good thing, but extremely high precision in engineering will lead to high cost and high sensitivity, which can be a bad thing. Hardness is a relative concept. When water is applied at a high speed (as in abrasive waterjet machining), the water flow can cut the hardest material.

The third point is that to achieve ideal solutions, the entity has to achieve a suitable balance of Yin and Yang factors. A typical engineering example is thermal management. In many engineering devices, we desire high power. High power means high productivity, but high power also leads to high heat generation. The device must be in a certain temperature window to function properly. Thus, there must be a suitable balance between heat generation and heat dissipation. Another example: human beings may consume as much natural resources as possible when it is economical in a short period of time, but such activities won't be sustainable if a sustainable cycle is not realized. For example, underground water has dropped to unusable levels due to being over-drawn over the years, resulting in serious issues of fresh water supply in many big cities. Fresh air and clean water are free gifts of nature, but when we pollute the environment beyond its self-cleaning capability, these basic needs of life become a memory in many major cities around the world. To solve these issues, we must learn to build balances.

With the fundamental mission of engineering and the philosophy of intelligent EFM defined, we will introduce the fundamental strategy of intelligent EFM, followed by a review of representative engineering methodologies.

3.2 THE GENERAL STRATEGY OF INTELLIGENT EFM

We have set lofty goals for intelligent EFM. To achieve these goals, we need a clear strategy. Strategic thinking is big picture thinking. A clear strategy enables us to see the big picture before working on the details. Too often people dive into detailed work without a clear strategy. It is like entering an unknown wild forest without a compass or driving into a big city without a map.

Here is the general strategy of intelligent EFM.

To implement intelligent EFM successfully, we should optimize the integration of dynamic M-PIE flows, optimize *the strategic values both for the customer and for the enterprise,* minimize *waste, and* maximize *the sustainability of the whole ecosystem.*

In short, we have the following *success equation of intelligent EFM*:

$$\text{Success} = (\text{optimizing the dynamic M-PIE flows}) + (\text{optimizing the strategic values})$$
$$+ (\text{minimizing waste}) + (\text{maximizing sustainability})$$

This strategy governs our tactics or the detailed operation plans and skills. This is a short and clear strategy standing out from all existing engineering methodologies, such as mass production, lean manufacturing, six sigma, TRIZ, and sustainable development. It is based on the study of historical engineering methodologies.

How did we get here? A review of existing engineering methodologies will help our understanding of this general strategy, which is a very important step in the implementation of intelligent EFM.

3.3 EVOLUTION OF REPRESENTATIVE ENGINEERING METHODOLOGIES

There are many textbooks on manufacturing processes, and there are many discussions of manufacturing methodologies. Table 3.1 is one effort to tidy them up to tell us how engineering methodologies evolved. The sequence of the methodologies and technical trends does not strictly follow the time they appeared, and this is by no means a complete list of all the important methodologies. We will give a high level overview of these methodologies to better understand the strategy of intelligent EFM.

3.3.1 INDIVIDUAL WORKSHOP AND CRAFTSMANSHIP

Since ancient times until the age of industrial revolutions, manufacturing was carried out in small-scale workshops, such as a shop for scissors, furniture, or bricks. The knowledge of the process was handed down from generation to generation. This is the style of craftsmanship. This style of manufacturing still exists today, such as a gold ring made by skilled workers. Due to the high reliance on the skills of individuals, the quality of production varies from person to person. The same category of products may not be interchangeable at all.

3.3.2 FACTORY AND MASS PRODUCTION

Entering the era of steam engines, factory mass production became the dominating production style when large-scale energy supply was not an issue anymore. The same design was used to make many parts on dedicated machines. Workers were managed to do repetitive assignments. The early days of the clothing business is a good example. Many workers were grouped in factories to operate weaving machines powered by steam engines. Production was divided into detailed processes,

---

TABLE 3.1
Evolution of Engineering Methodologies

Methodologies or Technical Trends	Examples
1 Individual workshop and craftsmanship	Craftsman made gold rings, artistic sculpture
2 Factory and mass production	Early clothing business
3 Standardization and streamlined assembly line	Ford's Model-T automobile
4 The rise of nontraditional manufacturing and hybrid processes → Energy field manufacturing	ECM/EDM, laser-assisted machining, EFM
5 Increased automation and digitalization	CNC, CAM/CIMS, RPM, concurrent engineering
6 Design for assembly, design for manufacturing → Lean manufacturing and six sigma	Toyota's business success
7 Systematic innovation	TRIZ
8 Interdisciplinary cooperation and bio-mimic engineering	MEMS, solar PV, nano, super-hydrophobicity
9 Sustainability challenge and the green movement	Recycle and alternative energy
10 Lean + quality control + sustainability + integration of dynamic M-PIE flows + systematic innovation	Intelligent EFM

---

greatly lowering the skill requirement of operators. This style helped increase the quality of production and reduce the production cost in the mean time. The management of production was mainly personal experience based. This ended the dominance of agriculture in the human economy. More and more complex systems were produced, calling for new methodologies of manufacturing and management.

3.3.3 STANDARDIZATION AND THE STREAMLINED ASSEMBLY LINE

The transportation revolution at the turn of the twentieth century and the two world wars pushed the quick progress of manufacturing processes, manufacturing methodologies, and engineering management.

To build the steam engine for locomotives, many metal parts needed to be machined or forged and finally be assembled together. Large volume automobile production forced manufacturing into the way of standardization. Parts made to standards were interchangeable, making it easier for cooperation between firms. Production in a single factory evolved into societal production systems. Factories could focus on a unit of a big system, optimize the process, and lower the cost.

Still, quality and robustness were bothering the assembled systems. Henry Ford solved the issues of automobiles in his time and eventually extended the car business into common American families. Henry Ford was the founder of the Ford Motor Company, the father of modern assembly lines used in mass production, and a pioneer of standardization and lean manufacturing.

Born in 1863, Henry Ford joined Edison's company in 1891 and became a chief engineer in 1893. In 1896, Henry Ford built his first prototype four-wheel gasoline-powered vehicle, the Ford Quadricycle. Ford resigned from Edison's company in 1899 and started his journey of automobile production. The products of his 1899 company suffered from low quality and high cost, and failed in 1901. He didn't give up, he didn't even wait a minute. He designed and produced a new car and started a new company the same year with new partners. Due to an unpleasant partnership, he left this company and formed Ford & Malcomson, Ltd. in 1902, which was renamed Ford Motor Company in 1903 with expanded investors.

Henry Ford was determined to make reliable and inexpensive automobiles. For reliability improvement, he used the latest material innovations. He used high salaries to attract and retain great talents. He also designed quality and reliability into his historical automobile, the Model-T. For example, the entire engine and transmission were enclosed, the four cylinders were cast in a solid block, and improved suspension was adopted. He simplified the user interface and fixed the color of the car to nothing but flat black.

Very importantly, he streamlined the assembly line to improve the efficiency of mass production. Figure 3.2 shows a picture of the Ford 1913 assembly line. He built a gigantic factory that shipped in raw materials and shipped out finished automobiles. With standardization and the movable assembly line, with the vertical integration strategy, and with a high-quality talent retainment policy, Henry Ford was able to continuously lower the price of the Model-T while maintaining high quality and high reliability. The price of the Model-T in 1916 was only $360, or equivalent to $7000 in 2008. Over 15 million Model-T automobiles were sold until 1927. It ran in cities and in the fields, affordable for many American families. This brought the automobile into a new era.

The power of Ford's mass production methodology was further manifested during World War II. The Ford Motor Company played a pivotal role in the Allied victory during World War II. When Europe was under siege, the Ford Company turned to mass production for the war effort. Ford examined the B-24 Liberator bomber. Before Ford and under optimal conditions, the military could produce only one B-24 bomber a day. With Ford's mass production and assembly strategy, the new plant produced one B-24 bomber an hour at a peak of 600 per month in 24 h shifts! The B-24 bomber became the most-produced Allied bomber in history, which quickly shifted the balance of power in World War II.

FIGURE 3.2   Ford assembly line, 1913.

Henry Ford greatly influenced how we do engineering today. When we talk about lean manufacturing, we should remember Henry's pioneering achievement in the beginning of the twentieth century. He not only cut the waste on the factory floor, he also cut the hidden cost (overhead) of manufacturing, paid attention to cutting the waste in talent loss, and designed quality into products. Ironically, the engineering industry in the United States renewed the values of Ford's wisdom from the Toyota way of production when the automobile industry in the United States faced increasing competition from abroad.

3.3.4   THE RISE OF NONTRADITIONAL MANUFACTURING AND HYBRID PROCESSES

Curiosity and creativity are inherent in human beings. In history, there were always some people who enthusiastically sought to uncover the secret of nature and the ways to conquer the limitations of nature. Thus came the use of fire, the birth of chemistry and materials science, the large-scale generation of electricity, the invention of lasers, and many other forms of energy generation or harnessing tools. Some people quickly found use in scientific progress, converting scientific knowledge into commercial applications. Our capacity of influencing nature increased quickly after entering the era of electricity and fossil-fuel-based economy.

Entering the twentieth century, the competition to fly in the sky and to enter space began. Aircrafts and rockets quickly evolved from a hobby into new weapon systems [2,3].

Many people, including the Wright brothers, improved the technologies of aircraft around 1900. World War I first tested the use of the aircraft as a weapon. Aircrafts initially served as mobile observation platforms, then as fighters when equipped with guns. After the war, the technology continued to improve. The first cross-Atlantic nonstop flight occurred in 1919. When the jet engine was developed in the 1930s, the aircraft achieved a new record of speed and agility. Aircrafts played a primary role in the Second World War.

Pressure and urgent needs are good accelerators of technological innovations. The Allied force and the Pivotal force competed to control the sky on the battlefield, while the foundation of competition was their national manufacturing system. New aircraft engines required the use of super alloys to withstand higher temperatures and maintain high strength, or the use of materials other than metals, such as ceramics and composites. Conventional machining based on mechanical force was improved in the steam engine and automobile industries, but mechanical machining ran into big challenges with the development of aircraft engines and rockets.

The reason is that these new materials are too hard or too brittle to machine with conventional machining. Traditional machining is most often based on removing material using tools harder than the workpiece. For example, polycrystalline diamond (PCD), which is almost as hard as natural diamond, cannot be effectively machined by the traditional machining process. One of the most commonly used conventional techniques is diamond grinding. In order to remove the material from a PCD blank, the diamond layer of the grinding wheel must be renewed by turning or dressing frequently. This resulted in rapid wear of the wheel, the G-ratio (the ratio of the workpiece volume removal rate to the grinding wheel volume wear rate) is 0.005–0.02. Thus, the grinding wheel wear rate is 50–200 times higher than the workpiece removal rate. Hence, classical grinding is suitable only to a limited extent for the production of PCD profile tools. The high costs associated with the machining of ceramics and composites and the damage generated during machining were the major impediments for the application of these materials. For example, the costs of machining structural ceramics (such as silicon nitride) often exceed 50% of the total production costs in the engine industry. High-quality ceramics machining can be done with diamond tools, but the process has to be well controlled and the machining rate is limited. In many cases, innovative techniques or modifications of existing methods are needed [4].

In addition to the advanced materials, stringent design requirements also pose new challenges to the manufacturing processes. More complex shapes (such as an aerofoil section of a turbine blade; complex cavities in dies and molds; noncircular, small, and curved holes), low rigidity structures, and components with tight tolerances and fine surface quality are needed. Traditional machining may be ineffective in machining these parts.

To meet these challenges, new processes need to be developed. The solutions were to (1) further improve the mechanical machining tools, (2) find new mechanisms of machining, and (3) develop hybrid processes. Today these are still the general strategies for machining difficult-to-machine materials.

For these reasons, the "nontraditional manufacturing processes" were invented and increasingly used. For example, electrochemical machining was first experimented in 1929, but didn't enter commercial application until 1959 [5]. The electro discharge machining process was invented by two Russian scientists, Dr. B.R. Lazarenko and Dr. N.I. Lazarenko, in 1943. Both processes used energy forms other than mechanical to machine conductive materials. With these machining mechanisms, hardness is not the limiting factor anymore. Waterjet machining, ultrasonic machining, laser machining, electron beam machining, etc., are similar examples. These processes break the limitations of contact and mechanical processes.

Not only were more energy forms explored, new combinations were tested. Abrasives were added to waterjets, thus the birth of the abrasive waterjet machining process [6]; ultrasonic energy was combined with mechanical machining, improving the machining of ductile materials [7]; plasma heating or laser heating was applied to the cutting zone in mechanical machining, lowering the cutting force and extending the tool life [8]; lasers were coupled into waterjets to combine the benefits of fast laser machining and strong water cooling [9]; and so on.

The above processes are called hybrid processes, which intend to integrate the benefits of individual energy forms while eliminating the weakness of some energy fields.

The concept of EFM was proposed by the author around 1995 (see Chapter 2) [10,11] to better reflect the technical trend of energy field integration in the manufacturing processes. The term of traditional and nontraditional manufacturing is a term of the twentieth century when human beings tried to overcome the difficulties in advanced material processing. The essence of manufacturing (or engineering in general) is to integrate energy fields under the control of information flow to convert materials into desired configurations. Thus, manufacturing should be called EFM rather than mechanical engineering.

Neither mechanical energy nor nonmechanical energy should be regarded as special energy. All energy forms have their uses. Nontraditional processes and hybrid processes should not be regarded as simply innovative processes. They actually guided us onto the road of energy field integration and optimization. This is an undeniable technical trend.

3.3.5   Increased Automation, Digitalization, and Intelligent Control

Energy field manufacturing also emphasized the integration of energy fields with information flow and material flow. Among myriads of technical trends, automation and digitalization are the most prominent.

Energy fields without control are either useless or harmful. In engineering, we inject our intelligence into the control of energy fields and their interactions with materials. Thus, any manufacturing has its information flows.

In the early stage, the human body acted as the controller of man-made systems. When the system got more and more complex, we tried all means to extend our ability to control energy. Levers, pulleys, switches, doors, etc., are such examples. Human beings are one loop, an important loop, in many engineering activities. Workers need to be trained to a certain level to do qualified work.

Humans have changing emotions and physical conditions. This leads to the variations of their performance from day to day or moment to moment. With increased automation, the variation of humans can be reduced.

Building more complex systems, such as the control of an electrical motor-powered lathe, the operation of the railway systems, the operation of aircrafts, the launching of satellites, etc., required more and more advanced control systems and strategies. With sensing and feedback control, we can control a dynamic system. With a proportional–integral–derivative (PID) controller, many complex machines were controlled with ease. Note that in the early history of automatic process control, the PID controller was implemented as a mechanical device, consisting of lever, spring, and mass, and was powered by compressed air. Later on, this changed into electronic analog controllers. Nowadays it has been replaced largely by digital controllers implemented with microcontrollers or FPGAs [12].

The request of scientific and engineering calculations increased exponentially with technological progress. The invention of modern electronic computers in the 1940s [13] and the invention of integrated circuits in the 1950s [14] marked the beginning of a new era of our civilization. Human beings now had the tools to easily inject intelligence into the physical world and then use this intelligence-integrated hardware to control the rest of world.

With the enhanced capability of computation and control, all branches of engineering embraced digital time. Computer-aided design (CAD) and computer-aided manufacturing (CAM) were developed in mechanical engineering, while various computer simulation and analysis tools were available for almost all branches of science and engineering. When I was an undergraduate student, several of my assignments were in computer programming of matrix transformation in Fortune language. Now, with mathematical software such as MATLAB®, Maple, Mathematica, etc., the same programming is just a single line command. My graduate thesis was based on MATLAB 4.0, using MATLAB to implement computer-aided complex control system design. That cut the system design time from several months to less than 10 min.

Looking around us, much of the equipment is computer controlled. Some of them can be remotely controlled. In our laser micromachining lab, we used computers to control a four-axis motion stage, a galvanometer scanner, laser firing, and the temperature-sensing system. We also used a design of experiments (DOE) tool to design our experiment. Once the programming was done, the remaining work was mainly clicking the button to run the programs. After experiments, the samples were analyzed with various tools, also computer driven.

It is difficult to capture all the implications of this technological progress. With computers, globalization, and the Internet, we entered the information age [15]. As pointed out in the Web-book

Beyond the Information Age [16], many of us drive back and forth from the office mainly to process information—many of us have become information processors in the information age.

For manufacturing, we entered the era of computer-controlled digitalized manufacturing. First was the arrival of computer numerical control (CNC) machining center [17]. Modern CNC systems are highly automated. CAD/CAM programs produce files that are interpreted by CNC systems to operate particular machines, to load the specified tools, and to finish multi-step processing in one machine. This saves tool changing time and part re-mounting times, resulting in improved productivity and consistency. NC (numerical control) technology replaced early day approach of automation, which was based on CAMs following. CAM-based automation relies on moving the tool using the input of a CAM follower riding the shape of the CAM. CAM-based automation is capable of write-once-read-many, but CAM has to be manually made first. In contrast, an NC system programs the tools to move to specified coordinates, being fully rewritable, changeable, and reusable. The flexibility of automation between the two is incomparable. If you read the story of how NC systems were developed, you would know that it was initially motivated by the blade production of helicopters. The complex geometry forced John T. Parsons and the MIT group to use a numerical approach. The first numerical machining system was built in 1952 by MIT [17]. It later used computer control and incorporated CAD and CAM. Isn't this a kind of virtual mold (see Chapter 2)?

Note that the CNC system is a general manufacturing methodology rather than tools dedicated to mechanical machining. Actually, CNC systems are used in laser-, ultrasonic-, or waterjet-based processes to digitally deliver various energy forms to the needed material-processing location.

There are many other technical directions related to increased automation and digitalization. Rapid prototype manufacturing (RPM) is one of those important directions. Complex 3D geometries are difficult even with five-axis CNC systems. With a 3D model of the part to make, and by taking a layered approach, complex solid objects can be built layer by layer. With the help of RPM systems, a new design can be transformed into prototypes in greatly reduced periods. Featuring a digital-driven and layered approach, RPM has evolved into a very active direction in manufacturing and is not limited to prototype anymore—some processes can directly build functional parts called rapid tooling or rapid manufacturing [18].

In a market-driven economy, the time to market is an important factor of success in intense competition. CAD/CAM/RPM/CNC, various simulation and modeling tools, computer aided engineering (CAE), along with Web communication tools enabled the implementation of concurrent engineering [19].

The concurrent engineering method is designed to optimize engineering design cycles. There are mainly two concepts. The first is that all elements of a product's life cycle should be carefully considered in the early design phases. The second concept is that the various design activities should occur at the same time, or concurrently, with the overall goal of significantly improving productivity and product quality, shortening the life cycle of new product development. This philosophy has the advantage of early feedback and error correction from related units of the product life cycle. The earlier the issues are identified and corrected, the less costly the development effort. This is in big contrast with the traditional sequential approach. Concurrent engineering has been used in the development of complex systems, such as new plane models, new space shuttles, etc. [19].

Advanced control requires several things: sensing and collecting information, processing the information based on certain control strategy, and using it to control the system. With the help of modern computers and sensing tools, such control can be readily integrated in the manufacturing system, thus increasing the intelligent level of the system. Open loop control, close loop control, adaptive control, real time control, and the introduction of neural networks, gene algorithms, fuzzy logic, system identification, etc. in manufacturing systems had changed manufacturing in many aspects. People have developed many other technologies, such as agile manufacturing, digital manufacturing, computer integrated manufacturing systems, etc. Robots had been used in automobile manufacturing many years ago. In the recent Iraq War on Terror, many robots were used on the battlefield. Even soldiers were equipped with many intelligence tools, such as global positioning systems and communication tools, night vision systems, etc. Nowadays, a cell phone has many functions, including TV, Internet, multimedia, camera, and games in addition to voice communication.

The progress of digitalized technology is dazzling and puzzling sometimes.

However, it is sufficient to point out that these new developments reflect the technical trend of manufacturing and engineering in general moving toward increased automation and digitization. In intelligent EFM terms, these are indications that the information flow is increasingly integrated into the total engineering activities. The more integration of information flows, energy flows, and material flows, the more advanced the technology.

The trouble in the information age is that there is too much information. The task of filtering out the useful information is becoming increasingly important. Information is like raw material, which may or may not be useful. What we actually need is intelligence and knowledge.

3.3.6 FROM DESIGN FOR ASSEMBLY AND DESIGN FOR MANUFACTURABILITY TO SIX SIGMA AND LEAN MANUFACTURING

If there is no market competition, you possibly can reinvent the wheel and still continue your business. This is never the case in reality. Whatever you do, there are possibly many competitors unless it is a brand new small market and you are temporarily the only player in the field. Once something is proven to be profitable, more social resources will enter the competition, following the rule of the forest—the most adaptable to survive.

Naturally, realizing functions, reducing cost, and improving quality and productivity have always been the top success factors in manufacturing. The pioneers of engineering explored ways of achieving these factors; we will review these methodologies, scrutinize them and absorb the essence of these brilliant solutions. Being inventive is not enough, you must not ignore these methodologies to be successful. These methodologies include design for assembly (DFA), design for manufacturability (DFM), six sigma quality control, lean manufacturing, etc.

A True Story—Two Minutes vs. Two Weeks

A friend asked me to pattern several small diameter tubes with my laser material processing system. He read that the fs laser could machine almost any material without much thermal effects (including polymer) and the laser process was very fast. So he showed me the long tubes and asked: "How much time do you need to finish the work?" He was expecting half a day or one day maximum. I answered: "At least 1 week if I work full time on this. Since this is urgent, I can finish the work in 2 weeks considering the needs of other projects."

Why? I needed to figure out a way to hold and rotate the tube while moving it on my stage. I didn't have a rotating stage at that point. I also needed to change my setup, redo the alignment and focusing process, and finish the programming work. Then I needed to do a series of iterative experiments, analyze the results, and work out the processing window for this unfamiliar material. Only after all those steps could we talk about the less than 2 min processing cycle time for his part. OK, this is how a 2 min process could extend into a 2 week process. I forgot one thing though. Before we do any experiments of new materials in the lab, we must get approval from our Environment, Health, and Safety (EHS) department. The final result was that I delivered the work in more than 1 month when I really pushed the progress.

Figure 3.3 shows the experimental setup of this work. It took me quite some time to hold the tube properly for processing. On one end it was held by the chuck on the rotary stage, while on the other end we mounted an L-shaped metal piece to prevent the tube from too much sagging. A fume extraction system was used. With this, I could finally do the laser processing work.

This is a typical situation in manufacturing. The takeaway of this story is that people without experience in manufacturing tend to underestimate the cost, time, and importance of the manufacturing processes. The other point is that proper fixturing and assembly could consume large quantities of manufacturing time, and there are hidden overhead costs related to any engineering activity.

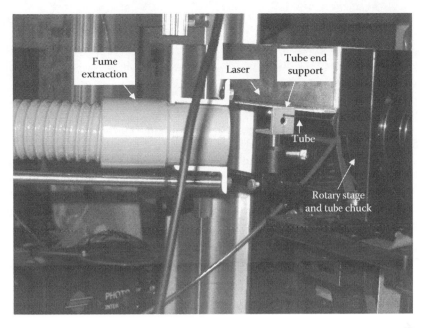

FIGURE 3.3   Experimental setup for laser tube patterning.

Historically, people have tried to improve the manufacturing processes starting from the more visible sources of cost and quality issues. Only in recent years have people addressed all sources of manufacturing issues with a large system approach.

Cut Part Numbers and Design for Fixturing, Assembly, and Manufacturability

Each part takes some time to fixture, process, and assemble, thus, cutting the number of parts is a straightforward and effective way of cutting manufacturing cost and improving overall quality. Sometimes, several parts can be replaced with an integrated structure. For example, a box made of multiple plates and put together with screws could be replaced with a one-body cast piece.

Be careful not to over-use this strategy. Some sub-division of structure is necessary. A one-body piece may lower the maintenance level of the system. When a small part of the piece malfunctions, it could be much more expensive to replace it than replacing a small part of it.

Cutting the number of parts is very visible. It is one of the low hanging fruits in cost-cut. Everyone should use this strategy, but it will soon reach its limit.

DFA is a process by which products are designed with ease of assembly in mind [20]. When parts are provided with features that make them easier to grasp, move, orient, and insert, the time and cost of assembly can be reduced, especially for the production of a system with many components, such as automobiles. In the 1960s and 1970s, various rules and recommendations were proposed to help designers consider assembly problems during the design process. Boothroyd studied how to estimate the time and cost of assembling products on an automatic assembly machine in the 1970s [21]. In the 1980s and 1990s, DFA became widely adopted in many big companies. There are many published examples of significant savings obtained through the use of DFA. The Ford Motor Company credited DFA for overall savings of around $1 billion in 1988 [22].

In our laser tube patterning example, we showed how bad small processing work could grow many times bigger than initially thought when the fixturing and assembly issues could not be solved immediately. Sometimes we need to self-design the fixtures for our special cases. But a good suggestion is that if your problem is a common problem, there are normally some professional assembly setups that are commercially available.

How do you hold a thin and flexible foil? Try a vacuum chuck.

How do you hold a circular component and adjust it in all freedoms? Try to find something in optical mounts.

Want to hold a tube with less than 400 μm OD? Learn from the people who do fiber coupling. Or simply check the Web to see how others do similar work. You can easily save time and money by going with professional fixturing and assembly.

Assembly is only part of the story. We need to process the part. The way a part is designed greatly affects the cost of manufacturing. Novice designers and experienced designers can give very different designs, resulting in very different manufacturing times and cost. For example, some designers may have the habit of specifying dimensional accuracy that is too high. High accuracy may require the use of expensive systems and high-end measurement equipment, lower productivity, and the yield of production. What is the best accuracy? It is the accuracy that is just enough for the function of the final system. Accuracy that is too high can be as bad as accuracy that is too low.

Some part geometry may be no trouble at all until you consider how to make it. If such issues are found in the manufacturing stage, the cost of correction is too high compared with a good design in the beginning. Some design may require special and expensive processes. Thus, along with DFA, we should consider DFM.

In DFM, we choose the right material, design the right geometry, and specify a suitable level of geometric accuracy considering the easiness of manufacturability. For example, putting holes initially on several planes onto one plane can save quite a bit of repositioning time in processing; a part with good reference planes for fixturing can help improve the repeatability of processing; avoid the processing of difficult-to-reach geometries; avoid processing with too many orientations; try to reduce the needed processing approaches, etc.

References [23,24] provide rich information related to DFA and DFM. For example, in reference [24], the University of Michigan offers a course dedicated to DFM. The course gives very systematic training on this topic, covering the whole process of developing a product.

In ASME, there is a DFM Committee [25]. The mission of the DFM committee in ASME is to disseminate practices, theories, and computational methods dealing with all areas of design and manufacturing integration among the engineering community as well as to encourage the growth and recognition of the value of design and manufacturing integration. This mission has expanded to a system level.

The readers are encouraged to find out more details on their own. In this section, we don't have space to explain all the details, since each topic can be a book in itself. We will review these methodologies and form a promising high level engineering philosophy along with workable procedures and principles to guide our interdisciplinary innovation work.

DFA and DFM could achieve very immediate and visible success. However, they are part of the mass production philosophy. Even after the implementation of DFA and DFM, it is still not good enough to ensure success in global competition. This is just what the U.S. automobile industry has gone through since the 1970s.

In 1973, members of the Organization of Arab Petroleum Exporting Countries (OAPEC) proclaimed an oil embargo against the Western countries, which triggered the 1973 Oil Crisis [26]. The economy of industrialized countries relied on crude oil. When the oil price jumped from $4 to $12 a barrel in a short time, many aspects of the economy were affected, including the automobile industry.

Japanese cars were known for their compact size, low cost, and high oil efficiency. On the other hand, large, heavy, and powerful cars were the standard in the United States before the oil crisis. When the Toyota cars entered the North American market, the compact cars were viewed as entry level or low-end products. U.S. car makers possibly felt they had better quality due to better technology and the strategy of a high-end market. The manufacturing methodology is mass production, design for manufacturing, computer-aided engineering, and concurrent engineering.

With a higher cost of gasoline, the demand for large cars dropped, and the Big Three (General Motors, Ford, and Chrysler) were forced to introduce smaller and fuel-efficient models for domestic sales. But Toyota, Honda, and Nissan had captured the market lead to a great degree with improved quality and efficiency. U.S. companies also lagged behind in the introduction of hybrid cars and electric cars. The status of the global competition in 2009 was that the Big Three of the United States were in deep trouble—Japanese and South Korean cars sold better than the American cars in general, with Toyota being the number one company in car production taking the honor from the American company, General Motors (GM).

This title change relates not only to strategic product line decisions and key technological innovations, but also to the manufacturing methodologies within these companies. Amazed by the success of Toyota, scholars studied the Toyota production systems and revealed their "secret" [27–32]. It is the methodology of lean manufacturing or lean production.

Lean manufacturing states that making products in mass production mode is not enough, one must focus on the values of the customer and cut the various sources of "waste."

Class activity: Watch the videos of James P. Womack: (1) *Key Lean Concepts* and (2) *Lean: A Fundamental Way of Thinking* in [30].

3.3.7 FROM MASS PRODUCTION TO LEAN MANUFACTURING

The automobile companies in the United States improved mass production, starting from the work of Henry Ford. Many people thought the key to mass production was the continuously moving assembly line, but this is the final superficial thing. The key to mass production is the *complete and consistent interchangeability of parts and the simplicity of attaching them to each other*. With greatly improved interchangeability, simplicity, and ease of attachment, Ford was able to lower the skill level of assembly workers. Workers remain in one spot to do repetitive work. Ford had not only created interchangeable parts, but interchangeable workers as well.

The assembly line might produce defects. Ford developed the rework specialist and general foreman of the assembly line to manage the errors of the "interchangeable" worker. Ford also streamlined the process to include the vertical integration of supplies.

Ford's mass production philosophy was adopted by other companies worldwide, including GM. Alfred Sloan, the long-time GM chairman, added the financial manager and marketing specialist to the mass production system [33]. Under Sloan, GM became famous for managing diverse operations with financial statistics. Sloan is also credited with establishing annual styling changes. With Ford's resistance to the change in the 1920s and with the more advanced manufacturing philosophy of Sloan, GM achieved sales leadership by the early 1930s, a position it retained until 2008.

We just witnessed the historical champion title change in 2008. Now Toyota is the largest and most profitable automobile company in the world. Now the whole world is learning from the Toyota production system. U.S. scholars termed it the lean production or lean manufacturing methodology [27,29].

3.3.7.1 The Lean Story of Toyota

Seeing the increased threat of abroad automobile competitors, researchers associated with the MIT International Motor Vehicle Program (IMVP) carried out a systematic study of the new system of manufacturing, especially the Toyota Production System in the 1980s. The book *The Machine That Changed the World* by Womack et al. summarized the IMVP research results in 1990 [29]. Lean production was a term coined by IMVP researcher John Krafcik [27]. The Toyota system is lean because it uses less of everything when compared with mass production—half of the manufacturing space, the investment in tools, the engineering hours to develop a new product, and the needed inventory on site. With less resources, it produces the amazing results of much fewer defects and a greater and ever-growing variety of products. Let's review the lean history of Toyota briefly.

Eiji Toyoda, Taiichi Ohno, and Shigeo Shingo are the key figures in this story. Taiichi Ohno was credited as the father of the Toyota production system and the pioneer of lean production.

Ohno said he learned from mainly three people, they were Eiji Toyoda who visited the Ford plants; Henry Ford, the founder of Ford Motor Company, and Shigeo Shingo, who was Toyota's primary consultant and teacher [34]. When young Eiji Toyoda visited Ford's River Rouge plant in Dearborn, Michigan, in the early 1950s, the Ford plant was manufacturing 8000 vehicles a day, while Toyota Motors was struggling and had just produced over 2500 automobiles for the first 13 years. Toyoda hoped to adopt the U.S. automobile mass production methods. After in-depth study, Toyoda went back to Japan. With the help of Taiichi Ohno and Shigeo Shingo, he concluded that mass production would never work in Japan for multiple reasons.

First, Toyota's domestic market was small and the demands included a wide range of vehicles—from luxury cars for executives, to large and small trucks for farmers and factories, to small cars for the crowded cities. Second, the Japanese workforce was not willing to be treated as interchangeable parts and Japan didn't have cheap immigrants like America. Third, Toyota didn't have enough capital to introduce the expensive automatic assembly lines in the 1950s. Finally, Toyota was eager to improve their quality and compete globally, which could only be realized through large-scale manufacturing with fierce waste cut and management innovations.

In a certain sense, history helped push Toyota onto the track of "lean production." Facing a deep business slump in the 1940s, Toyota was planning to fire one quarter of its workforce. The company reached an important compromise with the union: one quarter of the work force was let go, but the remaining employees received two guarantees. One was for lifetime employment, and the other was for pay by seniority rather than by job function and the pay was tied to company profitability. From this new company–worker agreement, workers established loyalty to the company and company enhanced management for its workforce.

Ohno combined the strength of craftsmanship and mass production in Toyota and gave birth to the Toyota Ways, which were termed lean production in the west. It is best illustrated with the comparison between mass production and lean production.

3.3.7.1.1   Assembly Teams vs. Assembly Lines

Ohno grouped the assembly workers into teams and the teams were responsible for all the tasks. In contrast, mass production leaves the jobs of tool repair and quality checking to specialists, while assembly workers only need to do dedicated and simplified tasks.

3.3.7.1.2   Defects Firefighting vs. Fire Prevention

In mass production, defect control was relying on quality check, a firefighting approach. Thus, once an error occurred, it went through the assembly line until it was detected. At this point, a batch of products would have defects and might be sent to rework.

Ohno forced Toyota into a new culture regarding defects. To reduce rework, the best way was to eliminate the need for rework. In Toyota, workers were asked to stop the whole assembly line immediately if a problem emerged and they could not fix it. The whole team then worked jointly to solve the problem. They would trace every error back to its ultimate cause and devise a fix so that it would never occur again in the future. With this mechanism in place, the amount of rework was minimized. The quality of cars was steadily improved.

Typically, 25% of the total work hours were spent on fixing mistakes in mass production, while today's Toyota assembly plants have almost no rework areas.

3.3.7.1.3   Long-Term vs. Short-Term Relationship with the Supply Chain

In mass production, a system is divided into many parts. These parts are made by suppliers. Usually the firms offering the lowest bids get the contract. This relation between the company and the suppliers is normally temporary; suppliers are normally not trusted enough to share business information to be actively involved in the quality improvement initiative of the company. To reduce business dynamics, surplus inventory is used, increasing the inventory cost.

Toyota forms long-term strategic relations with its supply chains and trusts them. Toyota maintains some ownership control over these suppliers. Necessary high-level information is shared between Toyota and the suppliers, so that suppliers can help improve the design process. The suppliers are intimately involved in Toyota's product development. They mutually share their destinies.

Mass production relies on sufficient inventory to maintain the stable operation of the system. Ohno developed a new way to coordinate the flow of parts within the supply system on a day-to-day basis, the famous just-in-time system.

3.3.7.1.4   Fine Division of Tasks vs. the Integration of Manufacturing and Design

Mass production divides a complex problem into fine tasks and assigns those tasks to specific specialists or outsourced suppliers. In this approach, if the subsystem integration is not optimized, the assembled system will reach a limit in meeting specifications. For example, if the parts of a car door were made by different suppliers, it was very difficult to ask the assembled system to meet strict tolerances, unless very tight tolerances were applied on individual parts. Applying tight tolerance increases the manufacturing cost.

In Toyota, team leaders are required to have experience in both manufacturing and design, taking a more integrated philosophy on engineering tasks. Suppliers are empowered to optimize their engineering design for both the super system and the subsystem. Taking a system approach rather than an overly fine division approach, Toyota was able to achieve higher quality than American cars with lower cost.

3.3.7.1.5   Customer Value Driven vs. Company Profit Driven

Mass production is driven by company profit. The big system is streamlined to improve productivity, performance, and quality and to lower the manufacturing cost and product cycle time. The voice of the customer is collected and used mainly in the design stage and the marketing stage.

In contrast, Toyota emphasizes customer value in all the processes. Anything not contributing to customer value is regarded as waste and is a target of elimination. This is a powerful and very different way of thinking relative to mass production. Procedures were designed to do value mapping and cut waste. Design for manufacturing is only part of the story. There are many other sources of waste revealed, such as unnecessary process steps, too many kinds of tools, distances between subtasks that are too long, idling time between processes, inventory, overhead, defects, etc.

With customer value at the heart of production, Toyota focused on cutting process waste and increasing customer values. This resulted in high-quality products, reduced costs, and high profit. Toyota entered the U.S. market in 1957. Its reliability, oil efficiency, competitive cost, and good quality won more and more customers. With the increased oil price and quicker pace of new product introduction as compared with their U.S. counterparts, along with the other troubles of the U.S. automobile industry, Toyota finally became the largest carmaker in the world in 2008. In this process, the Toyota production system, or what we call lean production today, played a vital role.

3.3.7.2   Reflection on the Philosophy of Lean Manufacturing

Today, lean production has been widely adopted worldwide, including the American automobile industry. The American automobile industry was actually the first in the west to study and implement the lean methodology. It is meaningful to think about why American companies still lost the competition after the adoption of lean manufacturing.

We could find some clues from the comment from Mr. Ohno himself. The Toyota production system, what we now call lean, is a manufacturing phenomenon that seeks to "maximize the work effort of a company's number one resource, *the People*." Lean is therefore a way of thinking to adapt to change, eliminate waste, and continuously improve [28].

As James P. Womack pointed out, lean manufacturing was not simply a program; it was a transformation of our philosophy, the philosophy of thinking, and the philosophy of doing things.

Unfortunately, the inertia of our mind is very powerful, only strong will or crisis can change it. People asked him how their company could afford such a big transformation. He replied "How could you afford not to make such a transformation in global competition?"

We also noticed that there are lots of unique cultural aspects tied to the Toyota production system, such as the life-long employment, long-term relationships with the supply chains, loyalty between the company and the employees, the attitude of work toward perfection, the integrated approach vs. the division approach, etc. Each culture has its unique strengths and weaknesses and each society has its own issues and solutions. The game in the automobile industry is not over yet. Among the crisis starting from 2008, we noticed some good moves toward next generation cars. The competition for technology lead in hybrid cars and all electric cars has just begun. Technological innovation plus proper methodology is the key to future success.

Lean is a manufacturing philosophy featuring the elimination of various sources of waste. In the beginning, the factory floor was the center stage for waste elimination. These sources of waste were visible and the resulting cost cut was immediate. These sources of waste are the low hanging fruits. The transformation of thinking requires us to go far beyond visible wastes. We need to redefine value, customer-based value. Anything not useful to the end customer is waste. Inventory, idling time, transition of working sites, overheads, inspection, documentation and paperwork, etc., are all sources of waste. Lean manufacturing may cut the inventory, the cycle time, the rework, etc., by half for an organization that hasn't implemented lean manufacturing. Lean manufacturing also drives for perfection in product quality rather than relying on inspections to "control" quality. That's why you can't afford not to implement lean manufacturing in global competition.

Another very important fact is that keeping lean only in manufacturing is a big mistake!

Lean is a system phenomena, an organization must change to succeed. Manufacturing is only one unit of an enterprise. The other units of the enterprise may ruin the efforts in a lean initiative. Thus, lean is the duty of everyone in a company. That is why people say that lean is not a program—it is a transformation in philosophy. Define the value and cut the waste. This philosophy has proven to be very powerful. Such a philosophy can be used in all aspects of our activity, including our daily life.

Another important fact: only 20% of the total waste can be eliminated if we limit the lean effort in manufacturing or production. The other 80% is decided in the design phase. Lean manufacturing is not the whole battlefield. The battle starts from the design or conceptual phase of an activity. Bart Huthwaite elaborated on the practical procedures to implement "Lean Design Solutions" [35]. He also shared his experience on how he arrived at this point.

Design for assembly and cut part numbers is straightforward, design for manufacturing is a good progress, lean manufacturing is a big leap, but the deciding factor is the integrated big system approach. Lean must flow up to the design stage and consider factors such as technological innovations and faster project feedback. Huthwaite gave an elegant universal equation for lean design: "Optimize strategic ilities, minimize evil ings," where ilities are the values that the customers and your organization desires and ings are the various potential sources of waste [35].

Lean has been extensively studied and applied. Many tools of quality control are used along with lean, such as combining six sigma methodology with lean [36]. Six sigma is a business management strategy originally developed by Motorola. Six sigma seeks to identify and remove the causes of defects and errors in manufacturing and business processes. In six sigma, defect is defined as anything that may lead to customer dissatisfaction. Like lean manufacturing, six sigma has been extensively studied and implemented to improve the quality of production and management [37]. When it is used as a philosophy of operation in a company, such as General Electric, it actually becomes the language of the company, deeply affecting the style of operation. All employees are trained in six sigma in General Electric. Various tools are used to report the six sigma projects.

It is worthwhile to point out some potential limitations of lean manufacturing and six sigma. Lean is focused on waste cut and customer value promotion, while six sigma is focused on quality improvement through defect analysis and control. Although efforts have been used to include the innovation elements into them, lean and six sigma are criticized as being weak in promoting innovation.

For example, lean manufacturing may favor mature technologies and avoid new technologies to reduce the development time—such a time is a waste in the standard of lean. This may be necessary for the short-term business interest, but may impede long-term new technology evolution if not well balanced. The long-term prosperity of a business is still decided by the major technological innovations. Cost cut and quality control are important. So is the choice of future strategy of technology. A good balance between lean and new technology development is critical.

In conclusion, lean production has surpassed mass production as the more competitive methodology of engineering. It is a transformation of our thinking, and it should be treated as a big system phenomena rather than a collection of tools and programs. Lean should be flowed up to the design phase. In the implementation of lean and other quality control methodologies, such as six sigma, we need to pay attention to the balance between technological innovation and technology improvement.

Finally, when we compare the philosophy of lean and six sigma with the philosophy of intelligent EFM, we notice that lean and six sigma are weak in dynamic M-PIE flow integration and in overall system sustainability. Intelligent EFM inherits the legacy of lean and other modern manufacturing methodologies, while pushing it to a new level.

3.3.8 INTERDISCIPLINARY COOPERATION AND LIFE MIMIC MANUFACTURING

Case Study: Finding a Suitable Pulsed Laser

I was asked to find a suitable laser for fluorescence spectroscopy with the following specs:

- Pulse duration <500 ps
- Wavelength between 532 and 740 nm
- Pulse energy >1 μJ
- Repetition rate >5 KHz

I am the laser micro/nano material processing guy in our research center and had "rich" contacts in the circle of laser vendors. I also worked on a Web-based laser material processing tutorial when I was at Columbia University [38]. So in the first round, I found that to satisfy these specs, we need a ps/fs laser that is >$70K. I was astonished when my colleague, whose research was on spectroscopy, said that he was using a system <$15K and the system was very close to the specs. How could such lasers exist? I mean with all the expensive Q-switching electronics and optics, it is very difficult to have a system with ps laser pulses at that price.

It turned out that not only are the short pulses achieved, such lasers are also very compact and only consume 4–10 W of electrical power to run!

I felt embarrassed. I assumed myself to be an expert in pulsed lasers, but my knowledge had been in the field of high power (≫1 W, >100 μJ) pulsed lasers, and I was totally ignorant of the laser systems used in spectroscopy. In laser spectroscopy, they need only tens of milliwatts of laser power. In the case of high power lasers, expensive electronics and optics are needed to modulate the pumping and timing processes. With much smaller pulse energy (<10 μJ), special pulsed lasers had been developed based on the passive Q-switch [39,40].

Put simply, passive Q-switch lasers rely on the following material–energy interaction. Diode lasers can convert electricity to laser energy at a very high conversion efficiency, but diode lasers have big divergence and poor beam quality. Thus, they are normally used as the pumping source in laser systems. Some nonlinear crystals can absorb the pumping diode laser with a unique property. It is opaque (absorptive) to the pumping laser initially, but this changes with the amount of absorbed photon energy. Photon energy excites the lattice of the crystal and irradiant energy is stored inside the crystal until a certain threshold is reached. At this moment, it suddenly changes into a state of highly transmissive, releasing the stored laser energy. By controlling the crystal design, one can control the pulse duration and the beam quality. So, by pumping such crystals with continuous diode lasers, the tiny crystal holds the energy for a while then emits it at another wavelength in less than one nanosecond. No complex electronics or optics are involved! The crystal does the majority of the pulsing work. That's why they can be so cheap compared with the lasers used in material processing.

The Interdisciplinary Trend in Modern Engineering R&D

The purpose of the above story is not to familiarize you with the working mechanism of lasers. I am trying to explain how valuable it is to share our gems (knowledge) from different disciplines. The tools or methods well known to one group of people (experts if they only know their area) may be totally new to the rest of the technical world, let alone the general public. In the process of fighting engineering challenges, individual discipline has established many valuable tools, instruments, and methods. As if mimicking the assembly line in automobile production, engineers become well trained in a narrow area and tend to be only concerned with dedicated tasks. For example, graduates from mechanical engineering are trained in how to design and make things in general. Even this is too broad an area, so people will specialize in further narrowed down directions, such as laser material processing, electrochemical machining, precision mechanical machining, casting, CAD/CAM, sensing and control, or fluid dynamics modeling. Engineers become experts in their area, but may not have a chance to systematically cross the borders of the individual directions within their major discipline and even less of a chance to cross the different disciplines.

In the real world, experts from different disciplines are teamed up to accomplish big engineering tasks. Like in an assembly line, each of us contributes a small part to the big system. The end products of any assembly line rely on the interchangeability of the parts (knowledge here) and the overall design of the big system (project management) and the assembly line (project team).

This mode is effective for many engineering tasks. However, we may waste a lot of human beings' collective wisdom if we don't exchange knowledge systematically. In other words, simply putting parts together won't be as effective as integrating the parts into useful systems.

If engineers and managers are trained as specialists, and the ability of integration is self-learned, would the above integration process be highly efficient? This really depends on what talent you have.

People in big research centers appreciate the invaluable resource of talents from many domains. The GE Global Research Center and many engineering schools are like this. As shown in the above example, things well known to one group of people may be totally unknown to another group. Here are two more examples: Suppose we want to know the hardness and strength of a thin coating and we don't want to damage the part after the analysis, what shall we do? Ask a material scientist, he may very quickly guide you to a nanoindentation testing setup close by and solve your issue in one day or so. Suppose we want to weld two pieces of metal together inside a controlled environment but we have no access to the location of work—what solution shall we use? Well, ask a laser expert and he will tell you that you can get the work done by passing the part under a long focused laser beam—laser energy can be applied through a transparent window.

In the GE Global Research Center, we had a laser forming project funded by NIST ATP [41]. Laser forming is a process to change the geometry of metal plates based on thermally induced plastic deformation. It has the advantages of noncontact and dieless forming. Our task was to control both the deformed geometry and the resulting material integrity. Once the project was decided, a team was formed. We had experts from geometric modeling, laser material processing, sensing and control, and a material scientist on the team. The bigger team also included Columbia University and three other companies. Columbia University did the modeling and fundamental research of the laser forming process, the industrial companies tried to bring the process to practical applications. So for this new process development, we had an integrated interdisciplinary team.

The trouble is that not all organizations have such a luxury of talents. Take start-ups of solar energy companies, for example. Small companies have to focus on monthly production progress.

Engineers are required to perform multiple roles for which they are not trained. They can learn, but this costs time and money. Sometimes, it can be as simple as a key word that may hold the progress of a project for a long time. For example, a solar team needs to figure out a way to measure the stress distribution of the layered structure. It may take a while before the engineer finds out that x-ray microdiffraction can be the solution [42]. X-ray stress measurement is well known, but it normally has a coarse resolution. X-ray microdiffraction can resolve the stress distribution in microns.

Solving the challenges of very complex systems requires the expertise in multi-disciplines. We are increasingly forced to cross the borders of disciplines. Interdisciplinary research has become an unavoidable trend in modern engineering activities, such as in the study of space exploration, micro engineering, alternative energy development, biomedical engineering, advanced manufacturing, nanotechnology, etc. ASME studied the trend of manufacturing in the twenty-first century; a list of the interdisciplinary research is in [43].

This interdisciplinary corporation is actually a manifest of the general trend of dynamic M-PIE flow integration in engineering. The information flow, the energy flow and the material flow must be properly integrated to achieve any engineering task. Today we see the trend of interdisciplinary corporation.

Crossing the disciplines gave rise to new research frontiers. For example, bio-mimic manufacturing was enlightened by many of the biological structures or processes in nature. The lotus leaf inspired people to develop self-cleaning surfaces, shark skin inspired new designs to lower the drag in water, etc. Life cycles over millions of years of natural selection have evolved or optimized themselves to adapt to the changes in nature. Lives are beautiful systems of M-PIE flow integration and role models of sustainability. Imagine the life cycle of grass. Its growth is more than a 3D process. The structure is active rather than passive. It can sense the environment and adapt to protect itself. It lives on solar energy and it is part of the natural cycle of the big ecological system. The photosynthesis process of plants is a process scientists are still trying to mimic. A recent article in *American Scientist* reported the quantum effect in photosynthesis [44].

Many universities have a department of bioengineering or biomedical engineering; some professors have roles in multiple departments, such as mechanical engineering or chemical engineering.

Engineering went through the process of branching stage after the Industrial Revolution. Engineering in the twenty-first century will see the increased trend of interdisciplinary integration. Such integration will be a scattered effort in the beginning, but will become more and more systematic with time. In this process, manufacturing is evolving quickly toward advanced integration of the dynamic M-PIE flows, with the future direction being life-mimic more than 3D manufacturing. This is part of the reason why this book was written. We want to address the interdisciplinary nature of modern engineering, with a focus on manufacturing processes.

3.3.9   SYSTEMATIC AND ORGANIZED INNOVATION

Class discussion: *Do you believe you can greatly improve your ability of technological innovation in one week? How could you achieve this?*

3.3.9.1   The Value of Technological Innovations

Technological innovations are the driving force of long-term economic waves. Innovations are also the differentiator in global competition. Thus, any means of improving the ability of innovation are critical to the success of an organization or a nation.

When this chapter was written (March 2009), the world economy was in a deep crisis. The Dow Jones Industrial stock index dropped from 14,000 in 2008 to less than 6,600 in March 2009, unemployment rate reached 8% nationwide in the United States. The American automobile giants were in big financial trouble, but there were at least two pieces of positive news. The first was the heavy investment in hybrid and all electric cars, which are the future of the automobile industry, and the other was converting some of the automobile production sites into the production lines of alternative energies, such as solar thermal electricity generation. In the latter case, mass production experience in the automobile industry could be quickly implanted in the solar business, whose runners were somewhat novices in mass production in 2009.

In the peak of economy, seemingly all players can survive in the field. In harsh environments, only the more adaptable can pass the test. This was the case of the solar business in 2009. Many solar PV startups appeared around 2008 when the oil price reached a historical high ($140 per barrel). In the crisis of 2009, many of them ran short of capital funding; the business was facing a big shake up. Big companies would acquire small companies with promising technologies, while less attractive companies would simply disappear.

Imagine we were in a big epidemic like the Severe Acute Respiratory Syndrome (SARS) crisis in 2003 [45]. All existing ways failed to stop the spread of SARS in 2002. Heavy social resources were put into developing new coping treatments. The most effective ones became the savior. Innovations, be it in technology, society, or politics, are the saviors of down times.

No doubt, technological innovation is important, especially in times of crisis. The next question is how we could better use our capability of innovation.

We were used to the stories of legendary scientific discoveries, such as Newton's discovery of the law of gravity, Alexander Fleming's "accidental" discovery of penicillin [46], etc. Many people have the impression that invention is a chaotic process, highly depending on personal inspirations or pure luck. Edison admitted that though he could improve machines, he could not improve men. Similarly, many people assume that an inventive mind is a gift at birth, so you cannot highly improve one's creativity through training.

Those views are understandable in the twentieth century, but people should have changed their minds in the twenty-first century.

3.3.9.2 Toward Organized and Systematic Innovation

In the scientific method [47], we first define the question, observe to gather information and resources, and then form a hypothesis. We do experiments and collect data. The data are analyzed and interpreted. Conclusions are drawn on the hypothesis. The results are published and retested by other scientists. This is an iterative process. We can call this the *hypothesis/test philosophy* for brevity.

Trial and error is a classical problem-solving methodology used by human beings and nature [48]. Biological evolution is a trial-and-error process. Through random mutations and sexual genetic variations (trials), the reproduced later generations accumulate the genomes with better potential of being adaptive to the environment. In engineering, we analyze the problem, prioritize the possible solutions or all of the possibilities, and then try them out. The earliest solution or the solution giving the best result will be chosen.

The trial-and-error approach is effective with simple problems or problems with no existing knowledge. It is widely used due to the following features:

1. It is solution-oriented. It does not attempt to discover why the solution works. Thus, you can find practical solutions to urgent tasks.
2. It needs little knowledge to start. You can proceed where there is little or no knowledge of the subject. This is not to say that it is totally random. Actually, you should make your best analysis of the problem and prioritize your test to be economical.
3. You may not reach optimal solutions and the solution may be problem-specific, but you do have the chance to find some valuable solutions. That's why it has traditionally been the main method for finding new drugs, synthesizing new polymers and alloys, etc.

This is a very effective methodology when used properly. In modern research and development, normally a team is formed to solve some engineering issues. Based on benchmark and experience,

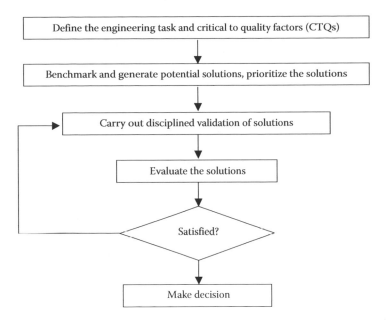

FIGURE 3.4   The modern trial-and-error methodology.

potential solutions are brainstormed, conjectured, and documented, then prioritized based on key critical to quality factors (CTQs). Then tests and experiments are planned out in a disciplined manner. Solutions are down-selected based on their performance. This is an iterative process as shown in Figure 3.4. Note that the iteration may need to go to the define phase if no good solutions were found in the first round. Each potential solution can be turned into a project, which may further include the development of some manufacturing processes. Methods such as design of experiments (DOE), six sigma tools, lean design solutions, concurrent engineering, etc., can be used in these projects.

The biggest limitation of the trial-and-error approach is that, for new problem solutions (innovations), it may not reach the optimal solutions cost effectively, given the limited time and resources. Edison invented the first practical light bulb after over 1000 trial and errors. Not everyone can survive such a long and expensive process.

Is there a better way out? In other words, can innovation be carried out in a systematic and more efficient way? The answer is positive.

Substantial studies had shown that there were rules to follow in innovations, and innovation could and should be an organized and systematic process [49–54]. Genrich Altshuller, a brilliant Soviet Union engineer and researcher, spent his life wondering not how to build better inventions but how to build better inventors. In this process, he brought TRIZ into the world [52]. TRIZ is the acronym for the Russian phrase "Теориярешения изобретательских задач," meaning "The theory of inventive problem solving."

TRIZ and intelligent EFM provide ways of organized and systematic innovation.

3.3.9.3   Introduction to TRIZ

After the Second World War, the Soviet Union assigned a team in the military to analyze the large quantities of technical information from Germany and the worldwide patents. Altshuller was on this team. From the late 1940s to 1969, Genrich Altshuller reviewed about 40,000 patents in order to find out in what ways innovations had taken place. As an engineer, an inventor, and a science fiction writer, Altshuller observed that *invention was mainly a process of removing technical contradictions with the help of knowledge and certain principles.* This vision finally led to the establishment of TRIZ, the theory of inventive problem solving. The major elements of TRIZ include the concept of technical contradictions, the concept of ideality, sufield analysis, the matrix of 40 principles of invention, laws of technical trend prediction, and knowledge-based engineering solutions. After the 1970s, TRIZ has become an important methodology for efficient inventive problem solving. Today, TRIZ has been used worldwide in both technical and nontechnical applications. Commercial software is available to assist the implementation of TRIZ.

3.3.9.3.1   The Key Difference between Compromise and Ideality

In a certain sense, conventional engineering solutions focus on how to achieve a good compromise between contradictory requirements. We define the problem, identify the key requirements, then try to find possible solutions. We rank these solutions based on how well they may satisfy the requirements.

The trap is that too often we rush to solve the problems without a deep understanding of the problem and without a good strategy to solve the problem.

The trial-and-error approach works, we are satisfied, so we stop at the un-optimized solutions. These solutions are the results of engineering compromises. Let's use the example of a car to illustrate this point.

We want the car to have good oil mileage and a high level of safety. To have good mileage, we should reduce the mass of the car and improve the efficiency of the engine. To run safely, the car must have a frame with enough mass and strength. So, the resulting design could be a compromise of the two requirements. Surplus mass is used to increase the safety of the riders. Today we all know a better solution. To improve the protection of life in an emergency, airbags are used. To reduce the weight of cars, composite and polymers are used. With the adoption of airbags and new materials in modern cars, the mass of modern cars can be much smaller than cars in the old times. This contributes to the mileage increase as well.

The first step of problem solving in TRIZ is to define the problem properly and formulate the ideal final result. Asking the right questions is the first step to success. In many situations, finding the right problem is the critical step for the whole process. To me, formulating the ideal final result is a very powerful tool in TRIZ and this distinguishes TRIZ from other methodologies.

As Ellen Domb pointed out [55], the goal of formulating the ideal final result is to eliminate rework—to solve the right problem the first time by addressing the root cause of the problem or customer need. It helps reach breakthrough solutions by thinking about the solution, not the intervening problems.

Suppose we need to develop the next generation motorized stage, and the current product is lead screw driven. The ideal final result can be a motorized stage with no friction in motion, no inertia to limit acceleration, no positioning error to affect accuracy, and scalable to unlimited range of travel if needed. This is very different from the approach we normally take. Normally, we apply our experience and knowledge into the goals in the beginning, forgetting to drill down what we really want to achieve in the end. Thus, what we can achieve is normally a compromise of specifications. For example, we may set our target of design to be: design a stage with <0.1 friction coefficient, 0.5 m travel, >4g accelerate, and 0.5 μm positioning accuracy. The idealized solution may immediately guide us to an air-bearing mechanism for next generation stages, while the normal approach may lead to an effort to improve the solid–solid contact, resulting in a product design lagging far behind the air-bearing stages.

A basic principle in TRIZ is that systems will evolve toward increased ideality, i.e., toward increasing benefits, decreasing costs, and decreasing harm. The ideal final result is thus a system with all the benefits, none of the harm, and none of the costs. The ideal system may occupy no space, have no weight, and require no labor and maintenance. In short, it is the solution to a technical problem without the constraints of the original problem. It delivers true customer values without the immediate constraints of the real world.

From here, one can bridge it to the real world constraints, making it practical. The human mind has big inertia. We tend to use what we are familiar with, although it may not be the best way to go.

The Ideal Final Result techniques help break the mind inertia, allow us to go out of the box of real (or current) world constraints, get closer to ideal solutions, and reject compromises whenever possible. Once the right direction is set, we have all the wisdom of human beings on hand to help solve the problem. That's the essence of TRIZ.

3.3.9.3.2   Some Important Concepts in TRIZ

Without going into details, we review some important concepts in TRIZ [51].

System Conflicts

Improving some system attributes leads to the deterioration of other system attributes; this situation has system conflicts. System conflicts present when (1) the useful action simultaneously causes a harmful effect or (2) the introduction (or intensification) of the useful action or elimination (or reduction) of the harmful action causes deterioration of the system.

Altshuller studied system conflicts and concluded that there were only around 1250 typical system conflicts. He developed the 40 typical techniques to solve the system conflicts and organized them in the form of a technical conflict matrix. Interested readers are encouraged to refer to [51,54] for more information.

Physical Contradictions

A situation in which one object needs to be in mutually exclusive physical states is called a *physical contradiction*. Physical contradictions are formulated in intensified ways in TRIZ to drive directly to the point of the issue. For example, part A needs to be cold and needs to be hot, or needs to be light (for good dynamics) and needs to be heavy (for strength), or needs to be in place A and not in place A.

Three generic methods are recommended in TRIZ to eliminate the physical contradictions rather than compromise them: (1) separate opposite properties in time, (2) separate opposite properties in space, and (3) separate opposite properties in different levels of a system. Traffic lights separate the physical contradiction in time, a bridge separates it in space. A stove in the kitchen solves the physical contradiction (be hot and not be hot) by isolating the conduction in different parts of the systems.

The solutions to system conflicts and physical contradictions are not bizarre things. Human beings have tried to solve them since ancient times. So, don't assume you have a "new" challenge, assume you have something someone already explored first. It could be a waste of time if one simply dives into "inventing" solutions to his engineering contradictions, neglecting the huge resource of human collective intelligence. This is not to deny the need of independent thinking. TRIZ emphasizes the value of knowledge base.

Mini- and Maxi-Problems

Solutions to a problem can occur at different levels. A mini-problem in TRIZ is the removal of undesired aspects without a major change to a system, while a maxi-problem is a solution without any constraints. Mini-problem solutions usually require less resources and are more economical than maxi-problem solutions, while maxi-problem solutions can be more dramatic and revolutionary.

For example, to reduce the air-pollution of automobiles, one can choose to improve the gas-based engine, try to find a mini-problem solution, or choose to develop all-electric cars, totally eliminating the source of air pollution. The latter case is a maxi-problem.

Substance-Field Analysis

A technical system performs certain functions. TRIZ treats technical systems as transformations between energy fields and substances. A process performed by a technical system consists of at least an energy source, a tool, and an article (substance). The term "field" in TRIZ is similar to but narrower than the General Energy Field concept in intelligent EFM. The field can be physical force fields (gravitation, electromagnetic fields, strong and weak forces) and distribution of physical fields (thermal fields, chemical medium field, etc.). The term "substance" has a broad meaning, which is the technical system with any possible level of complexities, such as a part (the tire of the wheel of a plane), an assembly, or the whole system (a space shuttle).

The substance-field structure of a technical system is called a sufield. A minimal technological system must have a field, a tool, and an article. The problem solving process can be treated as a process of completing and improving the sufield structure.

3.3.9.3.3 Using the Laws (Trends) of Technological System Evolution

There are significant, stable, and repeatable trends in the evolution of technological systems and in the interaction between the technological systems and the environment. They are called laws of technological system evolution, possibly for the reason of their general applicability to all engineering disciplines. These laws include the following:

1. All technological systems evolve toward an increased level of ideality. Ideality can be measured by the ratio of delivered benefits to required cost. A good example is the evolution of the cell phone. Its size shrunk, its functions multiplied, while its cost lowered to a commodity. The same thing can be said about computers, cars, the production of a metal cup, etc. This is also a trend of cross-function integration and cost reduction. With this law, one can immediately forecast some aspects of the future of a new product, and hopefully he acts properly to secure the opportunity. For example, the laser business is still a quickly developing business. Lasers should learn a lot from the histories of computers and automobiles.
2. Various parts of a system evolve at different rates and this leads to system conflicts. This may trigger chain reactions and the resulting effects could be positive or negative. For example, fuel-cell-based cars may have many subsystems of the car ready, but the quick charging system and distribution infrastructure are lagging behind. These system conflicts require inventive solutions.
3. Technological systems evolve from simple-functional systems toward multifunctional systems, from mono-systems to higher-level systems. You can find enough examples. This is a general technical trend, so you can safely expect the integration of functions. If such integration has not happened in an area, it can be your opportunity. For example, waterjet machining is conventionally used for through cutting. You can surely use it for hole drilling, 3D milling, turning, etc. Laser material processing is finding new applications every day and is increasingly integrated into sophisticated systems, such as a solar cell production system.
4. Technological systems evolve from rigid systems to more flexible and adaptive ones. To me, this should be replaced with "evolves toward increased intelligence."
5. Other laws include the shortening of the energy flow path and the transition from the macro to the micro level.

Using these laws can enhance the analysis of benchmarking, help forecast the future direction of products, and take proper actions. It is powerful in finding the focus of improvement in existing products, processes, and services. This leads to the immediate expansion of creativity.

Strong Support from the Knowledge Base of Engineering

Our world runs under the government of physical, chemical, biological, and mathematical laws. There are over 6000 physical effects described in scientific literature, while a normal engineer is only familiar with <200 of them. Purposely and systematically seeking support from this huge knowledge base can make a big difference in the efficiency of engineering activities. When an issue is becoming very difficult to solve in your familiar domain, it is possibly time to seek the help of the collective wisdom of human beings.

For example, could we create a very localized low temperature environment, such as −50°C over an area of far less than 1 mm²? Check out the cooling effects. We are familiar with phase transitions.

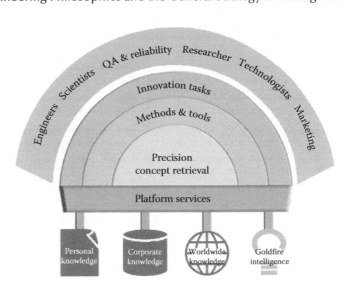

FIGURE 3.5   The structure of commercial software. (Courtesy of Invention Machine Co. Boston, MA.)

How about the Joule–Thompson effect, Rank effect, magnetocalorific effect, and thermo-electricity? In 2005, NIST produced a chip-based refrigeration system, merely $25\,\mu m \times 25\,\mu m$ [56,57].

On one hand, we are only aware of limited effects, on the other hand, we may not be skillful in using the known effects. Well, some effects may be critical to a specific engineering issue. This implies big potential for creativity improvement. To assist the application of scientific effects in engineering, various knowledge bases have been established. Now another issue stands out. Facing too much information, how can we effectively filter out the useful information? TRIZ developed tools and software to assist in this process.

Figure 3.5 shows the structure of the TRIZ implementation through the commercial software Goldfire of Invention Machine Co. [58]. It is a software to make the accumulated intelligence more valuable for engineering and management. The innovation platform includes personal, corporate, and worldwide knowledge. The software provides intelligent semantic indexing of information. TRIZ tools and algorithm are implemented, lowering the difficulty of operation.

Imagine the benefits of an internal knowledge base of a big organization. There are many departments in a big organization, such as GE, IBM, or NIST. Each department has its own talents and hardware assets. Talents have quite some dynamics (people join and leave), hardware may be used or put into storage. A knowledge base of available expertise and existing hardware can immediately result in big savings in time and materials cost if such a database is properly used.

3.3.9.4 Reflection on TRIZ and Technological Innovation

Historically, scientific inventions, innovations, and discoveries have been put on mysterious attires, ascribing too much to personal inspirations or heroism. Before the widespread accessibility of knowledge base, valuable information was only available to special groups. With the explosive accumulation of detailed knowledge in various disciplines, conventional education can only equip the educated with a narrow area of knowledge, without conveying a high level philosophy of engineering. This leads to unnecessary barriers in technological innovations.

TRIZ proved that creative problem solving had general rules to follow and was a capability that anyone could be trained to improve. To me, the main contributions of TRIZ are as follows:

1. Changing the engineering solutions from compromises to ideal final results. This can make a fundamental difference, since this can guide us to different directions. We say that technological innovations are important, high impact, and high risk. The right choice of research directions is the first step.

2. Systematic solution to engineering contradictions. The 40 principles, the laws of technology evolution, the concept of mini/maxi problems, and other tools can help engineers solve contradictions in a systematic way, rather than in random ways. TRIZ actually claims that a majority of the so-called inventions can be deduced from the TRIZ rules.

3. Making use of all potential resources. TRIZ gave many inspiring solutions that feature the clever use of existing resources. You might be surprised at how many resources you have neglected once you grasped the essence of resource analysis in TRIZ. This is one of the reasons that the substance-field analysis is a powerful tool. The general field and tool concept expand people's minds to wider resources. The resources are actually the collective wisdom of human beings, including personal, organization, worldwide, historical, and natural knowledge. Later advocators of TRIZ quickly found the market value of this aspect; commercial training centers and commercial software are spread worldwide. Accessing and filtering information are becoming much easier than before. Given the right information and being equipped with the right methodology, many people can make the right decisions. Global competition in technological innovations has entered a new era.

In short, TRIZ has triggered engineering into a new era of systematic and organized innovations.

We have emphasized the importance of technological innovations. Technological innovations can include process innovations, product innovations, service innovations, and strategic innovations. The success of innovation relates to the whole system, both internal and external.

In Chapter 1, we also pointed out the big difference between innovation and invention. Invention is just the early stage of innovation. Invention generates the creative ideas and potential technical implementations, while innovation has to carry the idea into commercialization.

Let's also differentiate engineering activities and innovations. Technological innovations are part of the engineering activities. Engineering activities also include the stable operation of matured technologies. All technologies have life cycles. While inventive problem solving requires breaking the thinking barriers, efficient engineering operation requires proven rules and standards. Thus, if not taken care of properly, the matured technology branches can be in collision with the innovation units—they may compete in resources, in exposure, in promotion, etc. Reference [50] elaborated on the difficulties and the processes for strategic innovations.

Now we are ready to discuss the limitations of TRIZ.

1. *Due to the values of TRIZ, people have developed TRIZ into quite complex tools. On one hand, TRIZ becomes a sophisticated package of tools; on the other hand, this may impede people's understanding of the general philosophy of engineering.*

2. It is important to treat TRIZ as a group of new engineering methodologies in addition to using the specific TRIZ tools and techniques. In this way, TRIZ can be seamlessly pieced together with other engineering methodologies and tools, such as lean manufacturing, six sigma, quality control, etc. Otherwise, people may run into the trap of justifying old ideas with TRIZ rather than improving creativity with TRIZ.

3. TRIZ is strong in creative idea generation. The idea generation stage is important in the whole process of innovation. Choosing the right idea may decide the future of an innovation. However, it must be understood that idea generation is only one of the beginning steps. Once a technical plan is decided, the execution of the plan requires disciplined engineering methodologies.

4. The effort to drive an ideal solution is good, but an ideal solution is dependent on what criteria you base your judgment. TRIZ defined ideality as the ratio of benefits to cost and harm. In engineering, we need more concrete guidance on optimization criteria. This is well addressed in intelligent EFM.

5. There are many rules, laws, and standards in TRIZ, but there is not enough emphasis on the philosophy of from nature to nature. In other words, it summarized the general rules

human beings tried to thrive in market driven economy. To adapt it to the future of technology, the fundamental philosophy of engineering should be added and used to guide all other aspects of technological innovations.

6. Despite the many rules, laws, and standards in TRIZ, there are some important principles of engineering that are missing. For example, information is treated as a knowledge base rather than a dynamic flow. In intelligent EFM, we introduced the concept of general intelligence, general energy field, and general logic functional materials, and emphasized the integration of dynamic M-PIE flows.

7. The use of TRIZ in manufacturing (or engineering in general) requires further adaptation. Many of the rules are abstract; one still needs to map the suggested ideas to real world solutions. So don't assume a machine can really help you do innovations; in the end, the human intelligence does it.

8. Although it may be correct to assume that your issue is an issue somebody had studied sometime somewhere, don't underestimate the value of personal inspirations and intuitions. TRIZ can be a useful tool to assist us, rather than a tool to govern us. Sometimes, it is useful to let people's minds run wild first to break the box, since the out-of-box thinking can come from people who don't know a box exists. For example, to improve laser drilling speed, my teammate suggested dividing a laser beam into multiple beams readily, while people familiar with the capability of nowadays lasers tend to eliminate this option.

9. Pay attention to the difference in mini-problems and maxi-problems. All innovations require resource assignment and can be costly. The most costly innovations are failed innovations. So there has to be a balance between new ideas and practiced ones. In short, tie TRIZ to reality and combine it with other engineering methodologies.

In summary, TRIZ and other innovation studies have advanced technological innovation to a new era of systematic and guided innovations. However, invention and R&D could never be a 100% mechanical process. The majority of innovation activities can be systematic, but we should not underestimate the importance of personal inspirations and intuitions, and we also need to pay attention to the limitations of TRIZ. It is important to use TRIZ and other innovation methodologies as valuable engineering philosophies and combine them with established ones. Intelligent EFM inherits the important essence of the existing methodologies and pushes them to a new stage.

3.3.10 SUSTAINABILITY AND INDUSTRIAL ECOLOGY

All the engineering methodologies we reviewed so far are efforts trying to maximize the profits of firms. To maximize the profits of firms, we need to add value to the products or services, do it efficiently (improve productivity), and minimize the cost.

This has been the dominating thought in the market-driven economy since the Industrial Revolution. In the twenty-first century, this kind of thinking must be changed.

In this thinking, human beings tactically assumed the following:

1. Natural resources are either free gifts to human beings or the commodities of human beings. Air and natural water were free until recently, when rules were passed in many governments to regulate air emission and water usage. Forests, islands, deserts, mountains, etc., were normally treated as realties owned by someone or some groups, and could be purchased. Minerals had a long history of being treated as commodities. Oil anywhere, be it underground or underwater, is a commodity that nations try to own and sell.

2. Human beings are the master of Earth, thus our engineering activities only optimize for the benefits of human beings or for the benefits of special groups. With this thinking, we took it for granted when human settlements devoured and continue devouring the habitats of other creatures on Earth, as long as the settlements had or will have contributions to

economy or someone can harvest money from it. Young mothers feel horrified when they read that small babies can catch strange immune system diseases after playing on lawns that were just sprayed with pesticides. In the United States, many people would spray pesticides simply to make the lawn look better. When human health is endangered, the society responses, but how much have we done to reverse the national scale pollution of the environment?

3. People also assumed that human activities were far from reaching the carrying capacity limit of nature, thus winners in market competition were and could be rewarded without considering their impact on the big ecosystem. This assumption will lead human society to a dead end if not revised. The global population increased from around 500 million in 1650 to 1 billion in 1850. The world population is currently 6.7 billion and may peak at 9 billion in 2070 [59]. The human impact on the environment is global and is creating serious consequences that require immediate actions.

In March 2009, more than 2500 climate experts from 80 countries at an emergency summit in Copenhagen said there is now no excuse for failing to act on global climate change. The Copenhagen Climate Council issued a strongly worded statement:

"The climate system is already moving beyond the patterns of natural variability within which our society and economy have developed and thrived. These parameters include global mean surface temperature, sea-level rise, ocean and ice sheet dynamics, ocean acidification, and extreme climatic events. There is a significant risk that many of the trends will accelerate, leading to an increasing risk of abrupt or irreversible climatic shifts."

The statement also added: "There is no excuse for inaction. We already have many tools and approaches—economic, technological, behavioral, management—to deal effectively with the climate change challenge. But they must be vigorously and widely implemented." [60]

With movies such as Al Gore's *The Inconvenient Truth* [61], the animated science fiction *WALL-E*, and with the dramatic climate disasters in recent years (the water crisis in California in the United States; the drought, flooding, ice rain, and sandstorms in China; the forest fire and flooding in Australia; the meltdown of the icecaps in the Arctic, Antarctic, Greenland, etc.), the general public's awareness of global sustainability has increased.

To the engineering group, we must *systematically* act on sustainability since we are the foundation and the long-term drivers of the global economy and we can influence the whole social system.

Thus, the general engineering methodology without in-depth consideration of sustainability is incomplete, no matter how efficient it may be in creating profits for firms.

So what is the definition of sustainability?

A widely accepted definition of sustainability by the World Commission on Environment and Development is *to meet the needs of the present without compromising the ability of the future generations to meet their own needs.* Another definition is *to improve the quality of human life while living within the carrying capacity of supporting ecosystems* [62].

Ecology is the scientific study of the distribution, abundance of life, and the interactions between organisms and their environment. An ecosystem is a natural unit consisting of all creatures (plants, animals, and microorganisms) in an area functioning together with all of the nonliving physical factors for the environment. An ecosystem, such as a desert ecosystem, the tropical reel ecosystem, etc., can be a completely independent unit of the interdependent organisms that share the same habitat [63]. One important feature of an ecosystem is its sustainability—it is able to maintain a state for a long time. The output of one part of the ecosystem is the input of the other part of the ecosystem, materials and energy flow in cycles, making the whole process sustainable.

In contrast, the modern economy features acquiring resources from nature, turning them into a commodity, consuming them, and leaving waste in all these stages. This is an open-loop process.

Inspired by the biological ecosystem, the concept of industrial ecology was first popularized in 1989 in *Scientific American* [64] and has grown quickly in recent years. Industrial ecology is an interdisciplinary study focusing on the sustainable combination of environment, technology, and economy, which is an analogy between natural systems and socio-technical systems. Industrial ecology tries to evolve our open-loop process to a closed-loop process where by-products of one unit in the economy can be used as inputs of other units, minimizing the waste generation and environmental impact of human activities.

In industrial ecology, industrial systems can be at different scales, such as a factory, a nation, or the whole world. The industrial systems are considered not as separate from the biosphere, but rather as cases of an ecosystem based on infrastructural capital. The research focuses of industrial ecology include the following:

- Industrial metabolism—material and energy flow studies
- Life-cycle planning, design, and assessment
- Eco-design—design for the environment
- Product stewardship—extended producer responsibility
- Analysis of eco-efficiency
- Dematerialization and decarbonization

The book *Natural Capitalism: Creating the Next Industrial Revolution* by Paul Hawken, Amory Lovins, and Hunter Lovins views the world economy as being within the larger economy of natural resources and the ecosystem than sustains us. While the traditional economy primarily recognizes the value of money and goods as capital, natural capitalism includes both natural capital and human capital. Natural capitalism is more than the three Rs—reduce, reuse, and recycle. It offers a big system view and gives many concrete suggestions. The authors believe that the next industrial revolution depends on the following four central strategies [65]:

- Conservation of resources through more effective manufacturing processes
- Reuse of materials as found in natural ecosystems
- Increase values more from quality than quantity
- Investing in natural capital, and restoring and sustaining natural resources

Sustainable development is a trend of our time. The requirements of sustainability present both challenges and opportunities to all parts of society. Green technology, environmental friendly technology, ecological development, green energy, etc., have become buzz words. The risk is that despite all the "green" activities, if we continue with the unsustainable levels of natural exploitation, or if we halt the trend too late, we may reach an irreversible stage in climate change and ecological system degradation.

Environmental protection is not the focus of this book. A good introduction on sustainability itself requires a dedicated book. Interested readers are encouraged to dig out more information on this subject. It is important to point out that modern engineering philosophy should accommodate the requirement of sustainability, and the criteria of engineering optimization should include sustainability. With this kind of thinking, we will design sustainability in engineering, creating a situation of more fire prevention than firefighting.

In this sense, we say, engineering must take systematic actions toward sustainability, working with the big system of human society.

3.4 BACK TO THE STRATEGY OF INTELLIGENT ENERGY FIELD MANUFACTURING

In the beginning of this chapter, we argued that the mission of engineering should be helping to *establish a market-driven sustainable economy through technological innovations*, and the fundamental philosophy of intelligent EFM is the combination of the theory of from-nature-to-nature and Tai-Chi. We also presented the general strategy of intelligent EFM:

To implement intelligent EFM successfully, we should optimize the integration of dynamic M-PIE flows, optimize the strategic values both for the customer and the enterprise, minimize waste and maximize the sustainability of the whole ecosystem. We have the following *success equation of intelligent EFM:*

$$\text{Success} = (\text{optimizing the dynamic M-PIE flows}) + (\text{optimizing the strategic values})$$
$$+ (\text{minimizing waste}) + (\text{maximizing sustainability})$$

It is clear that the above philosophy and strategy are not things coming from the empty sky, instead, they are rooted in the history of engineering and they are the new development of multiple important engineering methodologies. Intelligent EFM takes a holistic approach, absorbing the essence of mass production, lean manufacturing, intelligent manufacturing, quality control, theory of innovation, and sustainable development. This is a huge challenge, but once we realize the essence of engineering, it is not that involved. We regard engineering in general as a process of injecting intelligence into the integration of the dynamic M-PIE flows (material flow, information flow, energy flow, and project flow), thus the name of this book, intelligent energy field manufacturing.

We are facing critical challenges in the twenty-first century. The philosophy and methodology of engineering must be changed to meet the challenges. It is our wish that intelligent EFM can bring about such changes.

3.4.1 Discussion of the Sustainability and Prosperity Curve

Green and sustainability have become buzzwords nowadays. To bring out real changes to the way we do things, we need to know what actions are needed and act now, starting with each of us. Figure 3.6 is what I called the sustainability–prosperity (SP) curve of our society. We want to maintain high living standards, have plenty of desired things, be healthy, feel safe, be proud of our community, have a happy family life and satisfied social and spiritual life, and be confident in the future. Let's use the word prosperity to describe all these good things. The other side is the sustainability of the

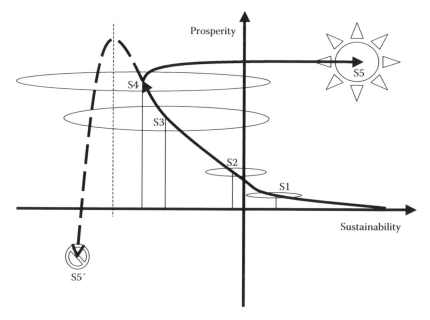

FIGURE 3.6 The S–P curve of human society.

big ecosystem. We share the Earth with all the animals, plants, microorganisms, rivers, mountains, oceans, the air, and the sunshine. Nature has many beautiful cycles, which run forever, like all the wild lives. There is no waste in nature; the output of one segment is the input of the other segments, while materials and energy flow and regenerate. We call this sustainability. Human beings are part of nature and rely on the sustainability of nature. Ideally, we want to have high prosperity and positive sustainability. Unfortunately, our record is not a proud one.

As shown in Figure 3.6, before the Industrial Revolution, human activities had low prosperity and positive sustainability on average. The agriculture in China 50 years ago is a good example. People harvested the plants by manpower and with animal assistance. Food was consumed and human and animal excretion and defecation were collected and mixed with plant leaves and stems. There were no garbage landfills in villages! All of the organic materials were transformed into organic fertilizers and sent back to nurture the plants and vegetables. No pesticides were used. River water was drinkable in many places and the air was always fresh. The field maintained fertility for a long time in this way, without degradation of productivity. The Chinese people had been farming in this way in the same field for over 4000 years! It was very natural for the Chinese people because the Chinese culture guided people to live in harmony with nature. This is the from cradle-to-cradle style of human society. The majority of human activity won't hurt sustainability on a large scale. Some activities protruded into the negative sustainability quadrant, such as the use of metals and ceramics. There were hunters, but the wild animals were plenty. Imagine the ancient people crossing the sea with wooden boats with the assistance of solar-powered wind. The life cycle of the wooden boat could easily become part of the material cycle in nature. The transportation system had zero net greenhouse emissions. Paint might be used, but a majority of them were biodegradable. We can call this stage S1, the stage before the Industrial Revolution.

Things changed very strikingly in the past 200 years. With the large-scale use of coal, a fossil fuel, Pandora's box was opened. Humans harnessed a condensed form of energy. Oil followed coal in the life change process of humans. With the invention of steam engines, locomotives, steam ships, and power plants, humans had a big jump in capability in all areas. We expanded our territory very quickly. We admired people who could build more chimneys that smoked and made fortune. Very few people considered the interest of the wild lives and the general ecosystems. Cities grew bigger and bigger. The environment in big industrialized cities, such as Detroit and Chicago in the 1920s, degraded to unbearable levels. Facing guns, no animal could compete with a human, including wolves, lions, and tigers. Many animals have become extinct due to killing by humans. With the development of chemical engineering and materials science, many synthetic materials were invented, such as plastics, composites, medicines, alloys, etc. The air emission was not a balanced carbon cycle anymore. Wastes could be freely dumped into rivers and oceans. The most dangerous thing was that humans started to produce and use toxic materials and consume materials very wastefully. In this period, we entered the era of negative sustainability. The global prosperity increased quite a bit. Many human activities, such as agriculture, were still sustainable. Our impact on the ecosystem was not at a global scale yet. This kept the sustainability negative but close to zero. This is stage S2.

With the rise of the oil industry, the invention of the gas-based internal combustion engine, and the convenience of electricity, human productivity advanced very quickly. More and more countries were industrialized—industry became more important than agriculture. So humans worried less about existence, since that problem was solved. We felt energetic. We broke many fronts—nuclear energy, highways, airplanes, space exploration, deep oil drilling, metropolitan cities, lasers, biology, etc. The quick progress in technology resulted in greatly affluent material supplies and two world wars. Then we entered the era of the market-driven economy. Synthetic fertilizers and pesticides were widely used to increase farming yield. The pollution and degradation of the environment finally triggered environmentalism. However, the volunteer work of environmentalists was too weak compared with the rich capitalists and national economy needs. In this period, we really dominated the planet. Even the whales may be extinguished without in-time legal restrictions. In this period, prosperity further increased while sustainability quickly slid into larger negatives. Think about the

Hudson River, a big beautiful river, from which the fish cannot be eaten. It is sad. This is stage S3, where regulations were loose and profit was the number one priority. With the developed countries evolved into S4, many developing countries are repeating the mistakes of S3. When economy development is number one in a nation, it is simply too difficult to protect the sustainability of our environment. The fertility of fields degraded quickly around the world in this period.

In stage S3, the consumption style was *from the cradle to the grave*, while the production was from door to door. People buy things from the market, consume them, then dump an alarmingly high percentage of technical materials as waste. Much of the waste ended their life in the landfill. Firms either get materials from others or directly from nature, make the products or deliver the service, and produce waste in many steps. Profits are harvested once the product or service is out of the gate—producers are not held responsible for recycling the materials they used. They may be required to pay a fee to handle the waste, but this fee is far from proportional to the long-term damage to sustainability. Our culture also changed in this period. With the luxury of cheap energy, people started living in a constant temperature environment. The automobile industry became the pillar of manufacturing in many developed nations. Global interconnection took shape in S3.

Many nations are in Stage 4, where prosperity is high and sustainability is even lower. People have realized the serious consequences of ecosystem degradation and many positive cultures had sprouted. More regulations were enforced to protect public interest. Government and international organizations act as guardians of our environment, while commercial organizations tried to adapt to the lowest cost way of obeying laws. With the lever of market to protect environment, green and sustainability became buzzwords. A product is called green simply because it is less "bad" in waste creation.

In the twenty-first century, we are facing many urgent challenges. Air and water pollution is prevalent in developing countries, toxic chemicals are used too much in developed countries, global climate change is approaching a turning point, many resources that are critical for continuous prosperity are becoming depleted.

The biggest mistake of our civilization is that we settle down with a slowdown of degradation but continue to remain on the road toward Stage S5′. In S5′, we have developed many advanced technologies, passed the peak of prosperity, but due to the breakdown of overall sustainability, human society enters a quick plummet of both prosperity and sustainability. S5′ is what has been depicted in many movies to warn us, including *WALL-E*, *The Day After Tomorrow*, *Happy Feet*, *The Inconvenient Truth*—several of my favorites. Think about nuclear and oil spill accidents; imagine a mad terrorist group owning large scale lethal weapons. You may have noticed many real-life examples. In March 2009, a truck carrying high density HF acid overturned; the acid spilled, forcing an evacuation of over 2000 people. In 2008, the big earthquake in Sichuan, China defeated the protection system of chemical plant containers. Unknown quantities of cancer-making chemicals might have been released into the water system. There is very little sign of life in the nuclear test field.

When we create danger and try to watch out to contain the danger, it is dangerous. Such a strategy is shortsighted. It is not a sustainable solution.

Can you imagine continuing in the current way (S4) for another 200 years? The firms will produce more high-risk products and the government will issue more regulations to contain the danger. We may slow down in degradation a little bit, but we will eventually approach the plummeting point.

It is very clear: we must change the route of our culture. Slow down is not the solution, change to positive sustainability is.

It is everyone's responsibility to change our track toward S5 instead of S5′. In stage S5, we maintain quite high prosperity; in the meantime, we live in harmony with nature rather than against nature.

Technology progress created the problems and opened Pandora's box in the past two centuries. Today's scientists and engineers are facing the challenge of solving the problems of sustainable development. We need to recover the air, water, and soil from the pollution of billions of tons of toxic waste; we need to quickly develop alternative energy to avert the trend of greenhouse gas emission; we need to establish a culture of sustainable development rather than being less bad and following regulations.

Could we have a way out?

Theoretically, yes. In reality, it will be a very tough struggle. This is part of the reason why this book was written. This is why we emphasize so much on the importance of understanding the mission of engineering, the fundamental philosophy of engineering, and the general strategy of engineering. The big direction has to be right before right things can be achieved.

3.4.2 Cradle-to-Cradle Design and Technical Nutrients

Sustainability has been widely studied in agriculture, architecture, biology, earth engineering and science, etc. It seems energy and manufacturing engineering is lagging behind. It is the task of this chapter to integrate the philosophy of sustainability into the conventional methodology of engineering. We have reviewed sustainability and introduced the concept of industrial ecology, natural capitalism, etc. In this section, we will introduce the cradle-to-cradle (C2C) design.

William Andrews McDonough and Michael Braungart published the book *Cradle to Cradle: Remaking the Way We Make Things* in 2002 and shed new light on sustainable industrial manufacturing [66]. C2C design is a bio-mimetic approach to the design of systems. The core concepts in C2C design are as follows:

1. Design to be sustainable, do not design to follow regulations. Human beings design things, make decisions. With our huge capability of influencing the ecosystem, we must be very conscious of our actions. Government and international organizations may issue regulations to protect the environment and public interest, but the focus is too much for human interest and may not be sufficient to avert the process of ecosystem degradation. As we said earlier, creating danger and trying to contain danger is dangerous. We must design and manage for sustainability.

2. Technical nutrients: limit the use of technical materials to those that can be treated as technical nutrients in the industrial ecosystem. Materials are divided into biological nutrients or technical nutrients. Biological nutrients are organic materials that can be disposed of in the natural environment and into the soil, providing food (nutrients) for other forms of life without negatively affecting the natural environment. Ancient agriculture is a good example of sustainable development based on the concept of biological nutrient circulation. In the C2C design, human processed technical materials are converted into or limited to technical nutrients, which are nonorganic technical materials that are nontoxic, nonharmful, have no negative effects on the natural environment, and can be used in continuous cycles as the same product without losing their integrity and quality [57]. By limiting the materials to nutrient materials, long-term sustainability can be achieved. There is no waste in this ideal situation. Waste equals food.

3. Ecological intelligence: The C2C design tries to achieve both high humane prosperity and high sustainability of the ecosystem. By eliminating the use of non-nutrient materials, by the design of our manufacturing processes and the management of material and energy flows, we can break out from the track of S5′ in Figure 3.6 and march toward S5.

4. A bright future: C2C design leads to a commercially productive, socially beneficial, and ecologically intelligent approach. This promises a positive and bright future for human beings. This is theoretically true. Some examples following C2C design partially proved this claim. For example, a carpet company used material that is 100% recyclable and healthy. This strategy lowered the production cost and led to big market success.

We say the struggle toward sustainable development is tough because we are in a market-driven economy and we would not accept lower prosperity unless out of choice. Unsustainable development creates bubbles and the bubbles will eventually explode, like the economic crisis that happened when this book was written. The financial bubble explosion on Wall Street triggered global economy deterioration. Few people expected such a big overturn in the world economy in such a short time. Well, that's a manifesto of the power of chain reactions in human society. We are part of the chain reactions in the big ecosystem, a still sustainable cycle at this point.

3.4.3 The Strategy of Intelligent EFM

Natural capitalism and C2C design, along with other sustainable methodologies, have pointed out the right directions for future engineering. These works are very inspiring. Unfortunately, the majority of the population is unaware of them. Our education system doesn't offer a sufficient introduction of these materials to students. People may have heard of them, but only a small percentage of people really spend time in understanding them. We are busy becoming more prosperous or contributing to the prosperity of groups. In short, we are trapped in a market-driven economy.

That's why it is urgent to change our philosophy of engineering. The Tai-Chi and from-nature-to-nature philosophy in this book is more general and more inclusive than the other philosophies. The C2C design is an idealized final solution for human society. With intelligent EFM, we take practical steps to bridge the gaps between idealism and reality.

In a market-driven economy, the door-to-door production without the responsibility of ecosystem protection normally will have lower production cost than strictly following the C2C design. In C2C design, people are required to use only the "nutrient" materials, but toxic and nutrient are relative terms depending on the situation in which they are used. Is carbon dioxide toxic? Is cadmium toxic? Shall we use them? There is no clear cut answer to many of these questions.

The task of engineering is to solve the contradictions and bring the material-energy system to the desired conditions. All material properties are useful if used properly, toxic materials are useful since they are an indispensable factor needed to balance the world. The Tai-Chi philosophy is also deeper than the concept of contradiction in TRIZ. The Tai-Chi philosophy indicates that the conflicting factors can be balanced to maintain a state or de-balanced to migrate to other states. Relativity, change, mutual inclusion and conversion, and balance and harmony are the essence of the Tai-Chi philosophy.

Natural capitalism and C2C design are still centered on the core interest of human beings. They show good progress from earlier methodologies, but they are still biased. We argue that *from-nature-to-nature* is a better term to summarize what sustainable development wants to achieve.

In Chapter 2, we presented the new criteria of engineering optimization (new CEO), and introduced the mechanism of practically evolving our economy from market driven toward market driven and sustainable. The from-nature-to-nature philosophy states that we are part of nature, we are supported by nature. Thus, we must consider the whole chain of material and energy flow from nature to nature, not from one human production point to another human production point.

In many ways, this is similar to what the C2C design tries to achieve. However, the from-nature-to-nature philosophy won't exclude the use of materials to only the "nutrient" materials—those materials are rare and there is no absolute nutrient material. Even water is not always good. Flooding or freezing rain may cause serious disasters. Some waste is necessary and does not need to be eliminated.

In reality, successful sustainable technologies have to survive the test of a market-driven economy. Relying on good intention or volunteered sacrifice is not reliable. The market has been an effective tool to drive technological progresses; it should also be the tool to drive our change of culture and economy. Of course, this requires the effort from all clans of our society.

Strategy governs tactics. The general strategy of intelligent EFM is a short and clear strategy standing out from all existing engineering methodologies, such as mass production, lean manufacturing, six sigma, TRIZ, digital and intelligent manufacturing, and sustainable development. It is based on the study of historical engineering methodologies.

In modern engineering, we want to achieve both high prosperity and positive sustainability. We have to be practical, and we have to be competitive in the global economy.

To achieve high prosperity and be competitive, we need to optimize the strategic values and minimize waste. This is the essence of design for lean solutions. We say optimize strategic values rather than maximize values because it can be very costly or impractical to maximize certain values. Overdoing can be as harmful as under-performance. Good enough is good enough. For example, the surface roughness of a surface can be as casted if it is not necessary to bring it to a higher surface finish. We also say strategic values, not all values, because one has to be practical to use his resources, picking the most critical groups. Adding more values may increase the cost and raise the system complexity, leading to other wastes in lean manufacturing.

This is a strategy, so it should be adopted from the very beginning, the concept phase of a project. We can thus design this strategy into our following activities. This is also aligned with many of the modern engineering methodologies.

Lean and six sigma are not enough to be successful in global competition. One must continue to innovate to be competitive. For this reason, our strategy includes the methodologies of innovation. This is reflected in the first term of our success equation—optimizing the dynamic M-PIE flows. Notice that we mentioned all four flows in engineering. The material flow and the energy flow have been studied widely, so has the process and project management flow, but not enough attention has been given to the information flows, although we claim we have entered the information age. In the optimization process, we learn from the biological processes and the TRIZ methodologies. We will introduce more principles in intelligent EFM in Chapter 4.

Additionally, we included sustainability in our general strategy. We view sustainability as the differentiator between the twenty-first century engineering methodologies and the older methodologies. We must change the way we evaluate things and bring our activities to positive contributions to the sustainability of our ecosystem. Again, we should design sustainability into our engineering activities. All of the valuable principles are applicable here, integrating with our new CEO and the from-nature-to-nature philosophy.

Finally, the four factors in the general strategy must work together to achieve both high prosperity and positive sustainability. Purely relying on an effective methodology for productivity enhancement may run into big risks of regulation constraints. For example, a milk can production process may be optimized for customer value and low manufacturing cost, but the government may forbid it due to the incompatibility with recycling rules. On the other hand, overly emphasizing ideal sustainability may add big cost to firms, lowering the global competitiveness of the same product. This is the complaint of many U.S. enterprises when they face competition from the developing countries, where environmental laws are not as strict as in the United States. It is better to stay in the playing field to be good rather than to criticize the bad ones while out of the game. Technology could not solve all the issues. We need a new culture and a new philosophy. This is part of the responsibility of our education system.

3.5 SUMMARY

In this chapter, we reviewed the evolution of representative engineering methodologies and philosophies, including mass production, standardization, the rise of nontraditional and hybrid processes, the digital and intelligent technologies, quality control methodologies, lean manufacturing, systematic technological innovation, and sustainable development. These are the grounds for modern manufacturing philosophies and methodologies. The Tai-Chi and from-nature-to-nature philosophies have rich meanings. They are inspired by nature and inherit the legacy of pioneers in history. The general strategy of intelligent EFM tries to absorb the essence of the earlier achievements and prepare modern manufacturing to meet the challenges of the twenty-first century, i.e., achieving high prosperity and positive sustainability simultaneously. Following regulations is not enough. We must change—change our philosophy and our culture. This requires an effort from all of us, especially the people in engineering.

QUESTIONS

Q.3.1 Find out more about the history of rockets, and discuss how aerospace exploration impacted manufacturing methodology.

Q.3.2 Read the story of numerical control. Why is the CNC system a big milestone in manufacturing?

Q.3.3 Form your database by searching the Web for "design for assembly."

Q.3.4 Investigate the reasons why the U.S. automobile industry struggled around 2008.

Q.3.5 Investigate why limiting the lean initiative to manufacturing is a mistake.

Q.3.6 Find out more about six sigma. Compare the methodologies of lean manufacturing, six sigma, and intelligent EFM.

Q.3.7 Do a case study of interdisciplinary research in your organization and report on the benefits and needed improvements.

Q.3.8 Form your database of TRIZ by searching the Web.

Q.3.9 Search for "sustainability" on the Web and write a report on your discoveries. Discuss why sustainable development presents both challenges and opportunities to future enterprises.

Q.3.10 What is the SP curve? Why is following regulations and being highly productive not enough?

Q.3.11 Explain the general strategy of intelligent EFM.

REFERENCES

1. Wu, L. C., *Fundamentals of Chinese Philosophy*, University Press of America, 1986.
2. Blatner, D., *The Flying Book: Everything You've Ever Wondered About Flying On Airplanes*, Walker & Co., 2005.
3. Goddard, R., *Rockets*, Dover Publications, New York, 2002.
4. Rajurkar, K. P. and Ross, R. F., The role of nontraditional manufacturing processes in future manufacturing industries, *ASME Manufacturing International*, 1992, 23–37.
5. McGeough, J. A., *Principles of Electrochemical Machining*, Chapman & Hall, 1974.
6. http://lyle.smu.edu/rcam/research/waterjet/int2.html
7. http://www.agfm.com/Cutting/Cutting.htm
8. Yang, B. and Lei, S., Laser assisted milling of silicon nitride ceramic: A machinability study, *International Journal of Mechatronics and Manufacturing Systems*, 1(1), 2008, 116–130.
9. http://www.synova.ch/english/synova.html
10. Zhang, W., Study on RPM, 3DM and energy field formics, in *Proceedings of International Conference of RPM (ICRPM98)*, Beijing, China, 1998, pp. 316–320.
11. Zhang, W., Yan, Y., Wang, W., and Feng, J., Method and system for 3D manufacturing based on energy field method, Chinese patent CN97104397.3, 1997.
12. Tan, K. K., Wang, Q.-G., Hang, C. C., *Advances in PID Control,* Springer-Verlag, London, 1999.
13. Aspray, W. and Campbell-Kelly, M., *Computer: A History Of The Information Machine (The Sloan Technology Series)*, Basic Books, 1997.
14. Rabaey, J. M., Chandrakasan, A., and Nikolic, B., *Digital Integrated Circuits*, 2nd edn., Prentice-Hall, New York, 2003.
15. Fang, I. E., *A History of Mass Communication: Six Information Revolutions*, Focal Press, Burlington, VT, 1997.
16. Ulmer, D., *Beyond the Information Age*, http://www.vias.org/beyinfoage/beyinfoage_03.html
17. Reintjes, J. F., *Numerical Control: Making a New Technology*, Oxford University Press, London, 1991.
18. Wright, P. K., *21st Century Manufacturing*. Prentice-Hall Inc., Upper Saddle River, NJ, 2001.

19. Kusiak, A., *Concurrent Engineering: Automation, Tools and Techniques*, John Wiley & Sons Inc., New York, 1993.
20. Boothroyd, G., Dewhurst, P., and Knight, W., *Product Design for Manufacture and Assembly*, 2nd edn., Marcel Dekker, New York, 2002.
21. Boothroyd, G., *Design for Assembly—A Designer's Handbook*, Department of Mechanical Engineering, University of Massachusetts, Amherst, MA, November 1980.
22. Henchy, L. W., American manufacturing fights back, *Business Solutions*, February 22, 1988, p. 10.
23. http://www.assemblymag.com/
24. http://gpd.engin.umich.edu/dfm.htm
25. http://www.isr.umd.edu/Labs/CIM/DFM/
26. Hammes, D. and Wills, D., Black gold: The end of Bretton Woods and the oil-price shocks of the 1970s, *The Independent Review*, v. IX, n. 4, Spring 2005.
27. http://www.autofieldguide.com/articles/010502.html
28. http://www.dau.mil/educdept/mm_dept_resources/navbar/lean/02tch-mtctw.asp
29. Womack, J. P., Jones, D. T., and Roos, D., *The Machine That Changed the World: The Story of Lean Production*, Harper Perennial, New York, 1991.
30. Womack videos available from Defense Acquisition University, http://www.dau.mil/educdept/mm_dept_resources/navbar/lean/womack_video1.asp
31. Womack, J. P. and Jones, D. T., *Lean Thinking: Banish Waste and Create Wealth in Your Corporation*, Simon & Schuster, New York, 1996.
32. Liker, J. K., *Becoming Lean: Inside Stories of U.S. Manufacturers*, Productivity Press, Portland, OR, 1998.
33. Pelfrey, W., *Billy, Alfred and General Motors,* Amacom Publishing, 2006.
34. http://www.strategosinc.com/_downloads/ohno_shingo.pdf
35. Huthwaite, B., *The Lean Design Solution*, Institute of Lean Design, Mackinac Island, MI, 2004.
36. George, M. L., *Lean Six Sigma: Combining Six Sigma Quality with Lean Speed*, McGraw-Hill, New York, 2002.
37. Stamatis, D. H., *Six Sigma Fundamentals: A Complete Guide to the System, Methods, and Tools*, Productivity Press, New York, 2004.
38. Zhang, W., Cheng, G., Yao, Y. L., et al., *Combined Research and Curriculum Development-Nontraditional Manufacturing* (web-based tutorial), available at http://www.mrl.columbia.edu/ntm/
39. Svelto, O., *Principles of Lasers*, Plum Press, New York, 1998, p. 311.
40. Svelto, O., *Principles of Lasers*, Plum Press, New York, 1998, p. 317.
41. General Electric Global Research, *Laser Forming of Complex Structures* (ATP project 00-00-5269), 2006, available at http://jazz.nist.gov/atpcf/prjbriefs/prjbrief.cfm?ProjectNumber=00-00-5269
42. Lagomarsino, S., Giannini, C., Guagliardi, A., Cedola, A., Scarinci, F., and Aruta, C., High spatial resolution strain analysis with x-ray microdiffraction, *Physica B*, 353, 104–110, 2004.
43. ASME report, Mechanical engineering in the 21st century: trends impacting the profession, ASME, New York, 1999.
44. Anderson, M., Entangled life, *Discover,* February 2009, pp. 59–63.
45. Thiel, V., *Coronaviruses: Molecular and Cellular Biology*, 1st edn., Caister Academic Press, 2007.
46. Hare, R. *The Birth of Penicillin*, Allen & Unwin, London, 1970.
47. Barrow, J., *Theories of Everything*, Oxford University Press, 1991.
48. Newell, A. and Simon, H. A., *Human Problem Solving*, Prentice-Hall, Englewood Cliffs, NJ, 1972.
49. Dunphy, S., Herbig, P. A., and Palumbo, F. A., Structure and innovation, in *The Innovation Challenge*, John Wiley & Sons, New York, 1997, p. 216.
50. Govindarajan, V. and Trimble, C., *10 Rules for Strategic Innovators—From Idea to Execution*, Harvard Business School Press, Boston, MA, 2005.
51. Fey, V. R. and Rivin, E. I., *The Science of Innovation*, TRIZ Group, Southfield, MI, 1997.
52. Ashtiani, M. et al., The Altshuller Institute for TRIZ Studies, see official website for TRIZ study, http://www.aitriz.org/
53. Rantanen, K. and Domb, E., *Simplified TRIZ, New Problem Solving Applications for Engineers & Manufacturing Professionals*, Times Mirror, London, U.K., 2002.
54. Altshuller, G., 40 principles of TRIZ, available online at http://triz40.com/
55. Domb, E., Ideal final result: Tutorial. Available at http://www.triz-journal.com/archives/1997/02/a/index.html
56. Science Daily, NIST produced chip scale refrigeration, April 22, 2005. Available at http://www.sciencedaily.com/releases/2005/04/050421211242.htm

57. Clark, A. M., Miller, N. A., Williams, A., Ruggiero, S. T., Hilton, G. C., Vale, L. R., Beall, J. A., Irwin, K. D., and Ullom, J. N., Cooling of bulk material by electron-tunneling refrigerators, *Applied Physics Letters*, April 25, 2005.

58. Invention Machine Co., Invention Machine Goldfire, software data sheet available at http://www. invention-machine.com/uploadedFiles/InventionMachineGoldfire_Datasheet.pdf

59. United Nations, Department of Economic and Social Affairs, Population Division 2007. *World Population Prospects: The 2006 Revision, Highlights*, Working Paper No. ESA/P/WP.202, 2007.

60. Adam, D., Stern attacks politicians over climate "devastation," *The Guardian*, March 13, 2009, available online at http://www.guardian.co.uk/environment/2009/mar/13/stern-attacks-politicians-climate-change

61. Al Gore, *An Inconvenient Truth: The Planetary Emergency of Global Warming and What We Can Do About It*, Rodale Press, Emmaus, 2006.

62. Daly, H. and Cobb, J., *For the Common Good: Redirecting the Economy Toward Community, the Environment and a Sustainable Future*, Beacon Press, Boston, MA, 1989.

63. Ulanowicz, R., *Ecology, the Ascendant Perspective*, Columbia University Press, 1997.

64. Frosch, R. A. and Gallopoulos, N. E., Strategies for manufacturing, *Scientific American* 261(3), 1989, 144–152.

65. Hawken, P., Lovins, A., and Lovins, L. H., *Natural Capitalism: Creating the Next Industrial Revolution*, Little, Brown & Company, Boston, MA, 1999.

66. McDonough, W. A. and Braungart, M., *Cradle to Cradle: Remaking the Way We Make Things*. North Point Press, New York, 2002.

4 Representative Principles and Techniques in Intelligent EFM

Wenwu Zhang

CONTENTS

Having discussed the philosophy of engineering and the general strategy of intelligent energy field manufacturing (EFM) in the previous chapter, we will introduce some representative principles of intelligent EFM. The discussion is still at the methodological level, with each principle illustrated with examples. The readers are encouraged to find more examples following these principles. The later part of this book presents specific examples that describe a technical direction in detail, such as laser material processing, electric-assisted machining, waterjet machining, etc. This book tries to enlighten the readers to think above the techniques. In this way, the true skills of process innovation can be gradually built up.

4.1　REPRESENTATIVE PRINCIPLES IN INTELLIGENT EFM

Principle #1: Make the from-nature-to-nature (N2N) philosophy the foundation of engineering.

From Chapters 1 through 3, we argued that the fundamental mission of engineering is to innovate for the long-term sustainable development of the big ecosystem. To reach this goal, we need to take a big system approach. Many of the engineering methodologies may have shown their values in promoting the progress in individual directions, but if they are not aligned with the N2N philosophy, they may lead us in wrong directions, intentionally or unintentionally. For example, mass production had improved productivity of fossil oil–based automobiles, but this improved productivity had created people reliance on a sustainability-degrading technology.

The big system approach is needed to consider the total effects of engineering activities. Currently, our economy is market driven, a from-door-to-door philosophy is prevalent. After reviewing the evolution of representative engineering methodologies, Chapter 3 talked about the general strategy of intelligent EFM, that is, we must simultaneously consider sustainability and other aspects of engineering, including optimizing the strategic values, minimizing the waste, and integrating the dynamic M-PIE flows. We will adopt this system approach in intelligent EFM.

Under this principle, society needs to change the culture of consumption, while engineering needs to change the criteria of engineering optimization. EFM and sustainable manufacturing have been two hot topics in the American Society of Mechanical Engineers/Manufacturing Science and Engineering Conference (ASME/MSEC) in recent years. The New Criteria of Engineering Optimization (New CEO) was introduced in Chapter 2. In ASME/MSEC 2009, a panel discussion was held at Purdue University. Invited experts discussed how engineering should change to meet the contemporary challenges. The key is the New CEO.

In a certain sense, engineering is about decisions. Once the decision is made, social resources and human intelligence are mobilized, and the ball rolls with all its consequences. Engineering activities are the frontline of the human–nature interaction. It is each engineer's responsibility, and each human's responsibility, to drive our economy toward both high prosperity and positive sustainability.

This principle is also the foundation of all engineering disciplines, thus, it is listed as Principle #1. It is extremely important to get the big direction right before initiating any engineering activities.

Principle #2: Solve engineering contradictions with the Tai-chi Yin-Yang philosophy.

As discussed in Chapter 3, engineering contradictions are pervasive and we should use the Tai-chi Yin-Yang philosophy when we try to solve these contradictions.

Contradiction is a condition that we want an entity to be in a state for one thing, but don't want it to be in that state for other purposes. Here are several examples. In an oven system, we want to achieve a high temperature inside the furnace and we want to keep the outside as close to room temperature as possible. We need big enough mass to make the structure of the car strong and spacious, while we desire to reduce the mass to improve mileage. We need to use huge amounts of energy to maintain a high standard of living, while we want to minimize the impact of human activities on the ecosystem.

The Yin and Yang factors in a system coexist, are relative in nature, and can convert mutually under suitable conditions. Harmony (or the ideal engineering solution) can be achieved only when the Yin and Yang factors are properly balanced.

It is important to realize that all engineering solutions have their Yin and Yang aspects. Sustainability can only be achieved through balanced cycles. This principle asks us to use balances. This is especially important in the twenty-first century when we are facing many sustainability challenges.

For example, an IT factory or chemical plant may consume large quantities of water. This may compete with civil water use and natural water use (rivers may be diverted from their natural flow pattern). Considerations must be used before the building of such plants.

Another example, many people advocate solar photovoltaic (PV) to be the future energy source. But the concentration of harmful elements is a potential risk. Imagine a power generation station based on CdTe solar cells. It may take tens of kilometers of land use. When everything is in place, this plant will generate human needed energy. On the other hand, we can expect two potential negative factors. First, "vacuuming" solar irradiation from the normal pattern of the earth surface to generate electricity on a large scale may change the weather pattern—just like water lakes, these are solar PV lakes. Second, the normal package of harmful materials in PV cells is only for normal conditions, it is not safe for big natural disasters or terror attacks. In case of wildfire or war, the high concentration of harmful materials may spread hundreds of miles. This is not to say that we should resist the development of some thin film solar cells. The point is to consider both the positive and negative (Yin-Yang) side of technological systems, be prepared, and try to achieve the ideal solution.

This is also aligned with the big system principle. The two should be used together.

Principle #3: Understand the role of process innovations.

The big system approach is implemented through processes. In Chapter 1, we discussed the fundamentals of technological innovations. Strategic innovations are organizational or system level innovations, product and service innovations are the cash-cows or cash-traps of organizations, while all of the three categories of innovations are riding on the myriad process innovations.

This is not to promote process innovations to unfairly high importance in engineering activities. What we want to emphasize here is to avoid the too low regard toward process innovations. Technological innovation is always a system project. It requires the concerted efforts from all customers and all relevant links to be successful and competitive. For many challenging tasks, process barriers, not the design barriers, are the final deciding factors toward final success. As we discussed in Chapter 3, we should design quality, manufacturability, lean, and sustainability into technological innovations. In the real world, the social status of process engineers could be so low that it can affect the efficiency of innovation.

For example, when a company decides to launch a new product, it may make the decision on leader's view, market study, inventor or major designer's expertise, etc. The process engineers may be called upon in the later stages of product development, when the design is almost decided. Once the process team is involved, they will provide feedback on the design. If it is too costly or too late to change the design, the process team has to use special alternatives to correct the drawbacks of the design. In this way, the project missed the opportunity of cutting cost by coupling process innovations with product or service innovations. When the process engineers are less powerful in the engineering decisions of a project, they could be forced to choose less suitable processing technologies and to develop new processes. For example, a small diameter metal rod is needed in a product. Above a certain outside diameter (OD), the material is rigid enough to support itself during turning, but below a certain OD, elastic yielding puts too much difficulty in direct mechanical turning, thus special processes have to be used, increasing the processing cost. If such a shaft is referenced by many other components and the design of these components has been finished, the shaft manufacturing team may have no choice but to find out ways to make it. Such a delicate shaft may create system robustness risks. Such a situation could have been avoided if the processing experts were involved at the conceptual stage of the project.

Here is another example. As shown in Figure 4.1, theoretically, with a cable long enough (from the Earth's surface to beyond the geostationary orbit height, >35,790 km) and strong enough, one could launch a mass into space without the need of rockets. This is the concept of a space tower or space elevator, first proposed by the Russian scientist, Konstantin Tsiolkovsky, in 1895 [1]. The concept evolved over time. In the twenty-first century, a new material, carbon nanotube, gives new hope to the implementation of the concept and NASA set up a space program to support related competitions and research. The property measured at a very short length of carbon nanotubes may be strong enough to build the cable needed in a space elevator, now someone has to overcome the engineering barriers of making long cables with carbon nanotubes. The focus has transferred from the design part to the manufacturing process part. This is one of the bottlenecks of the project, it is a bottleneck of manufacturing processes.

In summary, we need thinkers to break boundaries, we also need experts in processing to define boundaries and reduce a concept into reality. For technical innovations in market competition, the early involvement of process engineers and a good understanding of the importance of process innovations are important success factors. The bottom line is: all plans are realized in small processes. Manufacturing process innovations are important for many of the product or service innovations.

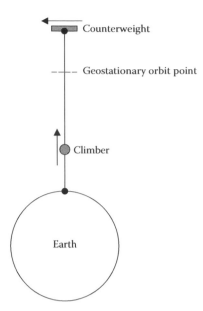

FIGURE 4.1   Illustration of the space elevator. A space elevator consists of a cable connecting the Earth surface to space. By attaching a counterweight at the end, centrifugal force stretches the cable taut, countering the gravitational pull, thus allowing the elevator to remain in geostationary orbit. Once raised beyond the gravitational midpoint, the carriage would be accelerated by the planet's rotation, all the way into space without the need of a rocket.

Principle #4: Use the generality of energy fields, logic functional materials, and intelligence.

In Chapter 2, we introduced the concepts of general energy fields, general logic functional materials, and general intelligence. These concepts are powerful tools to enhance process innovations and to understand technological trends. These concepts are the foundation of intelligent EFM. Readers are encouraged to read Chapter 2 once more to appreciate this principle. Here is a highlight of the key points.

Generality of energy field

- A general energy field is the spatial and temporal distribution of any entity (including but not limited to force, energy, mass, information and intelligence, social and financial resources, etc.) that contributes to the process of doing work.
- Doing work is the process of changing the state of a system.
- Materials are simply stabilized or slow variation energy fields. Treating the spatial and temporal distribution of material as an energy field is straightforward and very powerful. Yet, this is a point many people ignore in practice. People focused too much on the optimization of the familiar fields of forces.
- Information is a physical quantity, information can do work, and the distribution of information and intelligence is a kind of general energy field.
- The role of information flow in engineering is becoming increasingly important. We call it close loop control, intelligent manufacturing, digital manufacturing, smart products, lean manufacturing, production management, etc. However, it is rarely treated as an energy field.
- To be modest and safe, let's assume we cannot create information and intelligence, although we can collect and concentrate the field of information and intelligence (the **I–I Field**), just like we cannot create energy although we can convert or concentrate energy.

- In short, the definition of a general energy field relates to any factors that contribute to the change of state of a system, be it energy, force, mass, information and intelligence, or finance and society.
- This is a bold extension of the conventional definition of the energy field, which is limited to the distributions of various energy forms and forces.
- In engineering, our task is to transfer a system from a starting state to the final state in the (X, r, t) space, where X is the general energy field, r is the spatial dimension, and t is the time dimension. The more freedom we can manipulate, the higher chance of finding the optimal engineering solutions.

Generality of Logic Functional Materials

- Historically, materials used by human beings have evolved from natural, refined, and synthetic materials to functional and smart materials.
- Logic functional material is a concept originated from the study of three-dimensional (3D) dissociation forming. It is a material with three states: normal, excited, and the qualitatively changed state.
- All materials have properties that can be revealed under certain energy field conditions. All materials are "smart and intelligent" to certain degrees.
- *All materials are logic functional materials. We call this the generality of logic functional materials.*

Generality of Intelligence

- *General intelligence is the ability to drive (receive, process, and generate) the information flows through patterned responses, or the rule and logic-based responses, to the relevant energy fields.*
- Patterned response means repeatable response given the same states of energy fields.
- In this definition, we avoid saying learning, interpreting, reasoning, and inventing, which are too much tied to human beings or living organisms. According to this definition, intelligence is naturally not the privilege of advanced living organisms.
- Material and energy systems are by themselves receivers, processors, and actuators of information imparted to them.
- Physical laws governing energy–material interactions are part of the intelligence of an energy–material configuration. Fundamental particles have intelligence and we are built upon them to have advanced intelligence.
- Entities drive information flows in their ways. We couple our intelligence into their processes and guide them into our ways. We can work against their intelligence (rules, logic, or pattern) or work with their intelligence. This can make a huge difference in real life.
- The task of engineers is to capture the general intelligence of relevant entities, inject the intelligence of human beings into the entities, and drive the information flow in the directions of human beings.

To use these concepts, one should first forgo the conventional concepts of work, energy, energy fields, material, information, and intelligence. The knowledge base of these concepts affects the capability of process innovations. With a broader sense of energy fields, many engineering directions can be interpreted as energy field engineering. It is much easier to break the barriers between disciplines.

We also emphasize the degree of freedom of engineering optimization. Many times, difficult problems can be solved easily with increased dimension or in another dimension. For example, how do we create millions of nanometer scale holes in plastics? Such a structure can be used as membranes in seawater desalination or power plant green house gas sequestration. Suppose your expertise is in laser micro/nano processing. You may want to solve the issue with laser nano patterning. However, clever material property (material intelligence) utilization and a combination of thermal field, chemical etching, mechanical field, etc., enabled a much cheaper and faster process in use today than a laser-based process. Distribute and imbed nano-particles into the membrane structure first, etch out the nano-particles selectively, and you get the finished porous structure. It is a parallel hole-making process, scalable to large areas and high speed. Check the "membrane" processes in the Web.

Treating all materials as logic functional materials has immediate consequences. We should have confidence to find more of the logic functions of materials. Diamond was a precious crystal 100 years ago, because it could only be found naturally and it has amazing properties. Nowadays, the industrial application of diamond is pervasive. Think about the diamond turning and grinding processes. This is due to the fact that people uncovered the logic of synthesizing diamonds.

When we are facing the challenges of materials processing or product innovations, in addition to considering more energy fields, we can focus on unlocking the logic functionality of materials or alternative materials. For example, satellites need to limit their mass to control the cost of launching, but it requires sufficient energy to be functioning. We can put in certain high-density energy to keep the satellite working longer. However, this is not as good as the currently established solution, using solar panels, which can last much longer than any limited supply of fuel. Solar cells are materials with logic functions. You may be amazed by how many uses people have found in certain materials, such as bamboo, water, gas, powder, and bulk solid materials. Water can flow, has weight, and won't change volume much under normal pressure changes. So, it can be used as a flexible supporting material inside a container. By knowing more of the logic functionality of materials and using their intelligence instead of working against their intelligence, we can achieve unexpected success. If you find an unknown use for these functionalities, you become an inventor.

Treating the distribution of materials and information as kinds of energy fields is also a powerful tool. Chemists and materials scientists are playing with the medium field everyday. Material is a configuration of energy fields. In many organizations, a movement to improve the information field can immediately improve the overall efficiency of operation. For example, let's form rules to put instructions conveniently tied with all of the major equipment, such as using a Web-based database with all the standard of operations (SOPs). Substantial time can be saved. In my lab, there are many cabinets. People deposited many once-valuable scientific equipment or parts inside the cabinets. Some strange shaped parts are very interesting, but without proper instruction, many of them become dormant treasures, or become dumped waste when we are short of storage space.

General intelligence asks us to be more modest. Humans are not the only intelligent existence. All existences have a certain level of intelligence. Understanding this fact, building up our knowledge of such intelligence, and driving the information flow to our advantage or choosing the least resistance route of driving the information flow is what we should strive to achieve. Here, the guided and systematic innovation methodologies, such as TRIZ, should be adopted. Refer to Chapter 3 for detailed discussions.

Class Activity

Pick a specific process, such as metal forming, laser micromachining, rapid prototype manufacturing, or the production of solar PV cells; analyze the energy fields, materials, and intelligence involved; and try to extend the degree of freedom of engineering optimization through the use of the generalities of the concepts.

Principle #5: Drive towards a higher level integration of the dynamic M-PIE flows.

The dynamic M-PIE model is a high level model of engineering. Depending on its use, it can be used to analyze big projects or to improve specific manufacturing processes. Chapter 2 presented the background and the details of this model. There are three fundamental flows in any engineering activity: the material flow (M), the energy flow (E), and the information and intelligence flow (I). The task of engineering is to inject human intelligence into the process of material and energy interactions, in order to convert materials into desired configurations. Above the three fundamental flows is the project/program/process management flow (P). The P-flows package M–E–I flows as subsystems and interact with the rest of the world. These flows are dynamic in nature. Despite the various engineering disciplines, the dynamic M-PIE flows are the common feature of engineering, the gene of engineering, and the foundation that we can communicate between disciplines.

An important assumption in intelligent EFM is that engineering is evolving toward increasingly higher level integration of the dynamic M-PIE flows, and the more integrated the process or product, the more advanced they are.

People talk about advanced technology, innovative technology, or nontraditional technology, but advanced, innovative, or nontraditional are all relative terms. We suggest using the level of integration to evaluate the level of technology. This treatment has the potential of unifying various modern methodologies in engineering. It is straightforward and intuitive, and it enables a quick yet high-level view of any manufacturing processes.

Let's use several examples to illustrate this important principle.

Product example: Long-distance weapon systems

In the cold-weapon era, spears, catapults, and arrows were the long-distance weapons. These weapons relied on human body energy or the combination of human and machine energy. The launching was guided by a human estimation of targets. Once launched, the end results were basically decided, since there were no ways of correction after the launch. The effectiveness of these tools highly depends on personal skills. The level of M-PIE integration is low.

Modern missiles are quite different. The missile weapon system is a highly integrated system. The missile weapon system integrated the propulsion system (on-site energy system), the weapon head and structure (M), and the intelligent target tracking and signal process system (I) and it can receive corrections from the positioning systems (satellites and human directions). With such advanced integration, a missile can easily shoot down an airplane in high mobility. Soldiers can be trained through computer simulations. The importance of human body skills is much lower than in ancient times.

This is an example of product systems. Such a trend is seen in many other product systems. Computers, televisions, cameras, computer numerical control (CNC) systems, etc., have gone through similar processes. They were lab toys initially, requiring high personal skills to maintain and use. Only when they integrated the M-PIE flows to certain levels did they become commodities. The future trend is basically offering more powerful functions while keeping the skill requirement at low levels. In other words, increase the intelligent level of products and minimize the requirement on humans.

It is interesting to tie this trend to the evolvement of so-called nontraditional processes, such as laser material processing (Chapter 9), electromagnetic metal forming (Chapter 6), or ultrasonic machining. These processes are still used in niche applications. In my opinion, these processes should have gained much wider applications had big efforts been made to improve their dynamic M-PIE flow integrations. The complexity of technology may influence the cost of products, but there are other factors. Automobiles are complex systems. Compare their cost with that of laser systems. By the way, who can be 100% sure that laser material processing is more complex than mechanical turning? All energy fields are equal in intelligent EFM. The effectiveness of some fields depends on how we use them.

Process example: Three-dimensional manufacturing processes

There are many ways to make 3D objects. Sculpturing is one of them. An artist can use chisels to carve out the various 3D geometries, turning a dummy stone or wood into impressive art pieces. Again, the result of sculpturing depends a lot on the personal skill of the artist. A worker can duplicate many 3D art pieces if he is equipped with modern 5-axis CNC systems. With the computer model of the objects, the tool path can be automatically generated and the same geometry can be repeated on many materials easily. Here the CNC system knows where to do the machining (I-flow), it controls its energy with high precision (E-flow), and the tool is selected to be suitable to the materials to be machined (M-flow). The workers integrate these together, the plants organize workers together, and many complex 3D products are produced routinely (P-flow). Modern civilization has been built upon the high level integration of the dynamic M-PIE flows.

The development of modern additive processes pushed this integration to a new level. Compared with subtractive processes, additive processes, such as laser cladding and rapid prototype manufacturing, have very little material waste. With energy sources like laser beams or electric heating, a precision amount of energy can be applied in material–energy interaction. Taking a layer-by-layer approach under computer control, and making use of the properties of materials (general intelligence of materials and energy fields), very complex features can be built. The cycle time from design to market can be greatly reduced. Figure 4.2 shows a picture of the laser net-shape metal deposition of a hollow airfoil structure. Metal powders of high temperature alloys were delivered with four nozzles to the spot of laser focus. Laser energy melts the powders and deposits them layer-by-layer onto the existing structures. Un-used powders can be recycled. With the help of such a process, new airfoil designs can be validated in weeks instead of years. Once the processing window and program are determined, such manufacturing tasks can be easily transferred to a technician. 3D additive processes will be discussed in detail in Chapter 16 to give readers further appreciation of this process.

The technical trend can be used to forecast the future process/product/service innovations. For processes not fully developed in such integrations, there are opportunities for innovations and businesses.

This principle must be used in combination with the next principle, the New CEO. High-level integration may increase the complexity of systems, thus good judgment must be used to balance the critical requirements of a task.

Principle #6: Gain long-term success by competing with the New Criteria of Engineering Optimization (New CEO).

In Chapter 2, we introduced the New CEO. In Chapter 3, we introduced the sustainability and prosperity (SP) curve. Any engineering methodology excluding a good consideration of sustainability is biased. Each decision has consequences. Engineering decisions are made based on the criteria of engineering optimization. Conventionally, performance, quality, cost, cycle time, responsiveness, style, and esthetics, etc., are the basic requirements for a process, product, or service.

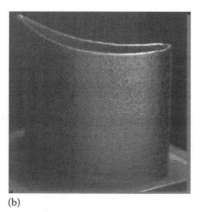

(a)   (b)

FIGURE 4.2 Laser net-shape metal deposition (GE Global Research): (a) nozzles deliver high temperature alloy powders to the laser focus and get deposited; and (b) the finished part.

Conventional engineering optimization tasks mainly focus on how to gain the enterprise the largest market competition advantages. In intelligent EFM, we strive for long-term sustainability in addition to gaining market competition advantages for specific enterprises. This book tries to encourage a new philosophy of engineering and a new culture of our society.

A practical way to achieve this lofty goal is to transform a market-driven economy into a market-driven sustainable economy. This requires effort from all classes of society. The fundamental thing engineering should do is to adopt the New CEO in education and engineering activities. Accordingly, the government and society should respond to this New CEO in order to give positive feedback to those who should win under the New CEO, and to prevent the unhealthy development of those who should be controlled or eliminated.

As discussed in Chapter 2, *the New CEO is used to maximize the index of sustainability and maximize the sustainability-adjusted value of engineering activities in a market-driven economy.* This can be achieved through maximizing the direct market value of processes/products/services and minimizing the nature economy value.

For convenience, here are the definitions of the core concepts:

The direct market value is the value of engineering activities measured in the current market.

The nature economy value is the sum of (1) the cost of energy and resources taken from nature, always negative in value; (2) the cost of the 100% recycling and environment-compatible treatment of the backflows into nature, always negative in value; and (3) societal compensation or penalty to balance the impact on the big ecosystem.

The sustainability-adjusted value is the direct market value plus the nature economy value.

The index of sustainability (IS) in this book is the nature economy value divided by the direct market value.

Currently, the technology cycles in a market-driven economy are missing a correcting mechanism to be sustainable theoretically. Introducing the New CEO can help establish the self-correcting mechanism to implement a market-driven sustainable economy, as shown in Figure 2.17.

The New CEO includes the requirements of current engineering optimization, but extends them two levels further. First, it extended the engineering optimization domain from technical measurements to the index of market performance. In Chapter 1, we defined technological innovation as the first commercialization of creative ideas. Extending the criteria of engineering optimization to market performance asks engineers to take a big system view of engineering activities, going beyond technical issues. Second, the New CEO extends engineering optimization to the big ecosystem, considering all the chains of the nature cycle. In Chapter 3, we reviewed the evolution of engineering philosophies and methodologies. We believe there are big risks tied to any engineering activities if they are not compatible with the sustainability of the big ecosystem. Consideration of sustainability brings both responsibilities and opportunities to enterprises. On the contrary, neglecting sustainability will bring about huge risks to both enterprises and the society. With more environmental regulations in place, enterprises producing pollution emissions face an increasingly higher cost of penalties, or may even be forced to stop functioning immediately.

Adopting the New CEO in all levels of engineering decisions is thus one of the fundamental principles in intelligent EFM. This will affect the long-term success of enterprises, districts, nations, and the world.

Additionally, the index of sustainability defined in this book can be used to measure a broad range of phenomena, such as accidents, natural disasters, etc., in addition to engineering activities. This helps bridge the communication between various disciplines.

In this book, we focus on using the New CEO for the optimization of manufacturing processes and for the fostering of future innovations. Think about the New CEO as an adjective and link it to a specific engineering activity; one can quickly find many ways to improve many of the existing products/processes/services.

For example, let's design a New CEO home heating system. We expect a home heating system that has a high index of sustainability, good performance, and affordability. Under this umbrella, many sustainable technologies can be brought into consideration and next generation products can be envisioned.

Another example, let's consider the process of New CEO thin metal plate patterning. When we use the conventional criteria, we may choose a mask-based chemical etching process. This may be beneficial to the enterprise temporarily, however, the post-processing of pollution may be a hidden bomb. The New CEO will guide us to choose a more environmentally friendly process, such as laser machining, electrical discharge machining (EDM), or waterjet machining. Furthermore, the sustainability requirement encourages the recycling of wasted materials. Thus, one can easily integrate multiple processes to raise the value of wasted materials. For instance, the particles in processing can be recycled. The micro-particles generated in laser machining can be nano-particles! This can be valuable for others. The recycling of water can cut overall cost and reduce environmental impacts.

Clearly, once we use the New CEO, many opportunities can be uncovered, giving new momentum to even stagnant processes/products/services. For example, think about a New CEO camping tent or New CEO chairs. One can form a business simply to replace used tents with new modern ones, harvesting the value of used items.

Class activity

Discuss the New CEO energy or transportation system in groups.

Principle #7: Integrate other engineering methodologies with intelligent EFM.

One of the roles of intelligent EFM is to bridge existing engineering methodologies to the twenty-first century. Intelligent EFM is not standing alone. It inherits and integrates with many other engineering methodologies, injecting sustainability, generality of the dynamic M-PIE flows, and New CEO into these methodologies. The most important engineering methodologies were reviewed in Chapter 3.

These engineering methodologies can be divided into four categories. First are those to improve productivity, such as mass production, design for manufacturing, lean manufacturing, nontraditional and hybrid processes, concurrent engineering, EFM, etc. Second are those to improve quality, such as various quality control tools, six sigma, standardization, etc. Third are those to improve the capability of technological innovations, such as brainstorming, TRIZ, etc. Finally are those to promote long-term sustainability, such as the concept of industrial ecosystem, design for sustainable engineering, natural economy, etc.

The above methodologies have shown success for specific engineering tasks. The established tools, skills, and procedures from these methodologies are valuable assets for the future success of engineering. However, these methodologies miss several critical points to be the engineering philosophy of the twenty-first century.

First, the existing productivity, quality control, and innovation methodologies don't have a systematic effort on sustainable development, while sustainable development is the top priority of the twenty-first century. As shown in the SP curve in Chapter 3, we can afford a low-prosperity society as long as it is sustainable, but we may run into disaster if we go for high prosperity but without sustainability.

Second, these existing methodologies scatter in various domains, such as mechanical engineering, industrial engineering, environmental engineering, engineering management, etc. This situation makes it difficult for people to gain a high-level philosophical view of engineering. Intelligent EFM tries to lower the communication barriers between disciplines by the following:

1. Establishing the common language of engineering based on the dynamic M-PIE model
2. Integrating the essence of existing methodologies into intelligent EFM to form a methodology that can help transit our economy towards both high prosperity and high sustainability

3. Preparing engineering with new tools, such as a knowledge base of general energy fields, general logic functional materials, general intelligence, and inheritable knowledge of process innovation, etc.

Clearly, this integration requires more effort from many people. The target is clear: we want to establish the engineering methodology for the twenty-first century, break the barriers of interdisciplinary communication, and enable a smooth transition of our economy from sustainability-numb to sustainability-conscious.

Principle #8: Systematically improve the network efficiency of talents and information.

What is the most important asset of a nation or a company in global competition? It is talents and information. We have entered the information age. The world economy has become closely connected. The borders between nations could no longer hold talents inside a nation, let alone inside a company; likewise the critical information of a nation or company.

This book is not about how to manage people in engineering. However, it must be understood that above all the other principles of intelligent EFM, the importance of human beings can never be underestimated. People are part of the engineering circle, they can produce value or waste and their overall value can be positive or negative. Talents in any organization are an important part of the organization. People are also part of the information flow in engineering. Once the critical talents are lost, the quality of the organization degrades.

We had discussed the importance of establishing the knowledge base of engineering. Here we emphasize the importance of systematically improving the network efficiency of talents and information. Improving the network efficiency of information is easy to understand. Why do we talk about the network efficiency of talents?

GE Global Research is famous for their career training. In a technical career seminar in 2009, the lecturer asked the audience: "What was the most important factor for the success of your career?" His answer was "network."

Everyone lives on networks all the time. If you have the right network, you get resources and relations that can help you solve many issues quickly; on the other hand, one can work very hard but get nowhere if he doesn't have a good network across an organization. Every event in our life is an event in the network of a bigger system. Some people call this soft-skills, communication skills, etc.

The success of an organization or a nation depends on how successful their talents can be. It is one of the deciding factors to retain talents. Rather than asking an individual to improve their skills of network interaction within an organization, the organization should have systematic efforts to improve the network efficiency of the organization, reducing the effort needed from individuals to be successful.

A good experience in GE Global Research is the mentoring system. New hires can be quite smart technically, but they may not be accustomed to the company culture. Employees at different stages of their careers need different guidance from more senior people. The mentoring system encourages employees to form close relations with more senior people. This is mutually beneficial for both the mentors and the mentees. Junior people are very creative, while senior people are very resourceful. This mentoring system thus improves the network efficiency within the organization. Imagine you have a very creative idea, such as using microfluidics for certain biomedical applications. To sell such an idea, you need people to help you build business cases; you also need to solve some challenging process issues, such as connecting capillary tubes into the PDMA polymer microfluidic system without leakage. With a good network of talents, especially in a big center like GE Global Research, one can quickly be guided to the right people to discuss a possible solution or form a team to solve the issues. For example, one can choose to drill tight through holes in PDMA with a femtosecond laser and insert capillary tubes into the holes in polymer, achieving a self-sealing connection.

On the other hand, if the person who initiated an idea doesn't have all the necessary skills to prove the concept by himself, and there is no mechanism to complement his skills, great innovations can die in the cradle. An organization needs to have mechanisms to foster the success of creative ideas. In GE Global Research, there is an annual cycle of new idea projects. In the first and second quarter of the year, applicants submit proposals for new ideas and seed money is allocated to enable jugular tests or proof of concept tests. These ideas are evaluated in technical committee meetings. The technical committee consists of senior talents of the center. Promising seed projects are promoted to the next stage, gaining higher technology development (TD) funding support. The TD progress is reported regularly to the leadership team. Mentors are normally assigned to these teams to build up business support and give feedback. These TD projects gain high exposure to the leadership team and the business. Some of them naturally evolve into next year's formal projects, entering the new stage of technological innovation and transition. The TD funding gives the team freedom outside their formal projects, which have to follow strict business needs and planned goals and objectives. Such an organism offers a channel for brilliant ideas to grow. In other words, it is a proven mechanism to improve the network efficiency of talents and information. Imagine an opposite case. If a talent doesn't have a channel to prove the ideas he is enthusiastic about, he may become unfocused in his assigned work, he may even choose to pursue his ideas elsewhere, creating disturbance in the talent base of the organization and lowering the efficiency of the information flow.

In short, the success of an organization lies in the use of its talents and information. Creating a culture to value talents and to systematically improve the network efficiency of talents and information should be the highest priority of engineering leaders, it is also one of the fundamental principles of intelligent EFM.

4.2   SOME TECHNIQUES IN INTELLIGENT EFM

In addition to the principles, there are many useful techniques in intelligent EFM. These techniques are boiled down from past innovations from all domains. The list of these techniques can grow with time, since it is an open system. It includes any proven techniques for technological innovations, such as techniques in TRIZ, lean six sigma, etc. This section will introduce several powerful techniques in intelligent EFM.

4.2.1   FOUR-ATTRIBUTE ANALYSIS AND MODULATION OF ENERGY FIELDS—TIME, SPACE, MAGNITUDE, AND FREQUENCY

Time, space, magnitude, and spectra are the four basic attributes of energy fields. To understand and to control the energy fields, one should do a four-attribute analysis of the energy field, and consider the modulation of the energy field in all aspects. In practice, we are frequently required to find out the processing windows of processes. Techniques of design of experiments (DOEs) are used. Limited by the specifications of the given system, we may constrain our mind on the degree of freedom offered by the system, while neglecting the full potential of the energy field. This technique asks us to consider the potential of all the attributes of the energy field. Understanding, manipulating, and exploiting the field from the four attributes point of view can help break our mind inertia, quickly finding unfilled areas in engineering and grabbing the innovation opportunities. In the following, laser energy is used to illustrate this technique.

Laser material interaction can be very complex, involving melting, vaporization, plasma, shock wave, thermal conduction, and fluid dynamics. One can get a relatively complete understanding of laser material processing following the *four-attribute analysis of energy fields*: time, space, magnitude, and frequency.

4.2.1.1   The Time Attribute

Time attribute is the temporal distribution and control of an energy field. Laser energy may be continuous wave (CW) or pulsed and can be modulated or synchronized with position and motion. For CW lasers, the average laser power covers a wide range, from several watts to over tens of kilowatts,

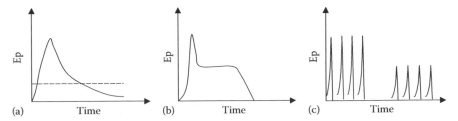

FIGURE 4.3 Temporal profile of laser beams: (a) CW vs. pulsed, (b) high peak followed by a low power average power, and (c) packets of short pulses.

but their peak power may be lower than pulsed lasers. There are many types of pulsed lasers, with pulse durations from milliseconds to femtoseconds. The major purpose of pulsating laser energy in laser material processing is to produce high peak laser power and to reduce thermal diffusion in processing.

Sometimes, it is meaningful to pay attention to the temporal beam profile. For example, in laser material processing, certain temporal profiles are desirable for certain applications. Saying the laser pulse duration is of 1 ms, 25 ns, 10 ps, or 50 fs is not enough. Energy with different temporal profiles has unique applications. Figure 4.3 shows three kinds of profiles. A laser pulse can be of a shape like that in Figure 4.3a, it has peak pulse energy much higher than a continuous power or long pulsed laser. In laser hole drilling, such high peak laser energy is needed to ablate materials in a short period of time. The profile in Figure 4.3b has a high initial peak followed by a lower pulse energy. Such a profile may be helpful in improve the drilling efficiency. The high peak laser intensity may heat material to the ablation point, but a continuous high peak power may suffer from plasma blocking effects. Thus, it can be desirable to use a lower but high enough laser energy immediately following a high initial peak to avoid such negative effects. Similarly, when a laser beam with huge pulse energy (such as a 1 ms and >10 J/pulse) is used, it can be very difficult to control the surface melting redeposition and reduce the sidewall heat affected zone. In practice, packets of short pulses (Figure 4.3c) are used to improve the drilling quality. Each packet has overall energy similar to a long pulse, but is divided into multiple pulses, such as 10 of the 50 μs pulses.

Case study

Timing is critical to the success of processes. Clever timing can lead to good inventions. Superpulse laser drilling is a good example [2]. When laser pulses about 5 ns are spaced far apart (>200 ns), they reach their depth limitation in micromachining at around 0.5 mm. But when the first pulse is followed by a very close-by (≪200 ns) second pulse, the plasma generated by the first pulse is still dense and not much has expanded when the second pulse arrives. The plasma strongly absorbs the second laser beam, ablates the substrate and then gains far more momentum than a single beam to bring out the removed material. This has been proven to have a faster machining rate than sequences of single 5 ns beams and results in better machining quality and much larger machining depth [2].

Another typical example in laser material processing is to compensate for laser firing delays or acceleration/deceleration of mirrors in galvanometer scanning. With a laser repetition rate of >50 KHz, even a slight mismatch in firing synchronization can lead to some undesired defects.

The enlightenment here is that we should be aware of the potential of time modulation of energy fields and purposely use it when conditions allow. Even for a matured energy source, such as EDM, if one can modulate the energy to very fine scales, the capability of the system changes. For example, EDM with ns or ps pulses can machine materials in very precise scales, making them good tools for micro/nano machining.

How about waterjet machining? Your impression on waterjet machining may still be a continuous waterjet machining system, but once we can control the precision pulsing of the waterjet energy, it can be used as a good milling tool. See Chapter 5 for more details.

The temporal modulation of an energy field provides many opportunities for new process innovations. The fundamental reason is that energy with different temporal profiles can induce different responses in materials. It has to be used in conjunction with other attributes.

4.2.1.2   The Spatial Attribute

A spatial attribute describes the spatial distribution and the propagation of an energy field. Along with other attributes, it can describe the isotropic or anisotropic interaction of energy fields with materials. A laser beam emanating from a cavity may have one or several transverse electromagnetic modes. For laser material processing, we are mainly concerned with the temporal and spatial distribution of the focused beam that affects the thermal field on the target. Laser intensity usually has a circular Gaussian beam distribution, but may take other shapes, such as a top-hat shape, in which the laser intensity at the center is uniform. In material processing, one can manipulate laser–material interactions by moving the beam while keeping the part fixed, moving the part on a stage while keeping the beam fixed, or by moving both.

In solar thin film scribing, a laser is used to remove a thin layer of material with the requirement of not affecting the underlying layers. When a circular Gaussian beam is used, the centerline has higher energy than the edge areas, resulting in nonuniform patterning tracks. When a rectangular and top hat beam is used, such an issue can be mitigated since the overlapped energy is uniform.

Laser milling can be used to machine out 3D structures in solids. To do a smooth milling, it is critical to use optimal overlap ratios; otherwise, a rough machined surface may degrade or even stop the continuous machine down of the structure. A common trick in laser raster scanning is to rotate angles in multi-layer scanning to avoid the accumulation effects of same direction machining.

Similarly, thinking about distributing the thermal effects in laser material processing may lead to huge differences. When a high power fiber laser was used to machine a thin metal piece, the piece burned out or totally distorted when it was machined line by line next to each other. By distributing the machining across the sample, delicate structures were machined out successfully.

The enlightenment here is that the spatial attribute of the energy field is just another freedom or opportunity we can use in process innovations. Don't be constrained by the normal boundaries of energy applications. Once you think about this freedom, you can possibly manage to realize it.

For example, a deep blind feature is difficult to observe by your conventional tool. How about changing the tool to reach a limited access area? Check out borescope, endoscope, or fiberscope. These tools can immediately solve some of the issues.

If one kind of energy field is not suitable for certain applications due to its spatial attributes, others might be a better choice. For example, optical energy can go through transparent windows or liquids, they can do machining or deposition inside a vacuum chamber.

Spatial modulation is a technique widely used. Space, time, and amplitude decide the intensity of energy fields. A mask, a focusing lens, physical boundaries, motion, tracking, mixing, layered structure, solutions, etc., are all examples of spatial modulation. Physics at the micro/nano scale can be very different from the macro scale. For example, powder from magnesium can react violently with water, while bulk magnesium only slowly releases hydrogen in room temperature water. Humans' exploration of the spatial attribute of the world seems to have just begun, although we have been studying the world for thousands of years.

Class activity

Discuss several interesting applications due to the clever manipulation of the spatial attribute of an energy field, such as super lens, superhydrophobicity, submarine, diamond, communication, etc.

4.2.1.3   The Magnitude Attribute

The magnitude attribute describes the amplitude, flux, or intensity of the energy field. Major magnitude parameters of laser energy are power (average power and peak power, units in W), pulse energy (J), and intensity (W/m^2 or W/cm^2). The average laser power might be low compared with other energy sources. A pulsed laser normally has an average power of less than 100 W. However, laser energy can have very high peak power (A 2 W fs laser can have a peak power of 40,000 MW; a 30 W 10 ns green laser can have a peak power of >100 kW). This results in high energy intensities, making laser suitable to process a wide variety of materials.

When the interaction between an energy field and target is not continuous, energy intensity is usually the deciding factor. Depending on the laser type, laser pulse energy can be varied from below 10^{-9} J to far over 1 J, the spot size can be varied from sub-microns to over 10 mm, and pulse duration can be varied from several fs (1 fs = 10^{-15} s) to over 1 s. The peak intensity of laser energy can be varied from 0 to 10^{22} W/cm^2. Lasers can be flexibly and accurately controlled. Optical filters, polarizers, attenuators, beam expanding, and focusing systems can be used to modulate laser energy distribution so that one can match the laser output to a specific application without disturbing the internal laser source. This makes lasers suitable for countless applications from thermal treatment and material removal to shock processing and fusion nuclear reactions.

Materials' response to energy at different magnitudes can be very different, linear relations are usually simplified relations, while nonlinear relations are more close to reality. For example, water is transparent to green laser light. This is correct only when laser intensity is low enough. When laser intensity is high enough, multiphoton absorption occurs, water may be very absorbent to the intense laser beam. Similarly, insulation materials insulate under certain voltage levels; once exceeding the threshold voltage, they may break out, such as the arcing of air.

The yield strength of metal is not a fixed value, it depends on how fast and at what level the load is applied. Although the yield strength of a metal remains constant at a low strain rate (<100 s^{-1}), at higher strain rates as happened in laser shock peening, the yield strength of copper becomes a function of strain rate, pressure as well as strain and temperature [3].

4.2.1.4 The Frequency Attribute

The characteristic frequency of an energy field is important because materials may respond very differently to energy fields of different frequencies. A well-known example is material heating by microwaves [4]. A microwave oven has a frequency of ~2.45 GHz. It passes microwave radiation energy to water, fats, and sugar contained in food through dielectric heating, thus through volume heating is realized. The characteristic frequency of the laser is its EM oscillation frequency or its wavelength. The frequency determines the individual photon energy of the laser beam. Lasers usually have very narrow spectral width, while other energy sources may have very broad and complex spectral distributions. The spatial resolution depends on the wavelength. For high precision applications, shorter wavelength lasers are preferred. Ultraviolet (UV) laser ablation of organic polymers can be very different in mechanisms compared with infrared or visible laser ablation. The infrared and visible laser ablation is mainly photo-thermal degradation, while UV laser ablation may involve the direct photochemical dissociation of the chemical bonds.

Materials show very different absorption properties at different wavelengths. Metals tend to have low absorption at far infrared (CO$_2$ laser 10.6 µm) while absorption increases with decreasing wavelength; nonmetals such as ceramics and liquids have strong absorption at far infrared, much decreased absorption at visible wavelengths, and increased absorption at UV. At deep UV, almost any material has very strong absorption. That's why different materials require the use of lasers at different wavelengths.

In general, the Four-Attribute Analysis can be applied to many other energy forms. From such an analysis, one can see the advantages and the limitations of a given process and can realize that many phenomena are relative rather than absolute. Thus, caution should be used when collecting material properties from literature. Material constitutive properties are highly temperature-, wavelength-, geometry-, and intensity-dependent in material processing.

TABLE 4.1
Keywords of Energy Field Analysis and Manipulation

Attributes of Energy Fields	Keywords	Typical Examples
Time	Duration, continuous vs. discrete, delay, synchronize, compensation, just in time, speed, machining rate, cycle time, impact, feedback control, real-time control	Pulsed laser or waterjet machining; thermal treatment of materials
Space	Volume, spot size, acting zone, energy-affected zone, spatial resolution, nano/micro/meso/macro, magnify, shrink, distributed, concentrated, floating, mixture, 1D/2D/3D	Microchannel or shaped hole cooling; rapid prototype 3D manufacturing
Magnitude	Power, pulse energy, intensity, fluence, level, pressure, increase/decrease, flow rate, dose	High pressure waterjet machining; impact forming
Frequency	Color, wavelength, repetition rate, resonance, purify, channel, tune, ringing	Cold machining of polymers with UV lasers; ultrasonic-assisted mechanical machining

Pure sciences such as physics and chemistry can map out the framework of a particular energy field, but it is the task of engineering to purposely instantiate and optimize the attributes of the energy field for a specific task.

Table 4.1 lists some keywords that can help us in the analysis and manipulation of energy fields. People might be amazed at how skillfully the world champions in tennis, golf, or table tennis play. The same thing can be seen in the manipulation of energy fields. The temporal, spatial, magnitude, and frequency modulation of an energy field is an art that can be trained and learned. When exploring an energy field, one should be aware of the full potential of energy field manipulation. This thinking can be assisted with modern tools of engineering, such as DOE, six sigma, lean manufacturing, etc.

4.2.2 Solve Engineering Challenges in Various Dimensions

A fish in a tank can explore the tank at maximum. An ant climbing around a piece of paper may think the paper is an endless world. When facing engineering challenges, one should first have a good understanding of what energy fields are involved, what resources are available, and whether the boundaries of engineering optimization can be extended to find out better results.

This technique can take many forms. Here are some examples.

Hybrid processes for better processing results

Conventional pulsed laser machining has a significant heat affected zone. To reduce the negative thermal effects, liquid assistance can be used. Liquid core fiber–based laser material processing was developed along this thought [5]. Liquids such as water can transmit laser intensity high enough ($>4\,GW/cm^2$ at 532 nm) to machine many materials, such as metals, ceramics, etc. Water cooling naturally limits the laser heat affected zone and helps in flushing the machined material. When a solid tube with a lower optical index than water is used to contain water, a liquid core fiber is formed. In practice, nanosecond or fs laser energy was focused into such a fiber, while water is pressurized to flow out of a coupling cavity, as shown in Figure 4.4. Far greater than the $1\,GW/cm^2$ laser energy was transmitted for the 25 ns 532 nm laser, and burr free machining was achieved with this process. As seen in Figure 4.4, a powerful 30 W pulsed green laser energy can be coupled into a tiny needle-like fiber, enabling solutions to many difficult tasks. For example, this process can enter a limited-access space to do laser marking, cleaning, or drilling without the worry of laser thermal effects.

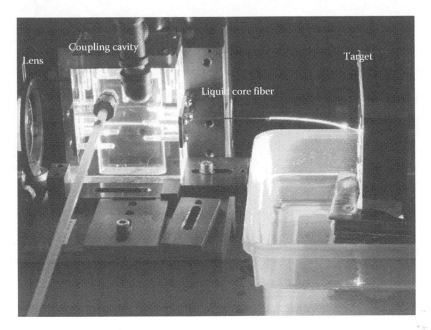

FIGURE 4.4 Experiments of liquid core fiber laser material processing.

There are many other well-known examples of hybrid processes, such as abrasive waterjet machining, electro-chemical discharge machining, ultrasonic-assisted composite machining, laser material deposition at elevated temperatures, spin coating, many other thin film processes, etc. The key point is that one should be prepared to integrate useful energy fields to use the strength of fields while controlling their negative effects. In intelligent EFM, we treat energy fields as additional engineering dimensions.

Use various dimensions

Problems difficult in certain dimensions may be easily solved in increased or other dimensions. Typically, the one-dimension (1D) process is good for two-dimensional (2D) manufacturing, while 2D techniques are good for 3D manufacturing. Using 1D technology to achieve 3D tasks can be slow and complex.

For example, many localized energy fields, such as a focused laser beam, a turning tool, etc., can do processing in a sequential manner. A point source has to be programmed in complex ways to achieve 3D forming. This is time consuming and sometimes may reach the limits of processing capability. On the other hand, a 2D process, such as layer-based manufacturing, can make very complex 3D geometries, as shown in Chapter 16.

On the other hand, super-systems can be divided into lower level systems to make processing easier. A lower dimension process does have the benefits of being easy to control. Thus, if we can divide an engineering task into suitable (manageable) dimensions, better processes can be developed. For example, a complex system can be first divided into a subunit, complex 3D objects can be assembled together by easy-to-make objects, and one can always localize or separate the 3D process from the rest of the part. In the aircraft engine business, expensive parts can be damaged in service. Instead of replacing the whole part, one can repair it on-site with various technologies. Lasers can be used to locally machine out the damaged zone, deposit a 3D metal geometry, and finally be EDM machined into the final dimension.

Are you ever curious about how some tiny tubes are made? They are very good examples of solving engineering tasks using different dimensions. A quartz capillary tube as tiny as 100 μm OD can be made, or more accurately, drawn, from a 5 mm diameter quartz tube in 10 s. Isn't this a big change in geometry? The trick is to heat up the quartz tube to a glass-molten state and then draw it manually or with a machine. The initial outside diameter/inside diameter (OD/ID) of the quartz tube naturally changes into the final capillary tube. No machining is involved, the tube surface can be very smooth, and the OD/ID can be controlled within several microns.

If we generalize this thinking, we can quickly arrive at the conclusion: many complex 3D structures can actually be implemented with lower dimension objects, such as tubes, rods, tiles, cables, layers, or blocks. We are already using this thinking in daily life. Our buildings are an integration of objects with different dimensions.

Here is a real example. One day I needed to have a fixture that can hold a tube and rotate. There was no suitable fixture at hand. I had a tube holder fixture, and I have a rotation component. The trouble was that the mounting holes of tube holder didn't match the mounting holes of the rotational stage. Initially, I thought about asking the shop to machine out an adapter plate out of aluminum. This would take several days to finish and was expensive. I soon had a good solution. I permanently glued the tube holder on the rotation component. It was a quick solution, but it was an elegant one. To generalize this example, I actually used a face combining process to replace a costly 3D process to integrate two components. Welding and gluing can be a good enough solution in many cases relative to mechanical assembly solutions. We use tape and glue a lot in daily life. We use bolts and nuts a lot in mechanical engineering. We need to remember that there are strong glues that can directly combine many materials.

How about hybrid dimensions in engineering?

Composites are actually hybrid dimension engineering structures. It is not simply 3D since it is not isotropic. The entangled fiber materials and the filling materials enhance each other, resulting in high toughness low weight materials, some time high temperature materials. The fibers are 1D materials. How about mixing some conducting materials with polymers? Polymers can become a conductive and flexible material.

How about embedding transparent fibers into concrete? Aha, you can produce transparent concretes!

How about coating a structure (1D growth) and then etching out the 2D patterns? This thinking gave birth to the information technology (IT) and micro electron mechanical system (MEMS) businesses.

The single axis rotation of mirrors can reflect a fixed light into dots or lines at high speed. Two such mirrors can be used as a 2D scanner to reach scanning speed of >5 m/s.

The spin-coating process is an elegant solution to a very challenging task—uniformly spread out the fluid-coating material on a surface. You simply need to drop the fluid on a rotating plate (1D motion)—after a while, the fluid spreads evenly under the centrifugal force and the surface tension forces. It is a parallel process.

What about integrating sensors in physical structures? We are already used to automatic doors and hand dryers. Imagine a wind tower on a remote island. One can choose to stop the power generation to check it by service personnel only several times annually or use sensors to monitor the health of the tower continuously without shutting down the tower and only stop it for big maintenances. Information flow is just another dimension that we should consider using to our advantage in engineering optimizations.

To summarize, we should consider the possibility of using different dimensions in engineering optimizations. We give dimension a broad meaning in intelligent EFM, which covers time, space, material, energy fields, and intelligence.

4.2.3 Explore the Value of Human Factors

An engineering project, such as the development of a process or product, may fail badly despite the extensive use of good technologies if the project neglects the human factors. Engineering activities are supposed to create values for customers. Customers usually face multiple choices. Facing similar products, processes, or services, those prevailing in human factors will win the market competition.

In engineering, using technology to implement the specified functions is usually the major task. But, it is equally important to "package" it nicely before it interacts with customers. Thus, in addition to the conventional technical requirements, such as cost, quality, performance, etc., we should also explore the value of human factors.

When a lady buys a wallet, many things other than the fundamental function (containing cards and money) are considered. How modern is the style? Is it from a famous brand? Does it match her dress at a party? Is it on sale from a high price? Can she get praise from peers?

My wife told me that our kids paid $3 to feed the birds in the park they were visiting. There are services in which you pay the farmers to do some of their farming work, such as picking up grapes or plowing the field.

Well, what is going on here? The customers are willing to pay you if you can provide satisfaction in their spiritual life. There is great value in human factors. This is common sense. Yet in engineering, many people blunder on this.

Some people make high-quality products without a good manual or a product with good functions but a poor user interface. Engineering design should consider human factors in the beginning. In the end, the customer may not need to know all the technical details of a product or process, but he directly feels the user interface of the technical system, experiencing the external functions of the technical systems. Few people really know how an LCD TV works, but people care about its performance, its outlook, its size, and its ability to connect to other media.

Similarly, people buying a laser micromachining system are interested in how to make high-quality products with minimal effort and with high safety. To achieve this, a vendor should have the product as an integrated system, ready to run in several clicks. Don't expect the workers to have the patience to tune the system day to day or to have the ability to add something critical in their work. It made me pretty mad when I ordered a laser micromachining head from a vendor. The laser head had good functions and was quite sophisticated. But when I tried to mount it, I found that the mounting holes were so long that none of our lab screws could be used. I had to use a special screw to mount it on a plate. Then I found that the three mounting holes were spaced at strange dimensions, while people are used to the 1″ or ½″ spaced optical table mounting structures in the laser world. The vendor could have made an adapter mounting plate and provided the special screw to me to make my lab work much easier.

The human factors in this section are the group of factors a technical system uses to interact with humans. An elegant user interface, a system with designed-in beauty, a system with some small things taken care of for the customers are part of this. How about a handy mark of operation procedures? How about error-proof switches? How about easy integration with other common technical systems? Personally, I think the limited use of nontraditional processes and technical systems are partially due to the cost of the system and partially due to the poor treatment of human factors in these systems. Good consideration of human factors can help shorten the gap between technology and people.

It is meaningful to discuss the three levels of human values. As shown in Figure 4.5, humans have fundamental needs (L1). To live, one has to eat, drink, dress, and sleep. Above this fundamental level, in L2, one wants to own something to show off or to maintain his or her value. In this regard, one wants to own some products, such as entertainment electronics, communication tools. Further up, people need better satisfaction in their spiritual world (L3). When one owns a house, one wants to make it a beautiful and lovely home. When one is doing something, one wants to be superior to competitors. When people have achieved many things, they will target their next challenges, such as higher quality products and higher speed transportation. When one is not worrying about material needs, one may be hungry for a more fulfilled family life or spiritual life.

The considerations of human values can affect our decision in engineering design. Note also that, L1 and L2 values tie to products or services that usually have a very broad consumer base. Intense

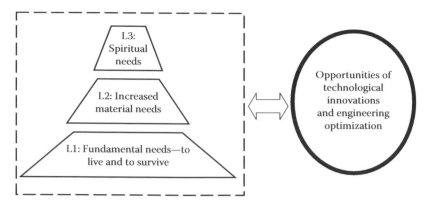

FIGURE 4.5   Three levels of human values.

competitions may exist in these areas. However, if you can identify an unmet need, you have the chance to quickly expand the market. For example, automobiles in the early days were the hobbies of some adventurers and scientists. Once they became suitable for daily transportation, they became one of the dominating economic powers in the twentieth century. The same thing can be said for cell phones, computers, the Internet, digital cameras, etc.

How about the so-called nontraditional manufacturing? Would you like to build your own 3D art pieces or customize your cakes? A 3D printing system for the home or food industry can meet this need. Nowadays many kids enjoy playing with legos. Sooner or later, they will play with digital solid modeling stuff and turn their wild dreams into sold objects.

The L3 human values, or the spiritual needs of human beings, are another source of process, product, or service innovations. High-quality sound systems, home theaters, green products, alternative energy, virtual reality chatting or pet caring, online video games, etc., are such examples.

To this point, we mainly discussed the human values at an individual level. Similar value levels can be ascribed to organizations, social groups, and whole human beings. Remember engineering activities are supposed to create values to be profitable in the market. If some values are not met, they are opportunities of innovation.

For example, many people feel it is really noisy when workers cut concrete to repair city roads. The mechanical cutting is not satisfactory in the regard of environmental quietness. This is an opportunity for new processes and new product development. Cities such as San Francisco are short of fresh water, although they may be very close to the sea. Using normal electricity to purify water is too expensive. This is an opportunity many big companies are trying to seize. Various desalination processes and products are being developed. Global warming is a crisis the whole world needs to face. Thus, various regulations were issued to lower green house gas emission. This triggers chain reactions in products, services, and processes. For example, to reduce CO_2 emission, engines need to burn fuel more efficiently. Low efficiency designs are facing retirement. High efficiency designs may require new processes to implement. Fuel injection hole processing thus becomes a front of competition among major players of cars and diesel engines.

Note also that human values can trigger technological innovations, while new technology can initiate new human values or needs. It is a two-way channel. The key is to remember how to connect new technologies to the different levels of human values. For example, remote conferencing is a new technology. One can adapt it for education, company-wide leadership meetings, working from home, etc. This technology has the potential to cut company travel cost or can help people meeting their family needs while working. These are new values that have been created in the information age. We are used to using dishwashers or sending greetings to friends through e-mails. These things save us time or labor. How about customizing the design of a product and processing it remotely with customers monitoring the process? This allows direct customer feedback and offers the spiritual satisfaction of good involvement in something important or fun to them. In GE Global

Research, we have open-house days for local students and families. The normal daily work we do in labs may just be equivalent to a visit to Disneyland for the kids, because they can see many cool technologies and products. There are human values involved in these activities. I bet one can open a business by training and allowing normal people to try some cool nontraditional processes.

When the thinking of human factors is combined with the other principles and techniques in intelligent EFM, we are better equipped in grabbing opportunities of technological innovations and engineering optimizations. In short, although it might be common sense in daily life, exploring the value of human factors of technological systems is a shortcut to success in the market economy, and this is quite an effective technique in intelligent EFM.

4.2.4 GRASP THE ART OF DC/AC

The history of electricity is quite interesting. Initially only static electricity and lightning were studied. With the invention of battery, small direct current (DC) was readily available for research. With DC circuits, one can vary its magnitude and testing materials. DC transmission theoretically has a large loss over a long distance. With the invention of generators, alternative current (AC) was widely used. The combination of DC/AC led human beings into the twentieth century. The art of DC/AC gave birth to many modern electronics. DC has limited information carrying capacity, while AC has much more room for variations and information carrying. The low loss of high frequency and high voltage electricity transmission is the foundation of modern power grids.

As shown in Figure 4.6, this DC/AC story in the electromagnetic world is a general innovation technique in the engineering world. If there is an energy field in which people can only exploit the power of DC or AC, we can immediately claim that there are opportunities in the DC/AC modulation for technological innovations. Both DC and AC energy fields are useful in their own ways. For example, continuous wave (CW) or pulsed laser energy has different applications. CW high power lasers can be used for thick section welding or cutting, while short pulse lasers are good for low thermal effect micromachining. In mechanical turning, a constant force may not always be the best processing condition. When the machining tool is in constant contact with a rotating workpiece, the tool may get overheated or wear out quickly. Suitable modulation of this tool-workpiece contact force has generated amazing machining results, such as in the case of ultrasonic-assisted machining cases (see Chapter 8).

When an energy field changes, it bears with it certain information and certain intelligence. This information can be used to control a process. Sonar obstacle detection is a well-known example. How about an engineering surface? Many people have the assumption that a flat or smooth surface is a good surface for engineering designs, such as a shaft, the skull of a high-speed structure, or the inside wall of a tube. But a smooth surface or a flat surface is just one of the countless possibilities of engineering surfaces; there is very high chance that a nonflat/smooth surface may have much better performances than flat or smooth surfaces. For example, one may want to use a flat surface for a solar cell; however, it has been found that a flat surface has much higher reflection loss than certain patterned antireflection surfaces [6]. For friction and drag reduction or heat transfer enhancement, specially patterned surfaces can have much better performance than flat surfaces. See the chapter on friction in this book for more in-depth discussions.

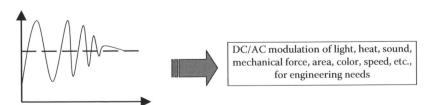

FIGURE 4.6   The art of DC/AC.

In short, the general methodology of DC/AC energy field modulation can be applied on any energy field. DC is a special case—it is just one of the many possible states of energy fields. The combination of DC/AC gives us more flexibility in energy field modulation and engineering optimization.

4.2.5 FORM THE TRADITION OF SYSTEMATIC USE OF PHYSICAL LAWS AND SCIENTIFIC KNOWLEDGE

For little kids, they think of something new and usually assume it to be new in the whole world. For young students, they learn a lot of things and may feel that almost all issues have been solved and very little is left for them. In higher education, people are trained to balance their independent thinking among the awe to human beings' prior achievements. In many cases, invention or innovation is the new way of integrating the existing scientific knowledge and physical laws. Due to the limited time of individuals, our personal knowledge is limited. Facing a challenging engineering task, one may try to find the best solution in the domain of his or her personal knowledge base. There are many ways to Rome. Those who know the best way of solving the engineering challenge can very possibly be the final winner in market competition. Thus, from the beginning, an organization or one person should form the tradition of systematically using the existing physical laws and scientific knowledge. In the information age, there is no excuse for complaining that one does not have enough resources to acquire the necessary information.

The sources of scientific knowledge lie in prior art and expertise. Consulting experts on a topic can quickly guide us to the door of useful information. This can be the mentioning of some keywords, some names, or company Web sites. One can always do a prior art search on a topic. Books and patents contain information that is at least several years old, journals contain information that is around one year old, while conferences and exhibitions reveal current work and some clue of future work. Of course, there is ongoing research and development that may not be reported for the sake of technology reservation. For those new to a domain, a good handbook can be a good start in understanding the prior art.

In many cases, we don't need to reinvent the wheel. Modern economy has formed a matrix of interactions. A company doing system integration does not need to make all parts by itself; it may reduce manufacturing cost by purchasing some of the sub-units. Balance is important. One has to have some core technologies to survive in market competition, while outsourcing the production of components may compromise its effort of cost control and cycle time control.

One also should balance his or her independent thinking in the sea of prior art. Sometimes those who don't know there is a boundary can easily break the boundaries. I would suggest you do your independent thinking first, then do the prior art search, improve your idea or borrow others' ideas, and implement the idea. Creativity is a very valuable characteristic. One should adhere to some new ideas, since any really new things will run into confliction with existing competing things.

As an engineer, one should have the capability of defining things. In many cases, once the issues are identified, the team can normally find good solutions, or the worldwide talent will help get it done. For example, I did a laser micromachining experiment. I complained to the team that the ablated material might be toxic, thus we should have 100% fume containment. I used a vacuum pump in air to collect the fumes, but I was not satisfied. So the issue of fine particle containment was raised. Very quickly, a senior colleague said that I could extract the fume into a fluid bath and the fluid would naturally contain and potentially recycle the ablated materials. If one searches the Web for "fume containment," you can find many potential solutions.

Physical laws and scientific knowledge are the foundation of technological innovations. A breakthrough in some frontiers of fundamental research can trigger revolutions in processes and products. For example, nuclear powered submarines changed the function of submarines strikingly, while the invention of the laser triggered many new applications of laser light.

In short, establishing the tradition of systematically using scientific knowledge and keeping a good balance between independent thinking and prior-art referencing are important for the success of individuals or organizations.

4.2.6 Find a Good Balance between Direct and Indirect Methods

My friends and I were hiking. I stood on top of a 3 m high rock, asking them to take a picture when I jump down. They thought I was joking. I said: "Let's bet I can safely jump down and the bet is dinner tonight." They agreed, so I jumped and won the bet. Well, I used a trick. I jumped in multiple steps with each step less than 1 m high. In this way, many people can win bets that they can safely jump down a cliff.

To solve a challenge, one has the choice of using direct methods or indirect methods. For example, to move rocks from one place to another, one may choose to directly pick it up by hand or with simple tools, or one can use a pickup truck to finish the work in a short period of time. Could you move 2 t of material 10 m in 5 min? You do that whenever you drive your car. In this case, the work done by direct human-power is reduced while the total work done is the same or even higher. We simply divide a challenge into multiple steps.

This seemingly mundane methodology gave rise to many marvelous engineering innovations. Could we develop a structural material that can survive the high temperature in combustion engines or aircraft engines? Well, people initially developed high temperature alloys to directly solve this challenge. But this is expensive and still not sufficient. So cooling holes were used to lower the thermal management load on the alloy. To further improve engine efficiency, even higher temperatures should be used. Here comes the "genius" use of the thermal barrier coatings (TBC) [7]. TBC is usually a thermally sprayed high temperature ceramics coating that can survive much higher direct temperatures than any metals. When TBC is used, it shields the high temperature from the rest of the metal structure. When this is used in combination with cooling holes, the temperature quickly drops to a level sustainable to the metal layer. Actually, less expensive metal alloys can be used in this case.

Well, don't we use this strategy everyday? Whenever you are cooking, you do not get hurt from the high temperature that can burn many tougher materials than the human body. We handle the thermal energy in indirect ways.

How about making nanometer scale structures over a large area? Many researchers manage to make nanometer scale structures over a very small area, say <5 mm, using various processes, such as ion beam etching, electron beam writing, atomic force tip patterning, etc. The serial direct process is slow and expensive. Here we are talking about an area of 0.5 m². An elegant solution was shown in Figure 4.7. Let's divide the task into two major steps. First let's use precision machining or the lithography method to make the desired patterns on mandrels (rollers used in embossing), then let's use two mandrels in a roll-to-roll (R2R) embossing process. When the workpiece such as polymer films or metal foils are fed through the rollers, the nanoscale features are copied to the workpiece in high speed and high fidelity. Such a process has been used in 3M to produce special optical films used in LCDs. The feature can be as small as 150 nm [8]. This is one of the reasons a high-quality LCD monitor can be so cheap.

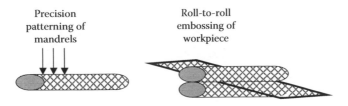

Precision
patterning of
mandrels

Roll-to-roll
embossing of
workpiece

FIGURE 4.7   Steps to make nanometer scale patterns over a large area.

Class discussion

What have you learned from this R2R process? List more than three points.

There are many other examples worth mentioning. To increase light transmission, one can choose to optimize the material itself, such as use high transmission materials or purify the workpiece, but a very effective indirect solution is to use antireflection coatings. In 3D manufacturing, one can choose to directly make the close to full density 3D structures, which requires high energy and complex process control, or one can form green parts by first "gluing" metallic powers together in a low-cost process, then sintering it in an oven to bring it to the desired strength and density. When we enjoy many of the cheap toys and gadgets, the molding process is behind the scenes; otherwise, direct machining of complex shapes are of higher cost.

Well, when we look at these processes, I hope we won't get lost in individual "innovations." In many of these processes, there is one common sense—achieve a good balance between direct and indirect solutions. Few people sleep outdoors, we stay inside houses to keep us warm. Many creative engineering solutions are just like that. Depending on your specific requirements, certain combinations of direct and indirect methods may be best suited for you.

4.2.7 Learn from Nature and Use Sustainable Cycles

I was traveling in Denver in June 2009. After working on solar thin film scribing for 5 days, I finally got the chance to climb the snow-covered Rocky Mountains. You can actually drive all the way above the snow lines. Amazed by the beautiful scene, I stopped on the roadside and took the picture shown in Figure 4.8. There are forests of pine trees on the mountain until the elevation is too high for trees. I still felt excited about this trip. What are you thinking about when you look at the trees, the highways, and the snow on mountaintops?

It is beautiful! It is exciting. The air is fresh. The trees and all the other plants are free and happy there. The living things are on their own and have been like that for millions of years. Over summers and winters, they are there. Nature offers them sustainable cycles. There is no waste. This cycle is beautiful. We love the wild possibly because there is something in our genes common to the rest of nature. We want to be vibrant and free, and we want to be in harmony with nature.

Back in my lab on the following Monday, I took safety glasses and gloves to handle the toxic materials. That's a very different feeling from the mountaintop.

We emphasized the importance of the from-nature-to-nature philosophy in the beginning of this chapter. Here we emphasize the effectiveness of learning from nature and using sustainable cycles.

FIGURE 4.8 Mountaintop.

People have learned a lot from nature and there are still many tricks we haven't grasped. The nature system is a big miracle. The growth of plants is an intelligent parallel manufacturing system. The plants sense the environment and choose to sprout and grow in spring. They carry the message to make fruits or seeds and prepare to propagate the following year. They are part of the ecosystem. They have evolved over a long period of time to adapt to a wide variety of environments, be it cold, drought, windy, or salty.

When we compare human beings' engineering system with the nature system, it is clear that we still have many things to learn from the art of nature. One of the key points is the use of sustainable cycles. In a market-driven economy, we normally optimize the solutions based on market values while the nature values are secondary. Thus, one can be in a business that is thriving in human interest, but might be creating serious environmental issues all the time. Gasoline-based automobiles are one such example. People also use fossil fuels to generate electricity, emitting huge amounts of toxic gases and green house gases. Policies and regulations are enforced in many places. However, our first priority is to thrive in market competition. Thus, those "less bad" may label themselves "green." This is not beautiful at all.

The key skill we should learn from nature is to use sustainable cycles to solve the engineering issues and side effects. In nature cycles, there is no waste. The output of one unit is the input of the other unit. Only with such cycles can the process be sustainable. If we build similar cycles in engineering solutions, we can be closer to high sustainability. Rather than asking for permission of emission, one should think about building cycles to use the emissions. In thermal power plants, coal is burned to generate steam. There is a lot of waste of generated heat. The emission also pollutes the atmosphere, while the burned dross of coals can be treated as garbage, cleaning it will require additional cost. If we get rid of the concept of "waste" in engineering, we can replace it with "potential resources," which covers both wasted energy and materials. All energy and materials are useful when used properly. For the example of the thermal power plant, what happens if we can capture CO_2 and store it in the form of construction materials? The dross of burned coal is already used as construction materials. In the emission of burned coal, there are rare earth elements such as cadmium. What happens if we can capture and reuse such elements?

Many companies in Japan have adopted the recycle strategy to cut the cost of production. Once a cycle is created, a continuous value chain is formed and the final waste can be greatly minimized.

The use of cycles is also a general trick used in many engineering innovations. It can be very expensive to build a large indoor pool with controlled flow speed. Yet, if you use the idea of cycles, you can design a small indoor pool with pump-controlled water flow. Such a system can be used to train professional swimmers. The roll-to-roll process is related to cycles. To make a 100 m long 2 ft wide aluminum foil, one can choose to make a huge dimension press or simply use the cyclic rotation of the rolls to roll out even longer batches of foils. Straight features have limits in length, while cyclic features can be unlimited in working length. The use of cycles can lead to very elegant engineering solutions. The internal combustion engine is such an example. Once ignited, the engine can continue the 4-piston process over a long time until the fuel is depleted. Imagine an automobile without such a cycle. It may be a "mobile" but may not be this "auto." An elevator is another example. There are two kinds of elevators: one that stops at different levels and one that runs continuously. The latter has higher carrying capacity and is widely used in airports and supermarkets.

Of course, there are many other things we can learn from nature. Super-hydrophobicity (SH) is a good example. Mimicking lotus leaves, people are trying to apply SH to many engineering surfaces to reduce drag or foliage. Some insects have very sensitive probes for air disturbances. One can mimic such functions by studying the structure of insect probes. Biological mimicry has become a hot topic in recent years.

In summary, use of cycles, especially the sustainable cycles, is an effective skill we can learn from nature. Instead of optimizing the engineering solutions to cut the idling time or to transfer the waste, we should think about cycles and potential resources. This is the art of nature, and this is beautiful.

4.2.8 BUILD AND MAKE GOOD USE OF YOUR INTELLIGENCE DATABASE

We have emphasized the importance of information flow in engineering. The implementation activity is the establishment, maintenance, and utilization of the intelligence database.

Recently, a colleague from another group in our center had a meeting with me. He is a materials scientist and wants to put fine grooves in a metal piece. He was very happy since I had the ready to go laser micromachining system to try many of their designs, and at the end of the discussion, several new approaches surfaced.

In medium to big organizations, one can immediately improve working efficiency by building an intelligence database, making the resources of all kinds knowable to individuals. Frequently, there are idling equipments or materials in one unit, but such things can be in urgent need in other units. Registering these resources in a database can save both time and money.

It is not rare that when you talk to a senior faculty regarding a new idea, he or she might recall that this has been investigated quite some time ago inside the company or elsewhere. Some organizations have documentation systems, but the documents may not be searchable to employees. In case there are no people in the team that are aware of earlier efforts in the organization, one may end up reinventing the wheel. For organizations with a long history, the internal technical reports are hidden treasures. For start-ups, forming a rule of documenting progress and effectively using these documents can improve their operation effectiveness. On the contrary, if all things are done in a hurry and not properly documented in a database, it can be costly to repeat some early results.

Talent dynamics is another aspect in which the intelligence database can help. Modern society has a high rate of talent in-outs. Talents carry the intelligence of their organizations. If the key talents of an organization leave, the know-how of the organization should not be disturbed too much. This is not the case in reality. For example, Lee was the sole operator of an advanced thin film deposition system. Thin film deposition is the key process in the business of the company. Lee left the company with very short notice. If the know-how of the process is not well documented, the company will face big loss in technology skill and in production time. Thus, a good intelligence documentation system can damper such unexpected changes.

For researchers, their treasure is the intelligence they grasped. A knowledgeable faculty is the bridge to resources both external and internal. One should always try to improve the efficiency of faculties, but one should also be prepared to secure the intelligence in long-lasting forms. This may add some cost to the organization, but when used properly, the payback is far greater than the added cost.

One big issue in the information age is that we are facing too much information. Every time I went to a big conference, I was bombarded with huge amounts of information. If the received information is not filtered in time or properly, their value is wasted. Conferences provide the latest intelligence in specific areas. A good trip report can be shared with wide audiences in an organization. This can connect the new information to people who didn't get a chance to attend the conference. In this way, the value of attending conferences is multiplied. I suggest that organizations deposit such reports as technical reports and one can standardize the format for easy information extraction.

Intelligence management affects the efficiency and culture of an organization. Many companies give awards to applied patents. To encourage effective intelligence build-up, how about setting up small awards for technical information deposition? Remember people have three levels of values. Feeling important and recognized is the L3 value.

Management of intelligence is a big topic—professional training is needed and standardization can be applied. It is believed that given sufficient information, many people can arrive at the same invention as many famous inventors. Thus, improving the effectiveness of intelligence collection and utilization should be of high priority on the agenda of an organization. Professional database management is widely used in big organizations in the Western countries. The situation in small companies can be quite different. However, if one pays attention to this high priority task, one can shape organization culture from the beginning. This will pay back in the long run.

4.2.9 USE THE FULL POTENTIAL OF YOUR RESOURCES

I was in TRIZ training in GE Global Research. One important skill in TRIZ is to use all of the available or convertible resources. The teacher used the example of the *Titanic*. She said: "The ship will sink in 2 h, the water is freezing cold, the nearest ship is 2 h away. What resources could people have used to save lives?"

We suggested putting wax and oil on clothes, making floating devices as soon as possible using any possible structures and materials. Finally, we found possibly the best solution: move people to the iceberg!

The sea was icy cold. People would die in 10 min or so in that kind of water. Safe rafts were too few to save thousands of people in a short period of time. But if people could stay alive for more than 2 h, nearby ships would have come to the rescue. As long as they were not in icy water, the cold temperature in the air won't kill people in 2 h, especially when people had their winter clothes during the trip.

As shown in Figure 4.9, imagine the following.

The *Titanic* bumped into the iceberg at night. The captain asked the crew to lower several rubber boats to climb up the iceberg. In 30 min, multiple bridges were ready for people to move from the deck to the top of the iceberg. The iceberg was big enough to float all the people there. In 1 hour, all people were evacuated and 30 min later, the *Titanic* sank. In another 30 min, the other ships arrived and saved all the people.

We felt really sad. Loss of life could have been totally avoided!

In the above case, we learned an important skill: use all possible resources, and sometimes you can convert the harmful factors into useful resources. The iceberg caused the accident, but it could have saved people had they thought of using it in time.

In engineering, there are many similar situations. Today people complain of energy shortage or high energy cost in many places. Well, a high percentage of the energy that we use on Earth is from the sun, and we receive a huge amount of solar energy every year. If we couldn't use solar energy to generate electricity immediately, we can at least use it immediately in other ways, such as solar drying, heating, lighting, etc. Lightning can be destructive. What happens if we could capture the huge amount of energy in lightning?

In engineering design, whenever we need to increase a function, we should think about whether we could use some parts of the existing system to help. An elevator is a good example. We need a machine to carry people up and down. When we use a balanced weight, we don't need to overcome

FIGURE 4.9   The iceberg could have been the savior in the *Titanic* accident.

the weight of the whole carriage, we only need to overcome the difference of the weights. If we have two carriages, they can mutually be the balance weight. Heat is generated in many systems and wasted. This heat usually negatively affects the environment. What about harvesting such heat and storing the energy in batteries and feeding it back to the system? Such ideas are under exploration worldwide.

The resource here takes a general meaning. What resources do you have at your house? In my house, I have an electric stove, a dishwasher, a refrigerator, a washing machine, a central AC, a hot water burner, a lawn mower, electric saws, two cars, and many others, such as water supply, electric supply, etc. These are the high energy sources. There are many other things, such as a wooden chair, curtain, etc. Depending on what you want to do, many things can change their original functions and be useful resources. On the other hand, seemingly useful things can be burdens if you could not use them properly to generate values. In lean manufacturing, anything not generating value is considered to be waste. Accordingly, anything that can generate value can be considered to be resources.

In summary, resources can be

1. Systems that can contribute values with their original functions
2. Systems that can contribute values with derived functions
3. Systems that initially are harmful but can be converted into useful contributors
4. Anything, including nature, nation, people, information, energy, and materials

If we adopt the general definition of resources and wastes, we can usually identify many places to improve a system.

4.3 CONCLUDING REMARKS

The principles and skills of intelligent EFM possibly should be a book by itself, since there are so many topics that should be discussed, and we should absorb the established methodologies and skills from other directions such as biology, materials science, civil engineering, TRIZ, DFSS, lean manufacturing, etc. For the balance of the book, we will stop here. But keep in mind that intelligent EFM is an open system.

We only give some brief discussions of interdisciplinary processes in the first four chapters of this book. The rest of the book will be dedicated to the technical discussion of specific directions and processes. I hope readers find them helpful in deepening their comprehension of the principles. You are encouraged to read the first four chapters again in the future.

QUESTIONS

Q.4.1 What is the N2N philosophy? Why should it be treated as the most fundamental principle of engineering?

Q.4.2 Find examples of engineering contradictions and try to solve the contradiction with the Tai-Chi Yin-Yang philosophy?

Q.4.3 What should be the role of process innovations?

Q.4.4 Explain the generality of energy fields, information and intelligence, and logic functional materials.

Q.4.5 Find more examples showing the trend of an increased level of integration of the dynamic M-PIE flows.

Q.4.6 Do you want to be a New CEO? Would you like to use the New CEO to gain long-term sustainability and prosperity? What are the major concepts in the New CEO?

Q.4.7 What are the seventh and eighth principles in intelligent EFM?

Q.4.8 Summarize all the skills discussed in this chapter.

Q.4.9   Choose an energy field, such as gravity or moisture, and explain how we can modulate its four attributes (time, space, magnitude, and frequency) and how these modulations can be useful.

Q.4.10 In the skill "solving engineering challenges in various dimensions," what is the meaning of dimension? Present your examples using this skill.

Q.4.11 What are the three levels of human values? Why is considering human factors so important in technological innovations?

Q.4.12 Using sustainable cycles is a very useful skill. Show several of your examples.

Q.4.13 What benefits can it bring when an effective intelligence is used?

Q.4.14 You need a good piece of equipment to continue your research, but you don't have enough funding to buy one. What alternative solutions can you think of? Suppose the equipment is a specific tool you need in your real work, such as a laser, an oven, a 3D microscope, etc.

Q.4.15 What other skills do you think we should add in this chapter?

REFERENCES

1. Swan, C. W. and Swan, P. A. (2006) Why we need a space elevator, *Space Policy*, 22 (2), 86–91.
2. Forsman A. C. et al. (2005) Double-pulse machining as a technique for the enhancement of material removal rates in laser machining of metals. *Journal of Applied Physics*, 033302.
3. Zhang, W. and Yao, Y. L. (2002) Micro scale laser shock processing of metallic components, *ASME Journal of Manufacturing Science and Engineering*, 124, 369–378.
4. Pozar, D. M. (1993) *Microwave Engineering*. Addison-Wesley Publishing Company, ISBN 0-201-50418-9.
5. Zhang, W. (2007) Photon energy material processing using liquid core waveguide and a computer program for controlling the same, U.S. patent 7,211,763.
6. Cid, M., Stem, N., Brunetti, C., Beloto, A. F., and Ramos, C. A. S. (1998) Improvements in anti-reflection coatings for high efficiency silicon solar cells, *Surface and Coatings Technology*, 106 (2–3), 117–120.
7. Gurrappa, I. and Sambasiva Rao, A. (2006) Thermal barrier coatings for enhanced efficiency of gas turbine engines, *Surface and Coating Technology*, 20 (6), 3016–3029.
8. Benson, Jr. O., Frey, C. M., Shusta, J. M., Nestegard, S. K., Lightle, V. L., Smith, K. L., and Bacon, Jr. C. A. (1997) Ultra-flexible retroreflective cube corner composite sheetings and methods of manufacture, U.S. patent 5,691,846.

Part II

Classic Nonmechanical Manufacturing Processes

5 Energy Fields in Waterjet Machining

Mohamed Hashish

CONTENTS

5.1 INTRODUCTION

It can be said that the harnessing and management of concentrated energy fields are behind the progress of mankind. Throughout history, we have been searching for and developing new tools to survive and prosper. The use of a sharp rock as a tool to hunt and cut was probably man's first innovation to focus energy for useful purposes. The energy of waterfalls and rivers has been known since creation. This energy—kinetic energy—is what alerted man to using water as a cutting tool. In ancient Egypt, river branches were diverted to wash out soil in search of gold and other minerals. This was possible because this excavation did not require high energy dense flows. Harder formations, such as coal, required the directing of river flow through pipes to focus the energy for washing out and carrying the ore. Early in this century, pumps were developed to obtain more energetic jets for profitable mining practices. The increase in pump pressures (>2 MPa) over time has been critical to modern waterjet cutting. Today, waterjets operate at 600 MPa and their energies are as dense as 100 kW/mm^2, which make them capable of cutting rocks and a wide range of nonmetals. Adding abrasives to waterjets provides another means of focusing the kinetic energy. In this case, the water kinetic energy is transferred to the abrasives, whose sharp tips do the cutting action, similar to man's first invention mentioned above. This has enabled the cutting of metals, brittle materials, composites, and hard rocks.

In the following sections, we will list the waterjet tool(s), systems, and processes. Then, we will discuss the associated energy fields before, during, and after jet formation. The control of the different energy fields requires knowledge of several disciplines, which will also be listed.

5.2 WATERJET TOOLS AND PROCESSES

In this chapter, a waterjet will be used as a general term for a wide range of energetic fluid, fluid/fluid, and fluid/solid jets. Waterjets or *waterjet tools* [1] can be classified as follows:

- Fluid jets
  Waterjets (WJs)—Plain water is used to form the jet.
  Pulsed waterjets (PJs)—The waterjet is pulsed to increase impact stresses.
  Polymer waterjets (PWJs)—Polymer is added to the water.
  Cavitating waterjets (CWJs)—Cavitation is formed at the impact zone mainly when submerged.
  Cryogenic and liquefied gas jets (CJs)—Cryogenic fluids are used to form the jet, which then becomes gaseous at ambient conditions.
- Fluid/fluid jets
  Gas waterjets (Fuzzy jets)—Air (or another gas) is entrained in waterjets.
  Liquid waterjets—A liquid is entrained in a waterjet.
- Fluid/solid jets
  Abrasive waterjets (AWJs)—Abrasives are entrained in a waterjet.
  Abrasive suspension jets (ASJs)—A premixed slurry is pumped through a nozzle.
  Ice waterjets (IWJs)—Ice is entrained into a waterjet or a waterjet is cooled to form ice.
  Abrasive cryogenic jets (ACJs)—Abrasives or vanishing abrasives are used in a CJ.

The above waterjet tools have been demonstrated for a wide range of material removal applications. The following is a list of applications where waterjets have been applied. While some of these applications are in commercial use today, other applications are still emerging.

- Kerf cutting—The jet is used to cut shapes or sever materials.
- Drilling (piercing)—The jet is used to drill a hole without trepanning.
- Turning—The jet is used to create a surface of revolution.
- Milling—The jet is used to remove material to a specific depth.
- Fragmentation—The jet is used to fragment the workpiece.
- Jet assist—The jet is used to assist other material removal processes such as cooling, lubrication, debris removal, and laser beam guiding.
- Surface modification—The jet is used to modify the surface, such as cleaning, rust removal, paint removal, peening, texturing, stripping, or polishing.
- Others—deburring, peeling, powder fabrication.

5.3 WATERJET ENERGY FIELDS

There are several energy fields in a waterjet system that ultimately affect its operation and machining results. These fields can be divided into the following processes:

- Pressure generation: This is the most upstream energy field and it starts from the water entry to the pump at ambient pressure until it exits the pump into the plumbing system at higher pressure.
- Pressure transport: High-pressure plumbing is used to transport the pressurized water to the jet-forming nozzle. This plumbing system may consist of tubing, hoses, fittings, swivel joints, and rotary swivels.
- Pressure release: Pressure is released using high-pressure valves either to the jet forming nozzle or to the atmosphere as a safety dump valve. A ruptured disc could also be used to release pressure when it exceeds a certain level.
- Jet formation: This is a most critical energy field in which the cutting jet is formed. In this field, the potential energy is converted to kinetic energy.
- Jet–material interaction: The jet energy is used in this field to remove material and achieve the desired machining results. A robotic system is used to either manipulate the jet or the workpiece to obtain the desired geometrical features.
- Energy dissipation: After the jet cuts through the material, its energy needs to be dissipated. A jet catcher is used for this reason, which may be of different sizes and shapes. Figure 5.1 shows a graphical representation of the energy fields.

In the following sections, we address these energy fields and their most critical factors.

5.3.1 PRESSURE GENERATION

In this section, we describe how the upstream conditions may affect the waterjet tool and its cutting capability.

5.3.1.1 Waterjet Pumps

Water compressibility at high pressures is an important factor in the design and generation of high pressures. The relationship between pressure, P, and density, ρ, of water can be obtained from

$$\frac{\rho}{\rho_o} = \left(1 + \frac{P}{L}\right)^n \tag{5.1}$$

Potential energy		Energy conversion		Kinetic energy	
Pressure generation (pump)	Pressure transport (plumbing)	Pressure release (valve)	Jet formation (nozzle)	Jet-manipulation (and interaction with workpiece)	Jet energy dissipation (catcher)

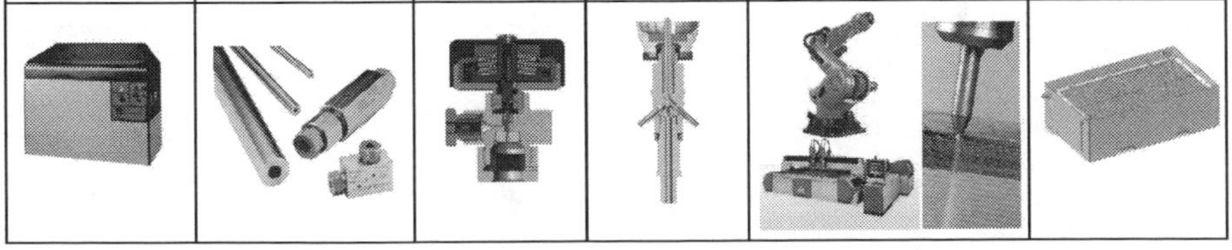

FIGURE 5.1 Waterjet cutting energy fields.

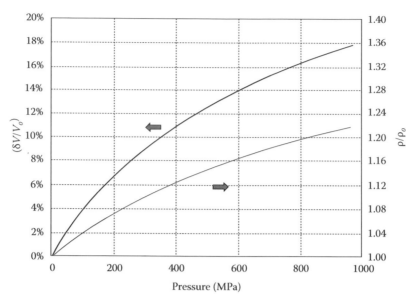

FIGURE 5.2 Water compressibility.

It was found that the above equation fits Bridgman's data [2] with $L = 300\,\text{MPa}$ and $n = 0.1368$ at 25°C. Figure 5.2 shows how the density and volume changes with pressure. This compressibility of water affects the overall energy performance of the system. For example, if the water compresses by 10% before it reaches the required waterjet cutting pressure, then at least 10% of the energy is lost due to compression. This is a critical factor in designing high-pressure pumps where the dead volume must be minimized to maximize the pump volumetric efficiency.

Two main classes of pumps are used in waterjet cutting: intensifier-type and direct drive pumps. Intensifier pumps (see Figure 5.3) work on the principle of pressure intensification. A low pressure acting on a large area results in a higher pressure acting on a smaller area. Conventional

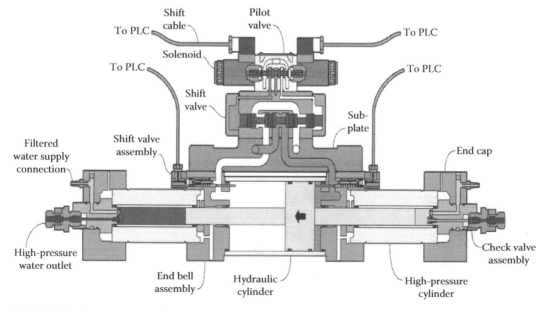

FIGURE 5.3 Pressure intensifier.

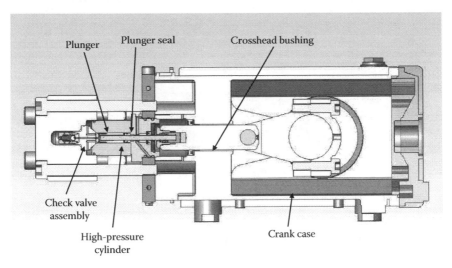

Plunger Plunger seal Crosshead bushing

Check valve
assembly

High-pressure
cylinder

Crank case

FIGURE 5.4 Direct drive pump (one of three cylinders shown).

variable output hydraulic pumps are used to generate pressures of about 20 MPa over a piston of an intensifier. A plunger with 20 times less area attached to that piston is used to transmit this hydraulic force to water, thus pressurizing it to 400 MPa. In double-acting intensifiers, two plungers are attached to one piston, so while high-pressure water is being discharged on one side, low-pressure water is charged on the other side. The typical pump frequency is 60 cycles/min. Significant energy is lost due to the flow and shifting of the hydraulic oil on the low-pressure side of the intensifier pump, which is converted to heat. An important component of the intensifier pump is a pressure attenuator, which is a pressure vessel used to store energy to compensate for a drop in pressure during intensifier position reversal.

In direct drive pumps, see Figure 5.4, plungers are connected to a crank shaft drive instead of using hydraulic pressure to affect pressure on water. Most pumps of this type use three plungers (triplex pumps). The rotational speeds (400–2200 rpm) of the crankshaft determine the flow rate and the required drive power. Current direct drive pumps are capable of 380 MPa operating pressure; however, higher reliability is obtained at lower pressures. These pumps are suitable for high flow rates as may be needed for cleaning or surface preparation.

5.3.2 Energy Transport: Plumbing

Ultrahigh-pressure (UHP) plumbing consists of tubing, hoses, fittings, and swivel joints. Small diameter (inside diameter [ID] = 2 mm; outside diameter [OD] = 6 mm) tubing is commonly used due to its relative flexibility in the form of long whips and coils. The pressure drop in this relatively small diameter tubing can be significant and thus its use must be limited to sections needing flexibility. An alternative option to achieving flexibility is to use swivel joints. In this case, larger tubing (less flexible) can be used with a number of swivels to provide the needed degrees of freedom. A drop in pressure can also be significant in swivels and thus the plumbing system components and tubing sizes must be carefully selected to minimize the pressure drop.

5.3.3 Pressure Release: On/Off Valve

Typical On/Off valves are naturally closed and pneumatically actuated. They consist of two principal parts: an actuator and a poppet/seat assembly. Standard shop air pressure (~0.4 MPa) acts upon the actuator piston against a spring to allow the lifting of the poppet off the seat (containing a hole)

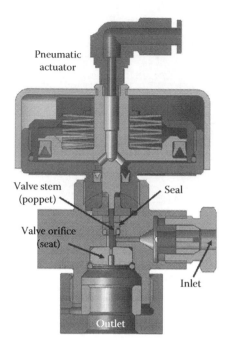

FIGURE 5.5   Typical pneumatically operated on/off valve.

and enables the flow to take place. When the air pressure is released, the spring force causes the poppet to contact the seat to shut off the flow. The poppet hole size needs to be carefully selected as a relatively large hole will require a larger actuator and/or higher shop pressure. Smaller diameter poppet holes will cause significant pressure drop. It is also desired that the valve shuts off within few or tens of milliseconds, especially when the cutting speeds are relatively high. Typically, the on/off valve is placed just upstream of the waterjet orifice so the upstream tubing is not subjected to pressure cycles and fatigue. In some valves, the orifice is used as a seat. This has a significant advantage in extending the fatigue lifetime of the upstream water body. Figure 5.5 shows a schematic of a typical on/off valve.

5.3.4   JET FORMATION

5.3.4.1   Waterjet

Typically, an ultrahigh-pressure waterjet is formed by an orifice made out of sapphire or diamond for more wear resistance. Figure 5.6 shows two methods of holding and sealing the orifice. Either plastic or metal seals are used to seal round orifices while sintering is used to mount and hold irregular shaped natural diamond orifices.

The jet velocity V_j (considering compressibility) and its kinetic power E can be expressed [3] as follows:

$$V_j = \sqrt{\frac{2L}{(1-n)\rho_o}\left[\left(1+\frac{P}{L}\right)^{1-n}-1\right]}. \tag{5.2}$$

$$E = \frac{\pi}{\sqrt{8\rho_o}}C_c C_v^3 \psi^3 d_n^2 P^{1.5}. \tag{5.3}$$

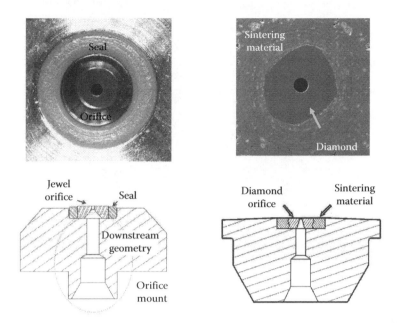

FIGURE 5.6 Waterjet orifices.

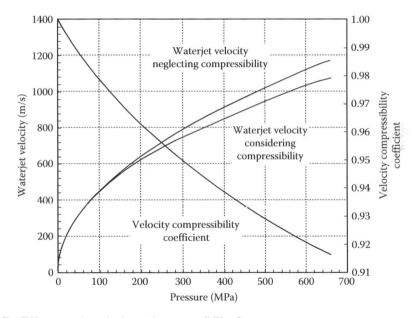

FIGURE 5.7 Effect waterjet velocity and compressibility factor.

where
 ψ is water velocity compressibility factor (less than 1) and is shown in Figure 5.7
 C_c is the jet contraction coefficient
 C_v is the coefficient of velocity

The orifice overall coefficient of discharge, C_d, includes these coefficients. Hashish [3] developed the following empirical formula for the dependency of the coefficient of discharge on pressure and orifice size:

FIGURE 5.8   Waterjets with different coherency levels at 350 MPa pressure.

$$C_d = 0.785 - 0.00014P - 0.197d_n. \tag{5.4}$$

It shows that increasing the jet size and pressure reduces the overall coefficient with more sensitivity to the orifice diameter. This suggests that, for a given jet power, it is more efficient (higher C_d) to increase the pressure than to increase the orifice diameter to convert the pressure's potential energy to kinetic energy.

An important feature of a plain water cutting jet is its coherency. Higher coherency increases the power density and it has been observed that coherent jets are more effective in cutting. Also, coherent jets can operate at longer stand-off distances. Figure 5.8 shows different coherency level waterjets. Yanaida [4] presented equations describing the waterjet structure and its velocity distribution. Basically, a high-pressure waterjet consists of an initial coherent waterjet core followed by a zone of droplet jet. The critical factors that affect the jet coherency (and its power density) include the following:

- Upstream tube diameter and length—The effects of the upstream tube diameter, d, and the level of turbulence above the orifice were studied. Results show that the jet coherency improves when the tube size reaches a certain critical size. Any further increase in the tube diameter does not further improve coherency. The upstream Reynolds number correlates well with the jet coherency length (defined as the length at which the jet diameter is doubled) as the laminar upstream flow results in more coherent jets. The addition of long chain polymers to the water was found to enhance the jet coherency due to drag reductions and upstream turbulence suppression. An upstream length of at least 20 tube diameters was found to be important in producing coherent jets.
- Orifice edge geometry and condition—A chipped edge on an orifice is the most important factor that affects jet coherency. Accordingly, water filters are typically used upstream of water pumps. Also, in abrasive waterjet nozzles, the migration of abrasives upstream of the orifice should be eliminated by proper sequencing of abrasive and waterjet shutdown. Also, controlling the pressure field in the mixing chamber is of critical importance.
- Downstream geometry of orifice holder—The downstream geometry should minimize the interaction of air with the jet. Figure 5.6 shows the geometry below the orifice. If the support area under the orifice is minimized to minimize the jet–wall interaction, then the orifice may fail due to bending stresses.

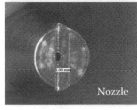
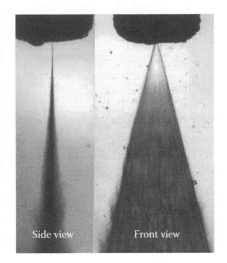

FIGURE 5.9   UHP fanjet and nozzle.

For cleaning and surface preparation applications, it was found that fan-type jets are more practical and more practical than round jets. In these jets, the energy is spread linearly in order to cover a wider zone. The power density is thus lower but only needs to exceed the material threshold for material removal. Shimizu [5] showed the effect of fanjet parameters on material erosion. Fanjet nozzles are typically made out of metal using special manufacturing techniques to obtain the correct geometry, surface finish, and the material pre-stress condition. The design and fabrication processes aim at evenly distributing the jet power and minimizing the edge "hot spots" where the jet tends to be more powerful. Figure 5.9 shows the shape of the fanjet orifice whose projected shape is elliptical while the front and side views of the fanjet [5] show an initial laminar zone followed by a turbulent and structured zone.

5.3.4.2   Abrasive Waterjet Formation

Several methods for forming an ultrahigh-pressure AWJ have been reported by Hashish [6]. In anticipation of component wear, a multiple-jet device was first used to have four waterjets converge at a focal point while abrasives were fed by gravity into the center of these jets. It was found, however, that jet alignment is difficult to achieve and repeat.

An alternative design was to use an annular jet. This was formed by placing an abrasive-feeding pin inside a relatively large sapphire orifice. It was observed that jet convergence can be achieved; however, the abrasive feed tube extending into the mixing chamber wore out quickly.

Figure 5.10 shows a schematic of the current AWJ nozzle. Typical waterjet diameters are 0.08–0.5 mm, and typical jet velocities are up to 900 m/s at 400 MPa. The flow of the high-velocity waterjet into the concentrically aligned mixing tube creates a vacuum, which is used to transport abrasives from a hopper to the nozzle abrasive chamber via a suction hose. A typical abrasive material is garnet, which has flow rates from a few grams per minute to 2 kg/min. Medium and fine abrasives (mesh 60 to mesh 200) are most commonly used for metal, glass, and resin composites. The abrasives are accelerated and axially oriented (focused) in the mixing tube, which has a length-to-diameter ratio from 50 to 100. Typical tube diameters are 0.5–1.3 mm with lengths up to 150 mm. A hard and tough material such as tungsten carbide is used as a mixing tube to resist erosion.

Based on the momentum and continuity equations of the liquid and solid flow, the following equation by Hashish [7] results for the particle velocity V_a at a distance x inside the mixing tube:

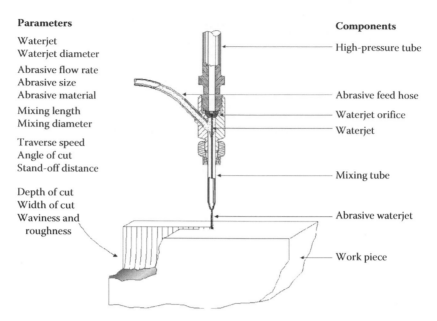

Parameters

Waterjet
Waterjet diameter

Abrasive flow rate
Abrasive size
Abrasive material

Mixing length
Mixing diameter

Traverse speed
Angle of cut
Stand-off distance

Depth of cut
Width of cut
Waviness and
   roughness

Components

High-pressure tube

Abrasive feed hose

Waterjet orifice
Waterjet

Mixing tube

Abrasive waterjet

Work piece

FIGURE 5.10   Abrasive waterjet nozzle and parameters.

$$x = \frac{1}{K}\left[\frac{\lambda}{\lambda-1} - \ln\left(\frac{1}{1-\lambda}\right)\right], \tag{5.5}$$

where

$$K = \frac{3C_D(1+r)^2}{4S_a d_p} \tag{5.6}$$

$$\lambda = \frac{V_a}{V_{a_{\max}}} = \frac{V_a}{V_j/(1+r)} \tag{5.7}$$

where $r = \dot{m}_a/\dot{m}_w$ is the abrasive loading ratio. The above equations suggest that larger particles require longer mixing tubes; note also that as the abrasive flow rate increases, shorter tubes can be used to attain the maximum velocity. For 100 mesh abrasives, for example, a mixing tube length of only 33 mm is required for an abrasive loading ratio of 0.12. The typical mixing tube length used in industry for this case, however, is about 76 mm. The additional length is used to collimate the jet and to raise the value of λ to about 0.95, as can be calculated from the above equations. Note that the additional 43 mm of mixing tube length contributes only 5% to the maximum possible velocity.

While the addition of abrasives to waterjets was found to significantly enhance its cutting power, the abrasive kinetic energy in the abrasives will not be more than 10%–15% of the waterjet. Moreover, this energy is distributed over the mixing nozzle cross section, which is typically 10 times more that the area of the waterjet. Accordingly, the power density of the abrasives in an AWJ is about 2 orders of magnitude less than that of a waterjet. However, this power is refocused through the tips of the abrasives on the workpiece to cause more local energy concentration and thus significantly more material removal than is possible with plain waterjets.

5.3.4.3 Air and Abrasive Entrainment

The entrainment of air in AWJ nozzles is a key process for cutting performance. The air is used as a carrier for the abrasives and, thus, must be of adequate momentum and velocity to perform this transport process effectively. Studies have been performed to characterize the jet pump performance of AWJ nozzles [3,8].

Figure 5.11 shows a graph of AWJ nozzle suction characteristics at different pressures; the curves are typical of jet pump performance. The airflow rate, Q_a, at standard ambient conditions can be approximated by the following equation:

$$\frac{P_a - P_v}{\left(P_a - P_v\right)_{max}} + \frac{Q_a}{Q_{a_{max}}} = 1, \tag{5.8}$$

where $(P_a - P_v)_{max}$ is the vacuum gauge pressure reading with no airflow. Theoretically, this should be equal to P_a if the mixing chamber is completely sealed. $Q_{a\,max}$ is the maximum airflow rate obtained when there are no restrictions to the intake flow.

The air entrainment characteristics in feed lines (hoses) can be obtained experimentally by measuring the airflow rate at different pressure differences between the ambient pressure and the suction pressure. A model for this airflow was developed [9,10] as shown in the following equation:

$$Q_a = A_h \sqrt{\frac{P_r P_a (2 - P_r)}{2\rho_a \left[\ln \dfrac{1}{(1 - P_r)} + \dfrac{fl_h}{2d_h}\right]}}, \tag{5.9}$$

where P_r is the pressure ratio $(P_a - P_v)/P_a$. Solving Equations 5.8 and 5.9 provides the airflow rate that satisfies both characteristics of the AWJ jet pump and the feed line.

Figure 5.12 shows a sample plot of hose and nozzle characteristics that can be used to determine the airflow rate. Selecting conditions with an adequate airflow rate is of importance for achieving

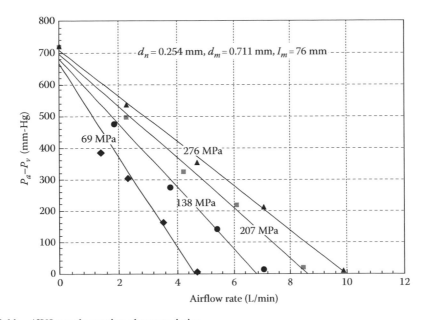

FIGURE 5.11 AWJ nozzle suction characteristics.

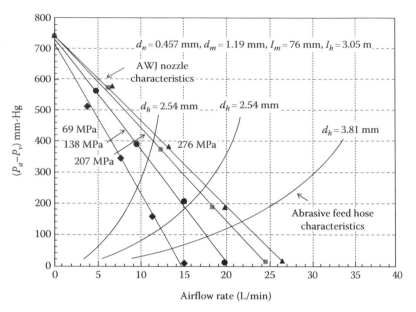

FIGURE 5.12 AWJ nozzle and suction hose characteristics.

a reliable abrasive feed process. However, it is important to realize here that the air velocity in the feed line is a parameter of equal importance. This velocity should exceed a certain threshold for stable flow.

The use of vacuum assist was introduced to allow jets with weak air entrainment performance to draw more air and, thus, provide a more effective abrasive-carrying capacity.

5.3.4.4 Slurry Jet Formation

To increase the power density of the abrasives in an AWJ, a premixed abrasive slurry (suspension) can be directly pumped through the nozzle [11,12]. This eliminates the need to mix the abrasives with the waterjet, which requires approximately 10 times more area in the waterjet. Accordingly, abrasive slurry jets (ASJs) will have an order of magnitude higher power density than AWJs at the same power levels. Also, the momentum transfer between the water and the abrasives is more efficient, resulting in enhanced power density. The power densities of AWJs and ASJs as a function of pressure are shown in Figure 5.13 in comparison with those for pure waterjets [1].

There are several methods of forming an ASJ. Hashish [6] suggested a fluidized bed bypass abrasive slurry system. Another approach is to use an isolator system [12] as shown in Figure 5.14.

Diamond orifice nozzles and tubular-type nozzles were used to form ASJs. Figure 5.15 shows jets at 241 MPa produced by these two types of nozzles. Observe that the jet produced by the diamond orifice spreads quicker than that produced by the tubular nozzle, which is contrary to the case of plain waterjets. It has also been observed by Hollinger and Mannheimer [11] that the rheological characteristics of the suspension fluid are of critical importance to the coherency of the ASJ. They found that the viscoelastic component of the jet controls the penetration depth and width.

5.3.4.5 Pulsed Waterjets

In general, pulsed waterjets are used to capitalize on the impact shock or water-hammer pressure ($p = \rho c V$, where ρ is the water density, c is the speed of sound in water, and V is the jet velocity) that results at the very early stages of impact. This pressure is higher than the hydrodynamic pressure by the ratio $2c/V$. By pulsating the jet, the material will be exposed to higher impact pressures but for shorter periods of time. If this pressure exceeds the threshold pressure for material removal, then

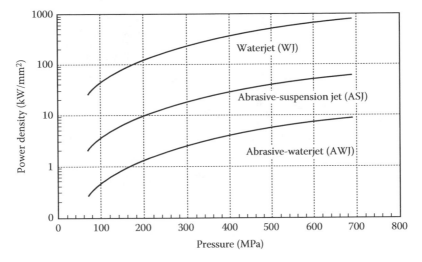

FIGURE 5.13 Power density.

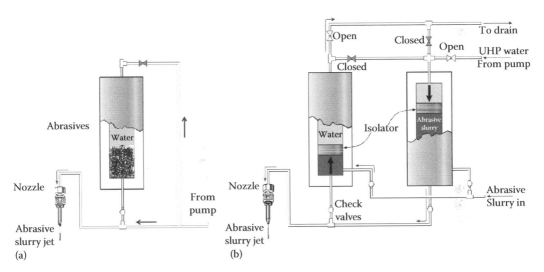

FIGURE 5.14 Concept for forming abrasive slurry (suspension) jet: (a) fluidized bed and (b) isolator concept.

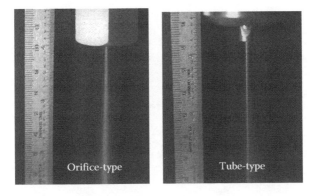

FIGURE 5.15 Abrasive suspension jets at 241 MPa.

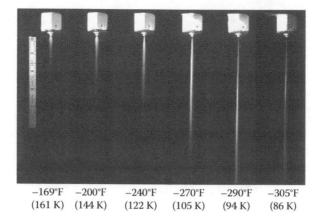

| −169°F | −200°F | −240°F | −270°F | −290°F | −305°F |
| (161 K) | (144 K) | (122 K) | (105 K) | (94 K) | (86 K) |

FIGURE 5.16   Effect of upstream temperature on nitrogen jet structure.

the pulsed jet will be effective. This method is particularly suitable when the steady-state pressure is less than the threshold pressure for material removal. The shape of the pulsed jet greatly affects its impact duration and is a critical parameter. A spherical shape is least preferred. Another important factor is the standoff distance. However, the steady-state waterjet effectiveness at larger standoff distances is attributed to the droplet impact. Vijay et al. [13] developed ultrasonically modulated jets with significantly enhanced material removal rates especially when relatively low-pressure jets are modulated. Yan et al. [14] applied this technique on the stripping of coatings.

5.3.4.6   Cryogenic Jets (CJs)

Pumping a cryogenic fluid, or more generally a liquefied gas, through a nozzle forms a CJ. The liquefied-gas jet will then evaporate to gas after performing the material removal task. To control and improve the performance of CJ tools, Dunsky and Hashish [15,16] indicated the need for accurate thermodynamic control of the cryogen at the upstream condition. Figure 5.16 shows the effect of the upstream temperature on the jet structure. It is noticed that the jet must be sub-cooled to about 100 K in order to get a liquid jet.

The addition of abrasives to CJs enhances their performance, as is the case with the AWJ. The entrainment of abrasives in a CJ, however, is more complex due to possible abrasive-feed-line freezing and plugging. The use of abrasives such as pelletized CO_2 in liquid-nitrogen jets may offer a zero-added-waste cutting and cleaning process.

5.3.5   MATERIAL REMOVAL

5.3.5.1   Cutting

The effects of the different parameters on the cutting speed or cutting depth is qualitatively shown in Figure 5.17. Among the most significant effects is the effect of pressure, which is the single factor that affects the jet's power density [17]. The power E of a jet shown in Equation 5.3 can be simplified as

$$E = KA_oP^{1.5} \tag{5.10}$$

where
   A_o is the orifice cross-sectional area
   K is a numerical constant
   P is the pressure

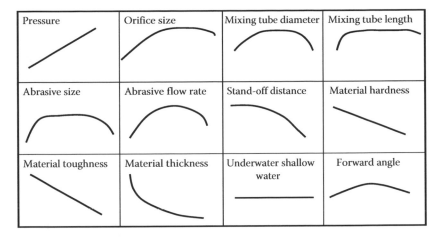

FIGURE 5.17 General trends of cutting with an abrasive waterjet.

From this equation, it can be seen that the power density, defined as the jet hydraulic power per unit area, is only a function of pressure

$$\frac{E}{A_o} = KP^{1.5} \tag{5.11}$$

For AWJs, the power density can be calculated by dividing the abrasive particle kinetic energy (E_a) by mixing the tube cross-sectional area (A_m). E_a can be expressed as

$$E_a = \frac{1}{2}\dot{m}_a V_a^2 \tag{5.12}$$

The following expressions can easily be deduced for the power density E_d:

$$E_d = \frac{E_a}{A_m} = K_1 P^{1.5} \tag{5.13}$$

K_1 in the above equation contains the loading ratio $r = \dot{m}_a/\dot{m}_w$, the momentum exchange relationship $V_a = \zeta_o V_j/(1+r)$, and also the area ratio of the orifice to the mixing tube A_n/A_m. If these ratios are kept unchanged, then the AWJ power density will only be a function of pressure. For $d_m/d_n = 2.5$, $r = 10\%$, and $\zeta_o = 0.9$, it can be calculated that increasing the pressure from 400 to 600 MPa results in an increase in the power density by 1.83 times. This is associated with a decrease in the water flow rate, and correspondingly, the abrasive flow rate by 33% (fixed loading ratio). Figure 5.18 shows data on the effect of pressure on cutting aluminum and steel [18].

5.3.5.2 Cutting Attributes

When jets (or any other beam cutting tool) cut through and separate the material, three phenomena are observed. The first is that the jet is deflected opposite to the direction of the motion [19–23]. This means that the exit of the jet from the material lags behind the point at the top of the material where the jet enters. The distance by which the exit lags the entrance is typically called the trailback, lag, or drag as shown in Figure 5.19. Crow and Hashish [3,24] developed a universal AWJ kerf equation by dividing the kerf zone into an upper direct impact zone and a more significant

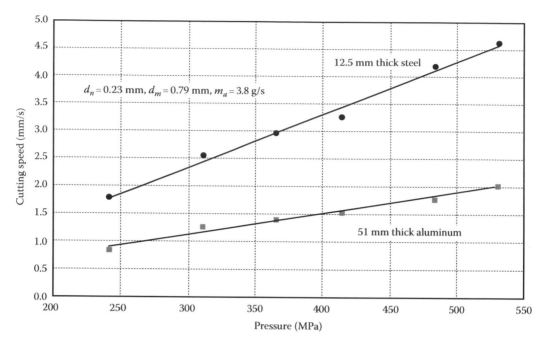

FIGURE 5.18   Effect of pressure on cutting speed.

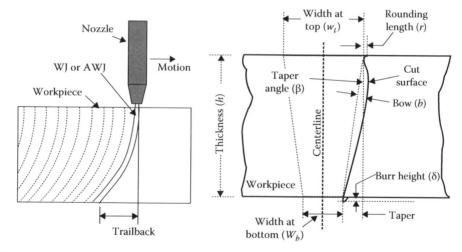

FIGURE 5.19   AWJ cut attributes.

centrifugal abrasion zone. Ignoring the upper zone, the trailback (t_b) at a depth (h) has been derived as follows:

$$\frac{t_b u}{v} = \ln\left[\sec\left(\frac{hu}{v}\right)\right] \tag{5.14}$$

where

$$v = \frac{\mu \dot{m}_a V_a^2}{\sigma_f d_m}. \tag{5.15}$$

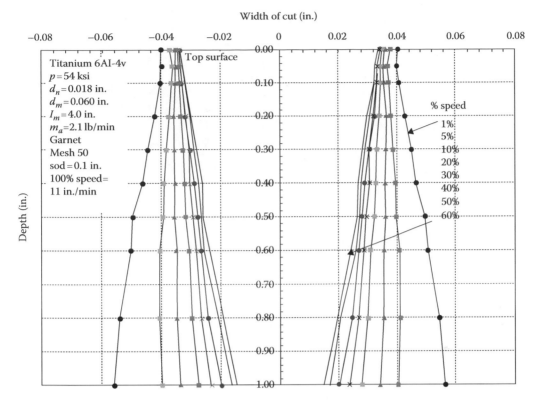

FIGURE 5.20 Width of cut profiles in 1 in. thick titanium.

While the above equation does not include the effect of particle velocity decay as depth increases, it shows that jet power density is an important factor in reducing the trailback.

The second phenomenon is that the width of the cut varies along the cut from top to bottom (see Figures 5.19 and 5.20). This difference in width is typically called the *taper* of the cut. A *taper* can be either positive or negative, that is, the width at the exit of the cut may either be smaller or larger than the width at the top. Typically, the kerf width at the exit side is smaller than that at the entry at practical cutting speeds. Hashish [3], based on a waterjet model [25], proposed a kerf width profile equation in the form

$$\frac{w_e}{d_m\sqrt{R}} = 0.335\sqrt{\frac{X}{X_c}}\left[1 - \sqrt{\frac{\pi d_m^2 \sigma_f}{8\dot{m}_a V_a^2}\frac{X}{X_c}}\right]^{2/3} \tag{5.16}$$

An example of the cut profile cross section is shown in Figure 5.20 at different cutting speeds as percentages of the maximum possible cut speed [26]. The cutting parameters are also shown in the same figure. Observe that the zero taper condition is somewhere between 5% and 10% cutting speeds. A striations-free cut is also observed to be at about a 30% cut speed from Figure 5.20 or about 1.4 mm/s, i.e., 4.1 times the zero-taper cut speed condition.

The kerf profiles suggest that tilting the jet to compensate for taper will be an advantage for increasing the cutting speed [26–28].

For shape cutting, the trailback and taper phenomena manifest themselves in distortions to the geometry of the cut at the exit side. The sketch in Figure 5.21 shows an undercut due to the trailback phenomena [26]. The picture in the same figure shows distorted square-shaped cuts at the bottom surface of the material due to trailback and taper.

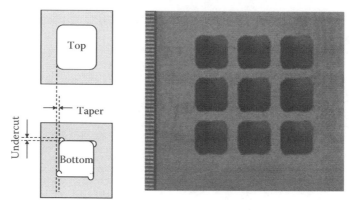

FIGURE 5.21   Undercutting at the bottom of cuts.

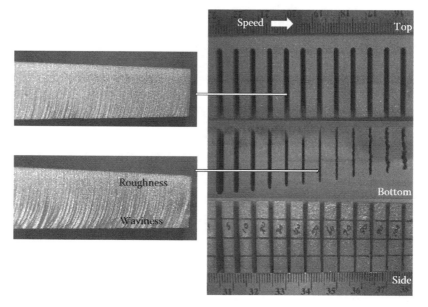

FIGURE 5.22   AWJ cuts showing kerf top, bottom, side, and surface morphology.

The third phenomenon is related to the surface waviness, which is the macro level surface finish of the cut. Figure 5.22 shows a typical striated (wavy) surface produced by an AWJ. Observe that the upper surface of the cut is free from waviness but still rough due to the abrasive erosion process (micro level material removal). The hypothesis of the waviness is that the jet/material interface is not steady. A step of material moves under the jet until it reaches the bottom of the workpiece. During this time, the jet traverses and its effective diameter is reduced as it penetrates deeper. Hashish [29] developed the following simplified expression based on this hypothesis:

$$\frac{2R_w}{d_j} = 1 - \left\{ 1 - (\pi/4)^2 \left[\frac{d_j hu}{0.5 m_a V_a^2 / \sigma_f} \right]^2 \right\}^{1/2} \tag{5.17}$$

This equation shows the effect of jet power density on the surface waviness.

Based on the above discussion, the basic strategies for managing the jet/material energy field can be defined to control the cut geometry. These strategies involve both fields of jet formation and kinematic manipulation:

- Jet formation
  - Parameter selection and optimization: The jet process parameters should be selected to cut the required depth at the required speed and surface finish. For a given jet power, the speed of cut can be determined based on the required surface finish. The parameter selection should also address the bow of the kerf especially when cutting thick materials in order to obtain flat surfaces.
  - Parameter steadiness: The surface finish of the cut surface is affected by the steadiness of the process dynamic parameters, which are pressure and abrasive flow rate.
- Kinematic issues
  - Motion: The surface finish is also sensitive to the quality of the motion system meeting accurate traverse rates with minimal deviations especially around corners. Also, in the case of 5-axis motion requiring angular changes, the angular velocities must be steady and free from vibrations or jerk.
  - Jet tilt parameters can be superimposed on the jet motion parameters to compensate for kerf taper and trailback. These angles are typically referred to as taper and lead angles, respectively.

The general trend of taper, trailback, and surface finish as functions of speed is illustrated in Figure 5.23. This figure shows general speed zones separated by four critical cutting speeds. The first critical cutting speed u_1 is the one at which zero taper occurs. Slower speeds than u_1 will result in divergent cuts with negative taper and no waviness. The second critical cutting speed u_2 characterizes the beginning of waviness formation. Increasing the speed beyond u_2 will continue to increase the taper to a maximum value at the third critical speed u_3. Beyond u_3, the taper will decrease and the

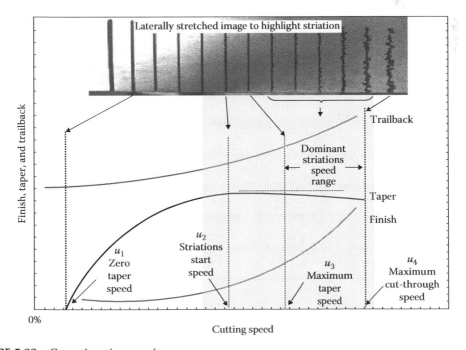

FIGURE 5.23 General cutting speed zones.

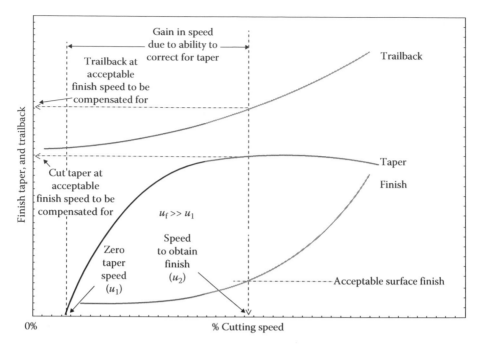

FIGURE 5.24 Gain in speed due to dynamic jet tilting.

surface will be highly wavy and irregular. At speed u_4, the jet will barely cut through the material not cut through completely.

The cut surface at speeds slightly below u_2 will produce a waviness-free surface similar to, but slightly rougher than, that obtained at speed u_1. Usually, u_2 is several times faster than u_1.

To capitalize on the dynamic waterjet angle tilting capability [28,29], the cutting speed can be maximized based on the required surface finish regardless of taper (and trailback). In this case, taper angles are used to obtain the required part accuracy by correcting the wall taper on the required side of the cut. Assume that an acceptable surface finish is Ra. This will identify a cutting speed of u_f. The taper obtained at this speed is then determined as shown in Figure 5.24. This will define the taper angle to be used. The same applies for trailback.

5.3.5.3 Small-Hole Drilling

Jets (or any energetic beam) do not necessarily produce straight-walled holes or holes with uniform diameters, like solid tool drills, due to the nature of a jet and its interaction mechanics with the material. Figure 5.25 shows different hole shapes that may result from waterjet piercing.

As the jet is penetrating through the material, the return flow is also exiting the hole and may cause secondary erosion. The strategy for controlling the qualitative and quantitative features of a hole, as well as reducing the drilling time, includes both before-breakthrough and after-breakthrough

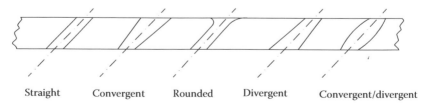

Straight   Convergent   Rounded   Divergent   Convergent/divergent

FIGURE 5.25 Hole geometries obtained with AWJ.

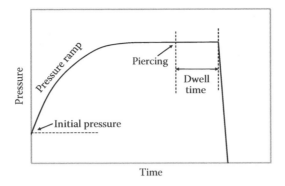

FIGURE 5.26 Pressure ramping.

techniques [30,31]. Some of the most effective techniques, which can be grouped into the categories of jet dynamic parameters and kinematic drilling parameters, are discussed below.

Jet dynamic parameters are those that can be changed during drilling, such as pressure, abrasive flow rate, and abrasive material (or particle size). Useful jet dynamic parameters include the following:

- Pressure ramping during drilling—This is particularly important when drilling sensitive materials that may chip or crack, such as glass, composites, and coated (or TBC) metals. The common strategy is to use a low enough pressure as not to cause initial impact that may damage the material. However, at low pressures, the abrasive entrainment to the jet may not be possible or reliable. To overcome this problem, a modified nozzle with vacuum assists was developed [32,33]. See Figure 5.26 for a schematic of pressure ramping. A side port, see Figure 5.27, in the nozzle chamber was added that attaches to a vacuum pump and draws abrasives through the mixing chamber before the waterjet starts. This allows the instantaneous formation of an AWJ, which then allows the drilling to start at higher pressures (still lower than full pressure in most cases). For high-volume hole drilling (tens of thousands), it was found that the periodic flushing of the mixing chamber with water is important to maintain process reliability. The flushing water removes the abrasives that have built up in the mixing chamber.
- Abrasive ramping during drilling—During pressure ramping, the abrasive flow rate may also be ramped. It is critical, however, for a wide range of materials to keep the abrasive flow rate above a certain critical limit [34].
- Pressure or abrasive flow rate alteration after breakthrough—The mode of erosion after breakthrough can be enhanced by altering the pressure or abrasive flow rate. This will change the structure of the jet and its spreading profile, which in turn will affect the hole shape.
- Abrasive material change after breakthrough—This process is used to improve the surface finish of holes drilled with relatively coarse abrasives. The abrasive flow rate may also be totally stopped to wash out the hole of any abrasives that may have been embedded into the wall.

The kinematic drilling parameters that can be varied to improve the drilling process and control hole shape include the following:

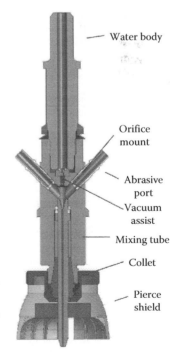

FIGURE 5.27 AWJ drilling nozzle with vacuum assist.

- Continuous angle change during drilling—This method is of importance when drilling at shallow angles [34], where the deflected jet may cause teardrop or shadow-hole anomalies. Starting at a relatively larger angle and then reducing the angle gradually at the correct rate may eliminate teardrop and shadow-hole formation. Highly accurate machines are needed to use this process to control the jet's focal point of impact.
- Trepanning while drilling—Trepanning while drilling can be achieved by rotation or orbital motion. Either the manipulator end effector, the workpiece, or the nozzle itself can be manipulated. Traversing the manipulator (in a circular path) is a simple and common approach [31]. Rotating the workpiece (under an eccentrically impacting jet) is only suitable for small round parts that can be easily rotated. Rotating the nozzle requires special devices such as swivels. Orbiting a nozzle (no need for swivels) requires an eccentric drive.
- Increase or decrease in standoff distance after breakthrough—This method is used to improve the hole taper and also to round the edges of the hole if needed.
- Dwelling after breakthrough—Dwelling after breakthrough applies to any of the above methods. It is used to affect hole taper. For example, hole taper may be reduced to achieve straight walls or reversed to produce a divergent hole shape. Dwelling, however, must occur while the jet is in the center of the hole to either maintain or improve hole roundness.

Advances have also been made in the AWJ process and in controls and inspection systems to allow the drilling of large numbers of holes reliably. For example, over 25,000 diamond-shaped holes (2×2 mm) were drilled in each of several titanium parts. Figure 5.28 shows a sample of the hole pattern.

Several analyses and modeling studies have been performed on AWJ drilling [35–37]; a simple analysis of the drilling process by Hashish [31] yields the following equation for the drilling time:

$$t = \frac{\sigma_f}{2E_d K_2}\left[e^{2K_2 h} - 1\right] \tag{5.18}$$

This equation shows the importance of increasing the power density of the abrasive particles in order to reduce the drilling time. It also shows that the drilling time is exponentially proportional to the depth. As the depth increases, the abrasive kinetic energy needs to be increased, and thus pressure ramping is important. Increasing the factor K_2 in the above equation will also increase the drilling time; thus, efforts should be made to reduce K_2. The factor K_2 from Equation 5.18 can be re-expressed as follows:

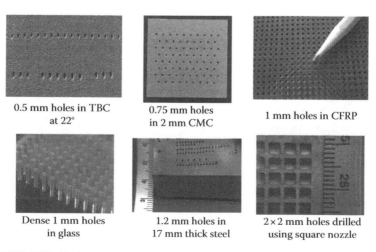

0.5 mm holes in TBC at 22°

0.75 mm holes in 2 mm CMC

1 mm holes in CFRP

Dense 1 mm holes in glass

1.2 mm holes in 17 mm thick steel

2×2 mm holes drilled using square nozzle

FIGURE 5.28   AWJ-drilled holes.

$$K_2 = \frac{3C_d}{4S_a d_p}, \tag{5.19}$$

where

S_a is the abrasive particle specific gravity

d_p is the abrasive particle diameter

Now, from this equation, it is obvious that reducing the drilling time requires reducing the drag coefficient as can be intuitively understood. This can be accomplished by reducing the water flow rate out of the hole (which is equal to the inlet flow rate), at least before breakthrough occurs. Trepanning, because it enlarges the hole and directs the return flow, reduces C_d significantly. This results in a dramatic improvement in drilling time. Increasing the abrasive specific gravity and size also helps to reduce the drilling time. Dense, large abrasives will carry their momentum deeper into the hole.

5.3.5.4 Controlled-Depth Milling

To control and limit the depth of cut by a waterjet, repeated passes are needed and only small amounts of material are removed per pass. This requires either the jet to be weak or moving it at relatively high traverse rates [38,39]. The latter approach involves the use of masks made out of more resistant materials because a high-speed contouring motion is difficult to achieve in practice. The following three methods are typically used for controlled depth milling [40]:

- Liner milling: In this method, Cartesian motion is used to scan the jet over the masked workpiece. Either the jet or the workpiece or both may be traversed and laterally indexed to expose the unmasked area to the jet and mill it to the required depth. To improve the milled corners, the jet may be tilted at different clocking angles.
- Radial milling: In radial milling, the jet moves radially over masked workpieces mounted on a rotating platter. To control the radial milling depth profile, both the rotational speed and the radial traverse speed can be selected as a function of the radial position of the jet. The jet angle can also be changed to improve the corners of the produced milled pockets.
- Cylindrical milling: Controlled depth milling can be achieved on both the outside and inside of cylindrical workpieces. In this case, the jet traverses axially while the workpiece is rotated. The inside or outside walls of a cylinder can also be used to mount samples and masks.

Figure 5.29 shows the radial and cylindrical milling concepts.

Masks can be cut with an AWJ or laser and mounted on the workpiece. This method has been used to mill isogrid patterns on the inside and outside of cylinders, cones, and domes [40,41]. Figure 5.30 shows an inconel plate about 150 mm × 80 mm in depth of 2 mm with an accuracy of 25 μm using radial milling where several plates were mounted on a rotary platter and simultaneously milled.

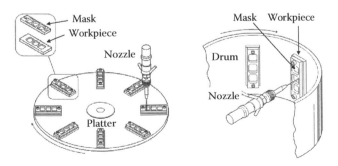

FIGURE 5.29 Milling methods using masks.

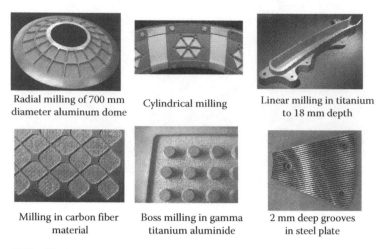

Radial milling of 700 mm diameter aluminum dome

Cylindrical milling

Linear milling in titanium to 18 mm depth

Milling in carbon fiber material

Boss milling in gamma titanium aluminide

2 mm deep grooves in steel plate

FIGURE 5.30   AWJ-milled parts.

The milling of variable-depth pockets can also be achieved by controlling the exposure time of the AWJ over the different areas to be milled, which can be accomplished by varying the traverse rate and the number of passes. Milling can also be used for multiple pattern cutting, such as holes, which can be milled through simultaneously by using a mask that has been predrilled with the hole pattern. The milling of thin, closely spaced slots can be accomplished with or without a mask as shown in Figure 5.30. To control the slot (or groove shape), both lead and taper angles may be used.

5.3.5.5   Turning

Waterjet turning is a relatively simple process where a workpiece is rotated while the AWJ is traversed axially and radially to produce the required turned surface [42]. Figure 5.31 shows AWJ turning methods. Work on AWJ turning has addressed the volume removal rate, surface-finish

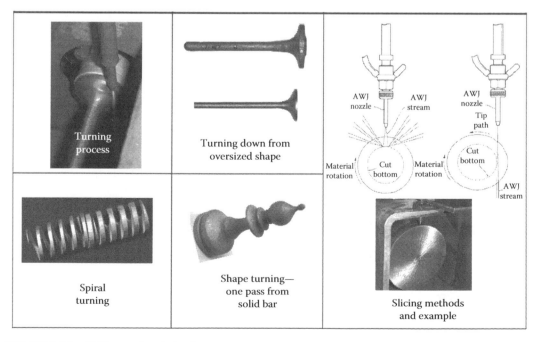

Turning process

Turning down from oversized shape

Spiral turning

Shape turning— one pass from solid bar

Slicing methods and example

FIGURE 5.31   AWJ turned and sliced parts.

control, visualization and modeling of the turning process [43–46], and the development of a hybrid AWJ/mechanical lathe [43].

The depth of cut, which is determined from the radial jet position, is a critical parameter for process optimization. Unlike conventional turning, AWJ turning is less sensitive to the original part shape. For example, a highly irregular geometry can be turned in one pass with a relatively large depth of cut to a surface of revolution. Also, AWJ turning is not sensitive to the length-to-diameter ratio of the workpiece. Long- and small-diameter parts have been turned to precise dimensions. Underwater turning has also been demonstrated to significantly reduce the noise on the AWJ.

A hybrid AWJ/mechanical lathe was built by modifying a conventional lathe to allow simultaneous turning with the AWJ and a solid tool, with the solid tool performing the finishing process. The AWJ nozzle can either be mounted on a separate manipulator for flexibility of pattern machining or mounted on the tool carriage when accurate synchronization between the rotational and axial motion is required. The AWJ was used to produce a diameter 0.25 mm greater than the required diameter. The machined surface was simultaneously finished using a solid single-point tool immediately behind the AWJ. This approach was found to be most efficient to obtain surface finishes better than 5 μm.

5.3.5.6 Multi-Process Machining

AWJs can be used for prototyping by cutting layers in the actual material and stacking them to form the required prototype geometry. Joining techniques can also be assisted by the AWJ machining process by providing holes or prepared surfaces for bonding.

Figure 5.32 shows the machining sequence used for a three-dimensional part made with an AWJ. This machining exercise demonstrates the flexibility of the AWJ process for cutting, turning, and drilling using the same setup and the same nozzle. A controller program was written to fully automate the machining process, which includes automatic quick-change nozzles. Also implemented was an intelligent manufacturing process involving parameter changes.

5.3.5.7 Hybrid Processes and Systems

There are two approaches to using waterjets as a hybrid tool with other traditional or nontraditional tools. These approaches are the hybrid process and the hybrid system.

5.3.5.7.1 Hybrid Process

In this method, the material removal energy field is modified either directly or indirectly by the waterjet. For example, the use of waterjets to assist drag bits was investigated by Hood [47] in mining tools and tunneling equipment is a hybrid process. The waterjet may modify the cutting field

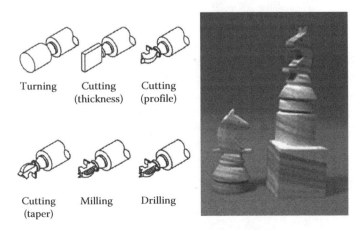

FIGURE 5.32 Three-dimensional AWJ machining of complex geometry.

by weakening the rock or creating free surfaces ahead of the tool tip such that less overall energy is consumed in the excavation. The waterjet may also act as a cooling tool to the otherwise excessively hot tool tip, thus elongating its service life and enhancing its cutting performance. In jet-assisted machining, the performance of cutting tools and grinding wheels are enhanced by using waterjets in different ways that may be by cooling tool tips or dressing grinding wheels. The cutting field itself is still primarily controlled by the solid tool process. The use of low-pressure waterjets to collimate laser beams [48] is another example of an indirect hybrid process as the waterjet is not used to assist in the material removal process but rather as a guide to the laser.

5.3.5.7.2 Hybrid System

In this method, two or more tools are integrated into one system to achieve technical and/or overall productivity objectives. Some of the commercially available hybrid systems are as follows:

- Waterjet–mechanical: In these systems, a waterjet is used to perform certain functions and the mechanical tools are used to either alter the produced surface or perform other operations in different locations on the workpiece. A waterjet composite machining center is used to trim composite structures with a waterjet tool and drill counter sunk holes using special drills. Figure 5.33 shows a typical dual mast hybrid waterjet–router system used for such parts as aircraft wings, tail fins, and some fuselage sections. In the stone and tile industry, a hybrid waterjet–saw machine is used where the waterjet is used to cut the interior shapes and the saw is used on external straight line cutting. Other machines may include edge finishing tools.
- Waterjet–EDM: A wire electrical discharge machine (EDM) cuts faster if the wire is partially engaged with the surface. For example, the cutting rate may double or triple if the wire is used on the surface to remove a depth equal to only one wire radius. A hybrid waterjet–EDM system is used to cut the part with a waterjet to within one wire radius tolerance. The wire is then used to finish the part. Figure 5.34 shows a picture of a hybrid waterjet–EDM machine.
- Waterjet–thermal methods: Waterjet band plasma has been combined with one machine. The plasma is used when accuracy or heat affected zones are not critical while waterjets are used to produce higher quality edges or when the material cannot be cut thermally. The two processes can be used on the same part to maximize productivity using special software that selected the waterjet or the plasma based on the edge specifications.

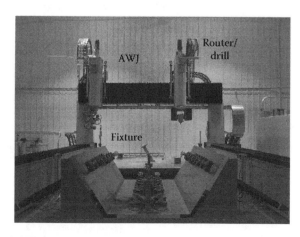

FIGURE 5.33 Trim and drill system.

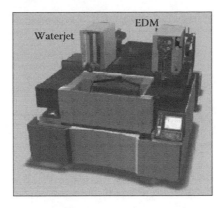

FIGURE 5.34   Hybrid waterjet–wire EDM.

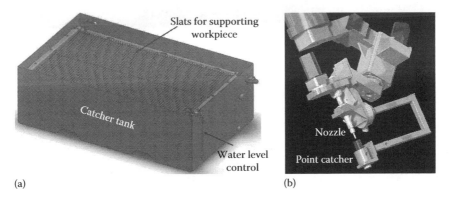

(a)   (b)

FIGURE 5.35   Tank and point catchers: (a) catcher tank and (b) point catcher.

5.3.6 ENERGY DISSIPATION

After the waterjet completes its cutting action, it exits the workpiece with a still significant amount of energy. Accordingly, several types of jet catchers are used. These can be classified as follows:

- Water catchers: In these catchers, water is used to dissipate the momentum of the jet. They can be in the form of a tank or a tube. Tank catchers are most commonly used with 2-axis, 3-axis, and 5-axis machines. A water height of about 1 m is commonly used. It has been noticed that when the tip of the nozzle is slightly submerged, less water can be used in the catcher tank. This is attributed to preventing the air from shrouding the jet, thus increasing its reach. Also, cutting under water reduces the aerodynamic noise. Accordingly, most tank catchers (see Figure 5.35) are equipped with water level control bladders. Slit catchers are narrow tanks and are used when the jet motion is restricted to a line rather than an area. Tube catchers are used when the jet is stationary. They are efficient in quickly dissipating the energy of the jet because the tube return flow imparts greater reverse momentum than in a catcher tank.
- Ball-filled catchers: Free moving steel balls may be used to absorb the momentum of the jet. This type of catcher is used when space is not available to use a water catcher. The steel balls may be placed in a tray with enough height (~0.1 m) or in a cup with drains at the opposite end. The steel balls will wear out and thus the catcher needs to be replenished.
- Carbide catchers: These are point catchers and are typically mounted on the end effector under the jet when edge trimming is needed. This catcher is typically made out of tungsten carbide material. A convergent opening is used to catch the jet and direct it to impinge on

another carbide rod. As the jet drills through the carbide rod, the reverse flow slows down its penetration. Eventually, the jet may penetrate through the rod if its length is not sufficient. Figure 5.35 shows a carbide point catcher mounted on the nozzle end effector.

5.4 FINAL REMARKS

There are several energy fields in waterjet technology. The most upstream energy field is obtained by increasing the potential energy of water by raising its pressure. This energy is, then, transferred and released to form energetic jets. The energy lost in pressure generation, transfer, release, and conversion to kinetic energy must be minimized. The consideration of water compressibility is of critical importance in pressure generation while fluid flow conditions are critical to efficient transfer and release. When the kinetic power density of the jet exceeds a certain threshold, the material can be removed and cut. The jet power density is a function of pressure only, which motivates pump manufacturers to continue raising the pump pressures. A limiting factor is compressibility. To increase the power density at the jet material interface, abrasive particles are used as agents for this power concentration. The energy of the waterjet is transferred to the abrasive particles. While this transfer is associated with power loss, the focus of the remaining power on the sharp tips of the abrasives significantly increase the cutting rates and material removal. To efficiently use the kinetic abrasive power for precise cutting, kinematic manipulation may be used to adjust and compensate for the anomalies associated with beam cutting. Advances in waterjet technology require the incorporation of several disciplines as follows:

- Solid mechanics (stress analysis, fatigue, fracture, etc.)
- Tribology (wear and erosion, friction, lubrication)
- Material technology (metallurgy, polymeric material, ceramics, etc.)
- Fluid mechanics (single- and multi-phase flows, compressible flow, rheology, etc.)
- Physics and mechanics (properties under pressure, material removal, impact, etc.)
- Mathematics (statistics, fuzzy logic, genetic algorithms, etc.)
- Electro-mechanics (kinematics, motion control systems, etc.)
- Software engineering (modeling, nesting, tabbing, etc.)
- Control systems (sensors, inspection, etc.)
- Reliability engineering

The multidisciplinary nature of waterjet technology continues to invite academia, industry, and the government to collaborate on the underlying sciences and technologies toward greater advances in all facets of the waterjet as a tool.

QUESTIONS

Q.5.1 What are the different energy fields in a waterjet system?

Q.5.2 Name three hard materials used to form an abrasive waterjet and where they are used.

Q.5.3 List five applications that can be performed with waterjets.

Q.5.4 Why can an abrasive waterjet cut metal while a plain waterjet cannot as effectively?

Q.5.5 Why do abrasive suspension jets have more power density than abrasive waterjets?

Q.5.6 What is the single parameter that affects the jet's power density?

Q.5.7 How can you apply the discipline of your study to waterjet technology?

Q.5.8 Develop an expression for the energy in a waterjet as a function of pressure.

Q.5.9 Develop an expression for the kinetic energy of abrasives in an abrasive waterjet.

Q.5.10 Develop an expression for the jet reaction force as a function of pressure.

ABOUT THE CONTRIBUTING AUTHOR

Dr. Mohamed Hashish is a senior vice president of technology at Flow International Corporation. He graduated from the Mechanical Engineering Department of Alexandria University, Egypt in 1970, obtained his PhD in mechanical engineering from Concordia University, Montreal, Canada, in 1977, and continued his post-doctoral work under a fellowship from the National Research Council (NRC) until 1979. Then, Dr. Hashish joined Flow Industries as a research scientist. In 1980, Dr. Hashish invented the abrasive waterjet process, revolutionizing the field of waterjet cutting technology.

Dr. Hashish continued to pioneer new applications for fluid jet technology. He pioneered new waterjet processes, such as precision cutting using tilt angles, 100 ksi cutting, side-fire nozzles used in aircraft composite trimming, drilling, milling, turning, polishing, and other processes. He patented new concepts in ultrahigh-pressure seals for pumps, high-speed rotary joints, and quick-change nozzles. He also led the Flow team in developing 100 ksi food sterilization systems. Most recently, Dr. Hashish and his team developed waterjet singulation systems for micro SD flash memory cards, which are in use today.

Currently Dr. Hashish is directing both internally and externally funded technology programs on many aspects of the waterjet technology covering cutting, pumps, and ancillary hardware such as sensors and recyclers. He has attracted over $30 million in contracts and grants to supplement his research work.

Dr. Hashish holds over 30 patents in the areas of jet cutting and high pressure. He has published more than 300 papers in many journals and conference proceedings. He edited several proceedings for the American Society of Mechanical Engineers (ASME) and the WaterJet Technology Association (WJTA). He was elected as an ASME fellow and was awarded both the technology and pioneer awards from the WJTA. Dr. Hashish also serves as an affiliate professor at the Mechanical Engineering Department, University of Washington, Seattle.

REFERENCES

1. Hashish, M. (1998) The waterjet as a tool, *Proceedings of the 14th International Water Jet Cutting Technology Conference*, BHR Group, Brugge, Belgium, September 1998.
2. Bridgman, P. W. (1970) *The Physics of High Pressure*, 1st edn., Dover Publications, Inc., New York.
3. Hashish, M. (2002) Waterjet cutting studies, *Proceedings of the 16th International Water Jetting Technology Conference*, BHR Group, Aix-en-Provence, France, October 16–18, 2002, pp. 13–48.
4. Yanaida, K. (1974) Flow characteristics of waterjets, *Proceedings of the Second International Symposium on Jet Cutting Technology*, BHR Group, The Fluid Engineering Centre, Canfield, U.K., Paper A2.
5. Shimizu, S. (2006) Structure and erosive characteristics of waterjets issuing from fanjet nozzle, *Proceedings of the 18th International Water Jetting Conference*, BHR Group, Gdansk, Poland, September 2006, pp. 337–345.
6. Hashish, M. (1982) Steel cutting with abrasive-waterjets, *Proceedings of the Sixth International Symposium on Jet Cutting Technology*, BHRA, Cranfield, U.K., April 1982, pp. 465–487.
7. Hashish, M. (2003) Inside AWJ nozzles, *Proceedings (CD Format) of the 2003 WaterJet Conference*, WJTA, Houston, TX, August 2003.
8. Hashish, M. (1989) Pressure effects in abrasive-waterjet machining, *ASME Transactions, Journal of Engineering Materials and Technology*, 111(3), 221–228.
9. Hashish, M. (1984) Suction characteristics of abrasive-waterjet nozzle experimental data, Flow Technical Report No. 319, Flow Research Company, Kent, WA, November 1984.
10. Hashish, M. (1984) Modeling of the air suction process in abrasive-waterjet nozzles, Flow Technical Report No. 315, Flow Research Company, Kent, WA, December 1984.

11. Hollinger, R. H. and Mannheimer, R. J. (1991) Rheological investigation of the abrasive suspension jet, *Proceedings of the Sixth American Waterjet Conference*, WJTA, Houston, TX, August 1991, pp. 515–528.

12. Hashish, M. (1991) Cutting with high pressure abrasive suspension jets, *Proceedings of the Sixth American Waterjet Conference*, WJTA, Houston, TX, August 1991, pp. 439–455.

13 Vijay, M. M., Bielaski, M., and Paquette, N. (1997) Generation of powerful pulsed waterjets with electric discharge: Fundamental study, *Proceedings of the Ninth American Water Jet Conference*, WJTA, Detroit, MI, August 1997, pp. 415–450.

14. Yan, W., Tieu, A., Ren, B., and Vijay, M. (January 2003) High-frequency forced pulsed waterjet technology for the removal of coatings, *Journal of Protective Coatings & Linings*, 20(1), 83–99.

15. Dunsky, C. M. and Hashish, M. (1994) Feasibility study of machining with high-pressure liquefied CO_2 jets, Manufacturing Science and Engineering, Book No. G0930A, PED-Vol. 68-1, the American Society of Mechanical Engineers, pp. 453–460.

16. Dunsky, C. M. and Hashish, M. (1996) Observations on cutting with abrasive-cryogenic jets (ACJ), *Proceedings of the 13th International Conference on Jetting Technology*, BHRA, Sardinia, Italy, October 1996, pp. 679–690.

17. Hashish, M. (2009) Trends and cost analysis of AWJ operation at 600 MPa pressure, *Transactions of the ASME, Journal of Pressure Vessel Technology*, 131, 1–7.

18 Hashish, M. (2000) 600 MPa waterjet technology development, *Transactions of the ASME, Journal of Pressure Vessel Technology*, 406, 135–140.

19. Hashish, M. (1984) A modeling study of metal cutting with abrasive-waterjets, *ASME Transactions, Journal of Engineering Material and Technology*, 106(1), 88–100.

20. Hashish, M. (1988) Visualization of the abrasive-waterjet cutting process, *Journal of Experimental Mechanics*, 28(2), 159–169.

21. Henning, A. and Anders, S. (1998) Cutting edge quality improvement through geometrical modeling, *Proceedings of the 14th International Conference on Water Jetting*, BHR, Brugge, Belgium, September 1998, pp. 321–328.

22. Henning, A., Goce, R., and Westkamper, E. (2002) Analysis and control of striations structure at the cutting edge of abrasive waterjet cutting, *Proceedings of the 16th International Conference on Water Jetting*, BHR, Aix-en-Province, France, October 16–18, 2002, pp. 173–191.

23. Henning, A. and Westkamper, E. (2000) Modeling of contour generation in abrasive waterjet cutting, *Proceedings of the 15th International Conference on Water Jetting*, BHR, Ronneby, Sweden, September 6–8, 2000, pp. 309–320.

24. Crow, S. and Hashish, M. (1989) Mechanics of abrasive jet cutting, *Flow Research Presentation*, Flow Research Company, Kent, WA, March 31.

25. Hashish, M. and DuPlessis, M. P. (1979) Prediction equations relating high velocity jet cutting performance to stand off distance and multi-passes, *ASME Transactions, Journal of Engineering for Industry*, 101(3), 311–318.

26. Hashish, M. (2007) Benefits of dynamic waterjet angle compensations, *Proceedings of the 2007 American Water Jet Conference*, Houston, TX, August 2007, Paper 1-H.

27. Knaupp, M., Meyer, A., Erichsen, G., Sahney, M., and Burnham, C. (2002) Dynamic compensation of abrasive water jet properties through 3-dimensional jet control, *Proceedings of the 16th International Conference on Water Jetting*, BHR, Aix-en-Provence, France, October 16–18, 2002, pp. 75–90.

28. Zeng, J., Olsen, J., Olsen, C., and Guglielmetti, B. (2005) Taper-free abrasive waterjet cutting with a tilting head, *Proceedings of the 2005 American Waterjet Conference*, Houston, TX, August 21–23, 2005.

29. Hashish, M. (1992) A modeling study of jet cutting surface finish, *ASME Proceedings on Precision Machining: Technology and Machine Development and Improvement*, Jouaneh, M. and Rangwala, S., eds., PED-58, Anaheim, CA, November 1992, pp. 151–167.

30. Hashish, M. and Whalen, J. (1993) Precision drilling of ceramic coated components with abrasive-waterjets, *ASME Transactions, Journal of Engineering for Gas Turbine and Power*, 115(1), 148–154.

31. Hashish M. (2002) Drilling small deep diameter holes using abrasive waterjet, *Proceedings of the 16th International Conference on Water Jetting*, BHR, Aix-en-Province, France, October 2002, pp. 33–49.

32. Hashish, M. and Craigen, S. (1990) Method and apparatus for drilling small-diameter holes in fragile material with a high-velocity liquid jet, U.S. Patent No. 4,955,164, September 1990.

33. Hashish, M. and Craigen, S. (1990) Abrasive-waterjet nozzle assembly for small hole drilling and thin kerf cutting, U.S. Patent No. 4,951,429.

34. Hashish, M. (1994) Drilling of small-diameter holes in sensitive materials, *Proceedings of the 12th International Water Jet Cutting Technology Conference*, BHRA Group, Rouen, France, October 1994, pp. 409–424.
35. Ohlsson, L. et al. (1992) Optimization of the piercing or drilling mechanisms of abrasive-waterjets, *Proceedings of the 11th International Symposium on Jet Cutting Technology*, BHR, St. Andrews, U.K., September 1992.
36. Guo, Z. and Ramulu, M. (1999) Simulation of displacement fields associated with abrasive waterjet drilling, *Proceedings of the 10th American Waterjet Conference*, WJTA, Houston, TX, August 1999, paper 19.
37. Ramulu, M., Posinasetti, P., and Hashish, M. (2005) Analysis of abrasive waterjet drilling process, *Proceedings of the 2005 American Waterjet Conference*, WJTA, Houston, TX, August 2005, paper 19.
38. Hashish, M. (1989) An investigation of milling with abrasive-waterjets, *ASME Transactions, Journal of Engineering for Industry*, 111(2), pp. 158–166.
39. Fowler, G., Shipway, P. H., and Pashby, I. R. (2005) Abrasive waterjet controlled depth milling of Ti6AL-4V alloy—An investigation of the role of jet workpiece traverse speed and abrasive grit size on the characteristics of the milled material, *Journal of Materials Processing and Technology*, 161, 407–414.
40. Hashish, M. (1994) Controlled-depth milling techniques using abrasive-waterjets, *Proceedings of the 12th International Water Jet Cutting Technology Conference*, BHR Group, Rouen, France, October 1994, pp. 449–462.
41. Hashish, M. (1998) Controlled depth milling of isogrid structures with AWJs, *ASME Transactions, Journal of Manufacturing Science and Engineering*, 120, 21–27.
42. Hashish, M. (1987) Turning with abrasive-waterjets—A first investigation, *ASME Transactions, Journal of Engineering for Industry*, 109(4), 281–296.
43. Ansari, A., Hashish, M., and Ohadi, M. (1992) Flow visualization study on macro mechanics of abrasive-waterjet turning, *Journal of Experimental Mechanics*, 32(4), 358–364.
44. Henning, A. (1999) Modeling of turning operation for abrasive waterjets, *Proceedings of the 10th American Waterjet Conference*, Houston, TX, August 1999, pp. 795–810.
45. Zeng, J., Wu, S., and Kim, T. J. (1994) Development of a parameter prediction model for abrasive-waterjet turning, *Proceedings of the 12th International Conference on Water Jetting*, BHR Group, Rouen, France, October 1994, pp. 601–617.
46. Manu, R. and Babu, R. N. (2008) Influence of jet impact angle on part geometry in abrasive waterjet turning of aluminum alloys, *International Journal of Machining and Machinability of Materials*, 3, 120–132.
47. Hood, M. (1977) A study of methods to improve the performance of drag bits used to cut hard rock, Chamber of Mines of South Africa Research Organization, Project GT2 NO2, Research Report Number 35177, August 1977.
48. Iscoff, R. (2003) On the cutting edge: Laser, water singulation bid for acceptance in a saw-diamond market, *Chip Scale Review*, August 2003, p. 45.

6 Electrical and Electrochemical Processes

Murali Meenakshi Sundaram and Kamlakar P. Rajurkar

CONTENTS

Electrical and electrochemical processes refer to a group of nontraditional manufacturing processes that primarily use electricity or the effect of electricity to produce desired features by material removal or material addition. The specific processes included in this chapter are electro-discharge machining (EDM), electrochemical machining (ECM), electrochemical discharge machining (ECDM), electroforming, and electroplating. All these processes are noncontact processes in which the tool never makes any physical contact with the workpiece. Hence, they can be successfully applied to any conductive material irrespective of its mechanical properties (such as high hardness). A historical view of the above-mentioned processes, their mechanism/working principle, the merits and limitations of individual processes, recent process improvements, and their environmental effects are discussed in this chapter.

6.1 ELECTRO-DISCHARGE MACHINING (EDM)

6.1.1 INTRODUCTION

EDM (also known as the spark erosion technique) is an electro-thermal material removal process that falls under the category of unconventional manufacturing methods because chips, in the traditional sense, produced by mechanical action, are absent in this process. In EDM, a series of randomly distributed discrete electric sparks are used to remove material from an electrically conductive workpiece immersed in a dielectric medium like kerosene. EDM is capable of producing intricate shapes on any electrically conductive material irrespective of its hardness. Hence, EDM is preferred for machining difficult-to-machine materials like tool steel and cemented carbides. Specific applications of EDM vary from well-known tasks like machining of hardened dies, fuel injector nozzles, and turbine blades to novel applications like the production of carbon nanotubes (CNT) and micro deburring [1,2]. Since EDM is a noncontact process, the mechanical forces are almost absent and hence this technique is an ideal choice for machining fragile components [3]. Materials with electrical resistivity below $300\,\Omega$ cm are machinable by EDM [4]. EDM provides good accuracy and repeatability, but the thermal damage of the machined surface due to high heat generated during the discharge pulses is a concern.

The history of EDM can be traced back to the experimental works of Joseph Priestly in the 1770s [5]. However, metal erosion by spark discharges was put to constructive use only in the 1940s by the Russian couple Lazarenkos [6]. From its inception in the world war era, EDM technology has undergone tremendous technological advancements in several disciplines such as power supply, control, and materials. From its humble beginning as a repair tool to remove broken drills and taps [7], EDM has emerged, over a period of time, as the fourth most popular machining process [8]. Today's commercial EDM machines with improved machine intelligence and better flushing are capable of the material removal rate (MRR) and surface finish on the order of $2500\,\text{mm}^3/\text{min}$ and $0.3\,\mu\text{m}$ Ra, respectively.

The two principal types of EDM processes are the die-sinking EDM (often simply mentioned as EDM) and the wire EDM (WEDM) processes shown schematically in Figure 6.1. Process variants such as EDM drilling and EDM milling are derived from the combination of the concept of die-sinking EDM and the simplicity in the tooling of WEDM. An important technological development is the application of EDM process variants for machining small feature sizes of dimensions less than $1000\,\mu\text{m}$. These EDM-based micromachining processes collectively are known as micro-electro-discharge machining (mentioned in the literature as micro-EDM, MEDM, or μEDM).

In die-sinking EDM, a tool with complex features is used to reproduce corresponding intricate shapes in the workpiece (historically *Die*) immersed in dielectric liquid. The two electrodes (workpiece and tool electrode) are initially separated by a dielectric medium. Potential is applied between the two electrodes, and the tool is gradually brought closer to the workpiece. When the gap between

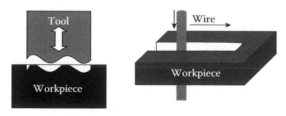

FIGURE 6.1   Schematic of two principal types of EDM processes.

the two electrodes reaches a critical value, the dielectric, which is initially nonconductive, breaks down and suddenly becomes conductive and energy is discharged as a spark. The sparking zone is subjected to a very high temperature as the temperature of the momentary local plasma column is estimated to be in the range of 10,000–60,000 K [9–11]. Experimental measurements using spectroscopic techniques show plasma column temperature between 8,000 and 10,000 K [12]. In the case of metals, material is removed from both the workpiece and the tool electrode by melting and vaporization [13–15]. In the case of conductive ceramics, which are characterized by a high melting point, low thermal conductivity, and high thermal expansion, the material removal is done by the flaking of material due to high temperature gradients—a phenomenon known as thermal spalling [16–18]. Besides these two typical material removal mechanisms, other mechanisms can occur, such as oxidation and dissolution of the base material [19].

WEDM is a special form of electrical discharge machining wherein the electrode is a continuously moving conductive wire. As the wire electrode is fed (from a spool) through the workpiece, material is removed by the erosive action of sparks between the wire and the workpiece. The computer numerical–controlled movement of the worktable allows intricate cutting and shaping of materials to nearly any complex three-dimensional ruled surfaces and shapes. As WEDM uses a thin wire as a single electrode, it is not necessary to make different shapes of tool electrodes to achieve the complex contours. However, prevention of wire breakage is critical to obtain a continuous machining process as the wire is subjected to electromagnetic force caused by direct and alternating components of the discharge current [20]. The optimal selection of wire properties, wire tension, and cutting speed would determine its final performance and success.

EDM milling is the process in which complex shapes are machined using simple-shaped electrodes. EDM contouring and planetary EDM are other techniques quite similar to EDM milling except that in the former cases, the electrode rotation is not present in some cases where electrodes other than cylindrical shapes are used. In EDM milling, simple-shaped electrodes are rotated at high speeds and follow specified paths in the workpiece like the end mills [21]. This technique is very useful as it makes EDM very versatile like the mechanical milling process. This process eliminates the problem of manufacturing accurate, complex-shaped electrodes that are required in the die-sinking of three-dimensional features. EDM milling improves flushing due to the high-speed electrode rotation. The electrode wear can be optimized in EDM milling because of the rotational and contouring motions of the electrode. Thus, unlike die-sinking EDM where several electrodes would be used to produce a part, one simple-shaped electrode can be used to machine different shapes in EDM milling. The main limitation in the EDM milling process is that complex shapes with sharp corners cannot be machined because of the rotating electrode.

Micro-electrical discharge machining (micro-EDM or μEDM) is a technique adapted from the conventional EDM machining process for the purpose of micro machining. For this, micro tools with diameters typically in the range of 50–300 μm are used in micro EDM. The working principle is the same for EDM and micro EDM [22]. However, the key difference lies in how, in micro-EDM, the discharge energy is reduced and electrode vibration is controlled [23]. In the case of micro-EDM, the process energy discharged per spark is precisely controlled to be on the order of 10^{-6} to 10^{-7} J [24]. Small unit removal (UR) is achieved in micro-EDM by reducing the discharge energy of each pulse [25]. For this purpose, the stray capacitance is kept to a minimum. The schematic configuration is shown in Figure 6.2 [26].

The traditional concept of mounting preformed tools in a machine spindle may lead to a coordinate shift—a serious difficulty in micromachining that includes problems such as tilting and off-centering. Additionally, damages such as tool bending and breakage may occur due to poor handling. These can be avoided by the on-the-machine tool making concept using a new technique called wire electrical discharge grinding (WEDG) to achieve common coordinates for the tool-making process and micromachining. The principle of the WEDG process is similar to wire EDM. A sacrificial wire, as shown in the upper right side of Figure 6.2, travels in a tungsten carbide wire guide immersed in a dielectric medium at a slow pace of 5–10 mm/min. The potential applied

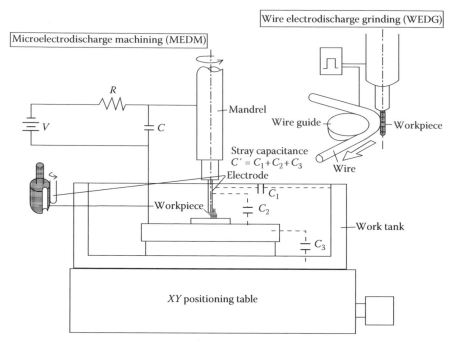

FIGURE 6.2 A typical relaxation circuit used in micro EDM. (From Rajurkar, K.P., Microelectrodischarge machining, Center for Nontraditional Manufacturing Research, University of Nebraska–Lincoln, http://www.unl.edu/nmrc/microEDM/medm3.htm)

between this wire and the tool electrode that is fed downward against the wire generates sparks to machine the tool to the desired dimension. Micro tools of different shapes can be machined in WEDG by numerical control of relative motion between the wire and the tool, which may be any conducting material including hard materials such as tungsten carbide alloys and sintered diamond. Hence, WEDG is a powerful method that is used to produce micro cutting tools. Based on the electrode being used, micro-EDM can also be classified into drilling, die-sinking, milling, WEDM, and WEDG [27,28]. An overview of the capabilities of micro-EDM is provided in Table 6.1 [29].

6.1.2 Process Description

The effect of electric sparks randomly occurring between the anode and the cathode erodes the material in electrical discharge machining. For this reason, the EDM process is also referred to as the "*spark erosion*" process. Spark erosion is a highly complex process and due to its stochastic

TABLE 6.1
Overview of the Micro EDM Capabilities

Micro EDM Variant	Geometric Complexity	Minimum Feature Size (µm)	Maximum Aspect Ratio	Surface Quality Ra (µm)
Drilling	2D	5	~25	0.05–0.3
Die-sinking	3D	~20	~15	0.05–0.3
Milling	3D	~20	~10	0.5–1
WEDM	2½D	~30	~100	0.1–0.2
WEDG	Axisymmetrical	3	30	0.8

Source: Rajurkar, K.P. et al., *Ann. CIRP*, 55, 643, 2006.

nature, it is very difficult to understand the mechanisms involved completely, even though it has commercially been used for more than half a century [30]. According to the *thermo-electric theory*, material removal in EDM operations takes place as a result of the generation of extremely high temperatures produced by the high intensity of the discharge current. Although well supported by experimental evidence, this theory alone cannot be considered as definite and complete because of the unknown contributions from other physical phenomena. The sequence of events leading to the electro-erosion is illustrated in Figure 6.3. The voltage applied between the electrode and the workpiece builds up an electric field throughout the space between the electrodes (Figure 6.3a). As a result of the power of the field and the geometrical characteristics of the surfaces, conductive particles suspended in the fluid concentrate at the point where the field is the strongest. This results in a bridge being formed (Figure 6.3b). At the same time, electrons are emitted from the negatively charged electrode (Figure 6.3c). They collide with neutral particles in the space between the electrodes and split them into positively and negatively charged particles. This process spreads at an explosive rate and is known as impact ionization. This development is encouraged by bridges of conductive particles.

Electrons and ions migrate to anodes and cathodes, respectively, at a very high current density. A column of vapor begins to form and the localized melting of the workpiece commences (Figure 6.3d). The discharge channel continues to expand along with a substantial increase in temperature and pressure (Figure 6.3e). When the power is switched off, the current drops (Figure 6.3f); no further heat is generated, and the discharge column collapses (Figure 6.3g). A portion of molten metal evaporates explosively and/or is ejected away from the electrode surface. With the sudden drop in temperature, the remaining molten and vaporized metal solidifies. A tiny crater is, thus, generated at the surface. The residual debris is flushed away along with the products of decomposition of the dielectric fluid (Figure 6.3h). The application of voltage initiates the next pulse and the next cycle starts. This EDM cycle may repeat up to 250,000 times per second, but only one cycle will occur at any given time. It is essential to understand this cycle to control the duration and intensity of the on/off pulses to optimize the EDM performance.

Compared with the vast literature available on experimental studies, the literature on analytical and numerical approaches to understand the EDM process is relatively scarce. The stochastic nature of the process and the complex plasma formation during discharge make it very difficult to fully

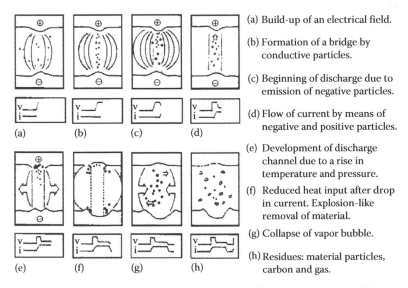

(a) Build-up of an electrical field.

(b) Formation of a bridge by conductive particles.

(c) Beginning of discharge due to emission of negative particles.

(d) Flow of current by means of negative and positive particles.

(e) Development of discharge channel due to a rise in temperature and pressure.

(f) Reduced heat input after drop in current. Explosion-like removal of material.

(g) Collapse of vapor bubble.

(h) Residues: material particles, carbon and gas.

FIGURE 6.3 Sequence of events during electro-discharge machining. A typical sparking cycle takes about 1–4 μs to complete. (From Petrofes, N.F., Shaping advanced ceramics with electrical discharge machining, PhD thesis, Texas A&M University, College Station, TX, 1989.)

understand the process. Almost all modeling efforts are based on Fourier's partial differential equation of heat conduction in solids [32] and a single-spark discharge is assumed [33]. The plasma fluid equation (similar to the Navier–Stokes equation of fluids) has been used to model the electrostatic force and stress distribution in the metal during discharge [34]. The heat transfer theory was used to simulate crater formation and to determine the relationship between the crater radius and the heat source radius [35,36]. A simple point heat source in which heat is partially dissipated is assumed in several works [10,37,38]. Other works assume a cylindrical heat source of a constant radius [39] or heat source with the radius expansion as a function of time, discharge current, and dielectric pressure [40–43]. As an additional heat input, Joule heating was also taken into account [44]. Joule heating occurs when an electrical current (I) is passed through a material and the material's resistivity (R) to the current causes heat generation. According to Joule's law, the heat generated for a time (t) is equal to I^2Rt. Almost all of these works assume isotropic material properties. Nonisotropic material properties were considered to study electrically conductive ceramics [42]. Surface integrity models have been proposed based on the metallurgical examinations of spark-eroded surfaces [45–48]. Recently, the finite element method (FEM) technique has been increasingly used to model the stress induced in spark-eroded surfaces [49–52]. Modeling the EDM process for multiple successive discharges is yet to yield satisfactory results.

As the traditional approaches are inadequate to model, simulate, and control the real-life random sparking behavior of the EDM process, other means like modern control concepts [53], such as adaptive control [54–58], statistical control [59,60], expert systems [61,62], fuzzy control [63–67], neural networks [68–71], and their combinations [72–74] have been used to study the process. The importance of process monitoring and control systems cannot be overemphasized, especially in micromachining, which must avoid or minimize the open circuit, arcing, and short circuits during machining. A stable gap control system enables better dimensional accuracy of the micromachined features by predicting the gap distance and offsetting tool position. The ignition delay time (t_d) is an important indicator of the isolation condition of the discharge gap. Larger gap widths cause longer ignition delays, resulting in a higher average voltage. The tool feed speed increases when the measured average gap voltage is higher than the preset servo reference voltage and vice versa [75]. Other than the average gap voltage, the average delay time can also be used to monitor the gap width [76]. In other attempts, gap monitoring circuits were developed to identify the states and ratios of gap open, normal discharge, transient arcing, harmful arcing, and short circuit [77,78]. These ratios were used as input parameters for online EDM control based on various control strategies.

The common behavior of EDM is that higher discharge energy will produce higher MRR, tool wear, and surface roughness [79–81]. Lowering the discharge energy, on the other hand, will improve the surface finish and reduce the tool wear but the MRR will also decrease. Hence, one needs to find an optimum balance based on the design requirements [68,82,83]. Apart from the discharge energy, several other process parameters such as dielectric medium, control systems, and tool and work material also influence the performance of EDM. These aspects are discussed in this section.

The dielectric medium in the gap acts as an insulator, cools the gap, and transports the debris. High flash point (the temperature at which the vapors of the fluid will ignite), high dielectric strength (the ability of the fluid to maintain high resistivity before spark discharge), low viscosity, and oxidation stability are some of the desired dielectric properties. Petroleum-based hydrocarbon mineral oils (such as kerosene) and deionized water are widely used as EDM dielectric fluids. Hydrocarbon-based dielectric is preferred because of its low conductivity (e.g., the conductivity of kerosene is $0.0017\,\mu S/cm$), which results in a small discharge gap and better accuracy [84]. However, a high temperature plasma channel decomposes hydrocarbon-based dielectric liquid and generates conductive carbon, which increases the debris concentration. This results in an unstable machining process. Therefore, deionized water with conductivity of $0.04\,\mu S/cm$ is proposed as a dielectric medium. Less debris and the large discharge gap obtained with deionized water lead to an aspect ratio of 10 in micro-hole drilling [85]. In addition, the MRR in micro-EDM using deionized water is much higher than that in mineral oil [86]. However, the disadvantage of deionized water is that the contaminated liquid increases the electrochemical reaction, thus resulting in either the misjudgment of short circuits by discharge detecting circuits or unexpected material removal. Another problem is that the plasma radius expansion will be more in case of deionized water due to its lower viscosity, which results in rougher surfaces.

Often, certain additives are added to kerosene to prevent gas bubbles and de-odoring. Adding surfactant to dielectric liquid to lower debris particle agglomeration has reduced the unstable concentrated discharges and improved the MRR by 40%–80% without sacrificing the surface finish [87]. It is possible to obtain a better surface by adding powders to the dielectric medium [88,89]. Certain debris concentration in the gap was reported to be helpful for micro-EDM [90].

In a recent development, high pressure (1 MPa) oxygen gas, instead of the dielectric liquid, was injected to get a higher MRR as a result of the thermally activated oxidation of the workpiece material [91,92]. The use of gas reduces the reaction forces and vibration, thus high-precision finishing is achieved [93].

Flushing is a useful procedure for removing debris from the discharge zone and the MRR in roughing is increased by flushing [94]. On the other hand, the most common cause of EDM failures is inadequate flushing [95]. Flushing is necessary to reduce the arcing tendency and to get good sparking. Arcing is an undesirable phenomenon caused by discharge concentration, leading to spark deterioration and an unstable process [96]. Arcing happens as a series of discharges strike the same spot repeatedly. Continuous and violent arcing will destroy both the tool and the workpiece. Since too much debris in a spark gap causes arcing, flushing is done to keep the sparking zone clean. Though flushing is predominantly achieved by mechanical means, alternate methods such as magnetic fields can also be effectively used [97]. Existing flushing methods are broadly based on the following concepts: (a) controlling the fluid flow (normal flow, reverse flow, jet flushing, and immersion flushing); (b) modifying the electrode shape; and (c) imparting relative movement between the workpiece and the tool electrode. Pressure as high as 5 MPa has been used to achieve higher MRR in fast EDM drilling [98]. An optimum fluid velocity (corresponding to MRR) needs to be used to minimize the conicity or taper of the drilled hole [41]. Dynamic flushing using moving nozzles to sweep along the sparking gap results in an even distribution of debris concentration, thus avoiding the undesirable accumulation of debris in the sparking zone [99]. Better jet flushing is feasible with the help of computational fluid dynamics (CFD) simulation and high-speed video camera observation [100].

With modified electrode shapes, such as mesh sheet electrodes [101], electrodes with multiple cavities [102], or notch [103], better flushing can be achieved [104]. Such shapes allow the quick escape of the gases and debris formed during the machining, facilitate high-speed EDM under lower electrode wear conditions, and have been found to be useful, especially in the machining of insulator ceramics [105].

Ultrasonic vibration is a well-known method for improving the performance of EDM [106–108]. It has been found to be a useful aid in exotic material machining [90,109–111]. The high-frequency pumping action of the vibrating surface accelerates the slurry circulation and reduces the machining time [112]. The vibrations can be imparted either to the workpiece [113,114] or the tool electrode [115,116]. These vibrations introduce acoustic streaming in the dielectric tank and particles move along this stream [117]. Apart from acoustic streaming, the debris particles are also subjected to ultrasonic irradiation, Bernoulli's attraction, and Stoke's force [118]. As a result of this induced fluid flow, better debris removal is achieved, which in turn results in higher aspect ratio machining [119] and better surface finish [120,121].

Electrode jumping is yet another flushing method. Die-sinking machines equipped with linear motors can provide a high-speed (1400 ipm) electrode jumping motion [8]. However, an electrode jump and debris exclusion model and experimental verification reveals that the electrode jump height, rather than the jump speed, plays more important roles in deep machining [122]. When the tool electrode [123,124] or workpiece [125,126] is rotated, better flushing is achieved by the

Taylor–Couette flow. Due to the accompanying centrifugal force, higher MRR is achieved. This is the most prominent method used in micro-EDM [127]. Additionally, the planetary motion of the tool electrode allows the dielectric liquid to flow in from one side and leave at the other side of the workpiece [128]. This technique is used in CNC machines for finish machining [129]. Micro-holes with an aspect ratio of 29 have been drilled using micro-EDM by combining ultrasonic vibration and the planetary movement of electrodes [130].

The EDM power supply basically consists of the rectifying circuit, high-frequency switching transistor, limiting resistor, and oscillating circuit, which form the pulsed power supply. Various types of functional circuits are incorporated in order to achieve stable machining characteristics in a wider range. The electro-discharge machining is performed by applying a voltage between the electrode and the workpiece and producing a pulsed discharge at the gap through the high-frequency switching of the power transistor of the main circuit. Pulse height (power setting I_p), pulse width (on time), and standby time (off time) (interval between the pulses) are the three main factors that determine the machining results (surface roughness, machining speed, electrode wear ratio, and clearance) from rough machining to finish machining. The pulse height is determined when the power setting is changed. The pulse width (on time) and standby time (off time) are set by the oscillator circuit. Since a few thousands to a few tens of thousands of pulses are generated in one second for a transistorized power supply, the duration of a pulse is very short, from a few hundred microseconds to a few tens of microseconds.

A micro-EDM power supply is provided by a relaxation-type circuit as well as a transistor-type circuit. In a relaxation-type circuit, the discharge pulse duration is dominated by the capacitance of the capacitor and the inductance of the wire connecting the capacitor to the workpiece and the tool [131]. The discharge energy is determined by the used capacitance and by the stray capacitance that exists (1) between the electric feeders, (2) between the tool electrode holder and the work table, and (3) between the tool electrode and the workpiece [75].

The transistor-type circuit has better controllability and improved capability to handle large currents with a fast response. Commercial transistor pulse generators can vary the pulse duration and duty factor and can also change the waveform of discharge pulse to reduce the tool wear and improve MRR. However, it is difficult to keep the constant discharge duration shorter than several tens of nanoseconds, using the transistor-type pulse generators [75]. By integrating the transistor-type isopulse generator with the servo feed control system, about a 24 times higher MRR has been obtained than that of the conventional RC pulse generator with a constant feed rate in both semi-finishing and finishing conditions [132]. Nevertheless, the relaxation-type pulse generators are still the better choice for finishing in micro-EDM because it is difficult to obtain a significantly short pulse duration with constant pulse energy using the transistor-type pulse generator. Other advantages of the relaxation circuit include simplicity and ruggedness, reliability at high frequencies, and low cost.

A summary of EDM process parameters and their effects is given in Table 6.2.

Additionally, interelectrode gap size and duty factor are also crucial for the stable operation of an EDM system. A recent report states that the machining rate was almost doubled with an adaptive control of gap states [136]. Gap sizes typically in the range of 10–50 μm are maintained by controlling the ram head or the quill movement by the servo control system with gap width sensors. Process parameters such as current, voltage, and the die-electric media affect the gap size, and, depending on the application, it is not uncommon to use gap sizes in the micrometer range. The duty factor is given by the ratio of the on time to the total time (on time + off time). A small duty factor yields a low machining rate due to a high off time. With a high duty factor off time, the flushing time is less and this might lead to the short circuit condition. Therefore, there has to be a compromise between the two depending on the tool used and the workpiece and the conditions prevailing. The duty factor most preferred is 0.5.

Extensive research using the commercial Panasonic MG-ED72W micro-EDM machine has revealed major micro-EDM process parameters and their influence [137]. The results show that the effect of capacitance, rather than voltage, is more pronounced on MRR. Additionally, an increase

TABLE 6.2
EDM Process Parameters and Their Influence

Parameter	Influence
Spark on-time	The time during which machining takes place.
	Increasing spark-on time will produce a larger crater, higher MRR, and roughness as more energy is discharged with longer spark-on time.
Spark off-time	The time during which the plasma channel collapses and molten material is removed.
	A longer spark-off time reduces the machining time and hence, less material is removed. A shorter spark-off time will make the process unstable. Hence, an optimum value needs to be used.
Voltage; current	Discharge energy is proportional to these factors.
	Spark gap increases with voltage. Higher current produces deeper surface damage. Hence, a lower voltage and current settings are preferable for finishing.
Material	Higher MRR can be achieved on materials having lower thermal conductivity and melting temperature.
Frequency	Roughing is done at a low frequency as longer spark-on time can be set. Finishing is done at a higher frequency.
Polarity	For wire EDM and micro EDM, it is preferable to have the tool electrode as the cathode as less tool wear is a critical requirement in avoiding tool breakage.
Dielectric liquid	Liquids with higher dielectric strength, lower viscosity, and specific gravity produce less arcing and hence are preferable for finishing operation.

Sources: Guu, Y. H. et al., *Mater. Sci. Eng. A*, 358, 37, 2003; Lee, H.-T. et al., *Mater. Trans.*, 44, 2718, 2003; Lee, H.T. and Tai, T.Y., 2003, *J. Mater. Process. Technol.*, 142, 676, 2003.

in feed gives a higher MRR. With proper understanding and control of the process parameters, it is possible to achieve a mean contour deviation within $\pm2\,\mu m$ [138].

6.1.2.1 Tool Materials

Copper and graphite are the two main types of electrode materials. Additionally, brass, zinc, tungsten, tungsten composites, and exotic materials (including tantalum, nickel, and molybdenum) are also used as EDM tool materials. In a recent study involving cemented tungsten carbide (WC), chemical vapor deposition diamond (CVDD), and polycrystalline diamond (PCD) in dry machining a copper-tungsten alloy, WC outperformed other materials to achieve lower tool wear and surface roughness [139]. Though technically tungsten and tungsten composites are the best material for the electrodes due to their high strength, hardness and high melting point, they are expensive and difficult to machine. In contrast, both copper and graphite are inexpensive and easily available. Graphite shows less wear than copper and also has superior fabrication capabilities. On the negative side, graphite pollution affects both personnel and equipment. In a recent study, it was noticed that copper-infiltrated-graphite (Poco EDM-C3) electrodes achieved a significantly higher MRR and lower electrode wear ratio than the graphite and copper electrodes in the electrical discharge machining of 95% pure alumina [140].

While simple EDM tool geometry can easily be designed by conventional methods, increasingly complex tool shapes, such as those required to make intricate injection mold shapes, need computer-aided design (CAD)/computer-aided manufacturing (CAM) systems for EDM tool design. It has been reported that CAD/CAM can improve the efficiency of the design process by at least 50% [141]. Desirable properties of wire materials include adequate tensile strength with high fracture toughness, high electrical conductivity, good flushability, a low melting point, and low-energy requirements to melt and vaporize. Wires are commercially available both as single component wires made of materials such as copper, brass, and molybdenum and multi-component wires like zinc-coated brass to improve the strength, conductivity, and flushability.

For micromachining, instead of complex-shaped expensive tools, simple cylindrical tools can be used to machine complex features by profiling [142]. It is possible to use hollow electrodes for drilling above 100 μm diameter holes. For smaller dimensions, solid tools from sintered carbide that have a diameter of around 50 μm are used [143]. Currently, WEDG is the widely accepted and commercialized method for fabricating micro tools [144,145]. Rapid prototyping, electroforming, and deposition are emerging tooling methods [146,147].

Using a single pulse discharge is an innovative technique for producing 20–40 μm diameter tungsten electrodes in hundreds of microseconds. The repeatability of tool shape after fabrication is hard-to-control in this rapid process [148–150]. Another method prepares a micro rod by using self-drilled holes [151]. This method does not need initial positioning of the rod with respect to the plate electrode, and the operation is easy and has good repeatability. A new hybrid technique combines WEDG technology with one-pulse electro-discharge (OPED) to fabricate multi-micro-spherical probes [152].

LIGA, a German acronym for Lithographie Galvanoformung Abformung (Lithography, Electroforming, and Molding) is yet another method used to fabricate the tool electrode for micro die-sinking EDM [127]. However, this process is limited by the availability of material suitable for the electroforming process.

A thin foil (thickness ~10 μm) has been used as a tool to machine long microgrooves [153]. Using graphite foil, a 20 mm long microgroove with an aspect ratio of about 2.3 was machined from tool steel. This process was further refined by using the gravitational effect for effective debris removal, which improved the aspect ratio to about 8 [154]. Using a wheel tool similar to the grinding wheel is another promising approach to machine microgrooves. Since the wheel rotates during machining, the removal of debris is improved [155].

Localized electrochemical deposition is yet another promising micro tool fabrication option [147]. Tungsten, which has a high melting point and tensile strength, is the predominant tool material in micro-EDM [156,157]. Tungsten carbide [158] and copper have also been used as tool materials [159]. Electrically conductive CVD diamond is a new entrant [160]. It has shown almost zero elec-trode wear, even at short pulse durations of 3 μs [161].

The mechanical properties and cutting performance of thin wires are a special concern in micro wire EDM [162]. In micro wire EDM, apart from tungsten, micro wires made of copper, brass, and molybdenum are also commonly used.

6.1.2.2 Tool Wear

The tool electrode wear is mainly influenced by the polarity and the thermal properties of elec-trode materials [75]. The energy dissipation distributed into the anode during discharging is always greater than that distributed into the cathode for both single discharges [163] and continuous pulse discharges [164]. The carbon layer deposited on the anode surface due to thermal dissociation of the hydrocarbon oil protects the anode surface from wear [165]. A thicker carbon layer leads to a smaller electrode wear ratio in macro EDM, where the tool is the anode. However, in micro-EDM, the tool is typically cathode and hence, deposition of the carbon layer is scarce. The effect of thermal properties on electrode wear was investigated in [166]. It was found that the boiling point in addition to the melt-ing point of the electrode material plays an important role in the wear of micro-EDM tools. It was found that the tool wear ratio reduces with the increase of the tool area [157]. Other factors affecting the tool wear ratio, like poor flushing conditions in a deep hole, are difficult to assess and control. This could easily result in the wrong estimation of the wear ratio and the produced depth [167]. The discharge current waveform is yet another factor affecting tool wear [75]. Graphite usually gives less wear at low frequencies, but the wear is very high during negative polarity and high-frequency applications. Different types of tool wear in EDM are shown schematically in Figure 6.4.

Compensating the tool wear is important especially in micromachining. Two tool wear compensation methods, namely, the linear compensation [169,170] and uniform wear method [167], have been used in micro-EDM. The linear compensation is to feed the tool toward the workpiece

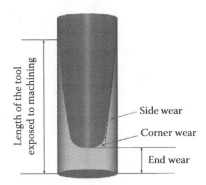

FIGURE 6.4 Types of tool wear in EDM. (From Sundaram, M.M. and Rajurkar, K.P., *Trans. North Am. Manuf. Res Inst. SME*, 36, 381, 2008. With permission.)

and compensate tool wear length after it moves along a certain distance. It is suitable to generate 3D cavities with straight side walls. The uniform wear method includes tool path design rules and tool wear compensation. Tool paths designed based on the uniform wear method can keep the tool wear uniform at the tool tip. This method has been verified by generating 3D microcavities with inclined side surfaces and spherical surfaces successfully as shown in Figure 6.5.

The EDM process produces a hard and cracked surface that contains a layer of recast material. Below this recast layer exists the heat-affected zone or the annealed layer, which has only been heated, not melted. The discharge energy used and the heat sinking ability of the material affect the depth of the recast and the heat-affected zone. A typical surface machined by EDM is shown in Figure 6.6.

Surface integrity is important as it affects the fatigue life of machined components. Surface integrity has two important parts, namely surface topography and surface metallurgy. In general, the depth of the heat-affected material in EDM roughing and finishing is on the order of 125 and 25 μm, respectively. Heat treatment, shot peening, metallurgical-type coating, low stress grinding, and chemical machining are some of the post-treatment processes recommended for restoring fatigue strength [173]. A recent study on the fatigue strength of electrodeposited nanocrystalline Ni reports a 50%–75% reduction in fatigue strength for a small depth of EDM-affected materials (~1% of width) [174]. Material properties also change from surface to core. A study on 36NiCrMo16 steel using

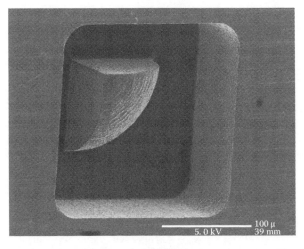

FIGURE 6.5 1/8 Ball in a square cavity. (From Narasimhan, J. et al., *Trans. NAMRI/SME*, 32, 151, 2004. With permission.)

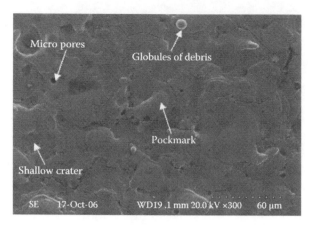

FIGURE 6.6 SEM image of SKD11 surface machined by EDM using pulse current of 4 A and pulse-on duration of 16 μs. (From Tai, T.Y. and Lu, S.J., *Int. J. Fatigue*, 31, 433, 2009. With permission.)

the photothermal deflection (PTD) technique suggests that the thermal conductivity and diffusivity would increase from the surface to the core [175]. In another study on a high-nickel-content superalloy (Inconel 718), depending on the process parameters used, a recast layer with an average thickness between 5 and 9 μm was observed [176]. The recast material was found to possess in-plane tensile residual stress contrary to compressive surface stress after grinding and polishing, as well as lower hardness and elastic modulus compared with the bulk material. Consequently, surfaces machined by EDM exhibit higher friction and wear compared with the ground and polished equivalents [177]. Considerable enhancement in the wear performance upon consecutive execution of gradually finer EDM regimes onto the WC-Co alloys has also been reported [178]. In micro-EDM, the generated surface has been studied with regard to topography, roughness, and craters in micro-drilling experiments [134,179,180]. Surface topography and roughness are largely determined by the eroded crater size relating to discharge pulse energy [181]. Smaller pulse durations result in the generation of smaller craters on the surface. The recast layer caused by micro-EDM can be reduced by changing the tool path or layer depth or can be removed using ECM or laser processes. The depth of the recast layer is influenced by the resistance and capacity in the circuit, both of which impact the discharge energy. Higher energy leads to thicker recast layers [89]. Low open-circuit voltage produces small craters and hence less surface roughness [145].

6.1.3 Advantages and Limitations of EDM

Advantages:

1. EDM can be used to machine any conductive material irrespective of its hardness. Hence, EDM is widely used in the die and mold industry.
2. EDM is a noncontact process that involves a negligible machining force. Hence, EDM is suitable for machining fragile parts.
3. EDM can be used in micromachining irrespective of the crystal orientation.
4. EDM can achieve near burr-free machining.

Limitations:

1. EDM induces thermal stress in machined surfaces.
2. Tool wear affects the accuracy of the machined features, especially in micromachining.
3. Batch processing possibilities are rather limited.
4. EDM is a relatively slower process when compared with its traditional counterparts.

6.1.4 RESEARCH ISSUES AND RECENT PROCESS IMPROVEMENTS

Recent technological advancements have extended the EDM capabilities to unprecedented horizons, especially in micromachining. It was found that cemented tungsten carbide micro rods (grain size = $0.4\,\mu m$) as small as $2.3\,\mu m$ in diameter could be obtained by micro-EDM even after a large number of repeated experiments [182–184]. Suggested reasons for this lower limit are the discharge crater size, residual stress, subsurface layer damages, and material structure of the workpiece. The minimum diameter of the micro rod was found to be almost the same whether the rod was used as an anode or cathode in WEDG [183]. This was probably because the machining time for the rod with reversed polarity was about eight times longer than that with normal polarity. By reducing the open circuit voltage to 20 V, a minimum rod diameter of $1\,\mu m$ was obtained [185]. Thin fins cut by wire EDM may bend due to the thermally induced residual stress when fin thickness is less than 0.1 mm in rough cutting [186,187]. The minimum machinable thickness of a micro wall was thinner when mono-crystal tungsten was used compared with poly-crystal tungsten [182]. However, since cracks were generated parallel to the (100) planes, it is not always true that mono-crystal is more suitable for miniaturization than poly-crystal.

A closed-loop wire tension control system has been reported in [188]. Compared with the open loop control, the developed wire tension control system contributes to better corner accuracy and vertical straightness [189]. Micromachining was done using tungsten ($\phi\,50\,\mu m$) and brass ($\phi\,70\,\mu m$) wires [190].

In micro-EDM drilling, the highest aspect ratio of micro-holes with a diameter of about $25\,\mu m$ is routinely obtained in the range of 15–18 [191]. The depth of hole drilling may be limited by the difficulties of ejecting generated gaseous bubbles and debris from the narrow discharge gap (several micrometers) during machining [86]. The tool is too small for internal flushing, and external flushing causes vibrations of the slim tool. The debris concentration results in abnormal discharges (arcs and/or short circuits) leading to unstable machining and excessive tool wear [192,99]. Several methods such as vibrating the tool, pre-drilling a hole to allow bubbles and debris to escape from the working area, the planetary movement of tools [86,157], and the rotation of tools [86,193] have been attempted to improve debris flow. The planetary movement of the tool has been used to machine noncircular blind holes [86]. By feeding the electrode at the desired angle, tapered holes with a diameter of $100\,\mu m$ at the electrode entry and a diameter of $160\,\mu m$ at the electrode exit were produced with a surface roughness of Ra under $0.3\,\mu m$ [127]. Micro-holes with internal grooves have been machined by a micro-EDM lathe [126].

Micro-EDM can machine not only conducting materials but also semiconducting materials like doped silicon and even insulating ceramics like Si_3N_4 using conductive coatings [188,194–198]. An electro-conductive layer was provided on the surface of single-crystal diamonds for initiating EDM machining [159]. The EDM of electrically conductive diamonds opens up a potential application for fabricating diamond-based micro dies with excellent mechanical properties. PCD tools have been made by micro-EDM [199].

A wire EDM machine that orients the wire horizontally is reported in [200]. Axisymmetric products can be machined using the micro wire EDM method with a rotating workpiece system [190]. The capability of micro wire EDM is fully exploited in the machining of a complex Chinese pagoda ($1.25\,mm \times 1.75\,mm$) shown in Figure 6.7.

6.2 ELECTROCHEMICAL MACHINING (ECM)

6.2.1 INTRODUCTION

The origin of ECM can be traced back to the principle of electrolysis discovered by Michael Faraday in 1833. However, the practical application of electrolysis for bulk material removal by anodic dissolution of metals evolved almost after a century [202]. It took another three decades for the appearance of commercial ECM machines in the 1960s. The reason for this rather slow pace

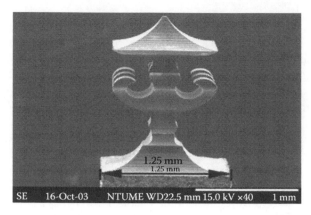

FIGURE 6.7 Pagoda machined by micro wire EDM. (From Liao, Y.-S. et al., *J. Micromech. Microeng.*, 15, 245, 2005. With permission.)

of development of ECM was probably due to the contemporary developments in EDM technology, which offered higher accuracy and less environmental pollution. An account of ECM developments during that period and references to early Russian literature can be found in [203]. ECM's unique ability to produce stress free and smooth surfaces of machined products and the capability of machining complex shapes without tool wear were too good to be ignored, especially by the aerospace industry for shaping and finishing operations, and the revival of ECM thus began. Soon other heavy industries followed to exploit the higher machining speed and MRR achievable in ECM.

6.2.2 ECM PROCESS DESCRIPTION

The ECM process is based on the principles of electrolysis [204]. Material removal occurs by the anodic dissolution in an electrochemical cell, as shown schematically in Figure 6.8.

The electrochemical cell consists of an anode (workpiece) and a cathode (tool) immersed in an electrolyte medium. The aqueous or molten electrolyte can be acidic, basic, or neutral in nature. The tool is positioned very close to the workpiece for the maximum amount of dissolution and minimum ohmic voltage drop between the two electrodes. With an electrical potential applied across the pair of electrodes, positively charged ions (cations) move toward the cathode while negatively charged ions (anions) move toward the anode. The energy required to separate the ions, and the increased concentration at the electrodes, is provided by an electrical power supply that maintains the potential difference across the electrodes. A high-amperage ($30–200\,A/cm^2$), low-voltage ($10–20\,V$) current

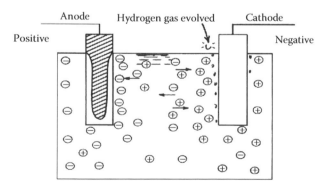

FIGURE 6.8 Schematic of an electrochemical cell. (From Wilson, J.F., *Practice and Theory of Electrochemical Machining*, Wiley-Interscience, New York, 1971. With permission.)

is generally used to dissolve and remove material from the electrically conductive workpiece. At the electrodes, electrons are absorbed or released by the ions, forming concentrations of the desired element or compound. For example, when water is electrolyzed, hydrogen gas (H_2) will form at the cathode, and oxygen gas (O_2) at the anode.

The electrolysis process is characterized by

- Electrolytic reactions involving the gain of electrons at the cathode and loss of electrons at the anode.
- The reactions at the cathode are redox (shorthand for *reduction-oxidation* reaction) as the positively charged ions (cations) receive electrons and at the anode, there is an oxidation reaction as the negative ions (anions) lose electrons.
- At the cathode, metals and hydrogen are released and at the anode, nonmetals are released.

The metal ions removed from the workpiece are taken away by the vigorously flowing electrolyte through the interelectrode gap and are separated from the electrolyte solution in the form of metal hydroxides by suitable methods. Both the electrolyte and the metal sludge can then be recycled.

The electrochemical reactions, for example, in the ECM of low carbon steel in a neutral salt solution of sodium chloride (NaCl), are given below.

With the passage of current in the electrolytic cell, the ionic dissociation reactions of electrolyte and water can be given by

$$NaCl \leftrightarrow Na^+ + Cl^-$$

$$H_2O \leftrightarrow H^+ + (OH)^-$$

The cations and anions move toward the tool and the workpiece, respectively. The positive hydrogen ions will take away electrons from the cathode (tool) forming hydrogen gas

$$2H^+ + 2e^- = H_2 \uparrow \text{ at cathode}$$

And the iron atoms will come out of the workpiece losing two electrons as

$$Fe = Fe^{++} + 2e^-$$

The positive iron ions combine with the other negatives in the electrolyte to form iron chloride and combine with the negative hydroxyl ions to precipitate as ferrous hydroxide. The ferrous hydroxide may react further with water and oxygen to form ferric hydroxide. The net reactions are shown below.

$$Fe + 2H_2O \rightarrow Fe(OH)_2(solid) + H_2$$

$$4Fe(OH)_2 + 2H_2O + O_2 \rightarrow 4Fe(OH)_3$$

As the workpiece gets machined, the removed material is swept away by the high-pressure electrolyte flow and gets precipitated as sludge. Only hydrogen gas is evolved at the cathode, so the tool electrode shape remains unaltered during the ECM process. This feature is perhaps the most relevant in the use of ECM as a metal shaping process. As the material removal is by atomic level dissociation, the machined surface is of excellent surface finish and is stress free. The rate of anodic metal dissolution that depends on the atomic weight and the ionic charge of the metallic element,

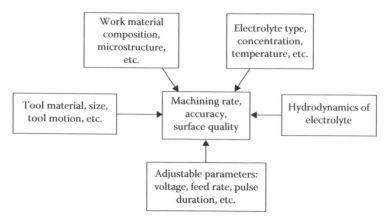

FIGURE 6.9   Important factors in ECM.

the current density, and the duration of the current flow is governed by Faraday's laws of electrolysis stated below.

First law: the amount of dissolution or deposition of material in electrolysis is directly proportional to the amount of electricity flowing through the circuit/cell.

Second law: the amount of material deposited or dissolved by the same quantity of electricity is proportional to the chemical equivalent weight of the material, which is a ratio of its atomic weight and valency.

The type of electrolyte, its concentration, temperature, and the rate of flow through the electrode gap affect the current density and in turn have great impact on the MRR, surface finish, and dimensional accuracy. The influence of the current density, current distribution, anodic reactions, and mass transport effects on material removal and accuracy are important considerations in ECM. Significant factors influencing the ECM machining performance are shown in Figure 6.9.

Major components of a typical ECM system and the process characteristics are given below.

Power supply: mainly two types of power supplies, namely, DC (full wave rectified) and pulsed DC are used in ECM. The power source adjusts the required voltage in either pulsed or continuous mode and provides current depending upon the selected current density (0.1–5 A/mm^2). A full wave rectified DC supplies continuous voltage where the current efficiency depends much more on the current density. The efficiency decreases gradually when the current density is reduced, whereas in pulsed voltage (duration of 1 ms and interval of 10 ms), the decrease is much more rapid. The accuracy of the form of the workpiece improves with decreasing current density.

Tools: the desired properties of an ECM tool include good corrosion resistance, inertness to chemical reactions, high electrical and thermal conductivity, sufficient stiffness to withstand the electrolyte pressure without vibration, good machinability, and availability. Copper, brass, bronze, copper-tungsten, stainless steel, platinum, tungsten, and titanium are some of the widely used ECM tool materials. With proper coating to prevent rapid erosion, graphite can also be used as a cathode in ECM. Since the shape of the machined surface in ECM is almost a mirror image of the tool used, the dimensional accuracy of the produced surface is linked with the shape and geometry of the tool along with other parameters, such as the tool feed rate, gap voltage, and cathodic tool insulation. It is essential to have proper electrolytic flow through the interelectrode gap for efficient machining that necessitates rigid fixture of the tool. To minimize the stray current effect, the tool should be properly insulated or coated. A proper insulation of the tool is a must for achieving high machining accuracy especially at the micro- and nano-levels.

Electrolytes: the electrolyte facilitates the desired electrochemical reactions that occur in an electrochemical cell and completes the electric circuit between the tool and workpiece. It carries away heat and reaction products from the machining zone. Therefore, the selection of proper parameters for electrolytes is critical for obtaining the desired dimensional accuracy and machining efficiency. Proper electrolytes with regard to their type, pH value, fluidic kinetic property, and concentration should be chosen depending on applications.

There are mainly two types of electrolytes used in ECM: passivating electrolytes containing oxidizing anions (e.g., sodium nitrate and sodium chlorate) and nonpassivating electrolytes containing relatively aggressive anions, such as sodium chloride. Generally, the MRR increases with an increase in electrolyte concentration. Due to the increase in electrolyte concentration, ions associated with the machining operation in the machining zone also increase. However, a higher concentration of ions reduces the localization effect of material removal by anodic dissolution. Machining accuracy can be improved by decreasing the electrolyte concentration. However, at very low ionic concentrations, the ion-content in the interelectrode gap is insufficient to supply the charge carriers necessary to complete the charging of the double layer capacitance. This depletion of ions prevents uniform dissolution and causes unstable machining [206]. The temperature and flushing conditions of electrolytes have a great impact on the machining accuracy and surface finish. The anodic dissolution rate can be significantly increased by using hot electrolytes, and the other effect is that the requirement on applied voltage is reduced [207].

MRR: in ECM, material removal takes place due to the atomic dissolution of work material. The machining rate is affected by many parameters, including current, the type of electrolyte used and its flow rate, and some of the workpiece properties. Current efficiency decreases with a rise in current density for some metals, for example nickel. The current efficiency of even the most easily electrochemically machinable metal is reduced when the rates of electrolytic flow are too low [208]. The inadequate electrolyte flow does not allow the products of machining to be promptly flushed away from the electrode gap and in turn reduces the machining efficiency and the MRR.

Dimensional accuracy: The current efficiency vs. current density characteristics of an electrolyte have a significant impact on the dimensional control of the produced surface in ECM. The current density controls the machining gap width and affects the accuracy of the workpiece. For sodium nitrate electrolytes, the current efficiency is greatest at the highest current densities and hence has the ability to help produce more accurate components than sodium chloride. Current efficiency remains steady at almost 100% for a wide range of current densities in the case of sodium chloride [209]. For high dimensional accuracy, a narrow interelectrode gap with a high feed rate using passivating electrolytes like sodium nitrate is recommended. Using a masked tool electrode or insulating the desired portion of the tool can minimize the stray current in ECM and substantially improve the accuracy of the component, especially in hole drilling operations.

Surface finish: surface roughness is greatly influenced by the grain size of the material on the surface; the insoluble inclusions in the material, e.g., graphite in cast iron; the overall composition of the workpiece material; and the precipitation of intermetallic compounds at grain boundaries. In ECM, the type of electrolyte used plays a significant role in determining the output surface finish. The production of an electrochemically polished surface is usually associated with the random removal of atoms from the anode workpiece, whose surface has become covered with an oxide film. This depends on the metal–electrolyte combination used. However, the detailed mechanism for controlling high current density electropolishing in ECM is still not completely understood. A surface finish on the order of $0.2\,\mu m$ has been reported for nimonic (a nickel alloy) machined in a saturated sodium chloride solution and $0.1\,\mu m$ when machining nickel-chromium steels in a sodium chlorate solution [210]. Oxide film formation on the metal surface sometimes reduces the efficiency of the ECM process and leads to poor surface finish. For instance, the ECM of titanium is very difficult in chloride and nitrate electrolytes because of the formed passive oxide film. Pitting is one phenomenon that arises from gas evolution at the anode; the gas bubbles rupture the oxide film. Process parameters also influence the surface finish. A higher current density can improve the finish of the machined surface and so does the increase in electrolyte velocity.

6.2.3 ADVANTAGES AND LIMITATIONS OF ECM

Advantages:

1. Can machine very hard electrically conductive materials.
2. Unlike other electrical machining processes, there is no tool wear and hence soft materials can be used as tools.
3. Nonrigid workpieces can be machined easily as there is no contact between the tool and workpiece.
4. Fragile and brittle materials that tend to develop cracks during machining can be machined easily by this process.
5. Complex geometrical shapes can be machined repeatedly and accurately.
6. Very high surface finish in the range of 0.1–1.25 μm (Ra) can be achieved.
7. No heat generation in the process, so no induced thermal stress in the workpiece or no thermally damaged machining surfaces produced.
8. Often faster than manual deburring processes.
9. No mechanical forces on the workpiece because of the noncontact nature of the process.
10. The process can be completely automated, hence low labor costs.
11. Suitable for mass production.

Limitations:

1. Limited to the machining of electrically conductive materials.
2. Tool design process is rather difficult.
3. Large amount of sludge generation per material removal.
4. High idle machining time and not economical for small lots.
5. Preparing, handling, and disposal of the electrolyte is cumbersome.
6. High energy consuming process compared with other traditional manufacturing techniques for the same material removal.
7. Slow rate of material removal.

6.2.4 RESEARCH ISSUES AND RECENT PROCESS IMPROVEMENTS

Passivating electrolytes generally give better machining precision [211]. This is due to the formation of oxide films and oxygen evolution in the stray current region. The pH value of the electrolyte solution is chosen to ensure the dissolution of the workpiece material during machining without affecting the tool. In ECM, it is typical to machine with sodium nitrate electrolyte solution (pH 7). To regulate the pH value, some chemicals can be added such as $NaHSO_4$ with specific concentrations that do not affect the process adversely. It was also found that hydrochloric acid solution is useful in fine-hole drilling because it dissolves the metal hydroxides produced from the electrochemical reactions. Recently, it was reported that less toxic and dilute electrolytes, 0.1 M H_2SO_4, can be applied for the machining of stainless steel 304 with ultrashort pulse voltage [212].

Pulsed electrochemical machining (PECM) is an ECM process for the fabricating of both parts and tooling with high-resolution features and well-defined edges. Pulsed voltage and pulsed current enable the recovery of the interelectrode gap condition during pulse-off time and, therefore, usually provides an improved machining status. PECM can be used for applications where high precision and three-dimensional contouring is required by a proper tool's position control with sensitive feedback schemes. PECM has more variable parameters and much better controllability than DC ECM. One can vary the pulse duration, shape of voltage pulse, and pulse duty-factor. These parameters provide a way to optimize the MRR and surface quality. The resolution of PECM can be further improved significantly by using ultrashort pulse voltage [206]. By using a pulse voltage

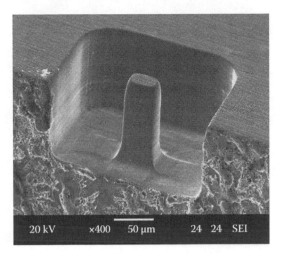

FIGURE 6.10   Micro column produced by micro ECM. (From Kim, B.H. et al., *Ann. CIRP*, 54, 191, 2005. With permission.)

of nano-second duration to the tool electrode, this method can lead to the strong spatial confinement of anodic dissolution with resolution down to nanometers. Micro-ECM is an electrolytic machining process applied to machine features typically in the size range 1–100 μm. Compared with other noncontact micromachining processes, such as dry etching techniques (ion beam milling) and wet chemical etching, micro-ECM features improved controllability and can use environmentally benign electrolytes and easily dispose of the waste products. ECM can be further downsized to the nanoscale due to its essence of atomic-level material removal. A micro column produced by micro ECM is shown in Figure 6.10.

6.2.4.1   Shaped Tube Electrolytic Machining

Shaped tube electrolytic machining (STEM), pioneered by workers at the General Electric Company, is a modified ECM process that uses an acid electrolyte so that the removed metal goes into a solution instead of forming a precipitate. In this process, holes are produced by the controlled deplating action in an electrolytic cell where the cathode is simply a metal tube of acid-resistant material such as titanium shaped to match the desired hole geometry. It is carefully straightened and insulated over the entire length except at the tip. The acid electrolyte under pressure is fed through the tube to the tip and returns via a narrow gap along the outside of the coated tube to the top of the workpiece. The electrode is given constant feed at a rate matching the dissolution rate of material [213]. Acid electrolytes such as sulfuric, nitric, and hydrochloric acids with 10%–25% concentration are preferred in STEM. However, neutral salt electrolytes (10%) with a small percentage of acid electrolytes (1%) have been used by researchers to minimize the sludge formation in the interelectrode gap. STEM is suitable for multiple hole drilling. Uniform wall thickness in repetitive production is achieved because of the noncontact nature of the process. Stress-free, high-integrity holes can be produced by the atom-by-atom dissolution of the material. The creation of uniform, good holes with aspect ratios of 11 has been reported in Inconel at a voltage and tool feed rate of 17 V and 1.0 mm/min, respectively [214].

6.3   ELECTROCHEMICAL DISCHARGE MACHINING (ECDM)

6.3.1   INTRODUCTION

ECDM is a hybrid process resulting from the combination of EDM and ECM [215]. Thus, the material is removed by anodic dissolution as well as by spark erosion. This concept is also published in literature as electrochemical arc machining (ECAM) [216], electrochemical arc cutting [217], and

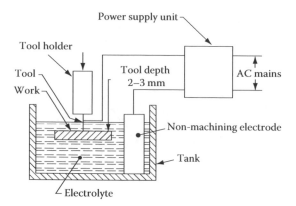

FIGURE 6.11 Schematic of ECDM process. (Adapted from Ghosh, A., *Sadhana*, 22, 435, 1997.)

spark-assisted chemical engraving (SACE) [218]. An important application of this process is in the machining of nonconductive ceramics [219,220]. ECDM has been used in hole drilling [221], die sinking [222], and cutting [217].

6.3.2 PROCESS DESCRIPTION

ECDM, like EDM and ECM, reproduces the form of the tool electrode on the workpiece. ECDM electrodes are of different sizes. The cathode (tool) is small and the anode is relatively large. Another interesting aspect is that unlike ECM, the *anode in ECDM is not the workpiece* to be machined but serves as a reference electrode only for the electrolysis generation. The actual workpiece is placed just below and very close to the tool as shown schematically in Figure 6.11. In the ECDM process, the workpiece, tool electrode (cathode), and auxiliary electrode (anode) are immersed in an electrolyte solution, such as sodium hydroxide or potassium hydroxide. The auxiliary electrode is far away from the tool electrode and has a larger surface than the tool electrode. When potential is applied between the two electrodes, an electrochemical reaction begins with the generation of the positively charged ions and gas bubbles, e.g., hydrogen at the anode and cathode, respectively. Furthermore, a higher current density due to smaller cathode size results in a rapid increase in temperature and boiling of the electrolyte in the vicinity of the tool. As a result of both of the effects, an insulating layer of gas bubbles and vapor accumulates across the interface of the tool and the workpiece. When the supply voltage is greater than the breakdown voltage of the insulating layer of the gas bubbles, a spark is initiated between the tool and the workpiece. This results in material removal by thermal melting in addition to chemical etching [223–225]. Since the material within the range of sparks is always heated, irrespective of its conductivity, ECDM is suitable for the machining of both conductive and nonconductive material. A detailed description of the ECDM mechanism is presented in [226].

6.3.3 ADVANTAGES AND LIMITATION OF ECDM

Advantages:

1. ECDM is suitable for the machining of both conductive and nonconductive materials [227,228].
2. To achieve a better quality of the machined surface and larger MRR than in ECM or EDM, surface irregularities caused by the electrical erosions on the machining surface are subjected to electrochemical smoothing by dissolution in ECDM. The productivity of ECDM can be 5–50 times greater than that of ECM or EDM [229,230].

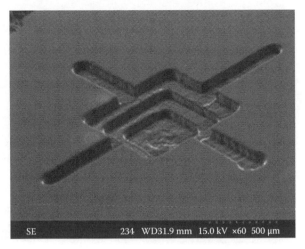

FIGURE 6.12 3-D microstructure formed by ECDM in Pyrex glass. (From Zheng, Z.-P. et al., *J. Micromech. Microeng.*, 17, 960, 2007. With permission.)

Limitation:

1. Due to the stray electrochemical attack, ECDM provides reduced accuracy compared with EDM.

6.3.4 RESEARCH ISSUES AND RECENT PROCESS IMPROVEMENTS

ECDM has been successfully applied in the micromachining of borosilicate glass [23]. The layer-by-layer machining concept has been used to demonstrate the great potential of ECDM for the 3D microstructuring of Pyrex glass shown in Figure 6.12 [231]. Fluidics interfacing through thin glass substrates and the formation of spherical microcavities by ECDM have been described in [232]. A side-insulated tool reduces stray electrolysis and minimizes the fluctuation of the peak current. This improved discharge current uniformity of the ECDM process results in improved geometric accuracy and surface roughness [224]. Higher machining efficiency is reported in another study [225] in which a different pulse voltage configuration called *offset pulse voltage* was designed to increase the gas film stability and enhance the discharge performance in the ECDM. Thermal damage of the microdrilled hole decreases in pulse current ECDM with higher pulse frequency and lower duty ratio [233]. A flat sidewall–flat front tool electrode has been used to reduce the taper phenomena due to the sidewall discharge [234]. Increasing the machining feed rate and depth of cut reduces the ECDM energy consumption [222,217]. ECDM using abrasive tools has been found to increase the machining ability in aluminum oxide drilling [235].

6.4 ELECTROPLATING AND ELECTROFORMING

6.4.1 INTRODUCTION

Electroplating, also known as electrodeposition, is a material addition process widely used as finishing operations in automotive, aerospace, electronics, and other engineering applications to deposit a metallic coating on a substrate for preventing corrosion, to provide wear resistance, and also for aesthetic reasons. For about a century and half, this process has been applied to enhance the beauty and durability of decorative arts and fine crafts. The plating material is typically a single metallic element, although occasionally metal alloys have also been electroplated. The engineering applications of electroplating include chrome plating on steel parts like crank and cam shafts for automobiles and zinc-coated nuts, bolts, and washers for hand tools.

The history of electrodeposition dates back to around 1800 when Luigi Brugnatelli, an Italian chemist and university professor, invented a process for coating gold by electroplating [236]. John Wright's innovation of using potassium cyanide plating baths for gold and silver electroplating was patented by George Elkington and Henry Elkington of England in 1840. Various types of nondecorative metal plating such as nickel, brass, tin, and zinc were developed by the 1850s. The copper electroplating of the printing press was invented by various researchers in Europe and Russia through the 1930s. Subsequently, safer acid plating baths instead of poisonous cyanide-based formulas came into existence. Regulatory laws for waste water emissions and disposal in the1970s set the guidelines for environmentally friendlier refinement of the modern electroplating industry.

Electroforming can be considered as a development of electroplating techniques for the fabrication of standalone products by electrodeposition. Electroforming has been practiced for almost two centuries to produce metal replicas of various shapes and textures. Though started as an art in the early days (around 1840), it has evolved over a period of time as a reliable and matured manufacturing process to produce stress-free products of complex configurations such as balloon ends, multiple lumens dies, guide pins, coins and banknotes, LP records, CD disks, and molded plastic holographic images [237,238]. Apart from the traditional applications such as molds, electronic components, jewelry, and aerospace and other industrial applications, the renewed interest in electroforming from academia and industry in recent days is due to its potential in the fabrication of micro- and nanoscale metallic devices and precision injection molds with micro- and nanoscale features for the production of nonmetalic micromolded objects [239]. An electroformed Ni-Co micro lens array mold is shown in Figure 6.13.

6.4.2 Process Description

Electroplating, like ECM seen earlier, works on the principle of electrolysis. It also involves a cathode, anode, and an electrolytic bath as shown schematically in Figure 6.14. However, in electroplating, the metal ions, acquired either by anodic dissolution or from metal salts, are allowed to flow through the bath to eventually plate on the cathode (workpiece).

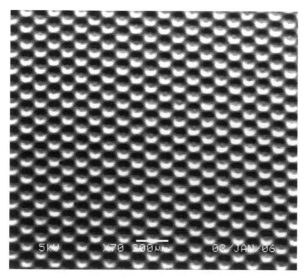

FIGURE 6.13   Micro lens array mold made by electroforming process. (From Lin, T.-H., et al., *J. Micromech. Microeng.*, 17, 419, 2007. With permission.)

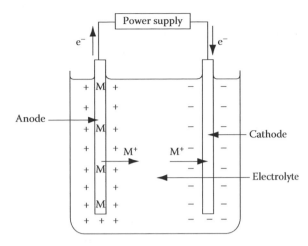

FIGURE 6.14 Schematic of electroplating process. (Adapted from Brown, R., Electroplating, in *RF/Microwave Hybrids*, Kluwer Academic Publishers, Dordrecht, the Netherlands, 2003, 169–184.)

Electroforming works on the principles of Faraday's laws of electrolysis as discussed earlier. A schematic of an electroforming process is shown in Figure 6.15.

Similar to electroplating, electroforming involves a cathode, anode, and an electrolytic bath. Electrolysis begins with the application of potential and metal ions get deposited on the cathode. However, an important difference between the two processes is that electroforming uses a mandrel as a cathode, which is preformed to the desired shape of the final product. At the end of the process, this mandrel is removed from the deposited material to leave a separate product.

Electroplating involves oxidation-reduction reactions, where anodes give up electrons (gets oxidized) and cathodes gain electrons (gets reduced). Ideally, the amount of metal deposited on the cathode is equal to the amount of metal dissolved at the anode. Theoretically, plating

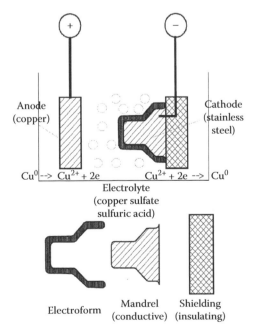

FIGURE 6.15 Schematic of electroforming process. (From McGeough, J.A. et al., *Ann. CIRP*, 50, 499, 2001. With permission.)

thickness is determined by the plating time and the amount of available metal ions in the bath relative to the current density. This means that workpieces with sharp corners and edges will have more deposits on the outside corners and fewer deposits in the recessed areas due to the difference in current density, which is denser around the outer edges than the less accessible recessed areas. The appropriate placement of the anodes to even the current density can result in uniform coating thickness. Metals such as gold, silver, nickel, palladium, platinum, ruthenium, and rhodium usually provide decorative finishes. Brass, cadmium, stainless steel, copper, gold, nickel, silver, tin, and zinc are widely used for engineering applications. Since electroplating makes preexisting surface imperfections even more prominent, removing any undesirable surface marks such as scratches, dents, or pits prior to the plating process is recommended. Electroplating process parameters include electrolyte pH, temperature, current density, bath composition, agitation, pulse duration, current density, current on- and off-time, polarity, and pulse height (in the case of pulse plating).

Electrolytes used in the process should posses (1) the ability to dissolve the salts of the metal to be deposited at a reasonable temperature and in adequate concentrations and (2) good electrical conductivity for uniform distribution of the plating material and to avoid excessive heating of the bath. Electrolytes are characterized by properties such as conductance, covering power (the ability of an electrolyte to cover the entire surface of an object being plated), and throwing power (the ability to deposit material uniformly) [242]. The main constituents of the electrolyte are the metal salt of the metal to be deposited and, in most cases, an acid or alkali to promote conduction. Certain additives may be needed to promote and/or to optimize the electrodeposition process. Baths can be acid (pH <3), neutral (pH = 7), or alkaline (pH >9). It is crucial to avoid significant pH changes during the process to ensure the reduction of the metal occurs before the reduction of hydrogen. Metal salts such as sulfates or chlorides, and in some cases phosphates or sulfamates, are commonly used in acid electrolytes. Acid optimizes electrical conductivity and minimizes pH fluctuations.

A typical example of an acid deposition bath is the *Watts bath* proposed by Watts in 1916 for nickel plating [243]. This bath is relatively easy to establish and is still used extensively. The Watts Ni bath was made of 40–60 g/L nickel chloride hexahydrate $NiCl_2 \cdot 6H_2O$, 240–300 g/L nickel sulfate hexahydrate $NiSO_4 \cdot 6H_2O$, and 25–40 g/L boric acid. H_3BO_3 was operated at temperatures from 40°C to 70°C and at a pH ranging from 2.5 to 4 with the boric acid acting as a buffer to maintain a constant pH.

The deposition current density is typically between 3 and 10 A/dm². Neutral electrolytes include systems that operate in the weak acid to weak alkaline range, pH range 7.5–8.8. These are not commonly used because of their poor electrical conductivity. Alkaline electrolytes can be of cyanide-containing or cyanide-free types like a variety of zinc baths used for rack or barrel plating. Leveling agents are added to the bath in some cases to smoothen preexisting irregularities in the surface, such as pits or scratches, by depositing a greater thickness of metal in the valleys than in the peaks [244].

Process variants of the direct current electroplating process include pulse and pulse-reverse electroplating [245] and brush plating. Pulse and pulse-reverse plating involve a series of pulses of direct currents of equal amplitude and duration, separated by periods of no current. During the pulse-on time, the metal ions next to the cathode are depleted and a layer rich in water molecules is left. During the pulse-off time, the metal ions from the bulk of the plating solution diffuse into the layer next to the cathode forming deposits. Pulse plating is carried out in constant current and constant voltage modes. Pulse reverse plating has an influence on the texture of the film and the anodic periods help to dissolve unwanted crystals, improve material distribution, and reduce the internal stress of the deposits [246].

Brush plating is a localized plating process. In brush plating, rather than dipping the entire workpiece in a bath, an electrolyte is applied to the targeted regions by the operator to achieve localized deposition [247,248].

6.4.3 Advantages and Limitations

Advantages of electroplating:

1. Engineering properties such as high wear strength, chemical resistance, and corrosion resistance can be incorporated into any metal by suitably coating it with another metal or alloy that has the desired properties.
2. Less expensive metals such as iron and aluminum can be coated with costly metals like silver and gold to give them a rich look as in the case of artificial jewelry.
3. A mirror-like surface finish and good thickness control (0.25 μm) are possible.
4. A high material deposition rate on the order of 0.010 mm/min is achievable.
5. The workpiece is subjected to low thermal load.

Advantages of electroforming:

1. Components with wall thicknesses as low as 0.001″ can be produced. Electroforming has the unique capability of producing thin cylinders without a joint line [249].
2. The replication of components with high dimensional precision (<5 μm), tolerances up to 1 μm, fine surface details (~0.01 μm), and surface finishes of 0.05 μm Ra are possible [239].
3. Complex shapes can be reproduced with minimum operations.
4. Components can be made of multiple materials with different properties.
5. High-volume production (size as well as quantity) is feasible with larger capacity baths.

Limitations of electroplating:

1. Holes and sharp angles are hard to plate.
2. It is difficult to control the uniformity of the plated structure.
3. High initial cost and long cycle time.

Limitations of electroforming:

1. Electroforming is a rather slow process and often produces components with nonuniform thickness.
2. Complex mandrels are required to replicate corresponding parts.
3. Careful process control and special techniques (such as the use of multiple anodes) are required to produce stress-free components.
4. Electroforming is successfully used on selected materials such as nickel, copper, iron, manganese, and nickel–cobalt alloys. However, application of this technology for a wide range of materials is not practical yet.

6.4.4 Research Issues and Recent Process Improvements

Certain chemical compounds (additives) are often added to the electrolyte in small amounts to produce smoother surfaces. By using additives in Ni electrolytes, properties such as hardness, internal stress, smoothness, texture, and structure can be altered considerably [250,251]. The synergistic effects of additives, such as polyoxyethylene nonyl phenyl ether, o-chloro benzyl aldehyde, and polyoxyethylene lauryl amine on smoothening the deposit surface, refining the grain size, and promoting the current efficiency of zinc electroplating in a wide range of current density has been reported in [252]. Using high current density, lotus-like super-hydrophobic surfaces were fabricated by electroplating without any chemical modification on hydrophilic copper substrates [253]. Plating coatings in a magnetic field increases the current efficiency of the alloy and the content of the electrochemically positive component in it, results in a finer structure, and leads to smoother surface relief and better quality of the coating [254].

The mass transport and charge transfer were enhanced by applying a magnetic field during the electrodeposition of the Ni/nano–Al_2O_3 composite coating. The alumina particles content and deposit rate increased with increasing magnetic flux density due to the magneto hydrodynamic effect [255]. Another study reports that the variation of the magnetic field vertical to the electrodes affected the deposition rate of the electroplated copper film, step coverage, and gap filling in trench. As the intensity of the magnetic field increased, the deposition rate of the copper film increased and the resistance of electromigration increased. However, the magnetic field did not affect the resistivity and the surface morphology of the electroplated Cu film [256]. Laser irradiation was used to suppress the generation and growth of whiskers on tin-electroplated film over copper [257]. Applying ultrasound sonication to the electroplating of iridium in aqueous hexabromoiridate(III) solution has decreased the defects including the cracks in the electrodeposits [258].

In the pulse electroplating of copper film, higher energy pulses caused stronger self-annealing effects on grain recrystallization and growth, thus leading to enhanced fiber textures, while lower energy pulses gave rise to more random microstructures in the deposits and rougher surface topography [259]. In the electroplating of nickel on carbon steel, when mechanical attrition was applied on the sample surface by the impact of glass balls with a special vibrating frequency, Ni coating with smooth surface morphology, refined grain size, pore-free increased microhardness, and excellent corrosion resistance was obtained [260].

LIGA is probably the most significant recent development of electroforming. LIGA is a fabrication technology especially suitable to create high aspect ratio microstructures [261]. Megasonic agitation improves high-aspect-ratio electroforming probably due to the enhancement of mass transport by resonant bubbles in the electrolyte [262]. Strong stirring effects, cavitation, and acoustic streaming caused by ultrasound effectively restrain the agglomeration and promote the uniform distribution of CeO_2 nanoparticles in the electroforming [263]. Conductive polymers are used as sacrificial parts in making high-aspect-ratio metallic microstructures by electroforming without hollow structure formation [264]. High-aspect-ratio metallic microparts of several hundred µms with localized deposition regions have been produced by periodically moving an insulating mask with patterned features along the metal growth direction in electroforming [265].

Better surface finish and a smaller grain size can be obtained by using a triangular wave in pulse current electroforming [266]. About a 28% improvement in hardness was reported when a ramp-down waveform was used instead of conventional rectangular waveform pulse current electroforming [267].

In a recent study, a shorter pulse-on time and larger duty cycle contributed to higher microhardness and strength of the Ni–Mn alloys. With an increase in average current density, there was an increase in the microhardness and strength of the alloys, a decrease in the elongation ratio because of the increment of Mn content of the alloys, and a decrement of the grain size of the alloys [268].

Inclusion of organic additives like imides helps in reducing the internal stress and increasing the hardness of electroformed products [269,270]. Microstructures and the mechanical properties of the electroformed product can be varied by the addition of brighteners to the electrolyte. A decrease in the grain size, degree of crystallization, and degree of preferred (002) orientation and an increase in the hardness, yield strength, and wear properties were observed when different combinations of brighteners were added [271]. An FEM-optimized anode led to nearly identical cathode uniformity of the current density, which was used to achieve a nickel thin-walled revolving part with a smooth, mirror-like outer surface and excellent thickness uniformity by electroforming without any organic additives [272].

6.5   ENVIRONMENTAL AND SAFETY ISSUES

Electrochemical processes in general operate at lower temperatures than their counterparts. The energy losses caused by nonhomogeneous current distribution, voltage drop, or side reactions can be minimized by the optimization of electrode structures and cell design. On the other hand, disadvantages arise from the fact that electrochemical reactions are heterogeneous and take place at the electrode–electrolyte interface in electrochemical reactors [20]. Therefore, the cell performance often suffers from mass-transport limitations and the size of the specific electrode area.

A crucial point is the durability and long-term stability of the cell components and electrode materials in contact with the aggressive media. It has been observed that harmful environmental impacts of the ECM process show up in the form of health hazards like eye and skin irritation, mainly due to electrolyte splashing, the release of toxic vapors during machining, and waste such as the ECM slurry produced in the process.

The use of passivating electrolytes like sodium nitrate in electrochemical deburring produces low surface roughness and a high machining rate, but because of nitrate reduction at the cathode, toxic chromate and ammonia are formed while machining metals contain chrome [273]. These harmful chemicals get absorbed in the ECM slurry in the form of metal hydroxides. So, extreme care is to be taken for its handling and disposal.

The addition of sorbitol as the complexing agent to an environmentally friendly noncyanide alkaline bath assists in the formation of fine-grained and highly adherent copper film on the stainless steel substrate [274].

Control of chromic acid mist is important for the health and safety of the plant personnel in the electroplating industry. The addition of mist suppressants and the removal of the gases and mist from above the surface of the plating tank by suitable extraction processes, in conjunction with ultrasonic irradiation, can help in safety and environmental control [275].

For the safe and efficient application of electroplating, several important aspects, including bath life, drag out, the reduction and the reuse of rinse water, the recycling of spent baths, the segregation of waste streams, and improving waste water treatment efficiency, need to be taken care of [276]. Using a packed column involving batch recirculation mode of operations, the nickel content of rinse water was reduced to less than 1 mg/L by the ion-exchange method and the harmful Ni(II) was recovered by desorption [277]. In another study, modified wood sawdust and sugarcane bagasse (the fibrous residue remaining after sugarcane or sorghum stalks are crushed to extract their juice) were used to remove Zn^{2+} from electroplating wastewater [278]. A designed two-stage electrodialysis system has been proposed to concentrate and purify chromate from a low-pH electroplating wastewater using monovalent selective electrodialysis membranes [279].

An ion exchanger could be one of the possibilities for recovering heavy metals in wastewater from electroplating plants as recent experimental results show that they are found to be effective for copper and chrome removal [280]. Compared to the raw sludge, vitrification of hazardous electroplating sludge displayed lower leaching concentrations for most metals, particularly chromium [281]. An x-ray fluorescence spectrometer is found to be effective for the rapid analysis and quantification of electroplating solutions [282]. In general, to minimize the effect of carcinogens like nickel dust, exceptional cleanliness and dust filtering is required especially in electrochemical processes.

6.6 CONCLUDING REMARKS

This chapter gives a detailed overview of the electro- and electrochemical processes. From the extensive literature, one may find many creative solutions to various engineering challenges. These processes have been used for many years. In other words, they are not "nontraditional" anymore; some of them are already part of the engineering history.

It is interesting to compare the strength and limitations of various energy fields. Mechanical energy requires direct contact to do work; it automatically provides a reaction force when it runs into something. Is this good or bad?

Electro-discharge and electrochemical processes are noncontact processes, but the gap between the tool and the workpiece is critical and has to be carefully controlled. Without a usable reacting force, people tried to use other signals—or the information in the energy-material interaction.

A very noticeable feature in the ECM/EDM/ECDM processes is the skillful use of the medium field. In processing, we simply use these physical or chemical fields, which are created by the interactions between the various materials and energy fields. Although this is a very pronounced feature of the ECM/EDM/ECDM processes, the use of medium fields is actually an indispensable part of any engineering processes. The world exists in an environment. All things happen under certain conditions. Interestingly, many people neglect the crucial functions of these medium fields. You do not care about the value of clean air until you have lost it.

Realizing this point, hopefully you can start digging into the treasure land of medium fields. For many processes, the potential of systematically optimizing the medium field has not been decently exploited. These are quick opportunities for process or product innovations. The electrical processes already set up a good role model for all the other processes.

Can these "matured" "nontraditional" processes be advanced processes? It depends on who uses them. The chemical processes are actually the truly atom-by-atom manufacturing processes, and they can reach extremely high resolutions. The terms traditional and nontraditional processes may create confusion. This is part of the reason this book is written.

One should also appreciate how the electrical processes evolved. With the availability of electricity, some scientific discoveries led to series of process innovations. More and more energy field combinations were tried to solve myriads of engineering challenges. Through all these convolutions, readers of this book should not get lost. No matter how "strange" these processes seem to be, they are just individual kinds of energy field manufacturing processes. They can be analyzed with the dynamic M-PIE model, which was discussed in the beginning of the book.

QUESTIONS

Q.6.1 Why are electrical and electrochemical machining processes considered to be nontraditional processes? When are they preferred over conventional machining processes?

Q.6.2 What are the two principal types of EDM processes? Describe the EDM process mechanism.

Q.6.3 Explain the different methods used to improve flushing in EDM.

Q.6.4 List the major EDM process parameters and their influence.

Q.6.5 What is micro-EDM? How does tooling in micro-EDM differ from EDM?

Q.6.6 Describe the ECM process mechanism. List the advantages and limitations of ECM.

Q.6.7 Explain the role of ECM and EDM in the ECDM process. When is ECDM preferred over ECM or EDM?

Q.6.8 How does electroforming differ from electroplating? What are the advantages and limitations of both the processes?

Q.6.9 Discuss some of the recent process improvements in electroplating and electroforming.

Q.6.10 Discuss the environmental concerns related to electrochemical processes. What are the different techniques used in the treatment of waste water from these processes?

ACKNOWLEDGMENTS

The financial support from the NSF under grant numbers CMMI-0728294, CMMI-0457346, CMMI-0423697, CMMI-0355380, and DMI-0331830 is acknowledged.

ABOUT THE CONTRIBUTING AUTHORS

Dr. Murali Meenakshi Sundaram is an assistant professor in mechanical engineering and director of the Micro and Nano Manufacturing Laboratory at the College of Engineering, University of Cincinnati, Ohio. He teaches design and manufacturing subjects. Dr. Sundaram received his PhD in mechanical engineering from Nanyang Technological University, Singapore. He has 10 years of industrial experience in aerospace manufacturing and 4 years of postdoctoral research experience in micro- and nano-manufacturing. Dr. Sundaram is a Co-PI in NSF-funded research projects and has co-advised 15 graduate students including three PhD students. He has authored over 25 refereed publications. Dr. Sundaram's research interests include nano-manufacturing, nontraditional machining, metrology, micromachining, hybrid machining, CAD/CAM, and process simulation.

Kamlakar P. Rajurkar, Distinguished Professor of Engineering at the College of Engineering University of Nebraska–Lincoln, received his MS and PhD degrees from Michigan Technological University, Houghton in 1978 and 1982, respectively.

Dr. Rajurkar is the founder and director of the Center for Nontraditional Manufacturing Research and professor of Industrial and Management Systems Engineering. He served as the interim chair of the Industrial and Management Systems Engineering department (January 2007–December 2008). He has also served as the interim associate dean for research of the College of Engineering (January 2005–December 2006). He served as program director of Manufacturing Machines and Equipment at the National Science Foundation (September 1999–November 2002). He was also chairman of the Manufacturing Systems Engineering Graduate Program of the College of Engineering (1988–1999).

Dr. Rajurkar is a fellow of ASME, SME, and the International Academy for Production Engineering (CIRP). Dr. Rajurkar served as a chairman of the Scientific Committee of the NAMRI/SME. He was President of the North American Manufacturing Research Institute of SME in 1998–1999. He also served as the ASME manufacturing technical group leader (previously called Vice president, manufacturing) for 3 years (2005–2008).

Dr. Rajurkar has more than 120 refereed publications and nearly 120 technically edited papers, which were published in conference proceedings. His research in macro-, micro-, and nanoscale machining has been supported by NSF, NIST/ATP, DoD, GEAE, Extrude Hone, Brush Wellman, Cummins Engines, NCMS, Mitsubishi Electric Corporation (Japan), Trans Tec Inc. (England), State of Nebraska, and other sponsors. He has received College of Engineering Awards for research, teaching, and service. He has also received the ASME Blackall Machine Tool and Gage Award for a paper on pulse electrochemical machining. He has received the 2005 Charles F. Carter Jr. Advancing Manufacturing Award from the Association of Manufacturing Technology. Recently, he received the 2009 International Honor, Gold Medal from the Society of Manufacturing Engineers. He is a co-inventor of a U.S. patent on cryogenically cooled tool machining.

REFERENCES

1. Tsai, Y. Y., Su, J. S., and Su, C. Y., 2008, A novel method to produce carbon nanotubes using EDM process, *International Journal of Machine Tools and Manufacture*, 48, 1653–1657.

2. Jeong, Y. H., HanYoo, B., Lee, H. U., Min, B.-K., Cho, D.-W., and Lee, S. J., 2009, Deburring microfeatures using micro-EDM, *Journal of Materials Processing Technology*, 209, 5399–5406.

3. Kohkonen, K. E., 2001, Manufacturing precision and delicate parts, *Proceedings of Electrical Insulation Conference and Electrical Manufacturing & Coil Winding Conference*, Cincinnati, OH. IEEE, Piscataway, NJ.

4. Puertas, I. and Luis, C. J., 2004, A study on the electrical discharge machining of conductive ceramics, *Journal of Materials Processing Technology*, 153–154, 1033–1038.

5. Priestley, J., 1778, Experiments and observations relating to the principle of acidity, the composition of water, and phlogiston, *Philosophical Transactions of the Royal Society of London*, 78, 147–157.

6. Lazarenko, B. R. and Lazarenko, N. I., 1948, Machining by erosion, *American Machinist*, 91, 120–121.

7. Jameson, E. C., 1983, *Electrical Discharge Machining: Tooling, Methods, and Applications*, Society of Manufacturing Engineers Marketing Services Division, Dearborn, MI.

8. Chapman, W. W., 2002, *Modern Machine Shop's Handbook for the Metalworking Industries*, Hanser Gardner Publications, Cincinnati, OH.

9. Kher, S. and Dua, A., 1999, EDM pulse control: A design approach, *Proceedings of the SICE 38th Annual Conference*, Morioka, Japan. IEEE, Piscataway, NJ.

10. DiBitonto, D. D., Eubank, P. T., Patel, M. R., and Barrufet, M. A., 1989, Theoretical models of the electrical discharge machining process. I. A simple cathode erosion model, *Journal of Applied Physics*, 66, 4095–4103.

11. Arunachalam, C., 1995, Modeling the electrical discharge machining process, PhD thesis, Texas A&M University, College Station, TX.

12. Albinski, K., Musiol, K., Miernikiewicz, A., Labuz, S., and Malota, M., 1995, Plasma temperature in electro-discharge machining, *Proceedings of 11th International Symposium for Electromachining*, Lausanne, Switzerland. Presses polytechniques et universitaires romandes, Lausanne, Switzerland, pp. 143–152.

13. Toren, M., Zvirin, Y., and Winograd, Y., 1975, Melting and evaporation phenomena during electrical erosion, *Transactions of the ASME, Journal of Heat Transfer*, 97 Series C, 576–581.

14. Koenig, W., Wertheim, R., Zvirin, Y., and Toren, M., 1975, Material removal and energy distribution in electrical discharge machining, *Annals of the CIRP*, 24, 95–100.

15. Erden, A., 1983, Effect of materials on the mechanism of electric discharge machining (EDM), *Transactions of the ASME, Journal of Engineering Materials and Technology*, 105, 132–138.

16. Gadalla, A. M. and Bozkurt, B., 1992, Expanding heat source model for thermal spalling of TiB_2 in electrical discharge machining, *Journal of Materials Research*, 7, 2852–2858.

17. Trueman, C. S. and Huddleston, J., 2000, Material removal by spalling during EDM of ceramics, *Journal of the European Ceramic Society*, 20, 1629–1635.

18. Gadalla, A. M., Bozkurt, B., and Faulk, N. M., 1991, Modeling of thermal spalling during electrical discharge machining of titanium diboride, *Journal of the American Ceramic Society*, 74, 801–806.

19. Lauwers, B., Kruth, J. P., Liu, W., Eeraerts, W., Schacht, B., and Bleys, P., 2004, Investigation of material removal mechanisms in EDM of composite ceramic materials, *Journal of Materials Processing Technology*, 149, 347–352.

20. Tomura, S. and Kunieda, M., 2009, Analysis of electromagnetic force in wire-EDM, *Precision Engineering*, 33, 255–262.

21. Liu, Y. H., Ji, R. J., Li, X. P., Yu, L. L., and Zhang, H. F., 2008, Electric discharge milling of insulating ceramics, *Proceedings of the Institution of Mechanical Engineers, Part B: Engineering Manufacture*, 222, 361–366.

22. Rajurkar, K. P. and Yu, Z. Y., 2000, Micro EDM can produce micro parts, *Manufacturing Engineering*, 125, 68–70.

23. Yang, C. T., Ho, S. S., and Yan, B. H., 2001, Micro hole machining of borosilicate glass through electrochemical discharge machining (ECDM), *Key Engineering Materials*, 196, 149–166.

24. Li, H. and Masaki, T., 1991, Micro-EDM, SME Technical Paper, MS91-485, pp. 1–7.

25. Masuzawa, T. and Toenshoff, H. K., 1997, Three-dimensional micromachining by machine tools, *Annals of the CIRP*, 46, 621–628.

26. Rajurkar, K. P., Microelectrodischarge machining, Center for Nontraditional Manufacturing Research, University of Nebraska–Lincoln, http://www.unl.edu/nmrc/microEDM/medm3.htm

27. Pham, D. T., Dimov, S. S., Bigot, S., Ivanov, A., and Popov, K., 2004, Micro-EDM—Recent developments and research issues, *Journal of Materials Processing Technology*, 149, 50–57.

28. Masuzawa, T., 2001, Micro-EDM, *Proceedings of the 13th International Symposium for Electromachining*, Bilbao, Spain, pp. 1–19.

29. Rajurkar, K. P., Levy, G., Malshe, A., Sundaram, M. M., McGeough, J., Hu, X., Resnick, R., and DeSilva, A., 2006, Micro and nano machining by electro-physical and chemical processes, *Annals of the CIRP*, 55, 643–666.

30. Schumacher, B. M., 2004, After 60 years of EDM the discharge process remains still disputed, *Journal of Materials Processing Technology*, 149, 376–381.

31. Petrofes, N. F., 1989, Shaping advanced ceramics with electrical discharge machining, PhD thesis, Texas A&M University, College Station, TX.

32. Myers, G. E., 1987, *Analytical Methods in Conduction Heat Transfer*, Genium Publishing Corporation, Schenectady, NY.

33. Erden, A., Arinc, F., and Kogmen, M., 1995, Comparison of mathematical models for electric discharge machining, *Journal of Materials Processing and Manufacturing Science*, 4, 163–176.

34. Singh, A. and Ghosh, A., 1999, A thermo-electric model of material removal during electric discharge machining, *International Journal of Machine Tools & Manufacture*, 39, 669–682.

35. Ikai, T., Fujita, I., and Hashiguchi, K., 1992, Heat input radius for crater formation in the electric discharge machining, *Transactions of the Institute of Electrical Engineers of Japan, Part D*, 112-D, 943–949.

36. Ikai, T. and Hashiguchi, K., 1995, Heat input for crater formation in EDM. *Proceedings of 11th International Symposium for Electromachining*, Lausanne, Switzerland. Presses polytechniques et universitaires romandes, Lausanne, Switzerland.

37. Patel, M. R., Barrufet, M. A., Eubank, P. T., and DiBitonto, D. D., 1989, Theoretical models of the electrical discharge machining process. II. The anode erosion model, *Journal of Applied Physics*, 66, 4104–4111.

38. Erden, A. and Kaftanoglu, B., 1981, Heat transfer modeling of electrical discharge machining, *Proceedings of the 21st International Machine Tool Design and Research Conference*, Swansea, U.K. Macmillan Press, Ltd, London, U.K.

39. Marty, C. C., 1977, Investigation of surface temperature in electro-discharge machining, *Transactions of the ASME, Journal of Engineering for Industry*, 99, 682–684.

40. Van Dijck, F. S. and Dutre, W. L., 1974, Heat conduction model for the calculation of the volume of molten metal in electric discharges [discharge machining], *Journal of Physics D (Applied Physics)*, 7, 899–910.

41. Heuvelman, C. J., 1969, Some aspects of the research on electro-erosion machining, *Annals of the CIRP*, 17, 195–199.

42. Spur, G. and Schoenbeck, J., 1993, Anode erosion in wire-EDM—A theoretical model, *Annals of the CIRP*, 42, 253–256.

43. Konig, W. and Wertheim, R., 1975, Material removal and energy distribution in EDM, *Annals of the CIRP*, 24, 95–100.

44. Jennes, M., Snoeys, R., and Dekeyser, W., 1984, Comparison of various approaches to model the thermal load on the EDM-wire electrode., *Annals of the CIRP*, 33, 93–98.

45. Roethel, F., Garbajs, V., and Kosec, L., 1975, Contribution to the micro-analysis of the spark eroded surfaces, *Annals of the CIRP*, 25, 135–140.

46. Rajurkar, K. P. and Pandit, S. M., 1984, Quantitative expressions for some aspects of surface integrity of electro-discharge machined components, *Transactions of the ASME, Journal of Engineering for Industry*, 106, 171–177.

47. Lee, L. C., Lim, L. C., Narayanan, V., and Venkatesh, V. C., 1988, Quantification of surface damage of tool steels after EDM, *International Journal of Machine Tools and Manufacture*, 28, 359–372.

48. Rebelo, J. C., Dias, A. M., Kremer, D., and Lebrun, J. L., 1998, Influence of EDM pulse energy on the surface integrity of martensitic steels, *Journal of Materials Processing Technology*, 84, 90–96.

49. Rebelo, J. C., Kornmeier, M., Batista, A. C., and Dias, A. M., 2002, Residual stress after EDM-FEM study and measurement results, *Materials Science Forum*, 404–407, 159–164.

50. Yadav, V., Jain, V. K., and Dixit, P. M., 2002, Thermal stresses due to electrical discharge machining, *International Journal of Machine Tools and Manufacture*, 42, 877–888.

51. Bhattacharya, R., Jain, V. K., and Ghoshdastidar, P. S., 1996, Numerical simulation of thermal erosion in EDM process, *IE(I) Journal-PR*, 77, 13–19.

52. Das, S., Klotz, M., and Klocke, F., 2003, EDM simulation: Finite element-based calculation of deformation, microstructure and residual stresses, *Journal of Materials Processing Technology*, 142, 434–451.

53. Rajurkar, K. P. and Wang, W. M., 1997, Advances in EDM monitoring and control systems using modern control concepts, *International Journal of Electrical Machining*, 2, 1–8.

54. Ahmed, M. S., 1988, Adaptive control of the EDM process, SME Technical Paper, MR88-222, pp. 1–11.

55. Wang, W. M. and Rajurkar, K. P., 1992, Modeling and adaptive control of EDM systems, *Journal of Manufacturing Systems*, 11, 334–345.

56. Rajurkar, K. P., Wang, W. M., and Zhao, W. S., 1997, WEDM—Adaptive control with a multiple input model for identification of workpiece height, *Annals of the CIRP*, 46, 147–150.

57. Weck, M. and Dehmer, J. M., 1992, Analysis and adaptive control of EDM sinking process using the ignition delay time and fall time as parameter, *Annals of the CIRP*, 41, 243–246.

58. Kruth, J. P., Snoeys, R., and Van Brussel, H., 1979, Adaptive control optimization of the EDM process using minicomputers, *Computers in Industry*, 1, 65–75.

59. Rajurkar, K. P., 1981, A physico-mathematical investigation of electrical discharge machining by Data Dependent Systems, PhD thesis, Michigan Technological University, Houghton, MI, p. 287.

60. Rajurkar, K. P. and Wang, W. M., 1990, Real-time stochastic model and control of EDM, *Annals of the CIRP*, 39, 187–190.

61. Dekeyser, W., Snoeys, R., and Jennes, M., 1988, Expert system for wire cutting EDM, based on pulse classification and thermal modeling, *Robotics and Computer-Integrated Manufacturing, Manufacturing Science, Technology and Systems of the Future*, 4, 219–224.

62. Filipic, B. and Junkar, M., 2000, Using inductive machine learning to support decision making in machining processes, *Computers in Industry*, 43, 31–41.

63. Kaneko, T. and Onodera, T., 2004, Improvement in machining performance of die-sinking EDM by using self-adjusting fuzzy control, *Journal of Materials Processing Technology*, 149, 204–211.

64. Boccadoro, M. and Dauw, D. F., 1995, About the application of fuzzy controllers in high-performance die-sinking EDM machines, *Annals of the CIRP*, 44, 147–150.

65. Tarng, Y. S., Tseng, C. M., and Chung, L. K., 1997, A fuzzy pulse discriminating system for electrical discharge machining, *International Journal of Machine Tools and Manufacture*, 37, 511–522.

66. Lin, J. L., Wang, K. S., Yan, B. H., and Tarng, Y. S., 2000, Optimization of the electrical discharge machining process based on the Taguchi method with fuzzy logics, *Journal of Materials Processing Technology*, 102, 48–55.

67. Lin, C.-T., Chung, I.-F., and Huang, S.-Y., 2001, Improvement of machining accuracy by fuzzy logic at corner parts for wire-EDM, *Fuzzy Sets and Systems*, 122, 499–511.

68. Su, J. C., Kao, J. Y., and Tarng, Y. S., 2004, Optimisation of the electrical discharge machining process using a GA-based neural network, *International Journal of Advanced Manufacturing Technology*, 24, 81–90.

69. Liao, Y. S., Yan, M. T., and Chang, C. C., 2002, A neural network approach for the on-line estimation of workpiece height in WEDM, *Journal of Materials Processing Technology*, 121, 252–258.

70. Tsai, K.-M. and Wang, P.-J., 2001, Comparisons of neural network models on material removal rate in electrical discharge machining, *Journal of Materials Processing Technology*, 117, 111–24.

71. Higuerey, E. E., 1998, Neural network modeling of process parameters for electrical discharge machining, PhD thesis, Lehigh University, Bethlehem, PA, p. 186.

72. Zhang, J. H., Zhang, H., Su, D. S., Qin, Y., Huo, M. Y., Zhang, Q. H., and Wang, L., 2002, Adaptive fuzzy control system of a servomechanism for electro-discharge machining combined with ultrasonic vibration, *Journal of Materials Processing Technology*, 129, 45–49.

73. Yuanfeng, L., Wansheng, Z., and Shichun, D., 2000, The detection of discharge state and neuro-fuzzy control in electrical discharge machining, *High Technology Letters*, 10, 72–74.

74. Huang, J. T. and Liao, Y. S., 2000, A wire-EDM maintenance and fault-diagnosis expert system integrated with an artificial neural network, *International Journal of Production Research*, 38, 1071–1082.

75. Kunieda, M., Lauwers, B., Rajurkar, K. P., and Schumacher, B. M., 2005, Advancing EDM through fundamental insight into the process, *Annals of the CIRP*, 54, 599–622.

76. Altpeter, F. and Perez, R., 2004, Relevant topics in wire electrical discharge machining control, *Journal of Materials Processing Technology*, 149, 147–151.

77. Rajurkar, K. P. and Wang, W. M., 1991, On-line monitor and control for wire breakage in WEDM, *Annals of the CIRP*, 40, 219–222.

78. Snoeys, R., Dauw, D., and Kruth, J. P., 1980, Improved adaptive control system for EDM processes, *Annals of the CIRP*, 29, 97–101.

79. Hocheng, H., Lei, W. T., and Hsu, H. S., 1997, Preliminary study of material removal in electrical-discharge machining of SiC/Al, *Journal of Materials Processing Technology*, 63, 813–818.
80. Kahng, C. H. and Rajurkar, K. P., 1977, Fundamental theories of the parameters of EDM process, *SME Technical Paper Series, MR for Midwest Engineering Conference*, Milwaukee, WI.
81. Lee, S. H. and Li, X. P., 2001, Study of the effect of machining parameters on the machining characteristics in electrical discharge machining of tungsten carbide, *Journal of Materials Processing Technology*, 115, 344–358.
82. Huang, J. T. and Liao, Y. S., 2003, Optimization of machining parameters of Wire-EDM based on Grey relational and statistical analyses, *International Journal of Production Research*, 41, 1707–1720.
83. Marafona, J. and Wykes, C., 2000, A new method of optimising material removal rate using EDM with copper-tungsten electrodes, *International Journal of Machine Tools & Manufacture*, 40, 153–164.
84. Klocke, F., Lung, D., Thomaidis, D., and Antonoglou, G., 2004, Using ultra thin electrodes to produce micro-parts with wire-EDM, *Journal of Materials Processing Technology*, 149, 579–584.
85. Masuzawa, T., Tsukamoto, J., and Fujino, M., 1989, Drilling of deep microholes by EDM, *Annals of the CIRP*, 38, 195–198.
86. Yu, Z. Y., Rajurkar, K. P., and Shen, H., 2002, High aspect ratio and complex shaped blind micro holes by micro EDM, *Annals of the CIRP*, 51, 359–362.
87. Wu, K. L., Yan, B. H., Lee, J.-W., and Ding, C. G., 2009, Study on the characteristics of electrical discharge machining using dielectric with surfactant, *Journal of Materials Processing Technology*, 209, 3783–3789.
88. Mohri, N., Saito, N., and Higashi, M., 1991, New process of finish machining on free surface by EDM methods, *Annals of the CIRP*, 40, 207–210.
89. Klocke, F., Lung, D., Antonoglou, G., and Thomaidis, D., 2004, The effects of powder suspended dielectrics on the thermal influenced zone by electrodischarge machining with small discharge energies, *Journal of Materials Processing Technology*, 149, 191–197.
90. Huang, H., Zhang, H., Zhou, L., and Zheng, H. Y., 2003, Ultrasonic vibration assisted electro-discharge machining of microholes in Nitinol, *Journal of Micromechanics and Microengineering*, 13, 693–700.
91. Kunieda, M. and Yoshida, M., 1997, Electrical discharge machining in gas, *Annals of the CIRP*, 46, 143–146.
92. Kunieda, M., Miyoshi, Y., Takaya, T., Nakajima, N., Bo, Y. Z., and Yoshida, M., 2003, High speed 3D milling by dry EDM, *Annals of the CIRP*, 52, 147–150.
93. Kunieda, M. and Furudate, C., 2001, High precision finish cutting by dry WEDM, *Annals of the CIRP*, 50, 121–124.
94. Lonardo, P. M. and Bruzzone, A. A., 1999, Effect of flushing and electrode material on die sinking EDM, *Annals of the CIRP*, 48, 123–126.
95. Bojorquez, B., Marloth, R. T., and Es-Said, O. S., 2002, Formation of a crater in the workpiece on an electrical discharge machine, *Engineering Failure Analysis*, 9, 93–97.
96. Luo, Y. F., 1997, The dependence of interspace discharge transitivity upon the gap debris in precision electrodischarge machining, *Journal of Materials Processing Technology*, 68, 121–131.
97. De Bruijn, H. E. and Pekelharing, A. J., 1978, Effect of a magnetic field on the gap cleaning in EDM, *Annals of the CIRP*, 27, 93–95.
98. Durgan, P. R., 1989, Fast EDM drilling, SME Technical Paper, EE89-815, pp. 1–10.
99. Masuzawa, T., Cui, X., and Taguchi, T., 1992, Improved jet flushing for EDM, *CIRP Annals-Manufacturing Technology*, 41, 239–242.
100. Okada, A., Uno, Y., Onoda, S., and Habib, S., 2009, Computational fluid dynamics analysis of working fluid flow and debris movement in wire EDMed kerf, *Annals of the CIRP*, 58, 209–212.
101. Lee, S., Minami, H., Masui, K., and Okami, M., 2003, EDM with mesh-sheet electrode, *International Journal of Electrical Machining*, 8, 45–50.
102. Sommer, C., 2000, *Non-Traditional Machining Handbook*, Advance Publishing, Houston, TX.
103. Yan, B. H., Huang, F. Y., Chow, H. M., and Tsai, J. Y., 1999, Micro-hole machining of carbide by electric discharge machining, *Journal of Materials Processing Technology*, 87, 139–145.
104. Batzinger, T. J. and Wei, B., 2002, Electromachining with perforated electrodes. U.S. Patent 6, 680, 454, General Electric Company, Niskayuna, NY.
105. Fukuzawa, Y., Tani, T., Ito, Y., Ichinose, Y., and Mohori, N., 1995, Electrical discharge machining (EDM) of insulator ceramics with a sheet of metal mesh. *Proceedings of 11th International Symposium for Electromachining*, Lausanne, Switzerland. Presses polytechniques et universitaires romandes, Lausanne, Switzerland.

106. Kremer, D., Bazine, G., Moisan, A., Bessaguet, L., Astier, A., and Thanh, N. K., 1983, Ultrasonic machining improves EDM technology. *Proceedings of the Seventh International Symposium in Electromachining*, Birmingham, U.K. IFS (Publ) Ltd., Kempston, U.K.

107. Komsa, E., 1974, Ultrasonic flushing in electric-discharge machining, *IBM Technical Disclosure Bulletin*, 16, 3429.

108. Sundaram, M. M., Pavalarajan, G., and Rajurkar, K., 2008, A study on process parameters of ultrasonic assisted micro EDM based on Taguchi method, *Journal of Materials Engineering and Performance*, 17, 210–215.

109. Wansheng, Z., Zhenlong, W., Shichun, D., Guanxin, C., and Hongyu, W., 2002, Ultrasonic and electric discharge machining to deep and small hole on titanium alloy, *Journal of Materials Processing Technology*, 120, 101–106.

110. Thoe, T. B., Aspinwall, D. K., and Killey, N., 1999, Combined ultrasonic and electrical discharge machining of ceramic coated nickel alloy, *Journal of Materials Processing Technology*, 93, 323–328.

111. Zhixin, J., Jianhua, Z., and Xing, A., 1995, Ultrasonic vibration pulse electro-discharge machining of holes in engineering ceramics, *Journal of Materials Processing Technology*, 53, 811–816.

112. Kremer, D., 1989, Effects of ultrasonic vibrations on the performances in EDM, *Annals of the CIRP*, 38, 199–202.

113. Gao, C. and Liu, Z., 2003, A study of ultrasonically aided micro-electrical-discharge machining by the application of workpiece vibration, *Journal of Materials Processing Technology*, 139, 226–228.

114. Zhang, Q. H., Zhang, J. H., Deng, J. X., Qin, Y., and Niu, Z. W., 2002, Ultrasonic vibration electrical discharge machining in gas, *Journal of Materials Processing Technology*, 129, 135–138.

115. Zhang, J. H., Lee, T. C., Lau, W. S., and Ai, X., 1997, Spark erosion with ultrasonic frequency, *Journal of Materials Processing Technology*, 68, 83–88.

116. Guo, Z. N., Lee, T. C., Yue, T. M., and Lau, W. S., 1997, A study of ultrasonic-aided wire electrical discharge machining, *Journal of Materials Processing Technology*, 63, 823–828.

117. Rozenberg, L. D., 1973, *Physical Principles of Ultrasonic Technology*, Plenum Press, New York.

118. Murti, V. S. R. and Philip, P. K., 1987, An analysis of the debris in ultrasonic-assisted electrical discharge machining, *Wear*, 117, 241–250.

119. Yeo, S. H. and Tan, L. K., 1999, Effects of ultrasonic vibrations in micro electro-discharge machining of microholes, *Journal of Micromechanics and Microengineering*, 9, 345–352.

120. Yeo, S. H. and Goh, K. M., 2001, The effect of ultrasound in micro electro discharge machining on surface roughness, *Proceedings of the Institution of Mechanical Engineers, Part B: Journal of Engineering Manufacture*, 215, 271–276.

121. Murali, M. and Yeo, S. H., 2004, Rapid biocompatible micro device fabrication by micro electro-discharge machining, *Biomedical Microdevices*, 6, 41–45.

122. Certin, S., Okada, A., and Uno, Y., 2004, Effect of debris accumulation on machining speed in EDM, *International Journal of Electrical Machining*, 9, 9–14.

123. Soni, J. S. and Chakraverti, G., 1994, Machining characteristics of titanium with rotary electro-discharge machining, *Wear*, 171, 51–58.

124. Sato, T., Mizutani, T., Yonemochi, K., and Kawata, K., 1986, The development of an electrodischarge machine for micro-hole boring, *Precision Engineering*, 8, 163–168.

125. Guu, Y. H. and Hocheng, H., 2001, Effects of workpiece rotation on machinability during electrical-discharge machining, *Materials and Manufacturing Processes*, 16, 91–101.

126. Masuzawa, T., Okajima, K., Taguchi, T., and Fujino, M., 2002, EDM-lathe for micromachining, *Annals of the CIRP*, 51, 355–358.

127. Ehrfeld, W., Lehr, H., Michel, F., Wolf, A., Gruber, H.-P., and Bertholds, A., 1996, Micro electro discharge machining as a technology in micromachining. *Proceedings of the SPIE—The International Society for Optical Engineering Micromachining and Microfabrication Process Technology II*, Austin, TX.

128. Masuzawa, T. and Heuvelman, C. J., 1983, Self flushing method with spark-erosion machining, *Annals of the CIRP*, 32, 109–111.

129. Altan, T., Lilly, B. W., Kruth, J. P., Koenig, W., Toenshoff, H. K., van Luttervelt, C. A., and Khairy, A. B., 1993, Advanced techniques for die and mold manufacturing, *Annals of the CIRP*, 42, 707–716.

130. Yu, Z. Y., Zhang, Y., Li, J., Luan, J., Zhao, F., and Guo, D., 2009, High aspect ratio micro-hole drilling aided with ultrasonic vibration and planetary movement of electrode by micro-EDM, *Annals of the CIRP*, 58, 213–216.

131. Lim, H. S., Wong, Y. S., Rahman, M., and Edwin Lee, M. K., 2003, A study on the machining of high-aspect ratio micro-structures using micro-EDM, *Journal of Materials Processing Technology*, 140, 318–325.

132. Han, F., Wachi, S., and Kunieda, M., 2004, Improvement of machining characteristics of micro-EDM using transistor type isopulse generator and servo feed control, *Precision Engineering*, 28, 378–385.

133. Guu, Y. H., Hocheng, H., Chou, C. Y., and Deng, C. S., 2003, Effect of electrical discharge machining on surface characteristics and machining damage of AISI D2 tool steel, *Materials Science and Engineering A*, 358, 37–43.

134. Lee, H.-T., Rehbach, W. P., Tai, T.-Y., and Hsu, F.-C., 2003, Surface integrity in micro-hole drilling using micro-electro discharge machining, *Materials Transactions*, 44, 2718–2722.

135. Lee, H. T. and Tai, T. Y., 2003, Relationship between EDM parameters and surface crack formation, *Journal of Materials Processing Technology*, 142, 676–683.

136. Zhou, M. and Han, F., 2009, Adaptive control for EDM process with a self-tuning regulator, *International Journal of Machine Tools and Manufacture*, 49, 462–469.

137. Yu, Z., Rajurkar, K. P., and Prabhuram, P. D., 2001, Study of contouring micro EDM characteristics. *10th International Conference Precision Engineering (ICPE)*, Yokohama, Japan, Euspen.

138. Michel, F., Ehrfeld, W., Koch, O., and Gruber, H. P., 2000, EDM for micro fabrication—Technology and applications. *Proceedings of the International Seminar on Precision Engineering and Micro Technology*, Aachen, Germany. Verlag Rhiem, Voerde.

139. Davim, J. P., Maranhão, C., Cabral, G., and Grácio, J., 2009, Performance of cutting tools in machining Cu/W alloys for application in EDM electrodes, *International Journal of Refractory Metals and Hard Materials*, 27, 676–682.

140. Muttamara, A., Fukuzawa, Y., Mohri, N., and Tani, T., 2009, Effect of electrode material on electrical discharge machining of alumina, *Journal of Materials Processing Technology*, 209, 2545–2552.

141. Lee, Y. H. and Li, C. L., Automation in the design of EDM electrodes, *Computer-Aided Design*, 41, 600–613.

142. Yu, Z. Y., Kozak, J., and Rajurkar, K. P., 2003, Modelling and simulation of micro EDM process, *Annals of the CIRP*, 52, 143–146.

143. IWF and IPT, 2002, Investigation of the International State of the Art of Micro Production Technology, MickroPRO.

144. Masuzawa, T., Fujino, M., Kobayashi, K., Suzuki, T., and Kinoshita, N., 1985, Wire electro-discharge grinding for micro-machining, *Annals of the CIRP*, 34, 431–434.

145. Masuzawa, T., Yamaguchi, M., and Fujino, M., 2005, Surface finishing of micropins produced by WEDG, *Annals of the CIRP*, 54, 171–174.

146. Monzón, M., Benítez, A. N., Marrero, M. D., Hernández, N., Hernández, P., and Aisa, J., 2008, Validation of electrical discharge machining electrodes made with rapid tooling technologies, *Journal of Materials Processing Technology*, 196, 109–114.

147. Habib, M. A., Gan, S. W., and Rahman, M., 2009, Fabrication of complex shape electrodes by localized electrochemical deposition, *Journal of Materials Processing Technology*, 209, 4453–4458.

148. Takezawa, H., Hamamatsu, H., Mohri, N., and Saito, N., 2004, Development of micro-EDM-center with rapidly sharpened electrode, *Journal of Materials Processing Technology*, 149, 112–116.

149. Takezawa, H., Itoh, N., and Mohri, N., 2001, The behavior of thin electrode wear in electrical discharge machining. *Proceedings of the 13th International Symposium for Electromachining*, Bilbao, Spain.

150. Takezawa, H., Mohri, N., and Furutani, K., 2001, Rapid production of a thin electrode by a single discharge machining. I. Machining phenomena and application of formed electrode, *Journal of the Japan Society of Precision Engineering*, 67, 1299–1303.

151. Yamazaki, M., Suzuki, T., Mori, N. and Kunieda, M., 2004, EDM of micro-rods by self-drilled holes, *Journal of Materials Processing Technology*, 149, 134–138.

152. Sheu, D.-Y., 2004, Multi-spherical probe machining by EDM: Combining WEDG technology with one-pulse electro-discharge, *Journal of Materials Processing Technology*, 149, 597–603.

153. Yeo, S. H. and Murali, M., 2003, A new technique using foil electrodes for the electro-discharge machining of micro grooves, *Journal of Micromechanics and Microengineering*, 13, N1–N5.

154. Murali, M. and Yeo, S. H., 2004, A novel spark erosion technique for the fabrication of high aspect ratio micro-grooves, *Microsystem Technologies*, 10, 628–632.

155. Ullmann, E., Doll, U., and Piltz, S., 2001, Electrical discharge grinding of microstructures, *International Journal of Electrical Machining*, 6, 41–46.

156. Song, X., Reynaerts, D., Meeusen, W., and Van Brussel, H., 1999, Investigation of micro-EDM for silicon microstructure fabrication, *Proceedings of SPIE—The International Society for Optical Engineering*, 3680, 792–799.

157. Yu, Z., Rajurkar, K. P., and Shen, H., 2002, Drilling of noncircular blind micro holes by micro EDM, *Transactions of the NAMRC/SME*, XXX, 263–270.

158. Yu, Z. Y., Masuzawa, T., and Fujino, M., 1998, 3D micro-EDM with simple shape electrode Part 1, *International Journal of Electrical Machining*, 3, 7–12.

159. Sharma, A., Iwai, M., Kawanaka, K., Suzuki, K., and Uematsu, T., 2004, Attempt at EDM of electrically conductive diamond and its application to miniature mold processing. *Seventh International Symposium on Advances in Abrasive Technology*, Bursa, Turkey.

160. Uhlmann, E. and Roehner, M., 2008, Investigations on reduction of tool electrode wear in micro-EDM using novel electrode materials, *CIRP Journal of Manufacturing Science and Technology*, 1, 92–96.

161. Sharma, A., Iwai, M., Suzuki, K., and Uematsu, T., 2004, Low-wear diamond electrode for micro-EDM of die-steel. *Seventh International Symposium on Advances in Abrasive Technology*, Bursa, Turkey.

162. Dauw, D. F., 1994, High-precision wire-EDM by online wire positioning control, *Annals of the CIRP*, 43, 193–197.

163. Xia, H., Kunieda, M., and Nishiwaki, N., 1996, Removal amount difference between anode and cathode in EDM process, *International Journal of Electrical Machining*, 1, 42–52.

164. Xia, H., Hashimoto, H., Kunieda, M., and Nishiwaki, N., 1996, Measurement of energy distribution in continuous EDM process, *Seimitsu Kogaku Kaishi/Journal of the Japan Society for Precision Engineering*, 62, 1141–1145.

165. Natsu, W., Kunieda, M., and Nishiwaki, N., 2004, Study on influence of inter-electrode atmosphere on carbon adhesion and removal amount, *International Journal of Electrical Machining*, 9, 43–50.

166. Tsai, Y.-Y. and Masuzawa, T., 2004, An index to evaluate the wear resistance of the electrode in micro-EDM, *Journal of Materials Processing Technology*, 149, 304–309.

167. Yu, Z. Y., Masuzawa, T., and Fujino, M., 1998, Micro-EDM for three-dimensional cavities—Development of uniform wear method, *Annals of the CIRP*, 47, 169–172.

168. Sundaram, M. M. and Rajurkar, K. P., 2008, Towards freeform machining by micro electro discharge machining process, *Transactions of the North American Manufacturing Research Institution of SME*, 36, 381–388.

169. Yuzawa, T., Magara, T., Imai, Y., and Sato, T., 1997, Micro electric discharge scanning using a mini-size cylindrical electrode, *Kata Gijutsu*, 12, 104–105.

170. Bleys, P., Kruth, J.-P., Lauwers, B., Zryd, A., Delpretti, R., and Tricarico, C., 2002, Real-time tool wear compensation in milling EDM, *Annals of the CIRP*, 51, 157–160.

171. Narasimhan, J., Yu, Z., and Rajurkar, K. P., 2004, Tool wear compensation and path generation in micro and macro EDM, *Transactions of NAMRI/SME*, 32, 151–158.

172. Tai, T. Y. and Lu, S. J., 2009, Improving the fatigue life of electro-discharge-machined SDK11 tool steel via the suppression of surface cracks, *International Journal of Fatigue*, 31, 433–438.

173. Machinability Data Center, 1980, *Machining Data Handbook*, 3rd edn., MDC, Cincinnati, OH.

174. Lai, L.-C., Chiou, W.-A., and Earthman, J. C., Influence of electrical discharged machining and surface defects on the fatigue strength of electrodeposited nanocrystalline Ni, *International Journal of Fatigue*, 32, 584–591.

175. Ghrib, T., Salem, S. B., and Noureddine, Y., 2009, EDM effects on the thermal properties of 36NiCrMo16 steel, *Tribology International*, 42, 391–396.

176. Newton, T. R., Melkote, S. N., Watkins, T. R., Trejo, R. M., and Reister, L., 2009, Investigation of the effect of process parameters on the formation and characteristics of recast layer in wire-EDM of Inconel 718, *Materials Science and Engineering: A*, 513–514, 208–215.

177. Bonny, K., De Baets, P., Quintelier, J., Vleugels, J., Jiang, D., Van der Biest, O., Lauwers, B., and Liu, W., Surface finishing: Impact on tribological characteristics of WC-Co hardmetals, *Tribology International*, 43, 40–54.

178. Bonny, K., De Baets, P., Ost, W., Vleugels, J., Huang, S., Lauwers, B., and Liu, W., 2009, Influence of electrical discharge machining on the reciprocating sliding wear response of WC-Co cemented carbides, *Wear*, 266, 84–95.

179. Kruth, J.-P., Stevens, L., Froyen, L., and Lauwers, B., 1995, Study of the white layer of a surface machined by die-sinking electro-discharge machining, *Annals of the CIRP*, 44, 169–172.

180. Liao, Y. S., Huang, J. T., and Chen, Y. H., 2004, A study to achieve a fine surface finish in Wire-EDM, *Journal of Materials Processing Technology*, 149, 165–171.

181. Yu, Z. Y., Rajurkar, K. P., and Narasimhan, J., 2003, Effect of machining parameters on machining performance of micro EDM and surface integrity. *Proceedings of the 18th Annual Meeting of American Society for Precision Engineering*, Portland, OR.

182. Kawakami, K. and Kunieda, M., 2005, Study on factors determining limits of minimum machinable size in micro EDM, *Annals of the CIRP*, 54, 167–170.

183. Han, F., Yamagata, Y., Kawakami, T., and Kunieda, M., 2003, Investigation on feasibility of sub-micrometer order manufacturing using micro-EDM, *Proceedings of the 18th Annual ASPE Meeting*, Portland, OR.

184. Han, F., Yamada, Y., Kawakami, T., and Kunieda, M., 2006, Experimental attempts of sub-micrometer order size machining using micro-EDM, *Precision Engineering*, 30, 123–131.

185. Egashira, K. and Mizutani, K., 2005, EDM at low open-circuit voltage, *International Journal of Electrical Machining*, 10, 21–26.

186. Kruth, J.-P. and Bleys, P., 2000, Measuring residual stress caused by Wire EDM of tool steel, *International Journal of Electrical Machining*, 5, 23–28.

187. Klocke, F. and Lung, D., 2001, Micro contouring by EDM with fine wires, *Proceedings of the 13th International Symposium for Electromachining*, Bilbao, Spain.

188. Yan, M.-T. and Huang, P.-H., 2004, Accuracy improvement of wire-EDM by real-time wire tension control, *International Journal of Machine Tools and Manufacture*, 44, 807–814.

189. Yan, M.-T., Huang, C.-W., Fang, C.-C., and Chang, C.-X., 2004, Development of a prototype Micro-Wire-EDM machine, *Journal of Materials Processing Technology*, 149, 99–105.

190. Tani, T., Fukuzawa, Y., Mohri, N., Saito, N., and Okada, M., 2004, Machining phenomena in WEDM of insulating ceramics, *Journal of Materials Processing Technology*, 149, 124–128.

191. Sheu, D.-Y., Masuzawa, T., and Fujino, M., 1997, Machining of deep micro holes by EDM, *Proceedings of the Annual Assembly of JSEME*, 105–108.

192. Koch, O., Ehrfeld, W., Michel, F., and Gruber, H. P., 2001, Recent progress in micro-electro discharge machining-part 1. *Proceedings of the 13th International Symposium for Electromachining*, Bilbao, Spain.

193. Masaki, T., Kawata, K., and Masuzawa, T., 1990, Micro electro-discharge machining and its applications. *Proceedings of the IEEE Micro Electro Mechanical Systems: An Investigation of Micro Structures, Sensors, Actuators, Machines and Robots*, Napa Valley, CA. IEEE, Piscataway, NJ.

194. Okada, A., Uno, Y., Okamoto, H., Tanaka, H., and Okada, S., 2001, A new micro EDM technique of monocrystalline silicon using fine triangular section electrode, *Proceedings of the 13th International Symposium for Electromachining*, Bilbao, Spain.

195. Song, X., Meeusen, W., Reynaerts, D., and Van Brussel, H., 2000, Experimental study of micro-EDM machining performances on silicon wafer. *Micromachining and Microfabrication Process Technology VI*, Santa Clara, CA. Society of Photo-Optical Instrumentation Engineers, Bellingham, WA.

196. Reynaerts, D., Meeusen, W., Song, X., Van Brussel, H., Reyntjens, S., De Bruyker, D., and Puers, R., 2000, Integrating electro-discharge machining and photolithography: Work in progress, *Journal of Micromechanics and Microengineering*, 10, 189–195.

197. Masuzawa, T., 1998, Recent trends in EDM/ECM technologies in Japan, *Proceedings of the 12th International Symposium for Electromachining*, Aachen, Germany.

198. Kawata, K., Fujimura, H., and Nishimur, K., 2004, Micro-electrodischarge machining of single-crystal diamond for micro-dies. *Seventh International Symposium on Advances in Abrasive Technology*, Bursa, Turkey.

199. Morgan, C. J., Vallance, R. R., and Marsh, E. R., 2004, Micro machining glass with polycrystalline diamond tools shaped by micro electro discharge machining, *Journal of Micromechanics and Microengineering*, 14, 1687–1692.

200. Albert, M., Wire EDM goes horizontal, *Modern Machine Shop*, November 2004.

201. Liao, Y.-S., Chen, S.-T., and Lin, C.-S., 2005, Development of a high precision tabletop versatile CNC wire-EDM for making intricate micro parts, *Journal of Micromechanics and Microengineering*, 15, 245–253.

202. Gusseff, W., 1930, Method and apparatus for the electrolytic treatment of metals, British Patent 335,003.

203. Bannard, J., 1977, Electrochemical machining, *Journal of Applied Electrochemistry*, 7, 1–29.

204. McGeough, J. A., 1974, *Principles of Electrochemical Machining*, Chapman & Hall; Halsted Press Division, Wiley, London, U.K.

205. Wilson, J. F., 1971, *Practice and Theory of Electrochemical Machining*, Wiley-Interscience, New York.

206. Schuster, R., Kirchner, V., Allongue, P., and Ertl, G., 2000, Electrochemical micromachining, *Science*, 289, 98–101.

207. Bhattacharyya, B. and Munda, J., 2003, Experimental investigation on the influence of electrochemical machining parameters on machining rate and accuracy in micromachining domain, *International Journal of Machine Tools and Manufacture*, 43, 1301–1310.

208. McGeough, J. A. and Thomson, J. R., 1986, Some effects of electrolyte flow in the electroforming of iron foil, *Proceedings of the Institution of Mechanical Engineers, Part C: Mechanical Engineering Science*, 200, 111–122.

209. McGeough, J. A. and Rasmussen, H., 1990, A theoretical analysis of electrochemical arc machining, *Proceedings of the Royal Society of London. Series A, Mathematical and Physical Sciences*, 429, 429–447.

210. De Silva, A. and McGeough, J. A., 1986, Surface effects on alloys drilled by electrochemical arc machining, *Proceedings of the Institution of Mechanical Engineers, Part B: Management and Engineering Manufacture*, 200, 237–246.

211. Bhattacharyya, B., Mitra, S., and Boro, A. K., 2002, Electrochemical machining: New possibilities for micromachining, *Robotics and Computer-Integrated Manufacturing*, 18, 283–289.

212. Kim, B. H., Na, C. W., Lee, Y. S., Choi, D. K., and Chi, C. N., 2005, Micro electrochemical machining of 3D micro structure using dilute sulfuric acid, *Annals of the CIRP*, 54, 191–194.

213. Sen, M. and Shan, H. S., 2005, A review of electrochemical macro- to micro-hole drilling processes, *International Journal of Machine Tools and Manufacture*, 45, 137–152.

214. Sharma, S., Jain, V. K., and Shekhar, R., 2002, Electrochemical drilling of inconel superalloy with acidified sodium chloride electrolyte, *International Journal of Advanced Manufacturing Technology*, 19, 492–500.

215. Basak, I. and Ghosh, A., 1997, Mechanism of material removal in electrochemical discharge machining: A theoretical model and experimental verification, *Journal of Materials Processing Technology*, 71, 350–359.

216. Crichton, I. M. and McGeough, J. A., 1984, Theoretical, experimental and computational aspects of the electrochemical arc machining process, *Annals of the CIRP*, 33, 429–432.

217. Kozak, J. and M, P., 1989, Electrochemical arc cutting, *Proceedings of ISEM*, 9, 103–106.

218. Fascio, V., Wuthrich, R., and Bleuler, H., 2004, Spark assisted chemical engraving in the light of electrochemistry, *Electrochimica Acta*, 49, 3997–4003.

219. Kita, Y. and Ajiro, H., 1986, Influence of electrochemical discharge phenomena on fine ceramics, *Journal of the Society of Materials Science, Japan*, 35, 1418–1424.

220. Bhattacharyya, B., Doloi, B. N., and Sorkhel, S. K., 1999, Experimental investigations into electrochemical discharge machining (ECDM) of non-conductive ceramic materials, *Journal of Materials Processing Technology*, 95, 145–154.

221. Kurafuji, H. and Suda, K., 1968, Electrical discharge drilling of glass—I, *Annals of the CIRP*, 16/4, 415.

222. De Silva, A. and McGeough, J., 1989, The production of full die shapes by electrochemical arc/electrochemical machining, *Proceedings of the ISEM*, 9, 107–110.

223. Sarkar, B. R., Doloi, B., and Bhattacharyya, B., 2006, Parametric analysis on electrochemical discharge machining of silicon nitride ceramics, *The International Journal of Advanced Manufacturing Technology*, 28, 873–881.

224. Han, M.-S., Min, B.-K., and Lee, S. J., 2008, Modeling gas film formation in electrochemical discharge machining processes using a side-insulated electrode, *Journal of Micromechanics and Microengineering*, 18, 1–8.

225. Zheng, Z.-P., Lin, J.-K., Huang, F.-Y., and Yan, B.-H., 2008, Improving the machining efficiency in electrochemical discharge machining (ECDM) microhole drilling by offset pulse voltage, *Journal of Micromechanics and Microengineering*, 18, 1–6.

226. Ghosh, A., 1997, Electrochemical discharge machining: Principle and possibilities, *Sadhana*, 22, 435–447.

227. Schopf, M., 2002, Dressing of metal-bond grinding wheels by ECDM, *Industrial Diamond Review*, 62, 82–85.

228. Peng, W. Y. and Liao, Y. S., 2004, Study of electrochemical discharge machining technology for slicing non-conductive brittle materials, *Journal of Materials Processing Technology*, 149, 363–369.

229. Drake, T. H. and McGeough, J. A., 1982, Aspects of drilling by electrochemical arc machining, *Proceedings of the 22nd International Machine Tool Design and Research Conference*, Manchester, U.K.

230. McGeough, J. A., Khayry, A. B. M., Munro, W., and Crookall, J. R., 1983, Theoretical and experimental investigation of the relative effects of spark erosion and electrochemical dissolution in electrochemical ARC machining, *Annals of the CIRP*, 32, 113–118.

231. Zheng, Z.-P., Cheng, W.-H., Huang, F.-Y., and Yan, B.-H., 2007, 3D microstructuring of Pyrex glass using the electrochemical discharge machining process, *Journal of Micromechanics and Microengineering*, 17, 960–966.

232. West, J. and Jadhav, A., 2007, ECDM methods for fluidic interfacing through thin glass substrates and the formation of spherical microcavities, *Journal of Micromechanics and Microengineering*, 17, 403–409.

233. Kim, D.-J., Ahn, Y., Lee, S.-H., and Kim, Y.-K., 2006, Voltage pulse frequency and duty ratio effects in an electrochemical discharge microdrilling process of Pyrex glass, *International Journal of Machine Tools and Manufacture*, 46, 1064–1067.

234. Zheng, Z.-P., Su, H.-C., Huang, F.-Y., and Yan, B.-H., 2007, The tool geometrical shape and pulse-off time of pulse voltage effects in a Pyrex glass electrochemical discharge microdrilling process, *Journal of Micromechanics and Microengineering*, 17, 265–272.

235. Chak, S. K. and Venkateswara Rao, P., 2008, The drilling of Al_2O_3 using a pulsed DC supply with a rotary abrasive electrode by the electrochemical discharge process, *International Journal of Advanced Manufacturing Technology*, 39, 633–641.

236. Cavallotti, P. L., 2005, Editorial, *Electrochimica Acta*, 50, 4523–4524.

237. Brown, J. A., 1991, *Modern Manufacturing Processes*, Industrial Press, New York.

238. Watson, S. A., 1989, Modern electroforming, *Transactions of the Institute of Metal Finishing*, 67, pt 4, 89–94.

239. McGeough, J. A., Leu, M. C., Rajurkar, K. P., De Silva, A. K. M., and Liu, Q., 2001, Electroforming process and application to micro/macro manufacturing, *Annals of the CIRP*, 50, 499–514.

240. Lin, T.-H., Hung, S.-Y., Yang, H., and Chao, C.-K., 2007, Fabrication of a microlens array electroformed mold with low roughness and high hardness, *Journal of Micromechanics and Microengineering*, 17, 419–425.

241. Brown, R., 2003, Electroplating, in *RF/Microwave Hybrids*, Kluwer Academic Publishers, Dordrecht, the Netherlands, pp. 169–184.

242. Cotell, C. M. et al., 1995, *ASM Handbook*, Volume 5: *Surface Engineering*, ASM International, Materials Park, OH.

243. Watts, O. P., 1916, Rapid nickel plating, *Transactions of American Electrochemical Society*, 29, 395.

244. Kruglikov, S. S., Kudriavtsev, N. T., Vorobiova, G. F., and Antonov, A. Y. A., 1965, On the mechanism of levelling by addition agents in electrodeposition of metals, *Electrochimica Acta*, 10, 253–261.

245. Mandich, N. V., 2002, Pulse and pulse-reverse electroplating, *Metal Finishing*, 100, 359–364.

246. Tang, P. T., Jaskula, M., Kubiczek, M., Mizushima, I., Pantleon, K., and Arentoft, M., 2009, Pulse reversal plating of nickel-cobalt alloys, *Transactions of the Institute of Metal Finishing*, 87, 72–77.

247. Anonymous, 1939, Brush plating, *Iron Age*, 143, 39.

248. Rubinstein, M., 1954, Brush plating now practical, *Materials and Methods*, 40, 98–101.

249. Sole, M. J., 1994, Electroforming: Methods, materials, and merchandise, *JOM*, 46, 29–35.

250. Tang, P.T., Dylmer, H., and Møller, P., 1995, Nickel coatings and electroforming using pulse reversal plating, *Metal Finishing AESF SUR/FIN*, 93, 529–536.

251. Amblard, J., Epelboin, I., Froment, M., and Maurin, G., 1979, Inhibition and nickel electrocrystallization, *Journal of Applied Electrochemistry*, 9, 233–242.

252. Hsieh, J.-C., Hu, C.-C., and Lee, T.-C., 2008, The synergistic effects of additives on improving the electroplating of zinc under high current densities, *Journal of the Electrochemical Society*, 155, D675–D681.

253. Xi, W., Qiao, Z., Zhu, C., Jia, A., and Li, M., 2009, The preparation of lotus-like super-hydrophobic copper surfaces by electroplating, *Applied Surface Science*, 255, 4836–4839.

254. Povetkin, V. V., Shibleva, T. G., and Zhitnikova, A. V., 2008, Electroplating lead-indium alloy from trilonate solutions in a magnetic field, *Protection of Metals*, 44, 487–489.

255. Wang, C., Zhong, Y.-B., Wang, J., Wang, Z.-q., Ren, W.-L., Lei, Z.-S., and Ren, Z.-M., 2009, Effect of magnetic field on electroplating Ni/nano-Al_2O_3 composite coating, *Journal of Electroanalytical Chemistry*, 630, 42–48.

256. Park, B.-N., Sohn, Y.-S., and Choi, S.-Y., 2008, Effects of a magnetic field on the copper metallization using the electroplating process, *Microelectronic Engineering*, 85, 308–314.

257. Mitooka, Y., Murakami, K., Hino, M., Takamizawa, M., and Takada, J., 2009, Effect of laser irradiation on generation and growth of whiskers in tin-electroplated film, *Nippon Kinzoku Gakkaishi/Journal of the Japan Institute of Metals*, 73, 226–233.

258. Ohsaka, T., Isaka, M., Hirano, K., and Ohishi, T., 2008, Effect of ultrasound sonication on electroplating of iridium, *Ultrasonics Sonochemistry*, 15, 283–288.

259. Zhang, X., Tu, K. N., Chen, Z., Tan, Y. K., Wong, C. C., Mhaisalkar, S. G., Li, X. M., Tung, C. H., and Cheng, C. K., 2008, Pulse electroplating of copper film: A study of process and microstructure, *Journal of Nanoscience and Nanotechnology*, 8, 2568–2574.

260. Ning, Z., He, Y. and Gao, W., 2008, Mechanical attrition enhanced Ni electroplating, *Surface and Coatings Technology*, 202, 2139–2146.

261. Saile, V., 2009, *LIGA and Its Applications*, Wiley-VCH, Weinheim, U.K.

262. Liu, G., Huang, X., Xiong, Y., and Tian, Y., 2008, Fabricating HARMS by using megasonic assisted electroforming, *Microsystem Technologies*, 14, 1223–1226.

263. Li, J.-S., Xue, Y.-J., Lan, M.-M., Liu, Y., and Yu, Y.-J., 2009, Effects of ultrasound on microstructures and properties of Ni-CeO$_2$ nanocomposite electroforming deposits, *Chinese Journal of Nonferrous Metals*, 19, 517–522.

264. Holstein, N., Schanz, G., Konys, J., Piotter, V., and Ruprecht, R., 2005, Metallic microstructures by electroforming from conducting polymer templates, *Microsystem Technologies*, 11, 179–185.

265. Zhu, D. and Zeng, Y. B., 2008, Micro electroforming of high-aspect-ratio metallic microstructures by using a movable mask, *Annals of the CIRP*, 57, 227–230.

266. Chan, K. C., Chan, W. K., and Qu, N. S., 1999, Effect of current waveform on the deposit quality of electroformed nickels, *Journal of Materials Processing Technology*, 89–90, 447–450.

267. Wong, K. P., Chan, K. C., and Yue, T. M., 2001, Study of hardness and grain size in pulse current electroforming of nickel using different shaped waveforms, *Journal of Applied Electrochemistry*, 31, 25–34.

268. Yang, J. M., Zhu, D., and Kim, D. H., 2008, Influence of high frequency pulse current on properties of electroformed nanocrystalline Ni-Mn alloys, *Transactions of the Institute of Metal Finishing*, 86, 98–102.

269. Huang, C.-H., 1993, Effect of organic additives on electroformed nickel alloys, *Metal Finishing*, 91, 107–110.

270. Huang, C.-H. O., 1999, Effect of imides on nickel-tungsten alloy electroforming, *Journal of Materials Science*, 34, 1373–1377.

271. Lim, J. H., Park, E. C., Joo, J., and Jung, S.-B., 2009, Effect of additives on microstructure and mechanical properties of nickel plate/mask fabricated by electroforming process, *Journal of the Electrochemical Society*, 156, D108–D112.

272. Zhu, Z.-W., Zhu, D., Qu, N.-S., Wang, K., and Yang, J.-M., 2008, Electroforming of revolving parts with near-polished surface and uniform thickness, *International Journal of Advanced Manufacturing Technology*, 39, 1164–1170.

273. Toenshoff, H. K., Egger, R., and Klocke, F., 1996, Environmental and safety aspects of electrophysical and electrochemical processes, *Annals of the CIRP*, 45, 553–568.

274. Hamid, Z. A. and Aal, A. A., 2009, New environmentally friendly noncyanide alkaline electrolyte for copper electroplating, *Surface and Coatings Technology*, 203, 1360–1365.

275. Mason, T. J., Lorimer, J. P., Saleem, S., and Paniwnyk, L., 2001, Controlling emissions from electroplating by the application of ultrasound, *Environmental Science and Technology*, 35, 3375–3377.

276. http://www.cdphe.state.co.us/el/library/metal/Electroplating.pdf

277. Priya, P. G., Basha, C. A., Ramamurthi, V., and Begum, S. N., 2009, Recovery and reuse of Ni(II) from rinsewater of electroplating industries, *Journal of Hazardous Materials*, 163, 899–909.

278. Pereira, F. V., Gurgel, L. V. A., De Aquino, S. F., and Gil, L. F., 2009, Removal of Zn2+ from electroplating wastewater using modified wood sawdust and sugarcane bagasse, *Journal of Environmental Engineering*, 135, 341–350.

279. Chen, S.-S., Li, C.-W., Hsu, H.-D., Lee, P.-C., Chang, Y.-M., and Yang, C.-H., 2009, Concentration and purification of chromate from electroplating wastewater by two-stage electrodialysis processes, *Journal of Hazardous Materials*, 161, 1075–1080.

280. Panayotova, T., Dimova-Todorova, M., and Dobrevsky, I., 2007, Purification and reuse of heavy metals containing wastewaters from electroplating plants, *Desalination*, 206, 135–140.

281. Li, C.-T., Lee, W.-J., Huang, K.-L., Fu, S.-F., and Lai, Y.-C., 2007, Vitrification of chromium electroplating sludge, *Environmental Science and Technology*, 41, 2950–2956.

282. Jung, S.-M., Cho, Y.-M., and Na, H.-G., 2007, Chemical analysis of zinc electroplating solutions by x-ray fluorescence spectrometry, *ISIJ International*, 47, 853–859.

7 Micro Electrical Discharge Machining of Spray Holes for Diesel Fuel Systems

Chen-Chun Kao and Albert Shih

CONTENTS

The goal of this chapter is to give readers a broad overview on the state-of-the-art manufacturing and metrology technologies for spray holes of diesel fuel systems. Spray hole geometry affects the fuel atomization and spray pattern and therefore it is a critical part for enhancing the combustion efficiency and reducing the particulate emission. This chapter starts with an introduction of micro-hole electrical discharge machining (micro-hole EDM) and then explains the current method employed to drill spray holes, EDM process monitoring, spray hole measurement, and EDM process control. Finally, it concludes with remarks on future work and the scope beyond. A section of questions and reflections is prepared for the interested readers to think deeper and out of the box.

7.1   INTRODUCTION

7.1.1   BACKGROUND

The spray hole geometry in the diesel fuel injector nozzle affects fuel atomization, spray penetration, as well as combustion and emission. Thus the spray hole is a critical feature for clean and energy-efficient diesel engines. Figure 7.1 illustrates the configuration of a diesel fuel injector nozzle. The needle, also called the plunger, is precisely controlled by a hydraulic, electromagnetic, or piezoelectric actuator to inject diesel fuel through spray holes into the cylinders. Conventional drilling using drill bits was the primary method for the fabrication of spray holes until the 1970s. Afterward, due to the need for higher fuel injection pressure, heat-treated high-strength steel was used as the base material of injector nozzles and the nozzle wall thickness was increased as well. The hardened surface, tough material property, and large hole depth made it very difficult to drill spray holes effectively using conventional drill bits [5]. Therefore, spray hole drilling by electrical discharge machining (EDM) was developed in the 1970s to replace the conventional drilling process.

A milestone for the early research of micro-hole EDM drilling was a successful experiment conducted in the Philips Research Laboratory in late 1960s. After these groundbreaking experiments, the micro-hole EDM has drawn increased attention and been extensively applied in industrial manufacturing, especially for the diesel fuel systems.

A schematic diagram of the EDM drilling process for injector spray holes is illustrated in Figure 7.2. A wire electrode is fed through the ceramic wire guide to the workpiece. Sparks are generated across the discharge gap at the tip of the wire electrode to remove the workpiece material and cause some electrode wear as well, which sharpens the electrode tip. Tungsten is the most common electrode material because of its high melting temperature (3370°C) and low tool wear rate. Tungsten wire electrodes for spray hole EDM are centerless ground [6] to achieve uniform size.

The specification of injector spray holes is 120–180 μm in diameter and 1.0–1.2 mm in depth. Figure 7.3 shows the scanning electron microscopy (SEM) micrographs of typical injector spray

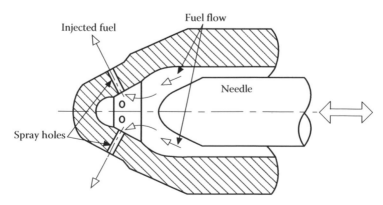

FIGURE 7.1   Illustration of diesel fuel spray holes in an injector nozzle, the needle movement, and diesel fuel flow.

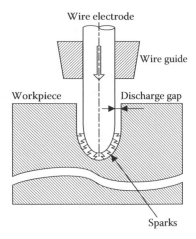

FIGURE 7.2   A schematic diagram of micro-hole EDM drilling.

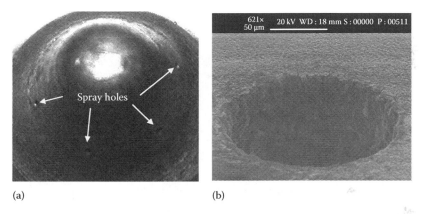

(a)   (b)

FIGURE 7.3   SEM micrographs of diesel engine injector spray holes: (a) overview and (b) close-up view.

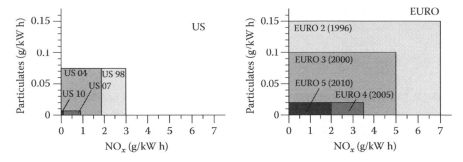

FIGURE 7.4   The emission standards of the United States and EU.

holes. The EDM process determines the dimension and geometry of spray holes. The dimension, geometry, flow rate, and surface roughness of the injector spray hole are critical to the spray pattern, fuel atomization, combustion, and emissions of diesel engines.

Smaller spray hole diameter (<100 μm) and higher injection pressure (>100 MPa) can generate more evenly distributed spray pattern and finer diesel fuel atomization [7,8], which can lead to lower nitrogen oxide (NO_x) and particulate emission, as required by the on-highway diesel emission standards (see Figure 7.4) set by the U.S. Environmental Protection Agency (EPA) and the European

Union. But the reduction of the spray hole size will result in a higher aspect ratio (hole depth vs. diameter), a major technical challenge for the micro-hole EDM. Negative tapered micro-holes that are truncated cone shaped with the diameter of the flow inlet larger than that of the flow outlet are reported to be able to minimize cavitation problems [9] that occur in the high-pressure diesel fuel injections. However, the EDM micro-hole drilling, using the configuration shown in Figure 7.2, typically generates positive tapered holes. These technical issues require an advanced study of the micro-hole EDM drilling technology.

7.1.2 MICRO-HOLE EDM DRILLING

One of the first experiments reported on micro-hole EDM drilling was performed by Van Ossenbruggen [10] in the Philips Research Laboratory in the late 1960s. High aspect ratio micro-hole EDM was studied by Masuzawa et al. [2,11], Takahata et al. [12], and Lim et al. [13]. Improvement of micro-hole quality could be obtained by lowering the level of the discharge energy [14]. The minimization of the stray capacitance between the electrode and workpiece for lower discharge energy was studied by Masuzawa [15]. The effects of polarity, electrode shape, and rotational speed of the electrode in the micro-hole EDM drilling of carbide were investigated by Yan et al. [16] and the experimental results showed that positive polarity must be used in micro-hole EDM drilling to reduce the tool wear and maintain hole accuracy.

Blind micro-hole EDM drilling is difficult because the debris concentrated at the bottom is hard to be completely flushed away and can lead to excessive electrode wear. A new approach employing planetary movement to produce self-flushing for blind micro-hole EDM was presented by Yu et al. [17]. Recent research on micro-hole EDM showed that very high aspect ratio (over 20) could be obtained [18].

7.1.3 PROCESS MONITORING OF EDM

Process monitoring provides the machining status that can be analyzed for the optimization of EDM process parameters. By monitoring the EDM process, the pulse types can be classified and information regarding the EDM machining condition can then be acquired.

Dauw et al. [19] developed an algorithm using the preset voltage and current threshold values to identify distinctive EDM pulses from the recorded pulse trains [20]. Pandit and Mueller [21] presented a method based on the data-dependent system to discriminate the pulse types. Ignition delay time of an EDM pulse was used as a discriminator for the pulse identification [22].

Artificial intelligence has been applied in the classification of EDM pulses. The application of fuzzy logic in EDM process monitoring was regarded as a reliable tool to discriminate EDM pulses [23]. Tarng et al. [24] developed a fuzzy pulse discriminating system for the EDM process. Kao and Tarng [25] and Liu and Tarng [26] conducted applications of neural networks and abductive networks, respectively, on the recognition of EDM pulses. To handle the transient non-stable signals generated from the EDM process, wavelet transform was applied to filter noise and extract waveform features of gap voltage and current [27].

7.1.4 MICRO-HOLE SHAPE MEASUREMENT

The geometry of diesel fuel injector spray holes is critical to the spray pattern and exhaust emissions. Micro-holes are difficult to characterize using optical microscopes and conventional coordinate measuring machines (CMM) due to the small diameter and high aspect ratio [1]. A common practice to study the spray hole shape uses either the destructive method [28] to cut a cross-section of the hole or the plastic molding method to make a replica of spray holes [29]. Both methods provide limited information of the shape and dimension of spray holes.

New technologies have been developed since the 1990s to enable the shape measurement of spray holes. Masuzawa et al. [1] developed a new technology called the vibroscanning method. This method utilized the vibration of a very thin conductive probe to detect and derive the distance between the probe and the hole surface. The measured hole profile was obtained by successive detections along the hole depth. Masuzawa et al. [30], Kim et al. [31,32], Yamamoto et al. [33], Pourciel et al. [34], Lebrasseur et al. [35], and Pourciel et al. [36] conducted advanced studies on the vibroscanning method.

A multi-sensor measurement technology, which combined the contact (tactile) and non-contact (optical) sensors into an integrated CMM, was developed by Werth GmbH [37]. The working principles, measurement procedures, as well as the gage repeatability and reproducibility (R&R) study of the Werth multi-sensor CMM will be presented in Section 7.3.

7.1.5 EDM PROCESS CONTROLLERS

The pulse generation during the EDM process is highly stochastic and complicated, which increases the difficulty of controller design and development. Several controllers have been investigated for the EDM process control for years and will be discussed in this section.

The proportional-integral-derivative (PID) controller is commonly used for EDM process control. The PID controller utilizes a predefined mathematical model to dynamically adjust the servo movement according to the sensor feedback. The difficulty of using a mathematical model to precisely describe the EDM process renders the PID controller less competitive in preventing the undesired arc and short circuit pulses [38,39].

Pulse width modulation (PWM) is another commonly used EDM process–control method. The PWM controller adjusts the servo position by varying the duty cycle of the input voltage pulses. The PWM controller, relying on the variation of pulse duty cycles, lacks the feedback of the EDM condition and operates with a severe non-linearity known as the deadband that degrades the controller's performance significantly [40].

Research on advanced control strategy for micro-hole EDM is needed to develop controllers that are more capable for the nonlinear, stochastic EDM process. Competent EDM process controllers should be able to distinguish the rapid changing EDM status to generate correct servo commands for the minimization of the undesired pulses and faster drilling cycle time. Fuzzy logic control has been extensively studied in EDM and is discussed in Section 7.4.

7.2 SUB-NANOSECOND MONITORING OF MICRO-HOLE EDM PULSES

7.2.1 IMPORTANCE OF MICRO-HOLE EDM MONITORING

The monitoring of EDM pulses using high-speed data acquisition is an essential part of the EDM process control and optimization [47]. The voltage and current across the gap between the electrode and workpiece greatly influence the material removal rate (MRR) in the micro-hole EDM. To enable efficient MRR in the micro-hole EDM, very short (less than $1\,\mu s$) pulse duration and high open-circuit voltage (above $200\,V$), as compared to conventional wire and die-sinking EDM, are typically utilized. The high repetition rate in the micro-hole EDM requires a high-speed data acquisition system to record the rapid change of gap voltage and gap current and to classify spark, arc, and short pulses, as shown in Figure 7.5.

At the end of discharging, due to the parasitic inductance and inherent capacitance of the EDM circuit, the voltage does not immediately reduce to the steady-state value. This has been recognized as the so-called ringing effect [41], as illustrated in Figure 7.5. For the micro-hole EDM, the ringing effect is significant. The voltage oscillates for a long period of time after discharging and slowly reduces to a steady-state value. Shortening the time duration of ringing may lead to more frequent EDM pulses and higher MRR.

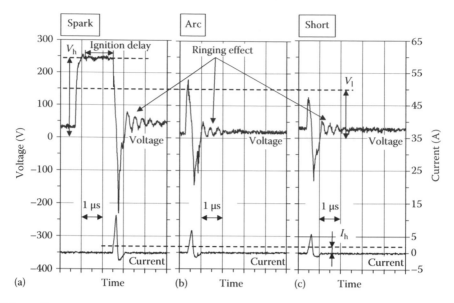

FIGURE 7.5 Characterization of EDM pulses: (a) spark, (b) arc, and (c) short. (From Kao, C.C. and Shih, A.J., *Int. J. Mach. Tools Manuf.*, 46, 1996, 2006.)

Another noteworthy phenomenon in the micro-hole EDM is the current flowing in the opposite direction near the end of the EDM pulse. The parasitic components induce a reverse current [29] that can lead to inferior surface roughness and excessive electrode wear [6,16]. The reverse current is also observed in both spark and arc pulses during this micro-hole EDM study.

7.2.2 EDM Process Monitoring

Five input or setup parameters are used in the micro-hole EDM process monitoring:

* *Polarity*: Polarity is the pole designation of the workpiece and electrode. Positive polarity sets the workpiece as the anode and the electrode as the cathode, and vice versa for negative polarity. The choice of polarity can greatly affect the wear of the wire electrodes. Electrode wear is expected to be high under negative polarity [6]. This property is used to blunt the tip of the electrode, a procedure known as electrode dressing. During drilling, to reduce the electrode wear, positive polarity should be used.
* *Drilling depth*: Drilling depth is the travel of the wire electrode from the initial contact with the workpiece. The drilling depth is controlled by the servomotor and affected by the electrode wear.
* *Open circuit voltage*, V_o: This is the system voltage when the EDM circuit is in the open state and the energy has been built up for discharging. An example of an open circuit voltage of 210 V is shown in Figure 7.6.
* *Pulse duration time*, T_D: As shown in Figure 7.6, T_D is the time duration of the positive discharging current in a single EDM pulse. The measured value of T_D is usually longer than the input value because of the time required for the current to rise and fall during discharging. The input value for T_D of the spark in Figure 7.6 was 0.1 μs. The actual value was about 0.3 μs.
* *Pulse off time*, T_{off}: This is the time from the end of one EDM pulse to the beginning of the next pulse, as shown in Figure 7.6. During the pulse off time, the pulse generator is in the off state and the current is zero.

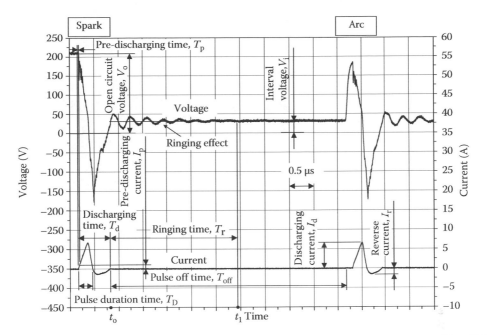

FIGURE 7.6   EDM process input and output parameters of sample spark and arc pulses. (From Kao, C.C. and Shih, A.J., *Int. J. Mach. Tools Manuf.*, 46, 1996, 2006.)

The EDM process used to drill a micro-hole suitable for a diesel engine injector is typically composed of three stages: electrode dressing, drilling, and penetration.

- *Stage 1, electrode dressing*: The tip of the tungsten wire electrode is sharpened during this micro-hole EDM process. Figure 7.7a shows an example of the sharpened electrode tip after completing the EDM drilling. The electrode dressing stage is implemented at the beginning of the contact between the electrode and the workpiece by using negative polarity to increase the electrode wear and blunt the tip. This has been proven to be important for maintaining the consistency of the micro-hole diameter in practical applications. Figure 7.7b shows the shape of an electrode tip after Stage 1. The electrode tip is blunted and covered by a thin recast layer.

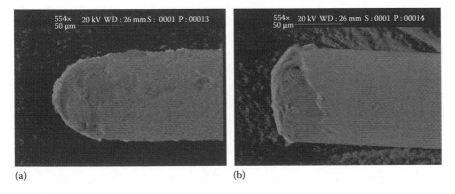

(a)   (b)

FIGURE 7.7   SEM micrographs of the tip of the tungsten wire electrode: (a) sharpened electrode tip after EDM Stage 3 and (b) dressed electrode tip after EDM Stage 1 (diameter of wire electrode: 125 μm). (From Kao, C.C. and Shih, A.J., *Int. J. Mach. Tools Manuf.*, 46, 1996, 2006.)

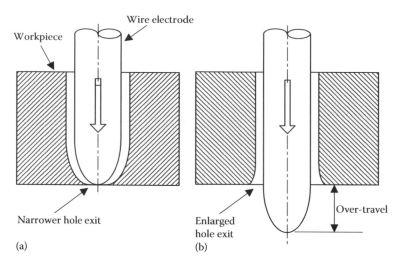

FIGURE 7.8 The effect of EDM Stage 3 penetration: (a) worn tip of wire electrode causes a narrower hole exit and (b) enlarged hole exit is generated due to the over-travel of the wire electrode. (From Kao, C.C. and Shih, A.J., *Int. J. Mach. Tools Manuf.*, 46, 1996, 2006.)

- *Stage 2, drilling*: Positive polarity was applied in Stage 2 to increase the drilling speed while maintaining reasonable debris size for efficient flushing.
- *Stage 3, penetration*: As shown in Figure 7.8a, at the time when the sharp electrode tip penetrates the other face of the workpiece, the electrode still needs to move forward by a set distance, called over-travel, to maintain a consistent diameter inside the hole. From the diesel engine emissions perspective, a larger hole diameter in the final penetration stage, as shown in Figure 7.8b, is preferred.

7.2.2.1 EDM Output Parameters

Seven major EDM output parameters can be measured from EDM pulse monitoring. All seven parameters are marked in Figure 7.6.

- *Discharging current, I_d*: This is the peak positive current during discharging.
- *Reverse current, I_r*: This is the peak negative current during discharging. The reverse current that has been reported by Hebbar [29] flows opposite to the discharging current due to the parasitic capacitance of the system [41,42].
- *Pre-discharging current, I_p*: Using the 2 GHz sampling rate, a unique phenomenon was discovered. The current starts to increase before the drop of voltage in discharging. The peak current before the voltage drop is defined as the pre-discharging current.
- *Pre-discharging time, T_p*: The time duration from the increase of current to the voltage drop in the beginning of discharging is defined as the pre-discharging time, T_p.
- *Discharging time, T_d*: This is the time duration that the discharging current, both positive and negative, exists.
- *Interval voltage, V_i*: This is the steady-state voltage after the ringing dissipates. The interval voltage is not constant for all pulses in an EDM pulse train. The value depends on the gap conditions, such as purity of dielectric fluid and concentration of debris.
- *Ringing time, T_r*: Analogous to the definition of settling time in the transient response analysis of dynamic systems [43], the ringing time is defined as the time duration from t_0, the instance when the negative current increases to zero, to t_1, when the voltage oscillation reaches and stays within the range of the 5% of the interval voltage V_i. Both t_0 and t_1 are marked in Figure 7.6.

7.2.3 Spark and Arc EDM Pulses

Using a 0.5 ns sampling interval, representative spark and arc in drilling (Stage 2) and electrode dressing (Stage 1) are presented in Figures 7.9 and 7.11, respectively. Six successive periods, denoted as Period I to VI, are identified in each sample spark and arc pulse.

- *Period I*: The voltage rises and stays at a specified open circuit voltage V_o while the current remains zero. In Period I, the wire electrode is gradually approaching the workpiece.
- *Period II*: The current rises and the voltage remains the same in this period. The pre-discharging current I_p occurs at the end of Period II. The time duration of Period II is the pre-discharging time T_p. In Period II, the parasitic capacitance [42] that is inherent in the EDM circuit begins to be charged.
- *Period III*: The voltage drops from the open circuit voltage V_o to zero in this period. In the beginning of Period III, a very rapid voltage drop occurs. The rate of voltage drop will be analyzed in Section 7.2.3.1. A high-frequency voltage oscillation due to the fast change of the gap impedance [44] occurs following the rapid voltage drop. The current typically rises to a peak value in Period III.
- *Period IV*: Period IV begins at the zero voltage and ends when the current reaches zero. The voltage becomes negative in this period.
- *Period V*: This is the period with the negative reverse current and its peak value is I_r.
- *Period VI*: This is the ringing period.

7.2.3.1 Spark and Arc Pulses at Drilling

A spark pulse from Stage 2, as is illustrated in Figure 7.9a, is more representative and, hence, discussed first. In Period I, the open circuit voltage V_o is 210 V and the current is zero. The pre-discharging time T_p is 30 ns and the pre-discharging current I_p is 1.1 A in Period II. A close-up view of the voltage drop, as marked by box A in Figure 7.9a, is shown in detail in Figure 7.10a. The rate of voltage drop is −37.6 V/ns. In Period III, the discharging current I_d is 6.3 A and the time duration is 230 ns. In Period V, the reverse current I_r reaches −1.2 A. The discharging time T_d that is the time duration from Period II to Period V is 652 ns. The interval voltage V_i is estimated to be 28 V in Period VI.

An arc pulse from Stage 2 is shown in Figure 7.9b. In Period I, the voltage gradually rises but does not stay at the V_o before the discharging. In Period II, the duration T_p is 23 ns and I_p is 1.2 A. In Period III, the rapid voltage drop begins at 172 V, which is 38 V lower than the V_o. The close-up view of the voltage drop, as marked by box B in Figure 7.9b, is shown in Figure 7.10b. The rate of voltage drop is −14.7 V/ns, which is lower than that in the spark pulse. Hebbar [29] has proposed to use the rate of voltage drop at discharge to differentiate between the spark and arc pulses. In Period III, the amplitude of the high-frequency voltage oscillation can also be seen. The current I_d is 6.5 A in Period III. The duration of Period IV is 69 ns. In Period V, the I_r is −1.5 A and the duration is 344 ns. The T_d is 690 ns, about 6% longer than that of the spark pulse.

The interval voltage V_i is estimated as 25 V in Period VI. In general, the duration of pre-discharging, the voltage at discharge, and the rate of voltage drop at the beginning of discharging are all lower in an arc than those in a spark.

7.2.3.2 Spark and Arc Pulses at Electrode Dressing

The negative polarity setup changes the shape of the waveform for both spark and arc pulses in Stage 1. A spark pulse is shown in Figure 7.11a. To make the shape consistent for the comparison, the measured signal was multiplied by −1. The voltage V_o is low, 170 V. The output parameters in Periods II and III are about the same as the spark in Stages 1 and 3: $T_p = 30$ ns, $I_p = 1.1$ A, $I_d = 8.3$ A, and rate of voltage drop is −20.5 V/ns.

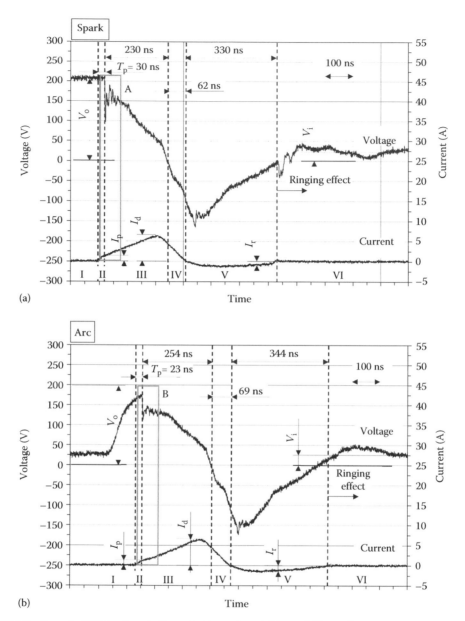

FIGURE 7.9 Sample EDM pulses in Stage 2: (a) spark and (b) arc (125 μm diameter wire electrode). (From Kao, C.C. and Shih, A.J., *Int. J. Mach. Tools Manuf.*, 46, 1996, 2006.)

The most significant impact of negative polarity occurs after the voltage drops below zero. Instead of continuing to reduce to −150 V in Stage 2 (Figure 7.9), the voltage reduces to only −70 V, as shown in Figure 7.11a, and starts to increase. The gradually increasing voltage results in a lower rate of current drop and elongates the time duration of Period IV, from less than 90 ns under positive polarity to 240 ns using the negative polarity. The effect of a polarity change on electrode wear has been studied [6]. The small debris generated is likely to be negatively charged and accelerated to impact and neutralize the positively charged tungsten electrode. This reduces the level of voltage below zero. The reverse current I_r is −1.7 A in Period V. The discharging time T_d is long, 1372 ns, more than twice of that in Stage 2. In Period VI, the interval voltage V_i is very low, only 8 V.

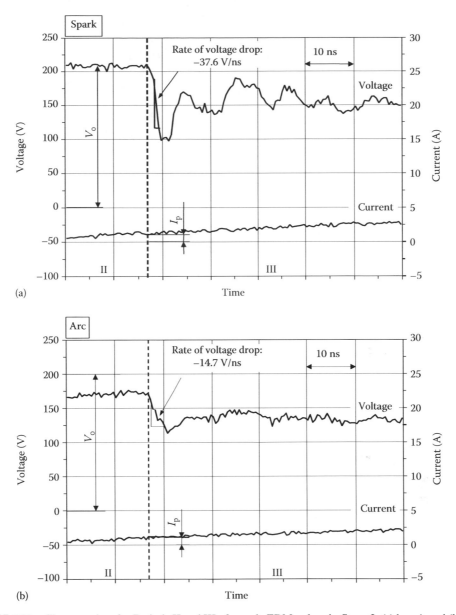

FIGURE 7.10   Close-up view for Periods II and III of sample EDM pulses in Stage 2: (a) box A and (b) box B. (From Kao, C.C. and Shih, A.J., *Int. J. Mach. Tools Manuf.*, 46, 1996, 2006.)

An arc pulse in Stage 1 is shown in Figure 7.11b. Two peaks of current are recognized. The voltage rises to about 70 V in the end of Period II. Both I_p (0.2 A) and T_p (16 ns) are very small. Although the initial voltage drop in Period III was slow, only −0.6 V/ns, the voltage drops quickly to zero. Period III is also short, only 125 ns. The peak current does not happen in Period III as in other arc and spark pulses in Stage 2. In Period IV, the current reaches its first peak (2.0 A), marked as I_{d1}, as the voltage continues to drop. After the first peak in current, the voltage rises to 50 V in 120 ns and starts another discharge. The magnitude of the second discharging current, denoted I_{d2}, is also about 2.0 A. The period of reverse current, Period V, still exists and I_r (−0.8 A) is very small. The time duration for an arc pulse with two discharges (T_d), is 1383 ns, about the same as that of the spark pulse in Stage I. In Period VI, the V_i is low, about 10 V.

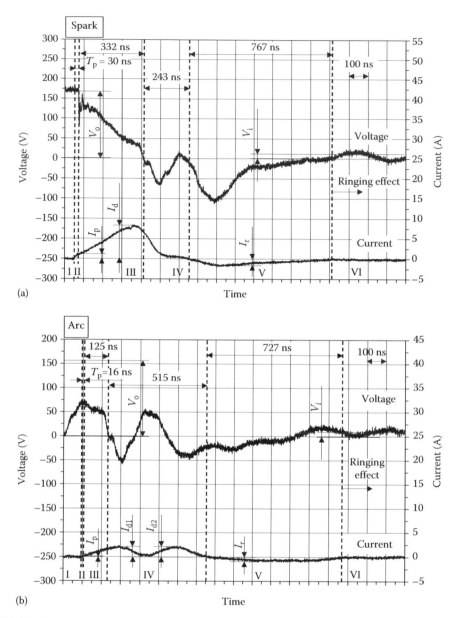

FIGURE 7.11 Sample EDM pulses in Stage 1: (a) spark and (b) arc (125 mm diameter wire electrode). (From Kao, C.C. and Shih, A.J., *Int. J. Mach. Tools Manuf.*, 46, 1996, 2006.)

In conclusion, this section demonstrates the capability of sub-ns monitoring of spray hole EDM and extracts distinguishing features of the spark, arc, and short EDM pulses.

7.3 MICRO-HOLE SHAPE MEASUREMENT

7.3.1 BACKGROUND

Accurate shape measurement of micro-holes is difficult due to the small diameter and high aspect ratio. For diesel engine industry, because of the stringent emission regulations [45,46], the spray hole geometry measurement has become an important quality control tool that synergizes efforts from product design engineers and manufacturing specialists to achieve this goal.

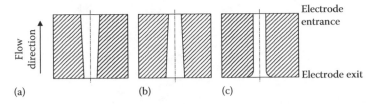

FIGURE 7.12 Classification of the shape of a micro-hole for the fuel injector: (a) positive taper, (b) negative taper, and (c) rounded flow entry. (From Kao, C.C. and Shih, A.J., *Meas. Sci. Technol.*, 18, 3603, 2007.)

Giedl et al. [3] has classified the shape of the micro-hole: negative taper, positive taper, and rounded flow entry (flow inlet), as shown in Figure 7.12. In EDM drilling, if the process parameters remain the same throughout the drilling, the hole will have positive taper due to the wear of the wire electrode. Advanced micro-hole EDM adjusts the process parameters in different stages to control the hole shape [47].

A flat-end electrode (see Figure 7.7b) can reduce the positive taper of the drilled hole. Near the end of the micro-hole EDM drilling, the gap voltage is increased to enlarge the gap width inside the micro-hole to enable the generation of negative taper, as shown in Figure 7.12b. The rounded flow entry, as shown in Figure 7.12c, can be achieved using the abrasive flow machining (AFM), which flows abrasive media through the micro-hole under high pressure to round the inlet edge of the micro-hole [48–50]. Both negative tapered hole and rounded flow entry have proven to improve the diesel fuel flow and atomization characteristics and can reduce engine exhaust emissions [3,28]. The form measurement using Werth CMM provides a quantitative method to specify and evaluate the effect of the AFM on micro-hole geometry [51].

7.3.2 Measurement

7.3.2.1 Machine Setup

Measurements of injector micro-holes were conducted with a Werth VideoCheck HA 400 CMM. Figure 7.13 shows the schematic diagram of the CMM measurement setup using a glass-fiber

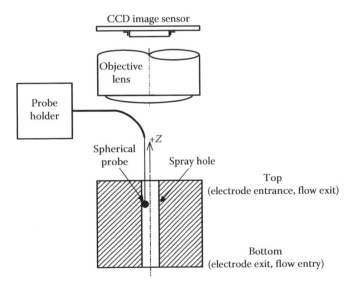

FIGURE 7.13 Micro-hole measurement using the combination of optical and contact sensors in a CMM. (From Kao, C.C. and Shih, A.J., *Meas. Sci. Technol.*, 18, 3603, 2007.)

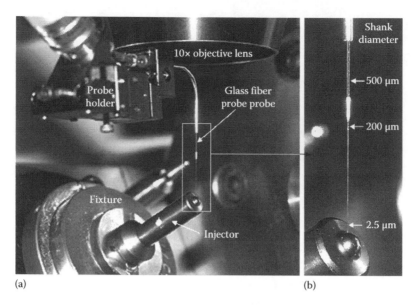

(a)   (b)

FIGURE 7.14   Setup of micro-hole measurement: (a) overview of the nozzle, probe, and optical lens and (b) close-up view for the glass fiber probe. (From Kao, C.C. and Shih, A.J., *Meas. Sci. Technol.*, 18, 3603, 2007.)

probe with a ball end tip and a charge-coupling device (CCD) optical sensor to measure the ball position.

An overview of the experimental setup is shown in Figure 7.14a. The probe is made of glass and its tip has a spherical ball of 74 μm diameter. A close-up view of the nozzle tip and the glass probe is shown in Figure 7.14b. Three light sources are available from the inside, top, and bottom of the micro-hole. The light can be transmitted through the glass to illuminate the inside of the micro-hole. On the top is the light around the objective lens. Inside the nozzle, a 1 mm diameter metal tube with a reflective mirror at the tip delivers the light to the bottom of the hole.

The CMM has three axes. The injector with the micro-holes is carried by the *X*- and *Y*-axes. The probe is moved up and down by the *Z*-axis. All axes have 0.1 μm resolution and are moved by either the manual or the computer-control mode. A specially designed fixture can orient the spray hole by three mutually perpendicular and computer controlled axes.

7.3.2.2   Determining the Coordinates of Contact Points

The contact point between the ball tip of the probe and the hole surface cannot be detected by the CCD image sensor. Figure 7.15a shows the image captured by the CCD when the probe is positioned near the center of the hole. When the probe is moved to contact the hole surface, as shown in Figure 7.15b, only a portion of the ball near the hole's center is visible. The image processing software is utilized to define a rectangular box, as shown in Figure 7.15b, to enclose the arc region of the ball to find its radius and center position.

Mathematically, as shown in Figure 7.16, point P_1 can be found using the CMM. The probe contacts with the hole wall surface at 12 locations, 30° apart from each other, under the same Z position. The centers of the probe tip in the next two contact locations are marked as P_2 and P_3. Sequentially, the probe center from P_4 to P_{12} can be measured. Using the information of P_1 to P_{12}, the center of the circle O can be found. The vector from the center O to the contact point C_1 is derived using the following equation:

$$\overrightarrow{OC_1} = \left(r + \left| \overrightarrow{OP_1} \right| \right) \frac{\overrightarrow{OP_1}}{\left| \overrightarrow{OP_1} \right|} \tag{7.1}$$

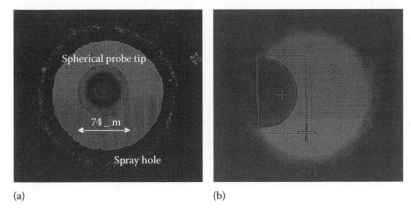

(a)   (b)

FIGURE 7.15 The measurement of a micro-hole by the spherical probe tip: (a) the spherical probe tip inside a micro-hole and (b) the spherical probe tip in contact with the micro-hole and the user-defined rectangle for arc identification. (From Kao, C.C. and Shih, A.J., *Meas. Sci. Technol.*, 18, 3603, 2007.)

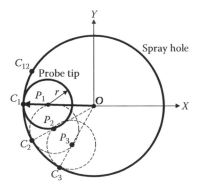

FIGURE 7.16 Determination of the contact point between the micro-hole surface and the probe tip. (From Kao, C.C. and Shih, A.J., *Meas. Sci. Technol.*, 18, 3603, 2007.)

where r is the radius of the probe tip. The contact points C_2–C_{12} can be found following the same procedure. The contact points on circles at different Z positions are measured to create an array of contact points to represent the micro-hole.

7.3.3 MICRO-HOLE MEASUREMENT PROCEDURE

As shown in Figure 7.17, the micro-hole was measured using 12 contact points in 12 sections in the Z-axis direction. The positions of these 12 contact points are illustrated in Figure 7.17a. As shown in Figure 7.17b, the spacing in the Z direction of the adjacent sections is about 30 μm in Stage 1 (electrode dressing). The spacing between adjacent sections increases to 130 μm in Stage 2 (drilling) and about 50 μm in Stage 3 (penetration). For each hole, 144 points were measured. These measured points were analyzed to find the form characteristics of the micro-holes. For gage R&R study, the experiment and data analysis were implemented based on standard procedures developed for measurement system analysis [52]. Numbers of sample, repeat, and operators are 3, 4, and 2, respectively.

7.3.4 CHARACTERISTICS OF MICRO-HOLES

Five features—cylindricity, diameter, roundness, taper, and straightness—are used to characterize the micro-holes.

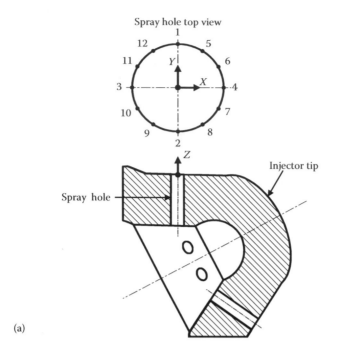

(a)

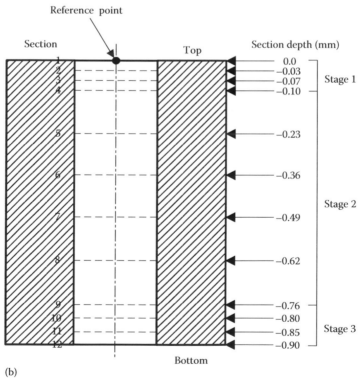

(b)

FIGURE 7.17   Shape measurement of micro-holes: (a) orientation of coordinate system and locations of selected measurement points and (b) depths of different sections according to the three EDM stages. (From Kao, C.C. and Shih, A.J., *Meas. Sci. Technol.*, 18, 3603, 2007.)

- *Cylindricity*: The cylindricity is defined as a tolerance zone bounded by two coaxial cylinders between which the measured cylinder must lie. The least-square (LS) fitting method [53] is used to calculate the cylindricity.
- *Diameter*: The diameter of the LS-fitted cylinder is used to represent the hole size.
- *Roundness*: The 12 points in a section are analyzed to calculate the roundness. The roundness vs. depth of circle in the micro-hole is investigated.
- *Straightness*: The measured points 1, 2, 3, and 4, as defined in Figure 7.17a, of the 12 sections are used to calculate the straightness. The LS fitted line of measurement points in 12 sections is used to represent the straightness at points 1, 2, 3, and 4 of the micro-hole.
- *Taper*: Two pairs of lines 1–2 and 3–4 determine two taper values of the micro-hole. Taper is the difference in distances between these two lines at the entrance and the exit ends of the micro-hole, that is, difference in size of the hole. Positive taper, as shown in Figure 7.12a, is defined to have the positive value of the taper.

As shown in Figure 7.17b, the spacing in the Z direction of the adjacent sections is about 30 μm in Stage 1 region, machined using negative polarity for the electrode dressing. The spacing between adjacent sections increases to 130 μm in Stage 2 and about 50 μm in Stage 3. For each hole, 144 points were measured. These measured points were analyzed to find the form characteristics of micro-holes.

7.3.5 Gage R&R

Definitions of standard deviation for repeatability, reproducibility, and gage R&R are based on the *Measurement Systems Analysis Reference Manual* [52].

The standard deviation for gage R&R, $\sigma_{R\&R}$, is

$$\sigma_{R\&R} = \sqrt{\sigma_{repeatability}^2 + \sigma_{reproducibility}^2} \tag{7.2}$$

The value of $5.15\sigma_{R\&R}$ is commonly used to represent 99% population of measurements for a normal distribution. The ratio $\sigma_{R\&R}/\sigma_{part}$ is used to assess the measurement system capability, where σ_{part} is the standard deviation for the part. For a capable measurement system, $\sigma_{R\&R}/\sigma_{part}$ needs to be smaller than 10% [52]. In practical applications, 30% of $\sigma_{R\&R}/\sigma_{part}$ could be acceptable. The part standard deviation under the 10% and 30% criteria can be calculated from the $\sigma_{R\&R}$.

The value of $5.15\sigma_{part}$ is defined as the capability tolerance of a measurement system. The capability tolerance is used to assess the capability of the multi-sensor CMM for micro-hole measurements. Values for capability tolerance are valuable for product engineers to specify the micro-hole tolerances that are measurable and for metrology engineers to quantify the measurement capability and further improve the capability of the gages.

7.3.6 Measurement Results

Table 7.1 summarizes the tilted angles and five measured characteristics of three spray holes. Diameter and roundness of 12 measured sections of three spray holes are shown in Figures 7.18 and 7.19, respectively.

7.3.6.1 Gage R&R for Micro-Hole Measurements

The $\sigma_{repeatability}$, $\sigma_{reproducibility}$, and $\sigma_{R\&R}$ of cylindricity, diameter, roundness, straightness, and taper measurements are summarized in Table 7.2.

The capability tolerance (CT) of the CMM for $\sigma_{R\&R}/\sigma_{part} < 30\%$ and 10% for each of the hole characteristic are summarized in Table 7.2. CT values specify the part tolerance that can be capably

TABLE 7.1
Measurements of Three Micro-Holes before and after Abrasive Flow Machining

		Before AFM			After AFM
Hole Number		1	2	3	1
Tilted angle (°)		0.43	0.45	0.59	0.52
Cylindricity (μm)		16.1	14.7	17.1	29.3
Diameter (μm)		160	157	158	178
Roundness (μm)	Stage 1	2.91	2.42	2.55	1.97
	Stage 2	4.14	4.98	4.50	4.02
	Stage 3	6.69	4.65	6.95	9.01
Straightness (μm)	Point 1	7.29	7.94	10.52	17.13
	Point 2	10.04	9.48	8.40	22.45
	Point 3	12.84	10.53	13.78	20.07
	Point 4	9.19	10.49	9.36	18.22
Taper (μm)	T_{12}	4.86	7.92	6.99	−21.9
	T_{34}	3.24	4.52	3.76	−25.9

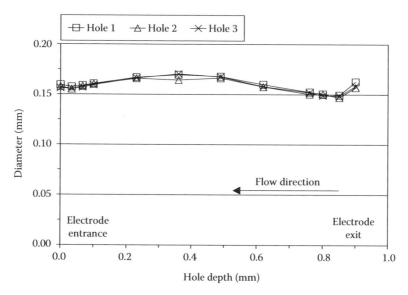

FIGURE 7.18 Diameter of 12 measurement sections in three micro-holes. (From Kao, C.C. and Shih, A.J., *Meas. Sci. Technol.*, 18, 3603, 2007.)

measured by the CMM. For cylindricity, the measurement capability is not very good: capable of only 70 μm and marginally capable of 23 μm. For diameter, the measurement is capable of 15 μm and marginally capable of 5 μm. Results in Table 7.2 provide a guideline for designers to specify tolerances for the diesel fuel injector spray holes and for metrology engineers to further improve the measurement capability.

7.3.6.2 Effects of AFM on Micro-Hole Geometry

AFM is a non-traditional machining process applied to improve the surface condition and round the flow entry (electrode exit) of the micro-hole [49]. The abrasive media, which is a mixture of abrasive

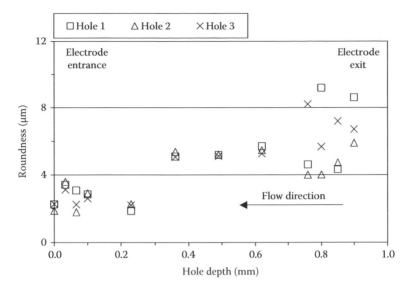

FIGURE 7.19 Roundness of 12 measurement sections in three micro-holes. (From Kao, C.C. and Shih, A.J., *Meas. Sci. Technol.*, 18, 3603, 2007.)

TABLE 7.2
Gage R&R and Capability Tolerance for CMM Micro-Hole Measurement

Dimension Type		$\sigma_{repeatability}$ (µm)	$\sigma_{reproducibility}$ (µm)	$\sigma_{R\&R}$ (µm)	Capability Tolerance (µm)	
					$\sigma_{R\&R}/\sigma_{part} < 0.3$	$\sigma_{R\&R}/\sigma_{part} < 0.1$
Cylindricity		1.36	0	1.36	23.4	70.2
Diameter		0.30	0	0.30	5.15	15.5
Roundness	Stage 1	0.60	0.12	0.63	10.8	32.4
	Stage 2	0.90	0.08	0.91	15.6	46.9
	Stage 3	1.44	0.04	1.45	24.9	74.7
Straightness		0.36	0.14	0.39	6.70	20.1
Taper		1.51	0.54	1.60	27.5	82.5

grit and a semi-solid carrier [50], flows through the micro-hole at high pressure to round the inlet or flow entry edge and improve the surface roughness. In this study, SiC abrasives with nominal size of 0.015 mm were used in an ExtrudeHone AFM machine. The flow pressure and the process cycle time were set at 27.6 MPa and 25 s, respectively.

The 3D wireframe representations of the measured points for Hole 1 before and after the AFM process are shown in Figure 7.20a and b, respectively. The fuel inlet (electrode exit) edge of the hole is enlarged and rounded after the AFM. The rounding is uneven around the circumference of the hole inside the edge. More material is removed at Point 3 that is away from the injector tip. The flow pattern of high-viscosity media in the AFM affects the material removal rate around the circumference of the hole edge. Such effect can be quantified using the micro-hole measurement technique introduced in this section.

The cylindricity, diameter, roundness, straightness, and taper of Hole 1 machined after the AFM are listed in the last column of Table 7.2. Due to the rounding of the hole inside the edge, the cylindricity, diameter, and straightness are all increased. The roundness is improved in Stages 1 and 2

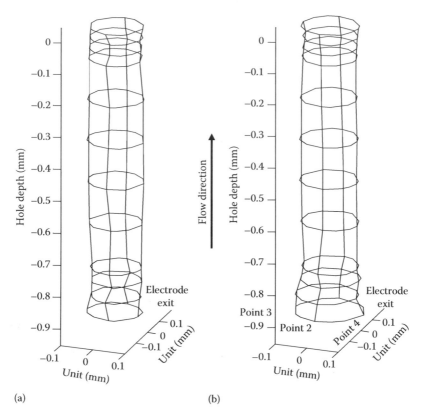

FIGURE 7.20 3D wireframe representation of Hole 1: (a) before AFM and (b) after AFM. (From Kao, C.C. and Shih, A.J., *Meas. Sci. Technol.*, 18, 3603, 2007.)

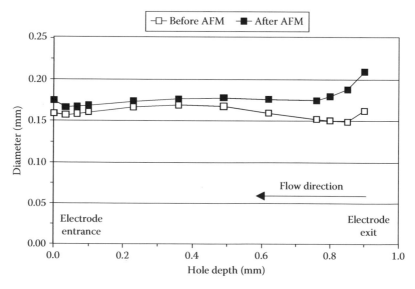

FIGURE 7.21 Effects of AFM on micro-hole diameter. (From Kao, C.C. and Shih, A.J., *Meas. Sci. Technol.*, 18, 3603, 2007.)

regions but worsened in Stage 3 region (near the hole inside the edge). The taper becomes negative, which demonstrates the effect of the AFM to alter the hole geometry.

The diameter of Hole 1 before and after the AFM for 12 sections in the Z-axis is shown in Figure 7.21. The diameter after the AFM is increased by about 10–25 μm in Stages 1 and 2 regions. In the Stage 3 region, the hole diameter is increased by about 35–50 μm.

7.4 FUZZY LOGIC CONTROLLER FOR MICRO-HOLE EDM

7.4.1 Review of the EDM Process Controller

The improvement of drilling cycle time is one of the major technical challenges for micro-hole EDM drilling. The key to faster drilling speed is dependent on the generation of normal discharges and the minimization of abnormal discharges. Spark pulse is the normal discharge for efficient material removal. Arc and short circuit pulses are abnormal discharges that increase the drilling time and are harmful to the machined surface [47].

The PID and the PWM have been used in the EDM process control for years. Due to their intrinsic limitations [38–40], the PID and the PWM may not be able to deliver the precision micro-hole EDM drilling process as required by the diesel engine industry. Artificial intelligence, such as neural networks and fuzzy logic, has been successfully applied to further advance the EDM process control [23,54]. Fuzzy logic control is more popular in EDM research because it responds faster and is highly stable [55] and can more accurately model real world events by allowing the existence of uncertainty to simulate human reasoning [56]. The fuzzy logic controller [57–62] has demonstrated the capability to adaptively control the EDM process by handling highly nonlinear processes with only qualitative knowledge [56,63] and to reduce the drilling time by minimizing the arc and short circuit pulses [39]. Gain scheduling that uses local linear controllers to collectively perform a global nonlinear process control by scheduling gains at different operating conditions [64] is also a common EDM process control method.

In addition to handling the nonlinearity, an ideal EDM process controller should be able to receive and analyze multiple input parameters from the feedback loop to synthesize more accurate and objective servo commands. The fuzzy logic control, well known for its capability of handling highly nonlinear processes with only qualitative knowledge available [56,63], is able to adaptively control the EDM process using multiple input parameters and prevent the undesired arc and short circuit pulses by actively responding to the gap conditions [39].

Fuzzy logic controllers with single and multiple input parameters have been investigated for die sinking and wire EDM. For example, the frequency of arc and short circuit pulses has been utilized as the single-input parameter [61]. Two-input fuzzy logic controllers have been investigated that use error of gap voltage and its change rate [58], servo position error and its change rate [60], pulse time ratio error and its change rate [62], or the ignition delay and percentage of abnormal discharges [57]. Three-input fuzzy logic controller for wire EDM has been developed using the spark frequency error and abnormal spark ratio error and its change rate [59]. The lack of information and limitations of the single-input fuzzy logic controller have been studied [65–67]. Multi-input fuzzy logic controller for micro-hole EDM drilling [65–67] will be briefly discussed in the following sections.

7.4.2 Fuzzy Logic Controller

Lotfi A. Zadeh, a professor of computer science at the University of California, Berkeley, proposed the theory of fuzzy logic in 1973. Since then, the concept of fuzzy logic has prevailed in many research fields, not confined to computer science anymore. Fuzzy logic control allows the existence of uncertainty in handling parameter values [56]. This is achieved by using linguistic variables that are associated with different levels of linguistic values to map a specific numerical value with

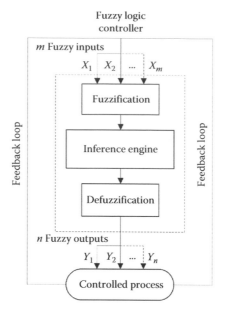

FIGURE 7.22   Schematic diagram of a typical fuzzy logic controller.

uncertainty, the so-called fuzziness. The schematic diagram of a typical fuzzy logic controller is shown in Figure 7.22. Three major parts of a fuzzy logic controller are fuzzifier, inference engine, and defuzzifier. The fuzzifier determines the fuzziness of each input parameter via the membership functions that map an input parameter from the universe of discourse (the input domain) to a fuzzy value between 0 and 1. Then, the inference engine, based on the input fuzzy values, triggers the If–Then inference rules and synthesizes output fuzzy values. The defuzzifier maps the output fuzzy values back to the output domain to generate the output parameters.

A fuzzy logic controller with three input parameters, the average gap voltage, V_g, deviation of spark ratio, ΔR_s, and change of deviation of spark ratio, $\delta(\Delta R_s)$, is presented in this section. ΔR_s and $\delta(\Delta R_s)$ are used to evaluate the current status and future trend of abnormal pulse occurrence. Mamdani's inference method [56], a commonly used fuzzy logic methodology, is utilized as the kernel of the fuzzy logic controller.

The fuzzy logic controller generates output parameters for servo motion commands that are synthesized through the inference engine based on the feedback information. The servomotor receives commands and feeds the electrode accordingly.

Three linguistic values, as shown in Figure 7.23, are associated with ΔR_s and $\delta(\Delta R_s)$: Positive (PO), Zero (ZE), and Negative (NE). Servo motion can be specified by the servo speed v and servo displacement d. There are five linguistic values associated with v: Forward Fast (FF), Forward (FO), Dwell (DW), Backward (BA), and Backward Fast (BF). The linguistic value, Dwell, is used to

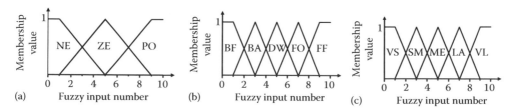

FIGURE 7.23   Membership functions for fuzzy logic parameters: (a) ΔR_s and $\delta(\Delta R_s)$, (b) v, and (c) d. (From Kao, C.C. and Shih, A.J., *ASME J. Manuf. Sci. Eng.*, 130, 064502-1, 2008. With permission.)

describe the transition between Forward and Backward. For d, there are also five linguistic values: Very Large (VL), Large (LA), Medium (ME), Small (SM), and Very Small (VS). Triangular membership functions are selected for the input and output parameters of the fuzzy logic controller, as shown in Figure 7.23.

7.4.3 Examples of Fuzzy Logic Control in EDM Drilling

7.4.3.1 Deep Micro-Hole EDM Drilling

The servo position and spark ratio of the micro-hole EDM drilling using the gain scheduling and fuzzy logic controllers are shown in Figure 7.24. The micro-hole depth is 2.28 mm. The gain-scheduling controller cannot complete the deep micro-hole drilling. After 131 s, the electrode has moved only 1.6 mm. The fuzzy logic controller can successfully complete the deep micro-hole drilling in 105 s. The failure of gain scheduling controller and the quite long drilling time under

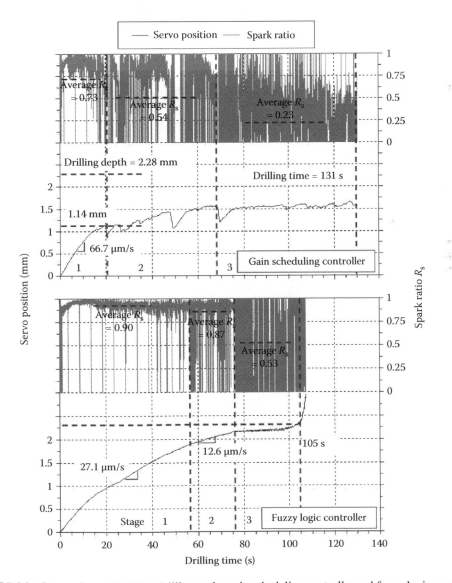

FIGURE 7.24 Deep micro-hole EDM drilling under gain scheduling controller and fuzzy logic controller.

fuzzy logic controller is primarily due to the difficulty of debris flushing, a challenge for deep micro-hole EDM drilling. This experiment shows that fuzzy logic controller is more adaptive to different micro-hole EDM drilling configurations.

7.4.3.2 Small Diameter Micro-Hole EDM Drilling

The servo position and spark ratio of small diameter micro-hole EDM drilling are shown in Figure 7.25. The diameter of the electrode is 75 μm, which significantly increases the drilling cycle time (compared with the standard 25–30 s cycle time for 150 μm electrode) because of the small size. Using the gain-scheduling controller, the drilling time is 165 s. For fuzzy logic controllers, the drilling time is 57 s, much shorter than that of the gain-scheduling controller. For both controllers, the drilling speed is reduced as the electrode advances deeper, due to the difficulty of flushing.

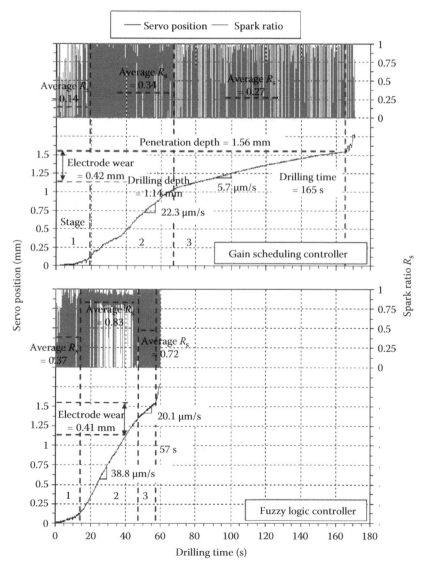

FIGURE 7.25 Small diameter micro-hole EDM drilling using 75 mm diameter electrode under gain scheduling controller and fuzzy logic controller.

7.5   CONCLUSIONS

The micro-hole EDM process monitoring and control were presented. Sub-ns EDM process monitoring technology demonstrated the capability of providing comprehensive views of voltage–current interactions. It also showed a future research on the combination of plasma physics and sub-ns EDM process monitoring to study the nanoscale discharging mechanisms. The fuzzy logic controller can be a foundation for the development of intelligent control systems as well as real-time expert systems for micro-hole EDM. Based on the current configuration, the fuzzy logic controller can be further expanded into an integrated system by adding online optimization modules to enable the self-learning ability, or by building a database to store various types of micro-hole EDM drilling process information for offline user queries.

The needs for clean and energy-efficient diesel engines continue to drive the innovative spray hole EDM and measurement technology. To effectively generate the reverse taper [51] for diesel fuel injector spray holes, special EDM mechanisms have been independently developed and commercialized, for example, the adjustable rotating wire guide by Posalux S. A. in Switzerland, and the dual-axis hole-shaping EDM head by Ann Arbor Machine Company in Michigan. Computed tomography using x-ray scanning has also been utilized to measure the spray hole geometry. The research and development in micro-hole EDM for diesel injector spray hole have been the enabling technology for diesel engine industry in the past four decades. Before other machining methods, such as pulsed laser micro-hole drilling, can prove to be effective in high-volume production with consistent quality, EDM will continue to be the dominating manufacturing process for micro-hole drilling of diesel fuel injector spray holes.

QUESTIONS

Q.7.1   Why small and tapered spray holes are advantageous to the high-pressure diesel fuel injection? There are many limitations, however, that constrain the smallest hole diameter and largest taper available on the current EDM drilling process. What are these limitations and how can they be technically resolved?

Q.7.2   According to Section 7.2.2, the negative polarity has some effects on the electrode tip shape. Is there any other application of this phenomenon in micro-EDM that the negative polarity can be possibly generated?

Q.7.3   Deep micro-hole EDM drilling and small diameter micro-hole EDM drilling are both very technically challenging due to the difficulties of debris flushing. The remaining debris floating in the discharge gap induces arcing as well as short circuits. Based on Figures 7.24 and 7.25, predict how the waveforms of gap voltage and current may look like in these two types of EDM drilling processes? Think about 20 or 30 consecutive EDM pulses in a time line and the possible distribution of spark, arc, and short pulses based on the spark ratios shown in Figures 7.24 and 7.25.

Q.7.4   Multi-input fuzzy logic EDM controller has shown its capability to reduce the cycle time and minimize the abnormal discharges, that is, arcing and short circuits. The input parameters used in this study are the average gap voltage and variation of spark ratio with time. What are the other EDM parameters that can be used for the fuzzy logic controller?

Q.7.5   Micro-EDM has a broad range of applications in industry. In addition to the diesel fuel injector spray holes, what are the other applications of micro EDM?

Q.7.6   There are many other non-traditional micro-hole drilling methods proposed, such as electrical chemical machining (ECM) [4], hybrid EDM/ECM method [68], combined method using EDM and ultra-sonic vibration [69,70], laser drilling [3], and magnetic field assisted EDM [71]. All of them show the potential for advanced diesel fuel spray hole drilling technology. Think about the advantages and weaknesses of each method and try to create a matrix for comparison.

ABOUT THE CONTRIBUTING AUTHORS

Dr. Chen-Chun Kao received his PhD degree in mechanical engineering from the University of Michigan in Ann Arbor in 2007. His research interests include non-traditional machining processes, fuzzy logic process control, and applications of x-ray computed tomography (CT) on micro-scale measurements. He is also a well-known expert in micro-hole electrical discharge machining (EDM) and has publications in referred journals and conferences. He is holding a pending U.S. patent on the development of advanced EDM drilling technology for diesel fuel injector spray holes. Dr. Kao is currently working as a senior engineer at Cummins Fuel Systems in Columbus, Indiana.

Dr. Albert Shih, professor, mechanical engineering, biomedical engineering, co-director of the SM Wu Manufacturing Research Center and associate director of the Medical Innovation Center, University of Michigan at Ann Arbor. After receiving his PhD from Purdue University, West Lafayette, Indiana in 1991, Dr. Shih worked at Cummins Inc. at Columbus, Indiana as a manufacturing engineer to advance the advanced engineering materials for a wide variety of diesel engines and fuel systems applications. From 1998 to 2002, he was associate professor in the Department of Mechanical and Aerospace Engineering, North Carolina State University at Raleigh. He joined the University of Michigan in 2003.

Dr. Shih's research and teaching interests are in manufacturing. He has conducted research in precision machining of advanced materials, precision machine design, precision optical measurements, electrical discharge machining (EDM), elastomer machining, infrared temperature measurements, manufacturing of permanent magnets, semiconductor ceramic machining, thermal management of power electronics, noncontact optical metrology, and friction stir joining. Currently, Professor Shih's research and teaching focus are in biomedical design and manufacturing—the application of advanced design and manufacturing technology to advance medical device, healthcare operations, and patient safety. He works closely with collaborators in the Medical School and is a founding member of the Medical Innovation Center at the University of Michigan. Professor Shih is a fellow of ASME and SME. He is the recipient of the 1999 ASME BOSS Award, 2000 NSF CAREER Award, 2004 SAE Ralph Teetor Education Award, and 2009 Fulbright Scholar. He served as the associate editor of the *ASME Journal of Manufacturing Science and Engineering* and was a guest editor of the Special Issue on Biomedical Manufacturing. He has 5 U.S. patents and over 180 publications, including over 100 journal papers.

REFERENCES

1. Masuzawa, T., Hamasaki, Y., and Fujino, M., Vibroscanning method for nondestructive measurement of small holes, *Annals of the CIRP*, 42, 589, 1993.
2. Masuzawa, T., Tsukamoto, J., and Fujino, M., Drilling of deep microholes by EDM, *Annals of the CIRP*, 38, 195, 1989.
3. Giedl, R. et al., Geometrical aspects of laser-drilled high precision holes for flow control applications, in *Proceedings of SPIE 5063, Fourth International Symposium on Laser Precision Microfabrication*, Bellingham, WA, 2003, p. 389.

4. Ahmed, M.S. and Duffield, A., Deep hole drilling using ECM, *SME Technical Paper Series*, MS89-816, 1989.
5. Iwata, K. and Moriwaki, T., Basic study of high speed micro deep drilling, *Annals of the CIRP*, 30, 27, 1981.
6. Her, M.G. and Weng, F.T., Micro-hole machining of copper using the electro-discharge machining process with a tungsten carbide electrode compared with a copper electrode, *International Journal of Advanced Manufacturing Technology*, 17, 715, 2001.
7. Nagasaka, K. et al., Development of fine atomization injector, *JSAE Review*, 21, 319, 2000.
8. Postrioti, L. and Ubertini, S., An integrated experimental-numerical study of HSDI diesel injection system and spray dynamics, *SAE Technical Paper Series*, 2006-01-1389, 2006.
9. Blessing, M. et al., Analysis of flow and cavitation phenomena in diesel injection nozzles and its effects on spray and mixture formation, *SAE Technical Paper Series*, 2003-01-1358, 2003.
10. Van Ossenbruggen, C., Micro-spark erosion, *Philips Technisch Tijdschrif*, 20, 200, 1969.
11. Masuzawa, T., Kuo, C.L., and Fujino, M., Drilling of deep microholes by EDM using additional capacity, *Bulletin of the Japan Society of Precision Engineering*, 24, 275, 1990.
12. Takahata, K., Shibaike, N., and Guckel, H., High-aspect-ratio WC-Co microstructure produced by the combination of LIGA and micro-EDM, *Microsystem Technology*, 6, 175, 2000.
13. Lim, H.S. et al., A study on the machining of high-aspect ratio micro-structures using micro-EDM, *Journal of Materials Processing Technology*, 140, 318, 2003.
14. Allen, D.M. and Lecheheb, A., Micro electro-discharge machining of ink jet nozzles: Optimum selection of material and machining parameters, *Journal of Materials Processing Technology*, 58, 53, 1996.
15. Masuzawa, T., State of the art of micromachining, *Annals of the CIRP*, 49, 473, 2000.
16. Yan, B.H. et al., Micro-hole machining of carbide by electric discharge machining, *Journal of Materials Processing Technology*, 87(1–3), 139, 1999.
17. Yu, Z.Y., Rajurkar, K.P., and Shen, H., High aspect ratio and complex shaped blind micro holes by micro EDM, *Annals of the CIRP*, 51, 359, 2002.
18. Kaminski, P.C. and Capuano, M.N., Micro hole machining by conventional penetration electrical discharge machine, *International Journal of Machine Tools and Manufacture*, 43, 1143, 2003.
19. Dauw, D.F., Snoeys, R., and Dekeyser, W., Advanced pulse discriminating system for EDM process analysis and control, *Annals of the CIRP*, 32, 541, 1983.
20. Gangadhar, A., Shunmugam, M.S., and Philip, P.K., Pulse train studies in EDM with controlled pulse relaxation, *International Journal of Machine Tools and Manufacture*, 32, 651, 1992.
21. Pandit, S.M. and Mueller, T.M., Verification of on-line computer control of EDM by data dependent systems, *Journal of Engineering for Industry*, 109, 117, 1987.
22. Weck, M. and Dehmer, J.M., Analysis and adaptive control of EDM sinking process using the ignition delay time and fall time as parameter, *Annals of the CIRP*, 41, 243, 1992.
23. Ho, K.H. and Newman, S.T., State of the art electrical discharge machining (EDM), *International Journal of Machine Tools and Manufacture*, 43, 1287, 2003.
24. Tarng, Y.S., Tseng, C.M., and Chung, L.K., A fuzzy pulse discriminating system for electrical discharge machining, *International Journal of Machine Tools and Manufacture*, 37, 511, 1997.
25. Kao, J.Y. and Tarng, Y.S., A neural-network approach for the on-line monitoring of the electrical discharge machining process, *Journal of Materials Processing Technology*, 69, 112, 1997.
26. Liu, H.S. and Tarng, Y.S., Monitoring of the electrical discharge machining process by abductive networks, *International Journal of Advanced Manufacturing Technology*, 13, 264, 1997.
27. Yu, S.F., Lee, B.Y., and Lin, W.S., Waveform monitoring of electric discharge machining by wavelet transform, *International Journal of Advanced Manufacturing Technology*, 17, 339, 2001.
28. Diver, C. et al., Micro-EDM drilling of tapered holes for industrial applications, *Journal of Materials Processing Technology*, 149, 296, 2004.
29. Hebbar, R.R., Micro-hole drilling by electrical discharge machining, PhD dissertation, Purdue University, West Lafayette, IN, 1992.
30. Masuzawa, T. et al., Twin-probe vibroscanning method for dimensional measurement of microholes, *Annals of the CIRP*, 46, 437, 1997.
31. Kim, B.J. et al., Dimensional measurement of microholes with silicon-based micro twin probes, in *Proceedings of the IEEE Micro Electro Mechanical Systems (MEMS)*, Heidelberg, Germany, 1998, p. 334.
32. Kim, B.J., Masuzawa, T., and Bourouina, T., The vibroscanning method for the measurement of micro-hole profiles, *Measurement Science and Technology*, 10, 697, 1999.

33. Yamamoto, M., Kanno, I., and Aoki, S., Profile measurement of high aspect ratio micro structures using tungsten carbide micro cantilever coated with PZT thin films, in *Proceedings of the IEEE Micro Electro Mechanical Systems (MEMS)*, Miyazaki, Japan, 2000, p. 217.

34. Pourciel, J.B. et al., Microsystem tool for Microsystems characterization profile measurement of high aspect-ratio microstructures, in *Proceedings of SPIE 4592, Device and Process Technologies for MEMS and Microelectronics II*, Adelaide, Australia, 2001, p. 244.

35. Lebrasseur, E. et al., A new characterization tool for vertical profile measurement of high-aspect-ratio microstructures, *Journal of Micromechanics and Microengineering*, 12, 280, 2002.

36. Pourciel, J.B., Jalabert, L., and Masuzawa, T., Profile and surface measurement tool for high aspect-ratio microstructures, *JSME International Journal Series C*, 46, 916, 2003.

37. Christoph, R., Multisensor coordinate metrology: Flexible measurements of form size and location, *VDI Berichte*, 1860,157, 2004.

38. Wang, W.M. and Rajurkar, K.P., Modeling and adaptive control of EDM systems, *Journal of Manufacturing Systems*, 11, 334, 1992.

39. Rajurkar, K.P. and Wang, W.M., Improvement of EDM performance with advanced monitoring and control systems, *ASME Journal of Manufacturing Science and Engineering*, 119, 770, 1997.

40. Ko, R.C. and Good, M.C., Improving contour accuracy of machine tools using an integral-gain scheduler, *Proceedings of the Institution of Mechanical Engineers – Part I: Journal of Systems and Control Engineering*, 219, 511, 2005.

41. Kuo, J.L. et al., Ringing effect analysis of the digital current pulse generator for the linear rail gun, *IEEE Industry Applications Society Annual Meeting*, 1, 176, 2002.

42. Takahata, K. and Gianchandani, Y.B., Batch mode micro-electro-discharge machining, *Journal of Microelectromechanical Systems,* 11, 102, 2002.

43. Ogata, K., *Modern Control Engineering*, 4th edn., Pearson Education Ltd., Singapore, 2004.

44. Bhattacharyya, S.K. and El-Menshawy, M.F., Monitoring the E.D.M. process by radio signals, *International Journal of Production Research*, 16, 353, 1978.

45. Johnson, T.V., Diesel emission control in review—The last 12 months, *SAE Technical Paper Series*, 2003-01-0039, 2003.

46. Holt, D.J., *The Diesel Engine*, SAE, Warrendale, PA, 2004.

47. Kao, C.C. and Shih, A.J., Sub-nanosecond monitoring of micro-hole electrical discharge machining pulses and modeling of discharge ringing, *International Journal of Machine Tools and Manufacture*, 46, 1996, 2006.

48. Perry, W.B., Abrasive flow machining principles and practices, in *Proceedings of Nontraditional Machining Conference*, Cincinnati, OH, 1986, p. 121.

49. Stackhouse, J., Abrasive flow machining, *SME Technical Paper Series*, MR93-148, 1993.

50. Loveless, T.R., Williams, R.E., and Rajurkar, K.P., Study of the effects of abrasive-flow finishing on various machined surfaces, *Journal of Materials Processing Technology*, 47, 133, 1994.

51. Kao, C.C. and Shih, A.J., Form measurements of micro-holes, *Measurement Science and Technology*, 18, 3603, 2007.

52. Automotive Industry Action Group, *Measurement Systems Analysis Reference Manual*, 1st edn., AIAG Publications, Southfield, MI, 1990.

53. Osborne, M.R., *Finite Algorithms in Optimization and Data Analysis*, John Wiley & Sons Ltd., Chichester, U.K., 1985.

54. Mediliyegedara, T.K.K.R. et al., An artificial neural network approach for the process control of electro discharge machining (EDM), in *Proceedings of the 36th ISCIE International Symposium on Stochastic Systems Theory and Its Applications*, Saitama, Japan, 2004, p. 303.

55. Karakuzu, C., An experimental comparison of fuzzy, neuro and classical control techniques, in *Proceedings of 21st IEEE Convention of the Electrical and Electronic Engineers in Israel*, Tel-Aviv, Israel, 2000, p. 160.

56. Reznik, L., *Fuzzy Controllers*, Newnes, Oxford, U.K., 1997.

57. Boccadoro, M. and Dauw, D., About the application of fuzzy controllers in high-performance die-sinking EDM machines, *Annals of the CIRP*, 44, 147, 1995.

58. Zhang, Y. et al., The research on the self adaptive fuzzy control system for electric discharge machining, in *Proceedings of the IEEE International Conference on Intelligent Processing Systems*, Beijing, China, 1997, p. 364.

59. Yan, M.T. and Liao, Y.S., Adaptive control of the WEDM process using the fuzzy control strategy, *Journal of Manufacturing Systems*, 17, 263, 1998.

60. Zhang, J.H. et al., Adaptive fuzzy control system of a servomechanism for electro-discharge machining combined with ultrasonic vibration, *Journal of Materials Processing Technology*, 129, 45, 2002.
61. Kaneko, T. and Onodera, T., Improvement in machining performance of die-sinking EDM by using self-adjusting fuzzy control, *Journal of Materials Processing Technology*, 149, 204, 2004.
62. Zhang, Y., The study in a new-type self-adaptive fuzzy logic control system in EDM process, in *Proceedings of the Fourth International Conference on Machine Learning and Cybernetics*, Guangzhou, China, 2005, p. 18.
63. Isermann, R., On fuzzy logic applications for automatic control, supervision, and fault diagnosis, *IEEE Transactions on Systems, Man, and Cybernetics—Part A: Systems and Humans*, 28, 221, 1998.
64. Shamma, J.S. and Athans, M., Gain scheduling: Potential hazards and possible remedies, *IEEE Control Systems Magazine*, 12, 101, 1992.
65. Kao, C.C., Miller, S.F., and Shih, A.J., Fuzzy logic control system for micro-hole electrical discharge machining, in *Proceedings of the ASME International Manufacturing Science and Engineering Conference*, Atlanta, GA, 2007, p. 793.
66. Kao, C.C., Shih, A.J., and Miller, S.F., Fuzzy logic control of microhole electrical discharge machining, *ASME Journal of Manufacturing Science and Engineering*, 130, 064502-1, 2008.
67. Kao, C.C. and Shih, A.J., Design and tuning of a fuzzy logic controller for micro-hole electrical discharge machining, *Journal of Manufacturing Processes*, 10(2), 61, 2008.
68. Masuzawa, T., Kuo, C.L., and Fujino, M., A combined electrical machining process for micronozzle fabrication, *Annals of the CIRP*, 43, 189, 1994.
69. Sun, X.Q., Masuzawa, T., and Fujino, M., Micro ultrasonic machining and its applications in MEMS, *Sensors and Actuators A*, 57, 159, 1996.
70. Wang, A.C et al., Use of micro ultrasonic vibration lapping to enhance the precision of microholes drilled by micro electro-discharge machining, *International Journal of Machine Tools and Manufacture*, 42, 915, 2002.
71. Yeo, S.H., Murali, M., and H.T. Cheah, Magnetic field assisted micro electro-discharge machining, *Journal of Micromechanics and Microengineering*, 14, 1526, 2004.

8 Ultrasonic Machining

Randy Gilmore

CONTENTS

8.1 HISTORY OF ULTRASONICS

The first documented attempts to produce ultrasonic vibrations were made more than one hundred years ago by Rudolph Koenig who sought to discover the highest pitched sound humans could hear. Mr. Koenig constructed various devices, such as tuning forks, for producing vibrations in air from 4 to 90 kHz.

Ultrasonics as a technology is believed to have begun during World War I in a laboratory in Toulon, France. There researchers were searching for ways to combat the submarine menace, which threatened France at the time. During this investigation, researchers built a high-power ultrasonic generator that used quartz crystals as active elements. With this equipment, researchers were able to produce cavitation in a tank of water. Small fish were killed when they swam into a beam of waves. Workers who placed their hands in the water near the source of the ultrasonic waves felt an intense pain and the feeling that the bones in their hands were being heated. In later experiments, it was found that this was exactly what was happening.

In the early 1900s, Professor Paul Langevin produced the first sandwiched ultrasonic transducer utilizing piezoelectric quartz crystals. Langevin developed several techniques for enhancing excitation of piezoelectric crystals so that high acoustic intensities could be obtained. One technique still used to this day was to vibrate the quartz at its resonant frequency. Another was to cement a thick metal plate to the front and back of the vibrating quartz crystal forming a sandwich, which vibrated at a lower and more practical frequency. The metal plate also acted to improve the coupling between the quartz and the water.

The next significant group of experiments in ultrasonics was performed by Woods and Loomis in 1927. These investigators observed agglomeration of particles, emulsification, dispersion of colloids, atomization of liquids, fragmentation of small fragile bodies, the destruction of red blood corpuscles, and various effects due to frictional heating [1].

They used quartz disks vibrating at resonance as sources of ultrasonic energy. The disks were submerged in oil and then excited by applying 50,000 V to them. The frequencies that they used ranged from 100 to 700 kHz.

In the years following the work of Woods and Loomis, many other workers investigated various aspects of ultrasonics and looked for practical applications of this new technology. Their efforts in terms of published technical papers amounted to more than 600 references. In spite of all this activity, ultrasonics as a processing tool remained a laboratory experiment until the mid-1940s when commercial equipment became available [2].

8.2 APPLICATIONS OF ULTRASONICS

The applications of ultrasonics can be divided into two groups. In the first group, the amount of power required is the most important factor. Ultrasonic energy is used in these applications to perform some kind of mechanical work or it is converted into heat at some distance from the sound source. Some examples are welding, friction reduction, drilling, and the processing of liquids. The acoustical power used in these applications ranges from just a few watts in simple laboratory units to several kilowatts in large industrial installations.

The power output of ultrasonic equipment belonging in the other category is usually in the range of microwatts or milliwatts, or essentially, it is insignificant. This equipment is used for producing and detecting an ultrasonic signal, which, in turn, is used for measuring a physical quantity, detecting flaws, or sensing the condition of some system.

8.3 THE ULTRASONIC MACHINING PROCESS

The ultrasonic machining (USM) process uses a nonabrasive form tool, called a sonotrode, attached to a transducer/booster combination, which vibrates at a frequency of approximately 20,000 cps. A machining action occurs as the sonotrode vibrates the fine abrasive particles flowing throughout the machining gap and propels them against the workpiece material. The form tool itself does not abrade the workpiece; the vibrating tool excites the abrasive grains in the flushing fluid, causing them to gently and uniformly wear away the material, leaving a precise reverse form of the sonotrode shape (Figure 8.1).

Unlike thermal and chemical processes, USM is a mechanical material removal process applicable to both conductive and nonconductive materials and particularly suited to the machining of brittle materials such as graphite, glass, carbide, and ceramics. The uniformity of the sonotrode vibration limits the process to forming small shapes typically under 6350 mm^2 (10 in^2) in the frontal area.

8.3.1 THE MACHINE

The ultrasonic machining system includes the machine tool, the generator, the slurry system, and the operator controls. Precision mounting areas on the machine table and on the upper platen accommodate the sonotrode tool assembly and the electrode holder or an optional positioning table. The sonotrode tool assembly consists of a transducer, booster, and sonotrode. The ultrasonic generator powers the transducer, creating impulses that occur at a range of 19,500–20,500 Hz, and automatically adjusts the output frequency to match the resonant frequency of the tool, which varies according to the sonotrode shape, length, and material. This adjustment occurs both at initial tuning and in process at high power to ensure the true resonance peak is continuously maintained. The transducer converts electrical pulses into vertical stroke. This vertical stroke is transferred to the booster, which may amplify or suppress the stroke amount. The modified stroke amount is, then, relayed to the sonotrode that contains the mirror image of the desired shape to be formed in the workpiece and serves as the actual tool in the process. The amplitude along the face of the tool typically falls

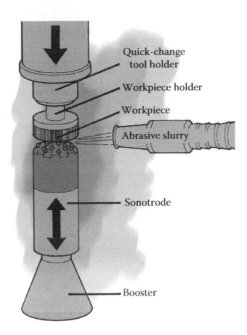

FIGURE 8.1 Ultrasonic machining process.

in a 25–50 µm (0.001–0.002 in.) range. The vibration amplitude is usually equal to two times the diameter of the abrasive grit used.

The slurry system supplies a mixture of water and abrasive grit, usually silicon carbide or boron carbide, to the cutting area. In addition to providing abrasive particles to the cut, the slurry also cools the sonotrode and removes particles and debris from the cutting area. The overcut produced with the USM is a function of the abrasive particle size (usually two times the particle diameter), as are the surface finish and material removal rates. Typical particle sizes range from 200 mesh, with an average particle size of 66 µm (0.0026 in.), to 800 mesh that has an average particle size of 11 µm (0.0004 in.). To maintain a constant temperature, an auxiliary chiller removes the residual heat from the slurry as it returns to the tank.

The operator controls provide inputs for manual or automatic sequencing of operations. Controls include variable cutting force, Z-axis position, speed control of ram movement, cycle timing, retract distance, and flush timing. These controls may be manually activated for set-up or operated in an automatic mode for part processing. Machining parameters for multiple workpieces can be pre-programmed. A built-in monitor displays operational parameters including machine settings, X, Y, and Z-axis positions, distance-to-go, and programming messages. In state of the art ultrasonic equipment, the feed of the Z-axis is adaptively controlled via a closed loop circuit linking static pressure and Z-axis feed rate. In this case, the operator programs a target static pressure threshold that is monitored real time by the system control. As the target pressure is approached, the system automatically slows the feed rate. If the target pressure is reached, the feed rate becomes zero and if the target pressure is exceeded, the feed rate becomes negative.

8.3.2 THE SONOTRODE

The sonotrode serves as a means of transmitting the vibratory energy from the transducer to the working surface and also usually acts as a transformer for the mechanical amplitude. It is normally a metal rod or bar with the form to be transferred either machined into or mounted on the tip or face of the tool. The length of the sonotrode normally corresponds to half a wavelength of the frequency (Figures 8.2 and 8.3) being used with the resonant length of the sonotrode in materials such as

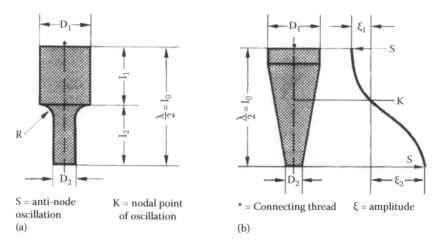

S = anti-node   K = nodal point   * = Connecting thread   ξ = amplitude
oscillation   of oscillation
(a)   (b)

FIGURE 8.2   Standard half wavelength sonotrode designs.

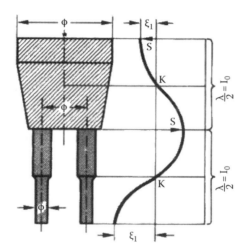

FIGURE 8.3   Full wavelength length sonotrode.

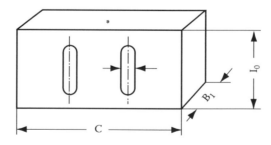

FIGURE 8.4   Large lateral dimension sonotrode with slots for vibration uniformity.

titanium, steel, and aluminum ranging between 115 and 140 mm (4.5 and 5.5 in.). Sonotrodes that have large lateral dimensions require slots to create uniform vibration along the working surface, as shown in Figure 8.4.

There are a number of ways to produce the initial sonotrode. Short run sonotrodes can be made by bonding a form (possibly a workpiece or photolithographic plate) to an ultrasonic horn (Figure 8.5).

FIGURE 8.5   The graphite electrode (right) was ultrasonically formed with the sonotrode (left) in 15 min with a 5 min redress cycle.

FIGURE 8.6   One sonotrode can form dozens of graphite EDM electrodes, and it can then be redressed by utilizing one of the EDM electrodes, the tool previously produced.

More commonly, the detail in the sonotrode is produced via conventional machining practices, such as milling, grinding, or turning, or via a non-conventional process, such as electrical discharge machining (EDM). Direct machining of the desired detail into the ultrasonic tool body is the preferred method since it does not rely on bonding of a secondary plate to the ultrasonic horn.

For the forming of graphite EDM electrodes by ultrasonic machining, an alternate approach of bonding graphite to a sonotrode blank can be utilized. In this approach, the bonded tool is vibrated into a form, such as a workpiece or mold, to produce a master electrode. The resultant electrode can then be used to EDM a sonotrode for forming and redressing a series of electrodes or components. A first run sonotrode or master electrode can also be produced by conventional machining methods, as described above.

To produce the sonotrode for forming the series of graphite EDM electrodes shown in Figure 8.6, an existing mold cavity was used as the master and graphite was ultrasonically machined into the mold. The initial electrode was then used to EDM a sonotrode for the manufacture of a subsequent series of production electrodes. Since the sonotrode does experience wear during the ultrasonic process (work-to-wear ratios typically range from 500:1 to 1500:1), it can easily be redressed (also by EDM, for example) with an electrode formed early in the sonotrode production cycle. Typically, a sonotrode can be redressed dozens of times before a new sonotrode must be machined.

To achieve maximum efficiency of ultrasonic forming of graphite EDM electrodes, a precision quick-change tooling system should be incorporated to facilitate the transfer of electrodes from the ultrasonic equipment to the EDM machine and back to the ultrasonic equipment. The reference surfaces of the tooling ensure the positioning accuracy and repeatability as the electrodes travel

FIGURE 8.7 This graphite electrode was ultrasonically formed.

from the USM after the initial forming, to the EDM for machining, and back again to the USM for redressing.

8.4 PROCESS CAPABILITIES

Machining speeds vary according to the vibration amplitude, abrasive grain size, and workpiece material. Typical ultrasonic abrading speeds in graphite range from 0.4 to 1.3 cm^3 (0.025–0.075 in.^{3}) per minute. Surface finishes for the ultrasonic process range from 0.2 to 1.5 μm (8–60 μin.). Accuracies of ±10 μm (0.0004 in.) are typical; tolerances of ±5 μm (0.0002 in.) can be obtained for specialized applications. Low machining forces permit the manufacture of fragile electrodes. The walls of the honeycomb structure in the electrode shown in Figure 8.7 have a width of 50 μm (0.002 in.). The electrode was manufactured by ultrasonic machining in less than 10 min.

Complex electrodes often comprise the most expensive part of an EDM manufacturing system. The ultrasonic machining process offers significant improvements in the manufacture of graphite electrodes over conventional techniques. The main advantage of ultrasonically forming graphite electrodes lies in the ability to quickly and conveniently redress worn electrodes, restoring the original detail for remachining. The convenience of electrode forming and redressing makes electrode wear less important, permitting higher machining rates and less EDM time. Additional benefits include graphite dust elimination, consistent accuracy of duplicate electrodes, the need for fewer electrodes, and lower graphite costs.

8.5 CAPABILITIES IN CERAMICS

Currently there is a major effort to develop new and enhanced ceramics, semiconductors and super-conductors for components requiring higher accuracy, better surface finish, lower degrees of surface degradation, and more complex geometry than ever before. Typical methods for processing ceramics today, including grinding, EDM, laser, and abrasive waterjet, have limitations: Grinding must be done with an expensive diamond wheel, a high concentration of microcracking often extending more than 250 μm deep may be produced and complex contours are not feasible. EDM is restricted to the machining of ceramics that have an electrical conductivity of 2×10^{-2} (Ω/cm) [1–3] and ceramics that have been machined by EDM exhibit lower fracture strength, a thermally altered zone up to 50 mm and microcracking. Laser machining requires extremely high power requirements that cause surface imperfections, and some composition changes or coatings may be needed to enable the material to be machined at all. Abrasive waterjet is limited to materials with a hardness less than 2000 kg/mm², making the machining of ceramics such as boron carbide (B_4C) and titanium diboride (TiB_2) impractical with this method [4].

Ultrasonic machining of ceramics has a number of important advantages over these methods: conductive or nonconductive materials can be machined, material hardness is relatively unimportant, and complex three-dimensional contours can be manufactured as quickly as the simple ones. The process produces no heat-affected zone and there are no chemical or electrical alterations on the workpiece surface, although the small impacts impart a shallow, compressive residual stress to the surface. It has been reported that this residual stress actually promotes an increase in the high cycle fatigue strength of the work material after ultrasonic machining [5].

The process is commonly used for drilling holes, slots, and irregular configurations in ceramics. Additional applications include machining of phased array radar components, cutting tool inserts, superconductors, biomedical devices, accelerometers, aerospace components, and wire draw dies.

Ultrasonic machining offers an alternative to diamond grinding with attributes that can be superior to diamond grinding in specific applications. Since ultrasonic machining is a process where the material is removed by a vertical motion into the workpiece, it is possible to install internal corners that are very nearly sharp (Figure 8.8). USM is also a very gentle machining method relative to diamond grinding, making it possible to produce very minute features in extremely fragile materials (Figure 8.9).

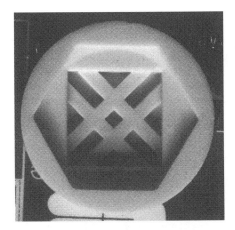

FIGURE 8.8　Ultrasonic machining offers the ability to create relatively sharp internal corners in virtually any ceramic material.

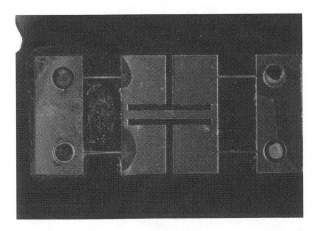

FIGURE 8.9　Ultrasonic machining is gentle enough to produce this accelerometer, which has features as small as 0.10 mm (0.004 in.) in width.

FIGURE 8.10   With ultrasonic contouring, intricate shapes can be produced with simple tools.

Abrasives used for machining ceramics include silicon carbide, which is used for the low-density ceramics such as silicon oxide (SiO_2) and aluminum oxide (Al_2O_3), and boron carbide, which is typically used for the higher density ceramics such as silicon nitride (Si_3N_4) and silicon carbide (SiC). Slurry circulation is important to remove the debris from the cut and to provide fresh abrasives to the machining gap. Penetration rates typically range from 0.25 mm (0.010 in.) per minute for high-density workpieces to 1.5 mm (0.060 in.) for low-density materials. Depending on the size of the abrasive grains used, material removal rates can range from 5 mm^3 (0.0003 $in.^3$) per minute to 65 mm^3 (0.004 $in.^3$) per minute. The process yields minimal radiusing on the face of the workpiece. Surface integrity depends on the workpiece material, abrasive size, and flushing conditions. Hard ceramics yield slightly better surface conditions than soft ceramics with finishes as low as 0.4 μm R_a (16 μin.) feasible.

CNC control allowing contour machining capabilities is a recent development. The glass piece shown in Figure 8.10 is used as a single point tool attached to a rotary ultrasonic tool body. The 2 mm (0.080 in.) diameter diamond-tipped tool was mounted on a rotary ultrasonic head, spinning continuously as the workpiece was driven along the programmed path. The program repeats continuously as the tool sinks into the workpiece to a typical depth of 0.005 mm (0.0002 in.) per pass.

Since current ultrasonic machining systems are predominantly CNC controlled, it is possible to create unique features in ceramic materials. Among these features are undercuts and threaded features. This type of geometry is produced by the utilization of planetary motion of the X- and Y-axes. An example of ultrasonic tapping can be seen in Figure 8.11.

Other current uses of multiple-axis contour machining include the manufacture of ceramic turbine blades. With the incorporation of a fourth and fifth action of motion, a standard three-axis ultrasonic machine can perform work very similar to a five-axis milling machine. One example is the creation of swept airfoil surfaces, such as the ones shown on the turbine wheel pictured in Figure 8.12.

Among the newest applications for ultrasonic machining is the creation of cooling holes, seal slots, and other features in ceramic matrix composite (CMC) materials. CMCs are emerging as a replacement for several aerospace turbine components historically manufactured from nickel alloys. Among the advantages of CMC are lower weight and higher temperature characteristics. These materials are very difficult to machine utilizing conventional machining techniques and cannot be machined by EDM, because they are not electrically conductive. Even CMC materials with a protective barrier coating can be successfully machined by ultrasonic machining. Figure 8.13 shows cooling holes installed in CMC aerospace turbine components. The holes are installed at a 30° angle to the surface to provide better cooling airflow across the airfoil.

FIGURE 8.11 An example of planetary ultrasonic machining and the tool used to create tapped features.

FIGURE 8.12 Silicon nitride auxiliary power unit turbine wheel shown before (right) and after (left) ultrasonic machining.

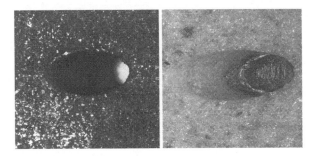

FIGURE 8.13 Cooling holes drilled in both uncoated (left) and coated (right) ceramic matrix composite material.

Since conventional machining of CMC materials is very difficult and causes extreme tool wear, USM is the only feasible approach for installing features less than 0.125 in. in width. In Figure 8.14, seal slots used for creating a sealing surface between adjacent vane structures are shown. These seal slots are only 0.060 in. in width and more than 0.100 in. in depth.

8.6 ULTRASONIC POLISHING

Proper finishing of molds and blind cavities has historically been a labor-intensive, time-consuming, and expensive procedure. Ultrasonic polishing techniques have now been developed for a variety of polishing applications including uniform surface improvement, deburring of delicate parts, and removal of recast layers left by thermal machining operations. By vibrating a brittle tool material

FIGURE 8.14   Seal slots produced in ceramic matrix composite material utilizing ultrasonic machining.

such as graphite or glass into the workpiece at ultrasonic frequencies and relatively low amplitudes, a polishing action occurs as the fine abrasive particles in the slurry abrade the high spots of the workpiece surface, typically removing 0.012 mm (0.0005 in.) of material or less.

The extent of polishing required is determined by the initial surface roughness of the workpiece and the finish required after polishing. Typical surface improvements range from 5:1 to 10:1; finishes as low as 0.3 µm R_a (12 µin.) can be achieved. A variety of materials including tool steels, carbides, and even ceramics can be successfully processed. Since the tool is not preshaped, but rather conforms to the workpiece configuration during the polishing process, polishing is uniform on all surfaces of the workpiece and repeatable from part to part, regardless of the complexity of the component. In addition, indexing and registration of the tool and workpiece is not required.

The ultrasonic polishing process offers automatic operation and is not dependent on manual polishing skills. The mold cavity shown before and after processing (Figure 8.15) is approximately 50 mm (2 in.) in diameter and 20 mm (0.8 in.) deep. An ultrasonic polishing cycle of approximately 10 min was used to remove the burrs and machining marks left by the CNC milling operation. In the illustration below, the ultrasonic polishing tool prior to the polishing cycle and after the polishing cycle is shown. As can be seen in the photograph, the polishing tool need not be pre-shaped, but rather assumes the shape of the part to be polished during the polishing cycle.

The ultrasonic polishing process is also capable of removing recast and heat-affected zones from components machined by EDM or other thermal removal processes (Figure 8.16). The recast area of a component manufactured by a thermal removal process is typically desired to be removed due to surface finish or mechanical strength concerns. This layer is extremely hard and commonly populated with a network of microcracks. The ability to automatically remove this unwanted layer provides a much faster surface treatment option to typical hand polishing, which must be done by highly skilled technicians.

FIGURE 8.15   Ultrasonically polished mold cavity and polishing tool shown before (left) and after (right) ultrasonic polishing.

FIGURE 8.16 A closure mold used for creating plastic caps for bottles shown before and after ultrasonic polishing.

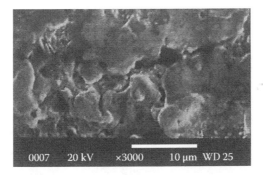

FIGURE 8.17 Photomicrograph at 3000× of a carbide surface machined by EDM.

FIGURE 8.18 Photomicrographs at 3000× show the effects of ultrasonic polishing.

The photomicrographs shown in Figures 8.17 and 8.18 were taken at a 3000× magnification. The original EDM surface finish of 0.75 μm R_a (30 μin.) on a carbide compacting die was ultrasonically polished to a final finish of 0.3 μm (13 μin.), removing only 0.005 mm (0.0002 in.) of the material.

Since the abradable tool automatically conforms to the workpiece geometry, the ultrasonic finishing method described offers a number of important benefits in finishing components with complex geometries, including

- No specially pre-shaped tools are required; consequently, even low-volume components are applicable.
- No precision alignment of the polishing tool to the workpiece is required.
- Work is uniform across the workpiece surface.
- Surface improvement is 3:1 or more on machined, EDM-ed and cast surfaces.
- Edges can be deburred and lightly radiused.
- No special operator skills are required.
- The system operates automatically without operator involvement.

A prototype ultrasonic finishing system was designed, built, and tested under a Machine Tool Initiative Program Research and Development Announcement sponsored by the Wright Laboratory of the U.S. Air Force. The equipment was designed in four subsystems including mechanical components, drive and controller systems, ultrasonic hardware, and slurry system. Selected parameter data were incorporated into a menu-driven display with prepackaged programs for automatic selection of machining parameters based on the depth of the area to be polished, beginning surface finish, and previous machining method.

In the Air Force program, process repeatability was examined based on the improvement of surface finish and amount of stock removal. Stock removal ranged from 0.003 to 0.013 mm (0.0001–0.0005 in.), dependent on the incoming surface roughness. In the study, it was proven that the stock can be removed accurately within 10% of the amount of stock removed; the stock can also be removed to a desired depth within 20% and is repeatable from workpiece to workpiece within 10%. Limitations observed are confined to vertical or near-vertical sidewalls where stock removal was as much as 50% less than that of the frontal surfaces.

Surface finish improvements are the greatest when incoming surface roughness is high. In this case, surface finish improvements as high as 10:1 were accomplished. On components with the beginning surface finishes in the range of 100 μin. (2.5 m) R_a, the improvement averaged 5:1. The repeatability of surface finish improvement measured ±10%.

Within the Air Force sponsored program, process parameters were optimized including slurry, feed/speeds, polishing tip material, machining pressures, and ultrasonic generator settings. Boron carbide is the best multi-purpose abrasive for ultrasonic polishing. With a 600-mesh abrasive, surface finishes as low as 8–12 in. (0.2–0.3 μm) R_a are possible; finishes as low as 14–18 μin. (0.36–0.46 μm) R_a can be produced using 320-mesh abrasives. Optimum abrasive concentration for most applications occurs between 18% and 22% by weight. Polishing tips of graphite show the most promise because they rapidly conform to the workpiece configuration and are of relatively low cost. Graphite, with particle sizes of 0.00004–0.0001 in (0.001–0.003 mm), performed the best. The optimum static pressure ranges between 1 and 1.5 pounds per square inch (0.5 and 0.7 kg/6.452 cm²).

8.7 ULTRASONICALLY ASSISTED MACHINING

Perhaps the most commonly used application of ultrasonically assisted machining is grinding. This technology is commonly referred to as rotary ultrasonic machining (RUM). Unlike conventional ultrasonic machining, RUM utilizes abrasives that are fixed to the ultrasonic tool rather than in a free-flowing slurry. RUM relies upon "hammer" machining, where the abrasive grains that are affixed to the tool are literally hammered into the workpiece surface. Although the surface and subsurface damage caused by the hammering action of RUM is more severe than that caused by conventional ultrasonic machining, it is superior in speed and geometric quality when it is used for the creation of round holes. The geometry is enhanced due to the rotational movement of the tool and material removal is commensurate with the combined removal rates of grinding and ultrasonic machining.

Ultrasonically assisted drilling has been found to aid in both deep hole drilling and drilling with greatly reduced use of coolant. The high-speed vibrations that are parallel to the vertical

FIGURE 8.19   Ultrasonic assistance in drilling can reduce built up edge. Ultrasonic assistance was used with the drill bit on the right, but not on the drill bit on the left.

drilling axis act as a chip breaker. When ultrasonic vibrations are not present, the drilled material is discarded from the cut area as a continuous ribbon. However, when ultrasonic assistance is incorporated, the material is ejected from the cutting area as a series of small chips. This chip-breaking characteristic not only permits drilling with minimal coolant, but also results in less built-up edge (BUE) or adherence of drilled material on the drill bit (Figure 8.19). BUE is one of the leading contributors to drill wear and drill breakage. Additionally, ultrasonically assisted drilling reduces the thermal characteristics of the drilling operation due to a reduction in friction. Finally, ultrasonically assisted drilling has been shown to reduce the size of burrs associated with the drilling process.

Ultrasonic assistance in cutting operations, such as textiles, honeycomb materials, Kevlar, plastics, carbon fiber, glass fiber, and even food products, offers advantages over conventional sawing or cutting operations. Ultrasonic assistance offers the very best in speed and accuracy of cut along with cleanliness. The amplitude of vibration of the cutting knives, whether in slitting or guillotining arrangements, creates a cutting blade that can slice through even the stickiest products with no residue and leave a clean and crisp edge along with superb dimensional stability. The rapidly reciprocating blade (typically 20–35 kHz) provides burn-free, smooth, and soft sealed edges.

Ultrasonic technology can also be applied to augment conventional machining technologies. The superimposition of ultrasonics to machining processes such as milling, grinding, drilling, EDM, and electrochemical machining (ECM) has been studied at various industrial and academic institutions. In the more non-traditional applications (i.e., EDM and ECM) it has been observed that by placing the electrode or cathode under the influence of ultrasonic vibrations, material removal rates can be increased and electrode wear decreased. These improvements are largely attributed to better removal of debris from the machining gap, which facilitates better machining efficiency.

8.8   CURRENT ULTRASONIC MACHINING EQUIPMENT

The current generation of ultrasonic machining systems are capable of multi-axis ultrasonic machining, rotary ultrasonic machining, and static ultrasonic machining. In addition, in the three to five axes of CNC movement, the output power level to the ultrasonic transducer is under computer numerical control and additional axes of manual movement are incorporated into the head of the machine to aid in the setup of the tool. These current models incorporate a number of features from previous generation machines, such as automatic resonance searching and following of the ultrasonic generator and the use of load cells in the machining head to sense machining pressure (Figure 8.20).

Other enhancements in the current generation of ultrasonic machining systems include menu driven software with enhanced graphical interface for ease of operation. Specialized NC codes are

FIGURE 8.20   Current ultrasonic machining system. (Courtesy of The Ex One Company, Irwin, PA.)

often incorporated into the software structure to create peck drilling cycles, planetary motion for tapping and undercuts, and automatic compensation for tool wear. In order to be as responsive as possible to customer demands, most current units have a remote communications software integral to the system, allowing technicians to remotely operate the equipment for troubleshooting, software upgrades, and programming assistance.

QUESTIONS

Q.8.1   Ultrasonic applications are divided into two groups, high power and low power. List some examples of each of these application groups.

Q.8.2   What frequency is used for most ultrasonic machining applications?

Q.8.3   What is the tool in ultrasonic machining called?

Q.8.4   What is the removal mechanism in ultrasonic machining?

Q.8.5   What is the name of the device that converts electrical pulses into mechanical stroke?

Q.8.6   What are typical abrasives used in ultrasonic machining?

Q.8.7   What are some common methods for producing initial sonotrodes?

Q.8.8   Why is ultrasonic machining more applicable to machining of ceramics than electrical discharge machining (EDM)?

Q.8.9   What is the typical range of abrasive sizes used in ultrasonic machining slurry?

Q.8.10 What are some reasons ultrasonic machining is the preferred method for machining ceramic matrix composite (CMC) materials?

Q.8.11 How does ultrasonic polishing differ from ultrasonic machining?

Q.8.12 What is the typical surface finish improvement resulting from ultrasonic polishing?

Q.8.13 Name some of the benefits of ultrasonic polishing.

Q.8.14 What is the most common ultrasonically assisted process?

Q.8.15 How does rotary ultrasonic machining (RUM) differ from standard ultrasonic machining?

Q.8.16 In what way does drilling benefit from ultrasonic assistance?

ABOUT THE CONTRIBUTING AUTHOR

Randy Gilmore currently serves as the director of technology for the Ex One Company, LLC in Irwin, Pennsylvania. Mr. Gilmore has 30 years of experience in the field of nontraditional machining techniques. He has acted as program manager on a number of federally sponsored research and development initiatives, has been a business unit manager for several divisions at both Extrude Hone and Ex One and acted as general manager for the micromachining division of Ex One. Prior to his employment at Extrude Hone and Ex One, Mr. Gilmore was the manager of applications for Elox, a world leader in electrical discharge machining technologies and equipment.

Mr. Gilmore was previously involved with the development, design, manufacture, and support of orbital, ultrasonic, and laser machining systems. He designed and built a new generation of ultrasonic machining systems as part of the Manufacturing Technology Development effort sponsored by the U.S. Air Force. He was also responsible for the development and design of orbital abrasive flow polishing systems. Among the commercial successes are a fully automated polishing cell for use at the United States Mint in San Francisco for the polishing of proof coin dies, a semi-automated system for the polishing of blow molds used in the manufacture of PET bottles, and a fully automated system for the polishing of prosthetic devices.

Most recently, Mr. Gilmore has served as the technical director of a program for the development of short pulse laser technologies for the Electro Optics Center and for an SBIR Air Force Program related to laser machining. At Ex One, Mr. Gilmore is currently involved as program manager for an Army program for the development of short pulse laser systems for diesel and aerospace military applications, another Army program for the development of light-weight castings for structural metal castings and powder metal parts, and a Navy program to support an RCT system at NUWC-Keyport and to fabricate demonstration castings representative of typical, low-quantity DOD metal castings required during legacy defense systems refurbishment and repair.

Mr. Gilmore is an alumnus of the University of North Carolina, Chapel Hill, NC where he majored in mathematics and computer science. He holds several patents and has numerous publications in his fields of expertise. He also has been an invited speaker at national and international symposiums on finishing and nontraditional machining and serves as an advisor to the National Science Foundation.

REFERENCES

1. Wood, R.W. and Loomis, A.L., *Physics Review 29*, 373 (1927).
2. Frederick, J.R., *Ultrasonic Engineering,* John Wiley & Sons Inc., New York, 1965.
3. Guitrau, E.B., *The EDM Handbook*, Hanser Gardner Publications, Cincinnati, OH, 1997.
4. Jones, F., Ryffel, H., Oberg, E., and McCauley, C., *Machinery's Handbook*, 26th edn., Industrial Press, New York, 2000.
5. *Machining Data Handbook,* 3rd edn., Vol. 2, Metcut Research Associates Inc., Cincinnati, OH, 1980, pp. 10–44.

9 Laser Material Processing

Andreas Ostendorf, Sabine Claußen, and Marshall G. Jones

CONTENTS

9.1 INTRODUCTION

Invented in 1960, the laser has become a very important tool in manufacturing. Lasers in the context of this book can be regarded as energy converters that convert common forms of energy (e.g., electrical discharge or conventional light emitted by flash lamps) into a unique kind of high-quality coherent energy fields. The laser active medium is pumped by conventional energy. Due to stimulated emission in a resonator setup, the generated photons are multiplied. As all photons provide the same wavelength and phase, the resulting electromagnetic wave is characterized by a high degree of coherence, a very narrow spectral distribution, and low divergence. This unique behavior results in very good focusing capabilities, and the resulting power densities in the focus of a high-power laser field can reach 10^{16} W/cm^2 easily. Those high intensities can lead to the melting and evaporation of almost all kind of materials, making laser sources an interesting tool in materials processing.

In this chapter, the most important laser-manufacturing processes are described. It will be shown how the laser radiation delivered and absorbed at the material site results in cutting, joining, surface treatment, or other material processes. In general, the different sections concentrate on the macroscopic processing of metallic materials. Other topics are addressed as well, but not in detail.

The first experiments in laser material processing began only 10 years after the invention of the laser itself in the 1970s, with cutting applications and expanded rapidly with the market launch of 1 kW CO_2 lasers in 1983. In the beginning of the 1970s, the first joining experiments using laser radiation were carried out. In the following years, the penetration in the industrial sector started with a series of productions in laser-based joining. Nd:YAG solid-state lasers in the kW range were invented in the early 1990s, followed by semiconductor lasers a few years later. Solid-state lasers rapidly succeeded because of their easy setups and their possibilities to couple the laser light in flexible guiding fibers. Today, diode lasers with powers up to 10 kW are commercially available.

Moreover, in the range of solid-state lasers, there have recently been promising developments in the field of fiber and disk lasers, which become increasingly important in laser material processing. The first fiber laser was developed in 1964, only 4 years after the invention of the laser. Initially, they were developed in the range of milliwatts with further improvements toward the end of the 1990s up to 100 W laser output power. In the beginning of the twenty-first century, new developments concerning fiber design were made and they led to fiber lasers in the kW-range. On the contrary, disk lasers with excellent beam quality were invented in the beginning of the 1990s [1]. Both fiber and disk lasers are superior to conventional solid-state lasers in terms of electrical to optical efficiency as well as in beam quality—both concepts deliver almost perfect Gaussian beams.

Nowadays, laser-based processes play a decisive role in industrial laser material processing. Despite the relatively low electrical to laser optical conversion efficiency and the high investment costs, laser-based processing has many advantages in laser material interaction compared to other processes. The conversion of laser energy into thermal energy at the workpiece enables processing with a much higher quality. In terms of energy field manufacturing, laser energy can be tailored to the material properties allowing a very flexible adjustment of the interaction that can lead to vaporization, melting, or just surface modification. There are not many alternative types of energy fields covering almost all of the important manufacturing processes over 8–10 orders of magnitude in dimension. With the further development of fiber and disk lasers, the efficiency of laser processing will be significantly increased.

9.1.1 LASER SOURCES IN INDUSTRIAL APPLICATIONS

Albert Einstein discovered in 1917 that photons could be generated and multiplied in a process called stimulated emission. In stimulated emission, one photon travels through the excited medium, causing the excited state to relax into the ground state while emitting a new photon. The new photon has the same photon energy, that is, optical wavelength as well as the same phase condition as the initial photon. The electrons of the matter in the lower energy state can be transferred into the excited state again by the application of external energy. This is called the pumping process. In 1960, Theodore Maiman used this principle and built the first laser. In his setup, a ruby crystal was used as the lasing medium. This was approximately 50 years ago. Since then, many different lasers have been developed covering a wide range of power, wavelength, energy, and beam quality. However, the basic principle of laser systems is still the conversion of pump energy, that is, the energy to bring matter into the excited state, into laser energy. Figure 9.1a shows the basic setup of laser systems.

The laser medium can be a gas, a liquid, or a solid-state material. Depending on the laser medium, the pumping process can be different. Gas lasers are often pumped by gas discharge, whereas a solid-state medium can be pumped by conventional arc light sources or more selectively by diode lasers. In the case of a semiconducting laser medium, that is, a diode laser, the pumping energy is supplied by electrical power. The resonator can be designed in many different ways in order to achieve an optimal gain factor of the photons. The energy difference between the upper level, that is, the excited state, and the lower level in the medium determines the photon energy and laser wavelength. Using various laser mediums, today's laser wavelength ranges from deep

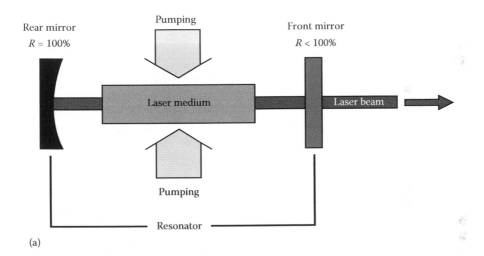

(a)

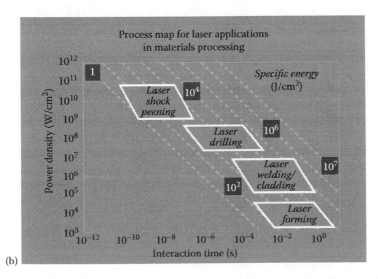

(b)

FIGURE 9.1 (a) Schematic setup of a laser and (b) process map for laser applications in materials processing.

UV, ultraviolet, visible, near infrared to the far infrared spectrum. Integrating suitable optical components and switches within the resonator, pulsed laser radiation can be generated, and the peak power of a pulsed laser can exceed the average laser output power by many orders of magnitude. Finally, good beam quality is necessary for optimal focusing. Gaussian-like beam profiles can be more sharply focused than any other mode. In order to achieve a perfect beam mode in solid-state lasers, the pumping energy that is not converted into photonic energy will show up as heat, and should be efficiently removed from the laser medium. One solution is to optimize the surface-to-volume ratio to enable the efficient cooling of the laser medium. This can be achieved by either stretching the medium in length, as in the case of a fiber laser, or reducing the laser medium length, as in the case of a disk laser. CO_2 lasers as well as diode-pumped solid-state lasers (DPSSL) are most widely used in material processing as they can be scaled up to several kilowatts of laser output power. Fiber lasers and direct diode lasers are becoming increasingly important in laser material processing due to their high conversion efficiency.

9.1.2 Understanding the Laser Fields

When considering the interaction of laser radiation energy with materials, there are key parameters that come into play when predicting what effects take place with the material of interest.

These parameters include laser radiation interaction time with the material; the amount of average laser power, in watts, delivered to the material interaction site when the laser is operating in a continuous wave (CW) mode; and the amount of laser energy, in joules, delivered to the material interaction site when the laser is operating in a pulsed mode. Figure 9.1b presents a guiding process map.

There have been CO_2 lasers with average powers up to 45 kW and fiber lasers with average powers up to 30 kW that have been or are being used for material processing. Peak powers in giga-watts can be achieved by the pulsed lasers when laser energy is delivered to the interaction site in micro-, nano-, or picoseconds pulse width or pulse length. The peak power level is determined by dividing the pulse energy by the pulse width (e.g., $10 J/10^{-9} s = 10^{10} W$).

Another important measure when assessing laser material processing is defined as power or energy density. The power density is the amount of laser power (average or peak) divided by the area of the focused laser beam on the surface of the material at the laser beam–material interaction site. The power density is typically given in W/cm^2 and the energy density is given in J/cm^2. It is usually the power density that is most often considered when determining which type of laser processing is possible. As a rule of thumb, the following power density ranges are associated with the corresponding material processes:

$<10^4 W/cm^2$	laser heat treatment, forming
$10^5–10^6 W/cm^2$	laser welding, soldering, brazing, cladding
$10^7–10^8 W/cm^2$	laser drilling, cutting, marking, scribing
$>10^9 W/cm^2$	laser shock peening, ablation

Other parameters that bare consideration in laser material processing are the wavelength of the lasers and the wavelength absorptivity of different materials. For example, the wavelength of CO_2 lasers is 10.6 μm, which is highly reflective with copper. For that reason, CO_2 lasers are not typically used when doing material processing with copper. Shorter wavelength lasers perform much better with copper. In some CO_2 laser systems, copper is actually used as mirrors in the system. The wavelengths of other lasers that are typically used for material processing are the Nd:YAG laser at 1.064 μm, the Yb-fiber laser at 1.07 μm, the frequency doubled Nd:YAG at 0.532 μm, and the excimer laser that can provide at least four UV wavelengths including 0.244 μm.

The remaining important laser parameter is laser beam quality, which is basically a measure of the focusability of the laser beam. The lower the beam quality number, the smaller the focus spot for a given lens. The other benefit of having a low beam quality number laser is that you can use longer focal length lenses that can increase the working distance in the material processing cell. The very high-power CO_2 lasers were not the best for key hole welding because their beam quality was not that good. Some lasers' beam quality would worsen as a function of increasing power. This is typical of a lamp-pumped Nd:YAG rod laser. The new disc and fiber lasers have very good beam quality. During 2009, a 10 kW fiber laser was introduced that was capable of operating in single mode; meaning a beam quality as good as it can possibly be. One beam quality measure is the beam parameter product (BPP). BPP is the product of the laser beam diameter times one-half the beam divergence angle. The BPP dimension is typically in mm-mrad.

Even though absorptivity, wavelength, laser spot size, and beam quality are not explicitly shown, the capabilities of laser material processing can be illustrated to the first order when the power density is plotted vs. time. With such a plotted process map, the operating regions or spaces for different laser processing regimes (forming, heat treating, welding, drilling, shock peening, etc.) can be estimated to the first order. This process map is depicted in Figure 9.1b. In addition to the material processing operating regimes, there are energy density curves that provide an estimate of the amount of energy deposition for a given process. In general, this process map can be a valuable aid for anyone who is thinking about using lasers for material processing.

9.2 LASER CUTTING

Regarding industrial applications, laser-based cutting is the most widely used technology. It allows faster processing along with better quality compared to alternative cutting processes. Laser cutting can be automated quite easily. Other advantages include the absence of tool wear or reduced noise at the workstation.

During laser cutting, the laser energy is absorbed at the surface of the workpiece and leads to the melting and/or vaporization of the material in the processing zone. Depending on the laser parameters, different cutting principles can be applied.

Nearly all materials can be cut by laser radiation, but there is a limitation in cutting depth depending on the laser power. Workpieces with large thickness cause a lower cut quality with rounded edges and extended heat-affected zones (HAZ). Surfaces with high reflectivity, low absorptivity and high thermal conductivities, such as copper or aluminum sheets, may complicate the laser cutting process, but this can be overcome with suitable processes and laser systems. In general, these materials require more powerful lasers. Another important variable in laser cutting is the cutting speed. It has to be adjusted to the thickness of the workpiece and the material itself. An inappropriate speed may cause burr formation or increased roughness of the cutting surfaces or even fusion penetrations. Figure 9.2 shows several interdependencies of laser cutting parameters [2]. The left

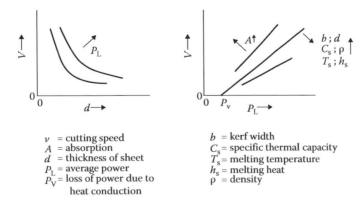

v = cutting speed
A = absorption
d = thickness of sheet
P_L = average power
P_V = loss of power due to
    heat conduction

b = kerf width
C_s = specific thermal capacity
T_s = melting temperature
h_s = melting heat
ρ = density

FIGURE 9.2 Processing parameters and their interdependencies. (From Hügel, H., *Strahlwerkzeug Laser*, Teubner Studienbücher, Stuttgart, Germany, 1992. With permission.)

field shows the dependency of the attainable cutting speed (v) vs. the thickness of the material (d). As mentioned before, a thicker workpiece requires lower cutting speeds. An increasing laser power (P_1) allows the processing of thicker workpiece with higher velocities.

In the right figure, the effects of specific material properties are shown. The energy losses due to heat conduction have to be resolved to permit the cutting of the material. A higher absorption (A) in the material leads to higher cutting speeds at constant average laser power. On the contrary, increased kerf width results in slower processes since more material has to be molten and blown out of the kerf during cutting. An increase of the quoted material properties, such as melting temperature, density, and the specific thermal capacity, cause a lower slope of the curve in Figure 9.2 and therefore a loss in cutting speed at constant laser power.

Absorptivity depends on the used material and its surface condition, and plays a significant role during laser processing since a sufficient amount of energy has to be absorbed to induce melting or vaporization of the material. In Figure 9.3, the absorptivity (A) of different metals is plotted against wavelength (λ). Absorption is strongly dependent on the wavelength [3]. For example, silver provides high absorption at UV wavelengths and shows increased reflection at longer wavelengths.

In Figure 9.4, a laser cutting head with a focusing mirror for CO_2 lasers (wavelength 10.6 μm) is depicted. Alternatively, processing heads using a lens as focusing optics can be used. However, mirrors usually possess a better thermal stability than transmissive focusing lenses since effective cooling during processing is possible. Lenses can be cooled as well, but the potential of effective cooling is limited.

During processing, a sharp gas jet (cutting gas: N_2, Ar, He, O_2, or compressed air) is targeted at the kerf to blow out the molten or vaporized material. An important function of the cutting gas is to protect the optics against fume and spatters. Depending on whether the material of the kerf is molten or vaporized, laser cutting can be subdivided into the following categories [4]:

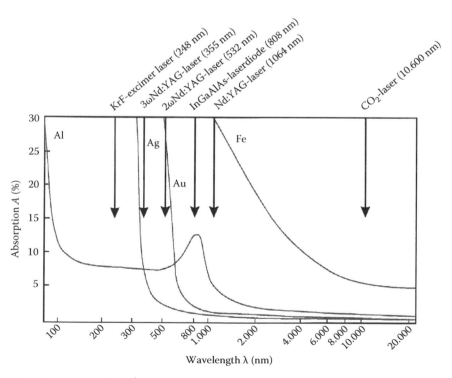

FIGURE 9.3 Absorption of different materials plotted against wavelength. (According to Killing, R., *Welding Processes and Thermal Cutting*, Verlag für Schweißen und verwandte Verfahren DVS-Verlag GmbH, Düsseldorf, Germany, 2001.)

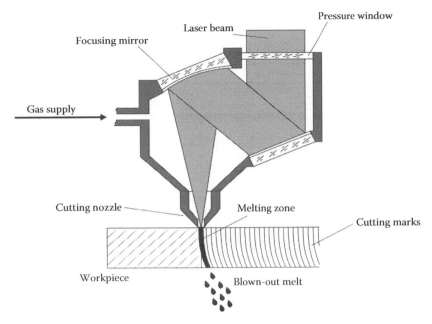

FIGURE 9.4 Schematic demonstration of a processing head with focusing mirror. (From Hügel, H., *Strahlwerkzeug Laser*, Teubner Studienbücher, Stuttgart, Germany, 1992. With permission.)

- Flame cutting (also known as oxygen cutting or combustion cutting)
- Fusion cutting
- Vaporization cutting

Flame cutting: Flame cutting is mostly carried out with oxygen as the cutting gas and leads to an exothermal reaction between the gas and the workpiece. Oxygen serves as an additional energy input supporting the laser cutting process. This enables faster cutting compared with fusion or vaporization cutting in which other gases are used. However, processing with oxygen results in cutting edges coated with an oxidation layer.

Fusion cutting: Compared to flame cutting, fusion cutting is carried out with inert processing gases. The function of the gas jet during fusion or vaporization cutting is to cool the cutting edge and shield it against oxygen. The absence of the exothermal reaction facilitates lower processing speeds relative to flame cutting but higher speed in comparison with vaporization cutting.

Vaporization cutting: Laser vaporization cutting is applied to materials without a distinctive liquid phase such as wood or paper, and to metals, ceramics, or plastics. The material in the processing zone is vaporized and melt formation is inhibited as far as possible. To vaporize the material, high-intensity laser radiation is required. This is why pulsed systems are applied in many cases. Due to the nonexistence of a liquid phase, there are normally minimum striations (since the mechanism can be mixed) on the cut surface as shown in flame or fusion cutting. In the case of vaporization cutting of materials with a liquid phase, the machinable thickness is decreased compared to materials without distinctive melt formation, since the generated vapor may condense on the cutting edges and seal the kerf.

The advantages of laser-based cutting compared to other thermal cutting techniques are the small HAZs, the possibility of cutting very complex shapes, clean-cut surfaces, narrow kerf widths, and very competitive cutting speeds. In contrast, there are high capital and operational cost. Compared with mechanical cutting processes, the laser-based process allows a wear free and, therefore, a reproducible process.

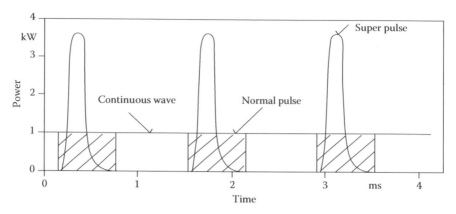

FIGURE 9.5   Different operating modes for laser cutting processes.

Laser cutting processes can be carried out with continuous or pulsed lasers (normal and super pulses). An example of these different operating modes is given in Figure 9.5. To process thick materials, pulsed laser is often advantageous because of reduced rounded kerf corners, side burning, and cut widths [5]. Pulsed laser cutting can achieve fine shapes due to limited heat conduction into the base material.

In industry, laser flame cutting has been applied to cut workpieces with a thickness of up to 40 mm depending on the material and the available laser power. However, such thickness requires a wide kerf to allow the effective removal of the melt. This results in relatively low cutting velocities, but the speed can still be very competitive compared with alternative methods. The actual processing parameters are 10 m/min for 1 mm, 3 m/min for 6 mm, and 1 m/min for 15 mm thick material, respectively, as reported in Ref [6].

Several criteria are considered to evaluate the quality of cutting. Important factors include sidewall striations, sidewall roughness, burr formation, erosion and pitting, taper, and the width of the kerf. Burrs are undesirable since an additional burr removal step needs to be used in production. Curved striations often appear during the processing of workpieces with increased cutting speeds, since the melt is blown out slightly behind the laser beam and not straight below. Generally, these curvatures as well as the kerf width do not affect the functionality of the workpiece. Precise and fine contours within a workpiece require narrow kerfs. On the other hand, erosion and pitting should be avoided since the cutting surface would be degraded functionally. Taper or perpendicularity describes the deviation of the cutting edge from a vertical cut.

The use of laser energy offers distinctive advantages in the field of carbon fiber reinforced plastics (CFRP). Conventional machining processes, such as mechanical milling and drilling, have reached their capability limits due to severe tool wear and high processing costs. The contact-free laser machining process is free of tool wear, and reduces the nonproductive time since tool changes are not as frequent as in mechanical machining. Additionally, the load on the workpiece during mechanical milling and drilling may cause delaminations. Controlled laser cutting processes can prevent this delamination of fibers from happening.

Fiber-reinforced materials can also be cut by abrasive water-jet energy that allows a wear-free machining as well as processing without heat affects of the workpiece but may lead to contamination because of the water or the abrasive material in the water jet. Figure 9.6 shows SEM pictures of UV laser cuts through a ply of a unidirectional carbon blanket with a thickness of 150 μm.

Figure 9.7 shows the results of UV-laser cutting of CFRP. The thickness is 2.5 mm in the left picture. The holes with a diameter of 5 mm and the fell-out material are shown in the right photo. The cutting edges exhibit smooth and crack-free surfaces with no visible HAZ or delamination.

  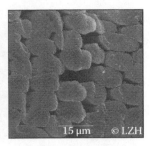

FIGURE 9.6 Cutting of dry carbon-fiber plies (SEM pictures) with UV-laser radiation. (From Denkena, B. et al., Novel UV-laser applications for carbon fiber reinforced plastics, in *Proceedings of Applied Production Technology APT'07*, Bremen, Germany, September 17–19, 2007.)

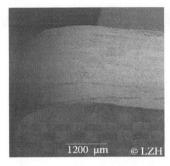 

FIGURE 9.7 UV-laser cut through CFRP (left) and cutting of holes (right). (From Denkena, B. et al., Novel UV-laser applications for carbon fiber reinforced plastics, in *Proceedings of Applied Production Technology APT'07*, Bremen, Germany, September 17–19, 2007; Völkermeyer, F. et al., Flexible and damage minimized machining of fiber plies and CFRP by UV-laser radiation, in *Proceedings of the CFK-Valley Stade Convention 2007*, Stade, Germany, June 13–14, 2007, CFK-Valley Stade, Stade, Germany, 2007.)

Laser cutting of wooden materials is possible. Figure 9.8 shows cuts through several types of wood with different thicknesses. For laser cutting of wood, CO_2 laser is the most suitable due to the high absorptivity at the wavelength of $10.6\,\mu m$.

A special application is the two-stage pre-notching and cutting process. This process allows the safe cutting of materials with a contaminated surface, as needed in the dismantling of nuclear installations. In this process, a bifocal optics is used to split up the laser beam into two less energetic beams. The first beam notches the top surface of the workpiece and an inert gas stream is used to blow out the contaminated melt and the fumes can be conveniently exhausted. The second laser beam follows the first beam and cuts through the cleaned surface. A schematic illustration of this process is given in Figure 9.9. On the right, the process with the two partial laser beams is depicted.

Fine cutting is mostly used in the electronics industry, such as solder masks in printed circuit board (PCB) manufacturing; or in the medical sector, where precise implants have to be cut. Cutting with conventional solid-state laser sources (e.g., Nd:YAG), which produce pulses in the temporal range of nanoseconds to milliseconds, has several limitations. First of all, the thermal load in the material due to heat conduction is relatively high, which results in large HAZs and melting. Melting leads to significant burr formation at the cutting edges, and may deposit solidified droplets on the surface. In many applications, both burrs and solidified droplets have to be removed via several subsequent treatments. Due to the thermal load in normal Nd:YAG laser cutting, the minimum producible structure size is limited. Moreover, a variety of materials cannot be structured using typical parameters from sheet metal manufacturing.

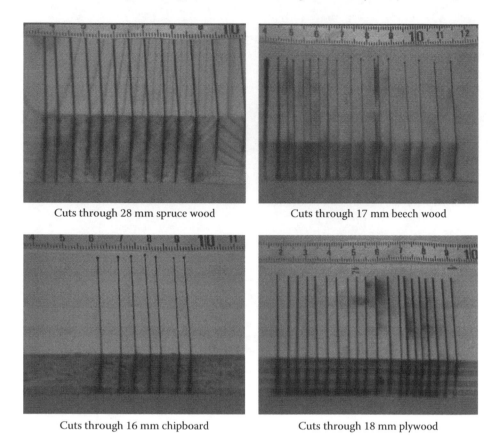

Cuts through 28 mm spruce wood

Cuts through 17 mm beech wood

Cuts through 16 mm chipboard

Cuts through 18 mm plywood

FIGURE 9.8 Laser cuts through several types of wood.

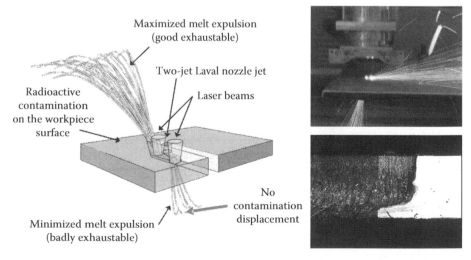

Maximized melt expulsion
(good exhaustable)

Two-jet Laval nozzle jet

Radioactive
contamination
on the workpiece
surface

Laser beams

No
contamination
displacement

Minimized melt expulsion
(badly exhaustable)

FIGURE 9.9 Schematic illustration of the two-stage pre-notching and cutting process (left) and picture of the process and the gradated cutting front (right). (From Hennigs, C. et al., Multifunctional hand-held laser processing device, in *25th International Congress on Applications of Lasers & Electro-Optics, ICALEO 2006*, Scottsdale, AZ, October 30–November 2, 2006.)

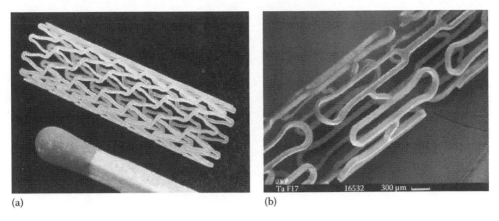

(a)   (b)

FIGURE 9.10   Femtosecond laser-machined cardiovascular implant.

Femtosecond (10^{-15} s) laser pulses can process almost all materials (metals, ceramics, glass, polymers, organic tissues, etc.) with minimum or no thermal damage. This is mainly due to the extremely short laser and material interaction time, although very high peak powers are used.

The energy per pulse and the average power of these femtosecond lasers are limited, making these lasers less cost-effective tools in the sheet metal industry than CW or longer pulse lasers. However, they can be used to process materials that are very difficult to process otherwise. Femtosecond laser cuts can be very smooth and completely burr-free, and free of depositions. The stent in Figure 9.10 has very smooth struts that cannot be machined using conventional laser techniques. The potential of femtosecond machining technique is even more pronounced in the case of extremely sensitive and delicate materials. At present, no other processes can cut and structure these materials with the achieved quality of femtosecond lasers. For many applications, the almost nonthermal nature of the femtosecond ablation process is necessary to ensure that material properties are maintained, including the area adjacent to the cut.

Structuring of medical implants, for example, coronary stents, is an emerging application for fine cutting. Coronary stents are used as a minimally invasive treatment of arteriosclerosis, and they can substitute for bypass surgeries. Since the requirements for medical implants (e.g., absence of burrs, x-ray opacity) are very strict, only a few materials are commonly used. Today, materials typically used for stents are stainless steel or shape memory alloys due to their super-elasticity. For these materials, chemical post-processing techniques have been developed to achieve the required properties. However, these materials are not optimal, concerning several medical aspects (risk of restenosis, limited biocompatibility, etc.). New approaches favor stents for temporary use only, which call for bio-resorbable and compatible materials such as Mg-based alloys, tantalum or special biopolymers (Figure 9.10). For these materials, no established post-processing technique is available. Furthermore, most of them react strongly to thermal load. Therefore, it is essential to avoid processing influences on the remaining material in order to retain the specific material properties. While CO_2 and excimer lasers show significant material modifications, femtosecond laser material processing meets the requirements of these sophisticated materials.

Due to their unique characteristics of minimal invasive ablation of material and tissue, femtosecond lasers offer new possibilities for medical treatment. In addition, fs laser material processing has become the popular tool of consideration when new and challenging materials are to be machined.

9.3   LASER DRILLING

Laser drilling is normally contact-free and without tool wear. Precision- and micro-drilling are employed in automotive industry, semiconductor manufacturing, and medical technology, for

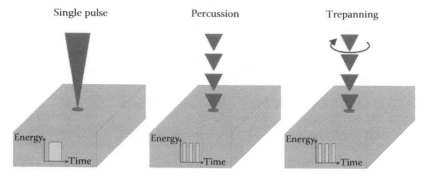

FIGURE 9.11 Schematic demonstration of the different drilling techniques.

example. Laser drilling can be melting-dominated drilling or sublimation-dominated drilling. Compared to melt drilling, sublimation drilling requires higher energy input due to the evaporation enthalpy and the less energy needed to blow out the molten phase by a gas jet. In most cases, pulsed systems are used to obtain suitable intensities to avoid melt deposition within the hole and to minimize the heat transferred into the bulk material.

As shown in Figure 9.11, laser drilling techniques include the following variants:

- Single-pulse drilling
- Percussion drilling
- Trepanning
- Helical drilling

Applications of laser drilling include the drilling of turbine blades for aircraft engines, fuel injectors in engines, wafers in solar cell manufacturing, and/or the perforation of packaging foils. In medical technology, laser drilling is used to fabricate surgery needles. Depending on the application, dimension and quality of the required hole, different methods of laser drilling are considered.

In single-pulse drilling, the hole is drilled within one single laser pulse, which is of high pulse energy. This method is mostly chosen if many holes have to be generated in a short period of time. However, the precision of these holes is somehow limited, and the reproducibility is strongly affected by the material homogeneity and/or the pulse-to-pulse stability of the laser system.

During percussion drilling, a sequence of pulses is used instead of one single pulse. There is no relative movement between the laser beam and the workpiece during processing. This drilling method allows the formation of deeper holes with a higher precision compared to single-pulse drilling. Smaller and more precise hole diameters can be realized, as well.

Trepanning is actually a combination of cutting and drilling. First, a through-hole is generated. Then the final diameter is cut from the drilled hole by a relative movement between the laser beam and the workpiece. The induced melt is mostly blown out of the hole at the backside via a gas flow that is coaxially guided around the laser beam. In principle, large holes of any diameter can be generated using laser-based trepanning. Cooling holes in turbine blades or the panels of combustion chambers can be realized by laser trepanning.

Trepanning can lead to excellent roundness of drilled holes. Since the focused beam moves in circles, the roundness of the hole is less dependent on the beam profile than when simply focusing the beam without rotation. Figure 9.12 compares the holes generated by percussion drilling and trepanning drilling. Trepanning hole drilling has better dimension control.

A variation of the trepanning process is helical drilling. In this case, the laser beam is moved on a circular path and runs along a helical path through the material, whereas the workpiece remains in position. In contrast to trepanning, no through-hole is generated in the beginning. With helical

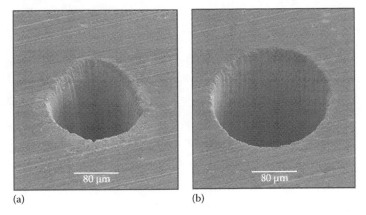

FIGURE 9.12 Femtosecond laser drilling in stainless steel using percussion (a) and trepanning (b) techniques, femtosecond laser pulses have been used to emphasize the difference.

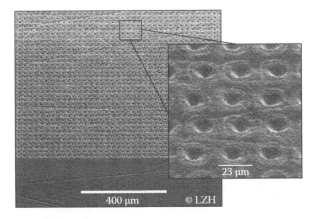

FIGURE 9.13 Laser structured metallic surface for tribological applications (blind holes). (From Siegel, F. et al., Laser-based surface modification—micro structures on macro areas, in *Proceedings of the Second Pacific International Conference on Application of Lasers and Optics 2006-PICALO*, Melbourne, Australia, April 3–5, 2006.)

drilling, holes with changing diameters, such as countersunk holes, can be drilled. Helical drilling is used to drill high-pressure injection nozzles that require holes without any melting redeposition.

To obtain holes with very high precision, shorter laser pulses should be used. Figure 9.13 shows a laser structured metal surface that is patterned using a picosecond laser (ps laser). Each blind hole in the figure used 600 pulses. The generated holes have a diameter of about 13 μm and a depth of 6 μm. These dimples on a surface are useful for friction reduction purposes [10].

It is meaningful to compare the results of drilling with different pulse durations. Figure 9.14 compares a hole made by nanosecond ablation and a hole drilled by a femtosecond laser (a titanium:sapphire laser with an emitting wavelength of 780 nm). The femtosecond laser shows much better drilling quality.

Femtosecond laser drilling leads to minimized HAZ, crack-free ablation and reproducible results. Further examples of fs laser drilling are given in Figure 9.15. Diamond, steel, tooth, silicon, and suprasil were drilled or milled. Clean and sharp features were achieved. Such results are difficult for other lasers.

For precise deep drilling of metals, which is necessary in many industrial applications, the most important criteria are the tolerances and quality. Similar to the cutting processes described above,

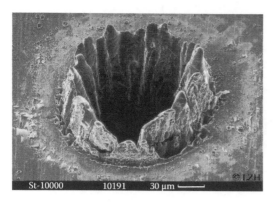

FIGURE 9.14   Comparison of a nanosecond (pulse duration 3.3 ns) in the left picture and a femtosecond (pulse duration 200 fs) drilling in the right picture.

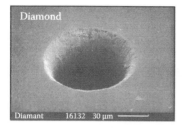
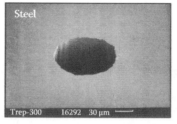

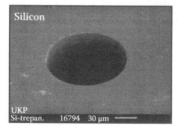
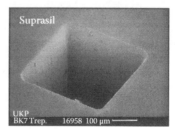

© LZH

FIGURE 9.15   Drillings and ablation in different materials with femtosecond titanium: sapphire laser.

ultra-short pulsed lasers are able to fabricate holes with special geometry, superior quality, and high reproducibility. These lasers have the potential to provide a simple processing technique that does not require any additional post-processing or special gas environment. Although extremely high intensities are applied, the thermal destruction is negligible. The high intensities are required to rapidly drill through. After the through-hole is drilled, the high intensity part of the laser pulse propagates directly through the hole without absorption. At the edge of the laser pulse, laser intensities are close to the ablation threshold, and only a small amount of material can be removed under such intensities. Starting from this point, the interaction of femtosecond laser pulses with the workpiece occurs in the low-intensity regime, bringing in some special benefits. It is responsible for the excellent hole quality and it actually becomes a low-intensity finishing process at this stage. The high-quality holes and their high reproducibility are depicted in Figure 9.16.

In conclusion, ultra-short pulsed laser material processing at high laser intensities is a practical tool for high-quality, deep drilling of metals. So far, however, industrial users have been reluctant to integrate femtosecond laser drilling in mass production, as the laser sources are known to be quite complex and expensive. But the market perception is changing, thanks to the significant advances in laser sources concerning the ease of use, the reliability, and the overall lifetime of the system.

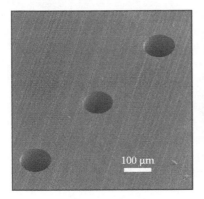

FIGURE 9.16 Scanning electron microscope image of holes drilled in 1 mm thick stainless steel plate using femtosecond laser pulses, and their replicas.

9.4 LASER-BASED JOINING

9.4.1 LASER WELDING

In laser welding, the absorption of laser radiation at the surface of the workpiece plays a decisive role as in all the other laser processes. The absorption of sufficient energy leads to the formation of a melting pool that is needed to join the workpieces. Laser welding can be performed with or without a filler material, and the laser radiation can be continuous wave (CW) or a pulsed mode. In the case of welding with a filler material, this additive material is typically a wire that is fed continuously into the processing zone. Burrs or the uneven cut edges of the mating plates could cause negative effects on the weld.

The use of a filler material in the welding process allows the joining of materials that are otherwise not weldable or may cause undesired properties like the formation of brittle intermetallic phases or an unwanted hardening of the welded material. In the same way, wider gaps can be bridged using additives resulting in reduced tolerances during seam preparation. Of course, using filler material requires additional energy to melt it, which leads to decreased processing velocities and increased cross section of the weld. Generally, the welding process can create a continuous welding seam, a quilting seam, or individual welding spots. Modifications of the weld seam geometry are often necessary in order to avoid distortion of the heated workpiece.

The advantages of laser welding over alternative joining techniques are

- Very narrow (few tenths of a millimeter or more, key hole welding) and deep (a couple of millimeters, deep penetration welding) weld seams, a depth-to-width ratio up to 10 can be generated (in extreme cases >20 have been reported using fiber or disk lasers)
- Localized heat input → minor shape distortion, fewer finishing processes
- Efficient operation with shielding (inert) gases due to the small weld seams
- Because of the small HAZ, the risk of corrosion is reduced
- Higher welding velocities
- Non-contact fabrication and therefore no tool wear
- Easily automatable
- Appropriate to lightweight construction

Compared to electron-beam welding (EBW), laser-beam welding induced a little larger heat input into the workpiece. The efficiency of EBW has improved. But due to the development of fiber and disk lasers, the efficiency of laser beam processes is still superior. Moreover, laser beam welding is often more practicable, especially for workpieces with large dimensions, where a vacuum chamber environment is too expensive or impractical. A variant of the EBW process is the non-va cuum-electron-beam-welding method where the electron beam, which is generated under vacuum, is guided into the atmosphere via several compression stages and nozzles, resulting in lower achievable welding depth.

Laser-welded joints can be continuous weld seams or single weld spots. Figure 9.17 depicts some of the common welding joint arrangements schematically. The arrows indicate the direction of the applied laser energy.

During laser welding, attention must be paid to the parameters shown in Figure 9.18. The absorption at the surface of the material at certain laser wavelength is an important criterion. Absorption depends on the chemical composition of the material and the surface condition of the workpiece among other factors. Likewise, the angle of incidence is important.

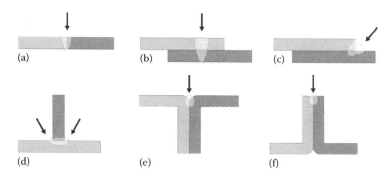

FIGURE 9.17   Some common welding joint arrangements: (a) butt joint, (b) lap joint, (c) fillet weld in a lap joint, (d) T-joint, (e) flare weld, and (f) flange joint.

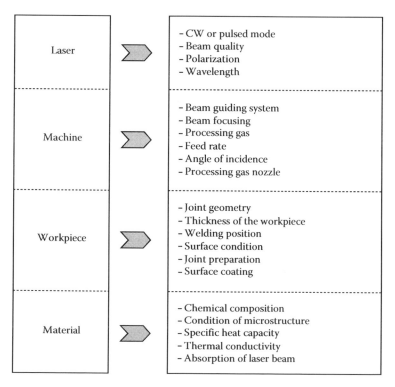

FIGURE 9.18   Parameters during laser welding.

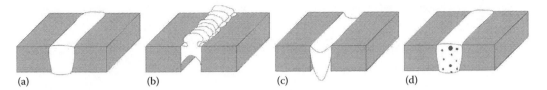

(a)   (b)   (c)   (d)

FIGURE 9.19   Weld shapes and non-conformities: (a) good seam, (b) humping, (c) drop out, and (d) pore formation.

As a result of disregarding one or more of the interacting parameters, inappropriate weld seams may occur. Possible non-conformities of laser-welded seams are pore formation, humping, drop out, or increased hardening of the weld (see Figure 9.19). Pore formation is one of the most common defects of laser-welded seams. Because of the high process velocities and high thermal gradients, the embedding of gas bubbles during solidification of the melt may take place. Humping, the formation of humps on the surface of the weld, often appears when welding velocities are not optimized. Excessive welding velocities can lead to lack of weld penetration, and insufficient welding velocities can lead to material drop out. Welding velocities in between could result in the humping effect on the weld crown and/or root. A stable condition occurs between the humping and the lack of penetration conditions.

In addition, cracks may appear during or after the welding process. The formation of cracks can be partially avoided by preheating and controlled cooling of the workpiece. Other factors influencing crack sensitivity include the inclusion of hydrogen in the workpiece and the aspect (width-to-depth) ratio of the welded seam.

The differences between continuous and pulsed laser-welded seams in steel H340LAD-sheets with a thickness of 1.25 mm are shown in Figure 9.20. The top figures compare the top views of the CW and pulsed laser welds, while the lower figures compare their cross sections. Since the parameters (such as spot size and welding velocity) in the CW and the pulsed laser welding are different, this only gives a qualitative evaluation.

Processing with pulsed laser power requires the control of additional parameters compared with continuous laser welding. Parameters to be considered include pulse duration, pulse energy, pulse frequency, overlapping, and pulse peak power. The installation costs of pulsed systems are normally cheaper. The drawbacks of pulsed welding processes are lower processing speeds and limitations in welding depths. The different heat input of continuous and pulsed laser welding is evident in Figure 9.20.

9.4.2   HEAT CONDUCTION WELDING

During heat conduction welding, laser radiation is absorbed at the surface of the workpiece and then transferred to the surrounding material via heat conduction that results in low welding depths. Therefore, this method is mainly suitable for joining of foils and thin sheets with thicknesses generally not exceeding 2 mm. The required intensity is sufficient to melt the mating components but not sufficient to evaporate them. The achieved welding depths depend on the thermal conductivity of the involved materials. A lower thermal conductivity leads to a slower heat diffusion and facilitates deeper welding seams and, respectively, higher welding velocities. A fusion of the molten materials takes place at the components interface and results in a welded joint seam after solidification. The welded joint seam commonly has a wider dimension than the laser spot size due to the high heat conduction from the weld-processing zone to the parent material.

Heat conduction welding is applied in order to join thin-walled workpieces with plane and rounded edge preparations and thus allows their line interface to be visual. An example of an industrial application is the heat conduction welding of stainless-steel sinks with diode lasers.

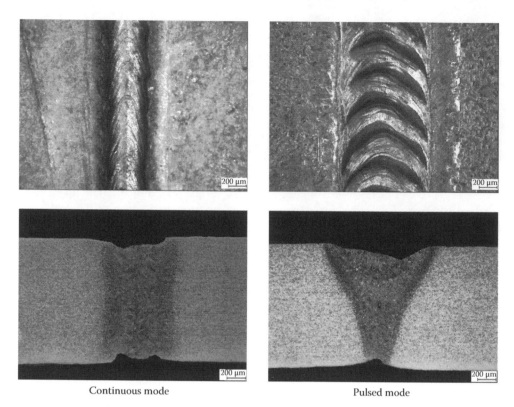

Continuous mode   Pulsed mode

FIGURE 9.20 Comparison of laser welding seams in Galvatite high-strength steel (H340LAD) in continuous and pulsed mode.

9.4.3 DEEP PENETRATION WELDING

The method of deep penetration welding requires very high power densities to induce the formation of metal vapor in addition to the melt. Due to the metal vapor, a pressure is generated in the melt resulting in a dynamic movement of the liquid phase. This incident results in the generation of a vapor capillary, a so-called keyhole. This keyhole is encircled by the melt during processing and solidifies on the backside of the keyhole depending on the welding direction. Thus, a very narrow weld seam with a high depth can be produced. Multiple reflections of the laser beam inside of the keyhole allow a higher absorption of radiation even while processing highly reflective materials. If the metal vapor is ionized by the laser beam due to a sufficient amount of metal vapor and intensity of the laser beam, a plasma will be induced that offers an additional energy input so that the welding velocity can be increased. Because of the higher velocities, the tendency of pore formation is raised compared to heat conduction welding. Otherwise, the produced gases may pass out of the processing zone through the generated keyhole whereby an important factor concerning pore formation is constricted. However, by passing a threshold energy density, a shielding of the laser beam occurs and causes a diminished intensity of the laser beam in the processing zone or a total abortion of the welding process. This effect of shielding is of higher interest during welding with CO_2 lasers compared to Nd:YAG lasers as the absorption of CO_2-laser radiation in the metal vapor plasma is, respectively, higher. In Figure 9.21, a laser-based deep penetration welding process of a tailored blank using a 6 kW CO_2 laser is shown.

A comparison of heat conduction welding by laser and deep penetration welding is given in Figure 9.22. In the left field, the low welding depth of heat conduction welding is apparitional whereas the right field demonstrates the formation of a keyhole and, therefore, increased welding depths.

FIGURE 9.21 Laser deep penetration welding process of a tailored blank using a 6 kW CO_2 laser.

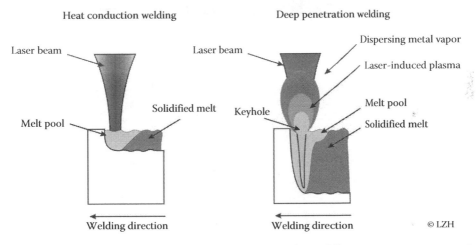

FIGURE 9.22 Comparison of heat conduction and deep penetration welding.

Feasible application fields for laser-based deep penetration welding are to be found in the construction of vessels, tanks, and containers, shipbuilding, and the automotive industry as well as in microtechnology. Adopted laser sources are CO_2 lasers, solid-state lasers, and diode lasers with sufficient power. Most recently, fiber and disk lasers have been used due their excellent beam quality that is predestined for deep penetration welding.

9.4.4 Hybrid Welding Techniques

Laser hybrid welding consists of a coupling of laser beam welding (CO_2 lasers, solid-state lasers, or diode lasers) with arc welding (gas tungsten arc [GTA], gas metal arc [GMA], or plasma welding) processes and is known since the end of the 1970s [11,12]. The technology of hybrid welding allows the utilization of the advantages of both processes in combination:

- Facilitated arc ignition
- Stabilization of the arc-welding process
- Higher penetration depths and welding speeds compared to the single processes

Figure 9.23 shows the difference between a combination and a coupling of a laser and an arc-welding process. The combination of both processes is associated with a distance *d* between the processing zone of the laser beam and that of the arc-welding torch. A following arc enables by way of example

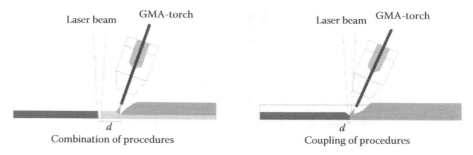

Combination of procedures   Coupling of procedures

FIGURE 9.23 Difference between a combination and a coupling of the two processes of laser beam welding and arc welding. (According to Dilthey, U. and Keller, H., Pilotstudie zum Einsatz des Laser-MSG-Hybridprozesses zum Hochleistungsschweißen von Stahl, Technical Report P 426/A 139, Stahl-Zentrum Düsseldorf, Germany, Studiengesellschaft Stahlanwendung, 2001. With permission.)

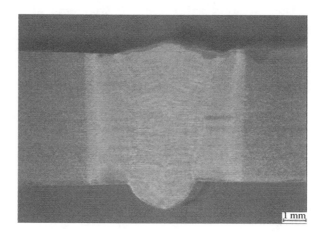

FIGURE 9.24 Cross section of a hybrid welding seam in a fine-grained steel.

an annealing of the laser welded seam and therefore a reduction of hardness and residual stresses. A diminishment of this distance that results in one shared processing zone causes a coupling of the two procedures and constitutes the hybrid welding technique as it can be seen in the right part of Figure 9.23. Another advantage of the hybrid welding process is the interaction between the coupled processes that involves a more stable process.

A cross section of a hybrid welded seam of a fine-grained steel is given in Figure 9.24. A sagging of the seam root at the bottom side that can be seen in this figure in moderate extent often necessitates finishing processes after welding and this can be diminished by correcting the energy input of the laser beam. However, a continuous weld seam through the whole workpiece of about 5 mm thickness has to be maintained.

9.4.5 Laser-Guided or Laser-Stabilized Welding

The main principle of laser-guided un-stabilized welding is the interaction between the electric arc and the laser beam. Depending on the adjustment of the welding parameters, there is a limit for the feeding speed of the GMA itself. Above this limit, the arc toe loses its stability and the welding seam is characterized by localized arc workpiece interaction. If a laser beam is used to guide the arc, feeding rates can be much higher. For this positive interaction, the power of the laser should range between 10% and 20% of the total power input. For this reason the heat influence and the continuous radiation of the laser is reduced to a minimum. The radiation of the laser is exclusively

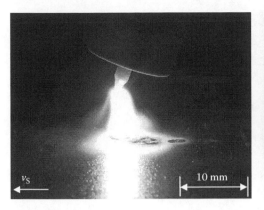

FIGURE 9.25 50 A-TIG-toe without laser (left) and a 50 A-TIG-toe stabilized by a 300 W Nd:YAG-laser (right).

used for the stabilization and guidance of plasma plumes of the electric arc [14]. Figure 9.25 shows pictures taken from the process without (on the left) and with a guiding laser beam (on the right).

9.4.6 REMOTE WELDING

During remote welding, the positioning movement and feed motion is realized by the deflection of the laser beam on moveable high-reflecting mirrors. The lightweight mirrors facilitate a very precise positioning of the laser beam with high velocities. Because of that, the nonproductive time of the process cycle can be reduced and thus the efficiency is enhanced. The large working distances existing in remote welding processes with focal lengths exceeding 1000 mm realize a rugged process. Due to the processing with scanning mirrors, the angle of incidence of the laser beam varies depending on the position of the working zone on the workpiece. Hence, the reachable welding depth is varying as well. Likewise, the nonexistence of a processing head close to the working zone poses an obstacle to a directed supply of processing gases. Today, remote welding is prevalently applied in automotive and component engineering. In many cases, the remote welding process is robot guided to obtain more flexibility concerning component geometries. A remote welding setup with a scanning system, which is mounted on an industrial robot, is shown in Figure 9.26 [15].

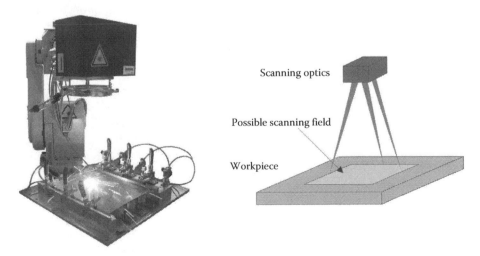

FIGURE 9.26 Remote welding setup with a scanning system mounted on an industrial robot (left) (From KeySysTech GmbH, Germany. With permission.) and principle of remote welding (right).

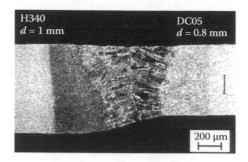

FIGURE 9.27   Cross section of a laser welded tailored hybrid blank. Materials: High-strength H340 (left) and mild steel DC05 (right).

9.4.7 TAILORED BLANKS

The proceeding concepts of lightweight construction in several industrial sectors afforded the demand for the joining of sheets with differing thickness, tailored blanks, or, respectively, joining of materials with differing quality and thickness, called tailored hybrid blanks. A cross section of a tailored hybrid blank made of H340 and DC05 is shown in Figure 9.27.

Deep drawing or comparable methods of plastic deformation, subsequently, shape the joined semi-finished parts. Tailored blanks can be joined in butt and overlap configuration [16]. While joining dissimilar materials attention must be paid to the formation of intermetallic phases in the processing zone. These phases are mostly brittle and deteriorate, as a result, the mechanical properties of the joint. The use of filler material or processing with a reduced heat input may inhibit or at least diminish the formation of these phases.

9.4.8 BRAZING

Brazing is a joining method that uses a filler material with a lower melting temperature compared to the joining partners. Therefore, in most cases only this filler material is molten during processing and flows in the joining gap where a diffusion-based bonding is realized. In addition to the joining partners and the filler material, a flux is often necessary to remove oxide layers from the surfaces of the joining partners. This flux even avoids an oxidation of the joining partners and the filler material but may cause a corrosive effect as well. A brazed seam exhibits a very plane and clean surface that renders finishing processes unnecessary. Brazing permits the joining of materials with a wide difference in the melting temperature where a welding process is complicated, respectively, impractical. Brazing is subdivided into soldering (also called soft brazing) up to temperatures about 450°C, brazing from temperatures of 450°C up to 900°C and high-temperature brazing at temperatures above 900°C. Brazing allows the joining of materials that are not suitable for any welding process. Because of the low-temperature melting filler material, the process can be carried out at lower temperatures compared to laser welding so that the HAZ is even smaller. Furthermore, laser-based brazing allows the generation of dense and nonporous seams. However, the achievable processing velocities are smaller than in laser welding processes. In the middle of the 1970s, Bohman [17] published the first research results concerning brazing with laser radiation. In Figure 9.28, an experimental setup of a laser brazing process with wire feeding is shown.

Using the example of hard metal cutting edges on base plates the laser brazing technique allows processing with lower heat input in the base plate compared to conventional flame- and induction-brazing processes. That way, the durability of the tools can be enhanced under a concurrent reduction of the required production time, since an overheating is avoided. An excessive heat input leads to

- A decrease of hardness in the HAZ resulting in a loss of strength inside of the saw tooth and its tearing during processing

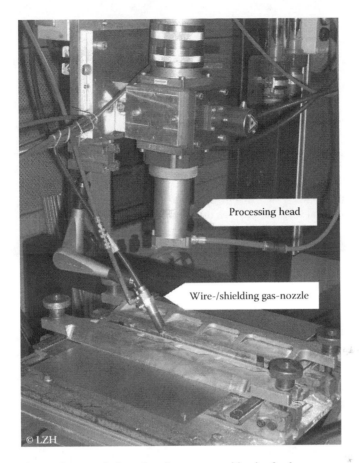

FIGURE 9.28 Experimental setup of a laser brazing process with wire feeder.

- Thermally induced residual stresses in the base plate and therefore to an increased wear of the cutting edges
- A distortion of the saw blade that necessitates a following flattening step

A picture of a laser-brazed saw blade with a diameter of 400 mm is given in Figure 9.29.

Another application example is the brazing of tailored hybrid blanks. These are tailored blanks as described above, consisting of different joining partners like steel and aluminum that are formed after the joining step. Thermal joining of these materials may result in a decreased ductility of the

FIGURE 9.29 Laser brazed saw blade of a diameter of 400 mm.

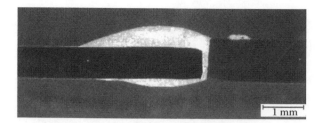

FIGURE 9.30 Example of a brazed tailored hybrid blank between DC05 and an Al-alloy AA6016 with a zinc-based brazing material.

seam because of the formation of brittle Fe_xAl_y-phases. Therefore, the formability of inappropriately joined tailored blanks and their adoption as lightweight construction materials is limited. An acceptable solution provides a laser beam brazing process during which the aluminum is not molten to avoid an embrittled seam. Previous studies with zinc as a braze metal showed that the amount of Fe_xAl_y-phases can be reduced but not impeded completely [18]. Other research studies handle with a combined welding and brazing technique where aluminum sheets were joined with zinc-coated steel sheets under the use of zinc- and an aluminum-based filler material [19]. An example of a laser-brazed tailored hybrid blank of a cold-rolled steel DC05 and an aluminum-alloy AA6016 with different thicknesses is given in Figure 9.30. The additional brazing material was a zinc-based type.

9.4.9 Laser Welding of Polymers

Regarding laser processing of polymeric components, laser welding is a comparatively new technology. Like laser processing of metallic workpieces, a contact-free machining and a precise energy input provide important benefits to the joining of polymers. Common techniques are laser penetration or transmission welding of thermoplastic polymers. In transmission welding, the laser beam passes through a laser-transparent component and is absorbed afterward at the joining partner with a higher absorptivity. The absorbing material plasticizes and via heat conduction, the transparent joining partner is plasticized, as well. A positive locking connection of the joining partners is a condition precedent to a successful joint. The technique of laser transmission welding allows the generation of high-quality and even optically hidden weld seams between the joining partners. A schematic demonstration of a laser transmission welding process is given in Figure 9.31.

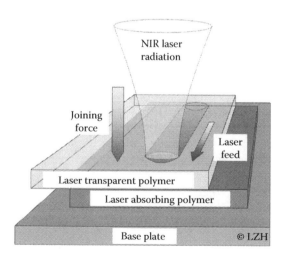

FIGURE 9.31 Schematic demonstration of transmission welding of thermoplastic polymers using near-infrared (NIR) laser radiation.

Due to the localized heat input, thermal damaging in other parts of the workpiece than the joining zone can be avoided. In particular, in mechanically and electronically high-sensitive micro-components, laser welding offers a new machining potential. A suitable process and assembly design moreover avoids the release of unhealthy by-products. Certainly, an essential drawback of transmission welding of polymers is the necessity of different optical properties of the joining partners that can be achieved by filler or additive materials to tailor the absorptivity.

Regarding the different laser sources, CO_2 lasers are particularly applied for welding of plastic foils, because of a low optical penetration depth of the CO_2-laser radiation in thermoplastic polymers. In the range of overlap joints, high-power diode lasers are preferred, since their emitting wavelengths in the near-infrared zone (NIR) generally pass through the upper workpiece.

Applied methods in laser welding of polymers are

- Contour welding
- Mask procedure
- Quasi-simultaneous processing
- Simultaneous processing

During contour welding, a relative movement between the workpiece and the laser beam has to be realized. This movement is arbitrarily programmable, thus, the welding seam contour can be designed with a very high flexibility.

As the name implies mask welding uses masks that are irradiated by the laser beam so that the contour of the mask is projected on the surface of the workpiece. Another method uses a mask that is placed in proximity to the joining zone and simply allows an admission of laser radiation in the designated area. This way, the generation of very precise welding seams is possible. Mask procedures are realized by a sequential tracking of the mask that is comparable to contour welding or by treating the whole mask with one single radiation pulse.

Quasi-simultaneous processing is based on scanning optics as mentioned before while describing the remote welding process. The high-reflecting mirrors allow high processing speeds and a focusing of the laser beam on the surface of the workpiece within the scanning field. Due to the high attainable speed, the laser beam is guided quasi-simultaneously across a two-dimensional welding seam contour.

On the other hand, simultaneous welding of a complete welding seam within one process step with a single laser radiation pulse is feasible because of the availability of high-power diode lasers. The advantage of this method is that processing of high quantities with very short processing periods is possible.

Compared to conventional polymer joining processes like bonding methods, laser-based transmission welding allows joining without idle periods. Additionally, the dispensing problem with the adhesive does not occur. Figure 9.32 shows an example of the application of laser transmission welding.

9.5   LASER SURFACE TREATMENT

Surface treatment under the adoption of laser-based processes offers the possibility of a selective treatment of the workpiece. The modification of the surface zones leads, for example, to an improved durability of the component that allows cost reduction and therefore an enhanced profitability. Surface technologies using laser energy are among other processes such as alloying and dispersing, hardening, and remelting as well as laser cladding in macro- and micro-dimensions in the field of repair welding, rapid prototyping, and rapid tooling. The different types of laser surface processes are presented in detail in the following sections. As rapid prototyping is of particular importance nowadays, there is a separate detailed section at the end of this chapter discussing the opportunities and limitations.

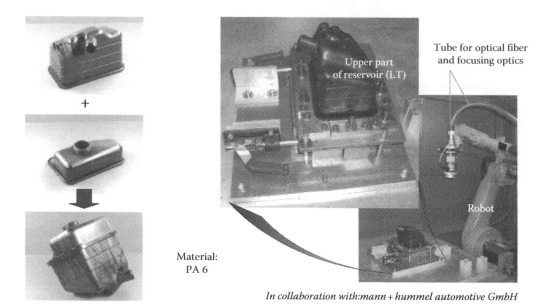

In collaboration with:mann + hummel automotive GmbH

FIGURE 9.32 Laser transmission welding of steering oil reservoirs made of PA 6. (From Haferkamp, H. et al., Welding of steering oil reservoirs using a high power diode laser system, in *Proceedings of the Second International WLT-Conference on Lasers in Manufacturing 2003*, WLT, Munich, Germany, June 2003, pp. 113–118. With permission.)

9.5.1 ALLOYING AND DISPERSING

As mentioned above laser-based processes allow selective alloying and dispersing of workpieces. Alloying elements that are inserted in the near-surface area by laser power are completely molten in the process and change the chemical composition and therefore the properties of the material. Thus, low-cost substrate materials can be alloyed to create surface zones with a high quality. Common substrate materials are cast iron, un- or low-alloyed steels as well as aluminum alloys.

On the contrary, dispersing materials are often ceramic particles like TiC, WC, and ZrO_2. These materials show higher melting temperatures and a high hardness and improve, if evenly spread on the base material, the wear behavior, and the strength of the workpiece.

The alloying and dispersing materials are often used in powder form but there is also the possibility to operate with pastes or wires. Figure 9.33 demonstrates how the alloying material can be applied in the process. On the left side, the simultaneous application of the additive material with an off-axial nozzle (single-step process) is shown whereas the second drawing demonstrates the previous deposition of this material on the surface of the substrate (two-step process). The simultaneous application of powder materials is realized by an off-axial (see Figure 9.33a) or a co-axial powder nozzle.

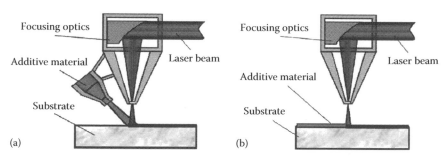

FIGURE 9.33 (a) Single-step and (b) two-step laser process with additive material.

Certainly, it is aspired to manage the single-step process because of saving one processing step and therefore useful time. However, there are many more process parameters to handle in the case of the single-step process. A very proper adjustment of several parameters has to be carried out to obtain feasible results. Parameters to be adjusted in both, the single- and the two-step, processes concerning mechanical handling are

- Type and quantity of shielding gas
- Geometry and position of the nozzle
- Focal position and focal diameter
- Preheating of the workpiece
- Traverse speed
- Exposure times
- Control of the temperature

Of course, these are not all of the influencing parameters that have to be noticed. The physical properties of the workpiece like melting temperature or the wettability behavior, for example, and the surface condition are important values, as well.

Processing the single-step method, the following parameters have to be considered additionally:

- Type and pressure of carrier gas
- Type of additive material
- Feeding rate and feeding method of the additive material
- Optical properties (i.e., absorption and scattering)

In Figure 9.34, the single-step laser dispersing process is schematically shown with an off-axial powder nozzle in longitudinal and cross section. The black spots in the dispersed layer represent the evenly spread ceramic particles. This dispersed layer is surrounded by an HAZ with a marginal extent as common in laser processing.

Compared to conventional wear-resistant layers, laser dispersing allows the generation of layers with a gradient concentration of the particles from the surface to the inner part of the workpiece.

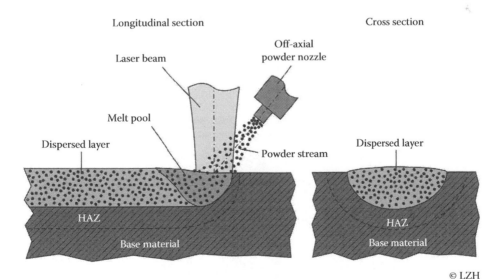

© LZH

FIGURE 9.34 Schematic illustration of single-step laser dispersing. (From Deutschmann, M., Lasergestützte Herstellung von Keramik-Stahl-Werkstoffverbundschichten für Werkzeuge der Blechumformung, Dissertation, University of Hannover, Hannover, Germany, 2007. With permission.)

This gradient dispersion leads to an improved support and adhesion of the wear-resistant layer by the base material and avoids flaking of the coating.

9.5.2 Hardening and Remelting

As mentioned before, the possibility to separate the function of surface and base material can be exploited for cost reduction using laser power. Hardening of the surface area of workpieces can lead to a decreased wear of the material. Heavy stressed tools for sheet metal forming of high strength steels, for example, are predestined for surface hardening to increase their durability. Generating a martensitic microstructure in hardenable steels leads to an increased hardness of the surface area of the workpiece depending on the thickness. In the martensitic structure, the carbon is dissolved under constraint in the tetragonal distorted lattice. To achieve this microstructure a specific temperature profile for each material has to be applied subsequently (time–temperature transformation [TTT] diagrams) with high cooling rates that is reached during laser processing (cooling rates between 100 and 1000 K/s).

In Figure 9.35, a schematic TTT-diagram of a heat-treatable steel with continuous cooling is shown with the temperature axis scaled logarithmically. The surface of the workpiece is heated above the austenite temperature and then cooled down continuously. Depending on the cooling rate, different microstructure conditions are passed through. Cooling curve number 1 is the one with the highest cooling rate and can be assigned with the process of laser hardening. This curve only passes the areas of austenite and martensite. Lower cooling rates, as demonstrated with the curves 2 and 3, can be reached in conventional hardening processes by quenching of the workpiece in water or oil after austenitizing.

There are even materials that are air-hardened. Cooling down following curve number 3, a ferritic–pearlitic microstructure is obtained by crossing these areas in the TTT-diagram whereas cooling rate number 2 leads to a bainitic martensitic structure. Every material has a specific TTT-diagram and even the time and temperature of austenitizing change the shape of the different sections.

Maximum hardening depths that can be reached by laser hardening are in the range of about 2 mm depending on the material and the workpiece geometry. The most important advantage of laser-based hardening over conventional processes is the potential of a local and flexible hardening so that finishing processes can be reduced.

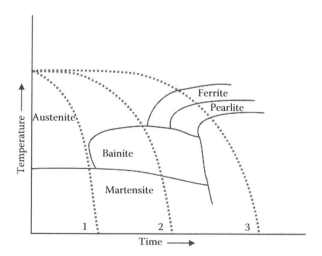

FIGURE 9.35   Schematic time–temperature transformation (TTT) diagram of heat-treatable steel.

9.5.3 LASER CLADDING

Laser cladding describes the buildup of layers on a substrate material under the utilization of an additive material. Several methods of depositing this additive have been described in the alloying and dispersing section whereas simultaneous powder feeding by a nozzle in a single-step process is proven as the most effective method [22]. As a matter of principle, the same assembly as shown for laser alloying and dispersing can be used for laser cladding. To avoid anisotropy of the built-up layer, coaxial powder-nozzles are usually chosen for powder feeding to the processing zone. An example for such a coaxial nozzle is given in Figure 9.36. Laser cladding is adopted in the range of generating wear-resistant coatings, like alloying, and dispersing, repair welding, and rapid prototyping as described in the following sections. Some metallic 3D-structures generated by laser cladding and made of Stellite 6 and Inconel 625 are shown in Figure 9.37.

9.5.4 REPAIR WELDING

Repair welding in the range of cladding is used for high-quality and expensive workpieces. Repairing of partially damaged components is often more economic than replacing the whole part. In the case of turbine blades, for example, the tip and the outer part of the blade can be rebuilt by laser cladding to allow for multiple use. Therefore, the damaged part is removed by a grinding step

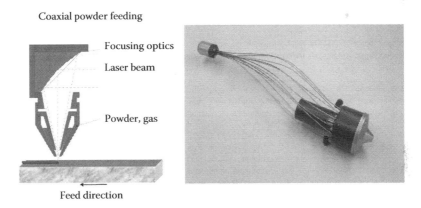

FIGURE 9.36 Schematic demonstration and a construction example for a coaxial powder nozzle.

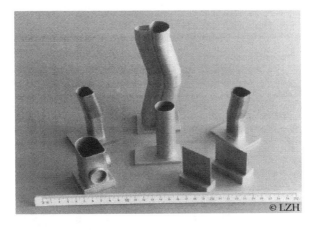

FIGURE 9.37 Metallic structures built up by laser cladding.

and rebuilt by a powder fed laser cladding process. A concluding grinding process obtains the final shape of the blade. A laser cladded tip of a turbine blade is shown in Figure 9.38. Within the scope of repair, welding extremely loaded forming dies are processed as well.

9.5.5 Laser Ablation

So far, in the range of laser surface treatment only generative methods have been mentioned. Therefore, this and the following paragraph will basically deal with surface ablation or modifying techniques. The ablated material may accrue in terms of oxide chips, melt, or slag and metal vapor. Different methods of laser ablation—chip removing, melt removing and removal by cutting—are shown in Figure 9.39. Laser ablation processes are mostly carried out with pulsed radiation to achieve high intensities and scanning optics for flexibility.

Using laser ablation, the surface of nearly all materials can be structured. Laser ablation is often applied to metals, ceramics, and plastics in order to generate microstructures. An application example is the machining of injection molds for micro-components. Also, in common tool and mold making, the laser is a suitable and often adopted processing device. In Figure 9.40, SEM-pictures of laser chip removing on a C45 tool steel using a 50 W diode laser can be seen. The left picture shows a bulking of the laser-treated section and its increasing ablation from the base material whereas the chip is already formed in the right picture.

FIGURE 9.38 Repair welded tip of a turbine blade. (From Weidlich, N. et al., Individual laser cladding for high pressure turbine blades, in *First EUCOMAS*, Berlin, Germany, May 26–27, 2008.)

An example of an application concerning selective laser ablation of fiber layers in carbon fiber reinforced plastics (CFRP) using UV-radiation is depicted in Figure 9.41. The number of scan cycles has been increased from the left to the right picture. Using scanning optics, any processing shape is possible.

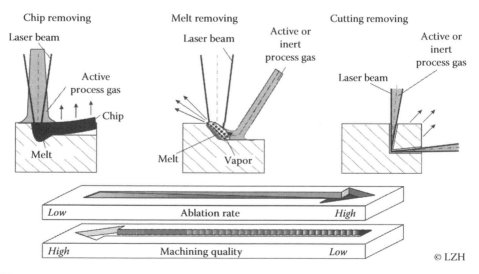

FIGURE 9.39 Principles of laser-based removal of geometric structures.

Bulking and ablation from   Chip formation
the base material

FIGURE 9.40 Laser chip removing of C45 carbon steel using a 50 W diode laser. The specific volume of the oxidized iron is higher compared to the bulk material resulting in surface layer stresses. The oxide material peels away like a chip in mechanical cutting (SEM photos).

FIGURE 9.41 Selective removal of fiber layers in a CFRP part by laser ablation with UV radiation, from left to right: increasing number of scan cycles. (From Denkena, B. et al., Novel UV-laser applications for carbon fiber reinforced plastics, in *Proceedings of Applied Production Technology APT'07*, Bremen, Germany, September 17–19, 2007.)

9.5.6 LASER MARKING

Laser marking is of high economic interest since it has great potential concerning resistance of the mark, high-contrast marking, and processing speed. Typical applications are computer keyboards, earmarks for animals, labeling of electronic devices, marking of cables and wires in airplanes, or marking of housings and control elements in the automotive sector. Further examples are quality control in manufacturing and writing of barcodes.

The process is based on an optical differentiation of the laser treated areas compared to the rest of the workpiece. The applied laser energy can be used in different forms and utilized mechanisms are

- Selective ablation
- Surface activation and modification
- Cutting of letters (technically considered, this is the process of laser cutting)

The marking itself can be realized by, for example, the generation of annealing colors on the surface, ablation of (different-colored) surface layers, that is, lacquer coats or by an activation of laser active materials. The latter technique facilitates processing without the disposal of smoke and renders a sealing of the surface that is unnecessary due to a missing ablation. In Figure 9.42, two application examples for laser marking with Nd:YAG laser radiation are given.

9.6 RAPID PROTOTYPING

Laser-based rapid prototyping is an example for solid freeform fabrication (SSF) and a layer-by-layer fabrication principle that does not require the use of any special mold or tool. Three-dimensional

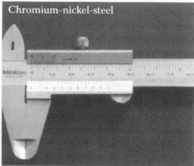

FIGURE 9.42 Examples for laser marking with Nd:YAG radiation.

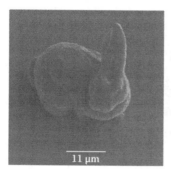
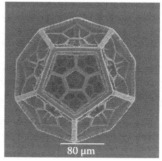

FIGURE 9.43 Structures created by two-photon laser lithography: bunny (left side) and a mathematical structure (cell in a cell in a cell) (right side) made of ORMOCER. (From Passinger, S., Two-photon polymerization and application to surface plasmon polaritons, Dissertation, 2008. With permission.)

model data can be generated with computer-aided design (CAD) software or three-dimensional scanning systems like computer tomography (CT). This data is virtually sliced in increments along the axis of the building direction. SFF may be used as a design tool during product development, an engineering field well known as rapid prototyping (RP).

This section attends to laser sintering, melting, and cladding of metallic and ceramic workpieces. Other essential fabrication methods that are based on photopolymerization, like the laser-based stereolithography and the two-photon laser lithography that allows the fabrication of micron-sized parts with a process resolution below 100 nm [24], are simply described as a general survey. In Figure 9.43, some structures that are created by two-photon laser lithography and made of ORMOCER®, inorganic-organic hybrid polymers are given.

Possible materials for processes based on sintering, melting, and cladding are theoretically all powder materials that form a liquid phase under the influence of laser radiation. But there are certain restrictions concerning the feeding behavior of the used powder, particularly the reproducibility and homogeneity of the powder deposition. Needless to say, there are limitations in part resolution due to the applied particle size of the powder.

9.6.1 FABRICATION BASED ON PHOTOPOLYMERIZATION

The fabrication methods based on photopolymerization achieve the highest resolution of all solid freeform fabrication processes and facilitate the generation of functional microparts. Both, the stereolithography and the two-photon laser lithography rely on the light induced linking of liquid monomers into a crosslinked polymer. While the stereolithography method utilizes the direct absorption of optical radiation using UV light sources and lasers the two-photon polymerization only

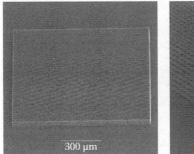

FIGURE 9.44   Photonic crystal created by two photon laser lithography in overview (left) and detail demonstration (right). (From Passinger, S., Two-photon polymerization and application to surface plasmon polaritons, Dissertation, 2008. With permission.)

takes place at high intensities, which are obtained in the focal region of a femtosecond laser beam. The former method only allows a processing at the surface whereas during two-photon polymerization a three-dimensional moving through the liquid matrix is possible. A limitation of these processes concerning the generation of microparts is the very slow processing speed. Figure 9.44 shows a photonic crystal as adopted in telecommunication applications, the details were generated by two-photon laser lithography.

9.6.2   Laser Sintering

The term laser sintering describes a process in which a powder material that is able to form a liquid phase is treated with laser radiation, leading to the surface fusing of the powder particles [26]. Contrasting the laser sintering with the laser melting process it has to be noticed that during laser melting, the powder material is completely molten so that the resulting workpiece densities are comparable to the theoretical ones. However, the meaning of laser cladding in literature is described as a laser coating process where a substrate material is coated with another, under laser radiation of completely molten material [22]. This coating material can be added in the form of powder, wire, pastes, or foils as already described in the previous section.

Employed alternatives of the laser sintering process are the selective, indirect, and direct laser sintering method that are described in the following paragraphs. An example of a machine setup suitable for the laser sintering processes is demonstrated in the schematic shown in Figure 9.45.

This figure shows a process chamber that allows processing in an inert atmosphere. Within this process chamber, a powder deposition system with a corresponding powder reservoir is installed to generate a plane powder bed for two-step processing as specified in the following paragraphs. The application of a scanning optic allows a focusing of the laser beam in the whole working zone with a minimal movement of the scanning mirrors. As described in Section 9.4.6, the use of a scanning optic facilitates fast positioning without moving the workpiece to the powder bed or a processing head. That way, the desired workpiece is generated in a layer-by-layer production.

9.6.3   Selective Laser Sintering

The process of selective laser sintering is very similar to the stereolithography technique. While three-dimensional workpieces using stereolithography are generated by the light induced curing of a liquid polymer, also referred to as resin, the fabrication of workpieces by applying selective laser sintering is usually realized by depositing a powder layer that is treated with laser radiation in *selected* areas. The applied powder materials must be able to form a liquid phase because of laser treatment in the laser sintering process. On the other hand, the term selective laser sintering

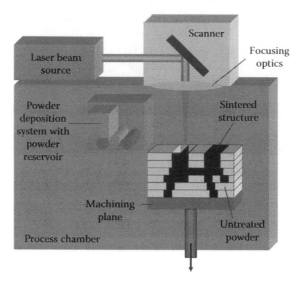

FIGURE 9.45 Machine setup for laser sintering processes.

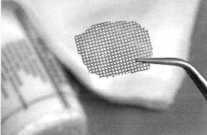

FIGURE 9.46 Metallic structures generated via selective laser melting. (From Höges, S., Fraunhofer ILT, Aachen, Germany. With permission.)

is also common for sintering processes handling with powders consisting of an alloy. In this case, one of the alloy partners is low-fusion melting or shows a much higher absorption of the used laser wavelength compared to the other components. Laser radiation treatment of such powders may lead to a selective melting of this component and therefore to a creation of sinter necks between the non-molten particles. In Figure 9.46, some metallic microstructures generated via selective laser melting are shown.

9.6.4 INDIRECT AND DIRECT LASER SINTERING

In the case of indirect laser sintering, the powder particles are coated with a polymer binder that is selectively molten in the laser process and thus connecting the adjacent, non-molten powder particles after solidification into a fragile green part. This induced polymer scaffold has to be removed during a following baking process. The obtained porous metal powder compound can be consolidated by infiltrating with a low-fusion melting metal to increase the low density.

In contrast, the realization of direct laser sintering allows the fabrication of close-to production prototypes without the need of baking or densifying steps after the sintering process. In the broadest sense, it is comparable to the above first described meaning of selective laser melting but in the course of several researches different denominations became apparent.

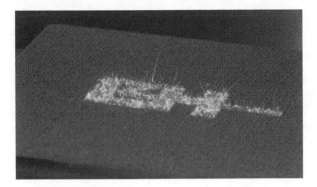

FIGURE 9.47   Laser sintering process of the Laser Zentrum Hannover-logo with Cu-powder.

9.6.5   SINGLE- OR TWO-STEP METHODS

Laser sintering processes can be carried out as single-step and two-step processes (see Section 9.5). The most applied method is the one with two processing steps. Before the application of laser power as the second step, the powder material is deposited in the working zone typically as a powder bed. After the fabrication of the first layer the powder bed is lowered by the thickness of one powder layer and a new layer of the powder bed is generated. The described processing steps are repeated until the complete workpiece is produced.

But the single-step method has to be mentioned as well, which needs less time because of the nonexistence of the second processing step. This method presumes the simultaneous feeding of the powder material during the application of laser power. In this case, a powder feeding system in combination with an adequate powder nozzle, co-axial or off-axial realization, has to be adapted to the sintering process as mentioned in Section 9.5. A picture of a two-step laser sintering process is given in Figure 9.47.

9.7   OTHER PROCESSES

Besides the rather conventional laser material processing technologies, there are many specialized processes whose existence is justified because there are no alternative methods. Lots of them are already applied in industry for a long time. This section deals with laser cleaning and hand-guided laser material processing. These methods are certainly just a diminutive selection of the wide range of possibilities in laser processing.

9.7.1   LASER CLEANING

Laser cleaning can be subdivided into two different methods. The first one deals with the removal of coatings, for example, the removal of paints from metals, where the chemical and physical composition of the coating is different from the substrate material. The laser parameters can be chosen for selectively removing only the upper surface layer or coating.

The second method is also called laser surface decontamination and deals with the removal of impurities that are localized near the surface of the substrate. In this case, the difference in chemical and physical composition between the lower and the upper part of the surface is very small. An example is the decontamination of certain radioactive metals under the adoption of laser radiation.

Compared to alternative cleaning methods, laser cleaning allows processing with lower hazards concerning safety and the environment. Certainly, the removed particles have to be discharged from the processing zone to avoid a recombination with the surface. Therefore, the surface should be treated with short-pulse laser radiation. Longer laser pulses may cause melting of the surface or the formation of larger droplets that cannot be discharged without problems. Depending on whether the ablated particles are posing a risk to health, or are even radioactive, a suitable exhaust system has to be installed. Other advantages over chemical cleaning or shot blasting for example are the possibility of automation or the reduced generation of toxic byproducts. In any case, the high initial capital costs of a laser system should be taken into account. Appropriate laser sources for laser cleaning depending on the various applications are CO_2, solid-state, and excimer lasers [28].

9.7.2 Handheld Laser Techniques

In the mid-1990s, the adoption of handheld laser devices started and has continuously been improved. Due to beam delivery via optical fibers and the availability of powerful and compact solid-state or diode lasers, new industrial applications have been rendered possible. Hand-guided systems cover the whole spectrum of laser material processing including cutting, ablation, drilling and hardening as well as welding, brazing, and cladding. This technique facilitates the processing of large and complex parts onsite, for instance, on construction sites. Another possible application is the dismantling of power plants or the disassembly of manufacturing plants. On the other hand, handheld laser processing is exacerbated by safety requirements. Different research groups dealt with this subject of designing suitable personal protective equipment [29].

In Figure 9.48, a schematic demonstration of a handheld laser device in longitudinal section and the functions of the implemented modules can be seen. By means of the double-focus-optic, for example, the creation of two focal points in the processing zone via the allocation of the raw laser beam is possible. The mirror adjustment adjacent to the double-focus-optic enables the adjustment of the nozzle and the focal distance as well as the intensity distribution to the two focuses.

As mentioned earlier, the safety aspect plays a significant role in processing with handheld laser devices. The presented system only allows a two-hand-processing so that an accidental encroachment on the laser beam by the operator is avoided. A temperature-control unit protects

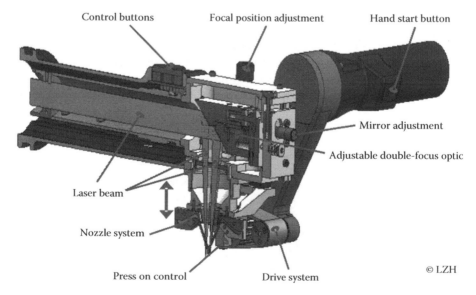

FIGURE 9.48 Schematic demonstration of a handheld laser device in longitudinal section. (From Hennigs, C. et al., Multifunctional hand-held laser processing device, in *25th International Congress on Applications of Lasers & Electro-Optics, ICALEO 2006*, Scottsdale, AZ, October 30–November 2, 2006.)

FIGURE 9.49 Illustration of a handheld laser device (left) and during cutting application (right). (From Hennigs, C. et al., Multifunctional hand-held laser processing device, in *25th International Congress on Applications of Lasers & Electro-Optics, ICALEO 2006*, Scottsdale, AZ, October 30–November 2, 2006.)

other modules from overheating and the press-on control on the bottom side will avoid an emission of laser radiation if the hand-held laser device has no contact to the workpiece. The control buttons on the upper side lead to the adjustment of the laser parameters and the feed rate whereas the drive system is realized by an outboard roller that is driven by a motor implemented in the grip. Figure 9.49 illustrates a current configuration of a handheld laser device and its handling during the cutting processes.

9.8 SUMMARY AND OUTLOOK

This chapter of laser materials processing was aimed at a general overview and better knowledge of present applied laser-based processing techniques. The application of high-power lasers allows processes without any tool wear. Processes can be roughly subdivided into material removal, joining, surface modifications, and generative technologies.

The recent development of laser sources with improved output power or beam quality, for example, allows an adoption of new processing routes or machining of materials whose handling has been problematic with prior laser techniques. Various laser materials processing methods are industrially used. In the future, further applications will be replaced and upgraded with laser devices due to their flexibility or a wear-free process requiring no necessary tool changes and thus does not have non-productive times. In other cases, laser technology allows the production of workpieces, especially in the range of microstructures, which were impossible to be fabricated in the past. This technology meets the requirements of structures with great accuracy for nearly any material. Moreover, a suitable and elaborated process positioning system for the laser head and workpiece facilitates processing without the need of additional finishing steps. The resulting shorter machining cycles often result in a significant cost reduction making laser processing even more attractive in future.

QUESTIONS

Q.9.1 What are the general advantages and disadvantages of material processing with laser radiation?

Q.9.2 Which are the processing parameters during laser cutting? Show their interdependency.

Q.9.3 Point out the difference between heat conduction and deep penetration welding.

Q.9.4 List the parameters that have to be adjusted in single- and two-step processes during surface processing.

Q.9.5 Define the difference of single- and two-step processing using powder as an additive.

ABOUT THE CONTRIBUTING AUTHORS

Andreas Ostendorf, born in 1968, studied electrical engineering at the University of Hannover, Germany. In 1994, he received his diploma (MSc) with some basic investigation on stability issues in Fuzzy Control. Starting in 1995 as a PhD student at the Laser Zentrum Hannover e.V. (LZH), he carried out research work in the field of micromachining using UV and ultrashort-pulsed lasers. Only 2 years later, in 1997, he became the head of the Department of Production & Systems at LZH. In 2000, he finished his PhD thesis on comparing the different interaction mechanisms at VUV wavelength lasers and ultrafast lasers. Also in 2000, he was appointed the CEO of Laser Zentrum Hannover. At that point of time, LZH had more than 130 scientists from physics chemistry, and engineering. In the following years, he was able to further expand the activities at LZH and make it one of the most powerful R&D centers in Europe in the field of laser research. In 2002, he was elected as the chairman of the board of directors at LZH. In 2003, he received the scientific award of the Gottlieb Daimler- and Carl Benz-Foundation for his achievements. Since 2004, he is a member of the Board of Directors at the Laser Institute of America (LIA) and in 2005, he was appointed to the Executive Committee of the LIA. In 2008, Dr. Andreas Ostendorf was elected president of the LIA. He is also a fellow of LIA and SPIE. In his scientific work, he recently focused on using laser pulses for generating structures much smaller than the diffraction limit. He has participated in many national and international projects in basic and applied research. Since October 2008, he is a full professor at the Ruhr-University Bochum, Germany, where he holds the chair of Laser Applications Technology. Dr. Ostendorf has published more than 40 peer-reviewed papers and 140 conference papers, he has contributed 5 book chapters, and edited 9 conference proceedings.

Mrs. Sabine Claußen started her studies in materials sciences with a major in the science of metals at the Clausthal University of Technology in Germany in 2002, and graduated in 2007. Since then she has been working at the Laser Zentrum Hannover e.V. as a scientist in the Department of Materials and Processes, as a member of the Surface Treatment Group.

Dr. Marshall G. Jones joined General Electric Global Research (GE GR) in 1974 after receiving his MS and PhD from the University of Massachusetts, Amherst, in mechanical engineering. He received his BS in the same field from the University of Michigan in 1965 and received his AAS degree in mechanical technology from Mohawk Valley Community College, Utica, New York, in 1962. He worked for 4 years (1965–1969) as development engineer at Brookhaven National Lab (BNL), Upton, New York, after completing his undergraduate studies in 1965. During his tenure at BNL, he functioned as an operational engineer on the largest hydrogen bubble chamber in the world and was an operational engineer on the largest superconducting magnet in the world. He also developed an advanced 70 mm film transport system that was the target for potential bubble chamber applications.

Dr. Jones, a principal engineer at GE GR, has performed research and development work for all the industrial business segments of GE. He has spent most of his GE career addressing laser material processing, laser device development, and fiber optics that has accorded him 49 U.S. patents, 31 foreign patents, and over 45 publications. His innovations include some of the earliest work in laser assistant machining, welding of dissimilar refractory conductors for lighting products, earliest successful laser to fiber optic coupling at high laser power levels, development of flux less soldering techniques for surface mount devices and fine pitch flex leads, and the development of thick section welding with high-power fiber lasers that have high brightness. Another of Dr. Jones' achievements is leading the transition of GE's invented slab or face pumped laser (FPL) from initial military applications to industrial applications, a dual use technology. Some of these innovations have been implemented in GE business' manufacturing plants such as lighting, aviation, and energy.

Dr. Jones is a GE-GR Coolidge fellow and is a member of the National Academy of Engineering (NAE), a fellow of ASME, and the Laser Institute of America (LIA). He serves or has served on both community and national boards including the Engineering Directorate for the National Science Foundation (NSF) and the LIA. Dr. Jones has received many community and technical awards including the GE Phillippe Award (community service) in 1992 and the 2007 Arthur Schawlow Award (LIA's highest achievement award). He also has written a children's book entitled *Never Give Up—the Marshall Jones Story.*

REFERENCES

1. A. Giesen, H. Hügel, A. Voss, K. Wittig, U. Brauch, and H. Opower, Scalable concept for diode-pumped high-power solid-state lasers, *Applied Physics B: Lasers and Optics* 58(5), May 1994, 365–372.
2. H. Hügel, *Strahlwerkzeug Laser*, Teubner Studienbücher, Stuttgart, Germany, 1992.
3. R. Killing, *Welding Processes and Thermal Cutting*, Verlag für Schweißen und verwandte Verfahren DVS-Verlag GmbH, Düsseldorf, Germany, 2001, ISBN 978-3-87155-790-3.
4. J. Powell, CO_2 *Laser Cutting*, 2nd edn., Springer, London, UK, 1998.
5. W. M. Steen, *Laser Materials Processing*, 3rd edn., Springer, London, 2003.
6. W. Rath, CO_2 laser—workhorse for industrial manufacturing, *Laser Technik Journal*, 6(3), 2009, 32–38.
7. B. Denkena, F. Völkermeyer, R. Kling, and J. Hermsdorf, Novel UV-laser applications for carbon fiber reinforced plastics, *Proceedings of Applied Production Technology APT'07*, Bremen, Germany, September 17–19, 2007.
8. F. Völkermeyer, R. Kling, and B. Denkena, Flexible and damage minimized machining of fiber plies and CFRP by UV-laser radiation, *Proceedings of the CFK-Valley Stade Convention 2007*, Stade, Germany, June 13–14, 2007. CFK-Valley Stade, Stade, Germany.
9. C. Hennigs, O. Meier, A. Ostendorf, and H. Haferkamp, Multifunctional hand-held laser processing device, *25th International Congress on Applications of Lasers & Electro-Optics, ICALEO 2006*, Scottsdale, AZ, October 30–November 2, 2006.
10. F. Siegel, A. Ostendorf, and U. Stute, Laser-based surface modification—micro structures on macro areas, *Proceedings of the Second Pacific International Conference on Application of Lasers and Optics 2006-PICALO*, Melbourne, Australia, April 3–5, 2006.
11. W. M. Steen, M. Eboo, and J. Clarke, Arc-augmented laser welding, *Fourth International Conference on Advances in Welding Processes*, Harrogate, U.K., May 1978, Paper No. 17, pp. 257–265.
12. E. J. Haas, Arc-augmented laser welding of aluminum, Master thesis, Oregon Health & Science University, Portland, OR, December 1986
13. U. Dilthey and H. Keller, Pilotstudie zum Einsatz des Laser-MSG-Hybridprozesses zum Hochleistungsschweißen von Stahl, Technical report P 426/A 139, Stahl-Zentrum Düsseldorf, Germany, Studiengesellschaft Stahlanwendung, 2001.
14. J. Hermsdorf, A. Ostendorf, C. Stahlhut, A. Barroi, F. Otte, and R. Kling, Guidance and stabilisation of electric arc welding using nd:yag laser radiation, *Third Pacific International Conference on Application of Lasers and Optics PICALO*, Beijing, China, April 16–18, 2008, paper 707.
15. KeySysTech GmbH, Germany. www.keysystech.de.

16. M. Kreimeyer and G. Sepold, Processing of laser joined aluminum-steel tailored blanks in overlap and butt joint configuration, *21st International Congress on Applications of Lasers & Electro-Optics ICALEO 2002*, Orlando, FL, 2002, ISBN: 0-912035-73-0.

17. C. F. Bohman, The laser and microsoldering, Society of Manufacturing Engineers, Tech. paper No. AD74-810, Mich. 481 28, 1974, p. 19.

18. L. Engelbrecht, O. Meier, A. Ostendorf, and H. Haferkamp, Laser beam brazing of steel aluminum tailored hybrid blanks, *25th International Congress on Applications of Lasers & Electro-Optics*, *ICALEO 2006*, Scottsdale, AZ, October 30–November 2, 2006, pp. 312–319.

19. H. Laukant, C. Wallmann, M. Müller, M. Korte, B. Stirn, H.-G. Haldenwanger, and U. Glatzel, Fluxless laser beam joining of aluminum with zinc coated steel, *Science and Technology of Welding and Joining* 10(2), 2005, 219–226.

20. H. Haferkamp, A. von Busse, M. Hustedt, and G. Regener, Welding of steering oil reservoirs using a high power diode laser system, *Proceedings of the Second International WLT-Conference on Lasers in Manufacturing 2003*, *WLT*, Munich, Germany, June 2003, pp. 113–118.

21. M. Deutschmann, Lasergestützte Herstellung von Keramik-Stahl-Werkstoffverbundschichten für Werkzeuge der Blechumformung, PhD Thesis, University of Hannover, Hannover, Germany, 2007, ISBN: 978-3-939026-47-1.

22. E. Toyserkani, A. Khajepour, and S. Corbin, *Laser Cladding*, CRC Press LLC, Boca Raton, FL, 2005.

23. N. Weidlich, A. Grüninger, O. Meier, K. Emiljanow, P. Stippler, and F. Seidel, Individual laser cladding for high pressure turbine blades, *First EUCOMAS*, Berlin, Germany, May 26–27, 2008.

24. J. Stampfl, S. Baudis, C. Heller, R. Liska, A. Neumeister, R. Kling, A. Ostendorf, M. Spitzbart, Photopolymers with tunable mechanical properties processed by laser-based high-resolution stereo-lithography. *Journal of Micromechanics and Microengineering*, 18(12), 2008, 125014+9.

25. S. Passinger, Two-photon polymerization and application to surface plasmon polaritons, PhD Thesis, University of Hannover, Hannover, Germany, 2008, ISBN: 978-3-86727-662-7.

26. P. K. Venuvinod and W. Ma, *Rapid Prototyping, Laser-Based and Other Technologies*, Springer, Berlin, Germany, 2003, ISBN: 978-1-4020-7577-3.

27. S. Höges, Fraunhofer ILT, Aachen, Germany.

28. J. F. Ready and D. F. Farson (Eds.), *LIA Handbook of Laser Materials Processing*, Laser Institute of America, Magnolia Publishing, Inc., Orlando, FL, 2001.

29. T. Püster, O. Meier, A. Ostendorf, H. Beier, and D. Wenzel, Qualification of personal protective equipment for the use of hand-held laser processing devices, *International Laser Safety Conference—ILSC*, San Francisco, CA, March 19–22, 2007, 91–100.

10 Laser-Induced Plastic Deformation and Its Applications

Gary J. Cheng, Wenwu Zhang, and Y. Lawrence Yao

CONTENTS

This chapter reviews two ways of laser-induced plastic deformation: deformation by thermal stress and deformation by shock wave. Several most important applications are discussed here: laser forming (LF), laser shock peening (LSP), laser peen forming, and 3D laser dynamic forming (LDF). The process mechanisms, microstructures, and mechanical properties of these processes are also discussed.

10.1 INTRODUCTION

10.1.1 BACKGROUND

The laser has become an indispensable tool for precision materials processing since it provides high localized energy intensity, high repeatability, fast setup, and flexibility. A laser can be used to generate plastic deformation in two ways: deformation by thermal stress and deformation by shock wave. Along these two lines, two most important applications have been developed in the last few decades: LF and LSP. LF has been applied in metals for 2D and 3D bending by introducing large temperature gradients along the thickness of metals and hence thermal stress. Microscale LF has been developed to modify and correct the shapes of micro-devices. LSP is used to improve the fatigue life of metals by generating work hardening and large compressive residual stress near surfaces. The applications of LSP have been extended to microscale LSP (μLSP), laser peen forming (LPF), and 3D LDF.

10.1.2 THERMAL-BASED LASER-INDUCED DEFORMATION

The LF of sheet metal components and tubes requires no hard tooling and no external forces and therefore is suited for die-less rapid prototyping and low-volume, high-variety production of sheet metal and tube components. A schematic view of the setup of a line scanning LF is shown in Figure 10.1. LF has potential applications in aerospace, shipbuilding, automobile, and other industries. It can also be used for correcting and repairing sheet metal components, such as pre-welding "fit-up" and post-welding "tweaking." Laser tube bending involves no wall thinning and little ovality, which makes it easier to work on high work-hardening materials such as titanium and nickel superalloys. LF offers the only promising die-less rapid prototyping (RP) method for sheet metal and tubes to date. MIT did a project on LF for the U.S. Navy in the 1980s. MIT, Boeing Company, and the shipbuilding industry did a project on LF in the 1990s. Germany (Prof. F. Vollertsen) and the United Kingdom (University of Liverpool) did many works on LF in the 1990s. With new laser sources, especially high-power direct diode lasers, and improved understanding of the process, both academic and industrial interests have been renewed in recent years.

10.1.3 SHOCK-BASED LASER-INDUCED DEFORMATION

Laser shock peening involves laser-induced, liquid- or solid-confined plasma, sending strong shock waves into the target, thus imparting compressive residual stress into the surface layer of components to improve their fatigue performance (see Figure 10.2). Compared with mechanical shot peening, LSP offers a deeper layer of compressive residual stress and is more flexible, especially for

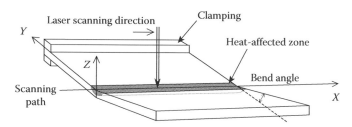

FIGURE 10.1   Schematic of straight-line laser forming.

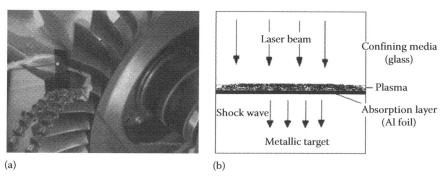

FIGURE 10.2   (a) Laser shock peening process on a turbine blade and (b) schematic view of laser shock peening process.

irregular shapes. It has been shown that LSP can improve fatigue life significantly. LSP was first systematically studied by Dr. A. Clauer of Battelle Institute in the 1960s. LSP requires powerful lasers to generate several to tens of GW/cm^2 laser intensity. Prof. Fabbro's group in France contributed to the theoretical understanding of LSP. General Electric Company industrialized this process and used it to improve the fatigue life of engine blades. To this point, LSP used a low repetition rate (<50 Hz) and large pulse energies (tens of joules) and the spot size was several millimeters. Columbia University carried out pioneering research on μLSP from 2000 to 2005 and developed prediction models and experimental results for μLSP, which used high repetition rate micron-sized laser beams while most LSP work used millimeter-sized beams. Nowadays, μLSP is studied in many sites around the world. μLSP has potential applications in improving the fatigue life of metallic micro-electro-mechanical systems (MEMS) components. More recently, LSP has been extended to LPF, in which the target is shaped and at the same time imparts desirable residual stress on both sides. The Lawrence Livermore National Laboratory (LLNL) has reported progress in LPF and Columbia University focuses on using a micron-sized laser beam to affect LPF. Most recently, LDF has been developed in Purdue to introduce micro/nanoscale 3D shape changes, which combines the advantages of LSP and high strain rate metal forming.

10.2 LASER FORMING (LF)

10.2.1 PHYSICAL PHENOMENA

LF involves heating sheet metal workpieces along a certain path with a laser beam directed normal to the surface, and with or without a jet stream of cold gas or water emulsion cooling the path after the beam passes. The temperature rise is typically below the melting temperature of the workpiece. During LF, a transient temperature field is caused by the irradiation and traveling of a laser beam. Consequently, thermal expansion and contraction take place, which give rise to the deformation of the workpiece.

LF has its roots in traditional sheet metal forming—flame bending, where the sheet metal is heated and deformed by an oxy-acetylene torch. LF has the following advantages over the traditional metal forming technologies. It requires no tools or external forces in the process. The cost of the forming process is greatly reduced compared with the traditional mechanical forming when small batch production and prototyping of sheet metal components are concerned. With the flexibility of the laser beam delivering and numerical control system, it is easier to incorporate LF into an automatic manufacturing system. Material degradation in LF is typically limited to a very tiny layer of the irradiated surface due to highly concentrated beam power and short interaction time, and therefore, LF may be used for materials that are sensitive to high temperature. Figure 10.3a shows

(a) (b)

FIGURE 10.3 (a) Experimental setup of laser forming at Columbia University; and (b) 3D shapes of thin plate after laser forming.

an experimental setup of LF at Columbia University. Figure 10.3b shows two 3D shapes (pillow and saddle) of the thin plate after LF.

10.2.2 PROCESS MECHANISMS

Efforts have been made to understand the mechanisms of LF. Vollertsen [1] suggested that three kinds of mechanisms exist in LF, namely, the temperature gradient mechanism (TGM), buckling mechanism (BM), and upsetting mechanism (UM). The TGM is dominant under conditions corresponding to a small Fourier number $F = \alpha t/s^2$, where α is thermal diffusivity, t is characteristic time, and s is sheet thickness or modified Fourier number $F = \alpha \cdot d/(s^2 v)$, where d is the beam diameter at the workpiece surface and v is the traveling velocity. The BM dominates for the case of a high Fourier number. Two characteristics of this mechanism are no steep temperature gradients along the sheet thickness direction and the extension of heated area compared with the sheet thickness. The UM is based on the increase of the sheet thickness and shortening of the sheet length. It is similar to the BM while the dimension of the heated area is much larger than that in the BM. Concaved LF can be achieved through BM. The buckling direction depends on the sheet surface and the pre-strain condition. Li and Yao [2] have proposed a new laser scanning scheme in which convex LF can be effectively insensitive to the above-mentioned disturbances.

In straight-line laser bending, variations of the bending angle along the bending edge have been observed. Understanding the transient stage is useful in analyzing the causes of the edge effects and realizing process control. Numerical and experimental attempts have been made by Bao and Yao [3] to study the edge effects. Both numerical and experimental results confirm that edge effects are characterized by a concave pattern in the bending angle variation along the scanning path and a bending edge curved away from the laser, when the forming process is dominated by the BM. The results from simulation and experiments also show that under the condition when the TGM dominates, the pattern of edge effects is different, which is characterized by a convex bending angle variation and the bending edge curved toward the laser. In the case of the BM-dominated bending operation, the X-axis contraction near the bottom surface is more significant than that near the top surface, resulting in a bending edge curved away from the laser, while the opposite is true in the case of the TGM dominated bending process. The observation of a convex or concave pattern of the bending angle variation is closely related with the direction of the curved bending edge.

10.2.3 MICROSTRUCTURE CHANGES AND THEIR EFFECTS ON MECHANICAL PROPERTIES

LF is a hot working process, during which the thermal distortion induced by a laser beam is made use of to shape material without hard tooling or external forces. Numerical and experimental investigations of LF processes have been carried out to better understand process mechanisms and the effects of key process parameters on dimension and mechanical properties of the formed parts [4–6]. Temperature and strain-rate-dependent material properties have been compiled and considered in the numerical models developed for concave, convex, and tube LF processes and nonlinear relationships including appropriate flow rule and yield criterion were specified for plastic deformation [7–9]. Experimental observations, however, have shown that the LF processes of metals are often accompanied by recovery, recrystallization, and phase transformation [10–13]. To enhance modeling accuracy and prediction capability, variations in material properties, such as flow stress, caused by microstructure changes need to be considered in the numerical simulation of the LF process. It is important to consider the microstructure changes in modeling the LF process because, first of all, for the high temperature experienced in the process, flow stress is more significantly influenced by the microstructure changes. Secondly, LF, like other hot working processes, is characterized by work hardening simultaneously relieved by dynamic softening processes. Apart from recovery, dynamic recrystallization is the primary softening mechanism determining the stress–strain relationship of a material and, hence its flow behavior. This is especially true for metals such as steels, which exhibit relatively low stacking fault energy. In particular, the simultaneous hardening and softening process repeats and its effects accumulate in multiscan LF, during which repeated heating and deforming take place to achieve the magnitude of deformation required for practical 3D LF.

For LF, the proposed mechanisms are TGM [15], BM [16], and UM [17] depending on operation conditions, material properties, and work piece geometry. So far, most thermomechanical research of the LF process has been focused on metals to understand process mechanisms and mechanical properties of the formed parts [4,5,18]. Cooling effects have been considered on multi-scan LF in finite element method (FEM) simulations [13]. Experiments have shown that the LF processes of metals are often accompanied by recovery, recrystallization, and phase transformation [10,11,13]. A microstructure integrated simulation of multiscan LF [14] has been presented considering the effects of strain hardening, dynamic recovery and recrystallization, superheating, and phase transformation. Figure 10.4 shows the grain and phase structure after LF [14]. As shown in Figure 10.5, the critical plastic strain for dynamic recrystallization and the levels of dynamic recrystallization of LF was predicted and found to be consistent with the SEM picture shown in Figure 10.4, where the size of grain refinement is close to the critical plastic zone.

(a)   (b)   (c)

FIGURE 10.4   SEM on the cross section perpendicular to the scanning path [14]. (a) Heat affected zone (dark region) and plastic deformation zone (white dash line); (b) microstructure (grain and phase structures) across the boundary of heat affected zone; and (c) microstructure within the heat affected zone. (From Cheng, J. and Yao, Y.L., *ASME Trans. J. Manuf. Sci. Eng.*, 124(2), 379, May 2002.)

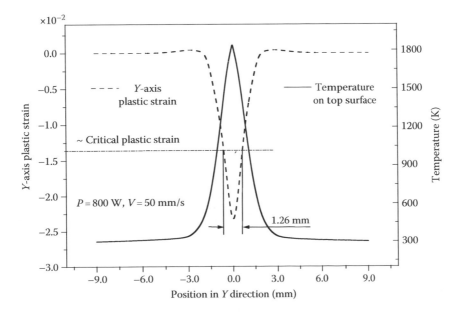

FIGURE 10.5   Predicted grain refined zone by the critical plastic strain in *Y*-axis.

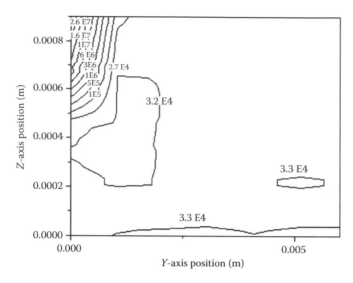

FIGURE 10.6 Predictions on fatigue life of AISI1010 sheet after laser forming: $P = 800\,W$, $V = 0.05\,m/s$. (From Steinberg, D.J. et al., *J. Appl. Phys.*, 51(3), 1498, 1980.)

Mechanical property prediction: For LF to become a practical process, however, it is crucial to predict the mechanical properties of materials after forming, such as fatigue life, fracture toughness, texture, etc. Systematical approaches have been developed to predict yield strength [14], fatigue life [19], and fracture toughness [20] after LF.

An integrated fatigue life modeling for the LF process of low-carbon steel has been presented in [19]. Incorporated with the FEM, mean stress, mean strain, and microstructure of material after LF have been considered in this model. Application of this model in the prediction of fatigue life has been validated through fatigue testing and microstructure analysis [19].

SEM analysis has also revealed some reasons why laser-formed low-carbon steel has a lower fatigue life than before LF. The tensile residual stress and nonhomogenous microstructure around the boundary of the heat-affected zone are the two most important reasons why the fatigue life of low-carbon steel after LF reduces. As shown in Figure 10.6, the fatigue life is lowest along the interface of the heat affected zone. The fatigue location and fatigue life complied with the results of fatigue testing after LF (as shown in Figure 10.7).

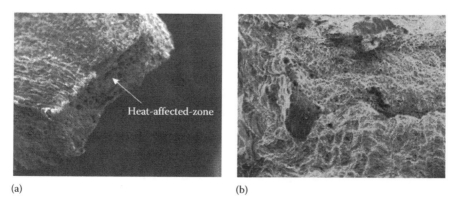

(a)   (b)

FIGURE 10.7 (a) SEM micrographs of cross section of laser formed AISI1010 steel sheet after fatigue testing. Dark zone is heat-affected zone; and (b) SEM micrographs around the boundary of HAZ of laser formed AISI1010 steel sheet after fatigue testing, revealing that a crack initiates on the boundary of HAZ and propagates along the boundary.

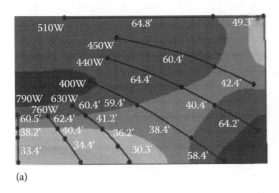

(a) (b)

FIGURE 10.8 (a) Designed paths and laser heating conditions and (b) resulted laser formed sheet steel. (From Cheng, J. and Yao, Y. L., *ASME Trans. J. Manuf. Sci. Eng.*, 126(2), 217, May 2004.)

10.2.4 PROCESS DESIGN AND SYNTHESIS FOR 2D AND 3D SHAPES

Process design: Sheet metal can be formed into a variety of shapes by manipulating the LF conditions. Shimizu [21] applied genetic algorithms (GAs) in LF. However, the results are not flexible and the program failed to obtain the heat conditions for a target shape. Yu et al. [22] presented algorithms for the development of doubly curved surfaces. However, no direct connections with LF were made. A process synthesis [23] has been developed based on GAs for the LF of a class of 2D-shaped sheet metal. For 3D target thin plates, a methodology was proposed to design laser scanning paths and heating conditions of LF [24]. The approach is effective in designing scanning paths and the heating condition for the LF process and the results agree with experimental measurements. Figure 10.8 shows the designed paths and processing conditions for a saddle target shape. The laser-formed shape was much closer to the target shape. So far, the process design of LF has only been applied to a single target function. However, it is important to consider multiple objectives in process design, such as mechanical properties, efficiency, time, etc.

Recent advances in LF for μLSP applications: A number of micro-adjustment applications for LF have been under development at Philips by Hoving and coworkers [25]. The laser adjustment of digital audio head mounting frames and the laser adjustment of reed switches are under development. At IBM, Tam et al. developed and implemented a laser curvature adjust technique (LCAT) system for adjusting the curvature of magnetic head sliders in disk drives by LF [26]. Gartner and coworkers reported the use of LF for the plastic reshaping of wet-etched silicon microscale structures [27,28]. Zhang and Xu [29] worked on the micro-bending of stainless steel and ceramics, which has proposed uses in MEMS. LF can be applied to accurately generate bending in brittle materials.

10.3 LASER SHOCK PEENING

10.3.1 PHYSICAL PHENOMENA

Laser shock peening has been studied since the 1960s [30–32]. Laser-generated shock waves in a confining medium have been used to improve the mechanical properties of various metals such as aluminum, steel, and copper. In particular, LSP can induce compressive residual stress in the target and improve its fatigue life (Figure 10.9).

Shock pressure generation: When a metallic target is irradiated by an intense laser pulse, the surface layer instantaneously vaporizes into a high-temperature and high-pressure plasma. This plasma induces shock waves during expansion from the irradiated surface and mechanical impulses are transferred to the target. If the plasma is not confined, i.e., in open air, the pressure

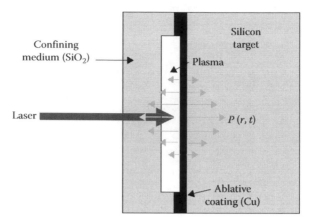

FIGURE 10.9   2D analytical model of laser shock peening process.

can only reach several tenths of one GPa. If it is confined by water or other media, the shock pressure can be magnified by a factor of 5 or more compared with the open-air condition [34]. At the same time, the shock pressure lasts 2–3 times longer than the laser pulse duration. In most LSPs, an ablative coating is used to protect the target from the thermal effects so that nearly pure mechanical effects are induced. The coating could be metallic foil, organic paints, or adhesives. These coatings can modify the surface loading transmitted to the substrate by acoustic impedance mismatch effects at the coating substrate interface and an additional 50% increase in the peak stress values can be achieved [35]. Pressures above 1 GPa are above the yield stress of most metals, thus plastic deformation can be induced. As a result, if the peak shock pressure is over the HEL (Hugoniot elastic limit) of the target material, compressive stress distribution in the irradiated volume can be formed [31].

Earlier modeling work on laser-induced shock waves was carried out by Clauer et al. [31]. Their model considered the nonlinearly coupled radiation and hydrodynamic equations governing pressure evolution at the metal surface during laser irradiation. Fabbro et al. [33] developed a model, which assumes that the laser irradiation is uniform, therefore, shock propagation in the confining medium and the target is 1D. This model was extended and analytical relationships for plastified depth and superficial residual stress were given [35]. The 1D assumption is appropriate when the size of the laser beam, which typically follows a Gaussian distribution, is relatively large. The spatial and temporal variation of laser energy and shock pressure must be considered when the laser beam is in the microscale. In the following, the modeling of µLSP will be presented.

When plasma is formed at the interface of the solid and confining medium, its volume expands, its pressure increases, and shock waves propagate into the sample and the confining medium. A portion of the incident laser intensity $I(t)$ is absorbed by the plasma as

$$I_p(t) = AP(t)I(t) \tag{10.1}$$

where
   $AP(t)$ is the absorption coefficient
   t is the time

Shock wave impedance is expressed as $Z_i = \rho_i D_i$, $i = 1, 2$, where ρ is the density and D is the shock propagation velocity. The subscripts 1 and 2 denote the solid target material and the confining medium, respectively. Defining $Z = 2/(1/Z_1 + 1/Z_2)$ and assuming a constant fraction of internal energy be used to increase the thermal energy of the plasma, the following relations between shock pressure $P(t)$ and plasma thickness $L(t)$ can be derived [33]:

$$\frac{dL(t)}{d(t)} = \frac{2P(t)}{Z} \tag{10.2}$$

$$P(t)\frac{dL(t)}{dt} + \frac{3}{2\alpha}\frac{d}{dt}[p(t)L(t)] = I_p(t) \tag{10.3}$$

If $I(t)$, $AP(t)$, and α are constants, shock pressure is found to be proportional to the square root of the laser intensity. If $I(t)$, $AP(t)$, and α are variables, the peak shock pressure is still proportional to the square root of the peak laser intensity and α. Thus, it is reasonable to assume that shock pressure follows a Gaussian spatial distribution with its $1/e^2$ radius proportional to the $1/e^2$ radius of the laser beam. In this way, the spatial nonuniformity of shock pressure is considered, which is needed when the laser spot size is small. The spatially uniform shock pressure $P(t)$ relates to the spatially nonuniform shock pressure as

$$P(r,t) = P(t)\exp\left(-\frac{r^2}{2r_0{}^2}\right) \tag{10.4}$$

where
  r is the radial distance from the center of the laser beam
  r_0 is the radius of the laser beam

$P(r,t)$ can be solved numerically from the above equations given the initial values of $P(t)$ and $L(t)$. The values of $P(r,t)$ are then used as a dynamic shock load in the stress analysis. The dependence of shock pressure on laser intensity is shown in Figure 10.1 where the laser intensity varies from 2 to 6 GW/cm^2 while α is kept as 0.2 and the pulse duration is kept as 50 ns in [41].

Rate-dependent material properties: In LSP, the target is subjected to very strong shock pressures (>1 GPa), the interaction time is very short (~100 ns), and the strain rate is very high (10^6–10^7 s^{-1}). A review of constitutive equations for such high strain rates was given by Meyer [36]. The simplest model to describe the work hardening behavior of metals is $Y = A + B\varepsilon^n$, where Y is the yield strength, n, A, and B are material constants, and ε is the equivalent plastic strain. The work hardening model was extended in Johnson's model [37] to include the influence of temperature and strain rate. Johnson's model is based on experiments with strain rates from 0 to 400 s^{-1} and it does not consider pressure effects, which are very important in laser shock peening. A constitutive model applicable to ultrahigh pressures was given by Steinberg et al. [28]. Steinberg's model did not consider rate-dependent effects; however, it was found that rate-dependent effects played a minor role at pressures above 10 GPa and their rate-independent model was verified to successfully reproduce shock experimental data in this range. But for shock pressures below 10 GPa, the rate-dependent effects cannot be neglected. In laser shock peening, the pressure involved is fairly high (>1 GPa), but normally less than 10 GPa. For laser shock peening, therefore, both the strain rate effects and ultrahigh-pressure effects on material yield stress need to be considered. Based on the above mentioned models and assuming that the material compression is negligible in the range of working pressure below 10 GPa, the following constitutive equations are suggested:

$$G = G_0\left[1 + \left(\frac{G_p'}{G_0}\right)P + \left(\frac{G_T'}{G_0}\right)(T - 300)\right] \tag{10.5}$$

$$Y = Y_0[1+C\ln\dot{\varepsilon}][1+B\varepsilon]^n\left[1+\left(\frac{Y_p'}{Y_0}\right)P+\left(\frac{G_T'}{G_0}\right)(T-300)\right] \quad (10.6)$$

$$G_p' = \frac{dG}{dP}, \quad G_T' = \frac{dG}{dT}, \quad Y_p' = \frac{dY}{dP}, \quad \frac{Y_p'}{Y_0} \approx \frac{G_p'}{G_0} \quad (10.7)$$

where

    G is the shear modulus

    P is the pressure

    T is the temperature

    Y_0 and G_0 are values at the reference state ($T=300$ K, $P=1$ atm, strain free)

    C is the logarithmic rate sensitivity at the strain rate s^{-1}

    ε is the strain

    $\dot{\varepsilon}$ is the strain rate

    B and n are material parameters describing the work hardening effect

Generation of plastic deformation: When the shock pressures are well above the HEL of target crystals, plastic deformation and compressive stress can be induced as the shock wave propagates through the sample. To understand the stress/strain distribution and residual stress induced by LSP, numerical simulations are needed to study the thermal-mechanical history of the process. Dynamic deformations of the target due to shock loading have been simulated using FEM for (a) the 1D case [31,32], (b) the axisymmetric case [39], and (c) the 3D case [40]. Compressive residual stress can be induced near the surface. The tensile stress at the edge of the laser beam can be changed into compressive stress through overlapping the laser shocks at a suitable spacing [41]. Since high strain rate deformation involves many uncertainties (e.g., work hardening rate, strain rate sensitivity at elevated temperatures), a multi-scale simulation approach [42,43] is used to investigate the deformation mechanisms of the materials' responses to shock waves, which merges the nano-microscale and the continuum scale. In the nano-microscale, dislocation dynamics (DD) analyses calculate the plasticity (e.g., work hardening and strain rate sensitivity). In the continuum scale, deformation is simulated with explicit finite element analysis (FEA) based on the laws of continuum mechanics. The result is a hybrid elasto-viscoplastic simulation model coupling discrete DD with FEA [44].

10.3.2 Microstructure and Mechanical Properties

Microstructure: During the last two decades, laser shock peening has become an efficient way to improve the surface hardness and fatigue strength of a variety of metals, such as aluminum alloys [31,45], titanium [46], and austenitic steel [31,47]. The microstructure changes during LSP and their effects on mechanical properties have been important research topics in many groups, such as hadfield manganese steel [48], copper [49–52], and austenitic stainless steels [53–55].

Generally, two types of linear substructures can be generated during the high strain rate deformation of metals [56,57]: deformation twins and microbands. Under high strain rate loading, the constitutive relationships and microstructure are different from those under low strain rate loading. Meyers [52] has studied the microstructure after LSP of single crystal copper. Murr and other researchers [50,58] investigated the formation mechanisms of deformation twins and microbands of metals under high strain rate loading. They found that the plane stress loading favors the formation of deformation twins, while the spherical one favored the formation of microbands. They also concluded that twins do not occur until the loading stress exceeds the critical twinning stress of materials. However, the critical twining stress is not a constant but is affected by many factors, such as stacking fault energy, dislocation structure, and point defects developed during shock loading. Mechanical twinning is favored in low stacking fault energy (SFE) single crystals (such as 304L steel, SFE = 21 mJ/m^2) while cross-slip starts in high SFE materials (such as copper, SFE = 78 mJ/m^2). The microstructures observed in shock pre-strained 304 stainless steel consist of deformation twinning and dislocation tangles for shock pressures above 15 GPa [59]. It has also been found that materials with low SFE deform by deformation twinning at very low pressures because they lack the ability to cross-slip [60,61].

The formation of microbands is related to the numerous dislocations generated on slip planes during the ultra fast strain rate forming. During shock loading, Frank–Read dislocation sources are activated since the magnitude of shear stress on the slip plane reaches critical shear stress. As a result, numerous dislocations are multiplied as the stress wave propagates through the sample. The duration of laser shock peening is on the order of nanoseconds; however, the velocity of dislocation propagation in the solid is on the scale of sound speeds. Therefore, large amounts of dislocations aggregate on {111} slip planes to form microbands due to the limited time for them to move out of the slip planes during laser shock peening. This observation is also consistent with the dislocation structure calculated after the laser shock peening of FCC metal—stainless steel (see Figure 10.5). The computational method follows the multiscale dislocation dynamic simulation [44] under conditions of shock pressure: 5.3 GPa, pulse width: 10 ns, SFE: 21 mJ/m, shear modulus (G): 73 GPa. It can be seen that the configuration of microbands is very similar to the findings in the TEM picture (see Figure 10.10b and c).

Mechanical property changes after LSP: LSP has been used in the aircraft industry to improve the fatigue resistance and stress corrosion resistance of metals. With the availability of cheaper laser systems, the application of LSP can be further extended to other industries.

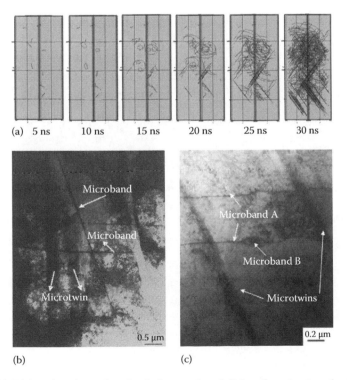

FIGURE 10.10 (a) Dislocation dynamics simulation results of dislocation structure (mobility = 0.052/pa s, strain rate = 5E6 s^{-1}, pulse width = 10 ns); TEM pictures of mixture of microtwins and microbands in 304L stainless steel after laser shock peening: (b) microtwins and two microbands with 70° angle; and (c) two pairs of parallel microtwin and microbands.

Fatigue resistance: The beneficial effects of LSP are primarily due to the near-surface compressive residual stress (which retards fatigue-crack propagation) and near surface work-hardening (which retards fatigue-crack initiation by limiting plastic deformation). There have been many studies of how a surface compressive layer induced by LSP can efficiently delay the development of fatigue cracking [31]. Laser shock peening successfully improved the fatigue behavior of Ti alloys, such as Ti6Al-4V, which are used in turbine compressors and Inconel superalloys, which are used in turbine hot sections [62,63]. Investigations on aluminum alloys [45] and steels [64,65] showed that laser peening could increase fatigue strength. The effects of μ-LSP on the fatigue life of small metallic components have been studied [39]. Fatigue life depends not only on the initial levels of residual stress and work hardening, but also on their stability under real-life service conditions [66].

Stress corrosion resistance: There has been limited study of the effects of LSP on stress corrosion cracking. LSP can improve the stress corrosion cracking resistance of stainless steel [67,68]. The effects of laser shock peening on the stress corrosion cracking behavior of two stainless steels (austenitic steel and martensitic steel) have been compared [69]—the effect on austenitic steels is more significant than on martensitic steels. In general, laser shock peening has been found to improve the stress corrosion cracking resistance because of the large compressive residual stress it produces.

The technical evolution of laser shock peening is quite interesting. In the beginning, only expensive lasers that could produce short (<30 ns) and huge laser pulses (>1 J) were used. Such lasers had a low repetition rate. Only high-end products such as aircraft engines could afford to use it. In recent years, more compact laser systems with a much high repetition rate (from <20 to ≫100 Hz) are commercially available. These lasers are cheaper and more robust than the conventional LSP lasers. With all the benefits discussed above, one can expect LSP to be used widely in many of the industrial products.

10.4 INTRODUCTION TO LASER-INDUCED 3D SHAPING

Recently, microforming has become a more important technology for manufacturing micro-metallic parts, in particular for bulk production, because of the quickly increasing requirement for many industrial products resulting from microtechnology. Laser microforming as a new technology has already been explored for a number of potential high-precision industrial applications, such as laser micro-bending of magnetic disk-drive components, adjustment of contact springs of miniature relays and reed-contacts, and the accurate bending of thin plate springs, etc. It can be seen that most of these applications encounter cyclic loadings. Thus, it is desirable to improve the fatigue life of micro-parts after the microforming process.

10.4.1 LASER PEEN FORMING

It is well known that laser shock peening (LSP) can induce desirable compressive residual stress in the target and improve its fatigue life. Recently, a new forming technology known as laser peen forming (LPF), which is developed from LSP has attracted more attention because of the inherent advantages of LPF. It is undesirable to form large structural components by bending or hydraulic pressing since these processes result in tensile residual stress on the surfaces subject to fatigue and stress corrosion cracking. As a result, mechanical shot peen forming was developed and has been successfully used for over 40 years to shape many of these large components. Laser peen forming can now compliment this traditional forming technology and has much greater depth of residual stress, which results in 3–8 times greater curvature than can be achieved with mechanical shot peen forming. Also, compared with shot peen forming, a desired shape may be achieved by laser peen forming with more accuracy, better repeatability, and faster. Figure 10.11 shows the schematic setup of the laser peen forming process.

Hackel et al. [70] employed a pulse laser beam of energy fluence 60–200 J/cm² with a spot size ranging from 3 mm × 3 mm to 6 mm × 6 mm for forming shapes and contours in metal strips.

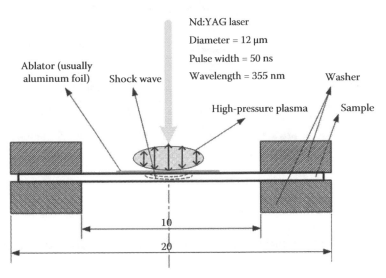

FIGURE 10.11 Schematic setup of laser peen forming.

They showed that the laser peen forming process can generate deep compressive stress without inducing unwanted tensile stress at the metal surface and is especially useful for thick (greater than 3/4 in. thick) material that is difficult to shape or contour. Zhou et al. [71] investigated the deformation mechanisms for a special configuration; that is, a sample clamped by two concentric washers. The laser spot size they used ranged from 3 to 10 mm in diameter. By using laser peen forming, metal sheet forming is realized without mold and the dimensions of the obtained shapes are determined by the boundary condition. Zhou et al. [72] further studied the response of various metal sheets to LPF and presented detailed deformation characterization. The investigation showed its potential to become a flexible manufacturing process with excellent properties and short manufacturing time.

Micro-laser peen forming (μLPF) has already attracted some attention. Compressive stress, like in the macro LPF process, can also be induced at the surface of both sides by μLPF. However, an important consideration in understanding the mechanics of μLPF at the microscale is that the laser spot size is likely to be in the same order of magnitude as the average grain size of the target so that the deformation is expected to occur predominately within a small number of grains. Thus, in order to accurately model the process, the material properties must be treated as anisotropic and heterogeneous rather than isotropic and homogeneous. In order to gain insight into the deformation and stress fields associated with μLPF, it is advantageous to first apply the process to single crystals so that the effects of the anisotropy can first be identified in a homogeneous material.

The effects of μLPF on single crystal aluminum have been studied by using both numerical and experimental methods [73]. Besides the residual stress distribution measured by x-ray microdiffraction, peen-formed crystal samples are also characterized by the profilometer curvature measurement and lattice rotation of electron backscatter diffraction (EBSD). In addition, a preliminary numerical simulation based on single crystal plasticity was presented to understand the process of μLPF, especially anisotropy and the effect of strain hardening. The anisotropic slip line theory was employed to compliment the numerical simulation to explain the stress and deformation state resulting from laser peen forming on a single crystal surface under plane strain conditions.

10.4.2 LASER DYNAMIC FORMING (LDF)

With the increasing application of MEMS in industry, its reliability and robustness calls for metallic micro-components with higher strength, longer fatigue life, and better precision. However, the

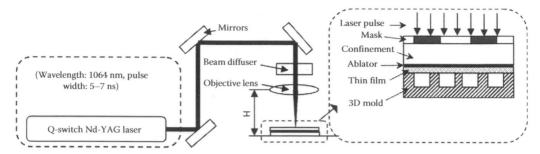

FIGURE 10.12   Schematic setup of laser dynamic forming.

quasi-3D microstructures by conventional photolithography could not satisfy the requirements in more complicated and integrated systems. There is a need to develop novel micro-manufacturing techniques to form 3D microstructures in metallic thin films with low cost and high efficiency. Microscale laser dynamic forming (μLDF) is a newly developed 3D forming process, which takes advantage of microscale laser shock peening and high strain rate metal forming [1,2]. LDF has great potential for direct 3D micro-nano-manufacturing since the laser provides high precision, high repeatability, fast setup, and superb flexibility. A schematic setup of the LDF process is shown in Figure 10.12.

The forming behaviors in LDF can be compared with those in some high energy rate forming (HERF) processes such as electromagnetic and explosive forming. The laser-induced shock wave is the energy source for the forming process. During LDF, as the laser penetrates the confinement layer and irradiates the ablative layer, the generated plasma is bounced back by the confinement producing a strong shock wave, which propagates into the thin film. Finally, the workpiece takes the 3D shape of the mold.

The LDF process has advantages over the current microforming techniques: (1) it is more suitable for micro- and mesoscale 3D forming, since laser energy is more localized and much more precisely controllable than other energy forms such as explosives or electromagnetic fields; (2) it can increase the ductility of brittle and hard-to-form materials, and this, in turn, leads to better formability, hence better ability to form complex shapes; (3) with the ultrahigh strain rates around 10^6–10^7s^{-1} during LDF, the inertia effects are particularly significant and contribute to the significant increase of ultimate plastic strain. Therefore, LDF results in much better formability than the static or quasi-static metal forming process.

The process mechanism of LDF has been experimentally studied [74]. As illustrated in Figure 10.1a, when the laser pulse irradiates onto the ablation coating on the thin film, the coating will be heated and will vaporize instantaneously. The vapor absorbs the remaining laser energy and gets ionized into plasma with ultrahigh pressure, which propagates into thin film and the confining media. The ablative coating prevents direct laser energy absorption in the thin film so that it deforms without overheating. The confining media not only magnifies the peak pressure, but also prolongs its shock wave pressure duration, so that the thin film acquires the momentum sufficient for plastic deformation. The thin film is deformed conformal to the underlying 3D molds. Figure 10.13a shows the optical profile image and SEM images of a dome-shaped aluminum foil after LDF using a microlens as a micro-mold. Figure 12.13b shows a SEM micrograph of an aluminum thin foil after LDF using a microgrid as a mold [76,77].

LDF takes places in the timescale of nanoseconds, which is too short to monitor the deformation behaviors in situ during the process. This transient deformation process has been studied numerically, with the modeling results verified by experiments [75]. Fabbro's model was used to simulate the laser-induced plasma shock wave pressure. Johnson–Cook's model was used to simulate the material's flow stress behavior considering work hardening and strain-rate sensitivity.

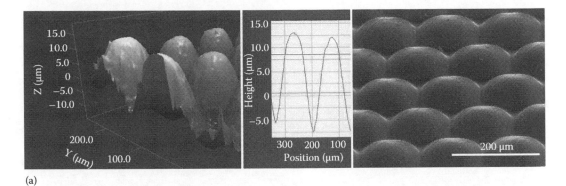

(a)

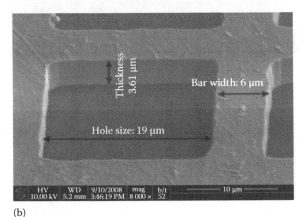

(b)

FIGURE 10.13 (a) An optical profilometer image and SEM image of a dome-shape aluminum foil after laser dynamic forming using microlens as a micro-mold; (b) a SEM micrograph of an aluminum thin foil after laser dynamic forming using a microgrid as a mold. (From Gao, H. and Cheng, G.J., *J. Microelectromech. Syst.*, Under revision, 2009.)

The Johnson–Cook failure criterion was used to predict damage initiation and evolution criteria. The effects of critical parameters on deformation behaviors were investigated systematically, including the fillet/film thickness ratio, aspect ratio of mold, as well as laser processing conditions. The simulated proportional relationship between laser pulse energy and the deformation depth has been verified with a series of LDF experiments.

10.5 CONCLUDING REMARKS

This chapter presents the development of two classes of laser-induced plastic deformation: deformation by thermal stress and deformation by shock wave. LF and LSP have been reviewed. Several more recent developments combining LSP and microforming have been introduced, such as laser peen forming and 3D LDF. The process mechanisms, microstructures, and mechanical properties of these processes have been discussed.

Laser-induced plastic deformation has the potential to be a key manufacturing process for 2D/3D forming and mechanical property enhancement of metallic components. Laser-based manufacturing is flexible. Laser energy can be delivered without mechanical contact and the laser spot size can be big or small, while the timing and intensity can be precisely controlled. The applied area can be very small or very big (through overlapping).

Over the years, one major obstacle in using laser-based processes has been the high capital cost. This situation has changed quite a bit in recent years. When one considers the total value chain of the competing processes, laser-based plastic deformation processes may already be cost-effective and more environment friendly. For example, mechanical shot peening consumes hard particles, and sometimes these particles can only be used once to ensure process consistency, and the used particles may need to be specially processed. On the other hand, laser shot peening generates less waste and can be more controllable and cost-effective.

The several processes discussed in this chapter might be "novel" processes to many people. They are not widely used yet. This doesn't mean they are not good enough processes. This means we need to bridge scientific discoveries and technical inventions with the efforts of commercialization. As said in Chapter 1, innovation is a long process. What is discussed in this chapter can be the seed for future major industrial innovations.

QUESTIONS

Q.10.1 What are the disadvantages and advantages of LF?

Q.10.2 What are the materials that can be formed to 2D/3D shapes by LF?

Q.10.3 What are the mechanical properties and microstructures as a result of LF in metals?

Q.10.4 What is the process mechanism of LSP?

Q.10.5 Why can LSP improve fatigue life in metals?

Q.10.6 What are the weaknesses of LSP in the real manufacturing process, and how would you think to improve the process?

Q.10.7 What are the differences between laser peening forming and LDF? What are the advantages of the two processes over the other metal microforming process?

Q.10.8 From this chapter and Chapter 9, we learn a wide variety of laser-related processes. Please summarize the features of the laser energy field, analyze how certain aspects of this energy field can be used to solve certain engineering challenges, and correlate these applications with laser parameters, including laser average power, repetition rate, pulse duration, wavelength, pulse duration, peak power, and spatial resolution. Think over whether this analysis shares some common features with other energy fields, such as ultrasonic machining or EDM/ECM machining.

Q.10.9 Work out the dynamic M-PIE model of the LF process, laser shot peening process, and LDF process.

ABOUT THE CONTRIBUTING AUTHORS

Dr. Gary J. Cheng is an associate professor in the School of Industrial Engineering at Purdue University, West Lafayette, Indiana. He received a PhD degree in mechanical engineering from Columbia University, New York in 2002. His research interests are laser materials processing, micro/nano manufacturing, and mechanical/physical property enhancement of materials. He has published more than 50 peer reviewed journal/conference papers. Dr. Cheng won the SME Award of Young Manufacturing Engineer, ONR Young Investigator Award, and the NSF Career Award.

Dr. Wenwu Zhang got his PhD degree from the Department of Mechanical Engineering of Columbia University, New York in 2002 and is currently a senior mechanical engineer in Laser and Metrology Systems Lab., Materials Systems Technologies of General Electric Global Research Center in Schenectady, New York. Dr. Zhang is a pioneer in the research of microscale laser

shock peening and energy field manufacturing and is the inventor of liquid core fiber laser material processing. He is currently leading the laser micro/nano R&D work in GE GRC. Dr. Zhang won the 2005 SME Robert A. Dougherty Outstanding Young Manufacturing Engineer Award, 2006 ASME Blackall Machine Tool and Gage Award, and the 2006/2008 ASME BOSS Award. Dr. Zhang is a member of the Optical Society of America, ASME, Sigma Xi, the Society of Manufacturing Engineers (SME) and Laser Institute of America. Dr. Zhang is a fruitful inventor as well as a science fiction writer.

Dr. Y. Lawrence Yao is professor and chair of the Department of Mechanical Engineering at Columbia University, New York. He received his PhD degree in mechanical engineering from the University of Wisconsin–Madison in 1988. Dr. Yao and his team in the Manufacturing Research Laboratory (MRL) are interested in multidisciplinary research in manufacturing and design; nontraditional manufacturing; laser materials processing; laser assisted material removal, shaping, and surface modification; laser applications in industry and art restoration; and robotics in industry and the health care industry. He currently serves as president of the North American Manufacturing Research Institute of SME and vice chair of the Manufacturing Engineering Division of ASME. He received many awards including the Blackall Award from ASME. He is a fellow of ASME, SME, and the Laser Institute of America.

REFERENCES

1. Vollertsen, F., 1994, An analytical model for laser bending, *Lasers Eng.*, 2, 261–276.
2. Li, W. and Yao, Y. L., 2001, Numerical and experimental investigation of convex laser forming process, *SME J. Manuf. Process.*, 3(2), 73–81.
3. Bao, J. and Yao, Y. L., 2001, Analysis and prediction of edge effects in laser bending, *ASME Trans. J. Manuf. Sci. Eng.*, 123(1), 53–61.
4. Hsiao, Y. et al., 1997, Finite element modeling of laser forming, in: *Proceedings of ICALEO'97*, Section A, San Diego, CA, pp. 31–40.
5. Magee, J., Watkins, K. G., and Steen, W. M., 1998, Advances in laser forming, *J. Laser Appl.*, 10(6), 235–246.
6. Bao, J. and Yao, Y. L., November 1999, Study of edge effects in laser bending, in: *Proceedings of 1999 ASME IMECE Symposium on Advanced in Metal Forming*, Nashville, TN, Vol. MED-10, pp. 941–948.
7. Li, W. and Yao, Y. L., 2000, Convex laser forming with high certainty, *Trans. North Am. Manuf. Res. Inst. SME*, XXVIII, 33–38.
8. Li, W. and Yao, Y. L., 2000, Numerical and experimental study of strain rate effects in laser forming, *ASME J. Manuf. Sci. Eng.*, 122, 445–451.
9. Li, W. and Yao, Y. L., 2001, Laser bending of tubes: Mechanism, analysis and prediction, *ASME J. Manuf. Sci. Eng.*, 123, 674–681.
10. Ramos, J. A. et al., 1998, Microstructure of laser bent aluminum alloy Alcad 2024-T3, in: *Proceedings of ICALEO'98*, Orlando, FL, Section E, pp. 178–185.
11. Maher, W. et al., 1998, Laser forming of titanium and other materials is useable within metallurgical constraints, in: *Proceedings of ICALEO'98*, Orlando, FL, Section E, pp. 121–129.
12. Li, W. and Yao, Y. L., 2001, Laser forming with constant line energy, *Int. J. Adv. Manuf. Technol.*, 17, 196–203.

13. Cheng, J. and Yao, Y. L., 2001, Cooling effects in multiscan laser forming, *J. Manuf. Process.*, 3(1), 60–72.

14. Cheng, J. and Yao, Y. L., May 2002, Microstructure integrated modeling of multiscan laser forming, *ASME Trans. J. Manuf. Sci. Eng.*, 124(2), 379–388.

15. Vollertsen, F., 1994, Mechanisms and models for laser forming, *Laser Assisted Net Shape Engineering, Proceedings of the LANE'94*, Meisenbach Bamberg, Germany, Vol. 1, pp. 345–360.

16. Arnet, H. and Vollertsen, F., 1995, Extending laser bending for the generation of convex shapes, *IMechE Part B J. Eng. Manuf.*, 209, 433–442.

17. Kraus, J., 1997, Basic process in laser bending of extrusion using the upsetting mechanism, *Laser Assisted Net shape Engineering 2, Proceedings of the LANE'97*, Meisenbach Bamberg, Germany, Vol. 2, pp. 431–438.

18. Rabier, J. and Demenet, J. L., 2000, Low temperature high stress plastic deformation of semiconductors: The silicon case, *Phys. Stat. Sol. B*, 222(63), 63–74.

19. Zhang, J., Pirzada, D., Chu, C. C., and Cheng, G. J., 2005, Fatigue life prediction of sheet metal after laser forming, *ASME Trans. J. Manuf. Sci. Eng.*, 127, 157–164.

20. Cheng, G. J., Zhang, J., and Zhe, J., 2005, Experimental study and computer simulation on fracture toughness of sheet metal after laser forming, *Int. J. Adv. Manuf. Technol.*, 26, 1222–1230.

21. Shimizu, H., 1997, A heating process algorithm for metal forming by a moving heat source, MS thesis, MIT, Cambridge, MA.

22. Yu, G., Patrikalakis, N. M., and Maekawa, T., 2000, Optimal development of doubly curved surfaces, *Comput. Aided Geometric Des.*, 17, 545–577.

23. Cheng, J. and Yao, Y. L., 2004, Process synthesis of laser forming by genetic algorithms, *Int. J. Mach. Tools Manuf., UK*, 44(15), 1619–1628.

24. Cheng, J. and Yao, Y. L., May 2004, Process design of laser forming for three dimensional thin plates, *ASME Trans. J. Manuf. Sci. Eng.*, 126(2), 217–225.

25. Verhoeven, C. M., der Bie, H. F. P., and Hoving, W., 2000, Laser adjustment of reed switches: Micron accuracy in mass production, in: *Proceeding of Laser Microfabrication, ICALEO 2000*, LIA, Orlando, FL, Vol. 90, p. B21.

26. Tam, A. C., Poon, C. C., and Crawforth, L., Laser bending of ceramics and applications to manufacture magnetic head sliders in disk drives, *Anal. Sci.*, 17, s419–s421.

27. Gartner, E., Fruhauf, J., and Jansch, E., 2001, Plastic reshaping of silicon microstructures: process, characterization and application, *Microsyst. Technol.*, 7, 155–160.

28. Gartner, E., Fruhauf, J., Loschner, U., and Exner, H., 2001, Reshaping of single-crystal silicon microstructures, *Microsyst. Technol.*, 7, 23–26.

29. Zhang, X. R. and Xu, X., August 2003, High precision microscale bending by pulsed and CW lasers, *Trans. ASME J. Manuf. Sci. Eng.*, 125(3), 512–518.

30. White, R. M., 1963, Elastic wave generation by electron bombardment or electromagnetic wave absorption, *J. Appl. Phys.*, 34, 2123–2124.

31. Clauer, A. H., Holbrook, J. H., and Fairand, B. P., 1981, Effects of laser induced shock waves on metals, in: Meyers, M. A. and Murr, L. E., eds., *Shock Waves and High Strain Phenomena in Metals-Concepts and Applications*, Plenum, New York, pp. 675–702.

32. Peyre, P., Scherpereel, X., Berthe, L., and Fabbro, R., 1998, Current trends in laser shock processing, *Surf. Eng.*, 14(5), 377–380.

33. Fabbro, R. et al., 1990, Physical study of laser-produced plasmas in confined geometry, *J. Appl. Phys.*, 68(2), 775–784.

34. Fox, J. A., 1974, Effect of water and paint coatings on laser-irradiated targets, *Appl. Phys. Lett.*, 24(10), 461–464.

35. Peyre, P., Fabbro, R., Berthe, L., and Dubouchet, C., 1996, Laser shock processing of materials, physical processes involved and examples of applications, *J. Laser Appl.*, 8, 135–141.

36. Meyer, L. W., 1992, Constitutive equations at high strain rates, in: *Shock-Wave and High-Strain-Rate Phenomena in Metals*, Meyers, M. A. and Murr, L .E. (Eds.), Marcel Dekker, Inc., New York, pp. 49–68.

37. Johnson, G. R., Hoegfeldt, J. M., Lindholm, U. S., and Nagy, A., 1983, Response of various metals to large torsional strain over a large range of strain rates, *J. Eng. Mater. Technol.*, 105, 42–53.

38. Steinberg, D. J., Cochran, S. G., and Guinan, M. W., 1980, A constitutive model for metals applicable at high-strain rate, *J. Appl. Phys.*, 51(3), 1498–1504.

39. Zhang, W. and Yao, Y. L., May 2002, Micro-scale laser shock processing of metallic components, *ASME Trans. J. Manuf. Sci. Eng.*, 124(2), 369–378.

40. Wu, B. and Shin, Y. C., 2005, A self-closed thermal model for laser shock peening under the water confinement regime configuration and comparisons to experiments, *J. Appl. Phys.*, 97(11), 113517.1–113517.11.

41. Zhang, W., Yao, Y. L., and Noyan, I. C., February 2004, Microscale laser shock peening of thin films, Part 1: Experiment, modeling and simulation, *ASME Trans. J. Manuf. Sci. Eng.*, 126, 10–17.

42. Cheng, G. J. and Shehadeh, M., 2005, Dislocation behavior in silicon crystal by laser shock peening: A multiscale simulation approach, *Scr. Mater.*, 53(9), 1013–1018.

43. Cheng, G. J. and Shehadeh M., 2005, Multiscale dislocation dynamics analyses of laser shock peening in silicon single crystals, *Int. J. Plast.*, 22(12), 2171–2194.

44. Zbib, H. M., Shehadeh, M., Khan, S. M., and Karami, G., 2003, Multiscale dislocation dynamics plasticity, *Int. J. Multiscale Comput. Eng.*, 1, 73–89.

45. Peyre, P., Fabbro, R., Merrien, P., Lieurade, H. P., 1996, Laser shock processing of aluminium alloys. Application to high cycle fatigue behaviour, *Mater. Sci. Eng.*, A210, 102.

46. Hintz, G. et al., 1996, XeCl-excimer laser-MOPA chain for shock hardening, *Proc. SPIE* 3092, 169.

47. Gerland, M. and Hallouin, M., 1991, Effects of laser induced shock waves on an austenitic stainless steel, in: Sudarshan, T. S., Bhat, D. G., and Jeandin, M., *Surface Modification Technologies IV*, The Minerals, Metals & Materials Society, Warrendale, PA, p. 713.

48. Chu, J. P. et al. 1995, Effects of laser-shock processing on the microstructure and surface mechanical properties of hadfield manganese steel, *Metall. Mater Trans. A*, 26a, 1507.

49. Andrade, U. R., Meyers, M. A. Vecchio, K. S., and Chokshi, A. H., 1994, Constitutive description of work- and chock-hardened copper, *Acta Mater.*, 42, 3183–3195.

50. Sanchez, J. C., Murr, L. E., and Staudhammer, K. P., 1997, Effect of grain size and pressure on twinning and microbanding in oblique shock loading of copper rods, *Acta Mater.*, 45, 3223–3235.

51. Li, G. A. et al., 2004, Study of deformed microstructures near the impact crater in pure copper targets, *Mater. Sci. Eng.*, 384A, 12–18.

52. Meyers, M. A. et al., 2003, Laser-induced shock compression of monocrystalline copper: Characterization and analysis, *Acta Mater.*, 51, 1211.

53. Murr, L. E. et al., 2002, Comparison of residual microstructures associated with impact craters in fcc stainless steel and bcc iron targets: The microtwin versus microband issue, *Acta Mater.*, 50, 121.

54. Chu, J. P. et al., 1999, Laser-shock processing effects on surface microstructure and mechanical properties of low carbon steel, *Mater. Sci. Eng.*, 260A, 260–268.

55. Meyers, M. A. et al., 2003, Microstructural evolution in adiabatic shear localization in stainless steel, *Acta Mater.*, 51, 1307–1325.

56. Bay, B. et al., 1992. Evolution of FCC deformation structures in polyslip, *Scr. Metal.*, 40(2), 205–219.

57. Huang, J. C. and Gray, G. T., 1989, Microband formation in shock-loaded and quasi-statically deformed metals, *Acta Metall.*, 37, 3335–3347.

58. Murr, L. E. et al., 1997, Shock-induced deformation twinning in tantalum, *Acta Mater.*, 45, 157–175.

59. Murr, L. E., 1981, Effects of peak pressure, pulse duration, and repeated loading on the microstructure and properties of shock deformed metals and alloys, in: Meyers, M. A. and Murr, L. E., eds., *Shock Waves and High Strain Rate Phenomena in Metals*, Plenum Press, New York, p. 753.

60. Nolder, R. L. and Thomas, G., Substructure of plastically deformed nickel, *Acta Metall.*, 12, 227, 1964.

61. Grace, F. I., Inman, M. C., and Murr, L. E., 1968, Shock-induced deformation faults in 70/30 copper-zinc alloy, *Br. J. Appl. Phys. (Ser 2)*, 1, 1437.

62. Brown, A. S., 1998, A shocking way to strengthen metal, *Aerospace Am.*, 21–23.

63. Ashley, S., 1998, Powerful laser means better peening, *Mech. Eng.*, 120, 12.

64. Banas, G., Elsayed-Ali, H. E., Lawrence, F. V., and Rigsbee, J. M., 1990, Laser shock-induced mechanical and microstructural modification of welded maraging steel, *J. Appl. Phys.*, 67, 2380–2384.

65. Peyre, P., Berthe, L., Scherpereel, X., and Fabbro, R., 1998, Laser-shock processing of aluminum coated 55C1 steel in water-confinement regime, characterization and application to high-cycle fatigue behavior, *J. Mater. Sci.*, 33, 1421–1429.

66. Prevéy, P., Hornbach, D., and Mason, P., 1998, Thermal residual stress relaxation and distortion in surface enhanced gas turbine engine components, in: Milam, D. L. et al., eds., *Proceedings of the 17th Heat Treating Society Conference and Exposition and the 1st International Induction Heat Treating Symposium*, ASM, Materials Park, OH, pp. 3–12.

67. Obata, M. et al., 1999, Effect of laser peening on residual stress and stress corrosion cracking for type 304 stainless steel, in: *Proceedings of International Conference on Shot Peening*, Warsaw, Poland, pp. 287–294.

68. Kritzler, J., 2002, Influence of shot peening on stress corrosion cracking in stainless steel, in: *Proceedings of Eighth International Conference on Shot Peening (ICSP)*, Garmisch-Partenkirchen, Germany. pp. 255–263.
69. Scherpereel, X. et al. 1997, Modifications of mechanical and electrochemical properties of stainless steels surfaces by laser shock processing, in: *Proceedings of SPIE-Europto Conference*, Munich, Germany.
70. Hackel, L. A., Halpin, J. M., and Harris F. B., 2002, Pre-loading of components during laser peen forming, U.S. Patent 6,670,578.
71. Zhou, M., Zhang, Y., and Cai, L., 2002, Laser shock forming on coated metal sheets characterized by ultrahigh-strain-rate plastic deformation, *J. Appl. Phys.*, 91(8), 5501–5503.
72. Zhou, M., Zhang, Y. K., and Cai, L., 2003, Ultrahigh-strain-rate plastic deformation of a stainless-steel sheet with TiN coatings driven by laser shock waves, *Appl. Phys. A Mater. Sci. Process.*, 77, 549–554.
73. Fan, Y., Wang, Y., Vukelic, S., and Yao, Y. L., 2005, Wave-solid interactions in shock induced deformation processes, *J. Appl. Phys.*, 98, 104904.
74. Pirzada, D., Cheng, G. J., and Field, D., 2007, Effect of film thickness and laser energy density on the structural characteristics of laser-annealed polycrystalline gallium arsenide films, *J. Appl. Phys.*, 102, 013519.
75. Gao, H., and Cheng, G. J., 2009, Deformation behaviors and critical parameters of microscale laser dynamic forming, *J. Manuf. Sci. Eng.*, 131(5), 051011, doi:10.1115/1.4000100. Oct. 2009.
76. Gao, H. and Cheng, G. J., 2010, Laser induced supperplastic 3D microforming of metallic thin films, *J. Microelectromech. Syst.*, doi:10.1109/JMEMS.2010.2040947.
77. Yu, C., Gao, H., Yu, H., Jiang, H., and Cheng, G. J., 2009, Laser dynamic forming of functional materials laminated composites on patterned three-dimensional surfaces with applications on flexible microelectromechanical systems, *Appl. Phys. Lett.*, 95, 091108; doi:10.1063/1.3222863.

11 Energy Field Interactions in Hybrid Laser/Nonlaser Manufacturing Processes

Lin Li

CONTENTS

11.1 INTRODUCTION

This chapter describes a number of hybrid laser and nonlaser manufacturing processes and energy field interactions in such processes. It is mainly based on the research work conducted by the author over the last 15 years in developing enhanced laser-assisted manufacturing technologies by combining laser energy and another form of energy/medium such as mechanical, chemical, and thermal systems. The processes described in this chapter will include abrasive laser machining (laser melting combined with grid blasting), laser-activated chemical machining, water-assisted laser machining, and laser-assisted flame spray coating. The purpose of introducing hybrid processes is to overcome some of the drawbacks of laser processing, such as heat-affected zone and recast generation, crack generation due to large thermal gradients, and uniformity of the surfaces.

11.2 ABRASIVE LASER MACHINING

Although lasers have been widely used in the industry for cutting, drilling, welding, and surface treatment applications, lasers are not used as often for bulk material macroscale milling and turning. This is partly because lasers generate a significant amount of heat in the material that can cause thermal distortions, microstructure variations, recast (solidified material after laser melting), spatter, and poor surface finish. The key challenge in laser macro-machining (to be distinguished from micro/nano-machining) is not melting and vaporizing the material but the effective removal of the molten materials once they are generated. Rapid vaporization is one route for material removal using a pulsed laser beam. Considerably more energy is required compared with melting the materials. For nonmetallic materials, this can be effective as the thermal conduction is low. In the "melt

319

and removal" process, a high-pressure (2–14 bar) gas jet is commonly used to remove the molten materials. This has been successful in some cases when the melt pool is relatively small. For a large melt pool (with a length or width greater than 3 mm), using gas alone is not sufficient for effective material removal. Therefore, there is a need for effective melt removal in laser macro-milling and macro-turning processes. By combining grid blasting with laser surface melting in situ, it has been shown that material removal is effective and surface finish is good (Figure 11.1) [1].

The material removal rates for three different materials for a CO_2 laser at different input powers are compared in Table 11.1. It is clear that the material removal rate is material dependent. For 304 stainless steel (thermal diffusivity = 3.9 m²/s) and Ti-5Al-4V titanium alloy (thermal diffusivity = 3 m²/s), much higher removal rates have been achieved compared with En3 mild steel (thermal diffusivity = 13.6 m²/s). Materials with lower thermal diffusivity would allow more energy to be used for melting the material rather than conducting away into the bulk. Although the material removal rates are much lower compared with mechanical milling, they are still higher than ultrasonic machining, electro-discharge machining, and electrochemical machining. An increase of laser power leads to an increase in the material removal rate, almost linearly. Compared with laser machining using a high pressure gas jet for material removal, abrasive laser machining offers over 100% improvement in the material removal rate for stainless steel and Ti-6Al-4V at 500 W laser power and a 7 mm/s scanning speed. For mild steel, a 20%–40% increase in the material removal rate has been found.

Neither laser melting alone nor grid blasting can achieve the high material removal rates compared with the hybrid process. Clearly, the benefit of the hybrid process is the use of the laser beam as a thermal energy source and grit blasting as the mechanical material removal means by

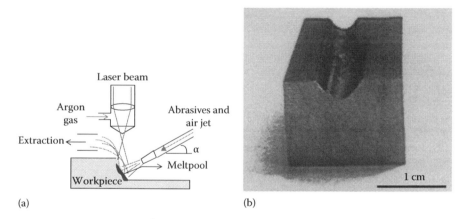

(a)   (b)

FIGURE 11.1 Abrasive laser machining: (a) experimental setup and (b) machined groove in stainless steel with two overlapping passes. (From Li, L. et al., *Ann. CIRP Manuf. Technol.*, 54(1), 183, 2005.)

TABLE 11.1
Material Removal Rate in Abrasive Laser Machining at Different Laser Power Levels

Materials	Removal Rate (mm³/min)	Laser Power (W)
Stainless steel (304)	4200	550
Ti-6Al-4V	2700	550
Mild steel (En3)	540	750

Source: Li, L. et al., *Ann. CIRP Manuf. Technol.*, 54(1), 183, 2005.

TABLE 11.2
Effect of Abrasive Jet Incident Angle (α) in the Hybrid Laser/Abrasive Machining Process

Workpiece	Injection Angle (°)			
Materials	15	25	35	45
Titanium alloy	√	√	√	X
Stainless steel	Δ	√	Δ	X
Mild steel	√	√	√	X

Source: Li, L. et al., *Ann. CIRP Manuf. Technol.,* 54(1), 183, 2005.

Note: X, completely failed; Δ, partially failed; and √, successful.

momentum transfer. Here the interaction of two different energy sources (thermal and mechanical) plays an important role in material removal. The geometry of the machined groove depends on the melt pool profile, which is a function of the laser processing parameters (spot size, beam geometry, beam intensity distribution, and interaction time). Only when the abrasive particle steams are delivered to the melt pool at certain angles can a successful hybrid abrasive machining process be carried out. Table 11.2 shows the effect of abrasive blasting angles on material removal.

Therefore, when two energy fields are interacting, certain rules that are governed by the specific requirements of each energy field and their interactions may need to be followed. This particular angle requirement for the abrasive jet is greatly influenced by the melt pool geometry at the leading end of the melt pool. What can be learned from here is that when manufacturing capability is limited by one energy source, the combination of two types of energy sources could be explored to extend their capabilities beyond those that can be achieved by individual energy sources.

The rapid removal of molten material by the grid blasting not only improves the material removal rate compared with gas jet–assisted machining but also allows significant benefits to be gained by improved surface finish (roughness) and reduced thermal distortion/heat affected zones. The typical surface roughness, Ra, values of the hybrid laser machined materials are 5–20 μm for Ti-6Al-4V, 15–42 μm for stainless steel, and 1.2–13 μm for the mild steel. These represent a 30%–60% improvement compared with gas jet–assisted melt removal laser machining. Increasing the laser traversing speed significantly improves the surface roughness.

Recast layers are almost completely eliminated in the abrasive laser machining. Heat-affected zones of the hybrid machined components, where microstructure changes are included, are 10%–15% less than in directly laser-machined (material removal by gas) components for the above three materials, indicating less heat input to the bulk material. However, the technique is still not competitive compared with mechanical machining in terms of the material removal rate and surface finish for common engineering materials. Some possible applications of the abrasive laser machining process may include the on-site repair (e.g., opening up cracks before metal deposition) of very large components where conventional mechanical machines are not suitable or have access difficulties or for machining very hard to machine materials. The multi-tasking capability of the lasers would allow machining, welding, and material deposition to be carried out using the same laser.

11.3 LASER-ACTIVATED CHEMICAL MACHINING

In laser micro-machining (e.g., grooving, drilling, cutting) of metallic materials, unless very short pulses (<10 ps) are delivered, recast (re-solidified layer) and heat-affected zones are always present.

When very short pulse lasers are used, the material removal rate is lower compared with longer pulsed lasers. To overcome the problems of heat-affected zones while maintaining high material removal rates, a hybrid laser and chemical machining process is introduced [2]. In this process, a laser beam heats up the workpiece (immersed in a NaCl salt solution or with a salt solution spray) to a high temperature. The high temperature in the workpiece locally generated by the laser beam heats up the surrounding liquid, which allows the chemical erosion process to take place at a much higher rate than that at room temperature. A comparison of micro-holes drilling in a stainless sheet (316) using a 30 ns pulsed diode pumped solid state (DPSS) laser alone and with the hybrid process is shown in Figure 11.2. Clearly, the hybrid process has considerably reduced the heat-affected zone. In addition, the machining time has been reduced (by up to 25%).

Figure 11.3 shows a hole drilled in a 2.5 mm thick stainless steel sheet using a standard Nd:YAG laser at 1 ms pulse length combined with the salt solution. The typical problems of spatter and recast layer formation have been removed. The advantage of the hybrid process is clearly demonstrated.

In this hybrid machining process, two forms of energy (thermal and chemical) interact with each other to enable a new capability in manufacturing to be demonstrated.

A possible energy flow for this laser activated reactive machining (LARM) is shown in Figure 11.4. Material removal is accelerated through a higher energy–enabled chemical process. Additional energy is added to the normal chemical corrosion/erosion process by laser heating of the materials in the interaction zone to enable a new route of energy flow. After the removal of the laser pulse,

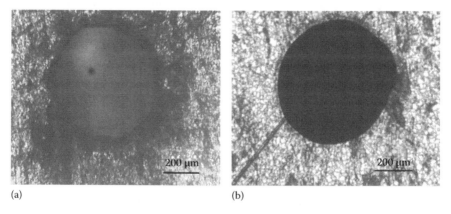

(a)   (b)

FIGURE 11.2   A comparison of holes drilled in stainless steel sheet with (a) a frequency doubled DPSS laser only, and (b) the DPSS + salt solution. (From Li, L. and Achara, C., *Ann. CIRP Manuf. Technol.*, 53(1), 175, 2004.)

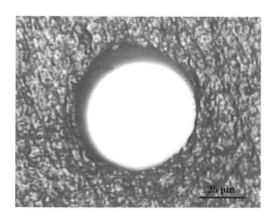

FIGURE 11.3   A hole drilling in 2.5 mm stainless steel using the hybrid Nd:YAG laser and salt solution. (From Li, L. and Achara, C., *Ann. CIRP Manuf. Technol.*, 53(1), 175, 2004.)

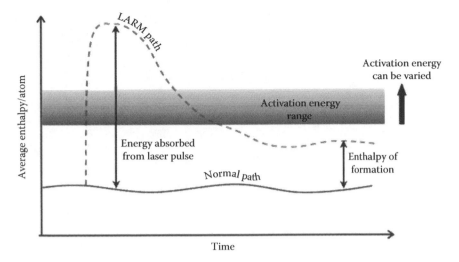

FIGURE 11.4 Energy flow of the laser activated reactive machining in the hybrid process. (From Li, L. and Achara, C., *Ann. CIRP Manuf. Technol.*, 53(1), 175, 2004.)

the local concentrated energy in the system is dissipated to other parts reducing the effectiveness between the laser pulses.

The chemical reaction process can be described as

$$3Fe + 6H_2O \Leftrightarrow 3Fe(OH)_2 + 3H_2. \tag{11.1}$$

The chloride ion is aggressive and helps break down passive compounds once a critical concentration is reached. The role of laser heating is to accelerate this process through [3]

$$K = \left(\frac{kT}{h}\right) e^{\Delta S^*/R} \cdot e^{-\Delta H^*/RT} \tag{11.2}$$

where
  K is the reaction rate
  T is the temperature
  h is the Plank's constant
  S is the entropy
  R is the gas constant
  H is the enthalpy

At higher temperatures, even below the melting point of the material, the chemical reaction rate could be increased by over 1000 times. In this process, the benefit of combining two forms of energy is to accelerate the material removal rate and to eliminate heat affected zones. The chemical used in this case is not environmentally harmful. Since alkali is generated as a by-product, a small amount of acid is needed to neutralize the alkali.

11.4 WATER-ASSISTED LASER MACHINING

Water-assisted laser processing has been widely used to produce nanoparticles and nanowires and for shock peening. For a material machining process, such as the laser ablation of partially sintered

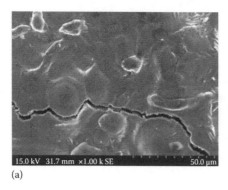
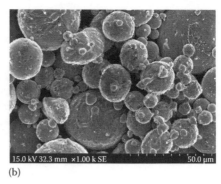

(a)
(b)

FIGURE 11.5 Excimer (248 nm wavelength) irradiated surface of partially sintered titanium powder, 10 pulses at a fluence of 1.8 J/cm^2. (a) Without water, no material removal, and (b) with water 0.19 mm removal depth.

Ti-alloys, water plays a critical role in the material removal process. Figure 11.5 shows an example of the excimer laser machining of partially sintered Ti-6Al-4V powders. Direct excimer laser radiation results in melting only, while water-assisted excimer laser ablation results in clean material removal without melting. As the particle surface temperature rises, the surrounding water is rapidly headed to a higher temperature and the boiling of water occurs close to the laser/workpiece interaction zone. The pressure from the water vapor helps dislodge the weakly bonded metallic particles and results in effective material removal.

In this case, two energy fields (thermal energy coming from the laser beam and mechanical energy coming from the water vapor pressure) interact with each other to create a condition for effective material removal.

11.5   LASER-ASSISTED FLAME COATING

The typical problems associated with the laser coating of ceramic materials are cracking (due to high thermal gradient) and low deposition rate. However, laser surface coating can normally result in high density (close to fully dense coatings) and fusion bonding between the coating and the substrates. Thermal spray coating, on the other hand, can produce coatings with a very high coverage rate and uniform coating thickness at a very low cost. However, flame spray normally produces coatings with a high proportion of porosity and there is a lack of fusion bonding to the substrates. By combining the two, their advantages can be enhanced and the weaknesses can be overcome. Figure 11.6 illustrates a hybrid laser/flame spray coating process configuration [4].

Figure 11.7 shows a comparison of Al$_2$O$_3$ coatings produced by flame spray and the hybrid laser/ flame spray processes. The flame spray was carried out using a standard thermal spray system (Sulzer-Metco 6P-II thermal spray gun) with an acetylene flow rate of 50 L/min at 1 bar pressure, an oxygen flow rate of 42 L/min at 2.1 bar pressure, and an spraying angle of 60° to the surface. The Al$_2$O$_3$ powders were fed at a rate of 12 g/min. The substrate was a 60% alumina-based refractory brick. The laser was a 1 kW Rofin-Sinar CO$_2$ laser, defocused on the substrate surface with a beam diameter of 16 mm. The scanning velocity was 20 mm/s.

Clearly, flame spray alone produces a porous and weakly bonded coating while the laser-assisted process produces a crack-free, porosity free, fully dense coating with a solidification structure.

The interaction between the two energy fields enabled the significant increase in material temperature reaching the substrate compared with the flame spray alone and the thermal gradient is significantly reduced compared with laser cladding as the flame spray system produces a larger pre-post heating area around the laser hot spot. As a result, the hybrid process can produce fully dense, crack-free, high-rate surface coatings of ceramics that are difficult to achieve with laser cladding or flame spray alone.

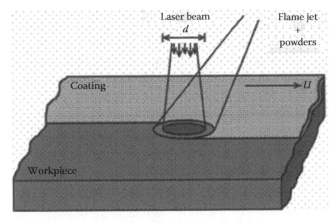

FIGURE 11.6 An illustration of a hybrid laser/flame spray coating process. (From Cotton, F.A. et al., *Basic Inorganic Chemistry*, 3rd edn., John Wiley & Sons, New York, 1976.)

(a)   (b)

FIGURE 11.7 Typical coating structure of Al_2O_3 by (a) flame spray coating and (b) hybrid laser and flame spray coating. (From Li, J.F. et al., *Surf. Coatings Technol.*, 180–181 C, 499, 2004.)

11.6 SUMMARY

Four hybrid laser/nonlaser manufacturing processes have been introduced to describe energy field interactions and the benefits of using hybrid processes to overcome the limits of each single-energy-field process. The abrasive laser machining combines grit blasting with laser melting enabling the effective removal of molten material to generate grooves in metallic materials. The laser-activated chemical machining process allows recasting of free features to be machined more effectively than the laser or the chemical process alone. The water-assisted laser ablation allows lower heat deposition into the bulk material whilst improving the material removal rate. The laser-assisted flame spray overcomes the problems of flame spray and laser cladding to allow fully dense, crack-free coatings to be produced at high surface coverage rates comparable to flame sprays. The energy field interactions in these hybrid processes are critical for the improved manufacturing processes.

The main disadvantage of the hybrid process is the increase in complexity and cost. This can sometimes be offset by the improved process efficiency, less material wastage, and better product quality or service life.

The general approach in developing and assessing a hybrid manufacturing process involving multiple energy fields is to (1) identify the strengths and weakness of individual energy fields, (2) integrate multiple energy fields to see whether it can enhance the process capability, (3) understand and optimize the new process, and (4) evaluate the process in terms of cost, processing time, material usage, energy usage, safety, and environmental impacts. Finally, the practical applications of any new process may require much further development to improve cost, productivity, repeatability, and reliability. Some examples of successful commercial applications of hybrid laser/non-laser manufacturing processes include laser shock peening (using water as a medium), hybrid TIG/laser welding, laser-assisted mechanical milling and turning of ceramics and hard alloys (heating up and softening the material by laser), and waterjet-guided laser machining. (The role of the water is to guide the laser beam, remove the debris, cool the workpiece, and reduce the beam divergence.)

QUESTIONS

Q.11.1 This chapter introduced several hybrid processes. Hybrid processes are designed to overcome the limitations of individual energy fields. This follows the general rule of increasing the degree of freedom when solving engineering challenges. Please analyze each of the four processes and describe the issues that are solved.

Q.11.2 Which engineering contradiction does abrasive laser machining focus on? Which rules are followed in energy field integration?

Q.11.3 What is the theoretical background for laser-activated chemical machining? Could you extend this phenomenon to other engineering applications?

Q.11.4 List examples of hybrid processes in your field.

Q.11.5 In this book, we talk about energy field manufacturing. Hybrid processes are examples of good energy field integration for specific problems. Actually, there is no "pure" process, which means there is no process that only uses one energy field. List as many energy fields as possible in one of the four hybrid processes, and think about what may contribute to solving the contradictions.

ACKNOWLEDGMENTS

The author would like to thank Dr. J.H. Kim, M.H.A. Shukor, C. Achara, Dr. J.F. Li, and Professor F.H. Stott for their original contributions to the work included in this chapter.

ABOUT THE CONTRIBUTING AUTHOR

Professor Lin Li holds a chair of Laser Engineering at The University of Manchester, United Kingdom. He started laser processing research in 1985 at Imperial College, London, United Kingdom. He is now the Director of the Laser Processing Research Centre at the University of Manchester. Professor Li is the author and co-author of over 400 publications in laser processing and 40 patents. He is a charted engineer and a fellow of the Institute of Engineering and Technology, Laser Institute of America, and International Academy of Production Engineering. He serves on the Board of Directors of Laser Institute of America and five international journal editorial boards. His research interests include laser cutting, welding, drilling, surface engineering, micro/nano-fabrication, and additive manufacturing.

REFERENCES

1. L. Li, J.H. Kim, and M.H.A. Shukor, Grit blast assisted laser milling/grooving of metallic alloys, *Annals of CIRP, Manufacturing Technology*, 54(1), 2005,183–186.
2. L. Li and C. Achara, Chemical assisted laser machining for the minimisation of recast and heat affected zone, *Annals of the CIRP, Manufacturing Technology*, 53(1), 2004, 175–178.

3. D.K.Y. Low, L. Li, and P.J. Byrd, Spatter-free laser percussion drilling of closely spaced array holes using anti-spatter composite coating for NIMONIC 263 alloy, *International Journal of Machine Tools and Manufacture,* 41(3), Feb. 2001, 361–377.

4. F.A. Cotton, G. Wilkinson, and P. Gaus, *Basic Inorganic Chemistry*, 3rd edn., John Wiley & Sons, New York, 1976.

5. J.F. Li, L. Li, and F.H. Stott, Multiple-surface coatings of refractory ceramics prepared by combined laser and flame spraying, *Surface and Coatings Technology*, 180–181 C, 2004, 499–504.

Part III

Interdisciplinary Process Innovations

12 Selected Process Innovations in Materials Science

Judson S. Marte

CONTENTS

12.1 HISTORICAL PERSPECTIVE

Materials have defined much of human history: the Stone Age, Bronze Age, and Iron Age. In each of these instances, the materials man used to make tools defined the era. Pre-historic man produced knives and arrowheads by chipping or grinding stone and bone. Toolmaking represents one of the earliest applications of intelligent energy field manufacturing. In this case, the energy field is mechanical and it is intelligent because it is applied with a purpose. The application of energy fields to materials, combined with an ability to accurately predict the response of the material, is one early version of materials science.

12.2 WHAT IS MATERIALS SCIENCE?

What is a material? A material is any solid matter; therefore, materials science is the study of matter. Materials science is often referred to as a multi-disciplinary field of study because it involves

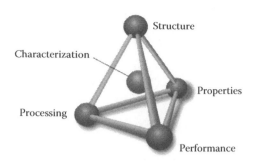

FIGURE 12.1 Tetrahedra of materials science. The tetrahedron symbolizes the interrelation of materials processing, structure, properties, and performance.

many fields of science and engineering: mechanics, chemistry, thermodynamics, physics, and electromagnetism. Because materials science touches all of these fields, it can act as a link among engineers and scientists.

In the terminology of intelligent energy field manufacturing, materials science is the study of a material's response to an applied energy field. Materials scientists study how to make materials (i.e., processing), what materials look like at all length scales (i.e., structure), and how materials behave (i.e., properties) so that humanity can make use of those materials (i.e., performance). The tetrahedra of materials science (Figure 12.1) is a symbolic representation of the inter-relationships between processing–structure–properties and performance. In the center of the tetrahedron is characterization and modeling. These are tools that allow the materials scientist to investigate and predict the response of a material to an applied energy field.

For millennia, materials were understood only on a phenomenological level. That is, the response of a material to a stimulus could be learned, but the mechanism behind that response was rarely known. This knowledge served as the basis for guilds of craftsmen who imparted the "secrets" of materials to the worthy. The study of materials remained more of a craft than a science, until recently. Materials science has evolved into a science that studies not only the response of materials to stimuli but also the atomic mechanisms responsible for that response. This is largely because of the invention of the electron microscope, which has allowed us to observe materials at a level where these mechanisms can be more easily inferred and, in some cases, directly observed.

Materials science studies the entire range of engineering materials. Engineering materials are often categorized according to what they are made of: metals, ceramics, polymers, or composites (i.e., collections of several materials). Metals are known for being ductile and shiny, with excellent electrical and thermal conductivity. For the most part, ceramics are extremely hard, with little ductility, and they are not electrically or thermally conductive under normal conditions. Polymers can be ductile or brittle, transparent or opaque, and they are nearly always nonconductive. Composites are combinations within, between, and among the other classes of materials. They often have a favorable combination of the properties of the materials from which they are composed. Table 12.1 lists some common applications of materials by material class in several industry segments.

In the introduction to intelligent energy field manufacturing, several energy states were described: the normal state, the excited state, and the qualitatively changing state. The normal energy state is similar to what materials scientists might call a metastable state. This is a local energy minima, but not necessarily the lowest possible energy state. The excited energy state is at the cusp between two energy minima and represents a local energy maxima. At this point, the material could move back to the previous normal state, or move on to another local minimum energy state, depending upon the smallest amount of energy added to, or subtracted from, the system. Intelligent energy field manufacturing calls the region between the normal state and the excited state the "qualitatively changing state."

TABLE 12.1
Types of Materials Used in Industrial Sectors

Energy	Class	Industrial	Class
Stainless steel steam turbine buckets	Metal	Silicone o-ring seals	Polymer
Superalloy gas turbine blades	Metal	Titanium valve bodies	Metal
Silicon-steel generator laminates	Metal	Stainless steel pipes	Metal
Alumina ceramic insulators	Ceramic	Ceramic catalyst substrates	Ceramic
Silicone high temperature wire insulation	Polymer	Glass sight gauges	Ceramic
Polymer matrix composite wind turbine blades	Composites	Powder coating paint	Polymer
Transportation		**Consumer goods**	
Plastic body panels	Polymer	Porcelain sink basin	Ceramic
Aluminum aircraft fuselage	Metal	Glass windows	Ceramic
Rubber tires	Polymer	Kevlar tennis rackets	Composites
Cast iron diesel engine blocks	Metal	Titanium golf club heads	Metal
Steel engine valves	Metal	Latex paint	Polymer
Ceramic permanent magnet motors	Ceramic	Copper wiring	Metal
Electronics		**Agricultural and Food**	
Polymer circuit board substrates	Polymer	Vinyl hose	Polymer
Aluminum heat sinks	Metal	Glass jars	Ceramic
Polycarbonate DVDs	Polymer	Steel harvester blades	Metal
Silicon microchips	Ceramic	Fluoropolymer fishing line	Polymer
Gold electrical contacts	Metal	Aluminum cans	Metal

These states are comparable to those that are used to describe different thermodynamic states of matter. The normal state is comparable to a material being in equilibrium with its environment. An equilibrium occurs when a material has no net mechanical, chemical, or thermal driving force acting upon it (Figure 12.2(a)). It represents the lowest energy state. Local equilibrium is a local minima in the energy that occurs under specific conditions (Figure 12.2(b)). These conditions include temperature, pressure, and chemistry (i.e., chemical composition). Materials are always struggling to reach an equilibrium, but engineers frequently apply driving forces (aka intelligent energy fields) that destabilize them. In these cases, a material can be transitioned from one local equilibrium condition to another by applying an energy field to overcome the activation energy barrier (Figure 12.2(c)). Activation energy barrier is the term that materials scientists use to describe the energy

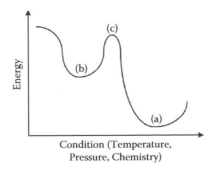

FIGURE 12.2   To reach a true minimum energy state at (a) from a locally minimum energy state (b), the activation energy barrier (c) must be overcome. This can be achieved through the application of an intelligent energy field, such as a heat or pressure.

level between the normal state and the excited state. Therefore, the activation energy is the energy barrier that is between two local minimum energy states.

12.3 STRUCTURE OF MATERIALS

12.3.1 BONDING AND CRYSTAL STRUCTURES

A basic understanding of how atoms behave is needed to appreciate the mechanisms responsible for the interaction of energy fields and materials. In solids, there are three types of strong bonds: covalent, ionic, and metallic. Covalent bonds involve the sharing of electrons between atoms. Ionic bonds form when positive and negative ions combine to produce a structure with a more neutral electronegativity. Metallic bonds involve the sharing of valence electrons among many atoms. The type of bonding has a strong influence on the ways that the atoms in the solid organize. For instance, ionically bonded atoms will form a crystal, in which the atoms are organized to maximize their packing, while maintaining the coordination, i.e., the number of nearest neighbor atoms, needed to minimize the charge state of the crystal. Covalently bonded atoms share electrons between two atoms. Polymers are long, chain-like molecules that are held together by covalent bonds. Metallic bonding involves the sharing of delocalized electrons among many atoms and gives rise to many of the unique properties of metals, such as ductility, high electrical and thermal conductivity, and reflectivity (i.e., metallic luster).

Most metals, many ceramics, and some polymers form crystals when in the solid state. A crystal is defined as a grouping of atoms that exhibit long-range periodicity. That is, they are put together in some repeating pattern. The reason for the pattern is tied to the energy minima, discussed above. The atoms find positions that minimize the energy of the atomic bonds in the material. The simplest crystals to understand are cubic. In these crystals, the three axes are orthogonal and have the same length. The three most common cubic crystals studied in materials science are body centered cubic (BCC), face centered cubic (FCC), and hexagonally close-packed (HCP). These structures are shown in Figure 12.3. The crystal structure of a material defines many of the engineering properties of the material. For instance, metals are more electrically conductive than ceramics, because the delocalized electron structure of metallic crystals allows the atoms to move throughout the metal, while maintaining the stability of the crystal. By contrast, a covalently bonded crystal tightly binds its electrons to the crystal, restricting the ability for these crystals to conduct electricity under normal conditions.

12.3.2 DEFECTS AND DISCONTINUITIES

Defects within the crystal structure also contribute to the properties of materials. Defects in crystals are defined by their dimensionality. One-dimensional defects, called point defects, involve extra, missing, or substituted atoms within an otherwise perfect crystal. Extra atoms can sometimes be squeezed into an existing crystal. These atoms are generally small in size and are referred to as interstitial atoms and cause an interstitial defect. When an atom is missing from the crystal, it forms a vacancy defect. When one atom is substituted for another on an existing atomic site, a

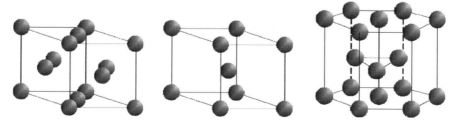

FIGURE 12.3 (a) Body centered cubic, (b) face centered cubic, and (c) hexagonal close packed crystal structures. These structures represent common ways in which atoms organize themselves into periodic arrays.

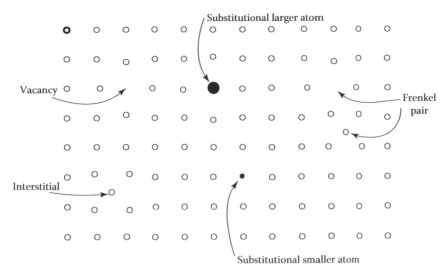

FIGURE 12.4 Point defects. A schematic showing one-dimensional (point) defects: (a) vacancy, (b) interstitial, and (c) substitution. These and other defects within the crystal structure control the properties of the material and can, to an extent, controlled by processing.

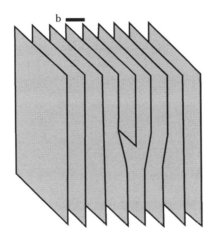

FIGURE 12.5 Edge dislocation. A schematic showing a two-dimensional (line) defect called an edge dislocation. Dislocations occur when a plane of atoms ends in the middle of the otherwise periodic array of crystal planes. Dislocations are important because their motion gives rise to plasticity in many important engineering materials.

substitutional defect is formed. These point defects are shown schematically in Figure 12.4. Point defects can contribute to the properties of a material by stretching the bonds between atoms. This stretching has the effect of locally raising the energy, which alters the response of the material surrounding the point defect to applied energy fields.

Two-dimensional defects, also called line defects or dislocations, affect a line of atoms and contribute to the response of materials to mechanical and other energy fields. There are three types of dislocations: edge, screw, and combined. Edge dislocations are the simplest to visualize, as in Figure 12.5. Edge dislocations occur when a plane of atoms stops in the middle of the periodic array of atomic planes. This causes a defect to form within the crystal, which follows a straight line, hence the name line defects. Edge dislocations can move within the crystal, under the influence of an external energy field. In Figure 12.6, a single edge dislocation moves from left to right, allowing

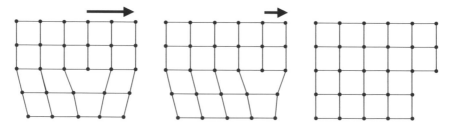

FIGURE 12.6 Dislocation motion. A schematic showing the movement of a dislocation across a crystal from left to right. The motion of the dislocation allows the atoms along the upper side of the crystal to slide past those along the bottom, giving rise to deformation.

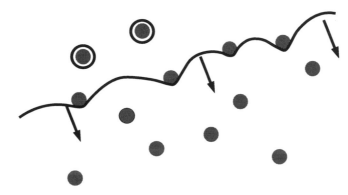

FIGURE 12.7 Precipitates. Precipitates and other crystal defects, can hinder the motion of a dislocation, giving rise to increase strength in metals.

the upper half of the crystal to slide past the bottom. This sliding of dislocations is what gives rise to plastic deformation in metals. The interaction of dislocations with other crystal defects and other dislocations affects the ease with which deformation occurs and controls the strength of the metal.

Three-dimensional defects include any defect that affects more than one plane of atoms. They include important microstructural features, like precipitates, voids, surfaces, and grain boundaries. The importance of these defects lies in the way they inhibit the dislocation motion. In Figure 12.7, a dislocation's motion is hindered by the presence of precipitates within the crystal. This causes the dislocation to bow around the precipitates, as shown in Figure 12.7. In order for a dislocation to move past a precipitate, it must either (a) deform the precipitate or (b) form a "dislocation loop" around the precipitate. Whether the precipitate is cut or forms a loop is determined by some critical size of the precipitate, as shown in Figure 12.8. The effect in either case is that the motion of the precipitate is hindered, making the material stronger. The potency of the dislocation–precipitate interaction is a function of the size and number of the precipitates. When the size of the precipitates is large, in the case of dislocations moving around the precipitates, the strength is proportional to

$$\sigma \propto \frac{G \cdot b}{L \cdot 2r} \qquad (12.1)$$

where
   G is the shear modulus
   b is the Burger's vector
   L is the distance between precipitates
   r is the radius of the precipitates

As the equation suggests, decreasing the size of the precipitates and the spacing between them leads to a higher strength. If, on the other hand, the precipitates are smaller than some critical value, then the dislocation will bow around the precipitate. When enough deformation energy is provided, the dislocation will move past the precipitate, leaving a dislocation loop in its wake, as shown in Figure 12.7. Under these conditions, the strength is proportional to

$$\sigma \propto \frac{r \cdot \gamma \cdot \pi}{b \cdot L} \qquad (12.2)$$

where

    r is the precipitate radius
    γ is the surface energy
    b is the Burger's vector
    L is, again, the distance between the precipitates

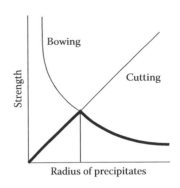

FIGURE 12.8 Particle strengthening. Precipitates (particles) interact with dislocations in two ways. When the precipitates are below some critical size, they can be sheared (cut) by the dislocations. When they are above that critical radius, the dislocation will move around the precipitate, first bowing, then eventually moving past, leaving a dislocation loop in its wake.

Because additional energy is required to move the dislocation around or through the precipitates, their presence retards their motion, leading to an increase in the metal's strength.

A similar behavior can be observed with the interaction of dislocations and grain boundaries. As the size of the grains in a metal decrease, the probability that a dislocation will interact with a grain boundary increases. Therefore, finer grained materials tend to be stronger than coarse grained materials. This behavior, known to materials scientists as Hall–Petch behavior, has the form

$$\sigma = \sigma_0 + \frac{k_y}{\sqrt{d}} \qquad (12.3)$$

where

    σ_0 is the Pierls stress
    k_y is a material-specific constant
    d is the grain size

The equation demonstrates the potency of grain boundaries in the strengthening of metals. This will be discussed in greater detail later in this chapter.

12.3.3 SUMMARY OF MATERIALS STRUCTURE

The structure of a material is determined by the composition of that material, which determines the type of bonding between atoms. The bonding between atoms often causes solids to form a regularly repeating pattern, called a crystal. The pattern of atoms in the crystal can help us to understand how materials behave under the application of an external energy field. Errors in the regular repeating pattern of the crystal are called "defects" and are categorized by their dimensionality. Defect structures in the crystal often determine important material properties, such as ductility and strength.

In some ways, each atomic bond in a material has its own energy field associated with it. All of the other bonds that surround the atom, as well as the application of external energy fields, influence that energy field. For instance, the introduction of a point defect, such as the substitution of a larger atom in a crystal of smaller ones, forces the other atoms to sit in non-ideal positions within the crystal. This leads to an increase in the overall energy of the system. In other words, the defect has imposed an energy field upon all of the other atoms in the system. One might refer to this as an "internal" energy field, i.e., an energy field that resides within the system of interest. Another example might involve the application of some mechanical force to the material. That mechanical force could be viewed as an "external" energy field being applied to the system. The response to that external energy field might be breaking the atomic bonds within the material, causing it to fracture. Materials science is largely devoted to studying the interactions of internal and external energy fields with the material structures and using those interactions to predict materials behavior.

12.4 PROPERTIES OF MATERIALS

All material properties can be categorized as either intensive or extensive. Intensive properties are those that do not depend upon the amount of material. Examples of intensive properties include density, specific heat capacity, and elasticity. Extensive properties do depend upon the amount of material present. Examples of extensive properties include mass, length, and electrical resistance. Materials scientists tend to favor intensive properties, since an extensive property for a material can always be derived if the "amount" of material is known. For instance, by knowing the density and volume of a material, it is easy to calculate the mass of the material.

In many cases, properties are treated as "isotropic." Isotropic materials behave in the same manner regardless of direction. While most engineers treat all materials as isotropic, they rarely are. Imagine that you have to push a car that has run out of gas. Obviously, it is much easier to push on the front or rear of the car to make it move than it would be to try to push it sideways. Many materials behave like the car…it is far easier to push on them in the easy direction. Materials that display directional behaviors are called "anisotropic." The reason for the anisotropy of the material is often related to the arrangement of the crystals in it. If there are many crystals, called grains, and they are randomly organized, the material may behave isotropically. If, however, there is some larger fraction of the crystals oriented in one direction or another, the material will likely behave anisotropically.

The properties of materials describe how a material responds to an applied energy field. Material properties are generally intensive in nature, as they do not depend on how much of the material is present. As stated earlier, these properties are derived from the material's structure, which is affected by the processes used to make the material. Properties are often categorized in terms of the energy field being applied to the material. For instance, mechanical properties describe how a material responds to a mechanical energy field. Materials can also cross categories of energy fields. For example, the application of a mechanical field gives rise to an electrical response.

An incomplete list of material properties that are of interest to engineers is shown in Table 12.2. There are many, many texts dedicated to the properties of materials. It would not be possible to do justice to them in this space, so the reader is directed to resources such as those shown at the end of the chapter.

12.5 PROCESSING OF MATERIALS

The processes used to manufacture materials are myriad, and a full description of them is beyond the scope of this chapter. Rather, the focus will be on the application of external energy fields to produce material structures with useful engineering properties.

There are many types of external energy fields that are used in materials manufacturing. In most manufacturing processes, combinations of energy fields are applied to the material. As discussed in the earlier section on materials structures, energy fields can be both internal and external. In terms of materials processing, internal energy fields might be residual stress energy, chemical gradients, etc. External energy fields are more plentiful and include mechanical, chemical, thermal, and electromagnetic energy fields, etc.

The manufacturing of most metallic objects involves numerous steps in which multiple external energy fields are used to produce and control material structures and internal energy fields. Let's take the production of a simple steel plate for an example. The production of the steel begins with

TABLE 12.2
Material Properties

Physical Properties

Crystal structure
Atomic weight
Density
Melting point
Boiling point
Vapor pressure
Viscosity
Porosity
Permeability
Reflectivity
Transparency
Optical properties
Dimensional stability

Electrical Properties

Conductivity
Resistivity
Dielectric constant
Hall coefficient (effect)
Superconducting temperature

Magnetic Properties

Magnetic susceptibility
Magnetic permeability
Coercive force
Saturation magnetization
Residual magnetization
Curie temperature
Magnetostriction

Thermal Properties

Conductivity
Specific heat
Coefficient of thermal expansion
Latent heat of fusion
Emissivity
Adsorptivity
Ablation rate
Fire resistance

Mechanical Properties

Hardness
Modulus of elasticity
Compression
Torsion
Bulk Modulus
Shear Modulus
Poisson's ratio
Stress–strain curve
Strain hardening exponent
Strain-rate sensitivity
Yield strength
Tension
compression
Shear
Ultimate strength
Tension
Shear
Bearing
Fatigue properties
Smooth
Notched
Corrosion fatigue
Rolling contact
Fretting
Charpy transition temperature
Fracture toughness (KIC)
High temperature behavior
Creep
Stress rupture
Damping properties
Wear properties
Galling
Abrasion
Erosion
Cavitation
Spalling
Ballistic impact

Optical Properties

Absorptivity
Reflectivity
Refractive index
Color
Photosensitivity
Transmittance
Photoluminosity

Chemical Properties

Position in electromotive series
Galvanic corrosion
Corrosion and degradation
Atmospheric
Fresh water
Salt water
Acids
Hot gases
ultraviolet
Oxidation
Thermal stability
Biological stability
Stress corrosion
Hydrogen embrittlement
Hydraulic permeability

Nuclear Properties

Half-life
Cross-section
Stability
Specific activity

Fabrication Properties

Castability
Heat treatability
Hardenability
Formability
Machinability
Weldability

Acoustic Properties

Acoustic absorption
Speed of sound

Note: This is an incomplete list of intensive and extensive physical, electrical, magnetic, thermal, mechanical, optical, chemical, nuclear, fabrication, and acoustic properties of materials.

a melting and casting process. During this process, the raw materials for the steel are heated (an external thermal energy field) and specific elements are added to produce both the desired chemistry and structure (an external chemical energy field). The molten metal is cast, perhaps in a continuous casting machine, to form a solid block of steel. The steel is cooled through the application of water, which provides another thermal and chemical energy field. The steel, still red hot on the inside, is then transferred to a hot rolling mill, where a mechanical energy field is applied, reducing its thickness and increasing its length. The steel may need to be reheated during this process to provide enough thermal energy for large amounts of deformation. Once the steel reaches the desired thickness, it is allowed to cool to room temperature. Later, another mechanical energy field is applied to the material as it is ground, to remove any surface oxidation. Finally, in order to achieve the desired mechanical properties, another thermal treatment is applied to rearrange the atoms to take advantage of an internal chemical energy field. The result is a steel plate with a microstructure made through the application of thermal, mechanical, chemical, and, in some cases, magnetic energy fields. The intelligent application of these energy fields is done to produce the desired set of material properties. The understanding of how these energy fields are applied, and how materials respond to them, is the essence of materials science.

Another example of the use of energy fields for material manufacturing can be found in the production of rare earth permanent magnets. The process begins with the application of a mechanical energy field to chunks of the various components to reduce them to powder. These powders are then blended to reach the desired chemistry, setting up the material for the desired internal chemical energy field, which will produce the desired microstructure. The blended powder is then pressed in a mechanical energy field, while a magnetic energy field is imposed. The magnetic field serves to align the powder particles into a preferred orientation (i.e., a preferred structure). The mechanical energy compresses the particles and temporarily holds them together. The "green" pressed powder is then transferred into a furnace where it is heated (i.e., an external thermal energy field is applied), providing an impetus for an internal chemical energy field to bond the particles together in a process called sintering. Because the powder particles are prone to oxidation, the atmosphere surrounding the block must be controlled during the sintering. This might be considered to be controlling another external chemical energy field. Finally, another magnetic energy field must be applied to the sintered block to align the magnetic domains within the block, thus magnetizing it.

12.6 SELECTED INNOVATIONS IN MATERIALS SCIENCE

In the following, examples of innovations in materials science and the processes of interest to the author will be described. The examples were selected because they emphasize the complexity of the interactions among energy fields during manufacturing. In each case, the intelligent application of energy fields results in a change in the material's structure. Keep in mind that changes to the structure of a material have an effect on the energy of the material system itself. Application of the material often takes further advantage of the energy field within the material. It is this complex interaction of energy fields that makes materials science and materials processes so interesting.

12.6.1 Nb$_3$Sn Superconducting Wire

Normally, if one wanted to make electrical conductors, i.e., wire, one would pull the wire through a series of smaller and smaller holes, reducing the diameter of the conductor while increasing its length. Kunzler et al. discovered the superconductivity of Nb$_3$Sn in the 1960s. The problem faced by Kunzler et al. [1] and others was that Nb$_3$Sn, an intermetallic compound, is brittle and couldn't be wire-drawn using the normal procedures. The bronze process overcomes this limitation by taking the basic elements of a Nb$_3$Sn superconducting wire (i.e., copper, niobium, and tin) and producing a wire from them. Niobium rods are placed into holes drilled into a bronze block. Bronze is an alloy of copper and tin. Unlike a copper block containing Nb$_3$Sn rods, the bronze containing niobium rods is ductile and can be drawn into wire. Once the fine wire is produced, the tin must be removed from the bronze and put into the niobium. This is done by heating the wire so that the overall thermodynamic energy of the system is decreased when the tin diffuses out of the bronze. This tin reacts with the niobium to form a thin layer of Nb$_3$Sn on each niobium filament. This transformation, or displacement of one chemical composition for another, is often employed by materials scientists.

FIGURE 12.9   Cu–Nb composite billet. A photograph of a copper–niobium composite billet. This billet has an outer diameter of approximately 90 mm and contains over 250 niobium rods.

The only problem with the bronze process, as it is described here, is that when more than about 13% tin is added to copper, the bronze loses its ductility and can't be efficiently drawn into wire. The result is that there is a limitation to how much niobium can be reacted to form the desired Nb_3Sn.

The solution to the limited tin that can be included in bronze has been to provide a reservoir of tin in addition to, or instead of, that which is alloyed within the bronze itself. One manifestation of this modified bronze process is called the mono-element internal tin process. In this process, a series of small holes are drilled into a cylindrical block of oxygen-free, high-conductivity copper. The holes are filled with small niobium alloy rods. A "lid" is placed over the niobium rods and sealed shut to prevent the contamination of the internal components. A photograph of a copper–niobium composite billet is shown in Figure 12.9. This cylinder, called a composite billet, is then heated in a furnace with a controlled atmosphere to prevent oxidation. The heated billet is then pushed through a small hole, in a process called extrusion. This reduces the diameter of the billet and increases its length. If the correct temperature, extrusion speed, and reduction in diameter have been chosen, the niobium filaments will be deformed in the same way as the overall billet. The extruded rod will still contain the niobium filaments, as shown in Figure 12.10. In this case, the original billet was 90 mm in diameter and it was extruded to make a rod that was less than 25 mm in diameter and over 12 times longer than when it began. A small, long hole was then gun-drilled down the center of this extruded rod. A tin rod was then placed in the gun-drilled hole and the rod was drawn to a wire less than 0.25 mm in diameter and over 125,000 times longer than the original composite billet. This wire still contains the small niobium filaments (now greatly reduced in size) that were in the original composite billet. The wire can now be transformed by heat treatment. However, the heat treatment is greatly complicated by the presence of the tin. At 231°C, tin melts and its volume increases by 10%. If the wire is heated to this temperature too quickly, the tin will quickly expand causing it to burst out of the wire before the reaction with niobium can occur. To combat this, the wire is slowly heated to a temperature just below the melting point of tin and it is allowed to equilibrate. The wire can then be heated causing all of the tin to melt at the same time, preventing tin-burst. As the temperature continues to increase, the tin diffuses through the copper, so that it can react with the niobium. This reaction creates the desired Nb_3Sn superconductor as a layer on the small niobium filaments within the copper matrix. A photo of the reacted wire, taken using backscatter electron imaging in a scanning electron

FIGURE 12.10 Extrudate. A photograph of the extruded copper–niobium composite billet. The extruded rod has a diameter of approximately 25 mm. All of the niobium filaments have been deformed so that they maintain the shape and configuration as in the original composite billet.

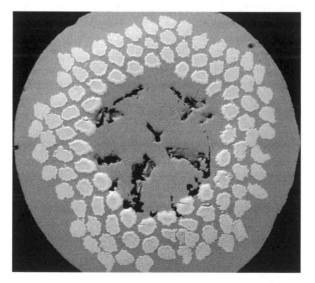

FIGURE 12.11 Nb_3Sn wire. A backscattered electron image of a reacted Nb_3Sn wire. The niobium filaments (light gray) are each coated with a layer of Nb_3Sn intermetallic (white). The filaments are contained in a matrix of copper. Note the residual porosity in the center of the wire that contained the tin reservoir prior to reaction heat treatment.

microscope, is shown in Figure 12.11. Once the reaction of the wire is complete, the wire must be handled with care, since it now contains the brittle Nb_3Sn intermetallic. If the wire is bent around too small a radius, the Nb_3Sn filaments can fracture, breaking the path for electrical conductivity.

The stamp of intelligent energy field manufacturing can be found in the example of bronze-processed Nb_3Sn superconducting wire. The process makes use of thermal and mechanical energy fields to produce the wire form. Then thermal energy is used with chemical energy to produce the

Nb_3Sn superconductor. Note that in both cases, thermal energy is used to enhance the effects of another energy field. Finally, one can look at how energy fields are used in the actual application of the superconducting wire. Superconductors are only superconducting within a narrow envelope of magnetic, electrical, and thermal fields. The inter-related critical temperature, critical field, and critical current define the conditions where the material has zero resistance. Outside of this envelope, the material has markedly higher electrical resistivity.

12.6.2 Nanostructured and Ultrafine-Grained Materials

Recent interest in ultrafine-grained and nanostructured materials has been driven by the enhanced properties that are often seen in these materials, compared with those of their conventional grain sized counterparts. While these properties show great promise, the ability to produce these materials at commercial scales continues to be a challenge. Three strategies for producing nanostructured and ultrafine grained materials have been attempted. These strategies, in broad terms, are (1) powder metallurgical, (2) severe plastic deformation, and (3) controlled crystallization.

12.6.2.1 Powder Metallurgy for Nanostructured Materials

Powder metallurgy (PM) represents a broad class of processes that include everything from the manufacturing of metallic powders or their precursors, to the consolidation of those powders into shapes and their use in real-world applications. Nanostructured and ultrafine grained products can be made using a PM approach, by producing nanostructured powders and consolidating them, without losing their characteristic nanostructured feature size. Several processes can be used to produce nanostructured powders. These might include precipitation from solution, comminution of coarser-grained powders, or the breaking up of larger particles by the introduction of a chemical that causes a volumetric expansion of the particles. These processes make use of chemical, thermal, and/or mechanical energy fields to produce the nanostructured particles. These particles must then be consolidated to make a useful shape. Normally, this is done using heat and pressure to make a solid shape. In terms of energy field manufacturing, this is the application of thermal and/or mechanical energy fields. Unfortunately, these processes rely upon the diffusion of atoms to accomplish the densification of the consolidated part. This diffusion often has the unfortunate side effect of causing the nanostructured grains to grow. If the grains grow too much, they will no longer have the advantageous properties that a nanostructure brings. For this reason, the PM approach must often employ the application of a thermal energy field to produce changes in the chemical energy field. For instance, by heat treating the consolidated powder in a temperature range where precipitation from a solid solution occurs, as in many nickel-based superalloys, it is possible to cause nanostructured particles to come out of the solid solution, adding greatly to the materials strength.

12.6.2.2 Severe Plastic Deformation

Another way to make nanostructured or ultrafine grained structured materials is through severe plastic deformation. Severe plastic deformation imparts large amounts of strain (i.e., deformation) to a workpiece without changing its overall shape. There are many severe plastic deformation processes. Among them, equal channel angular extrusion (ECAE) is the most mature.

ECAE involves the movement of a billet through two channels of approximately the same cross-section that intersect at an angle. As the billet moves from one channel into the next, it is sheared, imparting strain, but not substantially changing the geometry of the workpiece. The strain imparted during each pass is

$$\varepsilon = \frac{\cot\left(\dfrac{\Phi}{2} + \dfrac{\Psi}{2}\right) + \Psi \csc\left(\dfrac{\Phi}{2} + \dfrac{\Psi}{2}\right)}{\sqrt{3}} \tag{12.4}$$

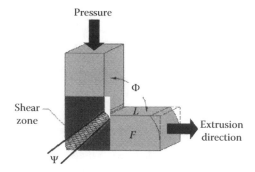

FIGURE 12.12   ECAE. A schematic of the intersecting channels in an equal channel angular extrusion die. The outer angle of the die, Ψ, represents the sharpness/smoothness of the channel intersection. The inner angle, Φ, is the angle between the two intersecting channels. (Courtesy of Bill Richardson, GRC Image C19864158 NASA Glenn Research Center June 1986.)

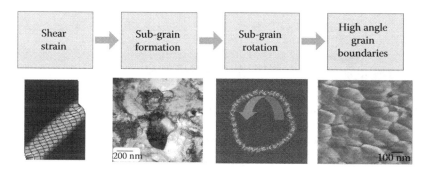

FIGURE 12.13   Grain refinement. The mechanism of grain refinement during equal channel angular extrusion. Shear strain leads to sub-grain formation and rotation. The result is the formation of high angle grain boundaries of substantially smaller size.

where

Φ is the outside corner angle of the die

Ψ is the inside corner angle

This is shown schematically in Figure 12.12. In a sharp-cornered 90° die, such as the one shown in Figure 12.12, the strain per pass is 115%. The workpiece does not change its cross-section despite this deformation, and so, the workpiece can be extruded multiple times with increasing deformation from each pass. Microstructurally, the grains in the workpiece are refined by a mechanism of sub-grain formation and rotation, resulting in ultrafine high angle grain boundaries, as shown in Figure 12.13. In terms of energy field manufacturing, the ECAE process uses mechanical energy to produce microstructural changes. These microstructural changes affect the properties of the material, because the manifestation of those properties requires more or less energy. One example of this is the yield strength of an ultrafine grained metal. Recall that the strength of a material is controlled by how easily dislocations can move through the material. Grain boundaries act as obstacles to dislocation motion, and an ultrafine grained structure naturally has a higher number of grain boundaries. The result is that an ultrafine grained material has higher strength than coarse grained metals. This is known as the Hall–Petch relationship. The strength of a material is inversely proportional to the square root of its grain size:

$$\sigma = \sigma_0 + \frac{k_y}{\sqrt{d}} \qquad (12.5)$$

12.6.2.3 Rapid Solidification and Nucleation of Nanostructures

The third processing method that can be used to produce nanostructured materials is controlled crystallization. This process is commercially practiced by several companies to make magnetic materials with low coercivity and high electrical resistivity. These nanostructured magnetic materials have low eddy current losses and, along with amorphous metals, are used commercially in electrical transformers. These nanostructured materials are made by first creating an amorphous metal sheet. This is accomplished by melting an alloy of a specific composition, typically containing a number of elements of widely differing atomic radii. The chemical composition is selected to take advantage of something called the confusion principle, where the elements are "confused" and don't know which crystal structure to form. The alloy is melted and squirted onto a large spinning copper disc, during a process called melt spinning (Figure 12.14), forming a very thin ribbon. The rapid solidification prevents the crystallization of the metal, making a metallic glass. The metallic glass is then heat treated in a very controlled way to cause the nucleation, but not the growth, of metallic crystals in the ribbon. Because the crystals, i.e., grains, are less than 100 nm in dimension when they nucleate and are not allowed to grow, the metal ribbon is nanostructured. This process uses many energy fields. An induction coil uses magnetic energy to cause heating (thermal energy) of the alloy to cause melting. The molten metal is moved through an orifice under the influence of a mechanical field (gas pressure), whereupon it is rapidly cooled (thermal energy). The rapid cooling manipulates the crystallization (chemical energy) until it is heat treated (thermal energy) to cause crystallization and the creation of the nanostructured material.

FIGURE 12.14 Melt spinning. A photograph of a melt spinning apparatus. The alloy is melted by an induction coil and is deposited onto a spinning copper wheel, where it is rapidly cooled, making a metallic glass.

QUESTIONS

Q.12.1 The tetrahedral of materials science (see Figure 12.1) accounts for how processing, structure, and properties influence performance. It also describes how processes, structures, and properties interact with one another. As with many other scientific endeavors, the computational modeling of materials has become very important in the past decade. Where does computational materials science fit in the tetrahedral of materials science? How could materials modeling produce a unified model of materials science?

Q.12.2 Most metals and ceramics have a distinct crystal structure that is a periodic arrangement of atoms in three dimensions. However, many polymers and all glasses lack this periodic arrangement of atoms. What impact might this structure, or lack thereof, have on the properties of these noncrystalline materials? How would "defects" manifest in this type of structure?

Q.12.3 Materials properties can be categorized as either intensive or extensive. Provide a list of intensive and extensive properties. Are any of these properties related? If so, how? If not, provide an example of the related intensive/extensive properties.

Q.12.4 Energy fields are used throughout the materials manufacturing processes. Describe, in detail, the manufacture of a simple object, such as a paperclip. List where energy fields are applied intelligently.

Q.12.5 Energy fields are the tools for materials scientists and chemists. This chapter gives many good examples of how energy fields are integrated or used alternatively to achieve engineering targets. Analyze the intelligence flows in these examples.

USEFUL MATERIALS RESOURCES

MATERIALS SCIENCE SOCIETIES

There are many professional societies associated with materials science. Below is a partial list of professional materials societies.

- Materials Research Society (MRS)
- The Minerals, Metals, and Materials Society (TMS)
- ASM International (ASM)
- Society of Advanced Materials and Processes (SAMPE)
- American Ceramics Society (ACerS)
- Institute of Materials, Minerals, and Mining (IOM^3)
- American Institute of Mining, Metallurgical, and Petroleum Engineers (AIME)

JOURNALS AND MAGAZINES

There are hundreds of journals, magazines, and e-zines that publish articles related to materials science. Below is a list of just a few:

- *Acta Crystallographica*
- *Advanced Materials and Processes*
- *Acta Materialia*
- *Ceramics International*
- *JOM*
- *Journal of the American Ceramic Society*
- *Journal of Materials Science*
- *Materials Chemistry and Physics*
- *Materials Research Bulletin*

- *Materials Science and Engineering*
- *Materials Science Forum*
- *Materials Today*
- *Metallurgical and Materials Transactions*
- *Nature Materials*
- *Philosophical Magazine*
- *Scripta Materialia*

HANDBOOKS

Materials science is a field that is very diverse. Because of its breadth, it is impossible for an individual to know everything about the field. To assist the materials scientist, there are many handbooks, including the following:

- *ASM Engineering Materials Reference Book*
- *ASM Metals Handbook,* 24 volumes
- *CRC Materials Science and Engineering Handbook, Third Edition*
- *Handbook of Advanced Ceramics: Materials, Applications, Processing and Properties*
- *Polymers Science*
- *Smithell's Metals Reference*

WEB SITES

In addition to the Web sites associated with the various professional organizations (listed above) there are several other Web sites that may be useful to the student. These include the following:

- www.azom.com
- www.forging.org
- www.matweb.com
- www.metallography.com
- www.mpif.org
- www.superconductors.org
- www.steelynx.net/
- ocw.mit.edu/OcwWeb/Materials-Science-and-Engineering/

ABOUT THE CONTRIBUTING AUTHOR

Dr. Judson S. Marte is the author or coauthor of over 20 technical articles on the thermo-mechanical processing of metals, intermetallics, and metal matrix composites. He works as a senior scientist at GE Global Research.

A native of Langhorne, PA, Dr. Marte earned his BS degree in materials science and engineering in 1993 from Johns Hopkins University in Baltimore, Maryland. He earned his MS degree in materials science and engineering in 1996 and his PhD in the same discipline in 1999, both from Virginia Polytechnic Institute and State University, Blacksburg, Virginia. Dr. Marte is currently a metallurgist at the GE Global Research Center. At GE, he leads programs to develop processes for the manufacturing of cost-effective, high-performance structural and functional materials. Of particular interest are the processes that combine multiple energy fields to affect the structure and properties of metals and composites. These include the processing of low temperature superconductors, soft and hard magnets, and high temperature structural materials.

Dr. Marte is very active in the local materials science community of upstate New York. He is a member of ASM International and The Minerals, Metals, and Materials Society (TMS). He was the chairman of the Eastern New York Chapter of ASM International in 2004 and 2005. He has worked with colleagues in the area to develop the Capital District Materials Camp and participated as a project mentor in 2007 and 2008.

REFERENCE

1. Kunzler, J. E., Buehler, E., Hsu, L., and Wernick, J. Superconductivity in Nb_3Sn at high current density in a magnetic field of 88 kgauss. *Phys. Rev. Lett.* 6, 89–91 (1961).

13 Processes and Methodologies in Nanotechnology

Tao Deng

CONTENTS

13.1 NANOTECHNOLOGY AS A REVOLUTION

As the technology revolution evolves into the twenty-first century, two key revolutions stand out: the revolution of micro- and nanotechnology and the revolution of information technology. Information technology is relying on the advancement of micro- and nanotechnology to provide the infrastructure for growth, while the rapid exchange of information also propels the development of next generation micro- and nanotechnology. The increased demand for capacity and speed in information technology really pushes the fabrication technologies to unprecedented complexity. In the middle of the twentieth century, when the microelectronic industry was just out of its infancy, microfabrication technology was the darling on the street and it provided devices and systems that met the need for the information industry. In the 1990s, as the Internet took off, the fast exchange of large amounts of information, for example a video game or movie, became a reality and necessity, which again calls for advancement in micro- and nanotechnology. With the market-driven demand, the revolution of nanotechnology experienced its best boom ever since Richard Feynman's famous lecture "There's Plenty of Room at the Bottom" in 1959.[1] Both fundamental research in academics and capital investment in industry have been growing at a fast pace in nanotechnology. Research centers dedicated to both the study of material properties and the engineering of material function at nanoscales are established everywhere, involving not only government funding but also private investment. Nanotechnology has touched almost all aspects of society (Figure 13.1)[2] from electronics to the environment and from revolutionary technologies to revolutionary products, which will reach a predicted market cap of ~$1 trillion between 2010 and 2015.[2] With nanotechnology, the direct economic impact on sustainability will be ~$45 billion/year worldwide.[2]

A measure of the revolution is the investment in technology. Figure 13.2 shows the comparison of worldwide research and development investment in nanotechnology for industrialized countries

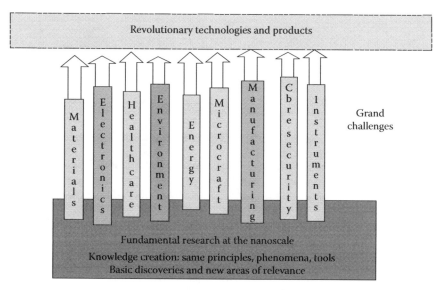

FIGURE 13.1   The nanotechnology revolution. (Courtesy of National Nanotechnology Initiative, Arlington, VA. www.nano.gov)

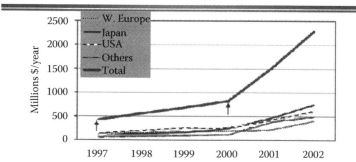

Note:
• U.S. begins FY in October, six month before EU & Japan in March/April
• U.S. does not have a commanding lead as it was for other S & T megatrends
   (such as BIO, IT, space exploration, nuclear)

FIGURE 13.2   Government investment in nanotechnology. (Courtesy of National Nanotechnology Initiative, Arlington, VA. www.nano.gov)

over the period between 1997 and 2002.[2] A jump in the upward swing clearly registered during this period and all the countries increased their funding to keep them competitive in the growing global market. In the United States, a new effort, the National Nanotechnology Initiative (NNI), was kicked off in 2001 and led the U.S. government's investment in driving fundamental research across U.S. academic institutes and industrial research and development laboratories. NNI is not the only funding agent, lots of other organizations also allocated nanotechnology research funding for the programs targeting their respective organizational needs (Figure 13.3).[2] Figure 13.4 shows the funding growth from 2001 to 2009 (predicted) for NNI.[3] Overall, the U.S. government investment in the nano-related area grew from ~$980 million in 2005 to an estimated $1.53 billion in 2009—a 50% jump in investment. Certainly the economic turmoil during the last several years has impacted every aspect of the society. The investment in nanotechnology also slows the pace,

Fiscal Year (Million $)	2000	2001 Appropr./eff./		2002 Appropr.	2003 Request
National Science Foundation	97	150	/150/	199	221
Department of Defense	70	110	/125/	180	201
Department of Energy	58	93	/88/	91.1	139.3
National Institutes of Health	32	39	/38.6/	40.8	43.2
NASA	6	20	/22/	48	51
NIST	8	10	/33.4/	37.6	43.8
Enviromental Protection Agency	-		/5.8/	5	5
Department of Transportation/FAA	-			2	2
Department of Agriculture	-		/1.5/	1.5	2.5
Department of Justice	-		/1.4/	1.4	1.4
Total	270.0	422.0	/464.7/	604.4	710.2

FIGURE 13.3 Funding by other U.S. government agents. (Courtesy of National Nanotechnology Initiative, Arlington, VA. www.nano.gov)

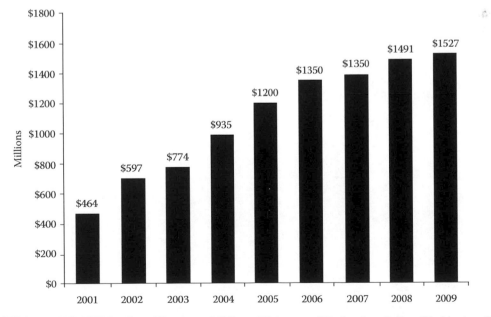

FIGURE 13.4 U.S. NNI funding. (Courtesy of Office of Science and Technology Policy, Washington, DC. www.ostp.gov)

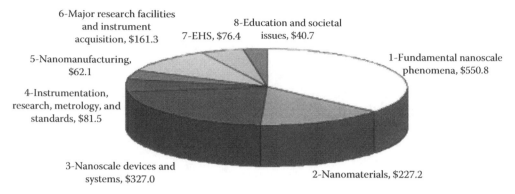

FIGURE 13.5 NNI funding distribution. (Courtesy of Office of Science and Technology Policy, Washington, DC. www.ostp.gov)

but still grows upwards. Figure 13.5 shows the NNI funding by programs—it clearly shows that fundamental research is still the largest piece of the pie and it indicates the relative infancy of the nanotechnology revolution.[3]

Another important indication of the nanotechnology revolution is the publication in this area. With the huge investment in research and development, there comes an increase in both published papers in fundamental research and published patents. Figure 13.6 shows the nanotechnology publications in the science citation index from 1990 to 2006 for major players worldwide.[3] The increase in the volume of publications matches well with the increase in investment over the same period.

The investment certainly generated lots of commercial interest as well. While research scientists in academic institutes file patents based on their serendipitous discoveries, industrial scientists focus more on using nanotechnology to meet the market needs. Many startup companies were formed, and numerous products based on nanotechnology were launched. Figure 13.7 shows the number of patents published worldwide and there is a similar upward increase, especially during the last decades.[3] The United States also leads the pack in terms of issued patents (Figure 13.8).[3]

Eventually, nanotechnology will mature as any other technology revolution, and there are implications of effort in the exploration of the post nanotechnology revolution. Purely scale-wise, angstrom is the next in line on the scale bar, and the appearance of an angstrom lab or research institute indicates effort toward this direction. Another trend in combining nanotechnology with other technologies, such as biotechnologies, also promises the next technology wave. Overall, both the market need and the demand for sustainability will drive the birth of the next technology revolution.

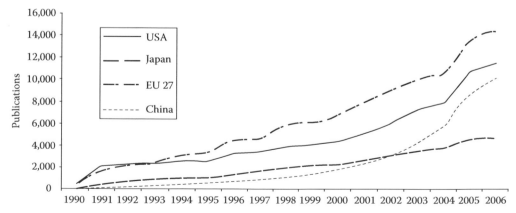

FIGURE 13.6 Nanotech publication. (Courtesy of Office of Science and Technology Policy, Washington, DC. www.ostp.gov)

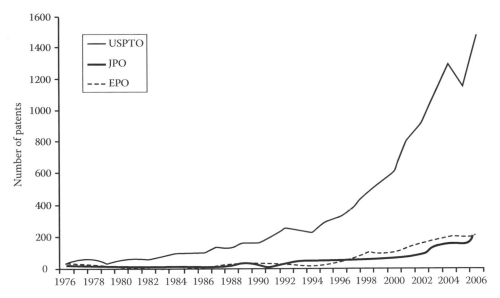

FIGURE 13.7 Patent growth in nanotechnology. (Courtesy of Office of Science and Technology Policy, Washington, DC. www.ostp.gov)

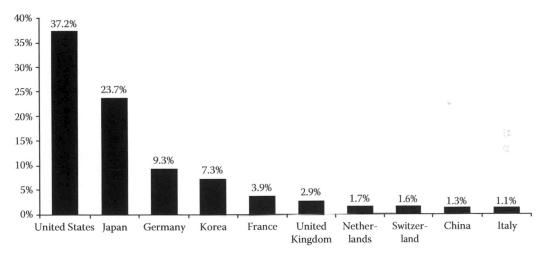

FIGURE 13.8 Issued patents in nanotechnology. (Courtesy of Office of Science and Technology Policy, Washington, DC. www.ostp.gov)

13.2 MATERIAL FABRICATION PROCESSES IN NANOTECHNOLOGY

13.2.1 THE NEED FOR MATERIAL FABRICATION PROCESS

There is a pyramidal structure embedded in the research of materials (Figure 13.9). It starts with the discovery of new materials. Engineering at the device level brings out the function of the new materials and integration of devices at the system level to benefit the society is the ultimate goal of using such materials.

The research of different materials can be traced back to millions of years ago when human beings first discovered the material for making fire. Figure 13.10a shows flint, a hard material consisting of silicate. Figure 13.10b shows the image of a modern lighter. Figure 13.10c shows the pyramid for the fire-making material. The discovery of flint, which has enough hardness to be used to break other materials, lays the foundation of the pyramid. Making a device out of this material to generate fire is at the next level of the pyramid, which requires the flint to be fabricated into a shape that can be fit into a human's hand and also has a sharp edge for the easiness of generating a spark. At the system level, the shaped flint and the operator are integrated together to make the flint fully functional in generating a spark. Combined with a flammable material, such as dry wood, the generated spark can ignite a fire. Currently, one of the most common hand-held fire-making devices is the lighter (Figure 13.10b), which also follows the same pyramid as that of flint: at the material level, there are materials inside the lighter that can generate a spark; at the device level, the shape of the lighter makes it easy to fit into an operator's hand and the shape of the plug inside the lighter makes it easy to generate a spark. The whole system consists of a lighter, an operator, and a flammable material. In both pyramids, material fabrication is the key to bringing out the functions of materials.

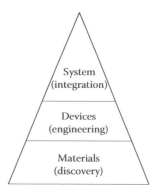

FIGURE 13.9 Materials research pyramid.

13.2.2 PHOTOLITHOGRAPHY

In the long history of using materials, there are many fabrication processes developed to shape materials into desired forms to express the function of the materials. The earliest fabrication example in human history is the use of mechanic force to shape stones, such as, the shaping of flint, the spear, and decorating stones, all of which involve fabrication processes using mechanic forces. Such early development in the fabrication processes probably is also the origin of the intrinsic bias towards mechanic tooling even in the modern manufacturing processes. As human beings evolve, the knowledge of materials increases and the development of fabrication processes also advances. After the Stone Age, humans learned to shape metals using high temperature casting or forging processes. In more recent times, lots of advanced technologies have been used to fabricate structures at different length scales. Processes at the macroscopic level include extruding, molding, and electroplating, etc. Processes at the microscopic level include microdrilling, embossing, photolithography, soft lithography, and other nonlithographic approaches. All these fabrication processes, whether at a large scale or a small scale, help to define the devices and bring function to new materials. Among

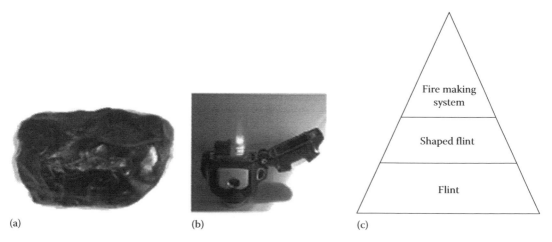

(a)   (b)   (c)

FIGURE 13.10 Fire making tools. (a) flint (Courtesy of Wikipedia); (b) lighter (Courtesy of Wikipedia); (c) pyramid of fire-making material.

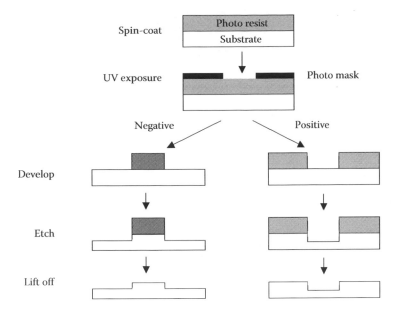

FIGURE 13.11 Standard photolithography process.

all these processes, photolithography is probably the most significant fabrication process that has revolutionized modern life. The manufacturing of modern gadgets, such as computers, digital TVs, radios, and cell phones, involves photolithography. Photolithography is applicable to lots of materials, such as silicon, silica, copper, etc. and can be used to integrate these materials into transistors, diodes, and circuits. The fabrication process makes these materials functional to support all the gadgets we are using.

Photolithography was invented in 1855 by a French chemist, Alphonse Louis Poitevin. Figure 13.11 shows the standard process of photolithography. The process, however, had not been used much for almost a century before it jumped onto the center stage when the microelectronics industry kicked off in the middle of the twentieth century. For the last several decades, photolithography has been following Moore's law: the number of transistors on the chip doubles every 2 years due to the shrinkage of the normalized sizes of transistors. The most recent breakthrough announced by Intel and IBM pushes the smallest feature size on a chip to ~30 nm (www.domino. watson.ibm.com). The increase of transistor density on chips makes the overall system smaller in size. For example, the size of the very first computer was as big as a room; now a power processor can be generated as small as a matchbox. As photolithography advances, the component size decreases, which also enhances the performance at the device and system level. When the density increases, the operation speed will be much faster, the energy loss will be much smaller, and the efficiency of the system will be much higher.

With the constant push for smaller and smaller sizes, photolithography also grows into new generations with more complex and expensive instruments required. As the demand of features approaches the optical limit, many different strategies are needed in the new photolithography process: the ultraviolet (UV) source with shorter wavelengths that lowers the smallest features generated through the process; the use of corresponding optical materials, such as deep UV lenses (CaF_2, for example) and photo resists; and the phase shift masks that can further scale down the features. All the complexities and extra requirements really push up the cost for the next generation lithography, such as the deep UV lithography.

Using the same principle as photolithography, e-beam lithography has been developed to generate primarily submicron structures. Figure 13.12 shows the process of e-beam lithography. The process uses focused electrons, rather than photons, to trigger the reaction in photoresist and generate either

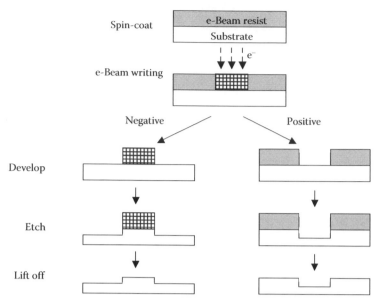

FIGURE 13.12 e-Beam lithography.

negative or positive patterns. Overall e-beam lithography is limited primarily for research use due to the nonparallel fashion of the fabrication process. The process generates one structure at a time, and it takes enormous operation time to generate nanostructures over large areas. Some other forms of these analog lithography processes are focused ion-beam lithography (Figure 13.13), in which a focused beam of ion is used to machine structures at the nanoscale, and dip-pen lithography (Figure 13.14) in which an atomic force microscope was modified to deposit/write nanostructures with a

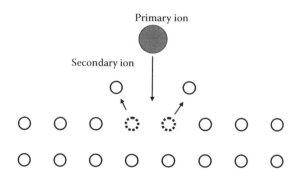

FIGURE 13.13 Focused ion beam lithography—surface atoms ejected due to the impact of the incoming primary ions.

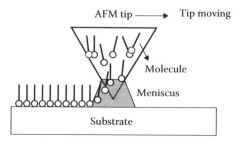

FIGURE 13.14 Dip pen lithography—AFM tip is used in writing molecues on the substrate.

super sharp tip. All of these analog processes have the characteristics of generating one structure at a time, and the output is limited due to the nonparallel nature of the processes. Recently, there has been development in pushing up the output of such processes, such as the multi-tip approach in dip-pen lithography, but challenges still remain for these processes to be adopted as the mainstream nanofabrication processes by major industries.

13.2.3 LIMITATIONS IN PHOTOLITHOGRAPHY AND THE DEVELOPMENT OF ALTERNATIVE NANOFABRICATION PROCESSES

Even though photolithography is the current workhorse in materials fabrication at micro- and nano-scales, in some cases it appears that the application of photolithography is limited. In the following sections, we will evaluate photolithography based on several aspects of the fabrication processes—feature sizes that can be generated, material compatibility, and shape adaptability. We will then discuss the development of alternative nanofabrication processes.

The smallest size that can be achieved varies with the fabrication processes. Different feature sizes require the invention of new fabrication processes or the improvement of the existing fabrication process. The feature size that can be fabricated by photolithography follows Moore's law and is defined by the infrastructure used in photolithography. As the feature size gets smaller, as we discussed above, the requirement for infrastructure gets prohibitively expensive and it is eventually impractical to continuously utilize photolithography in the fabrication of structures with dimensions less than tens of nanometers. The use of an optical field as the manufacturing field in photolithography has an ultimate limitation in the smallest feature sizes this field can handle and eventually it will be technically impossible to generate structures at nanometer scales. X-ray lithography, which uses x-ray as the manufacturing field that has better dimensional resolution than the optical field, can generate structures with smaller feature sizes than photolithography. The use of x-ray and the need to develop a similar set of instruments used in photolithography, however, limit the wide spread use of x-ray lithography.

Material compatibility is the second key element of the fabrication process. After the discovery of a new material, the present fabrication process might not be compatible with the new material. For example, before metals were discovered, people mostly used a simple mechanic fabrication process to deal with woods and stones. After metals were discovered, it was difficult to apply the same fabrication processes used for stones to metals. A high temperature process was then developed to fabricate structures of metallic materials. Photolithography was developed for semiconductor material fabrication. When dealing with some new materials, such as nanocrystal, sol-gel materials, and biological materials, there are limitations using photolithography due to the incompatibility of the chemicals used in the processing of these new materials. For example, the chemicals used in the rinse or lift-off process might have adverse effects to the functions of nanocrystals. The vacuum environment needed in the dry etching will certainly destroy the biological function of most biomaterials.

Shape adaptability is another aspect of the fabrication processes. Photolithography was developed for two-dimensional planar structures; it is challenging to apply photolithography in the fabrication of three-dimensional structures, especially on nonplanar surfaces. New fabrication processes that can make three-dimensional structures are thus needed to build more complex devices.

The above limitations in photolithography call for the development of alternative fabrication processes at micro- and nanoscales. There has been lots of development in this area over the last two decades, such as the development of soft lithography,[4–6] imprinting lithography,[7–9] dip-pen lithography,[10,11] block copolymer lithography,[12,13] and nano-sphere lithography.[14,15] Soft lithography is one of the alternative approaches that was developed in Professor George Whitesides's lab at Harvard.[4–6] In general, soft lithography consists of several molding and printing processes as indicated in Figure 13.15.

All the soft lithography processes involve a soft polymeric stamp or mold. The stamp or mold is made from polydimethylsiloxane (PDMS) through replicating against a master pattern. There are

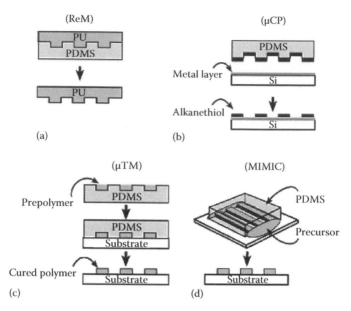

FIGURE 13.15 Soft lithography processes: (a) replica molding; (b) microcontact printing; (c) microtransfer molding; (d) micromolding in capillaries.

four main soft lithography processes (Figure 13.15). In replica molding (Figure 13.15a), a prepolymer is poured against the PDMS mold and then cured. A negative replica of the polymer is generated through this process. Microcontact printing (μCP, Figure 13.15b) probably is the best-known soft lithography process. In μCP, a PDMS stamp that is inked with certain materials is brought into contact with a substrate. After the transferring of the ink, the stamp is peeled off, and the ink is left on the surface of the substrate to form the patterns. In microtransfer molding (Figure 13.15c), a PDMS mold is filled with prepolymer (or any other desired liquid) and then the mold is brought into contact with the substrate. After curing the prepolymer, the mole was separated from the substrate and the polymer pattern is generated on the substrate. Compared with μCP, micromolding can generate patterns with taller features. In micromolding in capillary (Figure 13.15d), the PDMS mold is brought into contact with the substrate first to form continuous channels between the mold and the substrate. A liquid is then introduced to one of the open ends of the channels, and the channels are then filled with the liquid by capillary force. After the liquid is cured, the PDMS mold is peeled off to leave a patterned layer of the cured material.

Soft lithography has been complementary to photolithography in all three aspects of the fabrication process. The smallest size that soft lithography can generate is ~10 nm, which is much smaller than the feature size that photolithography can currently generate. Soft lithography is also compatible with a broader range of materials. It can be used to fabricate structures with materials beyond those used in the semiconductor industry, such as sol-gel materials, nanoparticles, biomaterials, and organic materials. Soft lithography has also been used in the fabrication of micro and nanostructures on nonplanar and three-dimensional surfaces.[4–6]

Imprinting lithography is another nonphotolithographic parallel fabrication process (Figure 13.16).[7–9] Professor Steven Chou's group at Princeton University pioneered the development of this process. In this process, an imprinting master is brought into intimate contact with the substrate and generates a negative replica of the structures in the molding materials. With the help of a releasing agent, the master can be separated from the substrate with both the master and the molded structures remaining intact. This process has been used to generate devices with feature sizes of ~10 nm. There are other forms of imprinting lithography, such as the flash-imprinting process (Figure 13.17),[16,17] in which a polymer precursor was cured during the imprinting process.

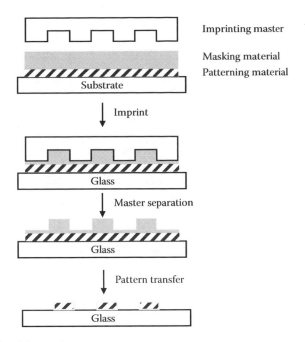

FIGURE 13.16 Imprint lithography.

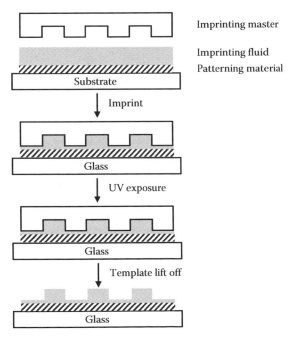

FIGURE 13.17 Flash imprint.

Both block copolymer lithography (Figure 13.18)[12,13] and nanosphere lithography (Figure 13.19)[14,15] are templated fabrication processes that use self-assembly processes to generate templates in the nanometer scales. The self-assembly process, which is the most important structure building process in nature, holds the promise of being the ultimate nanofabrication process; we will discuss it in detail in the later part of this chapter.

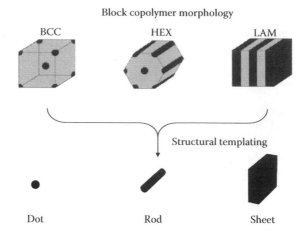

FIGURE 13.18 Block copolymer lithography through phase change of block copolymer.

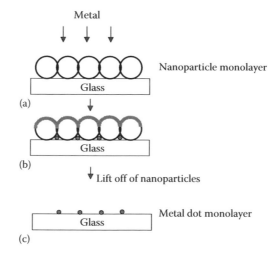

FIGURE 13.19 The use of nanoparticle template as deposition mask in nanosphere lithography.

Among all these nonphotolithography processes, soft lithography probably has the most widespread usage due to the simplicity and the broad material compatibility. In the following sections, we will discuss several applications of soft lithography.

13.3 APPLICATIONS OF SOFT LITHOGRAPHY

13.3.1 FABRICATE STRUCTURES OF MESOPOROUS SILICA

Mesoporous silica is a type of silica that has ordered pores with diameters ~2–50 nm. With large surface areas (~1000 m^2/g), mesoporous silica attracts lots of interest in the area of catalysts and material separation.[18] The tunability of the pore dimensions, geometry, and the composition of mesoporous silica also makes it an excellent host for optical and magnetic materials. To be fully functional, mesoporous silica needs to be fabricated into suitable forms, for example, micrometer-size lines used as on-chip waveguides and thin films used as catalyst support. Several processes, such as dip coating and spin coating, have been developed in shaping the mesoporous silica at the macroscopic level. At the micro- and nanoscale, soft lithography has been established as one of the most important tools in forming mesoporous silica structures. Various functional mesoporous silica–based devices were demonstrated through the application of soft lithography as a fabrication process.[19–22]

In processing mesoporous material, the primary soft lithography technique used is micromolding. The fabrication process follows a direct molding process[21] in which a sol precursor is placed on the substrate before the PDMS mold is pressed against the sol and the substrate. Micromolding in capillary can also be used to generate continuous structures in which the sol is placed at one of the open ends of the channels that are formed between the PDMS mold and the substrate.

In the demonstration of micromolding of mesoporous materials, a silica sol-gel precursor was used in a mixture containing amphiphilic block copolymers. The amphiphilic block copolymers were used as pore generating and structure directing agents. A typical composition of the mixture consists of poly(ethyleneoxide)-b-poly(propyleneoxide)-b-poly(ethyleneoxide) ($EO_nPO_mEO_n$), tetra-ethoxysilane (TEOS), ethanol (EtOH), HCl, and H_2O. The block copolymer, $EO_nPO_mEO_n$, is used in this particular system as a structure directing agent. For example, a hexagonal mesophase was generated when Pluronic P123 ($EO_{20}PO_{70}EO_{20}$) was used. The use of Pluronic F127 ($EO_{106}PO_{70}EO_{106}$) generated cubic mesophase instead of hexagonal mesophase.

Figure 13.20 shows SEM pictures of the waveguide structures of the mesoporous silica that was generated through micromolding.[22] The molding process can typically generate lines with lengths of several millimeters and cross-section dimensions ranging from tens of nanometers to hundreds of microns. The spacing between the lines can be easily tuned by changing the PDMS mold. The well-defined front faces in the cross-section SEM also demonstrated the high quality of the patterns by the shaping process, which enables the use of such structures in optical devices.

The use of P123 in the sol-gel precursor resulted in a hexagonal phase with channels parallel to the substrate (Figure 13.21). When micromolding in capillary was used, a strong capillary flow

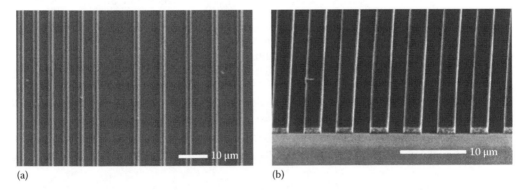

(a)   (b)

FIGURE 13.20 SEM image showing (a) mesostructured waveguide patterns with different spacing and (b) the waveguide ends after cleaving the substrate. (Reproduced with permission from the American Chemical Society.)

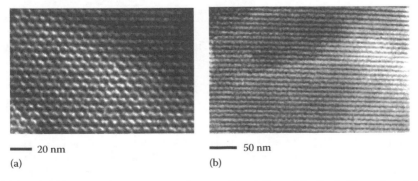

(a)   (b)

FIGURE 13.21 TEM images of mesostructured waveguides (after calcination). (Reproduced with permission from the American Chemical Society.)

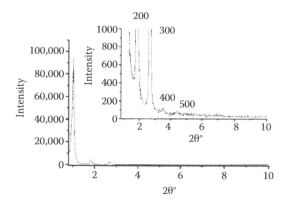

FIGURE 13.22 XRD patterns of waveguide arrays synthesized with P123. (Reproduced with permission from the American Chemical Society.)

aligned these channels not only parallel to the substrate, but also parallel to the flowing direction along the channels. The alignment of the mesopores along the channels would result in a better optical waveguide, when this material is used in a waveguide.

Figure 13.22 shows the XRD pattern of the P123 directed mesoporous material. Several (h00) reflections appeared in the XRD.[20] These reflection peaks showed the high degree of ordering of the mesoporous structure and the preferred parallel orientation of the pores along the substrate.

13.3.2 Functional Mesoporous Silica–Based Waveguide Shaped by Soft Lithography

An optically active waveguide was fabricated using the above process.[20,22] In the waveguide, a laser dye Rhodamine 6G was doped in the patterned mesoporous silica. Typically, the block copolymer–silica mesophase consists of the following components: P123: tetraethoxysilane (TEOS): EtOH: H_2O: HCl: R6G = 0.0172: 1.00: 22.15: 5.00: 0.02: 0.001–0.006 (in molar ratios). The five major components, P123, EtOH, H_2O, HCl, and TEOS, were mixed and put under refluxing for about 1 h. The dye, R6G, was added to the mixture after the solution was cooled to room temperature. The final solution was mixed and under stirring for one more hour before micromolding.

Figure 13.23 shows the procedure used for characterizing such waveguides. The air and a low refractive index underlay were served as a cap for the waveguide structures. The waveguide was

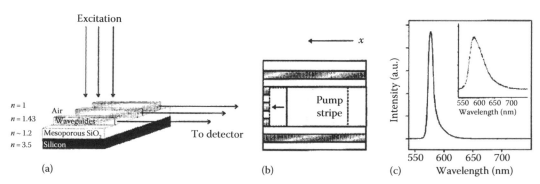

FIGURE 13.23 (a) Schematic representation of the mesoporous support—mesostructured waveguide architecture. The illustration also depicts the measuring geometry used. (b) Illustration of the loss measurements. The pump stripe is moved the distance x from the edge apart, and the output is monitored by the spectrometer. (c) Emission spectra of R6G-doped waveguides below and above the threshold. The inset shows the broad luminescence spectrum at low pumping energies (~1 kW · cm^{-2}). (Reproduced with permission from the American Chemical Society.)

pumped vertically and the detector was placed at the open end of the waveguide. Figure 13.23c shows the optical response as the structure was excited by a Nd:YAG laser (532 nm, ~10 nm pulse width, 10 Hz repetition rate) and it clearly shows an amplified spontaneous emission (ASE). The high quality of the patterns that are defined in shaping the mesoporous materials increases the application of soft lithography in building optical devices with nonconventional materials.

13.3.3 HIERARCHICALLY ORDERED MESOPOROUS SILICA GENERATED THROUGH SOFT LITHOGRAPHY

In the above mesoporous silica structures, two levels of ordering were generated—a microscale order defined by soft lithography and a mesoscale order defined by the structure-directing agent. To add another level of ordering between the micro- and mesoscale, microparticle templates could be used in the micromolding process.[21] Figure 13.24 shows the process of generating three levels of ordering using micromolding. The order at the largest scale is generated using micromolding, the medium scale ordering is generated using particle templates, and the order at the smallest scale is generated using structural directing agents such as amphiphilic block copolymers.

Figure 13.25 provides an example of such a hierarchically ordered system. The structure was made using micromolding in capillaries. The composition was similar to the one used in waveguides and F127 was used as a structural directing agent. The particles used were latex colloidal suspension (Bangs laboratory) containing polystyrene microspheres (200 nm in diameter and 10% in water). The images clearly show the ordering with discrete length scales at 10, 100, and 1000 nm.

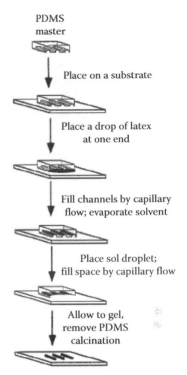

FIGURE 13.24 A sequential process for producing hierarchical ordering over three discrete and independent length scales. (Reproduced with permission from the American Association for the Advancement of Science.)

13.3.4 USING SOFT LITHOGRAPHY IN THE FABRICATION OF NANOCRYSTAL LASER

Besides mesoporous silica, inorganic nanocrystals also benefit from the fabrication process that involves soft lithography.[23] In one example, nanocrystals, together with a matrix, were processed into a device that was capable of generating multiple wavelength lasers. There are many applications that need the use of multiple wavelength lasers. The detection of different gas species simultaneously might require multi-wavelength laser sources. In telecommunication, the use of distributed feedback lasers (DFB) in wavelength division multiplexing (WDM) applications also promotes research on multi-wavelength lasers. Semiconductor materials have been investigated in these multi-wavelength applications due to the tenability of their gain profiles. The temperature sensitivity of the gain profile of semiconductor materials, however, makes them not the best candidate materials in such applications. It is thus necessary to develop alternative materials that have temperature independent gain profiles. One approach in pursuing these alternatives is to reduce the size of the bulk semiconductor material. When the size of the semiconductor materials decreases to the nanometer scale, Asada et al. predicted, theoretically, that the peak gain in zero dimensional structures, or nanocrystals, should be larger and also should be less temperature sensitive than the corresponding bulk materials.[24] Recent experimental studies showed the remarkable stability of the gain profiles of these nanocrystals over temperature.

The work by Sundar et al.[25] demonstrated the lasing from a sol-gel titania/nanocrystal composite optical gain system. In their system, the composite was coated on a single layer of DFB

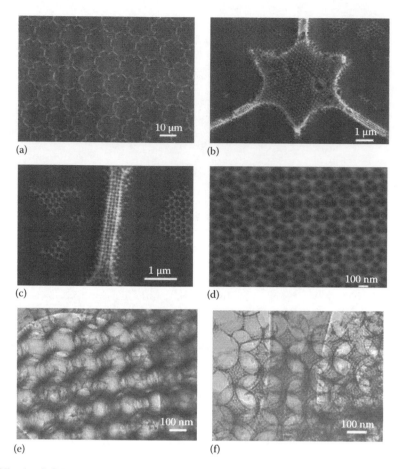

FIGURE 13.25 (a–d) SEM images of hierarchically ordered mesoporous silica displaying organization over three discrete characteristic dimensions at different magnifications. Note the excellent ordering both in the triangular regions and in the connecting bridges. A lattice of the macroporous framework skeleton is visible in (b) and (d). (e and f) TEM images of the same samples, showing that the framework of the macroporous skeleton is made up of ordered cubic mesoporous silica with an ordering length of ~11 nm. This sample was synthesized using Pluronic F127 block copolymer as the structure-directing agent. The sample was calcined at 450°C in air to remove the latex spheres and block copolymers. TEM images were recorded on a JEOL 2010 TEM operated at 200 keV. (Reproduced with permission from the American Association for the Advancement of Science.)

structures that was generated through photolithography. In another example,[26] Eisler et al. also demonstrated multiple color gain and amplified spontaneous emission in a multi-layer structure through repeated spin coating of structureless film. To generate multi-color lasing, however, the gains of the nanocrystals need to be coupled to the color-dependent periodicity of the DFB gratings in each layer. A multilayer titania/nanocrystal structure with DFB grating embedded in each layer was thus needed for the fabrication of the multiple wavelength laser devices. In the construction of such multiple DFB layer structures through photolithography, the nanocrystals would be exposed to different chemicals used in lift off and etching. The exposure would be detrimental to the optical properties of the nanocrystals. The use of soft lithography in the fabrication of such multiple layer structures, on the other hand, would be a better approach since nanocrystals would only be exposed to PDMS mold, which is inert and does not post any threat to the optical performance of the nanocrystals.

Figure 13.26 shows the process of using soft lithography to generate such multi-layer structures.[23] There are only two steps used in the soft lithography process to build each layer: molding and spin

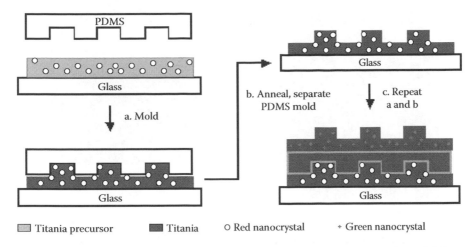

FIGURE 13.26 Fabrication of multilayer nanocrystal/titania composite structures using soft lithography.

coating. If photolithography is used, there would be five steps in the fabrication of each layer: spin coating, exposure to UV, developing, etching, and lift off. Soft lithography certainly offers the most straightforward approach in building such multi-layer films with embedded structures. The atomic force microscope images in Figure 13.27 shows the surface scan of the two-layer structures. The heterostructures consist of an initial 3.3 nm CdSe nanocrystal-tatania film (thickness: ~300 nm) molded with a 350 nm pitch PDMS grating. A buffer layer of neat titania (~600 nm, refractive index ~1.6) was used to planarize this structure. A final spin-coating of a 2.0 nm CdSe NC-titania film (thickness: ~300 nm) and molding with a 330 nm pitch PDMS grating completed the construction of the structure. The gratings were well defined, with the spacing matching the original mold. Due to the slight distortion of PDMS mold during heating, the gratings were a little rough. The purpose of using the neat titania layer in between two layers was twofold. First, it served as a planarization layer for the coating of the second DFB layer. Second, it served as a low refraction index buffer layer to cap both layers. The photons generated in each layer through the excitation of the nanocrystals were confined in the layer due to the capping.

Figure 13.28 shows the schematic of the optical measurement and the gains from the double layers in Figure 13.27. With the easy tuning of the sizes of nanocrystals, the emission wavelength of the nanocrystals can be readily manipulated. The insert in Figure 13.28 also shows the slight different angles of emissions from the first layer and the second layer. If the Bragg condition and the gain window of the nanocrystals are well matched, the emission should be normal to the surface. Due to the slight mismatch of the Bragg condition with the gain window of the nanocrystals, the emission had a small angle (3°–15°) of deviation from the normal. When the nanocrystal films were pumped above threshold, there was a dramatic change in line width (the full width at half maximum changed from ~30 nm to 1.4–3.2 nm). The successful coupling of the gain window with the gratings also reduced the lasing threshold, compared with amplified spontaneous emission threshold, by a factor of 3–5.

13.3.5 APPLICATION OF SOFT LITHOGRAPHY IN PROCESSING OTHER MATERIALS

The previous examples demonstrated the use of soft lithography in processing inorganic materials into functional devices. Research in other materials beyond inorganic materials also took advantage of the versatility of soft lithography. For example, the patterning of microtubules and centrosomes (Figure 13.29) demonstrated the application of soft lithography in dealing with fragile, bioactive materials.[27]

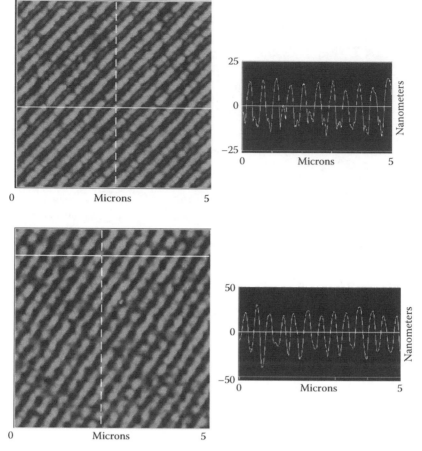

FIGURE 13.27   AFM images of the molded first layer with 350nm pitch size (top) and the second layer of 330nm pitch size (bottom). (Reproduced with permission from Wiley-VCH Verlag GmbH & Co.)

Besides processing materials through molding and printing, soft lithography can also generate micro and nano reactors for the synthesis of materials at nanometer sizes. Figure 13.30 shows such an application—the growing of organic acrylamide nanocrystals.[28,29] Such fine control of reaction at nanometer scales offers unprecedented power over the manipulation and synthesis of nanomaterials with desired properties. There are many other examples of such applications of soft lithography in non-inorganic materials. An extensive discussion of such examples is certainly out of the scope of this chapter.

Thousands of research articles have been published in the area of soft lithography since the invention of the technology at the end of the last century. There are many applications of soft lithography, especially in the areas of new materials or materials in unconventional applications. Soft lithography is still at its early stage; currently the widespread application of soft lithography is primarily in the research labs, both academic and industrial labs. The next big stage will be the commercialization of soft lithography, which will eventually bring its full potential impact to society. Several companies have started to invest in commercializing soft lithography, including some heavy weight players. The booming of nanotechnology and biotechnology will no doubt catalyze this commercialization process. The continued growth of soft lithography will make it mature and finally establish itself as a mainstream player, not only in the lab, but also in the market place—a stage that any mature technology will finally reach.

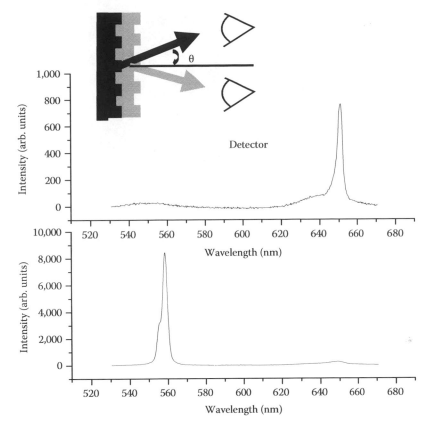

FIGURE 13.28 Above threshold spectra of lasing from structures shown in Figure 13.27. Simultaneous lasing is seen at two different spectral windows from two different angular orientations (top and bottom, respectively). Minor mismatch of the Bragg condition resulted in the laser beams being emitted in a noncollinear fashion (insert schematic). Hence the detector was moved to different angles with respect to the normal to collect and analyze the emitted beam. (Reproduced with permission from Wiley-VCH Verlag Gmbh & Co.)

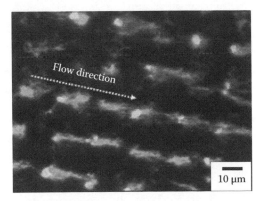

FIGURE 13.29 Directional assembly of microtubules using patterned centrosomes as template.

13.4 SELF-ASSEMBLY

In the development of manufacturing process for nanomaterials, several energy fields have been explored. Drilling or machining, which uses the traditional mechanical field, certainly is one of the early options for generating nanostructures. For example, focus ion beam drilling takes advantage

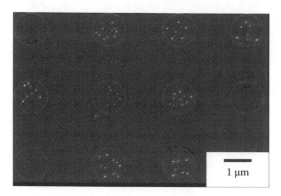

FIGURE 13.30   Acrylamide nanocrystals (white dots inside the circle) grown in microwells.

of the impact force of the high-speed ions. Diamond micromachining uses direct mechanic force to shape the underlying substrate, and there is a commercial instrument using diamond tools that can reach a resolution of sub-100 nm. Several forms of electromagnetic fields are used in nano-fabrication: the optical field in photolithography and the x-ray field in x-ray lithography. If we can categorize chemical interaction as a more general term of "chemical field," then dip-pen lithography, microcontact printing, and flash imprinting lithography certainly all take advantage of this chemical field. In molding and imprinting, the geometry field of the mold defines the shape of the final structures. The change of material property (for example, from a liquid state to a solid state) at the interface close to the mold surface fixes the final structure. Several fields are possibly involved at the interface: the chemical field, molecular force field (such as van der Waals force), and mechanical field.

In nature, the predominant way of generating structures with nanometer dimensions is self-assembly, which takes advantage of many different fields, such as the chemical field, biointeraction field, flow field, etc. In plants and animals, self-assembly processes build nanostructures all the time with unmatched shape and dimensional control. The trend in the nanofabrication process development involves more and more of self-assembly. For example, block copolymer lithography and nanosphere lithography mentioned in the previous section utilize self-assembly as a way of generating nanoscale templates for the subsequent pattern transfer. Self-assembly also evolves into a direct process for generating functional nanostructures. There are many self-assembly processes, such as the bio self-assembly process, chemical assisted self-assembly process, self-assembly process in evaporation processes, and wetting assisted self-assembly process. All these self-assembly process take advantage of preferential interaction between building blocks to generate specific structures. The self-assembly of nanoparticles is one of the most studied self-assembly processes in recent years.[30–39] There are also many important applications of self-assembled nanoparticle systems, such as opals or inverse opals for photonic applications, membranes for filtrations, sensors, and structure components for mechanical devices.

13.4.1   SELF-ASSEMBLY OF NANOPARTICLES

The self-assembly of nanoparticles primarily involves the interaction of particles dispersed in a liquid media. The ordering of the nanoparticles is a result of inter-particle force balance between attractive interactions and repulsive interactions. For example, colloid crystal assembly (CCA) can be formed within a concentrated polystyrene particle solution when the attractive force is balanced by the repulsive static force (Figure 13.31).

The polystyrene particles are well dispersed in the water solution, and the particles have Brownian motion in the solution due to the fluctuation of temperature. The charge groups on the surface of the particles form a shell of electric field around the particles, and the repulsive interaction between

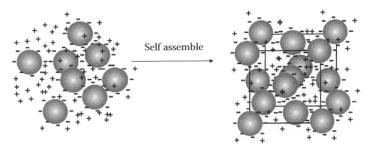

FIGURE 13.31 CCA forming process. Particles were charged in the solvent; the manipulation of surface charge interaction induces the forming of ordered colloidal crystals in solution.

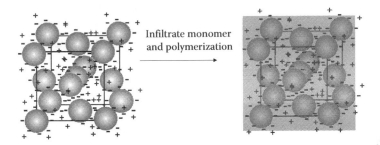

FIGURE 13.32 Hydrogel PCCA: cured hydrogel locks the ordered CCA structure.

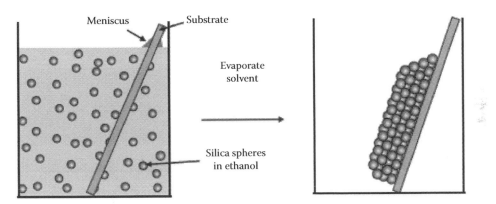

FIGURE 13.33 Evaporation assisted assembly process: ordered particle assembly can be formed on a substrate through slow evaporation of the solvent and the self-assembly of particles at the meniscus. (Reproduced with permission from Wiley-VCH Verlag Gmbh & Co.)

the neighboring electric fields the particles from sticking to each other. As the concentration of the particles gets higher, there is less room for each particle to move freely due to the electrostatic repulsion. If the electrostatic repulsion is not very strong, the concentration will reach a threshed (fluctuation threshold) and particles will stick to each other and precipitate out of the solution. If the electrostatic interaction is strong, the balance of the electrostatic repulsion and the dynamic force (the force that pushes particles together due to the Brownian motion) will induce a relatively stable colloidal crystal structure with order. Professor Sanford Asher's group has extensively studied the ordered colloidal crystal structure in hydrogel (Figure 13.32) as various types of sensors, such as chemical and temperature sensors.[40–45]

Besides forming ordered structures inside the solution, nanoparticles can also form ordered structures at the meniscus due to the evaporation of the solvent (Figure 13.33).[46]

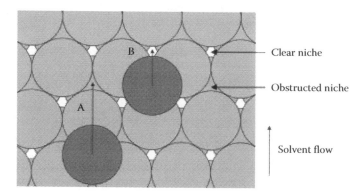

FIGURE 13.34 Flow-driven face centered closed packing: as liquid flows into the clean niche, the particles will preferentially select the site of clear niche due to the flow. (Reproduced with permission from Wiley-VCH Verlag GmbH & Co.)

The flow-driven assembly results in the preferential face centered closed packing for these nanoparticles (Figure 13.34).[46]

Xia, Colvin, Noris, Ozin, et al. have studied this evaporation assisted process and generated many beautiful structures.[38,46,47] Different layers of particles with different colors can be generated with such assembly processes.[37]

13.4.2 SELF-ASSEMBLY OF NONSPHERICAL PARTICLES

To date, most nanoparticle research has concentrated on the spherical particles, which are easily accessible due to the ample synthetic approaches developed. The use of nonspherical particles, however, is both fundamentally important and application-wise impactful. The study of nonspherical particles in blood flow will reveal the secret of geometry in the play of blood cell–related diseases.[48,49] The use of nonspherical particles in the opals and inverse opals will add another knob in tuning the optical response of such photonic structures.[50] The packing of nonspherical particles also promises other ways of increasing the packing density to generate mechanically more stable materials than materials of spherical particles.[51–56] Professor Paul Chaikin's group has theoretically predicted the aspect-ratio-dependent packing density of elongated particles.[54,55] They also tried use M&M candies to study the packing density and experimentally proved that the packing density indeed was high for nonspherical particles.

Research in the development of synthetic approaches and the study of assembling nonspherical particles are rising due to the impact of the nonspherical geometry.[57] In the development of synthetic approaches, many research groups have pioneered such study. Whitesides and his coworkers had developed an approach using microfluidic systems in the fabrication of particles with different shapes and sizes.[58] They also generated composite particles by injecting dopant during the synthesis of the particles. Their approaches are most effective in generating particles larger than 20 μm. It will be challenging to generate particles with sizes smaller than 1 μm using such an approach. Pine and his team also utilized droplets in generating nonspherical particles of polyhedrons.[53,56] In their method, they assembled a limited number of particles in a single droplet and then fixed the assembly as the building block for the next higher-level packing. The resulting polyhedrons are important for fundamental studies, for example, understanding the defects for nanomaterials packing. It is difficult though to extend such a strategy to generate particles other than polyhedrons. Dave Weitz and his team also reported recently in generating dimers and trimers by fusing the particles in the polymerization process.[59,60] Their approach offered flexibility in the generation of polymeric nonspherical particles, but it will be difficult to broaden it into other shapes as well. Most recently, Mitragotri's team at the University of California Santa Barbara (UCSB) demonstrated a versatile method for making particles of different shapes.[61,62] Xia's group used similar approaches

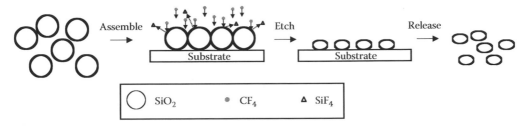

FIGURE 13.35 Fabrication of particles using anisoptrica etching field: a directional etching of monolayer of spherical particles resulted in nonspherical particles on the substrate.

in generating nonspherical polymer particles and used them in the fabrication of eggshell-type particles.[63] Both approaches used mechanical deformation, which is primarily applicable to polymeric particles. Obviously, methods for the generation of nonspherical nanoparticles that are applicable to both polymer and nonpolymeric materials are needed.

The author and his colleagues developed a different approach in generating nonspherical particles using a plasmonic field (Figure 13.35).[64]

In the process, monolayers or submonolayers of particles are first assembled on the substrate through a dip-coating process. A plasmonic field biased with an electric field that is perpendicular to the substrate surfaces is used to guide the etching of the particles.

In the case of silica nanoparticles, CF_4 is used as an etching gas and the ionic species are directed to impact the silica surface to generate SiF_4, which is volatile and diffuses away from the particle surface. Such an anisotropic field eventually will lower the top surface of the particles and will generate spheroid-like particles (Figure 13.36).

The etching rate can be controlled and maintained stable to generate the predicted geometry change of the particles (Figure 13.37).

After the etching, the particles can be lifted off the surface and redispersed in a solution for the subsequent study of the self assembly process. For spherical particles, there is no orientation order due to the spherical symmetry. For spheroid-like particles, there are several options regarding the orientation order (Figure 13.38) related to the long axis of the particles. As the orientation order moves from a parallel state to a tilt state to a perpendicular state, the center of gravity rises and the

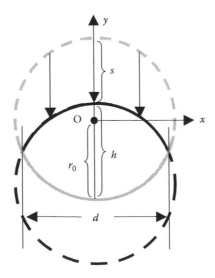

FIGURE 13.36 Schematic of the field assisted shape change: as the top profile moves down due to etching, the particles change from spherical shape to nonspherical shape.

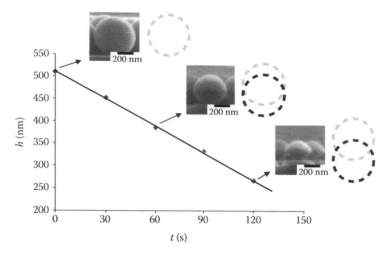

FIGURE 13.37   Etching rate plot and shape change (SEMs and schematic). (Reproduced with permission from the American Chemical Society.)

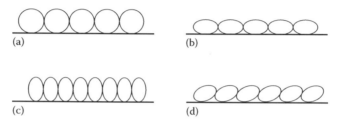

FIGURE 13.38   Orientational order: (a) spherical; (b–d) three possible order for nonspherical particles.

energy gets higher. The parallel state should thus be the most stable state. In the experiments, however, most of time the perpendicular state is the observed state (Figure 13.39). Further fundamental studies of the assembly forces are needed to explain the orientation orders these particles take during the self-assembly process.

Many challenges still remain for the use of the self-assembly process in nanostructure fabrication, for example, defect control, salability, materials compatibility, and shape adaptability. With increased effort in this area, however, it eventually will grow into the ultimate fabrication process for nanomaterials.

Lithography-based technologies dominate the nano-manufacturing processes currently in place. The industry that utilizes nano-manufacturing processes the most is the electronics industry. As the nano-manufacturing processes spread into the rest of the industries, the other nano-manufacturing processes, such as nanomachining, soft lithography, and self-assembly, will grow and meet the unique needs of the specific industry.

Nano-manufacturing as a technology area is still feeling its way amid the transition of many conventional manufacturing processes to micro- and nanoscales. The effort in this area, including both fundamental understanding and technology development, is rapidly growing and exploding. As manufacturing centers shift around the world, the development of advanced manufacturing processes, including nano-manufacturing processes, will give an edge in such a competitive global economy. Any development, however, needs to keep sustainability in mind. A process based on a high carbon footprint will not be sustainable and will not grow into mainstream technology road maps. The future of nano-manufacturing development will focus highly on precision, low cost, adaptability to various applications, and being environmentally friendly.

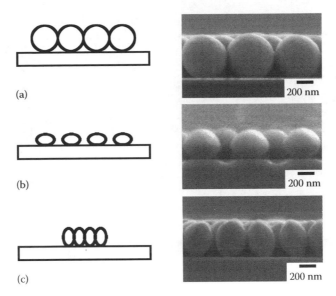

FIGURE 13.39   Orientational order of spherical particles and spheroid like particles observed in the experiment. (Reproduced with permission from the American Chemical Society.)

QUESTIONS

Q.13.1   Read general nanotechnology publications and discuss the nanotechnology revolution and trend.

Q.13.2   How much do you know about photolithography? Write a one-page summary of your understanding of photolithography.

Q.13.3   Summarize the application/impact of photolithography.

Q.13.4   Why is there a need to develop nonphotolithographic processes?

Q.13.5   List other important fabrication processes that are used in large-scale manufacturing.

Q.13.6   Summarize the different fields used in different nano/micro fabrication processes.

Q.13.7   How do you define a chemical field? Biofield?

Q.13.8   Read the literature on soft lithography and make a one-page presentation of your understanding of soft lithography.

Q.13.9   Summarize the applications of soft lithography and imprinting lithography.

Q.13.10 What is the self-assembly process? How does it fit into intelligent energy field manufacturing?

Q.13.11 What's the sustainability demand in nano/micro manufacturing? How does it drive the development of the nanotechnology revolution?

Q.13.12 What's next after the nanotechnology revolution? Can you predict using the intelligent energy field model?

Q.13.13 Where is the information flow in the development of nanomanufacturing processes?

Q.13.14 Discuss what the manufacturing community should do to catch/push the nanotechnology revolution in terms of intelligent energy field manufacturing.

ABOUT THE CONTRIBUTING AUTHOR

Tao Deng graduated with a bachelor's degree in the Department of Materials Science and Engineering with Professor Xingqin Liu at the University of Science and Technology of China, Hefei,

Anhui, China, in 1996. The focus of his undergraduate thesis was the synthesis and characterization of oxide sensory materials. He received a doctorate degree in the Department of Chemistry with Professor George Whitesides at Harvard University, Cambridge, Massachusetts, in 2001. His graduate work involved the development of soft lithography and the application of nonconventional lithographic processes in the fabrication of electronic, magnetic, and optical devices and systems. He joined the Department of Materials Science and Engineering, Massachusetts Institute of Technology, Cambridge, Massachusetts, as a postdoc scientist in 2001 and worked with Professor Edwin Thomas in the fundamental study and applications of block copolymer in microphotonics.

Since 2003, he has been with General Electric's Global Research Center at Niskayuna, New York. He is a member of the Nano Advanced Technology team and constantly promotes the use of nano-engineered surfaces for various GE industrial applications. His research areas include nano-structured surfaces, superhydrophobic and superhydrophilic surfaces, sensors, optical surfaces, self-assembly of nano objects, and industrial scale nanofabrication processes. He has seven issued patents, more than 20 filed patent applications, and more than 70 peer-reviewed external and General Electric internal publications.

REFERENCES

1. Feynman, R. P. *APS Annual Meeting*, Pasadena, CA, 1959.
2. http://www.nano.gov/html/res/NSETPresentations.html; M. C. Roca, *The Future of National Nanotechnology Initiative.*
3. http://www.ostp.gov/cs/pcast/documents_reports; The National Nanotechnology Initiative: Second Assessment and Recommendations of the National Nanotechnology Advisory Panel.
4. Xia, Y.; Whitesides, G. M. *Angewandte Chemie International Edition* 37, 1998, 550.
5. Xia, Y.; Whitesides, G. M. *Annual Review of Materials Science* 28, 1998, 153–184.
6. Zhao, X. M.; Xia, Y.; Whitesides, G. M. *Journal of Materials Chemistry* 7, 1997, 1069–1074.
7. Chou, S. Y.; Krauss, P. R.; Renstrom, P. J. *Science* 272, 1996, 85.
8. Chou, S. Y.; Krauss, P. R.; Renstrom, P. J. *Applied Physics Letters* 67, 1995, 3114.
9. Chou, S. Y.; Krauss, P. R.; Zhang, W.; Guo, L.; Zhuang, L. *Journal of Vacuum Science and Technology-Section B-Microelectronics Nanometer Structure* 15, 1997, 2897–2904.
10. Piner, R. D.; Zhu, J.; Xu, F.; Hong, S.; Mirkin, C. A. *Science* 283, 1999, 661.
11. Ginger, D. S.; Zhang, H.; Mirkin, C. A. *Angewandte Chemie International Edition* 43, 2004.
12. Park, M.; Harrison, C.; Chaikin, P. M.; Register, R. A.; Adamson, D. H. *Science* 276, 1997, 1401.
13. Harrison, C.; Park, M.; Chaikin, P. M.; Register, R. A.; Adamson, D. H. *Journal of Vacuum Science & Technology B: Microelectronics and Nanometer Structures* 16, 1998, 544.
14. Hulteen, J. C.; Duyne, R. P. V. *Journal of Vacuum Science and Technology-Section A-Vacuum Surfaces and Films* 13, 1995, 1553–1558.
15. Hulteen, J. C.; Treichel, D. A.; Smith, M. T.; Duval, M. L.; Jensen, T. R.; Van Duyne, R. P. *Journal of Physical Chemistry B* 103, 1999, 3854–3863.
16. Colburn, M.; Johnson, S.; Stewart, M.; Damle, S.; Bailey, T.; Choi, B.; Wedlake, M.; Michaelson, T.; Sreenivasan, S. V.; Ekerdt, J.; Willson, C. G. *Proceedings of SPIE* Vol. 3676, 1999; pp. 379–389.
17. Bailey, T.; Choi, B. J.; Colburn, M.; Meissl, M.; Shaya, S.; Ekerdt, J. G.; Sreenivasan, S. V.; Willson, C. G. *Journal of Vacuum Science & Technology B: Microelectronics and Nanometer Structures* 18, 2000, 3572.
18. Sayari, A.; Hamoudi, S. *Chemistry of Materials* 13, 2001, 3151–3168.
19. Schueller, O. J. A.; Whitesides, G. M.; Rogers, J. A.; Meier, M.; Dodabalapur, A. *Applied Optics* 38, 1999, 5799–5802.
20. Wirnsberger, G.; Yang, P.; Huang, H. C.; Scott, B.; Deng, T.; Whitesides, G. M.; Chmelka, B. F.; Stucky, G. D. *Journal of Physical Chemistry B* 105, 2001, 6307–6313.

21. Yang, P.; Deng, T.; Zhao, D.; Feng, P.; Pine, D.; Chmelka, B. F.; Whitesides, G. M.; Stucky, G. D. *Science* (*Washington, DC*) 282, 1998, 2244–2247.
22. Yang, P.; Wirnsberger, G.; Huang, H. C.; Cordero, S. R.; McGehee, M. D.; Scott, B.; Deng, T.; Whitesides, G. M.; Chmelka, B. F.; Buratto, S. K.; Stucky, G. D. *Science* (*Washington, DC*) 287, 2000, 465–467.
23. Sundar, V. C.; Eisler, H.-J.; Deng, T.; Chan, Y.; Thomas, E. L.; Bawendi, M. G. *Advanced Materials* (*Weinheim, Germany*) 16, 2004, 2137–2141.
24. Asada, M.; Miyamoto, Y.; Suematsu, Y. *IEEE J. Quantum Electronics* 22, 1986, 1915.
25. Sundar, V. C.; Eisler, H.-J.; Bawendi, M. G. *Advanced Materials* (*Weinheim, Germany*) 14, 2002, 739–743.
26. Eisler, H.-J.; Sundar, V. C.; Bawendi, M. G.; Walsh, M.; Smith, H. I.; Klimov, V. *Applied Physics Letters* 80, 2002, 4614–4616.
27. Shang, W.; Dordick, J. S.; Palazzo, R. E.; Siegel, R. W. *Biotechnology and Bioengineering* 94, 2006, 1012–1016.
28. Qin, D.; Xia, Y.; Xu, B.; Yang, H.; Zhu, C.; Whitesides, G. M. *Advanced Materials* (*Weinheim, Germany*) 11, 1999, 1433–1437.
29. Shang, W. Unpublished result, 2009.
30. Li, M.; Schnablegger, H.; Mann, S. *Nature* (*London*) 402, 1999, 393–395.
31. Rabani, E.; Reichman, D. R.; Geissler, P. L.; Brus, L. E. *Nature* 426, 2003, 271–274.
32. Shenton, W.; Davis, S. A.; Mann, S. *Advanced Materials* 11, 1999, 449–452.
33. Boal, A. K.; Ilhan, F.; DeRouchey, J. E.; Thurn-Albrecht, T.; Russell, T. P.; Rotello, V. M. *Nature* 404, 2000, 746–748.
34. Lu, Y.; Fan, H.; Stump, A.; Ward, T.; Rieker, T.; Brinker, C. J. *Nature* 398, 1999, 223–226.
35. Brinker, C. J.; Lu, Y.; Sellinger, A.; Fan, H. *Advanced Materials* 11, 1999, 579–585.
36. Velev, O. D.; Lenhoff, A. M.; Kaler, E. W. *Science* 287, 2000, 2240.
37. Jiang, P.; Bertone, J. F.; Hwang, K. S.; Colvin, V. L. *Chemistry of Materials* 11, 1999, 2132–2140.
38. Gates, B.; Qin, D.; Xia, Y. *Advanced Materials* 11, 1999, 466–469.
39. Vlasov, Y. A.; Yao, N.; Norris, D. J. *Applied Physics Letters* 62, 1993, 657.
40. Weissman, J. M.; Sunkara, H. B.; Tse, A. S.; Asher, S. A. *Science* 274, 1996, 959.
41. Pan, G.; Kesavamoorthy, R.; Asher, S. A. *Physical Review Letters* 78, 1997, 3860–3863.
42. Lee, K.; Asher, S. A. *Macromolecules* 27, 1994, 1446–1454.
43. Alexeev, V. L.; Sharma, A. C.; Goponenko, A. V.; Das, S.; Lednev, I. K.; Wilcox, C. S.; Finegold, D. N.; Asher, S. A. *Analytical Chemistry* (*Washington, DC*) 75, 2003, 2316–2323.
44. Holtz, J. H.; Holtz, J. S. W.; Munro, C. H.; Asher, S. A. *Analytical Chemistry* (*Washington, DC*) 70, 1998, 780–791.
45. Reese, C. E.; Baltusavich, M. E.; Keim, J. P.; Asher, S. A. *Interface Science* 232, 2000, 76.
46. Norris, D. J.; Arlinghaus, E. G.; Meng, L.; Heiny, R.; Scriven, L. E. *Advanced Materials* 16, 2004, 1393–1399.
47. Tetreault, N.; Miguez, H.; Ozin, G. A. *Advanced Materials* 16, 2004, 1471–1476.
48. Secomb, T. W.; El-Kareh, A. W. *Journal of Biomechanical Engineering, Transactions of the ASME* 116, 1994, 243–249.
49. Boryczko, K.; Dzwinel, W.; Yuen, D. A. *Computer Methods and Programs in Biomedicine* 75, 2004, 181–194.
50. Schmidt, H.; Oliveira, P. W.; Mennig, M.; *The International Society for Optical Engineering: Shanghai, China* 5061, 2002, 24–33.
51. Miyajima, T.; Yamamoto, K. I.; Sugimoto, M. *Advanced Powder Technology* 12, 2001, 117–134.
52. Cho, G.-C.; Dodds, J.; Santamarina, J. C. *Journal of Geotechnical and Geoenvironmental Engineering* 132, 2006, 591–602.
53. Manoharan, V. N.; Pine, D. J. *MRS Bulletin* 29, 2004, 91–95.
54. Donev, A.; Stillinger, F. H.; Chaikin, P. M.; Torquato, S. *Physical Review Letters* 92, 2004, 255506-1.
55. Donev, A.; Cisse, I.; Sachs, D.; Variano, E. A.; Stillinger, F. H.; Connelly, R.; Torquato, S.; Chaikin, P. M. *Science* 303, 2004, 990–993.
56. Manoharan, V. N.; Elsesser, M. T.; Pine, D. J. *Science* 301, 2003, 483–487.
57. Yang, S. M.; Kim, S. H.; Lim, J. M.; Yi, G. R. *Journal of Materials Chemistry* 18, 2008, 2177–2190.
58. Xu, S.; Nie, Z.; Seo, M.; Lewis, P.; Kumacheva, E.; Stone, H. A.; Garstecki, P.; Weibel, D. B.; Gitlin, I.; Whitesides, G. M. *Angewandte Chemie-International Edition* 44, 2005, 724–728.
59. Kim, J. W.; Larsen, R. J.; Weitz, D. A. *Journal of American Chemical Society* 128, 2006, 14374–14377.

60. Kim, J.-W.; Larsen, R. J.; Weitz, D. A. *Advanced Materials* 19, 2007, 2005–2009.
61. Champion, J.; Mitragotri, S. *American Institute of Chemical Engineers Annual Meeting,* New York, 2005, p. 4538.
62. Champion, J. A.; Katare, Y. K.; Mitragotri, S. *Proceedings of the National Academy of Sciences* 104, 2007, 11901–11904.
63. Lu, Y.; Yin, Y.; Xia, Y. *Advanced Materials* 13, 2001, 271–274.
64. Deng, T.; Cournoyer, J. R.; Schermerhorn, J. H.; Balch, J.; Du, Y.; Blohm, M. L. *Journal of American Chemical Society* 130, 2008, 14396–14397.

14 Nanofabrication and Nanocharacterization Using Near-Field Optics

Kaijun Yi and Yongfeng Lu

CONTENTS

In this chapter, techniques and platforms to fabricate and characterize nanostructures using optical approaches based on tip-enhanced near-field effects are introduced. The background of these techniques is reviewed. The details of a nano-Raman spectrometer and imaging system developed for characterizing materials and devices at nanoscales and a laser-assisted scanning tunneling microscope (LASTM) developed for performing surface nanostructuring are presented. Both the

377

nano-Raman spectrometer and LASTM processes feature spatial resolutions of ~30 nm, much beyond the optical diffraction limit.

14.1 INTRODUCTION

14.1.1 BACKGROUND

Nanoscience and nanotechnology, dealing with structures having dimensions of less than 100 nm, have attracted much attention in recent decades due to their potentiality and applications in various areas such as information processing, communications, biomedicine, chemistry, environmental science, tissue engineering, defense, and space exploration [1,2]. This scientific branch and technology were foreseen by Nobel Laureate Dr. Richard P. Feynman in the talk, "There is Plenty of Room at the Bottom" in 1959.

As highly multidisciplinary subjects, nanoscience and nanotechnology have significant impacts on other fields such as applied physics, materials science, colloidal science, device physics, supramolecular chemistry, biology, medicine, and mechanical and electrical engineering. For example, the 45 nm complementary metal oxide semiconductor (CMOS) technology has given rise to the emergence of small, light, long-lasting, and multifunctional smart personal digital assistants (PDA) and cell phones [3,4]. The discovery of giant magnetoresistance (GMR) and tunnel magnetoresistance (TMR), which use two ferromagnetic layers separated by a nonmagnetic layer with a thickness of several nanometers, has led to the invention of hard disk drives with high storage capacity and rapid access time [5,6]. The application of nanomaterials, e.g., gold (Au) nanoparticles, to biomedicine has enabled the invention of diagnostic devices, contrast agents, analytical tools, and drug-delivery vehicles for disease diagnosis and treatment [7,8]. As oil is depleting, nanotechnology is becoming more and more important in dealing with the energy issue by reducing power consumption, increasing the efficiency of energy production and building more environment friendly energy systems [9].

However, nanotechnology is far from reaching maturity, and there are still many areas to explore. For example, the need to further miniaturize high-performance and high-density devices is compelling us to develop quantum devices, based on moletronics that use molecules or nanoelectronics that use carbon nanotubes. These are superior to silicon (Si)-based microelectronics, which are severely constrained by the inherent tunneling effect when the device features are less than or comparable to the electron de Broglie wavelength, i.e., 100 nm [10–13]. We are currently approaching their size limitations.

To take advantage of nanoscience and nanotechnology, both the fabrication and the characterization techniques of nanostructures need to be significantly improved. In general, the fabrication methods are categorized into two types: "top-down" and "bottom-up" approaches. In the "top-down" approach, nanostructures are constructed from larger entities to smaller desired structures without atomic-level control. It evolves from conventional lithography. In the "bottom-up" approach, materials and devices are built from molecular components that assemble themselves chemically by principles of molecular recognition. In both approaches, however, without the precise characterization of the structures on an atomic or molecular level with the adequate tools, the controllable fabrication of such structures is nearly impossible. In fact, the impetus for nanotechnology comes from the invention of a new generation of characterization tools such as the scanning probe microscope (SPM) [14], which includes the scanning tunneling microscope (STM) and the atomic force microscope (AFM). When coupled with refined processes, such as electron beam lithography (EBL) and molecular beam epitaxy (MBE), these tools allow the deliberate manipulation of nanostructures and lead to the observation of novel phenomena [15]. The scanning electron microscope (SEM) is another analytical tool that images the sample surface by scanning it with a high-energy beam of electrons in a raster scan pattern. The electrons interact with the atoms of the sample and produce signals that contain information about the surface topography, composition, and other properties such as electrical conductivity.

Both the scanning probe microscope and the SEM are unable to identify chemicals, which provide vital information about the nanostructures. In order to have a cutting edge characterization tool, along with the capability of chemical identification, some other requirements must be fulfilled. These include (1) being able to operate in a versatile environment and (2) being fast enough to allow for dynamic observation on a fast timescale. Optical spectroscopic tools, e.g., Raman spectroscopy (RS), can meet these needs and are attracting more and more attention from scientists and engineers all over the world [16]. RS offers many unique advantages over its counterparts by identifying chemical structures based on light-matter interactions. For example, it is nondestructive and relatively inexpensive. In addition, it provides access to unique information under a wide range of operating conditions. Another appealing feature of this technique is that by coupling light with certain substances, surface nanostructuring, which is the basis of modern lithography, can be performed [17]. It is obvious that a versatile and multifunctional system for nanostructure fabrication and characterization using optical approaches should be developed.

14.1.2 Near-Field Optics: The Path to Achieving Nanometer Optical Resolutions

A wealth of light-matter interactions allows for a variety of highly selective spectroscopic techniques, providing information not only on the elemental composition of a sample but also on its chemical organization and structure. The classical far-field optics, however, have difficulty in confining optical fields to volumes sufficiently small for the purposes of characterization and modification at nanoscales. The smallest volume of a light spot generated by conventional optics is governed by the optical diffraction limit (see Figure 14.1a and d).

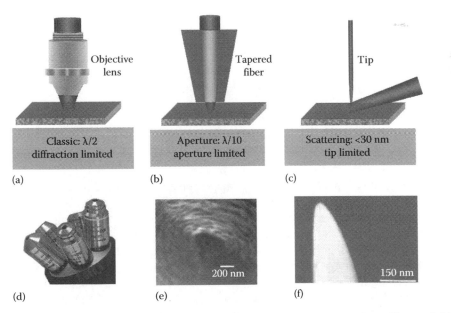

FIGURE 14.1 Comparison of approaches for (a) far-field optics using an objective lens, (b) near-field optics using a tapered fiber, and (c) near-field optics using a metal tip. (d) Practical objective lenses; (e) an aluminum (Al)-coated aperture probe prepared by pulling method; (e) (From Hecht, B. et al., *J. Chem. Phys.*, 112, 7761, 2000. With permission.) (f) a Au tip produced by AC-wet etching method. (From Bouhelier, A., *Microsc. Res. Tech.*, 69, 563, 2006. With permission.)

The optical diffraction limit, D, is defined as [18]

$$D = k \frac{\lambda}{\text{NA}}. \tag{14.1}$$

Here

λ and NA refer to the optical wavelength and the numerical aperture of a lens, respectively

k is a factor that depends on the incident beam profile onto the lens

For uniform illumination, $k = 0.61$ (Rayleigh criterion); for an optimized ring-shape illumination, k can be smaller, e.g., 0.36. With $k = 0.36$ and NA $= 0.9$, near-ultraviolet (UV) light ($\lambda = 400$ nm) can be focused down to $D \approx 140$ nm, a capability that is exploited in modern photolithography. With the same illumination, $D = 100$ nm can be achieved with the best commercially available solid-immersion objective lens (NA $= 1.4$). Clearly, classical optics is limited with respect to resolution right at the borderline of nanometer-sized dimensions.

The optical diffraction limit of the classical optics can be circumvented by near-field optics. Near-field optics is a branch of optics that considers configurations that depend on the passage of light to, from, through, or near an element with subwavelength features and the coupling of that light to a second element located a subwavelength distance from the first. This concept was first proposed by E. H. Synge in 1928, but was not realized experimentally until the 1950s when several researchers demonstrated the feasibility of subwavelength resolutions [21]. In 1984, D. Pohl at IBM in Zurich, Switzerland obtained, for the first time, a near-field optical scan trace recorded with an aperture-type probe at visible wavelengths [22].

As an evanescent wave, the optical fields from near-field optics exhibit exponential decay with distance along the optical propagation direction. Due to their rapid decay, they are of importance only very close to the interface. Using near-field optics, nanoscale resolutions could be attained (see Figure 14.1b, c, e, and f) [23]. For example, using an apertureless metallic tip, a resolution of less than 30 nm could be realized.

To understand the reason why such high resolutions could be achieved by the apertureless metallic tip, we need to introduce two concepts: plasmons and surface-plasmon polaritons. Plasmons are defined as electromagnetic excitations associated with the free charges of a conductive medium, and surface-plasmon polaritons are the plasmons bound to an interface of a dielectric medium by coupling with photons. The speed of these surface-plasmon polaritons is lower than the speed of light in the medium adjacent to the metal surface, so the electromagnetic field is evanescent. Principally, surface-plasmon polaritons can be excited in any kind of metal structure, but excitations in structures with feature sizes much smaller than the wavelength of the exciting light are attracting most interests due to the highly localized resonance property. These localized surface-plasmon polaritons can be easily excited in metallic nanostructures. The incident optical field is strongly enhanced in regions close to these nanostructures.

In tip-enhanced Raman spectroscopy (TERS), a sharp apertureless metallic tip (e.g., a silver [Ag] or gold [Au] STM tip) with a tip apex of tens of nanometers is used as the metallic nanostructure to generate the localized surface-plasmon polaritons. The electric field associated with the localized surface-plasmon polaritons at the apex of this metallic tip comprises an evanescent field. Since this enhanced field is highly localized around the tip apex, an extremely high spatial resolution can be achieved for RS by detecting the Raman scattering signals only from this field. It has been proved that these achievable resolutions are only determined by the size of a tip apex. Subsequently, a nanoscale resolution can be achieved in TERS.

Similarly, in a LASTM, which is another topic of this chapter, a sharp tungsten tip is used and serves as the field enhancer. Although this enhancement is not as strong as that induced by Ag or Au tips, it is still significant since the incident laser power in this application is much stronger than that used in RS. The enhanced field is also localized in a region that is well defined by the tungsten tip, and its intensity can be fine-tailored to a value by which a surface modification becomes possible. Therefore, only the surface within this region can be modified. As a result, nanoscale surface modification, or surface nanostructuring, becomes possible using LASTM.

14.1.3 Nano-Raman Spectroscopy: A Unique Tool for Identifying Chemicals at Nanoscales

Different types of radiation, including diffraction, scattering, photoluminescence, and fluorescence, occur when light interacts with matter. Based on this, different techniques and instruments have been developed and used to analyze compositions, chemical structures, and bandgaps of samples. There are two different types of scattering: Rayleigh scattering and Raman scattering. Rayleigh scattering is a two-photon process with a net effect of changing the direction of light while keeping its frequency constant, during which there is no exchange of energy. Therefore, Rayleigh scattering is elastic. Raman scattering, on the other hand, is a two-photon process with net effects of scattering photons and changing their frequency. This change of frequency, which is characteristic of inelastic scattering, provides a basis for the applications of RS [24].

Since traditional RS employs conventional objective lenses that follow the rules of classical optics, its spatial resolution is limited from a few micrometers to several hundred nanometers. To improve the resolution, the concept of near-field optics needs to be introduced. In this chapter, nano-Raman spectroscopy (n-RS), which is implemented by using tip-enhanced near-field effects, will be introduced as one example of nanoscale characterization methods using an optical approach. This technique, named TERS, is a combination of RS with a sharp metallic tip that can be controlled to be very close to the sample surfaces. Due to the highly localized and significantly enhanced optical field generated by the tip apex, very high spatial resolutions at nanoscales can be achieved [25–27].

In 1985, Wessel first proposed the idea of scanning a sample surface with an Ag nanoparticle, which was controlled by piezo crystals [28]. However, controlling the distance between the particle and the substrate was extremely difficult at that time. He then suggested that an STM might be used to control this distance and the scanning tip would serve as a light enhancer, which was validated later. The first TERS system was reported by Zenobi and Deckert who combined an AFM with an RS in 2000 [29]. A configuration in which the tip and the light were on the opposite sides of the sample, i.e., the bottom illumination optics, was used for this system. However, this configuration was only limited for transparent samples. To study opaque samples, Pettinger et al. developed a TERS system by employing an STM in combination with side illumination optics, in which the tip and the light were on the same side of the sample [30,31].

The origin of the enhanced electromagnetic field is a combination of two distinct physical effects: the "lightning rod" effect, which is a result of the geometric singularity and the localized surface-plasmon polaritons associated with sharply pointed metal structures. The excitation laser light is focused onto the very end of the sharp metal tip, which is kept in close distance to the surface. Surface-plasmon polaritons are excited at the tip apex or in the cavity between the tip and the sample, and only the molecules located directly underneath the tip in the enhanced field give rise to intense Raman scattering. The tip serves as an amplifying antenna for the incoming and outgoing electromagnetic waves, thus supporting extremely intense Raman scattering [20].

TERS can be applied in various applications, such as material analysis, cell identification, and device characterization. Using a Cr-coated quartz tip, Zhu et al. demonstrated that stressed Si lines with a width of 10 nm can be resolved by identifying a double band [32]. Hartschuh et al. obtained the Raman imaging of the individual single-wall carbon nanotubes (SWCNT) [33].

14.1.4 Surface Nanostructuring: The Way to Build Nanodevices

Nanostructures are crucial for nanodevices that can operate at very high speeds with extremely low power consumption. These devices play vital roles in future information storage and processing, communications, and biotechnology. The fact that nanostructures offer such a diverse range of potential application attributes to their capabilities of tailored electronic, optical, and magnetic properties. Surface nanostructuring techniques aim at building those kinds of nanostructures [34].

Conventional lithographic processes, which have been developed for the semiconductor industry for manufacturing microelectronic circuits and components, are incapable of fabricating the desired nanostructures, as they tend to approach their fundamental size limits. Therefore, novel cost-effective and technically feasible fabrication techniques are in high demand. The increased availability and quality of scanning probe microscope (SPM) instruments in recent years has propelled the development of a number of SPM-based surface nanostructuring techniques, which allow us to fabricate complex nanostructures from components as small as individual atoms. The advantage of using an SPM-based approach is that the structures can be fabricated, characterized, and modified by the same instrument, eliminating the contamination problem during transportation, which is unavoidable by other means. In addition, this approach is inexpensive, reliable, and easily operative [35–37]. The major challenge of using this approach is that the yield is relatively low, as only a single probe can be used at one time. Probe-tip arrays may overcome this problem [38].

An STM is an apparatus capable of obtaining surface images with atomic-scale resolutions, measuring tunneling spectroscopy that reveals a variety of surface electronic properties, and manipulating individual surface atoms and molecules with atomic-scale precision. The major purpose of using the STM is to characterize the electronic properties and surface morphologies of materials with conductive surfaces. It has been applied for various applications, such as adsorption and film growth, electrochemistry, and chemically selective imaging. Besides surface characterization, it can also be used to fabricate nanostructures on different substrates, including polymers, metals, semiconductors, and ceramic materials, in a precise location and with a predefined pattern. Compared with conventional lithography processes such as EBL and photolithography, nanostructures fabricated by STM feature more advantages, including ultrahigh spatial resolution, inexpensive equipment, and relatively convenient operation. Ultrahigh spatial resolution is achieved by the nanoscale electrostatic, electrochemical, and thermal energy sources induced by the STM tip through the tip–substrate gap, since the tunneling current can be treated as a miniature electron beam at a diminishing scale. Four types of surface nanostructuring can be performed: material modification, material deposition, material removal, and atomic manipulation. Without any assistance, however, application areas using STM only cannot be broadened due to the tiny tunneling current (generally around a nano Ampere) it generates.

On the other hand, laser microprocessing has been widely used to fabricate microstructures that have been employed in the semiconductor and data storage industries. Comparably, the fabrication of nanostructures using conventional laser microprocessing approaches is difficult due to the optical diffraction limit. LASTM came into being by combining the powerful fabrication capability of laser microprocessing with the high resolutions of the STM.

LASTM employs a metallic tip, such as a tungsten tip, to generate a highly localized and significantly enhanced optical field. This field could heat the substrate surface, induce phase change and modify the surface at nanoscales. As a direct consequence of nanoscale heating, a nanodomain of the sample experiences intense heating, phase change/explosion, stress development and propagation, and rapid structural evolution. Consequently, surface nanostructuring could be realized.

Compared with surface nanostructuring using STM alone, LASTM, with its additional energy source of lasers, is more robust by generating a thermo-voltage, inducing thermal expansion, exciting photoelectrons, or the combination of those effects. The ways to fabricate nanostructures include the following [39]:

1. Nanoindentation due to the mechanical contact between the STM tip and the sample surface
2. Nanoscale surface heating by the enhanced optical field attributed to laser-STM tip coupling

3. Chemical desorption of atoms from the sample surface by the increased tunneling current
4. Enhanced chemical interaction between the STM tip and the sample surface
5. Dissociation and deposition of gas species between the STM tip and the sample surface

Therefore, LASTM provides a wide variety of potential applications in nanoscience and nanotechnology. Examples of such applications include surface repairing, the fabrication and characterization of nanoelectronics and nanophotonics, and the machining and alignment of nanoparticles and nanotubes.

14.2 DEVELOPMENT OF A NANO-RAMAN SPECTROMETER USING TIP-ENHANCED NEAR-FIELD EFFECTS

14.2.1 NEAR FIELDS ACHIEVED BY METALLIC TIPS

Knowing the optical field distribution underneath the tip apex is critical for optimizing the experimental conditions under which the near-field optical intensity could be maximized. The influencing factors include the tip material (plasmonic material, lightning-rods, or dielectrics), the tip shape (spheroid, cone, ellipsoid, or pyramids), laser illumination (polarization, incident angle, and wavelength), laser type (continuous or pulsed), and the substrate type (dielectrics or metals) [20]. Different algorithms, including the discrete dipole approximation (DDA) method, the finite-different-time-domain (FDTD) method, the T-matrix method, the finite element (FE) method, the multiple multipole (MMP) method, and the modified long wavelength approximation (MLWA) method, have been developed to calculate the optical fields [40].

The electronic resonance, the lightning-rod effect, and the optical antennas are the three enhancement mechanisms involved in the simulations for tip enhancement. Denk and Pohl generalized the modeling of the tip by studying the eigenvalues of hyperboloids with boundaries stretching to infinity and the presence of a plane, and they claimed that an enhancement factor as high as 2000 can be achieved when wavelengths are close to the surface plasmon excitation wavelengths [41]. Zayats studied the influence from the substrates under the resonance conditions of an ellipsoid particle made of different metals. Their investigation revealed that the main resonance red-shifts when the particle-surface separation becomes smaller [42]. Martin and Girard calculated the field intensity underneath a tungsten tip on a glass substrate and found that an enhancement factor of 100 for certain excitation conditions could be achieved. Their findings indicate that the field enhancement is not necessarily limited to the metal surfaces on which surface plasmons would occur or to the tips that support plasmon resonance [43]. Notingher and Elfick integrated the samples and the substrates into their models and concluded that the electric properties of the samples and the substrate had a significant impact on the field enhancement at the tip apex. They suggested that Raman intensity could be optimized by tuning the geometry factors and electric properties of the materials. In general, they believe that the strongest enhancement occurs for the cases in which the metal is employed [40]. By calculating the surface charges localized along a Au tip for two different excitation polarizations using the MMP method, Novotny et al. demonstrated that if the polarization is aligned with the tip axis, the density of the surface charges is maximized at the tip apex, leading to an enhanced optical field [44]. Sun and Shen systematically studied the effects of a number of parameters, e.g., the polarization, the incident angle, the wavelength of the incident laser, the tip material, and the tip length, on the field enhancement. They concluded that those parameters have a significant influence on the enhancement and that care must be taken in the design of the experiment in order to maximize the near field [45]. Bohn et al. investigated the antenna effect of a Si tip, which is a material without resonance, by including the effect of a dielectric substrate very close to the apex of the Si ellipsoid. They found that if a certain geometric condition is satisfied, an enhancement factor of 250 could be achieved [46].

Among these algorithms, the FDTD method is frequently used to simulate optical wave effects including propagation, scattering, diffraction, reflection, and polarization. In addition, it can be

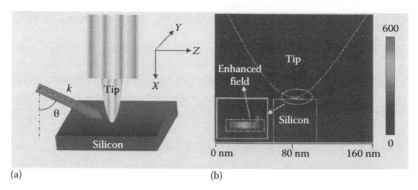

(a)   (b)

FIGURE 14.2   (a) Physical model for the simulation of electric field distributions underneath a Ag tip; (b) optical field distribution in the proximity of a Ag tip illuminated by a 514.5 nm laser beam. The inset is a zoomed view of the optical field distribution.

used to model material anisotropy, dispersion, and nonlinearities. In this chapter, the optical fields induced by two structures including metallic tips and metallic nanostructures, will be described by the simulation results using the FDTD algorithm based on the Lorentz–Drude model. Incorporating the Lorentz–Drude model solves the problem arising in conventional optics in which only the surface effect is considered. This approach includes the electromagnetic field effects on nanostructures by considering the optical penetration of the metal surfaces, which affect all the wave properties such as near field, far field, scattering, and diffraction.

The physical model is depicted in Figure 14.2a. The influence of the substrates on the field enhancement was taken into account for better accuracy. As illustrated in Figure 14.2a, a conical Ag tip is placed above a flat Si surface. The tip–surface gap is illuminated by a continuous-wave laser with a wavelength of 514.5 nm propagating along the k direction. The incident angle is 65° with respect to the tip axis (X axis). The tip apex radius is 50 nm. The tip–sample distance is 1 nm. The electric field amplitude of the beam is assumed to be 1 V m^{-1}. The mesh step is 0.5 nm, uniform for each axis. Figure 14.2b shows the calculated electric field distribution. It can be observed that an electric field is locally enhanced by as much as 600 times in the space between the tip apex and the substrate surface in a region of 14 nm along the Z axis. This indicates that a spatial resolution of 14 nm could be achieved.

14.2.2 SYSTEM SETUP OF A NANO-RAMAN SPECTROMETER

The schematic diagram of the developed TERS system is depicted in Figure 14.3. The system consists of an optical path, a spectral acquisition module (a spectrograph and a charge-coupled device [CCD] camera), an STM module, a motorized stage module, an optical microscope, and a computer. The optical path delivers the excitation laser beam onto the sample surfaces and collects the Raman signals from the surfaces. The spectral acquisition module is used to collect the scattered signals and convert the optical signals into electronic data. The tip–surface distance is controlled by the STM. The motorized stage module is used to position the samples. The optical microscope equipped with another CCD camera locates the regions of interest and monitors the optical alignment. The computer is used to control the STM. It also acquires, processes, and visualizes the Raman signals.

The instrument can accommodate both the AFM and the STM functionalities to combine with the Raman spectrometer. An SPM system (Agilent, SPM 5500) is used as a platform to build the TERS system. The tip is mounted on a piezoelectric transducer that has a stroke of 100 μm and closed-loop control. In addition, a two-axis nanopositioner (Mad City Labs, nano-H100) is placed beneath the sample holder, which is used for mapping the sample surfaces. The software used for mapping is developed using the LabView™ programming platform together with the instrumental drivers provided by the hardware suppliers.

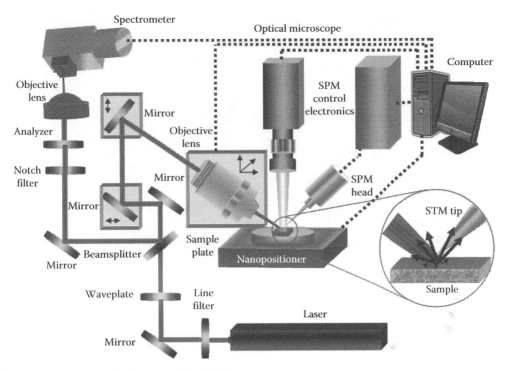

FIGURE 14.3 Schematic diagram of the TERS system.

In Figure 14.3, a laser beam from an Argon laser (Coherent, Innova 300) passes through the line filter (Newport), which only allows a wavelength of 514.5 nm to pass through. A half-wave plate (CVI) is used to control the polarization of the beam. The beam is split into two beams by a beam splitter. The transmitted beam is focused on the sample surface by an objective lens (Olympus, LWD 50×, NA = 0.45, WD = 15 mm) and the reflected beam is wasted because the same splitter will be used to reflect the collected signal to the spectrometer. The STM tip approaches the sample surface with the tip–surface distance in the tunneling range of 1 nm. The status of the tip, the laser beam, and the substrate surface are monitored by the optical microscope. The position of the objective lens is controlled by a motorized *XYZ* stage (shown as the box outside the objective) with a resolution of 3 µm. After successful optical alignment, the scattered Raman signals are collected by the objective lens. After passing through a notch filter (Kaiser, SuperNotch®, OD = 4.0) and an analyzer (Newport), the beam with Raman signals is focused by a lens (CVI) into the slit of the spectrograph (Acton Research, Spectro-2300i). A back-illuminated CCD camera (Princeton Instruments, PIXIS-400B) with a high quantum efficiency (>90% at 514.5 nm) is used to acquire the Raman spectra. The spectra are then analyzed, processed, and visualized by a computer.

To perform Raman mapping, a sample is placed on the nanopositioner that has a lateral resolution of 0.2 nm and can be controlled by the developed mapping software. Figure 14.4 shows a picture for the practical experimental setup. It can be seen that both the STM system and the motorized sample module are placed on a fixture frame.

14.2.3 PERFORMANCE OF THE DEVELOPED NANO-RAMAN SPECTROMETER

Contrast ratio, defined as the ratio of the near-field to the far-field Raman intensity, is used to evaluate the extent of the enhancement. The far-field Raman intensity is obtained when the laser beam is far away from the tip, ensuring no tip–beam interaction exists. Subtracting the far-field Raman intensity from the total Raman intensity obtained from a tip-enhanced Raman measurement gives

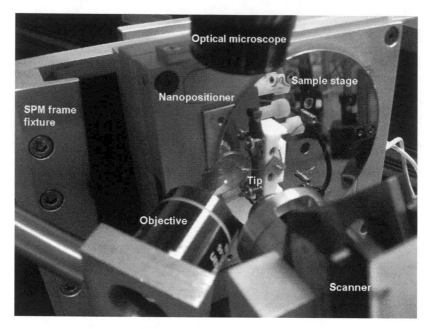

FIGURE 14.4   Experimental setup of the TERS system.

the near-field intensity when the laser beam is well aligned with the tip to achieve strong near-field effects. It is reported that the contrast ratio of the crystalline Si at 520 cm^{-1} is around 50% by Ag-coated W tips or Au tips [47].

Figure 14.5a shows the Raman spectra of the Si substrate with the Ag-coated W tip under different conditions. The total intensity in a well-aligned case, the far-field intensity, and the total intensity in a misaligned case in which the laser beam is partially blocked are indicated by Raman spectra 1, 2, and 3 in Figure 14.5a, respectively. In a well-aligned case, a contrast ratio of 52% can be observed (spectrum 1). When the beam is blocked by the tip, the intensity is significantly reduced (spectrum 3). Figure 14.5b shows the Raman spectra of the Si substrate with a Au tip. Similar to the case of using the Ag-coated W tip, a contrast ratio of 47% was obtained.

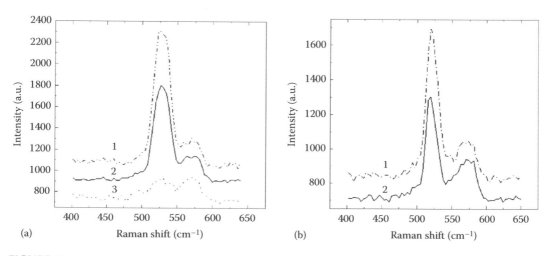

FIGURE 14.5   Raman spectra of the Si substrate using (a) a Ag-coated W tip and (b) a Au tip. Spectra 1, 2, and 3 were taken in the cases of good alignment, laser beam away from the tip, and laser beam on the tip but not in the gap, respectively.

14.2.4 Characterization of Single-Walled Carbon Nanotubes (SWCNTs)

Using the developed nano-Raman spectrometer, the Raman spectra of the individual single-walled carbon nanotubes (SWCNTs) dispersed on the Si substrate were obtained by measuring the Raman spectra with (near-field and far-field) and without (far-field only) the tip. With good alignment, the Raman enhancement for all vibration modes (RBM, D, G, and G′) can be found, as shown in Figure 14.6. This enhancement has a contrast ratio of about 180% for the G-band. The Raman peak of the Si substrate was observed, but not enhanced, because the tip was placed directly above the SWCNTs. Thus, the enhanced optical fields only enhanced the Raman signal of the SWCNTs but not the substrate.

By identifying different Raman intensities in different regions using a specific vibration mode as the fingerprint, Raman mapping can be performed to visualize the individual SWCNTs. Figure 14.7 shows the Raman image of SWCNTs dispersed on a Si substrate with the G-band as the fingerprint. As evidenced by the images, this instrument has a spatial resolution of approximately 30 nm.

This work of mapping the individual SWCNTs could pave a way for studying the overlapped optical, electronic, structural, chemical, mechanical, and topologic properties of the individual SWCNTs and reveal their relationship with the single instrument. A complete understanding of those

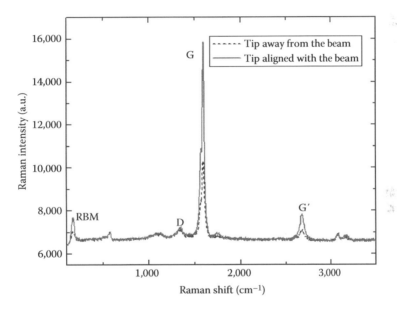

FIGURE 14.6   Raman spectra of the SWCNT with the tip away from (dotted curve) and aligned with (solid curve) the laser beam.

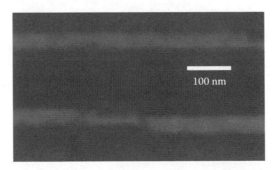

FIGURE 14.7   Raman image of the individual SWCNTs dispersed on a Si substrate.

properties and their relationship could assist us in designing and developing advanced devices using the SWCNTs for various applications. For example, to use a SWCNT to build the conductive channel of a MOS device, the chemical structure, tube diameter, electronic bandgap, and uniformity of the SWCNT must be well known before the desired performance of the device can be achieved. Using this nano-Raman system, the chemical structure could be identified with the function of RS; the tube diameter could be characterized with the function of STM and cross confirmed with the function of Raman spectroscopy; the electronic bandgap could be measured with the function of scanning tunneling spectroscopy (STS); and the uniformity could be obtained with the function of Raman mapping. Obviously, this nano-Raman system could guarantee a high level of characterization capability with which conventional instruments cannot compete, and could ensure a clean measurement environment by eliminating the need for transferring the device among different instruments.

14.3   ENHANCED RAMAN SPECTROSCOPY BY COMBINING A METALLIC TIP WITH SURFACE NANOSTRUCTURES

14.3.1   NEAR FIELDS ACHIEVED BY COMBINING A METALLIC TIP WITH SURFACE NANOSTRUCTURES

Using the FDTD algorithm for the above-mentioned Ag tip, the near-field distributions induced by combining a metallic tip with surface nanostructures were also calculated. The physical model is depicted in Figure 14.8a. This model includes a Si substrate. A conical Au tip with a diameter of 50 nm is placed 1 nm above the Si substrate. It is illuminated by a continuous-wave laser with a transverse Gaussian profile and a wavelength of 514.5 nm propagating along the k direction, which is 70° with respect to the substrate normal. The height of nanostructures is 30 nm with a 75 nm terminal distance. The mesh step for each axis is 0.5 nm in the FDTD simulation. Figure 14.8b and c shows the electric field distributions beneath the tip along the XY plane in the presence and in the absence of Ag nanostructures, respectively. In Figure 14.8b, the enhanced electric field is confined to a 1 nm range but spreads a distance of 20 nm along the X direction. A higher local electric field (1.5 times) is found in Figure 14.8b than in Figure 14.8c, indicating the strong influence of the nanostructures on the field enhancement.

The Raman intensity is approximately proportional to the fourth power of the field amplitude. Therefore, the presence of the nanostructures can further enhance the Raman signals by more than five times, similar to the enhancement from the Au sharp step [48].

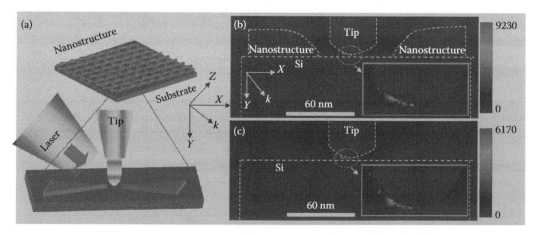

FIGURE 14.8   (a) Physical model to simulate the optical field beneath (a) Au tip under a laser irradiation. Electric field distribution beneath a Au tips (b) in the presence of the nanostructures and (c) in the absence of the nanostructures simulated by the FDTD algorithm. Insets in (b, c) show zoomed views of the electric field distributions.

14.3.2 Enhanced Raman Spectra of Silicon Substrates Covered with Nanostructures

Experimentally, the Ag nanostructures were fabricated by nanosphere lithography (NSL) [49,50], which typically includes five steps (Figure 14.9): (a) substrate cleaning, (b) solution dropping, (c) the self-assembly of nanoparticles, (d) the sputtering of the metallic layer, and (e) lift-off. Each NSL structure begins with the self-assembly of size-monodisperse nanospheres to form a two-dimensional (2D) colloidal crystal deposition mask. The methods of depositing a nanosphere suspension onto desired substrates include the spin coating, drop coating, and thermoelectrically cooled angle coating. All the deposition methods require that the nanospheres be able to freely diffuse across the substrate surface, so that a configuration of these nanospheres with the lowest energy can be formed. As the solvent (water) evaporates, the capillary forces draw the nanospheres together and the nanospheres form a hexagonally close-packed pattern on the substrate, serving as a nanosphere mask. Another material, such as a metal, is then deposited on the substrate over the nanosphere mask to a controlled thickness using direct current (DC) sputtering. After metal deposition, the nanosphere mask is removed by sonicating the entire sample in a solvent, leaving behind the deposited material on the substrate.

Different shapes of nanostructures were fabricated using the NSL technique. Figure 14.10a and b shows the SEM micrographs of the triangular Ag nanostructures on Si surfaces with different magnification. They have triangle geometries with a side length of around 200 nm and a tip-to-tip distance of around 100 nm. Figure 14.10c and d shows the AFM images of the ring-shaped Ag nanostructures with different magnifications. It can be seen that the minimum gap distance between two nanostructures is around 20 nm or less.

Using TERS, nanoscale residual strains in Si can be characterized by measuring the Raman shifts with respect to the vibration mode of $520\,cm^{-1}$ [51–54]. Under normal conditions, the enhancement contrast ratio, defined as the ratio of near-field to far-field Raman intensity, is usually around 50%, which is not sufficient for obtaining a clear Raman image. Suppressing the far-field signals by deparization is one way to increase the ratio. This ratio can also be increased by increasing the near-field signals using the nanostructure-assisted TERS technique. Figure 14.11 shows the Raman spectra of a Si substrate for three different cases: (1) signals from the far field only, (2) the sum of signals from the far field and the near field (total intensity) in the absence of the nanostructures, and (3) the sum of signals from the far field and the near field (total intensity) in the presence of the nanostructures. All spectra were accumulated for 1 s. Without the nanostructures, the contrast ratio is around 40%. If the nanostructures are present, the contrast ratio is increased up to 250%, which is more than five times higher. It was found that the strongest enhancement occurred when the tip was positioned in between the apexes of two adjacent nanostructures (see Figure 14.10b). The enhancement factor is generally defined as the actual signal enhancement per scattering volume, i.e., [55]

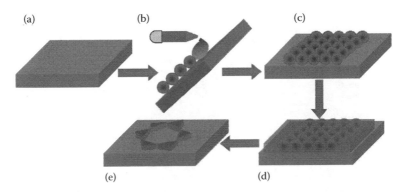

FIGURE 14.9 The process of nanosphere lithography. (a) Clean substrate, (b) drop solution, (c) self-assembly, (d) deposit metal, and (e) lift off.

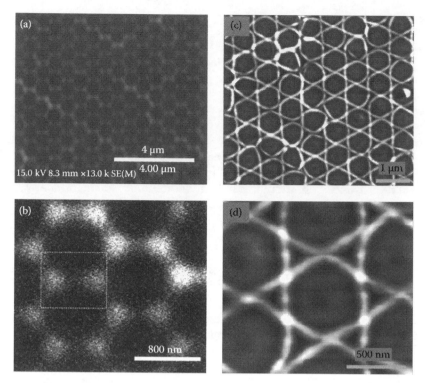

FIGURE 14.10 SEM micrographs of the triangular Ag nanostructures with (a) low and (b) high magnifications; AFM micrographs of the ring-shaped Ag nanostructures with (c) low and (d) high magnifications.

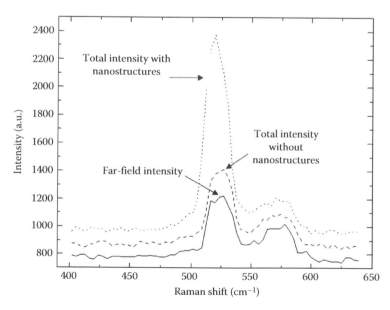

FIGURE 14.11 Raman spectra of the Si substrate under different cases: signals from far field only (solid trace), sum of signals from far field and near field in the absence of the nanostructures (dashed trace), and in the presence of the nanostructures (dotted trace).

$$EF = \frac{I_{near\text{-}field}}{I_{far\text{-}field}} \cdot \frac{V_{far\text{-}field}}{V_{near\text{-}field}} = \left(\frac{I_{total}}{I_{far\text{-}field}} - 1 \right) \cdot \frac{V_{far\text{-}field}}{V_{near\text{-}field}}. \qquad (14.2)$$

Here

$I_{near\text{-}field}$ and $I_{far\text{-}field}$ refer to the near-field and far-field Raman intensities

$V_{near\text{-}field}$ and $V_{far\text{-}field}$ refer to the near-field and far-field scattering volumes

$I_{total\text{-}field}$ refers to the total Raman intensity, which is the sum of the far- and near-field intensities

The term $I_{near\text{-}field}/I_{far\text{-}field}$ in Equation 14.2 is the enhancement contrast ratio, which is 250% in this case. For a Si sample, $V_{far\text{-}field}$ is defined as the surface area of the focused laser beam multiplied by the smallest one of three parameters: the sample thickness, the light penetration depth, or the focus depth. The light penetration depth, D, can be estimated using $D = \lambda/4\pi k$, with k being the extinction coefficient of Si at the wavelength λ. Using the value of $k = 0.06$ at $\lambda = 514.5\,nm$ (the wavelength of the excitation laser), the light penetration depth is estimated to be $0.68\,\mu m$. The focused beam diameter is estimated to be $3\,\mu m$. Subsequently, the far-field scattering volume is $1500^2 \times 680 \times \pi\ nm^3 \approx 4.8 \times 10^9\ nm^3$. The near-field scattering volume depends on the tip radius and the decay length of the enhanced optical fields. Because the radius of the enhanced optical fields is one half of the tip radius [56], it is estimated to be $12.5\,nm$. The decay length of the enhanced field is estimated to be about $2.5\,nm$. Then, the near-field scattering volume is $12.5^2 \times 2.5 \times \pi\ nm^3 \approx 1.23 \times 10^3\ nm^3$. As a result, the EF is calculated to be 9.75×10^6, which is more than one order higher than the reported results without the surface nanostructures.

14.4 FABRICATION OF NANOSTRUCTURES WITH HIGH ELECTRICAL CONDUCTIVITY ON SILICON SUBSTRATES USING A LASER-ASSISTED SCANNING TUNNELING MICROSCOPE (LASTM)

14.4.1 System Setup of a Laser-Assisted Scanning Tunneling Microscope (LASTM)

The schematic diagram of the experimental setup for surface nanostructuring using the LASTM is shown in Figure 14.12. The setup consists of an STM, a Q-switched pulsed Nd-YAG laser with a pulse width of 7 ns and a wavelength of 532 nm, and optics for delivering the laser beam. The STM can operate in both open-loop and closed-loop modes by changing different scanners. An on-axis optical microscope equipped with a CCD camera provides a zoomed view of the substrate surfaces for the beam alignment on the tip apex. The focused laser beam is delivered to the tip–surface gap from the front side of the probe head. Tungsten wires were used to fabricate the STM tips.

The alignment of the laser beam was performed under the optical microscope by observing the diffraction light from the tip. To make sure that the nanostructure was induced only from the near-field effects of the tip, the incident laser beam was adjusted to be completely perpendicular to the tip axis or in parallel with the sample surface. The STM scanner was placed in a transparent environment chamber to effectively suppress environmental noises. During nanofabrication, laser pulses were applied while the tip–surface distance was within the tunneling range. The incident laser beam was focused before reaching the tip–surface gap. The STM was kept in constant-current mode with a tunneling current of 3 nA for the fabrication and the imaging. The surface morphology was characterized in situ with the laser turned off. The STS was performed to measure the current–voltage (I–V) characteristics from which the differential conductance was derived.

14.4.2 Impact of Laser Fluence and Number of Pulses on Nanostructures

The pulse energy, the number of pulses, and the tip–surface distance strongly affect the formation of the nanostructures and their geometries. Figure 14.13 shows the STM images of the nanostructures

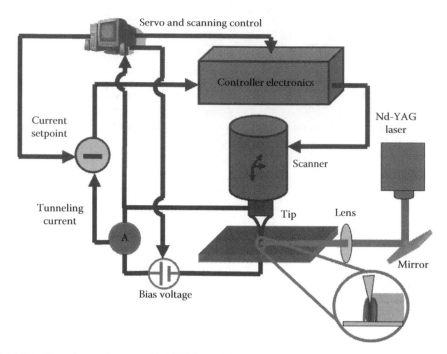

FIGURE 14.12 Experimental setup of LASTM used for surface nanostructuring.

fabricated using four different pulse numbers (15, 10, 5, and 3) under the same condition of a pulse energy of 15 mJ, a gap voltage of 1 V, and a tunneling current of 3 nA. Figure 14.13a and b shows two dots fabricated with 15 and 10 pulses, respectively. Figure 14.13c and d shows two dots fabricated with 5 and 3 pulses, respectively. It can be observed that these nanostructures have approximately circular shapes, due to the circular beam profile of the enhanced laser determined by the geometry of the W tip. The brighter regions in the figures correspond to the regions having higher electrical conductivities or higher tunneling probabilities. Therefore, the electronic properties of the fabricated structures are modified differently.

The dependencies of the diameters and the heights of the nanostructures in Figure 14.13 on the pulse numbers are shown in Figure 14.14. The diameters and heights increase almost linearly with the increasing number of pulses. The diameters of the nanostructure are in a range from 200 to 500 nm and their heights range from 4 to 16 nm, when the number of pulses is in the range from 3 to 15.

In order to fabricate smaller nanostructures, a more tightly focused and more highly enhanced optical field is required by tuning the gap voltage while keeping the same tunneling current. Different gap voltages (0.01, 0.1, and 1 V) and pulse energies (2, 5, and 10 mJ) at a constant current of 3 nA were utilized to fabricate smaller dots. As illustrated in Figure 14.15a, a dot with a diameter of 30 nm was fabricated when the gap voltage and pulse energy were 0.1 V and 5 mJ, respectively. A single line, as shown in Figure 14.15b, was also fabricated by scanning the STM tip on the Si surface at a speed of 10 nm s^{-1} while other parameters remained the same as those used in the fabrication of the 30 nm dot. This line has an average width of approximately 30 nm.

14.4.3 CHARACTERIZATION OF NANOSTRUCTURES USING SCANNING TUNNELING SPECTROSCOPY (STS)

The STS, a subsidiary function of an STM, provides vital information about the local density of states (LDOS) on the surfaces at an atomic or molecular scale [57]. H-passivated Si surfaces show an ideal surface electronic structure with a low surface state density [58]. The absence

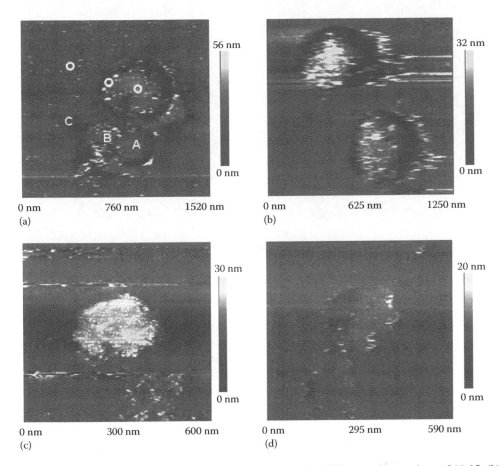

FIGURE 14.13 STM images of the nanostructures fabricated with different pulse numbers of (a) 15, (b) 10, (c) 5, and (d) 3 (laser energy = 15 mJ, gap voltage = 1 V, tunneling current = 3 nA).

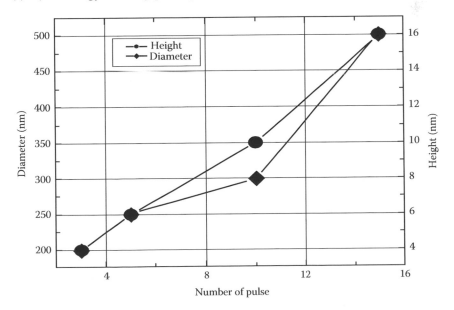

FIGURE 14.14 Nanostructure diameter (diamonds) and height (circles) as functions of number of pulses. (pulse energy = 15 mJ, gap voltage = 1 V, tunneling current = 3 nA).

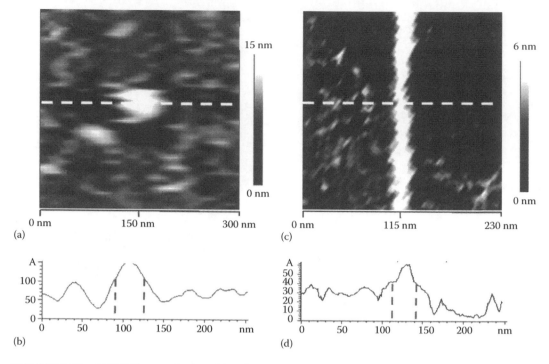

FIGURE 14.15 (a) STM image of a dot with a diameter of 30 nm, (b) profile along the line indicated in (a), (c) STM image of a line with a width of 30 nm, and (d) profile along the line indicated in (c) (gap voltage = 0.1 V, tunneling current = 3 nA).

of surface states and the Fermi level pinning enable direct observation of the Schottky barrier behavior. The *I–V* curves of the p-type Si surfaces indicate a typical Schottky behavior, exponential in forward bias and lower current in reverse bias. As the substrate voltage is tuned more positively, the conduction band moves up and more current will flow. In the reverse direction, as the voltage becomes more negative, the bands bend down and increase the potential barrier to the holes.

The STS was performed in different regions in situ after the nanostructuring. The tunneling currents as functions of the gap voltage at points A, B, and C indicated in Figure 14.13a are shown in Figure 14.16. The *I–V* characteristics are different in different regions. In the region far from the fabricated dot (curve C), a typical Schottky barrier behavior can be observed. At the edge of the dot (curve B), the tip–surface gap still demonstrates a Schottky behavior but with increased tunneling probability. Inside the dot (curve A), no obvious Schottky behavior can be seen. Furthermore, the curve becomes more symmetric and band bending disappears. The band gap can be estimated from the differential conductivity as a function of gap voltage, as shown in Figure 14.17. For the H-passivated surface without laser processing, the band gap is around 1.1 eV, which is a typical value for Si. The band gap of the Si surface decreases when the probe point approaches the dot and almost diminishes inside the dot, indicating the tendency of electronic structure change.

14.4.4 Mechanism Investigation of Surface Nanostructuring on Silicon Substrates

When the gap voltage is lower than the work functions of both the W tip and the p-type Si, the nanostructures with different differential conductances may arise from the surface states. These surface

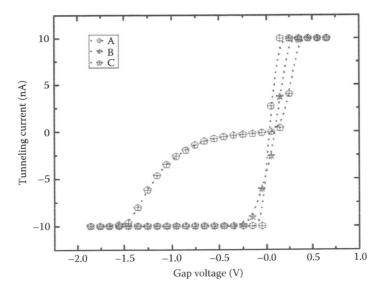

FIGURE 14.16 *I–V* characteristics as a function of gap voltage at points A, B, and C indicated in Figure 14.12a.

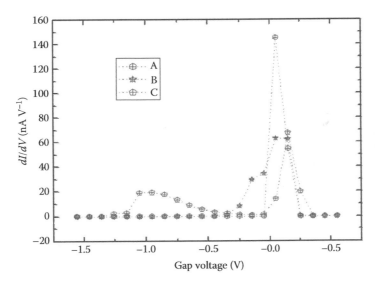

FIGURE 14.17 Differential conductance *dI/dV* as a function of gap voltage at points A, B, and C indicated in Figure 14.12a.

states are primarily associated with the critical points in the surface-projected bulk band structures or the surface dangling bonds. The mechanisms of the nanostructure fabrication on the Si surfaces using the LASTM might involve two individual steps: (1) the hydrogen atoms adsorbed on the Si surface for the passivation are thermally desorbed by the highly localized optical field, resulting in a clean and intrinsic Si surface; and (2) the dehydrogenized surface is further turned into the more conductive poly-Si through the thermal process by the enhanced field. The local temperature is of crucial importance in both steps. To better understand the mechanisms, a study of the temperature rising underneath the tip apex is necessary. It is assumed that the incident laser beam propagates in parallel with the sample surface. When a Gaussian distribution is used to describe the beam profile, the steady-state temperature on the Si surface underneath the W tip induced by a pulsed laser can be expressed by [59]

$$T_{max} = \frac{\sqrt{\pi}\alpha F(1-R_0)w_0}{8\kappa_T \tau}. \tag{14.3}$$

Here

T_{max} is the maximum temperature on the Si surface underneath the tip apex

α is the optical enhancement factor defined as the ratio of the enhanced electric field to the original electric field of the incident pulsed laser

F is the laser fluence

R_0 is the optical reflectance of the Si surface at the normal incidence

w_0 is the diameter of the tip apex

κ_T is the thermal conductivity of the Si

τ is the laser pulse duration

Obviously, the temperature induced by the LASTM is proportional to the laser intensity F/τ, the optical enhancement factor, α, and the tip radius. The laser wavelength also affects the temperature because the optical reflectance is a function of the laser wavelength. To increase the transient local temperature, we can use a pulsed laser with higher peak energy and a shorter pulse duration, select a lower gap voltage to maintain a shorter tip–surface distance that can lead to a higher optical enhancement factor, or employ a tip with a smaller tip radius. The temperature was calculated using the parameters obtained from both the practical experimental conditions and the simulation results, which are $R_0 = 0.37$ at 532 nm, $\alpha = 220$ at a gap distance of 1 nm, $w_0 = 30$ nm for the W tip, and $\kappa_T = 1.3$ W cm^{-1} C^{-1}. For a pulsed laser with peak energy of 5 mJ and a duration of 7 ns, the laser intensity F/τ was estimated to be 30 mW cm^{-2}. As a result, the temperature is calculated to be 1259°C, corresponding to a temperature at which the 30 nm dot was fabricated. This temperature is slightly lower than the melting point of Si (1412°C). The threshold pulse energy to form a nanostructure was found to be 2 mJ. Using the same parameters given above, the threshold temperature was calculated to be 503°C below which no nanostructure can be fabricated on the Si surface.

14.5   CONCLUSIONS

Using the FDTD method and the Lorentz–Drude model, the theoretical calculations of the enhanced optical near fields, induced by the metallic tips (Ag), and the combination of the Ag nanostructures with the Au tips were performed. The field enhancement by the metallic tips and the combination of the nanostructures and tips are based on the surface plasmon resonant excitation and the coupling of the excited surface plasmons of the nanostructures and tips.

A practical Raman spectrometer and imaging system capable of working at nanoscales was developed. Equipped with a back-illuminated CCD camera with an extremely high sensitivity and a notch filter with a high optical density at an excitation wavelength of 514.5 nm, this system features a resolution of around 30 nm. Thus, it allows us to characterize a variety of nanometer-sized samples, such as nanostructures on the Si substrate and individual SWCNTs.

Electronically active nanoscale dots and lines were fabricated on the p-type Si (110) substrates using the LASTM. Their feature sizes and heights are almost linearly dependent on the pulse numbers. The STS spectra indicate that their electronic structures were changed. The nanostructures are presumably polysilicon, which has a higher tunneling probability than the intrinsic Si. The thermal effect induced by the highly localized optical field is used to explain the fabrication mechanism and is believed to be the major driving force of this nanofabrication technique.

QUESTIONS

Q.14.1 What are the unique properties of near-field optics? Form teams and discuss the impact of nanofabrication and nanocharacterization using near-field optics.

Q.14.2 Nano-science and nano-technology can be mysterious to many people. After reading this chapter, what have you learned? Try to summarize your observations using the dynamic M-PIE analysis method. For example, which energy fields are used in the techniques discussed? What information or general intelligence can be derived from the material-energy interactions?

Q.14.3 To break through the conventional optical limit, new scientific phenomena can be used. In this chapter, what are the major scientific phenomena the authors used to bring the spatial resolution to ~30 nm?

Q.14.4 When we talk about a kind of energy, we may talk about point energy source, but actually they are all energy fields, having spatial and temporal distributions. Many nano-fabrication processes utilize localized and high-resolution energy fields. Playing with the scale of resolution in both the spatial and temporal dimensions can lead to a series of innovations. This chapter reported several good examples. Please find similar examples in other applications, such as electronics, telescope, etc., and try to appreciate the difference of energy fields at various scales.

Q.14.5 The authors are from the electrical engineering department. For students of mechanical engineering, what difference in methodology do you see between the two disciplines?

ACKNOWLEDGMENTS

The authors would like to thank Professor Natale J. Ianno in the Department of Electrical Engineering at the University of Nebraska–Lincoln for providing the sputtering system to fabricate tips and nanostructures and Dr. You Zhou at the Beadle Center of the Biology Department at the University of Nebraska–Lincoln for assistance in characterizing the tips and nanostructures using the scanning electron microscope. The authors would also like to acknowledge the efforts in design and fabrication of the fixtures by Mike Jenson at the Instrument Shop. The authors greatly appreciate Jesse Teas for his proofreading. This work was financially supported by the National Science Foundation (Nos. ECCS 0619553 and ECCS 0652905) and the Nebraska Research Initiative.

ABOUT THE CONTRIBUTING AUTHORS

Dr. Kaijun Yi obtained his PhD degree in 2009 from the Department of Electrical Engineering at the University of Nebraska–Lincoln (UNL) and is currently a post doctoral researcher in the same department. His current research interests include: (1) advanced techniques and platforms to fabricate and characterize nanomaterials, nanostructures, and nanodevices; (2) near-field optics; (3) nanoscale optical spectroscopic techniques (nano-Raman spectroscopy, coherent anti-stokes Raman spectroscopy, laser-induced breakdown spectroscopy, magneto-optical Kerr spectroscopy, etc.); (4) spectroscopic imaging; (5) plasmonic sensors; (6) laser-assisted scanning probe microscopes; and (7) carbon-based nanoelectronics and nanodevices. He has authored and co-authored over 20 peer-reviewed journal and conference papers and has presented his research findings in six international conferences. Prior to UNL, he served as an electronics design engineer at Philips Optical Storage, Singapore and then at Philips Consumer Electronics,

Singapore. Prior to joining Philips, he was a research and development engineer at Data Storage Institute, Singapore, where he began his professional career. He received his MEng and BEng degrees from the Department of Electronics Science and Technology at the Huazhong University of Science and Technology, Wuhan, China in 1996 and in 1999, respectively.

Prof. Yongfeng Lu received the bachelor, master, and Ph.D. degrees all in electrical engineering from Tsinghua University (Beijing, China) in 1984 and Osaka University (Japan) in 1988 and 1991, respectively. He is currently the Lott Chair Professor with the Department of Electrical Engineering at the University of Nebraska–Lincoln. He has nearly 20 years of research experience in the laser-based nano- and microprocessing of materials. Besides the fundamental research works that led to a large number of publications and a number of national and international awards, he also has successfully developed a number of laser nano- and microprocessing technologies and commercialized them in industries. His industrial experience also includes consultations with a number of companies on laser nano- and microprocessing in electronics and microelectronics. He became an SPIE fellow in 2008 and an LIA fellow in 2009.

Dr. Lu is widely recognized around the world as a leading researcher in the area of laser-based nano- and microscale material processing and characterization. He has made important contributions in the development of basic theories and industrial applications in the areas of laser-material interactions and laser microprocessing technologies. He is internationally known for his work on laser surface cleaning, pulsed laser deposition, real-time monitoring of laser interactions with materials, nanoprocessing and nano-characterization using laser-assisted scanning probe microscopes (SPM), laser-induced temperature rise and surface structures, and commercialization of laser microprocessing technologies.

Dr. Lu has received a number of national and international awards, including the *National Technology Award* (National Science and Technology Board, Singapore, 1998), *ASEAN Outstanding Achievement Engineering Award 99* (ASEAN Engineering Association, 1999), and *International Laser Award* (Berthold Leibinger Innovationspreis, Berthold Leibinger Stiftung, Germany, 2000). Dr. Lu has a strong track record of attracting external research funding. He has received more than $10 million in research funding over the past 6 years from NSF, AFOSR, ONR, DoE, and industry and private foundations. Dr. Lu has published more than 400 papers in leading journals and international conferences.

REFERENCES

1. H. J. Guntherodt and W. Meier, Nanoscience and nanotechnology, *Chimia*, 56, 483, 2002.
2. M. L. Cohen, Nanotubes, nanoscience and nanotechnology, *Mater. Sci. Eng. C*, 15, 1, 2001.
3. T. Sugii, High-performance bulk CMOS technology for 65/45 nm nodes, *Solid State Electron.*, 50, 2, 2006.
4. N. Tamura and Y. Shimamune, 45 nm CMOS technology with low temperature selective epitaxy of SiGe, *Appl. Surf. Sci.*, 254, 6067, 2008.
5. C. Z. Wang, Y. H. Rong, and T. Y. Hsu, Designs of higher tunnelling giant magnetoresistance in granular films, *Mater. Lett.*, 60, 379, 2006.
6. Z. P. Niu, Z. B. Feng, J. Yang, and D. Y. Xing, Tunneling magnetoresistance of double-barrier magnetic tunnel junctions in sequential and coherent regimes, *Phys. Rev. B*, 73, 014432, 2006.
7. M. Chen, S. Yamamuro, D. Farrell, and S. A. Majetich, Gold-coated iron nanoparticles for biomedical applications, *J. Appl. Phys.*, 93, 7551, 2003.

8. P. Eaton, G. Doria, E. Pereira, P. V. Baptista, and R. Franco, Imaging gold nanoparticles for DNA sequence recognition in biomedical applications, *IEEE Trans. Nanobiosci.*, 6, 282, 2007.
9. D. J. Schulte, Nanotechnology in environmental protection and pollution sustainable future, environmental cleanup and energy solutions, *Sci. Technol. Adv. Mater.*, 8, 11, 2007.
10. C. G. Smith, Low-dimensional quantum devices, *Rep. Prog. Phys.*, 59, 235, 1996.
11. R. Tsu, Applying the insight into superlattices and quantum wells for nanostructures: Low-dimensional structures and devices, *Microelectron. J.*, 38, 959, 2007.
12. B. Babic, M. Iqbal, and C. Schonenberger, Ambipolar field-effect transistor on as-grown single-wall carbon nanotubes, *Nanotechnology*, 14, 327, 2003.
13. H. Watanabe, C. Manabe, T. Shigematsu, K. Shimotani, and M. Shimizu, Single molecule DNA device measured with triple-probe atomic force microscope, *Appl. Phys. Lett.*, 79, 2462, 2001.
14. M. A. Poggi, E. D. Gadsby, L. A. Bottomley, W. P. King, E. Oroudjev, and H. Hansma, Scanning probe microscopy, *Anal. Chem.*, 76, 3429, 2004.
15. N. P. Kobayashi, T. R. Ramachandran, P. Chen, and A. Madhukar, In situ, atomic force microscope studies of the evolution of InAs three-dimensional islands on GaAs(001), *Appl. Phys. Lett.*, 68, 3299, 1996.
16. M. Moskovits, Surface-enhanced Raman spectroscopy: A brief perspective, *Top Appl. Phys.*, 103, 1, 2006.
17. P. Avouris, T. Hertel, and R. Martel, Atomic force microscope tip-induced local oxidation of silicon: Kinetics, mechanism, and nanofabrication, *Appl. Phys. Lett.*, 71, 285, 1997.
18. D. W. Pohl, Optics at the nanometre scale, *Philos. Trans. R. Soc. A*, 362, 701, 2004.
19. B. Hecht, B. Sick, U. P. Wild, V. Deckert, R. Zenobi, O. J. F. Martin, and D. W. Pohl, Scanning near-field optical microscopy with aperture probes: Fundamentals and applications, *J. Chem. Phys.*, 112, 7761, 2000.
20. A. Bouhelier, Field-enhanced scanning near-field optical microscopy, *Microsc. Res. Technol.*, 69, 563, 2006.
21. M. A. Paesler and P. J. Moyer, *Near-Field Optics: Theory, Instrumentation, and Applications*, New York: Wiley, 1996.
22. D. W. Pohl, W. Denk, and M. Lanz, Optical stethoscopy: Image recording with resolution lambda/20, *Appl. Phys. Lett.*, 44, 651, 1984.
23. D. W. Pohl and L. Novotny, Near-field optics-light for the world of nano, *J. Vac. Sci. Technol. B*, 12, 1441, 1994.
24. R. L. McCreery, *Raman Spectroscopy for Chemical Analysis*, New York: John Wiley & Sons, 2000.
25. B. Pettinger, G. Picardi, R. Schuster, and G. Ertl, Surface-enhanced and STM-tip-enhanced Raman spectroscopy at metal surfaces, *Single Mol.*, 3, 285, 2002.
26. D. Richards, R. G. Milner, F. Huang, and F. Festy, Tip-enhanced Raman microscopy: Practicalities and limitations, *J. Raman Spectrosc.*, 34, 663, 2003.
27. A. Hartschuh, M. R. Beversluis, A. Bouhelier, and L. Novotny, Tip-enhanced optical spectroscopy, *Philos. Trans. R. Soc. A*, 362, 807, 2004.
28. J. Wessel, Surface enhanced optical microscopy, *J. Opt. Soc. Am. B*, 2, 1538, 1985.
29. R. Zenobi and V. Deckert, Scanning near-field optical microscopy and spectroscopy as a tool for chemical analysis, *Angew. Chem. Int. Ed.*, 39, 1746, 2000.
30. B. Pettinger, Tip-enhanced Raman spectroscopy (TERS), *Top. Appl. Phys.*, 103, 217, 2006.
31. B. Pettinger, G. Picardi, R. Schuster, and G. Ertl, Surface-enhanced and STM tip-enhanced Raman spectroscopy of CN-ions at gold surfaces, *J. Electroanal. Chem.*, 554, 293, 2003.
32. L. Zhu, J. Atesang, P. Dudek, M. Hecker, J. Rinderknecht, Y. Ritz, H. Geisler, U. Herr, R. Greer, and E. Zschech, Experimental challenges for approaching local strain determination in silicon by nano-Raman spectroscopy, *Mater. Sci. (Poland)*, 25, 19, 2007.
33. A. Hartschuh, H. Qian, A. J. Meixner, N. Anderson, and L. Novotny, Nanoscale optical imaging of single-walled carbon nanotubes, *J. Lumin.*, 119, 204, 2006.
34. G. A. Baker and D. S. Moore, Progress in plasmonic engineering of surface-enhanced Raman-scattering substrates toward ultra-trace analysis, *Anal. Bioanal. Chem.*, 382, 1751, 2005.
35. P. M. Campbell, E. S. Snow, and P. J. McMarr, AFM-based fabrication of Si nanostructures, *Physica B*, 227, 315, 1996.
36. J. Schneir, J. A. Dagata, and H. H. Harary, Scanning tunneling microscope-based nanostructure fabrication system, *J. Vac. Sci. Technol., A*, 11, 754, 1993.
37. E. S. Snow and P. M. Campbell, Fabrication of Si nanostructures with an atomic-force microscope, *Appl. Phys. Lett.*, 64, 1932, 1994.

38. S. A. Miller, K. L. Turner, and N. C. MacDonald, Microelectromechanical scanning probe instruments for array architectures, *Rev. Sci. Instrum.*, 68, 4155, 1997.

39. S. Grafstrom, Photoassisted scanning tunneling microscopy, *J. Appl. Phys.*, 91, 1717, 2002.

40. I. Notingher and A. Elfick, Effect of sample and substrate electric properties on the electric field enhancement at the apex of SPM nanotips, *J. Phys. Chem. B*, 109, 15699, 2005.

41. W. Denk and D. W. Pohl, Near-field optics-microscopy with nanometer-size fields, *J. Vac. Sci. Technol. B*, 9, 510, 1991.

42. A. V. Zayats, Electromagnetic field enhancement in the context of apertureless near-field microscopy, *Opt. Commun.*, 161, 156, 1999.

43. O. J. F. Martin and C. Girard, Controlling and tuning strong optical field gradients at a local probe microscope tip apex, *Appl. Phys. Lett.*, 70, 705, 1997.

44. L. Novotny, R. X. Bian, and X. S. Xie, Theory of nanometric optical tweezers, *Phys. Rev. Lett.*, 79, 645, 1997.

45. W. X. Sun and Z. X. Shen, Optimizing the near field around silver tips, *J. Opt. Soc. Am. A*, 20, 2254, 2003.

46. J. L. Bohn, D. J. Nesbitt, and A. Gallagher, Field enhancement in apertureless near-field scanning optical microscopy, *J. Opt. Soc. Am. A*, 18, 2998, 2001.

47. W. X. Sun and Z. X. Shen, Near-field scanning Raman microscopy using apertureless probes, *J. Raman Spectrosc.*, 34, 668, 2003.

48. W. H. Zhang, X. D. Cui, B. S. Yeo, T. Schmid, C. Hafner, and R. Zenobi, Nanoscale roughness on metal surfaces can increase tip-enhanced Raman scattering by an order of magnitude, *Nano Lett.*, 7, 1401, 2007.

49. T. R. Jensen, G. C. Schatz, and R. P. Van Duyne, Nanosphere lithography: Surface plasmon resonance spectrum of a periodic array of silver nanoparticles by ultraviolet-visible extinction spectroscopy and electrodynamic modeling, *J. Phys. Chem. B*, 103, 2394, 1999.

50. H. Wang, C. S. Levin, and N. J. Halas, Nanosphere arrays with controlled sub-10-nm gaps as surface-enhanced Raman spectroscopy substrates, *J. Am. Chem. Soc.*, 127, 14992, 2005.

51. A. Tarun, N. Hayazawa, M. Motohashi, and S. Kawata, Highly efficient tip-enhanced Raman spectroscopy and microscopy of strained silicon, *Rev. Sci. Instrum.*, 79, 013706, 2008.

52. N. Hayazawa, M. Motohashi, Y. Saito, H. Ishitobi, A. Ono, T. Ichimura, P. Verma, and S. Kawata, Visualization of localized strain of a crystalline thin layer at the nanoscale by tip-enhanced Raman spectroscopy and microscopy, *J. Raman Spectrosc.*, 38, 684, 2007.

53. N. Hayazawa, M. Motohashi, Y. Saito, and S. Kawata, Highly sensitive strain detection in strained silicon by surface-enhanced Raman spectroscopy, *Appl. Phys. Lett.*, 86, 263114, 2005.

54. N. Lee, R. D. Hartschuh, D. Mehtani, A. Kisliuk, J. F. Maguire, M. Green, M. D. Foster, and A. P. Sokolov, High contrast scanning nano-Raman spectroscopy of silicon, *J. Raman Spectrosc.*, 38, 789, 2007.

55. W. H. Zhang, B. S. Yeo, T. Schmid, and R. Zenobi, Single molecule tip-enhanced Raman spectroscopy with silver tips, *J. Phys. Chem. C*, 111, 1733, 2007.

56. J. Steidtner and B. Pettinger, High-resolution microscope for tip-enhanced optical processes in ultrahigh vacuum, *Rev. Sci. Instrum.*, 78, 103104, 2007.

57. R. J. Hamers, Atomic-resolution surface spectroscopy with the scanning tunneling microscope, *Annu. Rev. Phys. Chem.*, 40, 531, 1989.

58. W. J. Kaiser, L. D. Bell, M. H. Hecht, and F. J. Grunthaner, Scanning tunneling microscopy characterization of the geometric and electronic structure of hydrogen-terminated silicon surfaces, *J. Vac. Sci. Technol. A*, 6, 519, 1988.

59. Z. H. Mai, Y. F. Lu, W. D. Song, and W. K. Chim, Nano-modification on hydrogen-passivated Si surfaces by a laser-assisted scanning tunneling microscope operating in air, *Appl Surf. Sci.*, 154, 360–364, 2000.

15 Coatings and Surface Technologies

Dalong Zhong

CONTENTS

15.1 INTRODUCTION

Historically, in the late 1950s, decorative coatings provided the initial thrust for surface-engineered products such as toys, textiles, etc. Since then, the uses of coatings and surface technologies in the areas of engineering and science have produced a dramatic growth in the sales of equipment and products produced, particularly in the last two decades. The semiconductor industry has changed its entire production line every 5–6 years. Suppliers of critical subsystems and components (e.g., vacuum subsystems and process power subsystems) to the semiconductor, flat panel display, and data storage industries reached $7.3 billion in 2000. The equipment suppliers for the semiconductor and related manufacturing industries have experienced a real rollercoaster ride in recent times. However, the photovoltaic cell manufacturing industry has rapidly emerged. Sales of critical subsystems for PV applications totaled less than $10 million in 2000 and have grown to a value of $600 million in 2008. The PV industry now accounts for over 11% of the total market for critical subsystems and is expected to reach about 25% by 2014. It is further estimated that only less than 20% of all the potential items that can benefit from coatings and surface technologies are being processed today. Components and devices manufactured with coating processes are estimated at hundreds of billion dollars annually.

The ever-increasing demands for the superior performance of components and tools in many important industries, including energy, aerospace, healthcare, bio-science, automotive, general manufacturing and engineering, optical and micro-electronics industries, have led to the constant development of new coatings and surface technologies for a wide range of applications. Because the interaction between a component and its surroundings usually occurs directly at the component surface, the properties of the component surface are crucial to the functionality which the component is desired to perform. For example, the superhydrophobic effect of lotus leaves is achieved through its unique surface structure. A wear-resistant component desires a surface with higher hardness, higher chemical stability, and lower coefficient of friction. These surface functionalities can all be achieved through coatings and surface technologies.

"Coatings" in this chapter broadly refers to the near-surface region of an object with properties differing from those of the bulk material (usually referred to as a substrate). A very large number of materials are used for coatings today. Numerous schemes can be devised to classify or categorize coatings, none of which are very satisfactory since several coatings will overlap different categories. With that said, coatings can be classified into two broad categories according to their applications:

1. As structural materials to protect the substrate against wear, oxidation, corrosion, and/or erosion to extend the life of the object or enable the usage of the object in harsh environments, e.g., thermal barrier coatings (TBC) used in gas turbines and hard coatings of engineering components and machine tools. Sometimes, a structural coating application may require other improved surface properties, for example, decorative coatings for hardware typically require different colors and appearances with scratch resistance.
2. As functional materials to achieve specific thermal, electrical, optical, micro-electronic, magnetic, and other physical and/or chemical functionalities, including thin films used in micro-batteries, semiconductors, electronic devices, superconductors, optical devices, magnetic media, sensors, and energy conversion and storage devices.

All coatings and thin films are formed by disposing or applying materials onto a substrate. For example, a coating is produced by depositing a layer onto a substrate by processes such as physical or chemical vapor deposition (CVD), electrodeposition, and spraying; by altering the surface material by diffusion or conversion processes (i.e., diffusion coating or chemical conversion coating); or by ion implantation of new material so that the surface layer now consists of both the parent and added materials. Coatings may also be formed by other processes such as melt/solidification (e.g., the laser glazing technique), by mechanical bonding of a surface layer (e.g., roll bonding or brazing), by mechanical deformation (e.g., shot peening), or other processes that improve the surface properties without changing the composition. Table 15.1 lists some common coatings, typical processes, and their applications. The aim of this chapter is to give the reader a perspective on several coating techniques used in critical or demanding (i.e., high technology) applications, with emphasis on the central concept of the book, *Intelligent Energy Field Manufacturing*. Consequently, some of the techniques such as painting, dip coating, or printing will not be emphasized. It is also realized that not all specific applications can be satisfied within this framework. Interested readers are encouraged to explore the references listed at the end of this chapter.

There are some generic areas common to coating technologies. The origin of coating materials may be a solid as powders in thermal spray or cold spray, a vapor as in physical vapor deposition (PVD) or in CVD, a fluid as in electrochemical deposition or in liquid phase epitaxy (LPE), or a plasma as in plasma-based vacuum deposition technologies. The disposing process to form a coating on a surface requires certain energy fields, including electrical fields, magnetic fields, chemical potential, and kinetic energy. For example, spin coating, a simple procedure to apply coatings to flat substrates, involves placing an excess amount of a solution on the substrate first and then spreading the fluid with centrifugal force by rotating the substrate at a high speed. The applied solvent is usually volatile and simultaneously evaporates. The thickness of the film depends on the angular speed of spinning and the concentration of the solution and the solvent.

TABLE 15.1

Examples of Common Coatings, Their Applications, and Processes

Category	Subcategory	Applications	Materials	Processes
Structural coatings	Hard coatings	Machining tools, metal or plastic forming tools, bearings, valves, pistons, solid particle or liquid droplet erosion resistant coatings, tribological coatings	Transition metal nitrides, carbides, and borides; diamond, diamond-like carbon (DLC); cubic boron nitride; MoS_2-metal	PVD, CVD, spray
	Decorative coatings	Watch bezels, bands, eyeglass frames, costume jewelry, household hardware	(Ti, Zr) N–gold, TiAlN—violet or bluish gray, DLC—metallic black	PVD, CVD
	Barrier coatings	Thermal barrier, environmental barrier	ZrO_2, Al_2O_3, SiO_2	PVD, CVD, spray
Functional coatings	Optical	Laser optics, architectural glazing, mirrors, reflective and anti-reflective coatings, optically absorbing coatings, selective solar absorbers	Oxides, multilayered oxides, fluorides, DLC	PVD, CVD, spray, solution-based deposition
	Electrical	Electrical conductors, electrical contacts, active solid-state devices, electrical insulators, solar cells	Metals, semiconductors, insulators	PVD, CVD, solution-based deposition
	Optical and electrical	Transparent conductors, solar cells	Oxides (e.g., ZnO:Al, ITO, SnO_2:F) semiconductors (e.g., Si, III–V or II–VI compounds)	PVD, CVD, MBE, solution-based deposition
	Thermal and electrical	Thermoelectric devices, thermal electric cooler, thermal electric generator	Alloys or semiconductors, e.g., ZnSb, PbS, $Bi_{1-x}Sb_x$, Bi_xTe_y,	PVD, CVD, MBE, solution-based deposition
	Chemical	Catalytic applications, fuel cells, photochemical devices, batteries	Metals (e.g., Pt, Pd), oxides (e.g., CeO_2, ZrO_2, TiO_2, WO_3), LiPON, $LiCoO_2$	PVD, CVD, spray, solution-based deposition
	Magnetic	Data storage, power electronics, generators	Metals and alloys (e.g., Fe, Co, FeCo)	PVD, CVD, electroplating

The more sophisticated coating technology, PVD processes, can be described in terms of the following three basic steps.

Step 1: Creation of a vapor phase species by evaporation, sputtering, or feeding chemical vapors and gases.

Step 2: Transportation of the vapor species from the source to substrate under line-of-sight or molecular flow-conditions (i.e., without many collisions between atoms and molecules). If the partial pressure of the vapor is high enough or some of the vapor species are ionized (by creating a plasma), there are many collisions in the vapor phase during the transport to the substrate.

Step 3: Film growth on the substrate by various nucleation and growth mechanisms. The film microstructure and composition can be tailored by controlling the chemistry (e.g., ion-to-neutral ratio) and physics (e.g., energy and ionization state of species) of the vapor or plasma and the substrate temperature.

Every PVD process can be usefully described and understood in terms of these steps. For example, sputtering is a momentum transfer vacuum coating process. A plasma will be generated in a sputtering process due to electrons emitted under the influence of an electric field. A plasma is a gas that contains charged species (e.g., electrons, positive ions, and negative ions) and neutral species (e.g., atoms and molecules) and is electrically neutral on the average. The primary particles for a magnetron sputtering process are supplied by the plasma formed in the process. The incident ions set off collision cascades in the target. When such cascades recoil and reach the target surface with energy above the surface binding energy, an atom can be ejected. Those ejected atoms will get into the plasma and a fraction of them will become ionized. Some of the species coming from the target will deposit onto the substrate surface or chamber walls, and thus a coating is formed. The electrical field applied between the target and the ground (e.g., chamber wall) can be direct current (DC), pulsed DC, or radio-frequency (RF) power. People who would like to learn more about sputtering are advised to read more in PVD textbooks.

From the intelligent energy field manufacturing point of view, all coating technologies can be described from three aspects: (1) what the building blocks are, (2) which energy fields are used, and (3) how the material and energy flow and how they control film growth in the process. Figure 15.1 illustrates energy, materials/resource, information, and process flows in magnetron sputtering using the energy field manufacturing methodology. Magnetron sputtering is a powerful and flexible technique that can be used to coat virtually any part with a wide range of materials—any solid metal or alloy and a variety of compounds. The building blocks for a coating are atoms from the target and gases fed into the chamber. The types of energies that can be found in sputtering are electrical power (electrical energy), magnetic field (electromagnetic energy), gas pressure (kinetic energy), and substrate temperature (thermal energy). These energy sources control the target processes during sputtering and they also affect how coatings grow on a surface. In other words, nucleation and coating growth are determined by the materials and energies involved in the sputtering process and the substrate (e.g., type of material, surface pattern, etc.).

It is a difficult task to predict the future trends of intelligent energy field manufacturing for coatings and surface technologies. What one might safely do is to try and extrapolate the currently available advanced manufacturing technologies and carefully suggest some innovative solutions with enough merit to be pursued.

In the following sections, we will review some of the state of the art on advanced coating systems and manufacturing technologies that should develop into commercial products in the short to medium term.

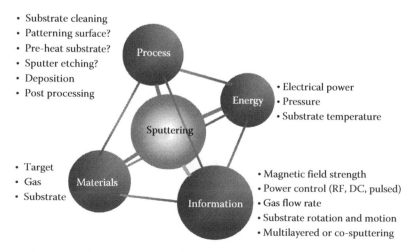

FIGURE 15.1 Illustration of magnetron sputtering using M-PIE model.

15.2 COATING ARCHITECTURES

In order to optimize the properties provided by the coating materials, one may have to engineer their deposition in such a way as to produce the desired characteristics in a cohesive, adherent film. This "tailoring" of coatings to meet specific application requirements may lead to the generation of complex micro-/nanostructures where the involved phases have controlled volume fraction, distribution, crystallite size, and crystallographic orientation. The clever assembly of these features may be described as the *coating architecture*.

The architecture of a successful coating may follow different approaches, such as the alternate deposition of selected materials (either in the micrometer or nanometer range), the co-deposition of immiscible phases yielding nanocomposites, or the deposition of materials with a gradient in composition. These architectures bring solutions to engineering issues (e.g., wear, oxidation, and corrosion) either by improving the intrinsic properties of the film or by providing the film with a self-healing capability. Most applications have complex requirements for the physical, chemical, and mechanical properties of the coating, and in most cases, a single-phase coating material cannot satisfy all these requirements. Coating materials with different properties and different coating concepts can all be engineered into an appropriate coating "architecture" to exhibit the best accommodation of the chemical, physical, and mechanical properties desired to achieve high performance and reliability for specific applications and to provide an optimum combination of functionality to meet the application requirements.

Two of the most promising approaches in thin film design, nanostructured films, and functionally graded or multilayered coatings are presented below.

15.2.1 NANOSTRUCTURED FILMS

Nanostructured coatings are structures that contain nano-sized features such as rods, tubes, precipitates, or film layers. In these cases, superior physical, chemical, or mechanical properties are feasible. The convergence of materials science with biology has begun to be seen at the nanoscale. Within the general denomination of nanostructured films one may include 1D nanostructures (e.g., nanorods, nanotubes, and nanowires), 2D nano-laminates, and 3D nano-composites such as nanocrystalline-amorphous systems (nc-a) and nanocrystalline biphase systems (nc–nc). Since 1D nanostructures are covered in another chapter of this book, the concept described here is focused on 2D and 3D nanostructures for superior properties or functionalities.

15.2.1.1 Nanolaminate or Superlattice Coatings

Nanolaminate or superlattice coatings consist of nanometer scale multilayers of two different alternating materials with a period (Λ) in the nanometer range (Figure 15.2a). Entirely new properties and characteristics not directly related to the individual layered constituents can be expected for superlattice coatings. For instance, the materials selected for this architecture should comprise the

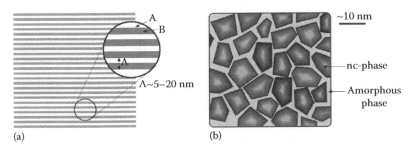

FIGURE 15.2 Schematic representation of nanostructures of (a) nano-laminates (superlattices) and (b) isotropic nano-composites.

following requirements for wear applications: (a) there should be a large difference in the shear moduli between the two materials, (b) the materials should be immiscible, and (c) the interfaces should be coherent with a small lattice mismatch between the constituent layers. Increased hardness and strength are generally observed for superlattice coatings. In the case of transition metal nitrides, superlattices have been effective in increasing hardness to over 50 GPa, well above the ~20–30 GPa values normally observed for either individual nitride and much higher than what would be predicted by the rule of mixtures [1,2]. The industrial scale manufacture of superlattice hard PVD coatings has been under investigation [3].

15.2.1.2 Isotropic 3D Nanostructured Coatings

Isotropic 3D nanostructured coatings differ from the superlattice coatings with respect to the distribution of the phases. In this case, instead of a 2D laminate structure, the nanocomposite will normally be formed by interwoven phases in three dimensions, see Figure 15.2b. For wear resistant coatings, the same principle of hardness enhancement by the suppression of the dislocation activity applies in this case as long as the frequency of occurrence of phase boundaries is the same in all directions. Moreover, the mechanisms of strengthening and toughening are likely active simultaneously, resulting in the simultaneous enhancement of strength and toughness.

These nanocomposites are usually formed from ternary or higher order systems and comprise two nanocrystalline (nc) phases or, more commonly, an amorphous phase surrounding nano-size crystallites of a secondary phase.

Some requirements to achieve the desired microstructure and properties are (a) materials must be immiscible at deposition conditions, (b) the cohesive energy at the interface between phases must be high, and (c) the second phase (amorphous or crystalline) must possess high structural flexibility in order to accommodate the coherency strain without forming dangling bonds, voids, or other flaws [4,5]. Additionally, for nanocrystalline-amorphous systems, according to Patscheider [6], to achieve the high hardness, the particle size of the crystalline phase should be less than 10 nm and the thickness of the amorphous phase separating the nanocrystals should be of only a few atomic bond lengths. In nanocrystalline-amorphous nanocomposites, the chemistry operating in these systems not only determines the constituent phases but also stabilizes their nanostructure and makes it fairly free of flaws. Therefore, there are critical values for the crystallite size and the relative amounts of nanocrystalline phases that control the interaction between the nanocrystallites and amorphous phase and thus determine the properties and performance of the nanocomposite.

15.2.2 Functionally Graded or Multilayered Coatings

In functionally graded coatings (FGCs), the composition and/or the microstructure of the coating system gradually changes over the volume (i.e., throughout its thickness) to produce different characteristics at each end without generating abrupt interfaces within the film, resulting in corresponding changes in the properties of the coating system. The most familiar FGC is compositionally graded from a refractory ceramic to a metal. Using such a FGC architecture, incompatible functions, such as high temperature, wear, and oxidation resistance of ceramics, can be combined with high toughness, high strength, and the bonding capability of metals without severe internal stress. Pores are also important structures in FGCs. Grading pore size and distribution from the interior to the surface of the coating can impart many properties, such as mechanical shock resistance, thermal insulation, catalytic efficiency, and the relaxation of stress. In general, this type of system is a multilayered multi-component coating system with a graded arrangement of phases to provide different synergistic properties and functions and is tailored for problem-specific applications.

The functionally graded coating architecture is very versatile and offers great potential with respect to the development of a coating system for protection and more favorable performance under extreme and complex environmental "loading" conditions. It is unlikely that a monolithic coating would provide the optimum system for any specific application that has complex requirements

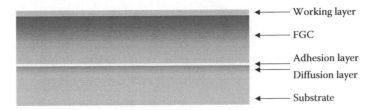

FIGURE 15.3 A schematic diagram of an optimized coating architecture for coating dies used for material forming applications, e.g., metal stamping, glass molding, aluminum pressure die casting.

for the physical, chemical, and mechanical properties of the coating material. The FGC concept is applicable to many fields. For instance, coating systems that incorporate specifically engineered, graded coating architectures have been developed by the Advanced Coatings and Surface Engineering Laboratory (ACSEL) at the Colorado School of Mines for coating dies used for material forming applications, e.g., metal stamping, glass molding, and aluminum pressure die casting [7–9]. A schematic diagram of such an optimized coating architecture is presented in Figure 15.3. The conceptual design of such an optimized "coating system" incorporates four sections of the total coating architecture from the substrate to the top layer as outlined below:

1. Surface modification of the substrate, e.g., plasma nitriding or ferritic nitro-carburizing
2. A thin (50–100 nm) adhesion interlayer, e.g., Ti or Cr, between the substrate (e.g., H13) and the coating system
3. An intermediate layer (compositionally graded) that facilitates the "accommodation" of thermal and residual stress as predicted using finite element modeling [10,11]
4. An exterior "working" layer that exhibits acceptable properties to meet the application requirements, such as low wettability with the material being formed, e.g., liquid Al or glass, coupled with high wear and corrosion/oxidation resistance

Thus, each layer of the coating's architecture will provide a specific function and the success of the coating system lies in the synergy of the properties and functions of each layer.

Many devices are built upon functionally multilayered coatings. As shown in Figure 15.4, a solar photovoltaic device consists of several functional layers, p-type and n-type semiconductor layers to form a semiconducting p-n junction, and a transparent conducting oxide (TCO) layer and a back contact layer to serve as two electrodes for collecting electricity. To maximize the light transmitted into the solar cell, the top encapsulation glass is typically coated with an anti-reflection (AR) coating. It is clearly shown that a functionally multilayered coating architecture is used in order to

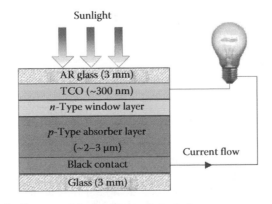

FIGURE 15.4 A schematic diagram of a solar photovoltaic device.

make a functional solar cell. Besides the requirement to meet the optical and electrical needs, the multilayered stack is also desired to be a compatible material system to meet other properties, such as adhesion. Thus, multilayered coating architectures seem to be the most versatile, powerful, and optimal solution in most cases. Note that a solar cell is a complex multilayered system and requires all components to work together and function properly. Issues at the interfaces could degrade the device performance significantly.

15.3   SMART COATINGS

Smart coatings are systems that respond in a selective way to external stimuli such as temperature, stress, strain, or environment [12]. "Smartness" results from the careful selection and combination of coating materials with distinctive intrinsic properties. The approach to produce such coatings may vary: one may achieve this goal by changing the composition of a single-phase material or by producing multilayered or graded films.

For instance, films that operate at moderate or high-temperature oxidation can be designed to have protective layers of either chromia or alumina depending on the environment. Alumina scales offer the best protection under high-temperature conditions while chromia is more resistant to hot corrosion [13]. TiAlN is an example of a very basic smart coating since it reacts to the presence of oxygen at high temperatures producing a protective Al_2O_3 layer that prevents further deterioration of the film by oxidation. Tribological coatings for aerospace applications may have to deal with broad ranges of environmental conditions such as air humidity, air pressure, and temperature while maintaining low friction and good wear resistance. Adaptative nanocomposite coatings, such as WC/diamond-like carbon (DLC)/WS_2 [14] and nc-YSZ in an amorphous YSZ/Au matrix [15], may present different friction mechanisms depending on the operating conditions. In the former case, for instance, a change in humidity triggers a reversible modification of the composition of the transfer film between WS_2 and graphite while in operation.

A more sophisticated type of coating that could be defined as "smart" is that of films with imbedded sensors and/or actuators. Little work has been conducted in this area, but the possible products of this research would be extremely valuable. The use of strain sensors alone could be beneficial for critical applications for monitoring stress in films. In the die casting industry, for instance, molds are subjected to wear, corrosion, and thermal and mechanical cycling. The ability to monitor stress in the film could help optimize casting parameters, mold design (geometry and placement of cooling channels), and provide a more efficient program for maintenance/replacement of the die.

An even more ambitious goal would be the use of both sensors and actuators, where the latter, triggered by the former, would relieve some of the stress imposed on the coating at peak conditions during the thermal or mechanical cycling of mechanical parts or tools.

Kim and Lee [16] investigated smart coatings containing piezoelectric materials and shape memory alloys. Their idea was to combine the fast response to strain provided by piezoelectricity with the large force-displacement produced by the martensite-to-austenite transformation in the TiNi shape memory alloy. The application of such systems may still be some years ahead, but the concept is certainly worth investigating.

15.4   NOVEL PROCESSES

CVD powered by thermal energy was the first process used commercially to produce coatings. Due to the high temperatures necessary, many materials could not be coated by such processes. The development of plasma-enhanced CVD and PVD processes, normally operating at much lower substrate temperatures, permitted the deposition of a variety of different coatings on several types of substrates, including ceramics, metals, and even plastic materials. While the fundamentals have not changed much, the industry has grown enormously in the last two decades. Cost-effectiveness and the major processing of technological advances have been the major driving forces for innovative process development. Because of the highly competitive nature of the coating industry, novel process improvements and inventions have generally been considered as proprietary information and kept secret. As such, it is a very difficult task to elaborate on many process innovations. Nevertheless, an outlook is given here on some technological advances based upon the current and potential developments of the coating processes reported in the literature. Although there are a wide variety of emerging processes, the criteria for the selection of an appropriate deposition process for a specific engineering application should be based on the ability of the coating method to do the following:

a. Deposit the required type of coating
b. Deposit it to the desired thickness
c. Not affect or impair the properties of the substrate or the layers beneath
d. Deposit the engineering components uniformly with respect to both size and shape
e. Be cost-effective in terms of coating techniques, with minimum equipment down-time and improved quality and performance of the coatings

Plasma-assisted vapor deposition processes are very important and extensively utilized for the deposition of compounds and novel technological materials, but its full potential has yet to be exploited. Over the past few decades, substantial progress has been made in the design, research, and development of various types of plasma sources to obtain coatings and surfaces with desired properties. Three important research and development areas are pulsed plasma processing, utilization of high-density plasma sources, and hybrid processes. These are also the topic of the following overview, with emphasis on pulsed plasma processing, which has the potential to revolutionize some of the established deposition processes. Note that other deposition processes, such as thermal or plasma spray, diffusion, laser ablating, and nonvacuum based CVD, are not covered in this chapter, but that does not mean that they do not have their advantages or are out of date. These techniques are still widely used in industry and are subjected to active research programs.

15.4.1 Pulsed Plasma Processing

Pulsed plasma processing is one of the state-of-the-art technologies implemented into PVD, CVD, surface modification, surface pretreatment, and cleaning processes such as pulsed magnetron sputtering [17], pulsed plasma-enhanced chemical vapor deposition (PECVD) [18,19], pulsed plasma nitriding [20,21], and pulsed ion beam treatment [22–24]. The recent development of pulsed plasma processing technologies opens an avenue for controlling plasma chemistry and physics [25–29], thus activating and energizing condensing particles. Activated or energized species will enhance surface mobility of adatoms, and, in turn, will improve the microstructure and properties of deposited films [30–32]. As an example, when pulsed power is applied to a growing dielectric film during magnetron sputtering, ion bombardment of the surface is enhanced thus promoting denser crystalline film growth at lower substrate temperatures [33,34]. Pulsed sputtering at very high rates and stable conditions over long production runs can be obtained for a variety of materials, such as Al_2O_3, SiO_2, and TiO_2.

Substrate bias has been used extensively in various plasma processes. It is usually of negative polarity to attract positive ions from the processing plasma to either pre-treat the substrate for coating or improve the coating by ion bombardment. In the last 10 years, pulsed biasing of the substrate has become popular for several reasons. One is the likelihood of the elimination of arcing and enhanced process stability when using a substrate bias with short pulses. Other reasons are associated with the nature of pulsing. Applying a pulsed bias at the substrate has a profound effect on the energy and flux of particles incident on the substrate [35]. This can both improve the effectiveness of pre-cleaning and the quality of subsequent coatings, which benefit from improved adhesion, uniformity, lower stress, and the suppression of columnar growth morphology.

Emerging technologies of pulsed plasma in PVD and CVD processes provide a new set of process parameters, such as pulse frequency, pulse duration, duty cycle, and pulse amplitude. Controlling these pulsing parameters, a pulsed plasma process can provide momentary high power, leading to more energetic and ionized particles in the plasma. High kinetic energy in pulsed processing allows film growth to occur much further from thermodynamic equilibrium than with continuous processing. The temporal development of the plasma composition in pulsed processing allows film growth under an environment unachievable in continuous processing. Consequently, superior films can be deposited. To illustrate the application of pulsed plasma processing, selected examples that have attracted increasing interest recently are briefly presented as follows.

Pulsed sputtering processes can be grouped into two categories: mid-frequency, "medium"-power pulsed sputtering; and low frequency, high-power pulsed sputtering. Mid-frequency, "medium"-power pulsed sputtering was developed in the 1990s. Pulsed DC was used either in unipolar or bipolar mode, at typical frequencies from 10 to 350 kHz and at high duty cycle. Typical pulsed power and current density are at the level below 100 W/cm^2 and 100 mA/cm^2, respectively. Mid-frequency, "medium"-power pulsed reactive sputtering has been successfully used to deposit various dielectric coatings such as Al_2O_3 [30,33]. Note that pulsed DC sputtering is used when a simple one-magnetron system is preferred, and unfortunately, such an approach does not solve the 'disappearing' anode problem suffered in the reactive sputtering of dielectrics. The problems of arcing and disappearing anodes are solved simultaneously using mid-frequency (20–100 kHz) AC powered dual-anode (DAS) or dual-magnetron (DMS) systems. In contrast, high-power pulsed sputtering, which was introduced in the late 1990s, uses low-frequency (<1 kHz), low-duty cycle, but very high peak power and current densities (at least one or two magnitudes higher). By increasing the power density to a magnetron by orders of magnitude (typically hundreds to thousands W/cm^2), significant ionization of the sputtered metal can be obtained [36]. A significant change in the slope of the voltage–current characteristics was observed by Ehiasarian et al. [25] when the current density at a chromium target exceeded 600 mA/cm^2, which is a sign for the transition to a fully ionized plasma. This method has been demonstrated as an extremely promising PVD technique suitable for both substrate pretreatment and coating deposition [37].

Pulsed glow discharge plasma immersion ion processing (PIIP) is a non-line-of-sight technique for depositing uniform coatings over large areas and can also be viewed as a type of pulsed plasma assisted CVD, which was proposed at Los Alamos National Laboratory (LANL) in the United States [38]. A pulsed plasma sheath is realized by applying a high negative pulsed bias to the substrate and is utilized to attract ions from the plasma for energetic condensation at the substrate surface and to accelerate secondary electrons in the sheath for plasma generation. Typical precursors utilized in PIIP are carbon-containing gases (e.g., C_2H_2) for DLC films [39], gas mixtures leading to doped DLC films [40], or metallo-organic precursors for TiCN coating [41].

Metal plasma immersion ion implantation and deposition (MePIIID) utilize both pulsed plasma and pulsed bias. The technique can be viewed as a hybrid system of ion plating with pulsed bias. Anders presented an excellent overview on the principles and trends of this method [42]. By synchronizing pulsed plasma generation and pulsed high-voltage bias, metal ion implantation without film formation is possible. This technique has been successfully demonstrated for wear-resistant surface modification [43], thin film deposition [44], and the minimization of intrinsic stress in coatings [45].

15.4.2 HIGH-DENSITY PLASMA SOURCES

High-density plasma sources with higher ionization efficiency are indispensable for large-scale plasma/ion beam technology. Such plasmas were generated in discharges excited at higher RF (including microwave), in discharges employing power coupling schemes that are more efficient than capacitive coupling, or by utilizing the confinement and resonant effects of a static magnetic field generated by external means (coils, permanent magnets). It is impossible to include in this chapter all of the novel high-density plasma sources that are suitable for low-pressure plasma processing. Two examples are given below.

Electron cyclotron resonance (ECR) plasma sources have proven to be very useful for plasma processing applications, primarily by exploiting their ability to operate at low pressures and high density (a few 10^{12} cm^{-3}) in a wave-supported electrodeless mode, which allows the plasma to be generated remotely from tool surfaces. It is widely used in sputtering [46] and CVD [47] for the fabrication of highly wear-durable coatings. The electron temperature of the ECR sources is generally higher than that of the other plasma sources due to its heating mechanism, resulting in plasma-induced substrate damage. Accordingly, much effort has been made to decrease the electron temperature, for instance, by a mirror magnetic field configuration [48] or pulse modulation [49].

Electron beam (e-beam) generated plasmas have several important attributes for processing applications including high plasma generation efficiency, independent control of ion and radical fluxes, decoupling of plasma production from chamber walls, scalability to a large area (square meters), and low electron temperature. The Naval Research Laboratory (NRL) has developed a number of hollow cathodes to generate sheets of electrons culminating in a "large area plasma processing system" (LAPPS) based on the electron-beam ionization process [50,51]. The LAPPS uses a sheet electron beam (2–10 kV, 10–20 mA/cm^2) confined by a 100–200 gauss magnetic field to produce a plasma with a density of ~10^{12} cm^{-3} in a neutral gas nearly independent of composition (oxygen, argon, neon, or other gas mixtures) over a few square meters of area within centimeters of the surface. It can be operated in pulsed or continuous mode depending on the application. This process has demonstrated considerable flexibility for materials processing applications, although only a few of the attributes have been investigated so far [52].

15.4.3 HYBRID PROCESSES

Another area subjected to extensive research and development that has shown great promise is that of hybrid processes. In these processes, different deposition techniques have been combined to extend the processing capabilities and overcome the limitations of each of the individual techniques. The general thrust is directed toward the following:

1. Separation of the various parts of the process so as to exert independent control over each part and avoid complications due to overlap between the parts
2. The use of substrate/film bombardment with different species of controlled energy as contrasted to a spectrum of energies

There are numerous examples of such hybrid techniques in the literature, such as ion beam assisted sputtering, microwave-/ECR-assisted PECVD, and hybrid systems of magnetron sputtering and filtered vacuum arc ion plating. These are generally innovation of effective plasma deposition processes by manipulating the processing discharges and then controlling the effects of ion bombardment on resultant coatings. For instance, plasma chemistry was successfully manipulated by the addition of helium in an ion beam–assisted reactive sputtering process. Helium was mixed with nitrogen in an inductively coupled plasma (ICP) reactor [53]. The addition of helium resulted in an approximately four times increase in the grain size, a 22% increase in surface smoothness, and a 44% increase in the hardness of the deposited TiN films [53].

Besides some hybrid processes discussed in the previous paragraph, novel hybrid processes of combining pulsed arc or pulsed laser ablation with molecular beam deposition are presented here as examples.

Laser-assisted molecular beam deposition (LAMBD) [54] and pulsed arc molecular beam deposition (PAMBD) [55] have been developed at AMBP Tech Corporation, U.S.A., a spin-off company of the State University of New York at Buffalo. These techniques are hybrid systems of pressurized chemical reactors and pulsed laser or arc ablation, which utilize a train of gas pulses to precisely control the chemistry and transport of species to be deposited on a substrate. These techniques have been used to grow a variety of films (e.g., oxides and DLC) on a variety of substrates.

Another example, hollow cathode plasma immersion ion processing, combines high density, hollow cathode plasma with plasma immersion ion deposition. Sub-One Technology Inc. [56] and Southwest Research Institute [57] have used this technology for coating interior surfaces. Using the hollow cathode discharge in PECVD, a high-density (over 10^{12} cm^{-3}) plasma can be generated within cylindrical substrates, such as pipes. In this case, the pipe itself is the vacuum chamber and high-density plasmas are generated and maintained by using asymmetric bipolar pulsed DC power. Very high deposition rates can be achieved on the order of $1\,\mu$m/min. A variety of applications can benefit from this novel process, such as industrial piping, offshore drilling, chemical delivery systems, gun barrels, and medical devices. Sub-One Technology Inc. deposited thick DLC coatings onto the inner surfaces of various lengths and diameters of pipes [58]. In general, these films have high corrosion, erosion, and wear resistant properties.

15.4.4 Nanostructure Growth by Glancing Angle Deposition

Nanomaterials are now no longer just a laboratory curiosity because new methods to synthesize, manipulate, order, and visualize materials at near atomic scales have quickly emerged over the past few years. Through the precise engineering of structures on the nanoscale, researchers can endow a material with remarkably different properties and open avenues for novel applications, such as nanoelectronics, spintronics, nano-optics, thermo-electrics, sensors, bio-imaging, and drug delivery. Recent advances in nanotechnology offer the possibility of a revolutionary approach to manufacturing. There are many different ways of generating nanostructures, such as lithography, laser micromachining, and template-directed electroplating. Lithography, especially e-beam lithography, is one way of generating metallic nanostructures, but the high aspect ratio of the metallic nanostructure needed will be difficult to achieve in the lithography process. Laser micromachining can generate metallic structures with high aspect ratios, but it is challenging to generate submicron-scale structures. Electrochemical processing is a promising approach for generating metallic nanostructures with high aspect ratios over large areas. However, it requires a post-deposition process to remove the template (e.g., using the wet chemical etching process). A good technique for fabricating sophisticated nanostructures (e.g., 3D nanosprings) should be capable of controlling the dimensions, uniformity, alignment, and interfacial property of the nanostructure. In addition, it is advantageous if the process is applicable to many materials and can directly grow the nanostructure without the need for a post-growth process step.

Glancing angle deposition (GLAD) coupled with typical PVD techniques is a nanostructure fabrication technique that can precisely engineer structures on the nanoscale in a single processing step and can offer the flexibility of producing complex nanostructures. Suitable systems for GLAD include electron-beam physical vapor deposition (EB-PVD), sputtering, and other PVD processes such as ion beam deposition (IBD), molecular beam epitaxy, and laser ablation. There are several in-depth reviews on nanostructure growth from the vapor phase using the GLAD approach [59–62]. The fundamental tool used in GLAD to control the arrangement of the atoms on a surface is atomic shadowing, which causes atoms to have a higher probability to land on the surface mound than in the valley that is shadowed. The surface morphological evolution of thin film growth can also be affected strongly by anisotropic surface diffusion, which refers to the fact that adatoms may move fast on certain crystalline planes but slow on others. Adatom mobility on the surface can be modified intentionally by using low-energy ion irradiation during deposition.

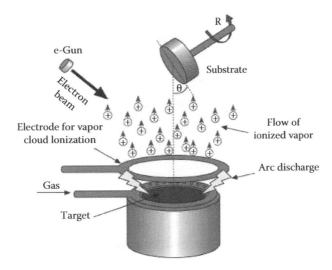

FIGURE 15.5 A schematic diagram of a GLAD process.

Figure 15.5 shows a schematic of the essential components for the growth of controlled 3D nanostructures using the GLAD process. Angle θ in Figure 15.5 is the oblique angle between the plane of the rotating substrate and the direction in which the vapor flux arrives. The oblique angle, θ, will be the primary factor in determining the tilt angle, α, of the growing nanostructure, as shown in Figure 15.6. The substrate rotation speed and pattern in both the polar and azimuthal directions will determine the geometry of the nanostructure. If the substrate rotates at 90° steps at fixed growth intervals, nanostructures as depicted in Figure 15.6a will be generated, while helical columns will grow if the substrate rotates continuously (Figure 15.6b). Using appropriate substrate movements, different types of sophisticated nanostructures have been demonstrated (e.g., vertical posts, helices, and zigzags) [60,61]. A uniform nanospring pattern with controlled spacing between features may be fabricated using a patterned seed layer on the substrate surface. The substrate may be pre-patterned with a periodic array of small protrusions or seeds (e.g., by block copolymer templates or colloidal self-assembly using polystyrene or silica nanospheres) to intercept vapor flux and control the nucleation location. This method offers a fully 3D control of the nanostructure with an additional capability of self-alignment. There is almost no limitation on the materials that can be fabricated into desired nanostructures. Besides nanorod arrays and nanospring arrays, people have demonstrated even more complex nanostructures, such as multilayer nanostructures by sequential deposition from multiple sources [62] and arrays of checkerboard nanorods by simultaneous deposition of multi-components with intermittent

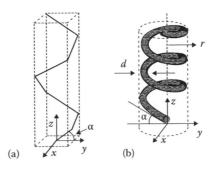

FIGURE 15.6 A schematic diagram of a 3D nanostructure growth using GLAD: (a) a square spiral with a π/2 rotation at discrete intervals, and (b) a helix with a constant and continuous rotation.

180° substrate rotations [63]. These nanostructured coatings exhibit tremendous potential applications due to tailored optical, thermal, mechanical, chemical, biomedical, catalytic, electrical, and magnetic properties.

15.5 SUMMARY

Coatings and surface technologies will continue in strong growth in the next decade because coating products can improve performance, reduce cost, and control surface properties relatively independent of the substrate, thus offering enormous potential due to the following:

- The creation of entirely new products or enabling new designs
- Solutions to previously unsolved engineering problems
- Improved functionality of existing products
- Conservation of scarce materials
- Ecological considerations—reduction of effluent output and power consumption

There have been some considerable advances in the development of coatings and surface technologies over the past decade. These advances have come through radical new ideas and approaches that have included coating architecture and design as well as advanced plasma processing concepts. The development of "adaptive" coatings that can respond to a changing or cyclic environment, such as high and low temperatures, humidity, and oxidizing atmospheres, is gaining considerable momentum, while the concept of " 'smart' coatings is a little farther into the future. Nevertheless, the time is rapidly approaching when both of these concepts will be combined to produce a "smart" and "adaptive" coating system that can first sense a problem in the coating, such as a sudden and critical increase in stress or corrosion/oxidation, and then respond to provide a cure for that problem.

There is no doubt that there will be considerable improvements and developments of new deposition processes in the near future that will provide increased control over the structure and properties of the coating and will provide more cost effective processing techniques. The most realizable of the current approaches will be further development in the use of controlled energy plasma sources and synergistic hybrid combinations of the currently used deposition systems.

There is an exciting future ahead for engineering radically new coating systems that will substantially improve the functionality of surfaces. Recent advances in nanotechnology offer the possibility of a revolutionary approach toward the advanced manufacturing of coatings and thin films.

QUESTIONS

Q.15.1 How are coatings classified into two broad categories according to their applications in this book?

Coatings are classified into two broad categories according to their applications: (1) as structural materials to protect the substrate and extend the life of the object and (2) as functional materials to provide specific thermal, electrical, optical, magnetic, and other physical functionalities.

Q.15.2 How many different states of matters are used as the source material for the deposition of a coating? Give an example of a process for each.

The different states of matter are solid, liquid, gas, and plasma/ions. Solid—targets used in PVD, liquid—electrochemical or solution deposition, gas—vapor deposition, and plasma/ions—plasma enhanced vapor deposition.

Q.15.3 What kinds of energies are used for coating processes? Give an example for each.

Examples: thermal—thermal evaporation, electron-beam—electron-beam evaporation or ionization processes, laser—pulsed laser deposition, arc—arc evaporation, microwave—microwave CVD or blazing.

Q.15.4   What is a plasma? How is a plasma started in sputtering?

In physics and chemistry, plasma is the fourth state of matter. It is a collective gas that contains charged species (e.g., electrons, positive ions, and negative ions) and neutral species (e.g., atoms and molecules) and is electrically neutral on average. The ability of the positive and negative charges to move somewhat independently makes the plasma electrically conductive so that it responds strongly to electromagnetic fields, thus it is used extensively in deposition technologies. A plasma is generated in a sputtering process due to the electrons emitted under the influence of an electric field.

Q.15.5   What is "coating architecture"? Give some examples.

"Coating architecture" refers to the assembly of coating materials in a structural way to meet specific application requirements. Examples: nanostructured films and functionally graded or multilayered coatings. Nanostructured films include nanorods, nanotubes, nanowires, 2D nano-laminates or superlattice, and 3D nano-composites.

Q.15.6   How has sputtering technology evolved?

DC/RF diode sputtering → DC/RF magnetron sputtering → pulsed DC magnetron sputtering → high power pulsed magnetron sputtering

Q.15.7   What is the future trend of coating processes? Give two or three example areas of development.

Examples: (1) processes to achieve high-quality coating at low substrate temperatures with high deposition rates, e.g., high-density plasma assisted vapor deposition; (2) non-line-of-sight coating technology for complex geometric objects or high aspect ratio interior surfaces, e.g., pulsed glow discharge plasma immersion ion processing; and (3) hybrid processes.

Q.15.8   What controls the nanostructure growth in GLAD? Which types of nanostructures can be grown using GLAD?

Nanostructure growth in GLAD is determined by atomic shadowing, anisotropic surface diffusion, and substrate rotation speed and pattern in both polar and azimuthal directions. GLAD can grow nanostructures as 1D (e.g., nanorods), 2D (e.g., zig-zag columns), and 3D (e.g., nanosprings).

ABOUT THE CONTRIBUTING AUTHOR

Dr. Dalong Zhong's research has been focused on the process, structure, and property of thin films/coatings with applications to energy conversion and storage, aerospace and aviation, healthcare, water, oil and gas, and related advanced manufacturing industries since 1996. After receiving his doctorate degree from Colorado School of Mines in 2001, Dalong worked at Golden, Colorado for the Colorado School of Mines, MVSystems Inc., and ACSEnT Inc. on (1) CleanTech, such as nc-Si/a-Si tandem junction solar cells, compound semiconductor thin films for energy conversion devices, lithium micro-batteries, and proton-conducting ceramic fuel cells; (2) nano-composite coatings for glass molding dies, metal die casting tools, and aerospace tribological applications; and (3) other functional coatings such as fluorinated DLC films for aerospace applications and a "smart coating" system for the monitoring of tools used in harsh environments. Dr. Zhong joined GE Global Research (GEGR) in 2005. At GEGR, Dr. Zhong has been leading various, multi-million dollar projects on thin films, coatings, and their advanced manufacturing and spearheading research on the direct growth of nanostructured films/coatings for GE businesses and for government-funded programs. Dr. Zhong has 18 peer-reviewed papers, 4 invited papers in leading international conferences, 2 book chapters, and over a dozen filed U.S. patents. He has actively participated in various international meetings, such as MRS Spring and Fall meetings, ICMCTF meetings, SVC annual meetings, TMS and ASM annual meetings and he has reviewed papers for various journals, such as *Thin Solid Films, Surface and Coatings Technology, Tribology International, Acta Materiala, Journal of Materials Synthesis and Processing, Journal of Thermal Spray Technology, Journal of Physics D: Applied Physics,* and *Journal of Physics: Condensed Matter.*

REFERENCES

1. P. Yashar, S.A. Barnett, J. Rechner, and W.D. Sproul (1998), Structure and mechanical properties of polycrystalline CrN/TiN superlattices, *Journal of Vacuum Science and Technology A* 16(5), 2913–2918.
2. W.D. Sproul (1996), Reactive sputter deposition of polycrystalline nitride and oxide superlattice coatings, *Surface and Coatings Technology* 86–87, 170–176.
3. W.D. Munz, D.B. Lewis, P.E. Hovsepian, C. Schonjahn, A. Ehiasarian, and I.J. Smith (2001), Industrial scale manufactured superlattice hard PVD coatings, *Surface Engineering* 17, 15–27.
4. S. Veprek (1999), The search for novel, superhard materials, *Journal of Vacuum Science and Technology A* 17(5), 2401–2420.
5. S. Veprek and A.S. Argon (2002), Towards the understanding of mechanical properties of super- and ultrahard nanocomposites, *Journal of Vacuum Science and Technology B* 20, 650–664.
6. J. Patscheider (2003), Nanocomposite hardcoatings for wear protection, *MRS Bulletin* 28(3), 180–183.
7. A.M. Peters, J.J. Moore, I. Reimanis, B. Mishra, and R. Weiss (1999), Cathodic arc evaporation of functionally graded chromium nitride thin films for wear resistant and forming applications, *Materials Science Forum* 308–311, 283–289.
8. D. Zhong, A.M. Peters, E. Hixson, B. Mishra, and J.J. Moore (2002), Processing, properties and modeling of graded thin films/coating systems. In: N. Chakraborti and U.K. Chatterjee (eds.), *Proceedings of the International Conference on Advances in Materials and Materials Processing*, Tata McGraw-Hill Publishing Company Limited, New Delhi, India, pp. 884–894.
9. D. Zhong, S. Carrera, A.M. Peters, O. Salas, B. Mishra, and J.J. Moore (2003), Development of surface engineered coatings for dies used in material processing. In: S. Seal et al. (eds.), *Surface Engineering in Materials Science II*, TMS, Warrendale, PA, pp. 183–194.
10. D. Zhong, G.G.W. Mustoe, J.J. Moore, and J. Disam (2001), Finite element analyses of a coating architecture for glass molding dies, *Surface and Coatings Technology* 146–147, 312–317.
11. S. Carrera, G.G.W. Mustoe, D. Zhong, and J.J. Moore (2002), Finite element modeling of coating architectures for aluminum die casting. In: T.S. Sudarshan, M. Jeandin, and J.J. Stiglich (eds.), *Surface Modification Technologies XV*, ASM International, Materials Park, OH, pp. 21–27.
12. J.R. Nichols (1996), Smart coatings: A bright future, *Materials World* 4(1), 19–21.
13. J.R. Nicholls, N.J. Simms, W.Y. Chan, and H.E. Evans (2002), Smart overlay coatings—concept and practice, *Surface and Coatings Technology* 149(2–3), 236–244.
14. A.A. Voevodin, J.P. O'Neill, and J.S. Zabinski (1999), Nanocomposite tribological coatings for aerospace applications, *Surface and Coatings Technology* 116–119, 36–45.
15. A.A. Voevodin, J.J. Hu, T.A. Fitz, and J.S. Zabinski (2001), Tribological properties of adaptive nanocomposite coatings made of yttria stabilized zirconia and gold, *Surface and Coatings Technology* 146–147, 351–356.
16. I.-J. Kim and H.-W. Lee (2001), Fabrication of TiNi/PZT heterostructure films for smart systems, *Scripta Materialia* 44(3), 525–530.
17. P.J. Kelly, and R.D. Arnell (2000), Magnetron sputtering: a review of recent developments and applications, *Vacuum* 56, 159–172.
18. U.K. Das, S. Morrison, and A. Madan (2002), Deposition of microcrystalline silicon solar cells via the pulsed PECVD technique, *Journal of Non-Crystalline Solids* 299–302, 79–82.
19. G. Fedosenko, A. Schwabedissen, J. Engemann, E. Braca, L. Valentini, and J.M. Kenny (2002), Pulsed PECVD deposition of diamond-like carbon films, *Diamond and Related Materials* 11, 1047–1052.
20. J.N. Feugeas, B.J. Gómez, G. Sánchez, J. Ferron, and A. Craievich (2003), Time evolution of Cr and N on AISI 304 steel surface during pulsed plasma ion nitriding, *Thin Solid Films* 424(1), 130–138.
21. P. Panjan, I. Urankar, B. Navinšek, M. Terčelj, R. Turk, M. Čekada, and V. Leskovšek (2002), Improvement of hot forging tools with duplex treatment, *Surface and Coatings Technology* 151–152, 505–509.

22. R.M. Bayazitov, I.B. Khaibullin, R.I. Batalov, R.M. Nurutdinov, L.Kh. Antonova, V.P. Aksenov, and G.N. Mikhailova (2003), Structure and photoluminescent properties of SiC layers on Si, synthesized by pulsed ion-beam treatment, *Nuclear Instruments and Methods in Physics Research Section B: Beam Interactions with Materials and Atoms* 206, 984–988.

23. A. Kondyurin, R. Khaybullin, N. Gavrilovd, and V. Popok (2002), Pulse and continuous ion beam treatment of polyethylene, *Vacuum* 68(4), 341–347.

24. H. Akamatsu, T. Ikeda, K. Azuma, E. Fujiwara, and M. Yatsuzuka (2001), Surface treatment of steel by short pulsed injection of high-power ion beam, *Surface and Coatings Technology* 136, 269–272.

25. A.P. Ehiasarian, R. New, W.-D. Münz, L. Hultman, U. Helmersson, and V. Kouznetsov (2002), Influence of high power densities on the composition of pulsed magnetron plasmas, *Vacuum* 65(2), 147–154.

26. J.T. Gudmundsson, J. Alami, and U. Helmersson (2002), Spatial and temporal behavior of the plasma parameters in a pulsed magnetron discharge, *Surface and Coatings Technology* 161, 249–256.

27. J.M. Schneider, A. Anders, I.G. Brown, B. Hjörvarsson, and L. Hultman (1999), Temporal development of the plasma composition of a pulsed aluminum plasma stream in the presence of oxygen, *Applied Physics Letters* 75(5), 612–614.

28. H. Sugai, K. Nakamura, Y. Hikosaka, and M. Nakamura (1995), Diagnostics and control of radicals in an inductively coupled etching reactor, *Journal of Vacuum Science and Technology A* 13(3), 887–893.

29. S. Samukawa and S. Furuoya (1993), Time-modulated electron cyclotron resonance plasma discharge for controlling generation of reactive species, *Applied Physics Letters* 63(15), 2044–2046.

30. R. Cremer, K. Reichert, D. Neuschutz, G. Erkens, and T. Leyendecker (2003), Sputter deposition of crystalline alumina coatings, *Surface and Coatings Technology* 163–164, 157–163.

31. F. Fenske, P. Reinig, B. Selle, and W. Fuhs (2003), Pulse-sputter deposition of highly <100>-oriented crystalline silicon films, *Surface and Coatings Technology* 174–175, 801–804.

32. H. Bartzsch, P. Frach, and K. Goedicke (2000), Anode effects on energetic particle bombardment of the substrate in pulsed magnetron sputtering, *Surface and Coatings Technology* 132, 244–250.

33. A. Schütze and D.T. Quinto (2003), Pulsed plasma-assisted PVD sputter-deposited alumina thin films, *Surface and Coatings Technology* 162, 174–182.

34. R. Cremer, M. Witthaut, D. Neuschutz, G. Erkens, T. Leyendecker, and M. Feldhege (1999), Comparative characterization of alumina coatings deposited by RF, DC and pulsed reactive magnetron sputtering, *Surface and Coatings Technology*, 120–121, 213–218.

35. P.J. Kelly, R. Hall, J. O'Brien, J.W. Bradley, P. Henderson, G. Roche, and R.D. Arnell (2001), Studies of mid-frequency pulsed dc biasing, *Journal of Vacuum Science and Technology A* 19, 2856–2865.

36. V. Kouznetsov, K. Macák, J.M. Schneider, U. Helmersson, and I. Petrov (1999), A novel pulsed magnetron sputter technique utilizing very high target power densities, *Surface and Coatings Technology* 122, 290–293.

37. A.P. Ehiasarian, W.-D. Münz, L. Hultman, U. Helmersson, and I. Petrov (2003), High power pulsed magnetron sputtered CrN_x films, *Surface and Coatings Technology* 163–164, 267–272.

38. M. Nastasi, X.-M. He, K.C. Walter, M. Hakovirta, and M. Trkula (2001), The use of plasma immersion ion processing in the synthesis of protective coatings for Al die casting, *Surface and Coatings Technology* 136, 162–167.

39. K.C. Walter and M. Nastasi (2002), Influence of ion and neutral flux on the properties of diamond-like carbon from pulsed glow discharges of acetylene, *Surface and Coatings Technology* 156, 306–310.

40. K.M. He, M. Hakovirta, A.M. Peters, B. Taylor, and M. Nastasi (2002), Fluorine and boron co-doped diamond-like carbon films deposited by pulsed glow discharge plasma immersion ion processing, *Journal of Vacuum Science and Technology A* 20(3), 638–642.

41. A.M. Peters and M. Nastasi (2002), Effect of carrier gas on the deposition of titanium carbo-nitride coatings by a novel organo-metallic plasma immersion ion processing technique, *Vacuum* 67(2), 169–175.

42. A. Anders (2002), From plasma immersion ion implantation to deposition: A historical perspective on principles and trends, *Surface and Coatings Technology* 156, 3–12.

43. S. Mändl, J.W. Gerlach, W. Assmann, and B. Rauschenbach (2003), Phase formation and diffusion after nitrogen PIII in molybdenum, *Surface and Coatings Technology* 174–175, 1238–1242.

44. P. Huber, D. Manova, S. Mandl, and B. Rauschenbach (2003), Formation of TiN, TiC and TiCN by metal plasma immersion ion implantation and deposition, *Surface and Coatings Technology* 174–175, 1243–1247.

45. S.H.N. Lim, D.G. McCulloch, M.M.M. Bilek, and D.R. McKenzie (2003), Minimisation of intrinsic stress in titanium nitride using a cathodic arc with plasma immersion ion implantation, *Surface and Coatings Technology* 174–175, 76–80.

46. T. Tokai, S. Umemura, S. Hirono, A. Imoto, and R. Kaneko (2003), Tribological characteristics of highly wear-durable ECR-sputtered silicon nitride films, *Surface and Coatings Technology* 169–170, 475–477.

47. K.Y. Li, Z.F. Zhou, C.Y. Chan, I. Bello, C.S. Lee, and S.T. Lee (2001), Mechanical and tribological properties of diamond-like carbon films prepared on steel by ECR-CVD process, *Diamond and Related Materials* 10(9–10), 1855–1861.

48. N. Itagaki, Y. Ueda, N. Ishii, and Y. Kawai (2001), Production of low electron temperature ECR plasma for plasma processing, *Thin Solid Films* 390, 202–207.

49. N. Itagaki, A. Fukuda, T. Yoshizawa, M. Shindo, Y. Ueda, and Y. Kawai (2000), Plasma parameter measurements and deposition of a-Si:H thin films in pulsed ECR plasma, *Surface and Coatings Technology* 131, 54–57.

50. R.A. Meger, D.D. Blackwell, R.F. Fernsler, M. Lampe, D. Leonhardt, W.M. Manheimer, D.P. Murphy, and S.G. Walton (2001), Beam-generated plasmas for processing applications, *Physics of Plasmas* 8(5), 2558–2564.

51. R.F. Fernsler, W.M. Manheimer, R.A. Meger, J. Mathew, D.P. Murphy, R.E. Pechacek, and J.A. Gregor (1998), Production of large-area plasmas by electron beams, *Physics of Plasmas* 5(5), 2137–2143.

52. D. Leonhardt, C. Muratore, S.G. Walton, D.D. Blackwell, R.F. Fernsler, and R.A. Meger (2004), Generation of electron-beam produced plasmas and applications to surface modification, *Surface and Coatings Technology* 177–178, 682–687.

53. C. Muratore, J.A. Rees, B. Mishra, and J.J. Moore (2003), The influence of particle energy distributions on the structure and properties of reactively sputtered titanium oxide and titanium nitride thin films. In: S. Seal et al. (eds.), *Surface Engineering in Materials Science II*, TMS, Warrendale, PA, pp. 251–261.

54. R.L. DeLeon, M.P. Joshi, E.R. Rexer, P.N. Prasad, and J.F. Garvey (1998), Progress in thin film formation by laser assisted molecular beam deposition (LAMBD), *Applied Surface Science* 127–129, 321–329.

55. E.F. Rexer, D.B. Wilbur, J.L. Mills, R.L. DeLeon, and J. F. Garvey (2000), Production of metal oxide thin films by pulsed arc molecular beam deposition, *Review of Scientific Instruments* 71, 2125–2130.

56. W.J. Boardman, A.W. Tudhope, and R.D. Mercado (2007), Method and system for coating internal surfaces of prefabricated process piping in the field, EP1619265 A1.

57. R. Wei, C. Rincon, and J.H. Arps (2009), Plasma immersion ion processing for coating of hollow substrates, EP2035596 A1.

58. D. Lusk, M. Gore, W. Boardmanx et al. (2008), Thick DLC films deposited by PECVD on the internal surface of cylindrical substrates, *Diamond and Related Materials* 17, 1613–1621.

59. D. Gall, (2005) Nanostructured transition-metal nitride layers. In: É.J. Knystautas (ed.), *Engineering Thin Films and Nanostructures with Ion Beams*, Series Optical Engineering, Vol. 95, CRC Press, Boca Raton, FL.

60. M. Hawkeye and M. Brett. (2007), Glancing angle deposition: Fabrication, properties, and applications of micro- and nanostructured thin films, *Journal of Vacuum Science and Technology* A25, 1317–1335.

61. J.J. Steele and M.J. Brett (2007), Nanostructure engineering in porous columnar thin films: Recent advances, *Journal of Material Science: Materials in Electronics* 18, 367–379.

62. Y.-P. Zhao, D.-X. Ye, G.-C. Wang, and T.-M. Lu (2003), Designing nanostructures by glancing angle deposition, *Proceedings of SPIE—The International Society for Optical Engineering*, 5219, p 59–73.

63. C.M. Zhou, H.F. Li, and D. Gall (2008), Multi-component nanostructure design by atomic shadowing, *Thin Solid Films* 517, 1214–1218.

16 Methodology and Process Innovations in Additive Fabrication

Lijue Xue

CONTENTS

16.1 INTRODUCTION

Additive fabrication refers to a group of technologies used for building physical models, prototypes, tooling, finished components, and production parts directly from 3D computer-aided design (CAD) data or 3D data generated using various measuring methods (such as coordinate measuring machine [CMM], computed tomography [CT], magnetic resonance imaging [MRI], or laser scans) [1]. Different from conventional material removal processes, additive fabrication produces parts through the adding and joining of liquid, powder, or wire materials layer by layer as per the cross sections of a 3D model. Due to the nature of the technology, additive fabrication provides the capability to produce complex 3D shapes with delicate features (such as an enclosed cavity, very small internal features, etc.) that are difficult or even impossible to manufacture using conventional manufacturing processes (such as machining, casting, forming, etc.).

Additive fabrication initially emerged in the 1980s. After the U.S. patent 4,575,330 "Apparatus for the production of three-dimensional objects by stereolithography" was granted, 3D Systems developed the first commercial "Stereo lithography apparatus (SLA)" system to produce demonstrative parts directly from photo-sensitive resins based on 3D CAD models [2]. Additive fabrication was initially used to produce models and prototypes for product development. Since then, many types of additive fabrication technologies have been developed. With the evolvement of technology, additive fabrication has been used to produce components made of polymers, metals, ceramics, composites, and other functional materials. Additive fabrication has evolved into an industry with estimated products and services of about $1.141 billion in 2007 [3].

With the globalization of economy, manufacturers face the increased challenge of delivering new products more quickly to meet customer demands. This results in the significant demands for the application of additive fabrication (also known as rapid prototyping, rapid manufacturing, etc.) technologies to produce demonstration parts, tooling, and even functional components directly from CAD models to significantly shorten the design and production cycle and reduce the manufacturing cost.

Additive fabrication technology consists of a family of different types of processes (such as printing-based processes, liquid- and powder-bed processes, and cladding-based processes). Each process has its unique advantages and drawbacks (Table 16.1). The following is a brief summary of some key processes.

TABLE 16.1
Comparison of Different Additive Fabrication Processes

Feature	3D Printing	SLA	SLS	Laser Cladding
Processing	Injection of liquid or paste	Liquid-bed	Powder-bed	Injection of powder or wire
Dimensional accuracy	Low to moderate	Moderate	Low	Moderate to high
Typical layer thickness	0.07–0.25 mm	0.05–0.15 mm	0.10 mm	0.10–1 mm
Minimum features	0.6 mm	0.25 mm	0.7–1.5 mm	0.65 mm
Material	Polymers, photopolymers, thermal plastics	Photopolymers	Polymers, ceramics, limited metals	Metals and ceramics
Strength	Low to moderate	Low	Moderate to high	High
Part quality	Low (e.g., fragile)	Low (e.g., fragile)	Low to moderate (e.g., porosity, cracking)	High
Support structure	Required	Required	Not required	Not required
Cost of system	Low	Moderate	Moderate	High

16.1.1 Printing-Based Processes

Three-dimensional printing is a group of technologies that are based on the printer concept to dispense small droplets of liquid or paste material onto a substrate as per the cross sections and connect successive layers to produce a 3D shape. One typical example is an inkjet printing system. Layers of fine powders are selectively bonded by "printing" adhesives from the inkjet print head following the shape of each cross section based on a CAD model (Figure 16.1) [4]. Based on the same concept, an inkjet printer can be used to dispense liquid (such as photopolymer) to form each layer of a part as per a CAD file, while a ultraviolet lamp can be used to cure each layer during deposition to form the part. As an alternative, fused deposition modeling (FDM) uses a nozzle to extrude molten thermoplastics, layer by layer, onto a platform as per the cross sections of a model and the part is built up from layers as the material hardens immediately after extrusion from the nozzle [4].

Aerosol jet is another alternative that first aerosolizes conductive photovoltaic inks or pastes and then forms an aerodynamically focused droplet stream of the material towards a substrate. The high velocity of the stream causes the particles to impact on the substrate. Thermal post-processing is usually required to sinter the particles together to adhere them to the substrate and/or to make them conductive [5].

Bioprinting is a biomedical application of rapid prototyping technology (or additive biofabrication) that involves the deposition of biological material for patterning and assembling tissue and next-level biological structures (such as organs) from cells, tissues, tissue segments, extracellular matrices, or biomimetic hydrogels [6,7]. Bioprinting is currently in the development stage and is primarily used as a scientific tool. The long-term goal is that the technology could be used to create replacement organs or even entire organisms from raw biological materials.

Printing-based technology is generally fast, affordable, and easy to use, making it suitable for making prototypes for visualization during the conceptual design stage when the dimensional accuracy and mechanical strength of the prototypes are less important. It is also used to produce molds and patterns for various applications (Figure 16.2).

16.1.2 Liquid- or Powder-Bed Technology

16.1.2.1 Stereolithography Apparatus [8]

Stereolithography utilizes a liquid bed of ultraviolet (UV) curable photopolymer "resin" and a UV laser to build a part, layer by layer. On each layer, the laser beam traces a cross-sectional pattern of the part on the surface of the liquid resin, cures (solidifies) the scanned regions, and adheres them to the underneath layer. After a layer is completed, the system descends the platform by one-layer thickness, while a resin-filled blade sweeps across the cured pattern to recoat it with fresh material. On the new liquid surface, the subsequent layer of pattern is traced and adheres to the previous layer. By repeating this process again and again, a 3D part is completed (Figure 16.3). The finished

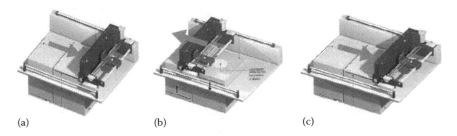

(a) (b) (c)

FIGURE 16.1 Illustration of 3D printing process, (a) spreading a layer of powder, (b) printing binder along desired cross section, and (c) spreading another layer of powder. (Courtesy of Z Corporation, Burlington, VT.)

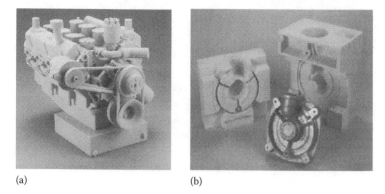

(a)   (b)

FIGURE 16.2   Parts produced by 3D printing process, (a) demonstration part, (b) 3D printed mold and a cast aluminum part. (Courtesy of Z Corporation, Burlington, VT.)

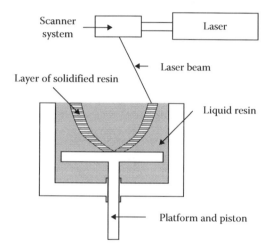

FIGURE 16.3   Illustration of stereolithography process.

SLA part will be immersed in a chemical bath to clean the excess resin and will then be finally cured in a UV oven.

SLA provides the capability to produce functional parts from photopolymers that posses certain strengths for demonstrations or even as master patterns for various molding and metal casting processes (Figure 16.4). With the liquid-bed design along with the use of support structures, SLA is capable to produce the shapes of parts with almost no limitations. The parts can be ready, depending on the size and complexity, from a few hours to a few days. However, SLA is really a "rapid modeling" technology since the objects generated from existing photo-sensitive resins or photo polymers do not have the physical, mechanical, or thermal properties typically required by production parts. In addition, the SLA process also has certain limitations associated with fabrication time, accuracy, durability, repeatability, toxic chemicals, and material cost.

16.1.2.2   Selective Laser Sintering

Selective laser sintering (SLS) is a powder-bed process, during which a layer of powder is dispensed onto a platform and leveled by a rolling device. Based on a sliced CAD model, a laser beam scans a two-dimensional pattern on the deposited powder layer to sinter the powder layer as well as to join it with the underneath sintered layer. After sintering a layer, the platform moves down by one-layer thickness and a new layer of powder is dispensed onto the previous layer and sintered in the same

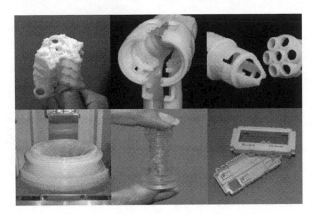

FIGURE 16.4   Parts produced by SLA process. (Courtesy of 3D Systems, Rock Hill, SC.)

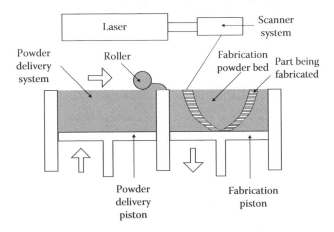

FIGURE 16.5   Illustration of SLS process.

manner. Successive powder dispensing and laser scanning gradually build up a complete 3D part based on the CAD model (Figure 16.5).

SLS was initially used to produce a durable nylon-based prototype and patterns for making molds. Although nylon-based parts can provide some level of functional evaluation and are economical for the production of limited quantities, it is necessary to have prototype parts made with the functional materials for rigorous testing and analysis. For that purpose, the SLS process has since been developed for plastics, plastic-coated metals, ceramics, and composite materials (Figure 16.6). A plastic or plastic-coated metal pattern can be initially made by the SLS process and then converted to a metallic component using a secondary process.

With the further evolution of the technology, some metallic functional components have been directly produced using SLS (or direct metal laser sintering [DMLS]) of stainless steel, nickel, titanium, and other alloys from 3D CAD data [9]. As a variation, an electron beam has also been used to replace lasers as the energy source for the powder-bed additive fabrication, which results in the "electron beam melting" process used to produce metallic parts from Ti, Ti-6Al-4V, and Co-Cr materials with high processing speeds (Figure 16.7) [10].

The metallic components directly produced using powder-bed processes demonstrated good mechanical properties and dimensional accuracy. The powder-bed processes provide support for building parts with enclosed features that are very difficult if not impossible to make using conventional manufacturing methods. The direct fabrication of functional metallic components from the CAD model using powder-bed processes provides a cost-effective alternative for low volume

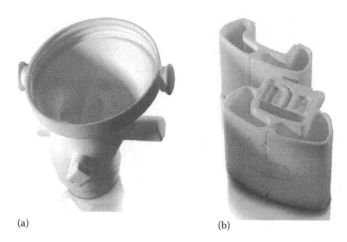

(a)   (b)

FIGURE 16.6 (a) Plastic part produced by SLS process. (Courtesy of EOS, Munich, Germany.) (b) Centrifuges with functional integration produced by SLS. (Courtesy of Andreas Hettich GmbH & Co. KG, Tuttlingen, Germany.)

FIGURE 16.7 A fully dense titanium part with integrated network structure for biomedical application, produced by the electron beam melting process. (Courtesy of Arcam AB, Mölndal, Sweden.)

production. However, the production speed for the powder-bed process is usually relatively slow and the surface finish of the buildup parts, especially underneath the surface, is usually a concern. Especially, the size of the part to be built is limited by the size of the powder bed, which restricts its application in building components of relatively small sizes. In addition, the powder-bed process will not be able to build features on existing components.

16.1.3 LASER CLADDING–BASED 3D MANUFACTURING

Laser cladding is a material deposition process by which powder (or wire) feedstock is melted by a focused laser beam and solidified to form a coating on a substrate or to fabricate a near-net shape part. The powder (or wire) is normally of a metallic nature and is typically injected through a coaxial or lateral nozzle into the molten pool created by the focused laser beam. When a relative movement between the laser beam and the substrate is applied through a computer

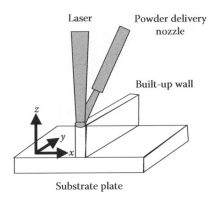

FIGURE 16.8 Illustration of laser cladding–based 3D manufacturing process.

numerically controlled (CNC) motion system or a robot, the deposited material solidifies quickly to produce a track of solid metal metallurgically bonded to the substrate. When the motion of the substrate is guided by a numerically controlled (NC) program that interpolates a CAD solid model into a set of tracks, a desired part will be built layer by layer following the trajectory (Figure 16.8).

Compared with the powder-bed direct metal laser sintering process, the laser cladding–based 3D manufacturing process has certain unique advantages, such as high deposition rate, large component size, easy change of materials, better mechanical properties, the capability to build features on large components and to repair damaged or undersized components, etc.

Laser cladding–based 3D manufacturing technology has been developed by various research organizations and different names have been used for a variety of processes, such as laser engineered net shaping (LENS), direct metal deposition (DMD), and laser consolidation (LC).

16.1.3.1 Laser Engineered Net Shaping

The LENS process was developed by Sandia National Laboratories and was commercialized by Optomec. The LENS system uses a Nd:YAG laser or fiber laser along with a multi-axis CNC motion system and offers a work envelop ranging from 300 mm × 300 mm × 300 mm to 900 mm × 1500 mm × 900 mm [5]. The LENS process is being intensively investigated for repairing components for aerospace and defense industries, although various case studies were also conducted for fabricating components for aerospace and medical devices applications (Figure 16.9). The U.S. Army's Anniston Army Depot has been using its LENS system to repair a number of Honeywell gas turbine engine components for the M1 Abrams Tank, while the U.S. Army has also selected LENS for its Mobile Parts Hospital to provide a real-time battlefield repair capability [5].

16.1.3.2 Direct Metal Deposition

The DMD process was developed by the University of Michigan and was commercialized by the POM group. The DMD process uses optical sensors for real-time closed-loop melt pool control during deposition. POM offers five-axis CNC as well as robotic-based DMD systems. The DMD process has been extensively investigated for repairing tooling (such as punch die and forging die) and parts as well as for fabricating molds with conformal cooling for automotive applications (Figure 16.10) [11].

16.1.3.3 Laser Consolidation

The LC process is being developed by the National Research Council Canada. As compared with other laser cladding–based 3D manufacturing technology that only produces near net-shape parts,

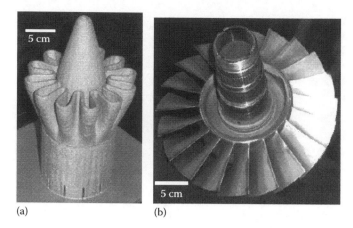

(a)   (b)

FIGURE 16.9 (a) Ti-6Al-4V gas thruster produced by, and (b) engine blisk repaired by LENS process. (Courtesy of Optomec, Albuquerque, NM.)

(a)   (b)

FIGURE 16.10 (a) Punch dies rebuilt, and (b) shapes produced by DMD process. (Courtesy of POM, Auburn Hills, MI.)

the LC process has the unique capability to produce net-shape functional components with excellent surface finish and dimensional accuracy [12]. The LC process can be used to build net-shape functional components with complex geometric features (or to build features on existing components) that would be difficult or even impossible to produce using conventional manufacturing methods (Figure 16.11). In collaboration with various industrial partners, the LC process has been investigated for many applications for aerospace, defense, and other applications [12–24]. Some examples will be described in the next section.

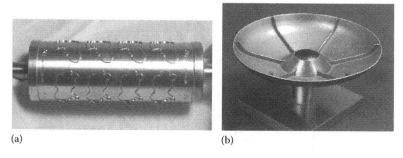

(a)   (b)

FIGURE 16.11 Functional net-shape metallic parts produced by laser consolidation process, (a) rotary cutting dies, and (b) demonstration piece.

16.2 SOME EXAMPLES OF POTENTIAL INDUSTRIAL APPLICATIONS OF LASER NET-SHAPE 3D MANUFACTURING

There have been myriad kinds of industrial applications of laser net-shape 3D manufacturing. In the following, the author's research area, laser consolidation–based applications will be used to show the benefits and development needs of 3D manufacturing.

16.2.1 STRUCTURAL COMPONENTS FOR THE ADVANCED ROBOTIC MECHATRONICS SYSTEM

The advanced robotic mechatronics system (ARMS) is a collaborative research project conducted by MD Robotics in collaboration with the Canadian space agency (CSA), National Research Council Canada, and several universities. Laser consolidation was used as a rapid functional prototyping method for making Ti-6Al-4V structural components [17].

The conventional design of a space robot manipulator generally consists of separate booms and joint housings that are connected to each other through a flanged interface, which substantially increases the weight and complexity. A one-piece integrated boom/housing design is preferable to reduce the weight and complexity and increase the interface stiffness of a typical robotic arm. However, it is extremely difficult or even impossible to make the one-piece integrated boom/housing using conventional manufacturing processes.

As a free-form fabrication process, the LC process allows the building of net-shape functional features on existing components and, therefore, offers the unique capability to build multi-functional boom on prebuilt housing to realize the innovative design for one-piece integrated boom/housing. Figure 16.12 shows an integrated boom/housing manufactured using laser consolidation of the Ti-6Al-4V alloy. The integrated LC Ti-6Al-4V boom/housing shows an as-consolidated surface finish, except for the contact surfaces that were initially machined for next stage final machining and assembling.

Each ARMS consists of four Ti-6Al-4V structural components:

- One multi-functional boom
- Two housings with interface
- One housing with integrated boom

The LC process was successfully used to build all of them from Ti-6Al-4V alloy powder. A prototype ARMS was assembled by MD Robotics using these structural components along with the other

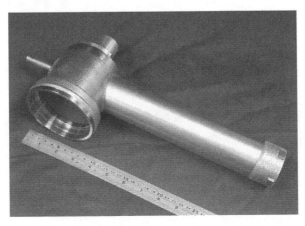

FIGURE 16.12 Integrated Ti-6Al-4V boom/housing made by laser consolidation process (initial machining only at the contact areas).

mechanical and electronic components. Figure 16.13a shows a close view of the assembled joint, while Figure 16.13b shows the ARMS with the required payload during laboratory testing. The real-time testing results demonstrated that the laser-consolidated components performed very well and all design requirements, such as low weight and high strength, were achieved.

Compared with the conventional cast or wrought material, the laser-consolidated Ti-6Al-4V also demonstrated excellent mechanical properties (Table 16.2). The average yield and ultimate tensile strengths of the as-consolidated Ti-6Al-4V are about 1062 and 1157 MPa, respectively, which is substantially higher than the as-cast/annealed cast Ti-6Al-4V and annealed wrought Ti-6Al-4V and is comparable to the wrought Ti-6Al-4V in the solution-treated plus aged condition. The elastic modulus of the as-consolidated Ti-6Al-4V (116 GPa) is about the same as the wrought material (110 GPa). However, the elongation of the as-consolidated Ti-6Al-4V material is about 6%, which is lower than the value of cast or wrought Ti-6Al-4V (8%–10%). An appropriate aging treatment may be needed to improve the elongation value [15].

LC Ti-6Al-4V demonstrated very good bonding strength to the wrought Ti-6Al-4V substrate. Under tensile testing, the as-consolidated Ti-6Al-4V shows a bond strength of about 1045 MPa in average with a standard deviation of 21 MPa. It is interesting to note that all bond test specimens failed inside the LC Ti-6Al-4V zone due to the stress concentration effect, instead of at the bond area, which indicates that the actual ultimate tensile strength at the bond area is high [15].

The preliminary fatigue test results are also very encouraging: the data of as-consolidated Ti-6Al-4V is at the high end of as-cast Ti-6Al-4V data [15]. The endurance limit of the LC

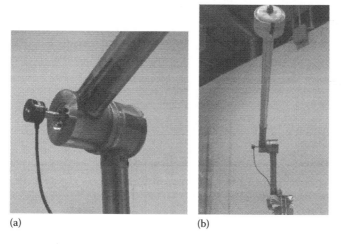

(a)   (b)

FIGURE 16.13   Assembled ARMS with LC Ti-6Al-4V structural components, (a) a close view, and (b) testing with payload.

TABLE 16.2
Tensile Properties of Laser-Consolidated (LC) Ti-6Al-4V Alloy

Materials	$\sigma_{0.2}$ (MPa)	σ_{UTS} (MPa)	E (GPa)	δ (%)
LC Ti-6Al-4V	1062±6	1157±11	116±8	6.2±0.9
Cast Ti-6Al-4V (as-cast or annealed)	890	1035	—	10
Wrought Ti-6Al-4V (annealed)	825	895	110	10
Wrought Ti-6Al-4V (solution-treated aged bar)	965	1035	110	8
Wrought Ti-6Al-4V (solution-treated + aged)	1103	1172	—	10

Ti-6Al-4V specimens is around 400 MPa, which is significantly higher than the 200 MPa achieved by the as-cast Ti-6Al-4V. The experimental results further demonstrated that the fatigue resistance of the LC Ti-6Al-4V alloy could be significantly improved through the optimization of the processing parameters. The preliminary results showed that the endurance limit of the LC Ti-6Al-4V material is in excess of 500 MPa, which is well within the upper scatter band of annealed wrought material [21].

16.2.2 AIRFOILS WITH EMBEDDED COOLING CHANNELS

The increase in thermal efficiency of the gas turbine is typically achieved by increasing its maximum operating temperature to as high as the first row of blades can withstand. The turbine blades with embedded cooling channels could significantly increase the turbine blade's resistance to higher temperatures. NRC worked with GE Global Research to demonstrate the potential of the LC process to produce advanced airfoil with embedded cooling channels, which is very difficult or even impossible to manufacture using other conventional manufacturing methods [18].

The airfoil has double walls with seven bridges connecting two walls to provide structural integrity. The wall thickness is about 0.762 mm, while the gap between the two walls is approximately 0.635 mm. The overall length of the airfoil is 33.02 mm and its overall width is around 10.16 mm. The height of the airfoil was set to 25.4 mm for the demonstration piece. The airfoil design contains several complex internal features such as sharp corners and relatively narrow gaps that extend throughout the body of the airfoil, which increases the technical challenge for manufacturing.

Using an optimized processing methodology and laser path planning strategy, laser consolidation successfully built the demonstration Ti-6Al-4V airfoils with embedded cooling channels. Figure 16.14 shows the "as-consolidated" airfoil without further surface operation except the grinding of the top surface. Measurement results reveal that the gap between the two walls is fairly uniform with an average value of 0.630 mm with a deviation of only 0.028 mm, while the design requirement of the gap is 0.635 mm [18].

FIGURE 16.14 A demonstration LC Ti-6Al-4V airfoil with embedded cooling channels.

The LC process demonstrated the unique manufacturing capability to produce functional net-shape components. The combination of the innovative design with the unique capability of the LC process can lead to the creation of components with highly complex geometry that is not possible with conventional means.

16.2.3 IMPELLERS BUILT ON PREMACHINED SUBSTRATE

Laser consolidation is a material addition process that can directly build functional features on an existing component to form integrated structures without the need for welding or brazing. Figure 16.15 shows an IN-718 impeller and its blades were directly built up on a premachined substrate using the LC process, which took about 5.5 h to complete [24]. It is evident that this novel process produces high quality, fairly complex shapes directly from a CAD model with good surface finishes in an as-consolidated condition without any further processing. The bond between the LC features and the existing component is metallurgically sound, without cracks and porosity. Compared with the conventional welding process, the heat input from the LC process to the substrate is minimal, resulting in a very small heat affected zone (several tens of micrometers). By using the LC process, more unique features can be added to the gas turbine components to provide additional functionality and to reduce manufacturing time and cost.

The laser-consolidated IN-718 shows very good mechanical properties. Table 16.3 compares the tensile properties of LC IN-718 with wrought IN-718 material. As expected, due to the difference in their microstructures, the as-consolidated IN-718 has relatively low tensile and yield strengths

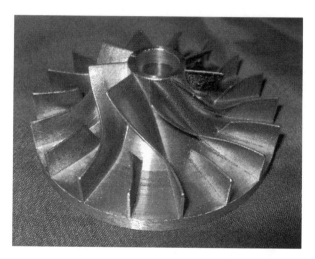

FIGURE 16.15 An IN-718 impeller with blades directly built up on premachined substrate using laser consolidation process.

TABLE 16.3
Comparison of Tensile Properties of IN-718 Alloy

Material	$\sigma_{0.2}$ (MPa)	σ_{UTS} (MPa)	δ (%)
LC IN-718	432±5	802±21	39±5
LC IN-718 (heat-treated)	1085±19	1238±12	21±2
Wrought IN-718 (heat-treated)	1036	1240	12
IN-718 Sheet (heat-treated)	1050	1280	22
IN-718 Bar (heat-treated)	1190	1430	21

(about 432 and 802 MPa, respectively) but much better elongation (about 39%) as compared with heat-treated wrought IN-718. After the standard heat treatment used for wrought IN-718 alloy, the yield and tensile strengths of the LC IN-718 are increased to about 1085 and 1238 MPa, respectively, and elongation is reduced to about 21%, which is fully comparable with the heat-treated wrought In-718 material [24].

16.2.4 SONAR SHELLS

The folded shell projector (FSP) is a compact flextensional sound source being developed by Defense Research and Development Canada (DRDC) for military low frequency sonar applications, including active towed arrays and sonobuoys. The FSP radiates sound from a thin-walled cylindrical shell driven by a piezoelectric or magneto-strictive motor. The shell is made of thin walls, high strength, and corrosion resistant metals (e.g., pure nickel, Ni-alloys, or Ti-alloys) and has superimposed corrugations, which creates a significant challenge for the existing manufacturing technology. Machining such thin wall parts is extremely difficult by conventional means, because without support, the thin wall will deflect under the load of the cutter, causing poor surface finish and loss of precision.

After considerable evaluation of alternative manufacturing technologies, laser consolidation was selected by the DRDC as the rapid manufacturing technology for the FSP and its newest variant, the shear mode projector (SMP). In collaboration with the DRDC, NRC investigated laser consolidation for manufacturing net-shape functional shells to support DRDC's design evolvements as well as for field testing of the developed sonar [22].

The LC process provides an excellent manufacturing method to accommodate rapid design evolution since it is a freeform fabrication technology without the needs of hard tooling. Figure 16.16 shows four types of shells manufactured by the LC process. FSP shell designs #1 and #2 were made of IN-625 alloy, while FSP shell design #3 and SMP shell design #1 were from Ti-6Al-4V alloy. Nondestructive inspections (including ultrasonic, dye penetrant, x-ray, and/or magnetic methods) reveal that laser-consolidated shells are metallurgically sound and free of cracks and porosity. As-consolidated shells show good surface finish and dimensional accuracy.

FSP shell designs #1 and #2 are about 130 mm in height and around 80 mm in diameter. The key difference is that design #2 has a reduction in the radius of curvature of the circular generator of the shell. For traditional manufacturing processes, such a change in design will require significant redesign and manufacturing of new tooling. Laser consolidation successfully built both types of shells from IN-625 alloy without requiring any change of expensive hard tooling. Therefore, it proved its effectiveness in supporting rapid design evolutions.

(a) (b) (c) (d)

FIGURE 16.16 Four types of sonar shells manufactured by laser consolidation process, (a) LC IN-625 FSP shell designs #1, (b) LC IN-625 FSP shell design #2, (c) LC Ti-6Al-4V FSP Shell Design #3, and (4) LC Ti-6Al-4V SMP shell design #1.

TABLE 16.4
Tensile Properties of LC IN-625 Alloy

Material		$\sigma_{0.2}$ (MPa)	σ_{UTS} (MPa)	δ (%)
LC IN-625	Horizontal	518 ± 9	797 ± 8	31 ± 2
	Vertical	477 ± 10	744 ± 20	48 ± 1
Cast IN-625		350	710	48
Wrought IN-625		490	855	50

Laser consolidated FSP shells demonstrated very good dimensional accuracy. The measurements on two LC IN-625 FSP shells (design #2) using a profile projector along the exterior fold crests revealed that the average absolute deviations from the CAD design are from 0.066 to 0.162 mm on one shell and from 0.106 to 0.128 mm on another shell. The wall thickness of the shells was inspected using a Brown & Sharpe CMM along 4 folds, 90° apart, at 3 elevations. The overall average wall thickness is 0.761 ± 0.014 mm for one shell and 0.779 ± 0.017 mm for the second shell. The excellent dimensional accuracy and wall uniformity of the LC shells successfully proved that rapid design evolutions could be readily accommodated by the novel LC process.

The LC IN-625 material exhibits very good mechanical properties. In general, the LC IN-625 material is stronger than the respective cast material and is comparable to wrought material (Table 16.4). The yield strength of LC IN-625 is around 518 and 477 MPa along the horizontal and vertical directions, respectively, which is significantly higher than cast IN-625 (350 MPa) and is comparable to wrought IN-625 (490 MPa). The tensile strength of LC IN-625 material also reveals the same trend.

The newest variant of the folded shell is the SMP [22], which looks superficially like the FSP, but with a twist applied to the shell. The SMP shell design #1 was developed as a test bed to explore shear mode projectors intended for wide band sonar and underwater communications. The SMP design requires precision, strength, corrosion resistance, and light weight. The most challenging component of the SMP is the thin-wall spiral corrugated titanium shell, which could not be readily made using conventional manufacturing techniques.

Laser consolidation was successfully used to build two SMP shells from Ti-6Al-4V alloy (Figure 16.16d). The shell is about 130 mm in height, around 82 mm in external diameter, and 0.84 mm in wall thickness. The surface finish of the LC Ti-6Al-4V shell was further improved. Visual inspection and acoustic testing reveals that these two shells are metallurgically sound, without any cracks or visible defects.

The ability to quickly and smoothly move the SMP from conceptual design to working prototype highlights the value that laser consolidation brings to product development and rapid manufacturing. For a conventional manufacturing technology, even if it could overcome the technical challenges to manufacture functional corrugated shells for FSPs, the change of design to the spiral shape needed for the SMP would introduce substantial technical challenges and cost for redeveloping manufacturing processes and tooling. The LC process readily adapted the significant design change for the SMP without any hard tooling requirements and no loss in precision or metallurgical soundness.

Although DRDC evaluated various conventional manufacturing technologies, such as electroforming, nickel vapor deposition, and hydroforming, at this moment, laser consolidation is the only proven rapid manufacturing method capable of building functional net-shape IN-625 and Ti-6Al-4V shells for the FSP and SMP.

16.2.5 SPHERICAL HOLLOW BALLS

Another demonstration piece for the LC process is a spherical hollow ball made of Ti-6Al-4V [23]. The ball requires an outside diameter of 51.501 mm with a deviation of ±0.254 mm and a wall thickness of about 0.7 mm. The ball is required to be fully sealed. The Ti-6Al-4V hollow ball is usually manufactured by forming two halves of the ball separately through machining, forming, or other appropriated methods and then welding them together, which involves many manufacturing steps. For high-end users, the thickness uniformity of the ball, especially at the weld area, is always a big concern. As a rapid manufacturing process, the LC process has the potential to directly build the spherical hollow ball.

With a five-axis motion system along with a specially developed processing procedure, laser consolidation was successfully used to build the entire functional net-shape Ti-6Al-4V spherical hollow ball from substrates of the same material in one step, which took about 1.5 h to complete. After laser consolidation, the ball was machined off from the substrate. Figure 16.17a shows a top view of the as-consolidated ball after removing the loose powder. It is evident that the as-consolidated Ti-6Al-4V ball shows good surface finish. It especially reveals that the LC process successfully closed the hollow ball without any noticeable defects. Figure 16.17b shows a bottom view of the ball, which reveals that a small portion of the substrate forms the part of the final ball after cutting off from the substrate.

The outside diameters of the LC Ti-6Al-4V ball were measured using a Mitutoyo Precision Height Gauge, 45° apart along both polar and equatorial orbits, and the data are listed in Table 16.5. The measurements were compared with the nominal diameter from the CAD model of the ball and

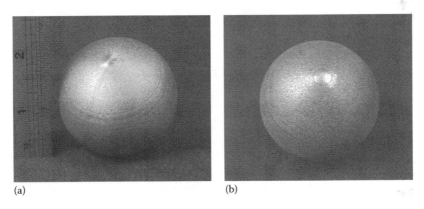

(a)   (b)

FIGURE 16.17 LC Ti-6Al-4V hollow ball, (a) top view, and (b) bottom view.

TABLE 16.5
Outsider Diameters of LC Ti-6Al-4V Ball (mm)

Angle	Measurement	CAD	Deviation
	Measured in Polar Orbit		
0°	51.575	51.501	+0.074
+45°	51.486	51.501	−0.015
+90°	51.397*	51.501	−0.104*
−45°	51.473	51.501	−0.028
	Measured in Equatorial Orbit		
0°	51.575	51.501	0.074
+45°	51.575	51.501	0.074
+90°	51.575	51.501	0.074
−45°	51.575	51.501	0.074

the deviation was calculated for each measurement. Based on the CAD data, the outside diameter of the ball should be 51.501 mm. The measurement of the LC Ti-6Al-4V ball along the polar orbit shows the maximum deviation of +0.074 mm and the minimum deviation of −0.104 mm. It should be noted that the minimum deviation was measured at the location where the ball was cut off from the substrate (Figure 16.17b), which contributes to a slightly large value of the deviation. Along the equatorial orbit, all the measurements are the same (51.575 mm in diameter), which demonstrates excellent uniformity of the ball.

It is obvious that the LC Ti-6Al-4V ball shows excellent dimensional accuracy. The deviation of the outsider diameter of the LC Ti-6Al-5V ball is in the range of +0.074 and −0.104 mm, which is significantly less than the required deviation of ±0.254 mm. If we disregard the (*) data that is affected by the cutoff process, the deviation is even smaller (+0.074 and −0.028 mm).

16.2.6 Tooling for Net-Shape Hot Isostatic Pressing

Hot isostatic pressing (HIP) of powder materials is widely used in industry for the production of critical components with high-level properties requirements. Net-shape HIP is a new technology that utilizes sacrificial tooling made from cheap low-grade steels, filled with powder, and HIP'ed to produce functional components. The low-grade steel tooling will then be etched away to "selectively" leave the net shape part. This technology provides unique advantages, such as low cost, short development cycles, and reduced design limitations of traditional manufacturing techniques. The capability to fabricate complex, monolithic shapes without welding provides an enhanced service life for the critical components in addition to the fabrication and inspection cost savings [25,26]. However, the removal of the sacrificial tooling after HIP possesses many technical challenges.

The LC process has the potential to produce net-shape tooling with the same material used for net-shape HIP. After HIP, the LC tooling will become the integrated part of the component and therefore will eliminate the need for removing tooling. In collaboration with Pratt and Whitney Rocketdyne (PWR), NRC investigated laser consolidation to build Ti-6Al-4V tooling for net-shape HIP for potential rocket engine components [23].

Initially, a cylindrical capsule was designed and manufactured to evaluate the feasibility of the LC process to make tooling for the net-shape HIP. Figure 16.18 shows a laser consolidated Ti-6Al-4V cylinder with a conical top built on a Ti-3Al-2.5V tubing with an inner diameter of 20.27 mm and a wall thickness of 1.59 mm. The total length of the LC Ti-6Al-4V portion is about 101.6 mm long. It is evident that the outside of the LC Ti-6Al-4V cylinder matches very well with the Ti-3Al-2.5V tubing substrate. The cross-sectional view (Figure 16.19a) shows that the LC Ti-6Al-4V cylinder has very uniform wall thickness and the conical top seals the tube completely. The wall of the LC Ti-6Al-4V cylinder matches well with the Ti-3Al-2.5V tubing substrate both internally and externally (Figure 16.19b). Metallurgical examination reveals that the bonding between the LC Ti-6Al-4V and the Ti-3Al-2.5V substrate is sound and laser consolidated Ti-6Al-4V portions do not have any porosity or cracks.

HIP testing of the LC Ti-6Al-4V cylindrical capsule was performed by PWR. The open end of the LC cylindrical capsule was seal welded to wrought titanium sheet stock and an evacuation tube was attached to support leak checks and hot vacuum degas of the titanium powder. HIP temperature, pressure, and time parameters were industry standards for the Ti-6Al-4V alloy and are generally conducted at a conservative margin below the beta transus temperature. Thermally induced porosity (TIP) testing, metallography, and density measurements were among the techniques used to verify the complete consolidation of the powder. These manufacturing technologies also verified the integrity of the LC structure with the HIP-consolidated powder and its ability to uniformly deform at high temperature and pressure.

Laser consolidation to build tooling for net-shape HIP was further investigated to manufacture segments of a Ti-6Al-4V impeller for demonstration. Figure 16.20 shows a CAD drawing of vane passages of an impeller. Because of the complex geometry of the component, as per PWR's

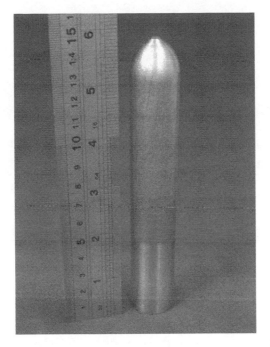

FIGURE 16.18 A laser consolidated Ti-6Al-4V cylinder with a conical top built on a Ti-3Al-2.5V tubing.

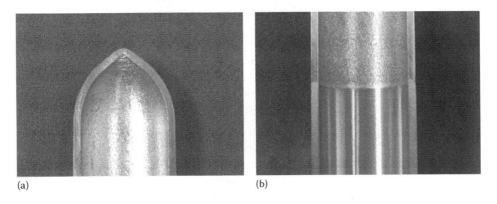

(a)   (b)

FIGURE 16.19 Cross section of LC Ti-6Al-4V capsule, (a) conical top, and (b) bonding to substrate.

FIGURE 16.20 A CAD drawing of vane passages of an impeller.

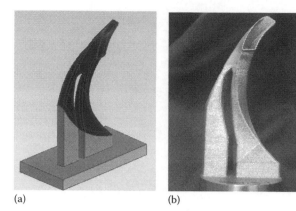

(a)   (b)

FIGURE 16.21   A vane passage segment of an impeller, (a) CAD drawing, and (b) LC Ti-6Al-4V part.

suggestion, one 1/6 section (highlighted in black) of the impeller was selected as sub-scale feature to be built using laser consolidation of the Ti-6Al-4V alloy. Based on the capability of the LC process, a CAD drawing of the vane passage segment was created (Figure 16.21a). The lower "leg" portions are supporting structures for the required vane passage (upper portion) and will be removed by CNC machining after laser consolidation. From the view of the LC process, the vane passage segment is quite complex and presents many technical challenges to build. For example, it requests a horizontal start to build the part from the substrate and finishes the building almost vertically.

The laser consolidation strategy was developed to build the vane passage segment using a five-axis motion system. It will start from the horizontal orientation to build two hollow "legs" from a Ti-6Al-4V disk in parallel. Two hollow legs will then be merged into one integrated cavity. With the help of the five-axis motion system, the cavity will be further built up gradually towards vertical direction. Figure 16.21b shows the as-consolidated Ti-6Al-4V vane passage segment. The part is about 127 mm high and 60 mm wide with a thickness of about 1.35 mm. It is evident that the LC Ti-6Al-4V part shows good quality. The preliminary results of this feasibility study have successfully proved that the LC process provides the potential to make tooling for net-shape HIP to produce complex components [23].

16.3 DISCUSSION OF METHODOLOGIES

Similar to other traditional and nontraditional technologies, additive fabrication is also an energy field manufacturing (EFM) technology that utilizes the energy field (such as laser or electron beam), information flow (such as CAD file and NC code), and material (such as metallic powder) to create the required components. A brief discussion on the methodologies relevant to the additive fabrication technology will help establish a better understanding and effective utilization of the general intelligence involved in the technology.

Specifically, additive fabrication is an interdisciplinary technology that involves design, materials, processing, sensing, control, measurements, CAD software, computer-aided manufacturing (CAM) software, simulation, etc. Therefore, it provides a unique opportunity for interdisciplinary innovations through optimizing information flow, inventing novel processes, minimizing energy consumption, and creating new materials/properties.

16.3.1 DESIGN INNOVATION—IMPROVED INFORMATION FLOW

The additive fabrication technology is a computer-aided rapid manufacturing process that minimizes the requirements for tooling or fixtures and therefore provides the flexibility to quickly change the design of the components. Thus, the lead-time to produce final parts could be reduced significantly. In addition, additive fabrication is a material addition process that is fundamentally different from traditional manufacturing processes (such as casting, forging, machining, etc.). The technology produces parts (or features on a part for certain processes) by adding material layer by layer to create the required shape. Because of the unique nature of the layered processing, this technology permits the innovative designs to create components that are difficult or even impossible to create using conventional manufacturing processes. For example, because of the layer-by-layer buildup along with the support provided by the powder-bed process, the selective laser sintering process can manufacture interconnected polymer parts (or metallic parts by the direct metal laser sintering process), such as a ball inside a ball or parts with internal passages.

Additive fabrication technology is relatively new to many people. Most engineers and designers are familiar with the conventional manufacturing technologies (such as casting, forging, and machining), but lack in-depth understanding of the capability, potential, and limitations of the additive fabrication technology. It will be critical to increase the effective communication/interactions (i.e., information flow) among designers/engineers from end users, and researchers/engineers from service providers of additive fabrication technology. Through the effective information flow, the end users will be able to create innovative designs to incorporate the advantages of the technology into their specific products.

As demonstrated in the previous section, NRC has been interacting with various organizations to explore the potential of laser consolidation technology for their specific applications through information exchange and design innovations. For example, through the collaboration, the designers from MD Robotics designed multi-function booms with internal rails to hold the electronic board inside the boom [17]. They also designed a one-piece integrated boom/housing structure (Figure 16.12) that eliminated a flanged interface, which reduces the weight and simplifies the structure. Both structures, very difficult if not impossible to manufacture using conventional methods, were successfully materialized using the LC process [17]. GE Global Research designed a demonstration airfoil with embedded cooling channels in their structural walls (Figure 16.14), which has the potential to significantly increase the turbine blade's resistance to high temperature while retaining its structural integrity. The LC process enables such an innovative structure, which is very difficult to make otherwise [18].

Researchers at DRDC worked closely with NRC to use the LC process as a rapid manufacturing technology for their design evolvement and innovation for low frequency sonar application [22]. They utilized their Mavart finite element software to design various shapes of the sonar shells and used the LC process to materialize these designs through producing net-shape functional metallic shells for evaluation and testing. Initially, they created a patented flextensional FSP that features a one-piece corrugated metal shell with an overall curvature that controls the mechanical transformer ratio. Laser consolidation successfully manufactured functional net-shape IN-625 shells for the FSP design (Figure 16.16a). In order to further reduce the resonance frequency of the FSP from 1400 to 1100 Hz, DRDC refined the outer profile of the FSP and LC process readily accommodating the design evolution (Figure 16.16b). The laser consolidated IN-625 shells perfectly yielded the resonance frequency predicted by the finite element analysis [22].

Based on the laser consolidation capability, DRDC researchers designed a Ti-6Al-4V version of the FSP with the reduced diameter to reduce the array storage volume and hydrodynamic drag (Figure 16.16c). Calibration tests successfully proved that the laser consolidated Ti-6Al-4V FSP met its design goals of resonant frequency, source level, and insensitivity to depth, showing identical resonant frequency of 1500 Hz as per the design requirement at depths of 15, 25, and 35 m [22].

DRDC researchers furthered their design innovation by creating a shear mode projector (SMP), which superficially looks like the FSP, but with a twist applied to the shell (Figure 16.16d). The most challenging component of the SMP is the thin-wall spiral corrugated titanium shell, which could not readily be made using conventional manufacturing techniques. The LC process quickly materialized the innovative design and the laser-consolidated Ti-6Al-4V shells enabled the field testing of the SMPs [22].

Although DRDC evaluated various conventional manufacturing technologies, such as electro-forming, nickel vapor deposition, and hydroforming, at this moment, laser consolidation is the only proven rapid manufacturing method capable of building functional net-shape IN-625 and Ti-6Al-4V shells for the FSP and SMP designs. The ability to quickly and smoothly move various types of sonars from conceptual design to a working prototype highlights the value that additive fabrication technology (laser consolidation in this case) brings to design innovation, product development, and rapid manufacturing.

16.3.2 Novel Processes—Interdisciplinary Innovation

The additive fabrication technology is a nontraditional manufacturing technology that provides a unique opportunity for interdisciplinary process innovation that is not readily available using the traditional technology. For example, the additive fabrication process enables the manufacturing of certain types of complex parts in one step, which otherwise requires multiple manufacturing processes.

The laser consolidation of spherical hollow balls is a good example [23]. The Ti-6Al-4V hollow ball is usually manufactured by forming two halves of the ball separately through machining, forming, or other appropriated methods and then welding them together, which involves many manufacturing steps. For high-end users, the thickness uniformity of the ball, especially at the weld area, is always a big concern. As a rapid manufacturing process, the LC process provides the capability of building the spherical hollow ball directly from a CAD file in one step (Figure 16.17a). The dimensional measurements show that the maximum deviation of the external diameter of the as-consolidated Ti-6Al-4V ball is significantly less than the allowable dimensional deviation of the design.

The laser consolidation of integrated tooling for powder materials processing is another example of process innovation [23]. Net-shape HIP utilizes sacrificial tooling made from cheap low-grade steels, filled with powder, and HIP'd to produce functional components. However, the low-grade steel tooling has to be etched away to leave the selective net-shape part, which possesses many technical challenges. Laser consolidation has been successfully demonstrated to produce net-shape tooling with the same material used for net-shape HIP (Figure 16.22). After HIP, the LC tooling will become the integrated part of the component and therefore will eliminate the need for removing tooling. The benefits of this process innovation take advantage of the uniformly laser-consolidated thin-walled structures to provide a simplification of the deformation modeling (rheology) currently employed to predict component shapes after HIP. This could eliminate both the difficult structural modeling and the requirement for bulk tooling to prevent part distortion during HIP consolidation.

16.3.3 Green Manufacturing—Minimized Material/Energy Consumption

Additive fabrication is a green manufacturing technology that creates complete net-shape or near-net-shape functional components (or features on existing parts for certain processes) by adding material layer by layer to create the required shape. Unlike conventional machining processes that

FIGURE 16.22 LC Ti-6Al-4V capsule after HIP.

create a significant amount of chips to produce parts, additive fabrication can produce net-shape or near-net-shape components directly from metallic feedstock materials (powder or wire). Therefore, it can significantly reduce the waste of materials through the reduction of machining chips. For expensive aerospace materials (such as titanium alloys or cobalt- or nickel-based superalloys), cost saving for the materials is especially important. Even further, the additive fabrication technology will also allow the reuse of waste materials (such as using metallic powder produced from machining chips of expensive materials).

The additive fabrication is also a one-step computer-aided rapid manufacturing technology that minimizes the requirements for tooling or fixtures and, therefore, significantly reduces the materials used to produce tooling in conventional manufacturing processes. For functional prototyping and low-volume components, the additive fabrication can provide significant cost savings for tooling and fixtures along with a shortened production cycle. As demonstrated in the previous section, laser consolidation successfully manufactured 17 pieces of sonar shells from four different designs and two totally different materials (IN-625 and Ti-6Al-4V) without any tooling. On the contrary, for a conventional manufacturing technology (such as hydroforming), even if it could overcome the technical challenges of manufacturing functional corrugated shells for FSPs, the redevelopment of the manufacturing process for each design and each material would induce substantially high tooling costs along with significant development time.

In addition, some additive fabrication processes (such as the laser cladding–based process) provide the unique capability to repair damaged or machining-undersized components and tooling that can not be repaired using conventional manufacturing technologies, which can significantly reduce waste materials for the manufacturing and maintenance of components/tooling for various industries [27].

As a green manufacturing technology, additive fabrication enables the significant reduction of material consumption through reducing or even eliminating machining, minimizing tooling, recovering machining under-sized components, repairing damaged components, and reusing waste materials, etc. The minimized material consumption will significantly reduce the energy consumed in the mining and raw material processing industry and will also reduce the impact to the environment created by these industry.

16.3.4 UNIQUE FUNCTIONALITY—ENGINEERED MICROSTRUCTURES AND MATERIALS

Additive fabrication technology utilizes lasers, electron beams, etc. as energy sources. For laser cladding–based 3D manufacturing processes, the rapid solidification inherent to the process will result in unique microstructures, which will in turn generate unique and favorable functionalities. These performance gains are particularly valuable in demanding aerospace environments like turbine engines, structural components, or landing gears.

IN-738 is a Ni-based precipitation hardening superalloy with excellent creep strength and hot corrosion resistance. It is being used for hot section turbine airfoils and for hot corrosion-prone applications, such as industrial and marine gas turbine engines. Laser-consolidated IN-738 demonstrated unique microstructures, such as directionally solidified microstructure (very fine columnar dendrites growing almost parallel to the direction of buildup), fine carbides (uniformly distributed at interdendritic regions), and fine γ' particle precipitation. These unique microstructural features effectively block the dislocation movement at elevated temperatures. As a result, laser consolidated IN-738 demonstrated excellent creep life. Stress rupture testing at 1010°C (1850°F) and 55 MPa (8 ksi) reveals that the average stress rupture life of LC IN-738 specimens is 423 h, which is about 150% longer than that of cast IN-738 baseline samples [14].

H13 is a hot work tool steel that has good resistance to thermal fatigue, erosion, and wear and is widely used for making molds and dies. Laser consolidated (LC) H13 tool steel shows a directionally solidified dendritic microstructure: layered features with very fine columnar dendrites (around 1–2 μm in dendritic arm spacing) aligned vertically. Cellular features have been observed along the transverse cross section. The unique microstructure of LC H13 contributes to its excellent mechanical properties. In addition to substantially higher tensile and yield strengths of the as-consolidated H13 than the annealed cast or wrought H13 and comparable to the hardened wrought H13, the sliding wear resistance of the as-consolidated H13 samples is also about three times better than the hardened wrought H13 and about one order of magnitude better than the annealed wrought H13 [20].

Functionally graded material is another example of engineered materials. Due to the nature of the layered buildup, additive fabrication enables the manufacturing of functionally graded materials: starting from material A and ending with material B with gradual composition transition in between. Functionally graded materials offer unique functionalities that are difficult to provide using conventional manufacturing technologies. For example, the laser cladding–based process has been used to fabricate load bearing Ti-6Al-4V implants with designed porosities up to 70 vol.% to reduce stress-shielding, while functionally graded Co–Cr–Mo coating was produced on porous Ti-6Al-4V to minimize the wear-induced osteolysis along with excellent bone cell–materials interactions [28].

The laser cladding–based process has also been used to produce tooling with functionally graded materials for plastic injection molding to speed up cooling, improve part quality, and cut cycle time. For example, LENS was used to build a copper core into a mold with a 420 stainless steel surface with gradual transitions between them. Copper has a coefficient of thermal expansion twice as much as tool steel. With the gradual material transitions, the thermal stress (and potential failure) at the interface between the two materials can be reduced significantly. During molding operations, the thermally conductive core rapidly conducts heat away from the steel surface and transfers it to conformal cooling channels. This functionally graded mold reduces cycle times by 15% [29].

16.4 FUTURE TREND—TOWARD INTELLIGENT 3D MANUFACTURING

As demonstrated in the previous sections, 3D additive fabrication technology provides the unique capability to produce net-shape functional components directly from CAD models without moulds or dies, which enables the rapid manufacturing of small quantity metallic components with short delivery times and reduced production cost. It is expected to make a huge impact on manufacturing in a wide range of industries, especially, for aerospace and deference applications. However, this technology is still emerging and much more developmental work is needed.

It is expected that more industrial materials will be investigated using 3D additive fabrication technology. A comprehensive database of materials, its microstructure, and its mechanical and other properties will be generated through collaboration among research organizations, universities, and industries. The processing capability of the technology will be further improved to build more complex net-shape functional components that can not be produced currently. The processing speed will be substantially improved to meet the industrial needs for production. The adoption of advanced sensing and control capability and software development will further improve the robustness of the process in the production environment.

With the increased interest from various industries, it is expected that more industrial applications will be developed, especially in the fields of aerospace, medical devices, and tooling. In combination with innovative design, 3D additive fabrication technology provides the unique capability to produce complex components with delicate details that are very difficult or even impossible to make using conventional manufacturing processes. In combination with other manufacturing technologies, laser cladding–based 3D manufacturing technology can also build integrated delicate features on components manufactured using conventional low cost methods. Compared with the traditional welding process, the heat input induced by the laser cladding process to the substrate is minimal. Therefore, the laser cladding–based 3D fabrication process also provides the possibility of building unique features on thin-walled components to provide additional functionality and reduce manufacturing time and cost.

With further evolvement, it is expected that a new generation of intelligent 3D manufacturing technology will be created, which may dramatically change the existing manufacturing process, procedure, and culture. The following is a snapshot of several potential situations.

16.4.1 GLOBAL MANUFACTURING NETWORK

Three-dimensional additive fabrication technology does not require any molds or dies. Therefore, it provides the possibility of designing and manufacturing components directly through the Internet to combine the real strength of virtual manufacturing and rapid manufacturing. It also provides the flexibility of changing the design quickly to make functional components that meet various industrial demands to produce customized components with significantly reduced lead-times.

It is expected that small quantity manufacturing will receive increasing demands and a global manufacturing network will be established to address such needs for customized rapid manufacturing. The network consists of many manufacturing hubs along with very powerful centralized service providers. Many huge manufacturing hubs with a variety of 3D additive fabrication systems and other conventional manufacturing technologies will be allocated in all major cities across North America, South America, Europe, Asia and the Pacific region, and Africa. An Internet-based service network will connect customers with these manufacturing hubs along with administration, design offices, CAD/CAM software, and service and maintenance providers.

When a company requires a certain amount of components, it can submit its request (quantity, materials, CAD models, etc.) to the central service of the network through the Internet. Based on the availability of the equipment and capacity, a work order will be automatically sent to specific 3D additive fabrication systems in the nearest manufacturing hubs. Service providers will create the required CAM programs based on the part design and selected manufacturing processes to process the work order. Parts will then be manufactured by the selected 3D additive fabrication systems following the schedule. After conducting the required post-processing (such as heat treatment, final machining, surface finishing, inspection) in the manufacturing hubs, the parts will be sent to the customer along with an automatically generated report that includes its identification, manufacturing information, and inspection results.

It is expected that such a manufacturing network will make a huge impact on the manufacturing practice in a wide range of industries in the not-so-distant future.

16.4.2 MANUFACTURING IN SPACE

Space exploration is the next frontier for finding required energy and resources for human beings to survive in the future. Manufacturing in space is a dream of human beings to support our space exploration activities. Space-based manufacturing mission concepts have focused on the development of terrestrial and space-based infrastructures to enable scientific and commercial missions in space, which includes terrestrial applications (Earth observation for resource management, global monitoring, and security); space exploration (extended duration missions); in situ resource utilization (ISRU) to provide a permanent, growing, autonomous manufacturing capability; and replicating systems. Although the development of space-based manufacturing capability has a long history, 3D additive fabrication technology has many unique advantages in this regard [30].

16.4.2.1 Materials Availability

Three-dimensional additive fabrication technology allows for the use of a broad range of feedstock materials (metals and alloys) in simple forms of feedstock (powder or wire). Operation in a vacuum is also a preferred environment. The NASA Apollo missions found that there are abundant metallic elements available in the soil of the Moon (including iron, aluminum, titanium, and magnesium) [31] and similar materials are available on Mars [32]. Various books and papers have been published to explore mining on the Moon and Mars [32,33]. In the near future, it is expected that mining on the Moon and then Mars will become a reality. Compared with other conventional metallic forms (such as ingot, plate, bar, and sheet) used by conventional manufacturing processes, powder may be the easiest form that can be produced on the Moon directly from lunar soil. Therefore, 3D additive fabrication processes have easy access to the metallic material readily available on the Moon and others planets.

16.4.2.2 No Hard Tooling Requirements

Conventional manufacturing processes (such as casting, molding, and deformation) require molds and dies to produce the required parts and many intermediate steps are required to support these processes. Although these conventional manufacturing processes are widely used in our daily earth living, it will be especially critical to minimize manufacturing steps in space. Three-dimensional additive fabrication technology is a free-form fabrication process that produces functional components in a layer-by-layer fashion directly based on CAD data without the need for molds or dies. The technology provides unprecedented flexibility and allows a single manufacturing station to produce an enormous variety of part geometries. Therefore, the technology is especially suitable for space manufacturing.

16.4.2.3 Minimized Waste Materials

The conventional machining process produces a part by removing materials from a feedstock (such as a plate, bar, or rod). For typical aerospace components manufactured by the machining process, the buy-to-fly ratio (the proportion of originally purchased material that is finally built into an aircraft) can be less than 10%, while the majority of the feedstock becomes waste material in the form of chips. In space manufacturing, material waste has to be minimized. The 3D additive fabrication process produces parts by material addition instead of material removal. It builds a part layer by layer exactly following the desired shape described by a CAD model, which produces almost no waste material, making it a very efficient method for space manufacturing.

16.4.3 Desktop Manufacturing

With the further involvement in the 3D additive fabrication technology, it is expected that more and more affordable desktop manufacturing systems will be available on the market in the near future. These desktop systems based on various 3D additive fabrication processes (such as 3D printing, SLS, and laser cladding) will be able to produce relatively small sizes and functional components made of plastics as well as metals. With relatively simple post-processing, these parts can be directly used for design verification, functional prototypes, and as real functional components.

It is expected that these desktop manufacturing systems will be widely used in various industry sectors. It will not be so surprising if someday people utilize these desktop manufacturing systems to make functional metallic parts to materialize their designs as today we use printers to print out our documents. The desktop manufacturing systems will be used not only in design offices but also on shop floors (or even in ships). It is expected that this CAM technology will make a huge impact on manufacturing as well as in our daily lives in the future.

16.5 FUTURE RESOURCES

- Annual International Congress on Applications of Lasers and Electro-Optics (ICALEO), organized by the Laser Institute of America (http://www.icaleo.org/)
- Annual International Solid Freeform Fabrication Symposium (https://utwired.engr.utexas.edu/lff/symposium/)
- Annual RAPID Conference & Exposition, organized by the Society of Manufacturing Engineers (http://www.sme.org/cgi-bin/get-event.pl?–001799-0)
- Annual Rapid Manufacturing Conference (http://www.rm-conference.com/)

- Annual Wohlers Report, Wohlers Associates (http://www.wohlersassociates.com/)
- TCT Magazine (http://www.tctmagazine.com/x/35.html)
- The Global Alliance of Rapid Prototyping Associations (GARPA) (http://www.garpa.org/main/)

QUESTIONS

Q.16.1 What are the major features of additive fabrication technology? What advantages can the additive fabrication technology offer over conventional manufacturing technologies?

Q.16.2 What are the major differences among printing-based processes, SLA, SLS, and laser cladding–based 3D manufacturing technologies? What are the advantages and drawbacks of each type of technology?

Q.16.3 Identify one or more potential applications of the additive fabrication technology based on your experience. Describe your rationales for your selection in terms of technical and cost considerations.

Q.16.4 According to Section 10.2, the LC process provides unique advantages of producing functional net-shape components or building features on existing components for various applications. Can you think of any application for which the LC process can be used to address your specific needs?

Q.16.5 Section 10.3 discussed the methodologies relevant to the additive fabrication technology (such as design innovation, novel processes, green manufacturing, and engineered functionality). Can you identify one specific example, based on your experience, in which the general intelligence involved in the additive fabrication technology can be effectively utilized to generate results that are difficult to obtain through conventional thinking?

Q.16.6 Additive fabrication technology has great potential in the future as briefly discussed in Section 10.4. Based on the information presented in this chapter and your experience, think about the future of the technology and list what you can do to help to achieve the goal.

Q.16.7 The additive fabrication processes rely on multiple energy fields (such as lasers or electron beams, heating, mechanical motion, chemical and physical reactions, etc.). Think about how these energy fields can be integrated and controlled intelligently in the future.

ABOUT THE CONTRIBUTING AUTHOR

Dr. Lijue Xue obtained his master of engineering degree in materials engineering from Shanghai Jiao Tong University (China) in 1987 and his PhD in mechanical engineering from Carleton University (Ottawa, Ontario, Canada) in 1994. He is currently a senior research officer with the Industrial Material Institute of National Research Council Canada (NRC-IMI) and has more than 27 years experience in conducting innovative R&D on materials and manufacturing processes. He is leading the development of a unique laser consolidation process to produce net-shape functional components. His research interests also include laser surface modification, the joining of dissimilar lightweight materials, materials development, characterization, and property evaluation. His team has established strong collaborations with industry and research organizations (including Pratt & Whitney

Canada, Pratt & Whitney Rocketdyne, Honeywell, Airbus, GE Global Research, MD Robotics, Messier-Dowty, DRDC, Canadian Space Agency, etc.). He has published over 75 papers in journals and conference proceedings and prepared more than 50 technical reports for various clients.

REFERENCES

1. Wohlers, T., Wohlers Report 2007—State of the Industry Annual Worldwide Progress Report, Wohlers Associates, Fort Collins, CO, 2007, 10 pp.
2. Jacobs, P., *A Brief History of Rapid Prototyping & Manufacturing: The Early Years, Metal Powder Deposition for Rapid Manufacturing*, D. Keicher, J. Sears, and J. Smugeresky, editors, Metal Powder Industries Federation, San Antonio, TX, 2002, pp. 1–4.
3. Wohlers, T., Wohlers Report 2008—State of the Industry Annual Worldwide Progress Report, Wohlers Associates, Fort Collins, CO, 2008, 39 pp.
4. L. M. Herman, 3D printers lead growth of rapid prototyping, *Plastics Technology*, 50 (8), 43–46, 2004.
5. Optomec Web site: http://www.optomec.com
6. Mironov, V., Kasyanov, V., and Markwald, R., Bioprinting: Directed tissue self-assembly, *Chemical Engineering Progress*, 103 (12), 2007.
7. Mironov, V., Reis, N., and Derby, B., Bioprinting: A beginning, *Tisssue Engineering*, 12 (4), 631–634, 2006.
8. 3D Systems Web site: http://www.3dsystems.com/company/index.asp
9. EOS Web site: http://www.eos.info
10. Arcam Web site: http://www.arcam.com/index.asp
11. POM Group Web site: http://www.pomgroup.com/
12. Xue, L. and Islam, M.U., Free-form laser consolidation for producing metallurgically sound and functional components, *Journal of Laser Applications*, 12 (4), 160–165, 2000.
13. Xue, L., Chen, J.-Y., and Islam, M.U., Functional properties of laser consolidated wear resistant stellite 6 alloy, in *Powder Metallurgy Alloys and Particulate Materials for Industrial Applications*, Alman, D.E. and Newkirk, J.W., editors, TMS, Warrendale, PA, 2000, pp. 65–74.
14. Xue, L., Chen, J.-Y., Islam, M.U., Pritchard, J., Manente, D., and Rush, S., Laser consolidation of IN-738 alloy for repairing cast IN-738 gas turbine blades, in *Proceedings of 20th ASM Heat Treating Society Conference*, St.-Louis, MO, October 8–12, 2000, pp. 1063–1071.
15. Xue, L., Chen, J.-Y., and Theriault, A., Laser consolidation of Ti-6Al-4V alloy for the manufacturing of net-shape functional components, in *Proceedings of 21st International Congress on Applications of Lasers and Electro-Optics (ICALEO 2002)*, Scottsdale, AZ, October 14–17, 2002, pp. 169–178.
16. Xue, L., Theriault, A., Chen, J.-Y., Islam, M.U., Wieczorek, A., and Draper, G., Laser consolidation of CPM-9V tool steel for manufacturing rotary cutting dies, in *Proceedings of 10th International Symposium on Processing and Fabrication of Advanced Materials, 2001 ASM Materials Solution Conference and Exposition*, Indianapolis, IN, November 5–8, 2002, pp. 361–376.
17. Xue, L., Theriault, A., Rubinger, B., Parry, D., Ranjbaran, F., and Doyon, M., Investigation of laser consolidation process for manufacturing structural components for advanced robotic mechatronics system, in *Proceedings of 22nd International Congress on Applications of Lasers and Electro-Optics (ICALEO 2003)*, Jacksonville, FL, October 13–16, 2003, pp. 134–143.
18. Xue, L., Theriault, A., Islam, M.U., Jones, M., and Wang, H.P., Laser consolidation of Ti-6Al-4V alloy to build functional net-shape airfoils with embedded cooling channels, in *Proceedings of the 23rd International Congress on Applications of Lasers and Electro-Optics (ICALEO' 2004)*, San Francisco, CA, October 4–7, 2004.
19. Xue, L., Chen, J.-Y., and Theriault, A., Laser consolidation of Al 4047 alloy, in *Proceedings of 24th International Congress on Applications of Lasers and Electro-Optics (ICALEO 2005)*, Miami, FL, October 31–November 3, 2005, pp. 344–351.
20. Xue, L., Theriault, A., Wang, S-H., and Chen, J.-Y., Laser consolidation of functional shell structure, in *Proceedings of the 5th International Workshop on Advanced Manufacturing Technologies (AMT 2005)*, London, Canada, May 16–18, 2005, pp. 231–236.
21. Theriault, A., Xue, L., and Chen, J., Laser consolidation of Ti-6Al-4V alloy, in *Proceedings of the 5th International Workshop on Advanced Manufacturing Technologies (AMT 2005)*, London, Canada, May 16–18, 2005, pp. 267–272.

22. Xue, L. and Purcell, C., Laser consolidation of net-shape shells for flextensional sonar projectors, in *Proceedings of the 25th International Congress on Applications of Lasers and Electro-Optics* (*ICALEO 2006*), Scottsdale, AZ, October 30–November 2, 2006, pp. 686–694.

23. Xue, L., Li, Y., Van Daam, T., and Bampton, C., Investigation of laser consolidation for manufacturing functional net-shape components for potential rocket engine applications, in *Proceedings of the 26th International Congress on Applications of Lasers and Electro-Optics* (*ICALEO 2007*), Orlando, FL, October 29–November 1, 2007, pp. 161–169.

24. Xue, L., Chen, J., Wang, S.-H., and Li, Y., Laser consolidation of Waspalloy and IN-718 alloys for making net-shape functional parts for gas turbine applications, in *Proceedings of 27th International Congress on Applications of Laser & Electro-Optics* (*ICALEO 2008*), Temecula, CA, October 20–23, 2008, pp. 255–263.

25. Bampton, C., Goodin, W., Van Daam, T., Creeger, G., and James, S., Net-shape HIP powder metallurgy components for rocket engines, in *Proceedings of 2005 International Conference on Hot Isostatic Pressing*, Paris, France, May 22–25, 2005, p. 53.

26. Samarov, V., HIP of net shape parts for critical applications from advanced powder materials, in *Proceedings of NATO AVT-139 Specialists, Meeting on Cost Effective Manufacture via Net Shape Processing*, Amsterdam, the Netherlands, May 15–19, 2006, pp. 4-1–4-8.

27. Hedges, M. and Calder, N., Near net shape rapid manufacture and repair by LENS®, in *Proceedings of NATO AVT-139 Specialists Meeting on Cost Effective Manufacture Via Net Shape Processing*, Amsterdam, the Netherlands, May 15–19, 2006.

28. Bandyopadhyay, A., Krishna, B.V., Xue, W., and Bose, S., Application of Laser Engineered Net Shaping (LENS) to manufacture porous and functionally graded structures for load bearing implants, *Journal of Materials Science: Materials in Medicine*, 20 (1), 29–34, 2009.

29. Grylls, R., Additive toolmaking process cuts cycle times, *Mold Making Technology*, January 2003.

30. Jessen, S. (MDA) and Xue, L., Development of a space manufacturing facility for in-situ fabrication of large space structures, in *Proceedings of 57th International Astronautical Congress* (*IAC 2006*), Valencia, Span, October 2–6, 2006.

31. Morrison, G.H. et al., Multielement analysis of lunar soil and rocks, *Science*, 167 (3918), 505–507, 1970.

32. Zubrin, R., *The Case for Mars*, Touchstone, New York, 1997, pp. 199–205.

33. David, L., Mining the Moon, the gateway to Mars, http://www.space.com/businesstechnology/technology/moon_mining_041110.html, posted: November 10, 2004.

17 Selected Topics in Biomedical Engineering

Ronald Xu

CONTENTS

17.1 BIOMEDICAL ENGINEERING

Biomedical engineering is an emerging field that integrates engineering advances with medical and biological sciences to help improve patient health care and quality of life. Biomedical engineers work closely with health care providers, such as physicians, nurses, therapists, and technicians, to analyze and solve problems at the interface of biology and medicine. Due to the diversity of clinical requirements and the complexity of biological systems, biomedical engineering projects typically require multidisciplinary teamwork involving clinical practitioners, engineers, scientists, and pharmacologists. To ensure the clinical safety and efficacy of biomedical systems, government regulations and industrial standards have to be considered at the early stage. Extensive benchtop research, cell/tissue validation, and preclinical studies are necessary before the successful translation of biomedical engineering techniques from the benchtop to the bedside. In the field of biomedical engineering, there is continuous change and the creation of new areas resulted from emerging clinical needs and technical advances. Therefore, an inclusive description of each research area is almost impossible. Recently, the National Research Council listed the following seven representative biomedical engineering domains: bioelectrical and neuroengineering; bioimaging and biomedical optics; biomaterials; biomechanics and biotransport; biomedical devices and instrumentation; molecular, cellular, and tissue engineering; and systems and integrative engineering. Each domain is briefly introduced below.

Bioelectrical and neuroengineering: Experiments, models, and simulations are used to elucidate the basic principles of excitable tissues, including the heart and brain, from the molecular to the systems level.

Bioimaging and biomedical optics: Knowledge of a unique physical phenomenon (sound, radiation, magnetism, photonics, etc.) is combined with high-speed electronic data processing, analysis, and display to generate an image for the identification of the structure and function of biological objects and for diagnosis and therapeutic purposes.

Biomaterials: Both living tissue and artificial materials are studied for clinical applications such as implantation. Biomaterials need to be nontoxic, noncarcinogenetic, chemically inert, stable, and mechanically strong enough to withstand the repeated forces of a lifetime.

Biomechanics and biotransport: Mechanical and transport principles, such as statics, dynamics, fluids, solids, thermodynamics, and continuum mechanics, are applied to living organisms in order to solve biological and medical problems.

Biomedical devices and instrumentation: Mechanical, optical, electrical, and measurement techniques are applied to develop medical devices and instruments that satisfy diagnostic and therapeutic needs with clinical safety and efficacy.

Molecular, cellular, and tissue engineering: Anatomy, biochemistry, and mechanics of cellular and subcellular structures are studied to understand disease processes and to intervene at the microscopic level.

Systems and integrative engineering: Engineering strategies, techniques, and tools are utilized to gain a comprehensive and integrated understanding of the function of living organisms ranging from bacteria to humans.

The above domains (or specialty areas) are not independent but are interwoven with each other to serve the common goal of improving health care and quality of life. For example, a biomedical imaging project typically involves the development of medical devices and instruments to characterize the structural and functional properties of biological systems at the molecular, cellular, and tissue levels and to enhance the safety and efficacy of a clinical diagnostic or therapeutic procedure. In this regard, a typical biomedical imaging project requires the seamless collaboration of an interdisciplinary team involving clinical practitioners, engineers, scientists, and regulatory personnel with diverse expertise and a professional background. Frequent information change and resource sharing are necessary within this interdisciplinary team in order to achieve common project milestones and deliverables. However, resource conflicts and knowledge barriers between individual team members may present major stumbling blocks for successful communication and efficient teamwork. Is there a systemic method for organizing the information flow, energy flow, material flow, and process flow involved in a multidisciplinary team for the optimal project outcome? This challenging question remains unsolved in the biomedical engineering community. The concept of an intelligent energy field may offer potential solutions to this challenging question. The objectives of this chapter are to introduce the state-of-art work in the field of biomedical engineering and to motivate the readers to seek systematic solutions from the viewpoint of an intelligent energy field. Since biomedical engineering is such a broad field, we do not intend to cover every discipline of the field. Instead, we will use diffuse optical imaging and its application in cancer detection to exemplify the multidisciplinary challenges that face the field of biomedical engineering and the need for innovative thinking of systemic methodologies.

17.2 BIOIMAGING

Bioimaging and biomedical optics is one representative discipline of biomedical engineering. Bioimaging represents the science and technology of developing and improving imaging platforms to identify the structure and function of biological objects and to serve diagnostic and therapeutic purposes. Major bioimaging modalities include computerized tomography (CT) [1], ultrasound (US) [2], magnetic resonance imaging (MRI) [3], positron emission tomography (PET) [4], single-photon emission computerized tomography (SPECT) [5], biomedical optical imaging and spectroscopy (BOIS) [6], and photoacoustic tomography (PAT) [7]. Each modality is briefly introduced below.

Computerized tomography (CT): CT is the technique of obtaining the depth-resolved spatial distribution of a biological object by visualizing the individual cross sections (Figure 17.1). In a CT system, the x-ray source and detector are rotated around the patient and the detector measures the amount of radiation absorbed in each rotational direction. The information is then used to reconstruct a sectional image of the plane that was exposed to radiation [8].

Ultrasound (US) imaging: US (also called diagnostic sonography or ultrasonography) uses an ultrasound transducer or a transducer array to send and detect acoustic energy in order to visualize subcutaneous body structures including tendons, muscles, joints, vessels, and internal organs for possible pathology or lesions (Figure 17.2). US is one of the most widely used clinical diagnostic tools and is relatively inexpensive and portable compared with other imaging modalities.

Magnetic resonance imaging (MRI): MRI (also called nuclear magnetic resonance imaging) detects the presence of hydrogens (protons) by subjecting them to a large magnetic field to partially align the nuclear spins, exciting the spins with radio frequency (RF) fields, and then detecting the rotating magnetic field induced by proton relaxation (Figure 17.3). The signal can be manipulated by additional magnetic fields to build up enough information to construct an image of the body with good contrast and high resolution.

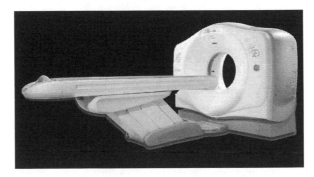

FIGURE 17.1   High-definition CT scanner. (Courtesy of GE Healthcare, Waukesha, WI.)

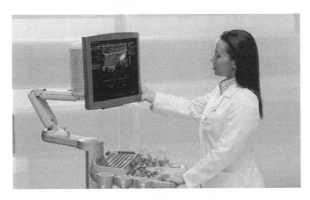

FIGURE 17.2   Acuson S2000 ultrasound system. (Courtesy of Siemens Healthcare, Malvern, PA.)

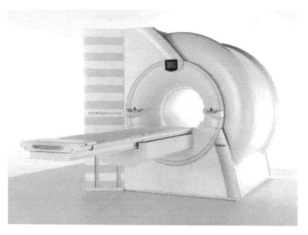

FIGURE 17.3   MAGNETOM Symphony 1.5T MRI system. (Courtesy of Siemens Healthcare, Malvern, PA.)

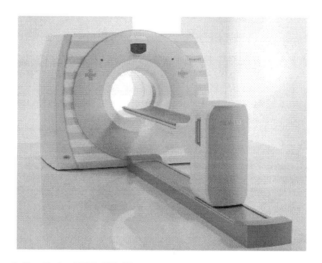

FIGURE 17.4   Biograph TruePoint PET-CT. (Courtesy of Siemens Healthcare, Malvern, PA.)

Positron emission tomography (PET): PET produces a 3D image of functional processes in the body by detecting pairs of gamma rays emitted indirectly by a positron-emitting tracer. It typically uses fluorine-18 radiolabeled fluorodeoxyglucose (18F-FDG) to detect the abnormal glucose uptake by high-glucose-using cells such as the brain, kidney, and cancer cells. PET has been integrated with CT in a hybrid system (PET–CT) for the simultaneous acquisition and fusion of the structural and functional images of the body (Figure 17.4).

Single-photon emission computerized tomography (SPECT): SPECT uses a gamma camera to acquire multiple 2D images of a radioactive isotope from different angles for the reconstruction of a 3D distribution of radioactive isotopes. SPECT is similar to PET in its use of radioactive tracer material and detection of gamma rays. However, the tracer used in SPECT emits gamma radiation that is measured directly, whereas the PET tracer emits positrons that annihilate with electrons up to a few millimeters away, causing two gamma photons to be emitted in opposite directions [9].

Biomedical optical imaging and spectroscopy (BOIS): BOIS uses the entire spectral range of electromagnetic radiation for screening, diagnosis, treatment, and therapeutic assessment of clinical anomalies (Figure 17.5). In recent years, BOIS has played more and more important roles in surgical

FIGURE 17.5 IVIS Lumina II optical imaging system. (Courtesy of Caliper LifeSciences, Hopkinton, MA.)

planning, intraoperative image guidance, and postoperative assessment as the field of image-guided intervention has developed at an impressive rate.

Photoacoustic tomography (PAT): PAT images the functional properties of biological tissue based on the photoacoustic effect. Tissue is first irradiated by a short-pulsed light beam. Part of the electromagnetic energy is absorbed by chromophores in the heterogeneous tissue and converted into heat. Heat-induced thermoelastic expansion generates a pressure wave that propagates in biologic tissue as a US wave. US transducers are used to detect this pressure wave and to reconstruct the absorptive heterogeneity.

The above bioimaging systems have different performance characteristics and different clinical applications. For example, CTs, USs, and MRIs are typically used for the structural imaging of anatomic characteristics; PETs, SPECTs, and BOISs are typically used for functional and molecular imaging. PAT combines the functional sensitivity of BOIS and the spatial resolution of US. Blood oxygen level-dependent (BOLD) MRIs and MRI spectroscopies add functional and molecular capabilities to MRIs [10]. The emergence of disease-specific microbubbles further extends the use of clinical US for functional and molecular imaging applications [11]. Since each of the above bioimaging modalities has advantages and limitations, not a single modality is able to satisfy all the clinical needs. Therefore, multimodal imaging modalities and hybrid systems are emphasized in the biomedical imaging community [12,13]. Multimodal imaging refers to multiple imaging modalities that are integrated within a single device for the simultaneous acquisition of anatomical, functional, and/or molecular composition information of biological tissue. Clinically available multimodal imaging systems include SPECT–CT [14] and PET–CT [15]. Other emerging multimodal techniques include MRI–PET [16], MRI–near infrared (NIR; a BOIS modality using near infrared light) [17], NIR–US [18], and mammography–NIR [19].

Despite the successful demonstration of technical feasibilities on animal models and human subjects, the extensive application of the above multimodal imaging systems in a clinical setting is hindered by the following technical limitations: (1) lack of image co-registration accuracy presents a major technical challenge for hardware design; (2) computational cost for image fusion presents another challenge for real-time image processing; (3) boundary mismatches between tissue structural and functional anomalies significantly affect image co-registration and image reconstruction accuracies; (4) many imaging modalities, especially optical imaging techniques, have to compromise between achievable image depth and spatial resolution; (5) biological interrelationships between biomarkers targeted by different modalities, such as FDG metabolism targeted by PET scan and oxy/deoxy-hemoglobin targeted by NIR, are not clear yet; (6) motion-induced imaging artifacts are difficult to correct due to the lack of information about tissue dynamic responses to external stimuli; and (7) typical multimodal imaging systems represent significantly added cost, equipment size, and complexity.

17.3   BIOMEDICAL OPTICAL IMAGING AND SPECTROSCOPY (BOIS)

BOIS is the field that involves generating and harnessing light and other forms of radiant energy for medical diagnostic and therapeutic applications [20]. The field of BOIS studies light–tissue interaction by using a wide variety of methods and technologies such as lasers and other light sources, fiber optics, electro-optical instrumentations, sophisticated microelectromechanical systems, and nanosystems. The major advantages of BOIS modalities include low cost, portability, no radiation hazard, molecular sensitivity, and real-time noninvasive measurements of multiple tissue parameters. However, clinical acceptance of optical imaging is hampered by the lack of calibration standards and validation techniques, the regulation hurdle, and toxic concerns about imaging contrast agents, the poor understanding of the interrelationship of multiple biologic parameters, the low spatial resolution for thick tissue imaging, and the inconsistent measurements associated with motion artifacts, test condition variations, physiologic changes, and inter-patient variations. To facilitate clinical translation from the benchtop to the bedside, it is important to develop quantitative methods and standards for the accurate detection of light–tissue interaction and appropriate interpretation of their biological and etiological correlations.

17.4   LIGHT–TISSUE INTERACTION

Absorption and scattering processes govern the light propagation in biological tissues. Absorption is the extraction of energy from light by molecular species. Major absorbing chromophores in biological tissue include oxygenated hemoglobin, deoxygenated hemoglobin, myoglobin, H_2O, eumelanin, pheomelanin, cytochrome oxidase, bilirubin, and lipids [21,22]. Absorption can be mediated by either an irradiative process where light energy is converted to thermal energy or a radiative process where absorption introduces the fluorescence emission of longer wavelengths (i.e., lower energy).

Scattering is the redirection of incident light over a range of angles due to tissue heterogeneities, such as the refractive index mismatch between subcellular organelles and the surrounding cytoplasm. Scattering events can be further classified into elastic scattering (i.e., the wavelength of the scattered light is the same as that of the incident light) and inelastic scattering (i.e., the wavelength of the scattered light is different from that of the incident light). Examples of elastic scattering include Mie scattering (i.e., the scattering particle size is comparable to the incident wavelength) and Rayleigh scattering (i.e., the scattering particle size is smaller than the incident wavelength). Examples of inelastic scattering include Brillouin scattering (i.e., emitting acoustic photons) and Raman scattering (i.e., emitting light of longer wavelengths). The level of inelastic scattering is much weaker as compared with that of elastic scattering.

17.5   NEAR-INFRARED DIFFUSE OPTICAL IMAGING AND SPECTROSCOPY (DOIS)

NIR light, in the spectral range from 650 to 1000 nm, is commonly utilized for biomedical imaging and spectroscopy. In this spectral region, light can penetrate up to several centimeters into biological tissues, enabling deep tissue imaging and tomography [23]. At shorter wavelengths, the absorption of major tissue chromophores, such as oxygenated hemoglobin and deoxygenated hemoglobin, is significantly higher [24], whereas, at longer wavelengths, the absorption of water is significantly higher [25]. The specific absorption coefficients of oxygenated hemoglobin and deoxygenated hemoglobin in the NIR wavelength range are shown in Figure 17.6.

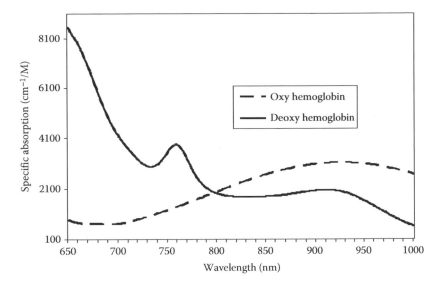

FIGURE 17.6 Absorption coefficients of oxygenated hemoglobin and deoxygenated hemoglobin in the near-infrared range of electromagnetic radiation.

Based on the Beer–Lambert law, light absorption under weak illumination follows an exponentially linear relationship with respect to the distance [26]. However, photon trajectories are much more complicated because of the highly scattering nature of most biological tissues. The complexity comes from the multiple scattering of light from a very heterogeneous tissue structure, as illustrated in Figure 17.7. The scattering mean free path (MFP) is used to represent the mean distance a photon propagates before being scattered. For biological tissues, the scattering MFP is typically on the order of 100 μm. At the maximum light intensity permissible on human tissues (at a wavelength of approximately 800 nm), the ballistic signal will fall below the short noise limited detection threshold after approximately 36 MFP, corresponding to a tissue depth of approximately 4 mm [27]. The depth of NIR diffuse optical measurement is roughly estimated by $\sqrt{2}r_{sd}/4$, where r_{sd} is the source detector separation [28]. The actual depth of measurement also depends on the contrast, size, and location of the embedded tissue heterogeneities [29]. It has been further estimated that the best

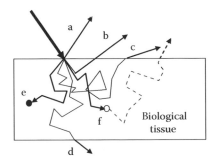

FIGURE 17.7 Typical trajectories of photons migrated in biological tissue. Photon "a" is bounced back directly on the surface of the biological tissue through "Fresnel reflection" (i.e., "specular reflection"). Photon "b" is back scattered by superficial portion of the biological tissue with minimal interaction (i.e., ballistic signal). Photon "c" experiences multiple scattering events and follows a complex photon trajectory before it reaches the original biological tissue boundary (i.e., diffuse reflectance). Photon "d" is multi-scattered and transmitted through the other side of the biological tissue. Photon "e" is absorbed by various tissue chromophores. The absorption of photon may trigger a second emission of longer wavelength in the case of fluorescence and inelastic scattering, as depicted by "f."

spatial resolution achievable when imaging through a scattering medium with diffuse light is given by approximately $0.2L$, where L is the thickness of the scattering medium [30].

17.6 DOIS DEVICES: CONTINUOUS WAVE, TIME DOMAIN, AND FREQUENCY DOMAIN

Commercially available NIR DOIS systems fall into the following three categories: (1) continuous wave (CW) [31], (2) time domain (TD) [32], and (3) frequency domain (FD) [33]. These three categories differ in the time dependence of the excitation source intensity.

A CW system illuminates the tissue with a steady or low-frequency light source. The optical properties are calculated based on the attenuation of the reflectance intensity. Most CW instruments assume a fixed reduced scattering coefficient (μ_s') or introduce a fixed differential path length factor (DPF) to account for tissue scattering [31]. Therefore, the result is a relative measurement of tissue chromophores absorption. Most fluorescence imaging systems are CW types. CW systems are relatively compact and less expensive. Many CW devices use low-frequency modulation/demodulation to effectively reject ambient noises. However, CW devices have two primary limitations. First, tissue scattering properties are not easily measurable. Second, it is hard to achieve depth resolution by using only the attenuation measurements under CW illumination.

TD systems were developed for tissue optical measurements with depth resolution. TD devices tap short laser pulses with durations from femto- to pico-seconds into biological tissue. The time-of-flight and the amplitude of photons are recorded by high-speed electronic instrumentations, such as streak cameras, fast avalanche photodiodes, and photomultipliers. By counting time-of-light and preferentially selecting the less scattered ballistic photons, one can obtain depth-resolved and quantitative measurements [34]. However, the resultant improvement in measurement accuracy and spatial resolution will be obtained at the expense of reduced signal intensity and the need for a more sophisticated detection system. Consequently, TD systems are typically very bulky and expensive.

An adequate balance between cost and performance can be achieved by FD systems, where the laser source is sinusoidally modulated at high frequency from 100 to 1000 MHz. By measuring the AC attenuation, DC attenuation, and phase shift of the resulting diffuse photon density wave (DPDW), one can retrieve both absorption and scattering properties of biological tissues with accuracy and resistance to ambient noises [33].

17.7 DOIS DEVICES: TISSUE OXIMETER, TISSUE TOMOGRAPHER, AND SURFACE IMAGER

Based on their functional applications, NIR DOIS devices can also be classified into the following three categories: (1) tissue oximeters, (2) tissue tomographers, and (3) surface imagers.

Tissue oximeters (also called "tissue spectrophotometers") utilize a limited number of light sources and detectors, and function to spot-check or continuously monitor functional parameters of the tissue underneath the sensor head. Most tissue spectrophotometers are CW or FD types, with multiple source-detector separations.

Tissue tomographers consist of a matrix of multiple light sources and detectors for the reconstruction of tissue heterogeneity. However, the achievable spatial resolution is very limited (in the order of one centimeter) due to the highly scattering nature of biological tissue. Proper gating of the multi-scattered light and advanced reconstruction algorithms are necessary for the improvement of the spatial resolution.

Surface imagers illuminate the region of interest with broadband excitation sources. The resulting fluorescence emission from the tissue is captured by a charge coupled device (CCD) or a scanning detector with installed filters. The image obtained by a surface imager is a 2D projection of diffuse reflected light in a 3D tissue structure without depth resolution. Fluorescent reflectance of the tissue captured by a surface imager under CW illumination is commonly used in cancer research [35–41].

Filtered broadband or laser light is diffused across the region of interest and excites the endogenous and/or exogenous fluorophores. The fluorescence emission is filtered and recorded with a cooled CCD. Multiple wavelengths (or channels) are used to characterize the spectral signatures of the lesions with single or multiple imaging probes.

17.8   CANCER BIOLOGY

Cancer is a class of malignant tumors that display an uncontrolled growth beyond the normal limits, invade and destroy adjacent tissues, and spread to other locations in the body. Carcinogenesis is the process of tumor development. It is initiated at the genomic level and is translated into aberrations in normal biological growth patterns at the molecular, protein, cellular, and tissue levels [42]. Hypoxia (i.e., a reduced level of tissue oxygenation) and angiogenesis (i.e., an increased number of microvessels) are two important characteristics of many solid tumors [43,44].

Tissue hypoxia results from the inadequate supply of oxygen that compromises biologic functions. Hypoxia can be caused by a number of factors such as hypoxic, anemic, diffusional, circulatory, and histotoxic [43,45]. Hypoxic hypoxia is caused by low oxygen partial pressure (pO2). Anemic hypoxia results from a decrease in hemoglobin concentration attended by a decline in oxygen-carrying capacity. Diffusional hypoxia is caused by the deterioration of the diffusion geometry, e.g., increased diffusion distances and concurrent versus countercurrent blood flow within microvessels. Circulatory hypoxia is also associated with normal pO2, whereas venous pO2 and tissue pO2 are reduced as a consequence of impaired tissue perfusion.

Tumor hypoxia is caused by an imbalance between oxygen delivery and oxygen consumption. Oxygen delivery is determined by perfusion, diffusional flux, and microvascular oxygen content. The oxygen consumption rate is determined by cell viability, respiration rate, temperature, and metabolic micromilieu substrate supply [46]. In the case of hypoxia, oxygen delivery is impaired by structural abnormalities present in the tumor vasculature. These structural abnormalities can cause numerous functional impairments, such as increased transcapillary permeability [47], increased vascular permeability, interstitial hypertension [48], and increased flow resistance. Altered tumor cell metabolism can also contribute to the presence of hypoxic regions within tumors [43].

17.9   APPLICATION OF DOIS IN CANCER DETECTION

Diffuse optical imaging is able to detect structural and functional abnormities mediated by tumor hypoxia [49]. Structural abnormalities, such as neovascularization, and functional abnormalities, such as perfusion and oxygenation, can produce early pathophysiologic changes in tumor tissue that may be detectable by NIR light, but may not otherwise be recognizable. These pathophysiologic changes have been studied successfully by measuring the intrinsic optical properties of normal and pathologic tissues [50,51] and by detecting molecular expressions of aberrant genes, proteins, and other pathophysiologic processes [52–54]. The reliable interpretation of these pathophysiologic changes requires the accurate detection of endogenous and exogenous biomarkers and the appropriate interpretation of light–tissue interactions [49].

17.10   ENDOGENOUS BIOMARKERS FOR DOIS

The endogenous biomarkers are intrinsic biomolecular markers that exist in biological tissue. Oxygenated hemoglobin and deoxygenated hemoglobin are two major endogenous biomarkers that can be detected by DOIS. The growth, progression, and eventual metastatic spread of malignant tumors requires neovascularization to provide both a source of nutrition and a route for tumor cell extravasations [44]. It has been demonstrated that NIR DOIS is able to assess changes in oxygenated and deoxygenated hemoglobin concentrations as a result of tumor-initiated angiogenesis. However, oxygenated hemoglobin and deoxygenated hemoglobin are not cancer-specific biomarkers. Tumors less than one cubic millimeter have their metabolic needs met by diffusion alone [55] and may not develop sufficient vascularity for NIR diffuse optical measurement of oxygenated and deoxygenated hemoglobin. Therefore, the current data are not sufficient to allow the development of definitive diagnostic criteria for differentiating malignant tissue from benign tissue at an early stage. In this regard, it is necessary to explore other biomarkers for the further characterization of disease-specific molecular signatures of malignant versus benign tissue.

In addition to oxygenated and deoxygenated hemoglobin, DOIS is also sensitive to various endogenous fluorophores. Fluorophores are fluorescence molecules that can emit the light of certain wavelengths when exposed to the excitation light. A large number of endogenous fluorophores exist within biological tissues. They are typically associated with the structural matrix of tissues (e.g., collagen [56] and elastin [57]) or with various cellular metabolic pathways (e.g., NAD and NADH [58]). Some of the most intense endogenous fluorophores that exist in humans and animals are involved in the process of cellular metabolism. The predominant species in this category include the reduced form of NADH, the various flavins, and the strongly fluorescent lipopigments. By measuring tissue autofluorescence, it is, in principle, possible to reveal the relative concentrations and redox states of these compounds, to learn the biochemical state of the tissue, and to detect various diseases, such as cancer, without requiring exogenous fluorescent markers or tags [59]. However, the clinical significance of endogenous fluorophores is limited by the overlap of the absorption spectrum with that of nonfluorescent chromophores, such as hemoglobin and melanin. Additionally, the excitation wavelength of autofluorescence spectroscopy is typically in the spectral range of ultraviolet (UV), which has a very limited penetration depth within biological tissues. In this regard, autofluorescence techniques are most suitable for the characterization of superficial tissue properties. Within the cancer arena, these techniques have been successfully implemented for the detection of skin cancer [35,36], cervical cancer [37,38], colon cancer [60,61], gastric cancer [39], and oral cancer [40,41].

17.11 EXOGENOUS BIOMARKERS FOR DOIS

Exogenous biomarkers are fluorophores or molecular beacons applied extrinsically to biological tissue to target and report specific molecular characteristics associated with tumor development and progression. In order to select proper exogenous fluorophores for more effective molecular imaging, one should consider the following factors: (1) fluorophores should possess an excitation/emission spectra with maxima in the NIR wavelengths (700–900) to allow for minimal absorption by endogenous chromophores, such as hemoglobin, lipids, and water; (2) fluorophores should possess a high quantum yield; (3) fluorophores should be chemically and optically stable; and (4) fluorophores should exhibit suitable pharmacological properties, including high specificity for the desired target and minimal nonspecific interactions [54].

Exogenous contrast agents for DOIS include various compounds such as indocyanine green dye and its derivatives, dye-antibody conjugates, dye-peptide conjugates, and molecular beacons [53]. Carbocyanine dyes, in particular indocyanine green dye (ICG) and its derivatives, are the most widely used exogenous contrast agents [62]. ICG is a water soluble tricarbocyanine dye approved by the Food and Drug Administration (FDA) for many medical imaging applications. Since its introduction, it has maintained a remarkably good safety profile. The basic structural framework of ICG consists of highly conjugated polymethine units flanked by symmetrical or nonsymmetrical aromatic groups. The homologation of carbocyanine probes results in compounds with a wide range of excitation and emission spectrum that are suitable for the multi-wavelength imaging of both superficial and deep biological tissues [63]. Because of their safety and broad excitation/emission spectrum, carbocyanine dyes, such as ICG, have been used for many contrast agent mediated optical imaging studies, such as retina and choroidal fluorescent angiography [62], cardiac output monitoring [64], and the assessment of hepatic function [65]. ICG was also used in multiple cancer imaging and therapeutic applications, such as NIR–MRI concurrent imaging of breast tumors [17], intraoperative surgical margin detection and sentinel lymph node mapping [66], and ICG-medicated phototherapy with enhanced cellular uptake and subcellular localization [67]. In general, ICG is highly bound to plasma proteins and, as a result, does not readily leak from the vasculature. Consequently, it is used primarily as a blood-pool agent for NIR diffuse optical imaging and spectroscopy [54].

The clinical use of carbocyanine dyes, such as ICG, in cancer imaging is hindered by several limitations. First, such dyes have a very limited tumor-to-normal tissue ratio for detecting cancer-related blood perfusion and vascular leakage. Second, these dyes have relatively low fluorescence quantum yields (<20%). Third, these dyes are nonspecific to cancer cells due to the lack of specific molecular information. Fourth, such dyes undergo an aggregation process in aqueous media, resulting in a wavelength shift of the main absorbance peak and a concurrent loss of absorption and fluorescence [68]. In the aqueous solution, the absorption and the emission peaks of ICG change with the dye concentration due to molecular aggregation [62]. In other solutions, such as plasma and human serum albumin, the nonspecific binding of ICG with both lipophilic and hydrophilic molecular species makes its absorption and emission spectra significantly different from those of the aqueous solution [62]. Recently, the noncovalent interaction of an anionic polymer, sodium polyaspartate, with ICG has produced an enhanced fluorescent stability in aqueous solution that retained strong fluorescence for 24 days under the same storage conditions [69]. PLGA nanoparticles or microbubble systems encapsulating ICG also demonstrated enhanced optical stability and an extended circulation lifetime [70,71]. Figure 17.8 shows the example of ICG encapsulated microbubbles fabricated by a modified double emulsion method [71,72]. Since ICG encapsulated microbubbles have simultaneous optical contrast and ultrasound contrast, they may be used for the concurrent imaging of tumor structural and functional characteristics [71]. Figure 17.9 shows NIR fluorescence images and B-mode ultrasound images acquired simultaneously as the suspension of ICG microbubbles flowed through different locations of a transparent tube. Microbubble flow rates that are calculated from NIR and ultrasound images are linearly correlated with actual flow rates. Highly linear correlations were observed between the normalized NIR and ultrasound image intensities and the actual MB concentrations, with repeatability errors less than 0.8%.

In order to increase the dye localization into tumors and to reduce its uptake within normal tissues, various tumor targeting conjugates have been developed. In this case, fluorescent probes are chemically conjugated with oligonucleotides or proteins to allow selective targeting of specific sites in cells. The high selectivity and affinity of receptor ligands results in a high signal-to-noise ratio and, therefore, enables the use of low concentrations of the fluorescent probes [73]. Typical tumor targeting conjugates can be classified into three groups [53,54,59,74]: (1) dye-antibody conjugate, (2) dye-peptide conjugate, and (3) molecular beacon. Dye-antibody conjugates target tumor-associated antigens by direct conjugation of cyanine dyes to monoclonal antibodies. One example is the conjugate

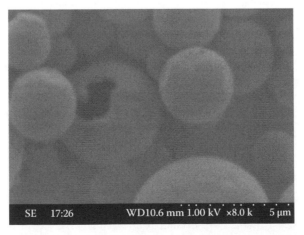

FIGURE 17.8   Scanning electron microscope (SEM) image of ICG encapsulated PLGA microbubbles.

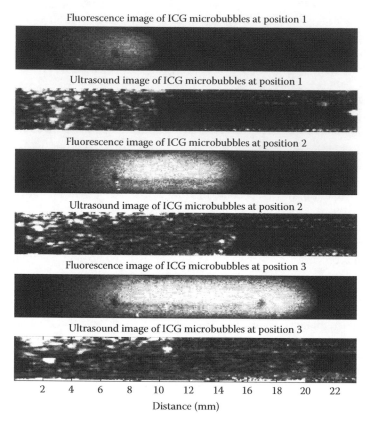

FIGURE 17.9 Concurrent NIR fluorescence imaging and ultrasound imaging of the ICG encapsulated microbubble suspension as it flows through a tumor simulator.

of the murine CC49 antibody (MuCC49) with Cy7 [75]. CC49 is an antibody targeting TAG-72, a human glycoprotein complex over-expressed in many epithelial-derived cancers, including colorectal, breast, ovarian, non-small cell lung, gastric, and pancreatic cancers [76]. TAG-72 is distributed in the extracellular space, therefore, the pharmacokinetics is slow and the biomarker can stay intact for weeks. Cy7 is an ICG derivative with an excitation peak wavelength of 747 nm and an emission peak wavelength of 776 nm. Compared with fluorescein and other fluorescence dyes with shorter emission peaks, Cy7 exhibits more favorable fluorescent signatures and has been shown to be more effective in visualizing the tumors, particularly in deeper tissues [75]. The CC49-Cy7 conjugate was injected into the LS174T colon cancer xenograft nude mouse model. Fluorescence images were acquired after 96 h (excitation: 730 nm, emission: 770 nm). MuCC49-Cy7 successfully targeted the tumor, as shown in Figure 17.10a. Fluorescence imaging of the dissected organs further confirmed the cancer-specific deposition, as shown in Figure 17.10b [75].

In addition to dye-antibody conjugates, dye-peptide conjugates have been synthesized such that the fluorescent dye is attached to a small bioactive peptide that specifically targets over-expressed tumor receptors. Recently, molecular beacons were also explored for changed physical properties after specific molecular interaction [77]. Tumor-targeting molecular beacons typically involve acid and enzyme cleavable conjugates of dye-biomolecules. Weissleder et al. conjugated Cy5.5 onto a graft copolymer consisting of a poly-L-lysine backbone and methoxypolyethylene glycol side chains to detect tumor-associated lysosomal protease activity in a xenograft mouse model [78]. Such a molecular beacon yielded a 12-fold increase in fluorescent signal and allowed for the detection of submillimeter-sized tumors [78]. Bremer et al. conjugated Cy5.5 to a matrix metalloproteinase (MMP) peptide substrate and a graft copolymer of methoxy-polyethylene-glycol-derivatized

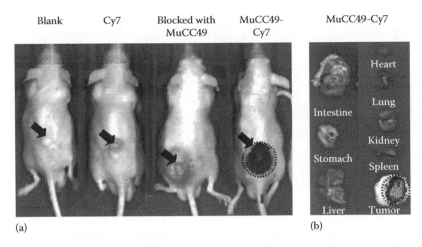

Blank   Cy7   Blocked with   MuCC49-   MuCC49-Cy7
MuCC49   Cy7

(a)   (b)

FIGURE 17.10 (a) Fluorescence images of the LS174T colon cancer xenograft nude mice 96 h after the injection of different contrast agents. "Blank" represents the control mouse without injecting anything; "Cy7" represents the mouse with IV injection of Cy7; "Blocked with MuCC49" represents the mouse blocked with MuCC49 followed by IV injection of MuCC49-Cy7; "MuCC49-Cy" represents the mouse with IV injection of MuCC49-Cy7. Solid arrows indicate the location of tumors. Dotted circles correspond to the areas of fluorescence emission. Fluorescence emission is only observed in the "MuCC49-Cy7" mouse. (b) Fluorescence images of the dissected tissues from the "MuCC49-Cy7" sample. Fluorescence emission is only observed on the tumor sample.

poly-L-lysine for the in vivo detection of MMP activities [79]. The probe fluoresced when MMP enzymes cleaved the peptide substrate and released previously quenched fluorophores [79]. In vivo imaging on mice with HT1080 human fibrosarcoma tumors demonstrated significantly less MMP-fluorescent signals in treated tumors compared with untreated tumors [79].

17.12 ADVANTAGES AND LIMITATIONS OF DOIS FOR CANCER DETECTION

Biomedical optical imaging is poised to become an emerging technology for the functional and molecular imaging of cancer. NIR DOIS is especially useful for imaging suspicious lesions in thick biological tissue. The technology will potentially provide new avenues and capabilities for image-guided cancer detection, staging, and therapy. The major advantages of NIR diffuse optical imaging and spectroscopy include the use of nonionizing radiation, real-time continuous data acquisition, low cost, portability, molecular sensitivity, and relatively large penetration depth (up to several centimeters) within biological tissues.

Like other imaging modalities, DOIS also has its technical limitations. These limitations and their potential solutions are briefly described below.

Limitation #1: NIR DOIS has low spatial resolution [80]. This is partially due to highly scattered and exponentially attenuated light transported in biological tissues [81] and is further complicated by the ill-posed nature of the inverse image reconstruction process [82].

Potential solution #1: Recognizing this limitation, as well as the potential advantages of integrating structural and functional modalities, has motivated many research groups to pursue the development of hybrid imaging systems. The structural information of tissues and organs that is obtained from conventional imaging tools, such as CT, MRI, and US, may provide highly useful co-registration and guidance that will ultimately improve the accuracy of NIR image reconstruction. One such example is an MRI and NIR hybrid system developed for the imaging of malignant and benign breast lesions [83]. In this imaging scheme, MRIs provide antecedent localization information for NIR diffuse optical tomography. In this clinical trial consisting of 14 patients with such an MRI–NIR imaging system, they demonstrated decreased oxygen saturation and higher blood concentration in malignant lesions. In addition to MRI, US has also been used to provide localization guidance for NIR diffuse optical imaging and spectroscopy. Another example is the hybrid system that integrates clinical US and NIR modalities. Tumor anatomic information collected by clinical US provides the a priori structural guidance for the improved NIR reconstruction of tumor functional properties. Portable imaging systems integrating NIR and US imaging modalities have been developed for differentiating benign and malignant breast lesions [84–87].

Limitation #2: NIR DOIS has relatively low reproducibility. The optical measurement is sensitive to the contact condition at the tissue-sensor interface. Improper contact of the sensor head to the tissue may not only introduce fluctuations in optical coupling, but may also alter local tissue vascular perfusion [88,89] and may result in changes in optical measurements. Besides, tissue heterogeneity and tissue property variations, in terms of age, race, hormone level, and menstrual phase, may also influence the measurement consistency. Previously, it has been demonstrated that the breast tissue of premenopausal women has approximately 2.5–3-fold higher absorption, 16%–28% greater scattering, 1.8-fold higher water concentration, and 50% lower lipid content than postmenopausal women [90]. Likewise, relative differences between symmetric breast locations within the same subject ranged from 18% to 30% for [Hbt], 10%–40% for adipose, 10%–25% for water, and 4%–9% for scattering [90]. Within the identical location of a given breast on the same subject, [Hbt], [StO2], water, and lipid contents varied by 54%, 15%, 43%, and 64%, respectively in different phases of the menstrual cycle [91]. Such temporal and spatial variations in the physiologic properties of breast tissue introduce additional inconsistency that may hinder the accurate differentiation between benign and malignant breast tissue.

Potential solution #2: To overcome the above limitation and enhance measurement consistency, a load cell was embedded into a tissue oximeter for continuous monitoring of the compression pressure [88,92]. The pressure-induced tissue deformation was also simulated numerically and was used to reconstruct images of tissue absorption and reduced scatter properties [93]. Changes in tissue optical and physiologic properties in response to external dynamic stimuli have been further studied. The external stimuli could be physiologic, chemical, or mechanical [88,89,92,94–98]. The simultaneous measurement of arterial hemoglobin oxygen saturation (SaO2), tumor vascular-oxygenated hemoglobin concentration (HbO2), and tumor oxygen tension (pO2) in response to a hyperoxic respiratory challenge has been previously studied in rat breast tumors and has demonstrated significant improvement in the tumor oxygen status [94]. Tumor physiologic changes in response to neoadjuvant chemotherapy have also been studied, demonstrating significant differences in total hemoglobin concentration, water fraction, tissue hemoglobin oxygen saturation, and lipid fraction before and after neoadjuvant chemotherapy [95]. A dynamic NIR optical tomographic (DYNOT) imaging system for simultaneous bilateral breast imaging was developed and evaluated on healthy subjects and breast cancer patients during the Valsalva maneuver, with demonstration of clinical feasibility and detectable differences between normal and malignant breast tissue [96]. A handheld dynamic imaging probe was also developed and tested in a 50-subject clinical trial for breast cancer characterization [99]. A possible correlation between the rate of pressure-induced [HbT] changes and malignant tumor characteristics was demonstrated by evaluating the data generated from human tumor clinical testing and from animal tumor model testing [98]. A major hypothesis for this dynamic imaging schema is that the altered neovascularization processes within cancer bearing tissues may significantly increase vascular resistance and cause an altered response of tissue oxygen and hemoglobin concentration during a dynamic compression stimulus.

Limitation #3: NIR diffuse optical imaging with intrinsic tissue contrast has relative low specificity in cancer detection. The intrinsic contrast can be derived from tissue chromophores, such as oxygenated hemoglobin, deoxygenated hemoglobin, myoglobin, H_2O, eumelanin, pheomelanin, cytochrome oxidase, bilirubin, and lipids [21,22]. Since hypoxia is a common characteristic of locally advanced solid tumors and can be associated with diminished therapeutic response and with disease progression [100], measurements of tissue oxygen saturation and hemoglobin concentration may provide useful information with regard to tumor characteristics. Clinical studies have shown that malignant tumors may be associated with increased hemoglobin concentration and reduced tissue oxygen saturation [50,59,83,101–105]. However, cancer diagnosis based on tissue oxygen and hemoglobin changes have low specificity. Similarly, exogenous blood flow tracers such as ICG are not cancer specific.

Potential solution #3: Recently developed nanoparticle contrast agents offer new avenues for cancer-specific imaging. However, clinical use of these nanoparticles is hindered by toxicity concerns and regulatory restrictions. Extensive preclinical research and toxicity studies are necessary before nanoparticles can be safely used on patients for cancer imaging and therapeutic applications.

17.13   FUTURE TRENDS

Future trends in biomedical optical imaging: Biomedical optical imaging modalities such as DOIS are currently at a stage of transition from that of basic laboratory research and animal modeling to that of potential adjuvant clinical applications. The next few years will witness major progress and advancements in the clinical applications of this technology in the arena of cancer detection and therapy. With regard to instrumentation, hybrid imaging systems are being developed that integrate optical imaging with various other established imaging modalities, such as CT, MRI, US, PET, and SPECT. Technological and computational advances will enable image reconstruction with higher spatial resolution and sensitivity. With regard to exogenous contrast agents, it is expected that concerns with toxicity issues will be fully addressed. Similarly, many regulatory hurdles will be overcome as toxicity issues are resolved and will ultimately enable the clinical use of exogenous contrast agents in human subjects. In addition, more effective drug delivery techniques and more sensitive fluorescence detection techniques are being investigated in order to minimize the necessary drug dosages for maintaining contrasts between tumor and normal tissues.

Future trends in biomedical engineering: Biomedical engineering is poised to become a new leading engineering discipline of the twenty-first century. It is one of the fastest growing engineering fields and opens new avenues for many exciting scientific research and technical development works at the interface of engineering and medicine. Future biomedical engineering research will revolutionize clinical practice by providing personalized health care with improved safety and clinical efficacy. Personalized health care represents the integrated practice of medicine and individual support based upon a patient's unique biological, behavioral, and environmental characteristics. Examples of biomedical engineering research areas that support personalized medicine are gene research, bioinformatics, and nanomedicine. In the arena of gene research, human gene expression and regulation and its relationship to genetic variation are studied to identify genetic changes associated with common human diseases, such as heart disease, cancer, diabetes, asthma, and stroke. In the future, a patient's gene expression profile will be used to tailor medical care to an individual's needs. In the arena of bioinformatics, advanced computing techniques and large databases will be used to process, analyze, and present genomic information for better understanding of diseases and accurate identification of new drug targets. Web-based biomedical information systems will provide scientists, engineers, and clinicians with more convenient access to biological information for the advancement of scientific, engineering, and clinical innovations. As an offshoot of nanotechnology, nanomedicine provides highly specific medical intervention at the molecular scale for curing diseases or repairing damaged tissues, such as bone, muscle, or nerve. Nanotechnology also offers the ability to detect hidden or metastatic diseases at early stages. The recent emphasis of theranostics combines the diagnostic and the therapeutic agents to appropriate sites with high specificity and in adequate concentrations to realize the promise of combined diagnosis and treatment of diseases in a single sitting. The multifunctional theranostic agents may allow for a more precise diagnosis, assessment of the effectiveness of therapeutic interventions, prediction of individual susceptibility/responsiveness to therapeutic intervention, and optimized personalized therapeutic strategies.

QUESTIONS

Q.17.1 Please list the commonly used biomedical imaging tools and explain their differences in terms of performance characteristics and clinical applications.

Q.17.2 What are the major limitations of multimodal biomedical imaging systems?

Q.17.3 What are the processes that govern the light propagation in biological tissues?

Q.17.4 Why is near-infrared (NIR) light utilized for diffuse optical imaging and spectroscopy (DOIS)?

Q.17.5 What are the advantages and disadvantages of continuous-wave (CW), time-domain (TD), and frequency-domain (FD) diffuse optical imaging devices?

Q.17.6 What are the major differences among tissue oximeters, tissue tomographers, and surface imagers?

Q.17.7 Why can DOIS devices be used for cancer detection?

Q.17.8 What are the differences between endogenous and exogenous biomarkers?

Q.17.9 Explain the advantages and limitations of using DOIS for cancer detection.

Q.17.10 What are the major challenges for multidisciplinary biomedical research?

Q.17.11 How do you think the concept of an intelligent energy field may help to overcome the major challenges in multidisciplinary biomedical research?

ABOUT THE CONTRIBUTING AUTHOR

Professor Ronald Xu received his PhD degree in mechanical engineering from Massachusetts Institute of Technology, Cambridge, followed by postdoctoral training at Stanford University, Palo Alto, California. Before joining OSU as an assistant professor of biomedical engineering, Dr. Xu had been director of technology development in a medical device company for 5 years and led the company efforts in technology development, benchtop tests, animal validation, and FDA 510k application. Dr. Xu's teaching and research interests include medical device design and innovation, bioinstrumentation, multimodal imaging, and theranostic agent fabrication. He is currently teaching two graduate courses: "Medical Device Design" and "Biomedical Optics." Dr. Xu has led the development and validation of various biomedical imaging devices for multiple clinical applications, such as breast cancer detection, peripheral vascular disease diagnosis, and wound assessment. His lab recently developed the multifunctional microbubble and nanobubble agents for cancer targeting, multimodal imaging, and integrated theranostics. Dr. Xu is inventor/co-inventor of five published patents, two pending patents, and over ten provisional patents in the field of biomedical imaging (http://www.bme.ohio-state.edu/bmeweb3/Xu/).

REFERENCES

1. G. Salamon and J. M. Caill, *Computerized Tomography*. Berlin, Germany/New York: Springer-Verlag, 1980.

2. H. B. Meire, *Abdominal and General Ultrasound*, 2nd edn. London, U.K./New York: Churchill Livingstone, 2001.

3. E. Feigenbaum, *Magnetic Resonance Imaging (MRI)*, Rockville, MD: U.S. Department of Health and Human Services, Public Health Service, National Center for Health Services Research and Health Care Technology Assessment, 1985.

4. P. E. Valk, *Positron Emission Tomography: Basic Science and Clinical Practice*, London, U.K./New York: Springer, 2003.

5. H. N. Wagner, Z. Szabo, and J. Buchanan, *Principles of Nuclear Medicine*, 2nd edn. Philadelphia, PA: Saunders, 1995.

6. T. Vo-Dinh, *Biomedical Photonics Handbook*, Boca Raton, FL: CRC Press, 2003.

7. X. Wang, Y. Xu, M. Xu, S. Yokoo, E. S. Fry, and L. V. Wang, Photoacoustic tomography of biological tissues with high cross-section resolution: Reconstruction and experiment, *Med. Phys.*, 29, 2799–2805, 2002.

8. Computed tomography, http://en.wikipedia.org/wiki/Computed_tomography.

9. Single photon emission computed tomography, http://en.wikipedia.org/wiki/SPECT.

10. J. D. Van Horn and R. A. Poldrack, Functional MRI at the crossroads, *Int. J. Psychophysiol.*, 73(1), 3–9, 2009.

11. F. S. Villanueva, W. R. Wagner, M. A. Vannan, and J. Narula, Targeted ultrasound imaging using microbubbles, *Cardiol. Clin.*, 22, 283–298, vii, May 2004.

12. D. W. Townsend, Multimodality imaging of structure and function, *Phys. Med. Biol.*, 53, R1–R39, February 2008.

13. W. R. Hendee and G. S. Gazelle, Biomedical Imaging Research Opportunities Workshop III: A white paper, *Ann. Biomed. Eng.,* 34, 188–198, February 2006.

14. T. F. Lang, B. H. Hasegawa, S. C. Liew, J. K. Brown, S. C. Blankespoor, S. M. Reilly, E. L. Gingold, and C. E. Cann, Description of a prototype emission-transmission computed tomography imaging system, *J. Nucl. Med.,* 33, 1881–1887, October 1992.

15. T. Beyer, D. W. Townsend, T. Brun, P. E. Kinahan, M. Charron, R. Roddy, J. Jerin, J. Young, L. Byars, and R. Nutt, A combined PET/CT scanner for clinical oncology, *J. Nucl. Med.*, 41, 1369–1379, August 2000.

16. B. E. Hammer, N. L. Christensen, and B. G. Heil, Use of a magnetic field to increase the spatial resolution of positron emission tomography, *Med. Phys.*, 21, 1917–1920, December 1994.

17. V. Ntziachristos, A. G. Yodh, M. Schnall, and B. Chance, Concurrent MRI and diffuse optical tomography of breast after indocyanine green enhancement, *Proc. Natl. Acad. Sci. USA*, 97, 2767–2772, March 2000.

18. Q. Zhu, E. Conant, and B. Chance, Optical imaging as an adjunct to sonograph in differentiating benign from malignant breast lesions, *J. Biomed. Opt.*, 5, 229–236, April 2000.

19. A. Li, E. L. Miller, M. E. Kilmer, T. J. Brukilacchio, T. Chaves, J. Stott, Q. Zhang et al., Tomographic optical breast imaging guided by three-dimensional mammography, *Appl. Opt.*, 42, 5181–5190, September 2003.

20. T. Vo-Dinh, Biomedical photonics: a revolution and the interface of science and technology, in *Biomedical Photonics Handbook*. Boca Raton, FL: CRC, 2003, pp. 28-1–28-50.

21. R. R. Anderson and J. A. Parrish, Optical properties of human skin, in *The Science of Photomedicine*, J. D. Regan and J. A. Parrish, Eds. New York: Plenum Press, 1982, pp. 147–194.

22. K. Meszaros, B. Chance, and H. Holtzer, Myoglobin and cytochrome oxidase in the myocardium of the developing chick, *J. Mol. Cell. Cardiol.*, 12, 965–975, October 1980.

23. A. Yodh and B. Chance, Spectroscopy and imaging with diffusing light, *Phys. Today*, 48, 34–40, 1995.

24. M. Cope, *The Development of a Near-Infrared Spectroscopy System and Its Application for Non-Invasive Monitoring of Cerebral Blood and Tissue Oxygenation in the Newborn Infant*. London, U.K.: University College London, 1991.

25. S. Prahl, Optical properties spectra, http://omlc.ogi.edu/spectra/index.html.

26. L. V. Wang and H.-i. Wu, *Biomedical Optics: Principles and Imaging*. Hoboken, NJ: Wiley-Interscience, 2007.

27. C. Dunsby and P. M. W. French, Techniques for depth-resolved imaging through turbid media including coherence-gated imaging, *J. Phys. D: Appl. Phys.*, 36, R207–R227, 2003.

28. S. Feng, F.-A. Zeng, and B. Chance, Photon migration in the presence of a single defect: A perturbation analysis, *Appl. Opt.*, 34, 3826, 1995.

29. R. Xu, B. Qiang, and J. Mao, Near infrared imaging of tissue heterogeneity: Probe design and sensitivity analysis, *Proceedings of IEEE Engineering in Medicine and Biology Society*, 1385, 278–281, 2005.

30. J. A. Moon, R. Mahon, D. M. D., and R. J., Resolution limits for imaging through turbid media with diffuse light, *Opt. Lett.*, 18, 1591–1593, 1993.

31. B. Chance, Q. Luo, S. Nioka, D. C. Alsop, and J. A. Detre, Optical investigations of physiology: A study of intrinsic and extrinsic biomedical contrast, *Philos. Trans. R Soc. Lond. B Biol. Sci.*, 352, 707–716, June 1997.

32. B. Chance, S. Nioka, J. Kent, K. McCully, M. Fountain, R. Greenfeld, and G. Holtom, Time-resolved spectroscopy of hemoglobin and myoglobin in resting and ischemic muscle, *Anal. Biochem.*, 174, 698–707, November 1988.

33. S. Fantini, M. A. Fanceschini, J. S. Maier, S. A. Walker, B. Barbieri, and E. Gratton, Frequency-domain multichannel optical detector for non-invasive tissue spectroscopy and oximetry, *Opt. Eng.*, 34, 32–42, 1995.

34. J. C. Hebden, Evaluating the spatial resolution performance of a time-resolved optical imaging system, *Med. Phys.*, 19, 1081–1087, July–August 1992.

35. H. J. Sterenborg, S. Thomsen, S. L. Jacques, and M. Motamedi, In vivo autofluorescence of an unpigmented melanoma in mice. Correlation of spectroscopic properties to microscopic structure, *Melanoma Res*, 5, 211–216, August 1995.

36. P. Diagaradjane, M. A. Yaseen, J. Yu, M. S. Wong, and B. Anvari, Autofluorescence characterization for the early diagnosis of neoplastic changes in DMBA/TPA-induced mouse skin carcinogenesis, *Lasers Surg. Med.*, 37, 382–395, December 2005.

37. R. Richards-Kortum, M. F. Mitchell, N. Ramanujam, A. Mahadevan, and S. Thomsen, In vivo fluorescence spectroscopy: Potential for non-invasive, automated diagnosis of cervical intraepithelial neoplasia and use as a surrogate endpoint biomarker, *J. Cell. Biochem. Suppl.*, 19, 111–119, 1994.

38. I. Pavlova, K. Sokolov, R. Drezek, A. Malpica, M. Follen, and R. Richards-Kortum, Microanatomical and biochemical origins of normal and precancerous cervical autofluorescence using laser-scanning fluorescence confocal microscopy, *Photochem. Photobiol.*, 77, 550–555, May 2003.

39. B. W. Chwirot, S. Chwirot, W. Jedrzejczyk, M. Jackowski, A. M. Raczynska, J. Winczakiewicz, and J. Dobber, Ultraviolet laser-induced fluorescence of human stomach tissues: Detection of cancer tissues by imaging techniques, *Lasers Surg. Med.*, 21, 149–158, 1997.

40. D. C. de Veld, M. Skurichina, M. J. Witjes, R. P. Duin, H. J. Sterenborg, and J. L. Roodenburg, Autofluorescence and diffuse reflectance spectroscopy for oral oncology, *Lasers Surg. Med.*, 36, 356–364, June 2005.

41. D. R. Ingrams, J. K. Dhingra, K. Roy, D. F. Perrault, Jr., I. D. Bottrill, S. Kabani, E. E. Rebeiz, et al., Autofluorescence characteristics of oral mucosa, *Head Neck*, 19, pp. 27–32, January 1997.

42. A. Sarasin, An overview of the mechanisms of mutagenesis and carcinogenesis, *Mutat. Res.*, 544, 99–106, November 2003.

43. M. Hockel and P. Vaupel, Tumor hypoxia: Definitions and current clinical, biologic, and molecular aspects, *J. Natl. Cancer Inst.*, 93, 266–276, February 2001.

44. J. Folkman, Tumor angiogenesis and tissue factor, *Nat. Med.*, 2, 167–168, February 1996.

45. K. M. Scheufler, Tissue oxygenation and capacity to deliver O2 do the two go together? *Transfus Apher. Sci.*, 31, 45–54, August 2004.

46. P. Vaupel, D. K. Kelleher, and O. Thews, Modulation of tumor oxygenation, *Int. J. Radiat. Oncol. Biol. Phys.*, 42, 843–848, November 1998.

47. D. M. McDonald and P. Baluk, Significance of blood vessel leakiness in cancer, *Cancer Res.*, 62, 5381–5385, September 2002.

48. Y. Boucher, M. Leunig, and R. K. Jain, Tumor angiogenesis and interstitial hypertension, *Cancer Res.*, 56, 4264–4266, September 1996.

49. R. X. Xu and S. P. Povoski, Diffuse optical imaging and spectroscopy for cancer, *Expert Rev. Med. Devices*, 4, 83–95, January 2007.

50. B. J. Tromberg, N. Shah, R. Lanning, A. Cerussi, J. Espinoza, T. Pham, L. Svaasand, and J. Butler, Non-invasive in vivo characterization of breast tumors using photon migration spectroscopy, *Neoplasia*, 2, 26–40, January-April 2000.

51. A. Garcia-Uribe, N. Kehtarnavaz, G. Marquez, V. Prieto, M. Duvic, and L. V. Wang, Skin cancer detection by spectroscopic oblique-incidence reflectometry: Classification and physiological origins, *Appl. Opt.*, 43, 2643–2650, May 2004.

52. U. Mahmood and R. Weissleder, Near-infrared optical imaging of proteases in cancer, *Mol. Cancer Ther.*, 2, 489–496, May 2003.

53. S. Achilefu, Lighting up tumors with receptor-specific optical molecular probes, *Technol. Cancer Res. Treat.*, 3, 393–409, August 2004.

54. M. Gurfinkel, S. Ke, X. Wen, C. Li, and E. M. Sevick-Muraca, Near-infrared fluorescence optical imaging and tomography, *Dis. Markers*, 19, 107–121, 2003.

55. D. Darland and P. D'Amore, Endothelial cells and pericytes in tumor vasculature, in *Tumor Angiogenesis and Microcirculation*, E. Voest and P. D'Amore, Eds. New York: Marcel Dekker, Inc., 2001, pp. 1–8.

56. D. Fujimoto, Isolation and characterization of a fluorescent material in bovine achilles tendon collagen, *Biochem. Biophys. Res. Commun.*, 76, 1124–1129, June 1977.

57. D. H. Tinker, R. B. Rucker, and A. L. Tappel, Variation of elastin fluorescence with method of preparation: Determination of the major fluorophore of fibrillar elastin, *Connect. Tissue Res.*, 11, 299–308, 1983.

58. G. A. Wagnieres, W. M. Star, and B. C. Wilson, In vivo fluorescence spectroscopy and imaging for oncological applications, *Photochem. Photobiol.*, 68, 603–632, November 1998.

59. I. J. Bigio and S. G. Bown, Spectroscopic sensing of cancer and cancer therapy: Current status of translational research, *Cancer Biol. Ther.*, 3, 259–267, March 2004.

60. T. J. Romer, M. Fitzmaurice, R. M. Cothren, R. Richards-Kortum, R. Petras, M. V. Sivak, Jr., and J. R. Kramer, Jr., Laser-induced fluorescence microscopy of normal colon and dysplasia in colonic adenomas: Implications for spectroscopic diagnosis, *Am. J. Gastroenterol.*, 90, 81–87, January 1995.

61. B. H. Li and S. S. Xie, Autofluorescence excitation-emission matrices for diagnosis of colonic cancer, *World J. Gastroenterol.*, 11, 3931–3934, July 2005.

62. T. Desmettre, J. M. Devoisselle, and S. Mordon, Fluorescence properties and metabolic features of indocyanine green (ICG) as related to angiography, *Surv. Ophthalmol.*, 45, 15–27, July-August 2000.

63. K. Licha, B. Riefke, V. Ntziachristos, A. Becker, B. Chance, and W. Semmler, Hydrophilic cyanine dyes as contrast agents for near-infrared tumor imaging: Synthesis, photophysical properties and spectroscopic in vivo characterization, *Photochem. Photobiol.*, 72, 392–398, September 2000.

64. V. V. Kupriyanov, S. Nighswander-Rempel, and B. Xiang, Mapping regional oxygenation and flow in pig hearts in vivo using near-infrared spectroscopic imaging, *J. Mol. Cell. Cardiol.*, 37, 947–957, November 2004.

65. J. Caesar, S. Shaldon, L. Chiandussi, L. Guevara, and S. Sherlock, The use of indocyanine green in the measurement of hepatic blood flow and as a test of hepatic function, *Clin. Sci.*, 21, 43–57, August 1961.

66. A. M. De Grand and J. V. Frangioni, An operational near-infrared fluorescence imaging system prototype for large animal surgery, *Technol. Cancer Res. Treat.*, 2, 553–562, December 2003.

67. C. Abels, S. Fickweiler, P. Weiderer, W. Baumler, F. Hofstadter, M. Landthaler, and R. M. Szeimies, Indocyanine green (ICG) and laser irradiation induce photooxidation, *Arch. Dermatol. Res.*, 292, 404–411, August 2000.

68. J. F. Zhou, M. P. Chin, and S. A. Schafer, Aggregation and degradation of indocyanine green, *SPIE Proc.*, 2128, 495–505, 1994.

69. R. Rajagopalan, P. Uetrecht, J. E. Bugaj, S. A. Achilefu, and R. B. Dorshow, Stabilization of the optical tracer agent indocyanine green using noncovalent interactions, *Photochem. Photobiol.*, 71, 347–350, March 2000.

70. V. Saxena, M. Sadoqi, and J. Shao, Indocyanine green-loaded biodegradable nanoparticles: Preparation, physicochemical characterization and in vitro release, *Int. J. Pharm.*, 278, 293–301, July 2004.

71. R. Xu, J. Huang, J. Xu, D. Sun, G. Hinkle, J. Edward, W. Martin, and S. Povoski, Fabrication of indocyanine green encapsulated biodegradable microbubbles for structural and functional imaging of cancer, *J. Biomed. Opt.*, 14(3), 034020, May/June 2009.

72. J. S. Xu, J. Huang, R. Qin, G. H. Hinkle, S. P. Povoski, E. W. Martin and R. X. Xu, Synthesizing and binding dual-mode poly (lactic-co-glycolic acid) (PLGA) nanobubbles for cancer targeting and imaging, *Biomaterials*, 31(7), 1716–1722, March 2010.

73. A. Becker, C. Hessenius, K. Licha, B. Ebert, U. Sukowski, W. Semmler, B. Wiedenmann, and C. Grotzinger, Receptor-targeted optical imaging of tumors with near-infrared fluorescent ligands, *Nat. Biotechnol.*, 19, 327–331, April 2001.

74. J. V. Frangioni, In vivo near-infrared fluorescence imaging, *Curr. Opin. Chem. Biol.*, 7, 626–634, October 2003.

75. P. Zou, S. Xu, A. Wang, S. Povoski, M. Johnson, E. Martin, V. Subramaniam, J. Schlom, R. Xu, and D. Sun, Near-infrared fluorescence labeled anti-TAG-72 monoclonal antibodies for tumor imaging in colorectal cancer xenograft mice, *Mol. Pharm.*, 6, 428–440, February 2009.

76. V. G. Johnson, J. Schlom, A. J. Paterson, J. Bennett, J. L. Magnani, and D. Colcher, Analysis of a human tumor-associated glycoprotein (TAG-72) identified by monoclonal antibody B72.3, *Cancer Res.*, 46, 850–857, February 1986.

77. S. Tyagi and F. R. Kramer, Molecular beacons: Probes that fluoresce upon hybridization, *Nat. Biotechnol.*, 14, 303–308, March 1996.

78. R. Weissleder, C. H. Tung, U. Mahmood, and A. Bogdanov, Jr., In vivo imaging of tumors with protease-activated near-infrared fluorescent probes, *Nat. Biotechnol.*, 17, 375–378, April 1999.

79. C. Bremer, C. H. Tung, and R. Weissleder, In vivo molecular target assessment of matrix metalloproteinase inhibition, *Nat. Med.*, 7, 743–748, June 2001.

80. M. Ferrari, L. Mottola, and V. Quaresima, Principles, techniques, and limitations of near infrared spectroscopy, *Can. J. Appl. Physiol.*, 29, 463–487, August 2004.

81. D. A. Boas, A. M. Dale, and M. A. Franceschini, Diffuse optical imaging of brain activation: Approaches to optimizing image sensitivity, resolution, and accuracy, *Neuroimage*, 23 (Suppl 1), S275–S288, 2004.

82. S. R. Arridge, Optical tomography in medical imaging, *Inverse Probl.*, 15, R41–R93, 1999.

83. V. Ntziachristos, A. G. Yodh, M. D. Schnall, and B. Chance, MRI-guided diffuse optical spectroscopy of malignant and benign breast lesions, *Neoplasia*, 4, 347–354, July–August 2002.

84. Q. Zhu, M. Huang, N. Chen, K. Zarfos, B. Jagjivan, M. Kane, P. Hedge, and S. H. Kurtzman, Ultrasound-guided optical tomographic imaging of malignant and benign breast lesions: Initial clinical results of 19 cases, *Neoplasia*, 5, 379–388, September–October 2003.

85. R. X. Xu, B. Qiang, J. O. Olsen, S. P. Povoski, L. D. Yee, and J. Mao, Localization and functional parameter reconstruction of suspicious breast lesions by near infrared/ultrasound dual modal imaging, in *IEEE EMBC'05*, Shanghai, China, 2005, pp. 4473–4476.

86. Q. Zhu, N. Chen, and S. H. Kurtzman, Imaging tumor angiogenesis by use of combined near-infrared diffusive light and ultrasound, *Opt. Lett.*, 28, 337–339, March 2003.

87. R. X. Xu, J. Ewing, H. El-Dahdah, B. Wang, and S. P. Povoski, Design and benchtop validation of a handheld integrated dynamic breast imaging system for noninvasive characterization of suspicious breast lesions, *Technol. Cancer Res. Treat.*, 7, 471–482, December 2008.

88. X. Cheng and X. Xu, Study of the pressure effect in near infrared spectroscopy, *Proc. SPIE*, 4955, 397–406, 2003.

89. S. Jiang, B. W. Pogue, K. D. Paulsen, C. Kogel, and S. Poplack, In vivo near-infrared spectral detection of pressure-induced changes in breast tissue, *Opt. Lett.*, 28, 1212–1214, 2003.

90. N. Shah, A. E. Cerussi, D. Jakubowski, D. Hsiang, J. Butler, and B. J. Tromberg, Spatial variations in optical and physiological properties of healthy breast tissue, *J. Biomed. Opt.*, 9, 534–540, May–June 2004.

91. R. Cubeddu, C. D'Andrea, A. Pifferi, P. Taroni, A. Torricelli, and G. Valentini, Effects of the menstrual cycle on the red and near-infrared optical properties of the human breast, *Photochem. Photobiol.*, 72, 383–391, September 2000.

92. R. X. Xu, B. Qiang, J. J. Mao, and S. P. Povoski, Development of a handheld near-infrared imager for dynamic characterization of in vivo biological tissue systems, *Appl. Opt.*, 46, 7442–7451, October 2007.

93. H. Dehghani, M. M. Doyley, B. W. Pogue, S. Jiang, J. Geng, and K. D. Paulsen, Breast deformation modelling for image reconstruction in near infrared optical tomography, *Phys. Med. Biol.*, 49, 1131–1145, April 2004.

94. Y. Gu, V. A. Bourke, J. G. Kim, A. Constantinescu, R. P. Mason, and H. Liu, Dynamic response of breast tumor oxygenation to hyperoxic respiratory challenge monitored with three oxygen-sensitive parameters, *Appl. Opt.*, 42, 2960–2967, June 2003.

95. D. B. Jakubowski, A. E. Cerussi, F. Bevilacqua, N. Shah, D. Hsiang, J. Butler, and B. J. Tromberg, Monitoring neoadjuvant chemotherapy in breast cancer using quantitative diffuse optical spectroscopy: A case study, *J. Biomed. Opt.*, 9, 230–238, January–February 2004.

96. C. H. Schmitz, D. P. Klemer, R. Hardin, M. S. Katz, Y. Pei, H. L. Graber, M. B. Levin et al., Design and implementation of dynamic near-infrared optical tomographic imaging instrumentation for simultaneous dual-breast measurements, *Appl. Opt.*, 44, 2140–2153, April 2005.

97. R. Xu and A. Rana, Dynamic near infrared imaging with ultrasound guidance (dNIRUS): Analytical model and benchtop validation on multi-layer tissue simulating phantoms, *Proc. SPIE*, 6086, 353–364, 2006.

98. B. Wang, S. P. Povoski, X. Cao, D. Sun, and R. X. Xu, Dynamic schema for near infrared detection of pressure-induced changes in solid tumors, *Appl. Opt.*, 47, 3053–3063, June 2008.

99. R. X. Xu, D. C. Young, J. J. Mao, and S. P. Povoski, A prospective pilot clinical trial evaluating the utility of a dynamic near-infrared imaging device for characterizing suspicious breast lesions, *Breast Cancer Res.*, 9, R88, 2007.

100. P. Vaupel, The role of hypoxia-induced factors in tumor progression, *Oncologist*, 9 (Suppl 5), 10–17, 2004.

101. X. Cheng, X. Xu, S. Zhou, L. Wang, B. Hu, F. Li, M. Wang, C. Zhou, H. Li, and H. Zhang, A novel optical scanning system for breast cancer imaging, *Proc. SPIE*, 4244, 468–473, 2001.
102. M. J. Holboke, B. J. Tromberg, X. Li, N. Shah, J. Fishkin, D. Kidney, J. Butler, B. Chance, and A. G. Yodh, Three-dimensional diffuse optical mammography with ultrasound localization in a human subject, *J. Biomed. Opt.*, 5, 237–247, 2000.
103. B. W. Pogue, S. Jiang, H. Dehghani, C. Kogel, S. Soho, S. Srinivasan, X. Song, T. D. Tosteson, S. P. Poplack, and K. D. Paulsen, Characterization of hemoglobin, water, and NIR scattering in breast tissue: analysis of intersubject variability and menstrual cycle changes, *J. Biomed. Opt.*, 9, 541–552, May-June 2004.
104. T. O. McBride, B. W. Pogue, S. Poplack, S. Soho, W. A. Wells, S. Jiang, U. L. Osterberg, and K. D. Paulsen, Multispectral near-infrared tomography: A case study in compensating for water and lipid content in hemoglobin imaging of the breast, *J. Biomed. Opt.*, 7, 72–79, January 2002.
105. X. Cheng, J.-m. Mao, R. Bush, D. B. Kopans, R. H. Moore, and M. Chorlton, Breast cancer detection by mapping hemoglobin concentration and oxygen saturation, *Appl. Opt.*, 42, 6412–6421, 2003.

Part IV

Selected Innovative Processes

18 Energy Field Methods and Electromagnetic Sheet Metal Forming

Glenn S. Daehn

CONTENTS

Sheet metal forming has seen few changes in its practice over the past 50 years, yet it suffers from a number of long-standing problems, including difficulties in fabricating lightweight components from high strength materials, distortion due to elastic strains that remain in the forming process, and expensive inflexible forming equipment. An alternate approach to sheet metal forming is presented in this chapter. Electromagnetic forming is a process that can accelerate sheet metal to high velocities using a pulsed current and the resulting transient magnetic field. This method can easily drive conductive sheet metal components to velocities over 100 m/s without contact and in a controlled and reproducible manner. This chapter gives a background of this process by first discussing how this impulse-based forming method has distinctly different characteristics from those of traditional quasi-static formability. Inertia and speed can inhibit instabilities, improve formability, and improve dimensional conformation. Next, the electrodynamics that provides impulse is presented in a simplified quantitative manner. Important problems are placed in three classes, two of which can be analyzed using relatively simple computational methods. Finally, a number of important processes in forming, joining, and cutting are broadly discussed and are related to the computational methods that are sufficient to analyze the problems. Overall, the case is made that this method allows a direct control of the forming energy distribution and this can provide a powerful complement to the stamping methods that are currently practiced.

18.1   HIGH-VELOCITY FORMING AND ITS BENEFITS

The earliest days of metal forming used hammers to shape billets and sheets into useful shapes. Since the industrial revolution, fixed tools such as stamping dies have become the standard for metal forming. While fixed tools are highly productive, they do not offer the local control of energy density that was provided by hammer forming. Even today, difficult components in prototype production and aerospace manufacture are commonly produced with the aid of manually operated hammers. High-velocity metalforming methods in general, and electromagnetic metal forming techniques in particular, are more akin to hammers than presswork as the local impulse is controlled rather than forces or displacements. Also, like hammer work, when properly implemented, one can control the spatial distribution of the impulse.

High-velocity forming methods include techniques such as explosive forming, electromagnetic forming, and impact forming. These techniques are distinct from most other metal forming methods. In a common implementation, the explosive or electromagnetic force first accelerates the workpiece to a high velocity, giving the workpiece significant kinetic energy. The sheet metal workpiece, then, changes shape as it either strikes a die or is decelerated by plastic deformation in flight. Sometimes, both means of shape change are important. These methods can provide robust methods of performing metal forming operations that may be quite difficult by conventional means. Although these methods have been known for over 100 years and saw significant developments years ago (particularly in the 1960s), these methods have not been developed or taught in such a way that they can be routinely used to their potential. However, because these methods work well on even hard-to-form materials and because manufacturing systems are generally very simple and can be established quickly, there is a strong resurgence in interest in these methods.

The major sections of this chapter will first discuss how high-velocity impulse-driven metal forming is fundamentally different from quasi-static metal forming. These differences offer possibilities for treating long-standing problems in sheet metal forming. Next, the analytical foundation for understanding and designing electromagnetic metal forming processes is developed. Three classes of problems are considered. One important lesson that can be learned from this exercise is that relatively simple analyses are often sufficient for the design of electromagnetic forming systems. Section 18.2 will show how these analyses can be used to support several diverse electromagnetic forming problems. Discrete examples are shown from problems of shaping, cutting, and joining. In particular, a hybrid electromagnetically assisted stamping approach is described, where a careful control of the spatial distribution of plastic work is the key to giving an acceptable strain distribution to a part to obtain the required shape. Section 18.3 discusses how electromagnetic sheet metal forming can be an enabling technology for the agile manufacturing of sheet metal components.

18.1.1   HISTORICAL DEVELOPMENT OF HIGH-VELOCITY METAL FORMING METHODS

High-velocity forming methods were discovered in the late 1800s in the form of explosive forming and saw some application in forming thick plates in the 1930s. Between about 1950 and the early 1970s, the U.S. government funded numerous studies on the development and application of high-velocity forming. The U.S. Defense Advanced Projects Agency (DARPA) invested heavily in explosive forming technology in the mid 1960s. DARPA [1] cites explosive forming as "a cost-effective process for forming a variety of metals and alloys that results in remarkably high reproducibility (~0.5%) for complex, large metal structures. The method was used extensively in DoD projects; the applications included making afterburner rings for the SR-71, jet engine diffusers, Titan 'manhole' covers, rocket engine seals, P-3 Orion aircraft skin, tactical missile domes, jet engine sound suppressors, and heat shields for turbine engines." Also in the same period, significant research investments were made in the allied technologies of electromagnetic forming and electrohydraulic forming. These methods are based on storing the forming energy in a capacitor bank instead of as chemical energy, as detailed later.

Explosive metal forming still sees limited commercial use (i.e., Exploform of The Netherlands and PA&E Company of Washington State). It is a very simple process, usually done by placing a sheet metal on a form die, evacuating air behind it, immersing the die and sheet in water, and using an explosively driven shock to form the metal into the cavity. This process is simple and remarkably robust. It represents one of the simplest possible ways to make a first part, but dies often do not last long enough to make a long production run.

Electrohydraulic forming is in operation quite similar to explosive forming but is usually limited to discharge energies on the order of 100 kJ, while explosive forming operations can expend MJs of energy in a single event. In electrohydraulic forming, an underwater spark gap converts electrical energy to mechanical work by forming and heating plasma in the spark gap. This produces a shock wave in the water that can do useful mechanical work. Forming using shock waves in fluids (as both electrohydraulic and explosive forming do) can be almost magical with respect to the complex parts that can be produced with very simple dies. The shock waves enable much higher pressures than quasi-static approaches such as hydroforming. Also, because there are very high deformation rates and because forming is often done in compression between a high pressure fluid and the workpiece being struck, it often results in much better formability, which allows very aggressive shapes to be formed. Lastly, as discussed later in this chapter, die impact gives access to remarkable detail and accuracy. The example shown in Figure 18.1 shows how deep draw depths can be created from a very simple electrohydraulic forming operation and formability can be greatly extended. This method saw extensive commercial development in the United States and in the former Soviet Union in the 1960s and 1970s. In the United States, a commercial manufacturing system known as the Cincinnati Electroshape was developed. This process has not become very common in industry. The author has sought out many practitioners to talk about this process; many of the "old timers" who have worked with electrohydraulic forming enthusiastically attest that it can often form parts impossible to fabricate by any other process. But the process has also suffered from a reputation of

FIGURE 18.1 Electrohydraulically formed rectangular impression formed in a single operation in electrohydraulic forming of 2024-T4 aluminum, with a very simple single sided die. Plane strain plastic extensions of over 70% are seen on the vertical walls. There is virtually no draw-in and the part is made entirely by stretching. The depression is about 1 cm deep.

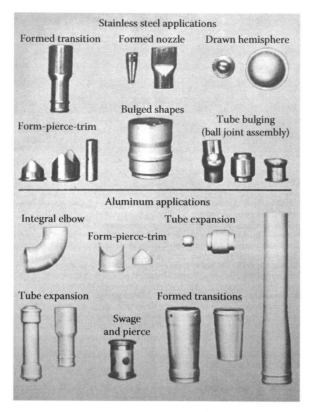

FIGURE 18.2 Typical applications of electrohydraulic forming as practiced at the General Dynamics Fort Worth Plant in the early 1960s. (From Bruno, E.J. ed., *High Velocity Forming of Metals*, ASTME, Dearborn, MI, 1968.)

being inconsistent and hard to control. Members of the Society of Manufacturing Engineers (SME) can easily access many papers from the early development of this at the SME Web site. A Web site maintained at Ohio State at www.osu.edu/hyperplasticity also has much archival information on this and the related process. The out-of-print book *High Velocity Forming of Metals* that was produced in 1968 by the American Society of Tool and Manufacturing Engineers gives possibly the best overview of electrohydraulic forming and allied impulse forming methods. Figure 18.2, from that book, gives a range of examples of how electrohydraulic forming was used in aircraft manufacture.

Electromagnetic forming has been used primarily in industry as an assembly technique. It is the focus of this chapter and will be described in depth later and arguments will be made to extend its range of application. It is only directly applicable to materials with high electrical conductivity. However, conductive aluminum or copper drivers can often be used to form materials with low intrinsic conductivity. General Atomics and Maxwell Laboratories of San Diego developed the commercial Magneform machine in the early 1960s and equipment is still available from the Magneform company. Presently, there are several small commercial competitors producing equipment (capacitor banks, coils, etc.) for electromagnetic forming. They include IAP of Dayton, Ohio; Poynting of Dortmund, Germany; PST Products of Alzenau, Germany; and Pulsar of Yavne, Israel.

18.1.2 LIMITATIONS OF CONVENTIONAL SHEET METAL STAMPING

The central point of this chapter is to show how impulse-based high-velocity metal forming as driven by electromagnetic forming can be effective in eliminating or working around some of the "problems" related to traditional sheet metal forming or stamping as it is known.

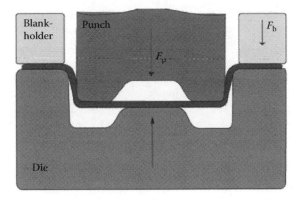

FIGURE 18.3 A conventional metal stamping process.

Stamping is a term that refers to a number of sheet metal forming manufacturing procedures. The prototypical stamping press is shown schematically in Figure 18.3. With appropriate tooling, stamping can accomplish a number of operations including bending; stretching; drawing (which is similar to stretching, but additional metal is brought into the die cavity by sliding in from the blank-holder region of Figure 18.3, as the punch moves downward); coining or embossing (putting intricate or useful shapes into a metallic surface, using high pressure); flanging (bending metal edges up, which will cause compressive or tensile strains for concave or convex curvatures); and cutting, which can be referred to as blanking, piercing, or shearing. A detailed review of these processes is well beyond the scope of this chapter, but useful summaries can be found in the literature [2–4]. While stamping has been a very low-cost and robust technology for mass production, it does have a number of long-term drawbacks that can be partially overcome by electromagnetic metal forming. These problems and solutions will be discussed in detail later in this chapter.

There are three chronic limitations in sheet metal stamping that are expanded upon in the remainder of this section, and proposed solutions based on electromagnetic forming are provided in Section 18.3 of this chapter. The limitations are as follows:

1. *Slow and expensive system setup and changes*: Figure 18.3 schematically shows the metal stamping system. The press and dies are all massive and the die set has blank holders, punches, and dies cut specifically for the shape to be made as well as for the specific metal thickness. If the part to be produced has significant complexity, it is usually made from a progression of dies, each adding shape or features to the part. When a new sheet metal part is commissioned, it is a significant undertaking. The dies are expensive and take a long time to produce. Tuning the dies to make the part is also required—adding or removing small amounts of material strategically to make sure the metal forms properly. With this two-sided die arrangement, clearance issues between the two halves often set the final required tolerance. Later sections of this chapter will show that electromagnetic impulse forming can allow the use of lighter presses and one-sided dies and can allow more to be done and/or tuned in a single press station, allowing more part complexity, fewer die stations, or both.

2. *Forming difficult shapes and materials*: It is a general rule of thumb that as materials become stronger, they become less ductile; that is, the strain to failure decreases. In this case, the complexity of the shapes that can be created by stretching and/or drawing significantly decreases. Often, a compressive strain is required to shape metal to a needed shape. Compression (particularly of strong, thin metals) will cause buckling. This chapter will show that high-velocity forming can avoid these problems as well as tearing caused by necking and wrinkling, which are all examples of instabilities that can be overcome by inertial effects at high velocity.

3. *Springback*: Again referring to Figure 18.3, if the sheet metal is drawn into the die and the punch is fully extended into the die, a shape will be created in the sheet metal wherein the walls are perpendicular to the floor of the pan. When the punch is withdrawn and the blankholder is disengaged, there is no guarantee that the part will have the shape it had when it was constrained. Generally, the shape will be different due to the elastic strains that existed at the end of the process. These cause distortion of the part. There is extensive literature on springback as well as best practices to avoid it, but again this requires further die tuning or additional operations. This chapter will show that electromagnetic impulse forming can be used to control springback as well.

18.1.3 Fundamental Characteristics of High-Velocity Forming

When metal is formed at a high velocity using impulsive loading, the forming characteristics can be much different from those under quasi-static loading. The physics of this unusual mode of non-contact direct electromagnetic launch will be considered in detail in Section 18.2. But the most important general aspects of high-velocity metal forming are considered briefly in the following section. Because the fundamental processes are quite different from quasi-static forming, high-velocity forming offers some complementary practical capabilities. The practical implications of this will become clear in the examples in Section 18.3, where applications are discussed.

18.1.3.1 Impact and Pressure

Commonly, electromagnetic forming works by accelerating a sheet metal sample to a high velocity (over 100 m/s) and striking a hardened metal die. Very high transient impact pressures are produced on impact. A detailed understanding of this requires study of the wave behavior in solids and is beyond the scope of this chapter. Johnson's book [5] gives an excellent treatment of the background. The impact pressure, P, which is developed when two semi-infinite elastic bodies labeled 1 and 2 collide at an impact velocity V_i is given as

$$P = \frac{\rho_1 \rho_2 C_1 C_2}{\rho_1 C_1 + \rho_2 C_2} V_i \tag{18.1}$$

Here

for each material, ρ represents density
C is the longitudinal wave speed, which can be expressed as

$$C = \sqrt{\frac{3K(1-\nu)}{\rho(1+\nu)}} \tag{18.2}$$

where

K represents the bulk modulus
ν is Poisson's ratio

Longitudinal wave speeds are on the order of about 7000 m/s for most structural metals. Equation 18.1 shows that for a modest impact velocity of 50 m/s, impact pressures of 500 MPa and 1.4 GPa are generated for aluminum-to-aluminum and steel-to-steel impacts, respectively. This analysis shows it is relatively easy to develop pressures large enough to produce extensive plastic deformation at the interface. More sophisticated analyses are required to study the deformation of the interface once plastic deformation takes place. Such large pressures can be used to put great surface details into materials or to perform embossing or coining-like operations, even with very lightweight forming systems. In contrast, if coining is to be carried out over significant areas using traditional methods, one must exceed the flow stress of the material being formed by a factor of 2–3. As a result, very large presses are required for parts of significant size. Examples shown in Section 18.3.5 illustrate the principle that high workpiece speeds generated in small lightweight forming systems can allow them to replace large equipment such as coining presses.

18.1.3.2 Formability and Strength

Several studies have shown that at high-velocity and high strain rates, stretch forming can be carried out at strains in considerable excess of the usual quasi-static forming limits (which are usually represented on the forming limit diagram) [6–14]. In fact, depending upon the exact conditions of loading, impact, and so forth, the forming limits may increase or decrease. Specifically, metals with strong ductile-to-brittle transition characteristics may see decreased formability in high-strain-rate forming, due to brittle failure ensuing. Also, forming over a discontinuity such as a ledge perpendicular to the forming direction can encourage strain localization at high strain rates, reducing formability versus quasi-static loading.

There are many factors that can improve formability in high-velocity and high-strain-rate conditions. These include: (1) the fact that the strain rate sensitivity of most metals increases significantly at strain rates above 10^3 s^{-1}, (2) the observation that impact can produce through-thickness compressive stress that causes the material to deform in a manner more like forging or ironing than stretching, and (3) inertial forces tend to deter the formation of necks and slow their growth. All these factors can give much improved tensile ductility. *The important practical outcome of all of this is that in high-velocity forming, the conventional forming limit diagram does not set material forming limits.* Strains can exceed these nominal values by better than a factor of two, but the extent of improvement (or penalty) depends upon many factors including the details of the velocity distribution, material thickness, failure mode, and so forth. A comprehensive design approach for high-velocity forming does not yet exist, but it is clear that practical use can be made of ductility enhancement at high speeds and strain rates.

18.1.3.3 Buckling

High-velocity forming can have similar benefits to compressive forming as it does in tension. The dominant instability modes are stabilized by inertia. Often, in sheet metal forming, it is desirable to form a material in compression. Examples of this include tube shrinking, shrink flanging, corner drawing, and other common sheet metal forming operations. Electromagnetic forming gives the ability to perform such operations with rapid acceleration to high velocities. In high-velocity forming, inertial terms become dominant and the material tends to continue its motion along its launch vector. Hence, if, for example, a tube can be brought very quickly to a high velocity in compression, it will tend to resist buckling and large reductions in diameter are possible. There is considerable literature on this effect [15–19], and it remains to be seen if this effect can be harnessed to inhibit buckling in sheet metal forming.

18.2 ANALYSIS AND MODELING OF ELECTROMAGNETIC FORMING

18.2.1 Physical Introduction to the Process

Electromagnetic forming (EMF) is a noncontact technique where large forces can be imparted to any electrically conductive workpiece by a purely electromagnetic interaction. The physics of this interaction is covered well elsewhere [20–24]. Only a short description is included here, where the present goal is to give a technically fluent person interested in using this process an appropriate and simple approach to the design of an electromagnetic forming process. A general schematic diagram is provided in Figure 18.4. The primary inductance–resistance–capacitance (LRC) circuit denoted as 1 in the figure drives the process. A significant energy (usually between 1 and 100 kJ) is stored in a large capacitor, or a bank of capacitors, by charging to a high voltage (usually between 2,000 and

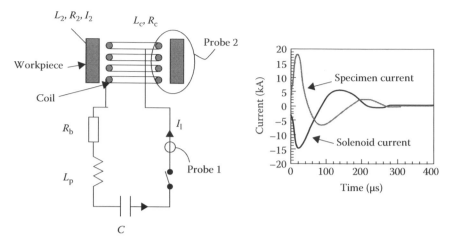

FIGURE 18.4 Circuit schematic (left) and typical current traces (right) for an electromagnetic forming operation. The solenoid can take on a variety of shapes to affect the pressure distribution, and the probes measure the currents.

20,000 V). The charge is switched through low inductance conductive buswork and carried through an engineered path near the workpiece that is presently referred to as an actuator. The current history for such a discharge takes the form of a damped sine wave and can be understood as a ringing LRC circuit. The peak current is typically between about 10^4 and 10^6 amperes and the time to peak current is on the order of a few microseconds to tens of microseconds. This high current and short pulse duration generate an extremely strong transient magnetic field in the vicinity of the coil. The change in magnetic field induces eddy currents in any conductive materials nearby (demonstrating both Faraday's law of induction and Lenz's law). Hence, any nearby metallic workpiece will carry induced currents and these will necessarily be opposite in direction to the primary current. The magnetic fields established by the currents in the coil and workpiece thus interact so as to produce a repulsive force on both conductors (according to the Lorentz force law). This electromagnetic force can produce stress in the workpiece that is several times larger than the material flow stress. If the workpiece is free to move, it may accelerate it at a velocity up to hundreds of meters per second.

Developing a relatively high system ringing frequency is important, which will become clear from the quantitative presentation that follows as well as the prior literature. If the electrical oscillation frequency is too low, intense eddy currents are not induced in the workpiece and the force developed is low. The ringing frequency is directly related to the circuit LRC characteristics, where low capacitance and low inductance favor a high ringing frequency. Materials of lower conductivity demand higher ringing frequency for effective forming. For these reasons, metals with high conductivity such as aluminum and copper are very well suited to electromagnetic forming. Carbon steel can be formed, but a sufficiently high oscillation frequency of the primary circuit is very important to the system efficiency. Metals with relatively low electrical conductivity, such as titanium and austenitic stainless steels, are almost impossible to directly form by electromagnetic forming (but they can be formed with the aid of a more conductive driver plate or similar impulse-based techniques).

In general, the highest electromagnetic efficiency can be generated when the working surface of the coil of the primary actuator, carrying a high current density, is directly coupled with the workpiece. Hence, large pressures can often be created with single turn coils and large primary current densities. Often larger magnetic fields can be created with multi-turn actuators, because multiple windings multiply the effective current density. Often these multi-wound coils are not mechanically robust enough for industrial service. Field-shaper based coils can be used to develop high electromagnetic pressure without coil failure while being able to increase and tailor the inductance of the coil. The principle of a field shaper coil is illustrated in Figure 18.5. A typical multi-turn coil

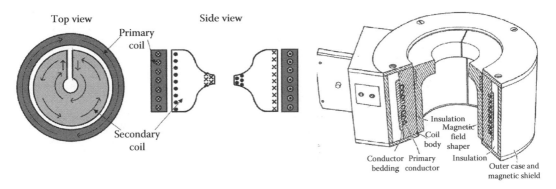

FIGURE 18.5   (a) The behavior of a field-shaper based coil looking at the coil cross section. The many turns of the primary coil induce a current in the shaper. (b) The situation looking down the axis of the shaper. A slot in the coil forces the current to the bore of the coil concentrating current there. (c) The practical construction of a field-shaper based coil.

is used as the primary inductor to create a magnetic field. This is coupled with a secondary inductor or field shaper. If properly designed, nearly the entire magnetomotive force (amp•turns) that is created by the primary circuit can be transferred to the bore of the shaper. The advantage of a coil of this type is that it can concentrate large magnetic pressures in a desired area. This is done at the cost of being less efficient than what a well-designed single-stage coil would be (the difference is typically due to having two separate transformer circuits in the field shaper case, producing more leakage inductance; however, this is convoluted with the fact that ringing frequencies will usually be different too when comparing single stage to field shaper). Using this approach, a single coil with a number of varied inserts can be applied to a wide range of compressive forming operations. This basic approach has been creatively applied to making coils of many types for many operations [25]. These can be dealt with in a straightforward analytical manner as is discussed elsewhere [26], and that approach can be used to extend the electromagnetic forming classes that follow.

In the following, the basic physical elements that must be taken into account in the analysis of the process are described quantitatively, then, limiting cases are described that allow significant simplification of the analysis. These can be divided into three basic classes, depending on the degree of the complexity of the motion of the system. Then in Section 18.3, several operations are described that show the application of this forming method. These are grouped based on the analysis classes developed here.

18.2.2   Quantitative Introduction to Electromagnetic Metal Forming

The mathematical and analytical modeling and simulation of electromagnetic metal forming has been covered in depth elsewhere [20–24]. The goal of this presentation is simply to show that in many or most cases simplified analysis methods are appropriate and comprehensive methods, such as finite element analysis, are not often required to develop engineering insight to a given problem. The problem can be considered as three coupled phenomena: the discharge of a capacitor bank to create a current pulse, the action of the magnetic field that is developed to produce eddy currents in the workpiece, and the interaction of the primary and induced currents to produce an electromagnetic repulsion.

From an electromagnetic point of view, the system is modeled as two LRC circuits coupled through their mutual inductance. The mutual inductance, M, will generally decrease as the workpiece and coil separate due to deformation. The primary, or driven, circuit is given a subscript 1 and is coupled with a workpiece that is the induced circuit, labeled as 2, and the governing equation for the primary circuit is given by

$$\frac{d}{dt}\left(L_1 I_1 + M I_2\right) + R_1 I_1 + \frac{Q_1}{C_1} = 0 \tag{18.3}$$

Here

    L is the inductance
    I is the current
    R is the resistance
    Q is the charge
    C represents the capacitance

The workpiece is also modeled as a resistive and inductive circuit (without capacitance). Its governing equation is

$$\frac{d}{dt}\left(L_2 I_2 + M I_1\right) + R_2 I_2 = 0 \tag{18.4}$$

Currents running in opposite directions in the workpiece and actuator give rise to an electromagnetic repulsion between the two elements. There are three consistent, but somewhat different, ways to calculate this electromagnetic force or pressure.

18.2.2.1 Pressure Calculation Method 1—Direct

Maxwell's equations can be used in a direct manner starting from the fundamental equation

$$\Delta \mathbf{F} = \mathbf{I} \times \mathbf{B}\, \Delta L \tag{18.5}$$

where $\Delta \mathbf{F}$ is the force acting on some wire length ΔL where the wire (or element of the material that can be idealized as a wire) carries a current $\mathbf{I}$ and is surrounded by a magnetic field, $\mathbf{B}$. This approach is easily applied if the vector current in the workpiece and the field from the actuator are known.

18.2.2.2 Pressure Calculation Method 2—Spatial Derivative of M

There are generally two components to magnetic pressure. The first is due to the interaction between the workpiece and the actuator and is given by

$$P_m = \frac{1}{A} \frac{dM}{dx} I_1 I_2 \tag{18.6}$$

where

    A is the area of the workpiece adjacent to the actuator
    x represents the distance

As the distance between the workpiece and actuator increase, dM/dx will decrease, accounting for the effect that distance has on reducing the magnetic force at the given current levels.

The second component of magnetic pressure is the self-force that results from the current running in a curved object such as a tube. This force always tends to expand a tube with a current running circumferentially, and this can be represented as

$$P_{m,s} = \frac{1}{2} \frac{1}{A} \frac{dL_2}{dr} I_2^2 \tag{18.7}$$

These magnetic pressures can be summed to give the total magnetic pressure acting on the body.

18.2.2.3 Pressure Calculation Method 3—Magnetic Pressure

The magnetic pressure, P_m, can also be seen as being related directly due to the difference between the magnetic field, B, on the inside, B_i, and outside, B_o, of the conductive workpiece (due to the existing field and current) as

$$P_m = \frac{B_i^2 - B_o^2}{2\mu_0} \qquad (18.8)$$

For a workpiece with no internal current, the field will be the same on both sides, hence there will be no magnetic pressure. Current alters the magnetic field; the location of the current within the finite thickness of the workpiece can also be important. For diamagnetic metals of finite conductivity that are subject to eddy currents with a given frequency, the current will tend to ride predominantly on the surface and decay exponentially with depth into the conductor (skin effect). The effective depth of the skin it will ride upon, δ, is given by

$$\delta = \sqrt{\frac{2\rho}{\mu_0\omega}} \qquad (18.9)$$

where ρ represents material resistivity. The excitation frequency given by ω is

$$\omega = \sqrt{\frac{1}{LC}} \qquad (18.10)$$

where

L is the effective system inductance (which usually increases as the workpiece moves from the actuator)
C is the bank capacitance

The system inductance L can take on two limiting values. At very high frequencies, the skin depth approaches zero and no field penetrates the workpiece. In this case, the magnetic field on the outside in Equation 18.8 is zero and the maximum possible magnetic pressure is obtained. Conversely, if the workpiece resistivity is high or the system ringing frequency is low, the field will fully penetrate the workpiece (i.e., the skin depth is much greater than the workpiece thickness) and the magnetic pressure in this case tends to be zero.

Another similar frequency dependent effect, known as the proximity effect, exists and becomes pronounced in the case of closely spaced conductors. Currents will tend to flow so as to minimize the energy stored in the corresponding magnetic field. This results in increased current density on the inside faces of opposed, closely spaced conductors. This offers another insight into the desire to not only operate at as high a frequency as possible, but also to closely couple EMF actuators and workpieces in space, thus increasing the electromagnetic coupling and efficiency by maximizing the magnetic pressure.

18.2.2.4 Role of Mutual Inductance

The mutual inductance between the primary and secondary (workpiece) circuits is key in understanding their interaction. Feynman's classic text has an excellent section on this as well [22]. To a large extent, the mutual inductance represents the coupling between the primary and the secondary and is also bounded by energy considerations and the relationship

$$M \le \sqrt{L_1 L_2} \qquad (18.11)$$

Thus, mutual inductance is often expressed as

$$M = k\sqrt{L_1 L_2} \qquad\qquad (18.12)$$

where k must take on a value between 0 and 1 (the difference between unity and k corresponds to the leakage inductance). Unity is referred to as tight coupling (the goal in an ideal transformer) and zero is no coupling, or the primary and secondary are far removed from each other. Generally, the mutual inductance is related to the volume enclosed between the primary and secondary circuits and decreases as that volume increases. Mutual inductance and its spatial gradient are also clearly important in determining the mutual force between a primary and secondary circuit, as shown in Equation 2.4. In short, mutual inductance both mathematically and physically provides the linkage between the primary circuit, the secondary or workpiece, and the forces or pressures between them.

18.2.3   LIMITING CASES WITH REGARD TO ELECTROMAGNETIC METAL FORMING ANALYSIS

The practically important cases for electromagnetic metal forming can be broken into three distinct limiting classes that are based primarily on how the mutual inductance changes with time during the metal forming event. It will become clear that these are limiting cases, and ultimately the choice may depend on what aspects of the problem are important. The goal of this is to avoid over-analysis of a problem. The classes are briefly defined below.

Class I—Essentially no motion of the workpiece during the experiment
Here, mutual inductance does not change during the experiment, allowing very simple analytical tools to be used.

Class II—Simple one-dimensional motion of the workpiece during the experiment
Here, motion of the sample can cause mutual inductance to change. The motion of the sample is simple enough to allow force, pressure, or stress to be calculated using coupled differential equations.

Class III—Complex motion of the workpiece
In some cases, there are strong spatial and temporal variations in magnetic pressure to be captured and mutual inductance changes during time. In these cases, full numerical solutions are required.
   Further details on each of these classes are provided below.

Class I—Essentially no motion of the workpiece during the experiment
Surprisingly, there are a number of cases in which electromagnetic metal forming can provide substantial benefit even if there is no gross motion of the workpiece. The precise conformation of a component to set dimensions, what is referred to as "calibration," is a prominent example of this, as is detailed in Section 18.3. In some cases, there may be some motion of the part, but this may be after the time scale that is important in the experiment. There can be considerable design freedom even in this case.
   Let us consider what is termed the "path actuator" in a concrete example. The path actuator has been used in a number of recent studies at Ohio State University [27–29]. The path actuator is simply a conductive nearly closed circuit that carries a current pulse in the proximity of a workpiece, as illustrated schematically in Figure 18.6. A long-life path actuator can be created from a monolithic block of a high-strength, high-conductivity alloy or single-use actuators can be created on-demand from conductive metal sheets. The current enters and exits, and the same current must flow through the full section of the coil. Note that the actuator has varied width (and thickness) along its location, giving changing values of pressure at each location and the inductance per unit length is a function of coil width. At the high frequencies used in electromagnetic forming, in accord with the proximity effect, the current in the coil will tend to concentrate near the workpiece surface, generally being

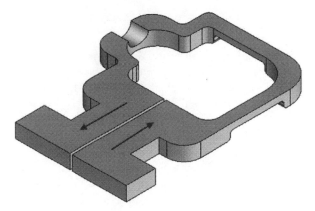

FIGURE 18.6 A typical path actuator. Current is fixed through the coil, but current density varies with the actuator width. Because of skin depth considerations, the current will tend to ride on the surface near the workpiece, provided the gap to the actuator is small.

within one skin depth of the sample. This must be accounted for in the calculations of inductance and magnetic pressure.

Imagine that this coil is used to induce a current and produce a force on a fixed sample. In this case, as the mutual inductance does not change, the primary circuit can be modeled as a simple LRC circuit. This LRC circuit would have a fixed effective primary circuit inductance, $L_{1,\text{eff}}$, which is made up of the sum of the system inductance plus the coil inductance, properly accounting for the coupled workpiece path. It could also be calculated as the piecewise sum of the inductance of each segment of a given width and length as derived from a handbook. Alternatively, a physical meter could be used to measure the system inductance and resistance at the estimated discharge frequency. This approach is actually relatively simple, as the coils can be created inexpensively. With LRC values, the current with time can be calculated as

$$I_p(t) = \frac{V_0 \sqrt{C/L_{\text{sys}}}}{\sqrt{1-\xi^2}} e^{-\xi\omega t} \sin(\omega t) \tag{18.13}$$

where
  V_0 is the system voltage at discharge
  ξ is the damping factor
  ω is the ringing frequency and the term

$$\xi = \frac{1}{2} R_{\text{sys}} \sqrt{\frac{C}{L_{\text{sys}}}} \tag{18.14}$$

where R_{sys} is the resistance of the system

With this current-time profile known, the pressure with time can be estimated at each location (based on the known coil width and effective actuator-workpiece gap, accounting for skin depth) according to Equation 18.5, 18.6, or 18.7. In all cases, these equations give the result that the pressure will scale with the inverse of the actuator width squared. Hence, narrow locations give higher force per length and much higher pressure (and higher local inductance) and wider regions give much reduced pressure. This path actuator approach allows great latitude in tailoring the pressure distribution (and location). The magnitude of the peak pressure is easily changed by changing the magnitude of the peak current. This is increased by increasing the charging voltage and reducing the fixed system inductance.

Class II—Simple one-dimensional motion of the workpiece during the experiment

Class I has mutual inductance that does not change during the experiment, and it can be modeled as a simple LRC primary circuit that does not change with time. In many cases, the workpiece will move significantly, causing changes in mutual inductance during the experiment. Because the equations are coupled, the ringing frequency in the primary and secondary can change substantially during the experiment. If a single displacement can express the displacement between the workpiece and the actuator, then Equations 18.3 and 18.4 can be coupled with the equations for force and resistance; and coupled differential equations can be used to fully solve for currents, pressure, velocity, and displacement with time. Some geometry satisfies the requirement of simple motion. These include: ring expansion or compression by a helical actuator, expansion or compression of a tube that remains cylindrical, or launch of a plate that remains planar. Many examples of such solutions exist in the literature [6,20,21,30] and are very accurate when the constitutive behavior of the material being deformed or accelerated is well known.

Class III—Complex motion of the workpiece

In other cases, the motion of the workpiece is not simple, requiring full numerical solutions to simultaneously track the changes in mutual inductance as well as the local pressure, deformation, resistance, and so on. There is a rich body of literature that has been developed in this area, and there have been several approaches that work well. One of the earliest comprehensive studies was of a two-dimensional problem of this type carried out by Takatsu and now general packages exist that are based on a number of base packages including CALE [30], LS-DYNA [31], and other commercial solutions where the deformation solution is coupled to an electromagnetic solution [32]. While a complete review of this is beyond the scope of this paper, the methods have become quite accurate and robust, however, they still tend to suffer from long computation times for truly three-dimensional analyses.

18.3 MOTIVATION AND EXAMPLES OF POSSIBLE OPERATIONS

The short time duration and high speeds in electromagnetic forming give related manufacturing operations very different characteristics than quasi-static forming. In terms of the equipment used, the comparison is like using a press versus a hammer to do useful work. Traditionally, if large pressures are to be obtained, large presses are needed. Alternatively, large forces can be obtained by the rapid-deceleration of a hammer-head upon impact. However, even after the free fall of a hammer from 3 m high, the velocity is about 7.5 m/s. The operations considered here typically produce velocities around 100 m/s or higher.

This has many characteristic advantages. In addition to providing lightweight forming equipment, the high strain rates and velocities can inhibit the formation of instabilities. Necking can be de-stabilized by both velocity and possible changes in the material constitutive behavior at high strain rates. Also, wrinkling is stabilized and can be virtually eliminated at high velocities. Figure 18.7 provides a cartoon that illustrates how increasing velocity can change the fundamental response of an impact problem [33]. Here, simulations are shown where a hypothetical piece of aluminum with a flow stress of 15 MPa is impacted with an ideally nondeformable block. As the impact speed increases, deformation becomes more and more severe and the impact pressures rise. These fundamental aspects of how strain rate and velocity affect formability and wrinkling have been discussed elsewhere [3,7,34,35].

This section will show that there are a number of different ways that electromagnetic forming can be used to effectively treat metal forming problems as an agile sheet metal forming method. These are divided roughly according to the class of analysis required as shown above. Here, a simple introduction to the several methods is presented, rather than a deep analysis. The following

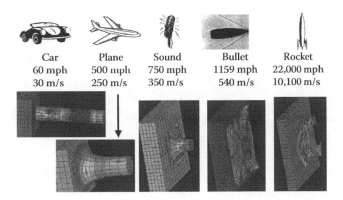

FIGURE 18.7 Results of a computed impact between aluminum rod with a flow stress of 15 MPa impacting a rigid wall. The bottom five simulation results correspond to the top cartoons of speed, respectively. Simulations were carried out by Peihui Zhang using the AUTODYN code.

is intended as a broad review with references listed to provide more extensive detail. The practice of hybrid forming using both traditional and electromagnetic forming makes more explicit integration of the energy fields, and for this reason, additional emphasis is placed on these techniques.

18.3.1 Class I—Springback Control and Shape Calibration

Springback is a very common problem in sheet metal forming wherein the component is shaped (or stamped) between two fixed tools. After removal from the toolset, inhomogeneous elastic strains cause distortion of the part. Springback is particularly prominent and troublesome for thin components stamped from strong alloys with low elastic moduli. One relatively surprising, but effective, use of electromagnetic forming has been in the precision dimensional calibration of shapes in sheet and thick metal structures [36–38].

In each case, the procedure used to effectively correct springback resembles that shown in Figure 18.8. The part with unwanted distortions is fixed such that there is a die with the desired net-shape on one side; the system forces the part into near conformity with the die and an electromagnetic actuator on the other side, wherein the current path runs near the regions that need the most correction. Upon pulsing, the shape becomes very close to what is in the net-shape die.

This basic approach has been demonstrated for simple corners and more complex examples. Two prominent examples are shown in Figures 18.9 and 18.10. The first case is that of a rocket nozzle that was re-shaped with several electromagnetic impulses onto a net-shape die [36]. Further details on this process are available in the figures and the related publication. What is significant here is that this method permits one to precisely calibrate a three-dimensional item to a net-shape without any compensation for springback. The other industrially common solutions

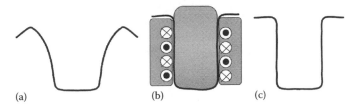

FIGURE 18.8 General approach used in springback control. (a) The part after forming where upon removal from the tool it has large dimensional errors. (b) Placement of the part into a fixed tool with the net shape. The circles indicate the actuator path with outbound and inbound currents. Without a current pulse, the part would return to the original shape. With appropriate pulsing, the desired shape can be obtained as shown in (c).

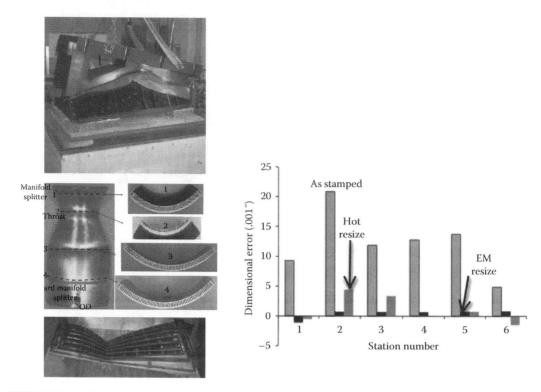

FIGURE 18.9 From top to bottom photos on the left show (top) the Aerojet stamped rocket nozzle, (middle) the electromagnetic actuator used in electromagnetic calibration of the liner, (bottom) the urethane-potted liner with the nozzle segment above it and the form die to be calibrated to on the top. In the process, several modest electromagnetic impulses are used to provide the nozzle with an impulse that throws it against the die. If the impulse is too small, there are no changes in dimension. If it is too large, it can cause collapse of the cooling channels. The best results are obtained when several modest impulses are used. The graph on the right shows the dimensional error along the minor radii of curvature moving from the nozzle (station 1) to barrel end (station 2 being at the throat). Dimensional error is always less than 0.0015″ after EM resizing, considerably better and more reproducible than those when hot resized. EM calibrated panels have been welded together and test fired successfully.

FIGURE 18.10 Two TRIP 700 U-channels. The upper sample has been calibrated by means of a 24 kJ impulse at each sidewall. The lower profile is as deep drawn after removal from the die.

to this problem are using rubber hammers and check-gages to calibrate the part or using combinations of simulations and iterative approaches to guide the development of tools that properly overbend the part to compensate for springback. Other advantages of electromagnetic (EM)-sized panels in comparison with hot-sized panels are: (1) evidence of less channel deformation, resulting in more uniform flow in the channels, less pressure drop, and greater and more uniform flow admittance and (2) dramatic cost and schedule improvements—calibration cost is reduced by 98% and operation time is reduced by 95%.

The second example is shown in Figure 18.10. The drawn of a high-strength steel frame suffered die wall curl and springback during the normal forming operation. The use of a system very much like that shown in Figure 18.10 was able to nearly fully correct the springback [37,38]. The actuator had a geometry very much like that schematically illustrated in Figure 18.8, with a serpentine shape.

At this point, the mechanisms for springback correction are not fully understood, but they are effective. It may be that impact with the net-shape die is able to impart small plastic strains that remove the deleterious elastic strains. If this were the case, sheet-die impact velocity should be very important, however, the standoff between the die and the sheet does not seem to be really important. Another possible explanation is that the current running through the material itself softens it. With both approaches, there is a great amount of local stress and power densities that are somehow finding a way to overcome the local inhomogeneous elastic strain. Even without complete understanding, it is clear that this technique is effective, even if the mechanisms are still debatable [39]. This is a fruitful area for further study and application. The analysis methods for Class I forming can also assist in basic process design because there is no significant motion of the actuator or workpiece over the duration of the operation.

18.3.2 Class I—Electromagnetically Augmented Stamping

The next example will be developed in greater detail, as it is directly related to the issue of controlling the energy distribution in sheet metal forming. Electromagnetic actuators can be embedded in dies and can provide local impulses that can be small and repeated. Each one of the pulses may move the material a small distance and provide small plastic strains, but if they are repeated, they can deposit plastic strain energy as required. In this way, plastic strain distributions are directly controlled and largely conform to Class I analysis. If motion is significant, Class II or III analysis may be required. This is a very good example of how energy field integration can help overcome engineering difficulties.

Sheet metal forming is almost always carried out in one of two fundamentally different ways based on the number of parts required. In *mass production* (such as automotive production), typically 10^4–10^7 parts per year are required. Here the process must be highly reliable and must require no significant human intervention. High costs for fixed tooling can be amortized over many parts. The forming process in mass production known as stamping was illustrated in Figure 18.3. In designing this metal forming process, one walks on a fine line. Varied proportions of stretching and drawing create the increase in surface area from the initial blank to the final part. Although one must stretch the component to make the part, if it stretches too far or the stress becomes locally too large, it will tear (as is usually described by the forming limit diagram [40–42]). One can allow more metal to be drawn into the die by reducing the binder force, but this will encourage wrinkling. Hence, it is often impossible to form a required geometry from a given material in a single press operation.

The short run production of sheet metal articles is very common in the aerospace industry as well as in the production of heavy trucks and other military, specialty, or prototype vehicles. Here, production volumes may be measured in the dozens to few thousands per year. The fabrication of fixed matched tools (which can start at tens of thousands of dollars per set) can be prohibitively expensive. Therefore, single-sided tooling solutions are sought. Hydroforming and superplastic forming are common examples of this. Figure 18.11 schematically shows sheet hydroforming as is commonly practiced on a Vearson-Wheelon or ABB flexform press. While tooling costs are minimized using this production method, it suffers from high cost per-part. Again, the goal is to hold the part firmly enough with the binder (and developing the proper spatial distribution of restraint) so that the part resists wrinkling. Then the punch will push the sheet into the die cavity,

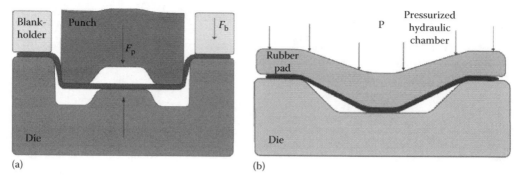

FIGURE 18.11   (a) Generic stamping process used in mass production; (b) hydro-forming process that is common in aircraft production.

drawing material into the cavity and stretching it. The key in making a good part is reducing the localization of strain (or energy dissipation) so that the material does not reach its local failure criterion.

There are several solutions commonly used to treat difficult-to-form materials or component geometries. These are all essentially designed to distribute the forming energy over a larger area, in order to reduce the propensity for energy and strain localization. However, the great advantage of a net-shape forming process is that the fixed tooling precisely constrains the shape of the part to be formed. This important advantage will be maintained in the new approach, while allowing local control of forming energy.

In improving the ability to make a given part, the majority of the conventional solutions are focused on devising ways to better distribute the forming energy over the part (i.e., reducing localization). These approaches include the following.

Forming lubricants [43–45]: using forming lubricants is one of the obvious methods of spreading energy over the part. As the coefficient of friction between the part and die approaches zero, for simple systems, the stress distribution of the sheet inside the blankholder tends to become uniform, eliminating localization. There are limitations to this approach. The system geometry retains complex stress distributions and lubricants are often environmentally objectionable, and their removal produces environmental and economic costs.

Material properties [46–49]: several approaches can be used to improve intrinsic materials formability, including increasing the amount of strain hardening through the use of softer materials; using superplastic materials; using tougher or more fracture resistant materials, which are usually more expensive; and tailoring anisotropy to encourage draw in. These solutions limit material selection and favor lower strength materials or high temperature forming, adding cost and/or reducing structure performance.

Intermediate annealing: intermediate annealing is commonly used in the aerospace industry to make complex parts from a single die. This essentially relies on simply reducing the value of $d\sigma/d\varepsilon$ at chosen intervals. This process is quite costly and environmentally deleterious, as many cycles of lubrication annealing and cleaning are typical.

Multiple operations [50]: in mass production, if a component cannot be made in one press operation, it is common to fabricate it in multiple press operations in separate tools. This is the main concept behind progressive dies or transfer presses for larger parts. The typical concept is to initially draw and stretch to approximately develop the required surface area of material into the die cavity and then use a later operation to form the detailed shape. The obvious drawback of this is the increased cost and complexity of using multiple hard tools and multiple press stands. In low-volume production, the tooling cost cannot be justified.

Active blankholders [51–55]: in order to draw more material into the die cavity without wrinkling, there has been significant research on spatially and temporally changing the blankholder force during the forming stroke. This can increase the ability to make a part, but obviously adds to the complexity of the forming press and tooling system. Note that this process is passive in the sense that forming energy is not added to the system.

Improved die design [56–58]: while this is useful, there are limitations on the existing process. For example, the limits by which one can draw a simple shape such as a cup are well known [59]. Such limits also exist for more complex component shapes, but are more difficult to compute.

Other approaches, such as viscous pressure forming [60–62] and selectively heated tools, have been used to modify the forming energy distribution in the part.

All of the approaches described above essentially conform to the paradigm of remotely applying the forming energy from the mechanism driving the press and modifying the details related to the system to enable the production of a given part. The hybrid forming approach in the following is based on the idea of delivering the deformation energy where it is required in the part. The key enabler of this is the incorporation of electromagnetic actuators within otherwise relatively conventional stamping tools.

Figure 18.12 shows how electromagnetic actuators can be added to tooling in short-run production with rubber-pad hydroforming and in mass production with electromagnetically augmented stamping systems as illustrated in Figure 18.13. Figure 18.12 provides a cross section through an arbitrarily shaped part where this section is nearly in plane strain. Without any electromagnetic augmentation, the incremental local of plastic work deposited into the part is plotted as a function of position and is shown below the sketch. It shows that the part will fail in the unsupported region. Increasing the effectiveness of the lubricant will aid in the drawing of the blank off the blankholder surface and may allow some further deformation of the flat region at the bottom of the tool, which is in contact with the die. Electromagnetic actuators (like the path type) can be embedded in the

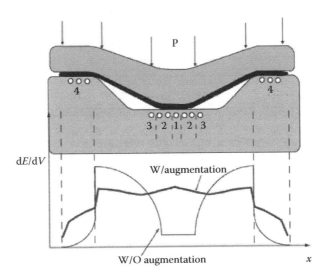

FIGURE 18.12 An electromagnetically augmented rubber-pad hydroform process. In concept it is simple to independently control the energy provided to the actuators in regions 1, 2, 3, and 4. These increase the energy available for deformation in regions that would not otherwise form. This will enable much more aggressive parts to be fabricated in a single press operation.

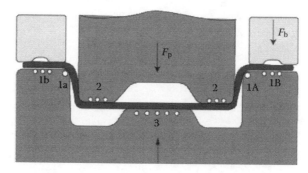

FIGURE 18.13 Mass production hybrid approach-electromagnetic actuators are used to assist draw and stretch. Actuators located in regions like 1a can be used to put material in compression in corners to assist the forming of pan corners without wrinkling.

blank holder surface and at the bottom of the tool, as shown in the figure. When one of these actuators in the blankholder region is coupled with a capacitor bank and pulsed, pressures exceeding the fluid pressure by several times can be easily created. This will develop a stretched "bubble" of material above the coil and cause plastic extension of the material (i.e., plastic work will be done in the regions above the coils and nominally in a direction that will allow the material to feed the open cavity). This bubble will respond to the unbalanced forces and material will slide to feed the open volume in the die. It is a simple matter of controlling which actuators will be energized and the energy level that will be provided with each shot. The fixed dimensions of the die cavity assure dimensional accuracy in the usual way. While using this technique may enable the fabrication of aerospace components in a single augmented press cycle, it would represent an important cost savings if three to four process cycles (lubricate, form, clean, anneal, and repeat) could be reduced to one or two.

The approach to augment stamping tools for mass production is conceptually similar to that described for hydroforming and is illustrated in Figure 18.13. The primary goal is to provide forming energy to regions that are receiving too little strain or draw-in. Starting with the blankholder, the function of actuator 1b is to draw in the outer region of the blankholder surface taking advantage of the unbalanced forces over the surface and possibly to extend the material in the x direction. The "bubble" that 1b could form can be actively pulled into the die cavity using actuator 1a. These can be run sequentially to create what is in essence a controllable electromagnetic pump that can feed material into the die cavity. The function of actuators 2 and 3 is to stretch these regions essentially in the x-direction to provide the increase in line length needed to create the part.

In all of these cases, both with respect to mass and short-run production, the key idea is to energize the actuator many times where a small work increment is done in each cycle. There are many advantages to this approach. It is not difficult to cycle the capacitor bank many times in a second or less. Capacitor bank systems often have many discrete capacitors that once charged can be individually switched. It is a simple matter to charge all the capacitors while loading a part and then closing the switches for the capacitors (all connected to a common bus line) sequentially as the punch moves over a time period that may be a small fraction of a second. Total part-to-part cycle time on the order of 5 or less seconds can be easily achieved (as is typical in auto body panel stamping).

Electromagnetically augmented stamping was carried out in the PhD project of Jianhui Shang, and further details are available in this document [63]. His approach was to use many relatively small electromagnetic impulses to improve the draw depth of a pan-like component. An existing tool set originally developed by Hasida and Wagoner [64] was modified for these experiments. A replicate punch was formed from a polymer-metal composite known as STAMP. Then this STAMP

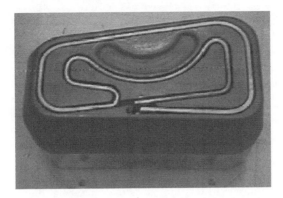

FIGURE 18.14 Electromagnetically augmented metal forming punch. The brass pathway provides pulsed EM pressure.

punch was machined to accommodate an actuator that was based on a brass bar bent to the appropriate configuration and inserted into the punch. Figure 18.14 is a photo of the punch inserted with an actuator.

A coaxial cable connects the capacitor bank to the actuator embedded into the punch. The capacitor bank supplies the current pulse to the actuator. During the stamping process, the blankholder load is fixed and the punch is advanced a prescribed amount between discharges. The punch is then advanced incrementally with each discharge until failure.

The basic procedure is that the punch is advanced and then a small impulse is driven trough the coil. This puts plastic strain across the face of the punch. The augmented punch is then advanced again. A typical result of forming is shown in Figure 18.15. It shows that the hybrid method can greatly improve the draw depth of the parts without any reliance on lubricants by using electromagnetically assisted sheet forming. The draw depth increased by 44% (from 4.4 to 6.35 cm). This punch and coil has suffered no damage after well over 1000 shots.

This experiment clearly shows that the approach of placing forming energy where it is required can dramatically increase the ability to make aggressive sheet metal parts. One primary effect of the embedded electromagnetic coil is to produce tensile strain across the top surface of the part, and the tooling system constrains the part shape in the usual way. A detailed analysis of this [63] shows that the increased draw depth is produced both by providing additional strain across the face of the part, but also by encouraging additional draw-in of metal into the die cavity. Together these effects are quite potent. Shang's PhD thesis also demonstrates examples wherein adding an electromagnetic actuator to the blankholder surface (as shown in Figure 18.13 as region 1) can also aid drawing dramatically.

The progress with the stretched and drawn pan shows that this method does work and the process enables the forming of less-formable higher-strength materials without a reliance on lubricants. Forming depths are increased by broadly moving deformation energy to areas that may otherwise not deform. Also, more challenging or complex shapes can likely be fabricated in a single operation. Moreover, this approach is easily modified because instead of relying on hard tool design, the

FIGURE 18.15 6111-T4 aluminum formed conventionally (left) and using 22–5.4 kJ pulses, without lubrication. (Draw depth increased from 4.4 to 6.35 cm. Clamp load 35 klb.)

energy-time schedule (software) greatly affects the strain distribution in the part. The adoption of such methods can dramatically reduce the amount of time required for die development.

In the example shown, the coil configuration used was chosen based largely on convenience, rather than a systematic analysis of the required energies or strains. It would be desirable to develop a formal design method based on the concept of mapping the required deformation energy onto the blank and then designing coils and schedules of punch depth and energy for each pulse that will optimize the ability to make the part.

18.3.3 Class I—Punchless EM Shearing

As schematically sketched in Figure 18.6, the path actuator can provide a force along a length and is ideally suited for a number of operations such as shearing, flanging, hemming, and beading, where energy is to be deposited along a line or path-like region. In its simplest form, punchless shearing is a Class I operation. If enough pressure is deposited along a line, this can cause shearing over a tool on the opposing side. The path actuator can take the form of virtually any nearly closed path so long as it does not cross itself. The primary discharge current is provided by the capacitor bank and induces an opposing secondary current of nearly equal magnitude in the nearby sheet.

Figure 18.16 schematically shows how the path actuator interacts with the workpiece and die. This simple system can be used to flange or shear complex shapes in the plane of the workpiece sheet. For shearing, the best results are obtained by centering the coil over a cutting edge and using sufficient electromagnetic pressure.

The electromagnetic pressure is controlled by a few important degrees of freedom; in particular, the magnetic pressure in this case basically scales with the square of the current density. Therefore, for a fixed peak current in the pulse, pressure scales with the inverse of the actuator coil width squared. As the current must be fixed around the length of the actuator, larger pressures will be produced where the actuator is constricted.

Another important degree of freedom is the coil position in respect to the forming features. When it is centered over the forming edge, shearing is encouraged. If it moves outbound to the sheet edge as shown in Figure 18.16, flanging is encouraged.

Electromagnetic shearing also lends itself greatly to agile forming by the relative ease involved in creating tooling. In a traditional mechanical press, a precisely matched punch and die are used on both sides of a sheet to create a sheared edge and having tightly controlled punch-die clearance is essential in reducing the burr on the workpiece. Costly fine-tuning of these dies must be done to achieve the desired performance. Only one die is required for electromagnetic shearing, which immediately reduces the tooling cost and time. Complex path coils, dies, and insulating backings can be created by quick and relatively inexpensive means such as those illustrated in Figure 18.17 from the MS thesis of Golowin [27]. A path actuator with a smooth shape and a width of about 6 mm was water-jet cut

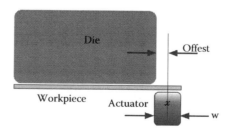

FIGURE 18.16 Cross section of a path actuator interacting with a workpiece and a die. While the current that runs through the actuator is the same at all locations, local coil width sets the current density and pressure scales approximately with width squared. The local die radius, offset of the actuator and discharge energy determine if there will be local bending, flanging or shearing.

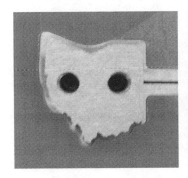

FIGURE 18.17   At left, the shearing die of Ohio State contour prepared by wire EDM; laid over the smooth water-jet cut copper coil is within 3 mm of the die's cutting edge. At right, a successfully sheared sample.

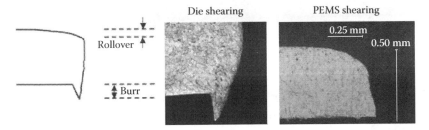

FIGURE 18.18   Conventional press shearing causes burr and rollover as illustrated on the left, and photographed in the center. In order to avoid burrs and slivers, tight tolerance between the punch and die is typically required. The results of many punchless electromagnetic Shearing (PEMS) experiments on copper, aluminum, and steel alloys have robustly shown a cut profile as shown on the right. Modest rollover is present, but no burr. The tolerance is set by the single tool that sits on top of the sheet and path coil.

from copper and a simple shearing die was created by wire EDM machining from low carbon steel. This coil and die pair was tuned to shear thin sheets of aluminum, copper, silver, and magnesium sheets by adjusting only the capacitor bank discharge energy. Materials with a low conductivity, such as austenitic stainless steel, can be sheared by the use of a more conductive driver sheet.

Many experiments have indicated [27] that the edges created by electromagnetic shearing show almost no burrs and do not suffer from slivering, as indicated in Figure 18.18. A traditional mechanical die would require very tight clearances to achieve comparable results, and this produces considerable expense in the production of the toolset. The ability of electromagnetic shearing to create high quality edges and hold tight dimensional tolerances is ideal for low volume, specialized production runs associated with agile forming. A clearance tight enough to minimize burrs and slivers can lead to binding and rapid tool wear, which degrades the edge quality over time. Water-jet and laser cutting are capable of relatively tight tolerances, but suffer from a high cost per part as well as large capital costs. Water-jet and laser cutting also produce serrated part edges that are subject to fatigue crack initiation. The edges produced in electromagnetic shearing are smooth. Furthermore, electromagnetic shearing and forming could be integrated with stamping, as another example of hybrid processes to yield additional savings.

18.3.4   Class II—Conformal Impact Joining

Conformal impact joining or impulse-based interference-assembly is by far the most common industrial application of electromagnetic forming. It is used to collapse a tube or ring onto nominally axisymmetric mating elements and has received some attention in the engineering literature [3,65–71]. In many cases, the strength of the joint is not of utmost importance, for example,

(a)   (b)   (c)

FIGURE 18.19 Prototype of torque carrying tube for aircraft application. The steel tube end with recesses for strength are shown in (a). The tube after destructive torque testing is shown in (b) and a cross section of the crimped tube is shown in (c). High-velocity forming produces a natural interference fit such that the joint is slop-free, and even though there is no metallurgical bonding between the tube and the end, the parent strength of the tube is maintained. (Courtesy of Gunter Zittel, Elmag Inc., San Diego, CA.)

where an electromagnetic crimp is used to seal a mechanism such as the casing over a fuel pump. However, in cases where high strength and low mass is paramount, such as in torque-tubes for Boeing 777 aircraft (see an example of electromagnetic conformal joint from a similar component in Figure 18.19), the process can provide joints with the strength at or near the strength of the weakest component in the joint and still meets the safety and fatigue life requirements [71]. In addition to higher strength than available in welding, high performance conformal impact joints made by EMF have several advantages over traditional joints. The joints are distortion free and dimensionally accurate because of less heat input and a very small, if any, heat-affected zone and no distortion. High dimensional accuracy is possible with proper fixtures. This is a definite advantage over fusion joints. Excellent structural joints are much easier to produce as compared with solid-state electromagnetically driven impact *welding* (which is discussed in Section 18.3.6). In conformal joining, much lower impact velocities can be used, whereas impact speeds in excess of 300 m/s are typically required in impact welding.

Three examples of pulsed electromagnetic conformal interference joints are shown in Figure 18.20. The typical physical process consists of an axisymmetric tube that has an electromagnetic actuator placed outside of it (often using a field shaper to concentrate the field in the appropriate region) and a mandrel on the inside of the tube. The discharge of the capacitor bank accelerates the

FIGURE 18.20 Example of three possible crimping operations. The component on the left is a simple aluminum tube with nuts crimped on both ends. The center example shows a copper tube expanded with an expandable coil into an aluminum plate. The example on the right is an archival example circa 1960 inherited by the author that nicely demonstrates the capabilities of electromagnetic joining. It is an aluminum tube with a compression crimp to a complex hexagonal shape on the bottom and a wrap-around onto a ring. In all cases high strength, slop-free joints can be created.

tube inward onto the mandrel. When the tube strikes the mandrel, deformation takes place, causing the two surfaces to conform, and almost invariably there is a remnant interference fit between the two components. This is likely caused by the relaxation of the shock front, which induces the elastic interference. The tube that is launched is described well by the approaches described in the Class II analysis, but, upon impact, the analysis may depart from the axisymmetric assumptions. A detailed analysis of this part of the process may require finite element analysis. If the impact velocity is known, a full electromagnetic analysis may not be required. Instead once velocities or pressures are known, a fully mechanical impact analysis can be carried out.

Alternatively, an expansion actuator may be placed inside a tube, driving the tube into a cavity to develop a crimp joint.

There are three different mechanisms that give mechanical strength to conformal impact joints that have been emphasized by different studies. First, the dynamics of the tube impact on the mandrel and the remnant interference strain were studied in the PhD work of Zhang [33]. Rings were crimped onto mandrels of various materials. The level of interference strain was measured by electromagnetically crimping simple tubes onto round mandrels, then putting strain gauges on the rings, cutting them, and measuring the elastic strain released. The results demonstrated that beyond a critical impact velocity, the level of interference strain increases with increasing impact velocity. In addition, the results showed that the tensile hoop stress in the ring could approach the flow stress of the material at high impact velocities. Golovashenko [72] came to similar conclusions for the case of a ring being expanded into a pipe-like region.

The second important effect that gives strength to tube-based conformal impact joints is the mechanical conformation between the tube and the mandrel. This results in significant torsional and axial strength. Recent studies for joints similar to the torque tube in Figure 18.19 by Park et al. demonstrate that the strength of these types of joints can be accurately predicted by conventional finite element analysis. Lastly, even small undulations of the mandrel surface can give excellent joint strength. The study of Eguia et al. [65] shows that impacting onto knurled or threaded surfaces with axial grooves can give excellent combinations of axial and/or torsional strength. These grooves can have thicknesses that are comparable to, or less than, the tube thickness. It is possible to use interference, macroscopic undulations, and small-scale surface relief together to make very effective joints at low levels of electromagnetic energy. This is a fruitful area for future study.

In general, this method is ideally suited for forming frame structures using tubular members because it is distortion-free, easily and inexpensively implemented, and produces very high joint strength.

18.3.5 CLASS II—THE UNIFORM PRESSURE ACTUATOR

A variant on the simple tube expansion actuator shown in Figure 18.4 is the uniform pressure actuator (UPA). This is a device whose primary purpose is to accelerate a piece of sheet metal to a high velocity (typically 100–300 m/s) over a distance of a few millimeters. The construction and application of this device have been published earlier [73–77], and in terms of analysis, it is relatively similar to tube expansion with the exception that only part of the secondary circuit is able to move. A UPA can be treated similarly to a tube expansion through a number of coupled differential equations as has been shown in the PhD thesis of Kamal [74].

The UPA is a tool ideal for agile sheet metal forming in at least three ways. First, it uses one-sided tools. Second, it uses very light tooling. Third, it provides opportunities to perform multiple functions in a single operation. The construction of the UPA is schematically shown in Figure 18.21. A current pulse from a capacitor bank runs through the central primary coil. This is coupled with the outer structural conductor that has a conductive workpiece or driver that bridges the open ends. The magnetic field created by the induced current in the workpiece is in opposition to the magnetic field created by the primary current, repelling the part. Typical current traces and velocities during acceleration are shown in Figure 18.22 [78]. The velocity is dissipated by impact with a tool surface.

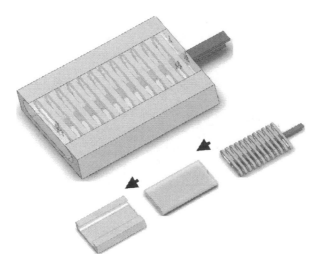

FIGURE 18.21 Three-dimensional schematic of the construction of the uniform pressure actuator. The inner coil represents the primary circuit. The secondary is created by the combination of the outer shell and workpiece, or driver, that completes the circuit.

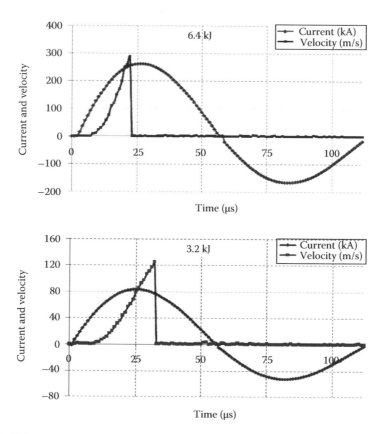

FIGURE 18.22 Primary current and measured velocity in the launch of a 0.006″ thick copper driver and 0.003″ thick grade-3 titanium sheet from a uniform pressure actuator at launch energies of 6.4 and 3.2 kJ, respectively. Standoff distance was 2 mm, at the end of which the sheet strikes a die face and comes to zero velocity. The fact that primary (and secondary) currents are running at the moment of impact maintains a pressure on the sheet after tool impact, and this reduces the rebound effects.

Impact produces forming, embossing, or cutting. Because of the high strike velocities, it is easy to produce very high impact pressures (and hence highly detailed surfaces) over a wide region of the material. If the pressures exceed about three times the flow strength of the workpiece, it will conform fully to the die surface. As such, it can be used for forming, embossing, or micro-embossing over large areas. An example of this is shown in Figure 18.23. Here optical diffraction gratings were reproduced onto copper and aluminum sheets by impact with an initial grating that was formed in hard electroformed nickel. Large surface areas can be micro-embossed in this manner with very simple and light tooling.

Forming can also be carried out over longer length scales. Because the impact velocity is high (>100 m/s), strain rates upon impact can be on the order of 10^{-5} s^{-1} or greater (smaller feature sizes cause higher strain rates at the same impact velocity) and there is a large net hydrostatic compressive stress component upon impact. Therefore, the usual rules that govern quasi-static forming limits do not apply. It is not yet clear how much benefit this may afford in the ability to make parts. An example of recent experiments at Ohio State [79], seen in Figure 18.24, shows some preliminary

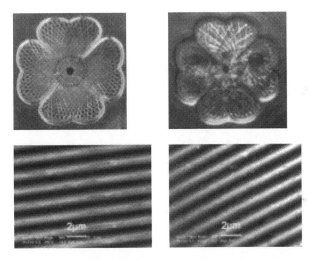

FIGURE 18.23 Examples of micro-embossing onto sheet from a coin with a holographic optical diffraction grating image (about 2.5 cm across) into 0.13 mm thick copper sheet (left) and 0.25 mm thick 5052-H32 sheet. Both experiments had rough (several torr) vacuum on both sides of the sheet and a standoff of 2.32 mm. The copper was formed at 2.4 kJ; the aluminum was formed at 4.0 kJ. SEM images compare the original holographic surface (electroformed nickel) and the pattern embossed in the copper (right). In both cases the entire area formed was about 100 × 75 mm.

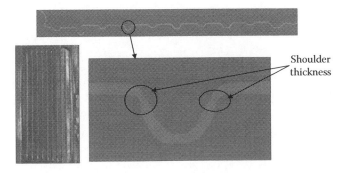

FIGURE 18.24 Overall plan view of a copper workpiece that is driven into a die with a series of grooves with varied depth, wall angles and entry radii. This tool is being used to assess formability in impact forming.

work. A forming die was developed to assess the forming limits when forming shallow channels in nominally flat plates. In general, very high local strains can be developed without tearing in high-velocity impact forming.

18.3.6 CLASS III—IMPACT WELDING

There are a number of processes that can be idealized as two-dimensional, but they may require a complete analysis of the motion of the workpiece in order to produce a suitable engineering analysis of the process. Flanging, hemming, and welding are three similar processes that lend themselves well to electromagnetic forming, but they require a detailed analysis of the shape the workpiece takes over time during forming in order to optimize the process. While these processes have some technical similarities in the way workpiece motion is developed, only electromagnetic impact welding will be considered here.

The solid-state process referred to as magnetic pulse welding (MPW) has gained considerable attention recently in the welding community [80–88] because it is able to join both similar and dissimilar metals. Dissimilar metals are often a special problem. Extensive brittle intermetallic phases in fusion welding generate brittle interfaces that are not suitable for structural applications.

Unlike the fusion welding process, MPW does not need filler metals or shielding gases. The most important advantage is that it eliminates the heat-affected zone and fusion zones [85,87–89]. Based on these practical advantages, MPW technology has great potential in a variety of applications [90–92]. For example, aluminum alloys and steels are both widely used in the automotive industry, but they cannot be fusion welded because the joint has very poor strength due to the formation of iron-aluminum intermetallics.

MPW is a variant of several solid-state impact welding processes. The oldest and most established of these is explosive welding [93–98]. Explosive welding is usually carried out by separating a cladding layer by a few millimeters from a base metal plate, spreading explosives on top of the flyer (clad) layer, and detonating the explosive from one corner. This causes the flyer to impact the base at a high velocity (usually on the order of 1 km/s). If the impact angle is appropriate (usually under 15°), a jet of material is formed that cleanses both materials' surfaces prior to creating a true metallurgical bond. There has been extensive research in this area, and this helps in our understanding of MPW.

MPW is thematically similar in that two metal surfaces must collide at a sufficiently high velocity (usually over 200 m/s) and at a proper impact angle. If these conditions are met, a metallurgical bond will form. Because the required speeds are more difficult to obtain using magnetic means versus explosives, surface cleanliness and reaching proper impact conditions are more critical in MPW.

There is a significant history in the MPW of tubes onto mandrels [85,89,99–102]. This is similar to the crimping process described earlier, but higher impact velocities are sought and the impact angle must be controlled. Companies such as Maxwell-Magneform and Pulsar have commercialized tube-implosion MPW-based processes. Typically, energies of over 20 kJ are required, and the production of long-life actuators may prove challenging for this application. Driving material more than a few millimeters, high pressures, and current densities are required that may produce fatigue in the highly stressed regions of the actuators.

Recently, a few groups have practiced magnetic pulse seam welding [87,88,102–105]. This allows lap welds along long lengths of sheet metal and most of these experiments have been done with small and efficient capacitor banks with 5 kJ or less of stored energy. Typically, this can produce several cms of weld length between 1 mm thick aluminum and base materials such as aluminum, steel, or other metals. Longer lengths and curved joints should be an easy extension of this work, but this has not been demonstrated to date. The basic approach used at Ohio State is shown in Figure 18.25. The high current density on the narrow side of the "U" shaped actuator gives a high local pressure to the end of the flyer sheet, accelerating it towards the fixed target sheet. If the velocity

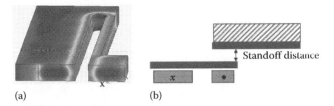

(a)   (b)

FIGURE 18.25 (a) Shows the physical configuration of the out and back coil typically used in EM pulse seam welding. The magnet current density simulation without a workpiece interaction is superposed. The workpiece covers both the out and back regions of the coil and captures an eddy current. This produces a large electromagnetic pressure on the exposed end shown in cross section in (b). This end must strike the counter face at a high-velocity and at an appropriate angle to produce welding.

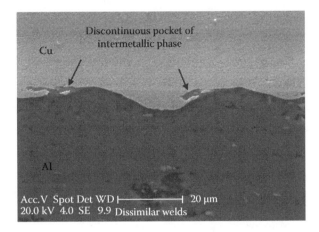

FIGURE 18.26 Typical interface structure of a dissimilar magnetic pulse seam weld between copper and aluminum. Most of the interface is discrete with no intermetallic formation. However, small isolated pockets are often present.

and impact angle are within the critical windows, a metallurgical bond is formed upon impact, as is illustrated in the cross section shown in Figure 18.26. These joints can have excellent mechanical properties, which normally exceed the strength of the weaker base metals. Typically, the interface has regions where there is native bonding between the base metals with no intermetallic formation, as well as small isolated regions where intermetallics have formed. Overall, these interfaces have proven to have an excellent combination of mechanical properties, but there are not many systematic investigations. MPW interfaces generally have much flatter interfaces than those produced at the higher energy densities in explosive welding.

For explosive welding, there has been much empirical development by practitioners at companies such as Dynamic Materials of Mount Braddock, PA. Their general experience indicates that virtually any pair of metals is explosively weldable, with two caveats: the conditions must be dialed in properly with respect to impact conditions and both metals must be able to sustain the impact without extensive fracture [106]. Because electromagnetic impulse seam welding can be operated very efficiently and is able to produce metallurgical welds between dissimilar metals, it has a bright future for commercialization.

18.4 CONCLUDING REMARKS

Electromagnetic metal forming allows precise temporal and spatial control of pressure on a sheet metal component. This pressure can be large enough to cause a high-velocity launch. The method is remarkably simple and often involves no moving parts. For these reasons, it can be easily used in a hybrid manner with traditional metal forming. Electromagnetic forming is easily analyzed by energy field methods. The examples shown here just scratch the surface of what can be done. With some diligence, creativity, an understanding of the process basics, and a need to expand the boundaries of traditional metal forming, this approach will yield some surprising and very different ways to approach manufacturing. High-velocity electromagnetic forming offers great potential for manufacturing many high value parts in an agile and cost-effective manner, precisely due to their unique advantages relative to the conventional forming processes.

QUESTIONS

Q.18.1 The discharge frequency in the capacitor bank circuit can be easily varied. How is this varied? What are the practical effects of very high and very low discharge frequencies? Is there an optimum?

Q.18.2 The operations carried out by impact embossing produce results much like those from a coining operation. What are the relative advantages of each?

Q.18.3 List the ways one can tailor the force distribution along the length of a path actuator.

Q.18.4 Consider how the force distribution can be tailored for a multi-turn actuator and sketch designs that can produce pressure and energy distributions that can be useful for specific metal forming problems.

Q.18.5 This chapter was largely confined to electromagnetic forming as the driving technique for impulse-based metal forming. Consider what other impulse techniques might be available. How might these be used in conjunction with existing metal forming processes?

Q.18.6 Consider the two techniques that are available for joining frames from tubular sections: crimping and magnetic pulse welding of tubes. What are the relative merits of each technique? In what cases would you favor each?

Q.18.7 Consider the distribution of plastic strain energy in a simple stamping. Show how lubrication, softening the material, and the use of embedded electromagnetic actuators can change this distribution.

ACKNOWLEDGMENTS

This work is an overview of the activities of members of my research group and outside collaborators over the past 15 years or more. Original papers are cited in each of the areas that give credit to the people who did the hard work and organizations that funded it. Significant long-term external collaborations with General Motors, American Trim, and Applied Research Associates have improved the quality and relevance of our work. Several people from my current team helped put this document in final form. Kathy Babusci, Jason Johnson, Yuan Zhang, Huimin Wang, Anupam Vivek, Shekar Srinivasan, and Brad Kabert deserve thanks.

ABOUT THE CONTRIBUTING AUTHOR

Glenn Daehn is a professor in the Department of Materials Science and Engineering at Ohio State University, Columbus and has been designated as the Mars G. Fontana Professor of Metallurgical Engineering. His research group has been actively developing electromagnetic metal forming technologies for the past 20 years. He is one of the founding members and the vice chair of the International Impulse Forming Group. He also has research interests in the high temperature creep deformation of solids and the reactive processing of ceramics to create ceramic-metal composites. One long-term theme in his work is using unusual mechanics or reactions to open new paths to manufacturing processes. For example, strain mismatch in pressure cycling has been shown as an exceptional way of improving the densification of composites. Daehn was named a National Young Investigator by NSF and was the recipient of the Hardy Medal of TMS (1992). Daehn's academic training includes a BS degree from Northwestern University (1983) and MS and PhD degrees from Stanford University, Palo Alto, California (1987), all in materials science and engineering. A more complete background can be found at mse.osu.edu/~daehn.

REFERENCES

1. DARPA, DARPA technology transition report 2002. http://www.darpa.mil/Docs/transition.pdf, 2002.
2. S. Kalpakjian and S. R. Schmid, *Manufacturing Engineering and Technology*, 4th edn. Prentice Hall, Upper Saddle River, NJ, 2001.
3. G. S. Daehn, High velocity metal forming, in *Metalworking: Sheet Forming*, vol. 14B. ASM International, Cleveland, OH, 2006.
4. C. Wick, *Tool and Manufacturing Engineers Handbook*, vol. 2, *Forming*, 4th edn. Society of Manufacturing Engineers, Dearborn, MI, 1984.
5. W. Johnson, *Impact Strength of Materials*. Edward Arnold, London, U.K., 1970.
6. M. Altynova et al., Increased ductility in high velocity electromagnetic ring expansion, *Metallurgical and Materials Transactions A-Physical Metallurgy and Materials Science*, 27, 1837–1844, July 1996.
7. V. S. Balanethiram et al., Hyperplasticity—Enhanced formability at high-rates, *Journal of Materials Processing Technology*, 45, 595–600, September 1994.
8. G. S. Daehn et al., Hyperplasticity—A competitor to superplastic sheet forming, in *TMS: Superplasticity and Superplastic Forming*. TMS, Warrendae, PA, 1995, pp. 11–16.
9. B. M. Dariani et al., Experimental investigation of sheet metal formability under various strain rates, *Institution of Mechanical Engineers Part B-Journal of Engineering Manufacture*, 223 (6), 703–712, 2009.
10. X. Y. Hu et al., The effect of inertia on tensile ductility, *Metallurgical and Materials Transactions A-Physical Metallurgy and Materials Science*, 25, 2723–2735, December 1994.
11. X. Y. Hu and G. S. Daehn, Effect of velocity on flow localization in tension, *Acta Materialia*, 44, 1021–1033, March 1996.
12. J. D. Thomas et al., Forming limits for electromagnetically expanded aluminum alloy tubes: Theory and experiment, *Acta Materialia*, 55, 2863–2873, May 2007.
13. J. D. Thomas and N. Triantafyllidis, Theory of necking localization in unconstrained electromagnetic expansion of thin sheets, *International Journal of Solids and Structures*, 44, 6744–6767, October 2007.
14. M. Zadra et al., Tensile ductility of AA5182: Effect of impact loading, *Materials Science and Technology*, 24, 1259–1264, October 2008.
15. H. E. Lindberg, Buckling of a very thin cylindrical shell due to an impulsive pressure, *Journal of Applied Mechanics*, 6, 267–272, 1964.
16. S. T. S. Al-Hassani, The plastic buckling of thin walled tubes subject to magnetomotive forces, *Journal of Mechanical Engineering Science*, 16 (2), 59–70, 1974.
17. V. Psyk et al., Electromagnetic forming as preforming operation for tubular hydroforming parts, in *International Conference on High Speed Forming 1*, March 31–April 1, 2004, Dortmund, Germany, 2004, pp. 171–180.
18. A. V. Florence, Dynamic plastic flow buckling of short cylindrical shells due to impulsive loading, *International Journal of Solids and Structures*, 4 (8), 741–756, 1968.
19. M. Kleiner et al., Analysis of process parameters and forming mechanisms within the electromagnetic forming process, *CIRP Annals-Manufacturing Technology*, 54, 225–228, 2005.
20. W. H. Gourdin, Analysis and assessment of electromagnetic ring expansion as a high-strain-rate test, *Journal of Applied Physics*, 65 (2), 411–422, 1989.
21. J. Jablonski and R. Winkler, Analysis of the electromagnetic forming process, *International Journal of Mechanical Science*, 20, 315–325, 1978.
22. R. P. Feynman et al., *The Feynman Lectures on Physics*. Addison Wesley, Reading, MA, 1964.
23. F. C. Moon, *Magneto-Solid Mechanics*. John Wiley & Sons, New York, 1984.
24. P. L'Eplattenier et al., Introduction of an electromagnetism module in LS-DYNA for coupled mechanical-thermal-electromagnetic simulations, in *3rd International Conference on High Speed Forming*, March 11–12, 2008, Dortmund, Germany, 2008.

25. I. V. Belyy et al., *Electromagnetic Metal Forming Handbook*. Vischa Shkola, Kharkov State University, Kharkov, Ukraine, 1977.

26. P. Zhang et al., Analysis of the electromagnetic impulse joining processes with a field concentrator, in *Materials Processing and Design: Modeling, Simulation and Applications—NUMIFORM 2004—Proceedings of the 8th International Conference on Numerical Methods in Industrial Forming Processes. AIP Conference Proceedings*, vol. 712. 2004, pp. 1253–1258.

27. S. M. Golowin, Path actuators for magnetic pulse assisted forming and punch-less electro-magnetic shearing [3 32 mb]. Available: http://rave.ohiolink.edu/etdc/view?acc%5Fnum=osu1211986132, 2008.

28. S. Woodward, Senior thesis, Ohio State University, Columbus, OH, 2009.

29. C. Weddeling, Dortmund University, Dortmund, Germany, 2009.

30. G. K. Fenton and G. S. Daehn, Modeling of electromagnetically formed sheet metal, *Journal of Materials Processing Technology*, 75, 6–16, March 1998.

31. P. L'Eplattenier et al., Introduction of an electromagnetism module in LS-DYNA for coupled mechanical-thermal-electromagnetic simulations, *Steel Research International*, 80, 351–358, May 2009.

32. A. Brosius et al., Finite-element modeling and simulation of material behavior during electromagnetic metal forming, in *Proceedings of the 6th International ESAFORM Conference on Material Forming, ESAFORM 2003*, April 28–30, 2003, Salerno, Italy, 2003, pp. 971–974.

33. P. Zhang, Joining enabled by high velocity deformation. Available: http://rave.ohiolink.edu/etdc/view?acc_num=osu1061233577, 2003.

34. V. S. Balanethiram and G. S. Daehn, Hyperplasticity-increased forming limits at high workpiece velocities, *Scripta Metallurgica Et Materialia*, 30, 515–520, February 1994.

35. M. Seth et al., Formability of steel sheet in high velocity impact, *Journal of Materials Processing Technology*, 168, 390–400, October 2005.

36. E. Iriondo et al., Electromagnetic springback reshaping, in *Proceedings from 2nd Annual International Conference on High Speed Forming (ICHSF) 2006*, March 21–22, 2006, Dortmund, Germany, 2006, p. 153.

37. E. Iriondo et al., Experimental study of springback elimination using electromagnetic field. International Deep Drawing Research Group, in *IDDRG 2006*, June 19–21, 2006, paper no 44.

38. E. Iriondo et al., New approach for HSS Springback correction: Electromagnetic pulses, in *International Conference New Developments in Sheet Metal Forming*, Stuttgart, Germany, May 10–12, 2006.

39. S. F. Golovaschenko, Springback calibration using pulsed electromagnetic field, in *6th International Conference and Workshop on Numerical Simulation of 3D Sheet Metal Forming Process*, 2005.

40. S. P. Keeler and W. A. Backhofen, Plastic instability and fracture in sheet stretched over rigid punches, *AMS Transactions Quarterly*, 56, 25–48, 1964.

41. G. M. Goodwin, Applications of strain analysis to sheet metal forming in the press shop, *SAE Transactions*, 77, 380–387, 1968.

42. T. B. Stoughton, A general forming limit criterion for sheet metal forming, *International Journal of Mechanical Sciences*, 42, 1–27, January 2000.

43. P. Carlsson et al., Tribological performance of thin organic permanent coatings deposited on 55% Al-Zn coated steel—Influence of coating composition and thickness on friction and wear, *Wear*, 251, 1075–1084, 2001.

44. H. Yoshimura et al., Improvement of the Erichsen values of titanium sheets using various cereal flour lubricants, *Tetsu to Hagane-Journal of the Iron and Steel Institute of Japan*, 87, 474–479, July 2001.

45. C. G. Wall, The laboratory evaluation of sheet-metal forming lubricants, *Lubrication Engineering*, 40, 139–147, 1984.

46. J. Cao et al., Prediction of localized thinning in sheet metal using a general anisotropic yield criterion, *International Journal of Plasticity*, 16, 1105–1129, 2000.

47. H. Yao and J. Cao, Prediction of forming limit curves using an anisotropic yield function with prestrain induced backstress, *International Journal of Plasticity*, 18, 1013–1038, 2002.

48. R. J. Comstock et al., Simulation of axisymmetric sheet forming tests, *Journal of Materials Processing Technology*, 117, 153–168, November 2001.

49. S. Murty et al., Instability criteria for hot deformation of materials, *International Materials Reviews*, 45, 15–26, 2000.

50. D. A. Peterson, *Progressive Dies: Principles and Practices of Design and Construction*. SME, Dearborne, MI, 1994.

51. K. J. Pahl, New developments in multi-point die-cushion technology, *Journal of Materials Processing Technology*, 71 (1), 168–173, 1997.

52. J. R. Michler et al., A strip-drawing simulator with computer-controlled drawbead penetration and blank-holder pressure, *Journal of Materials Processing Technology*, 43, 177–194, June 1994.

53. D. E. Hardt and R. C. Fenn, Real-time control of sheet stability during forming, *Journal of Engineering for Industry-Transactions of the ASME*, 115, 299–308, August 1993.

54. G. Liu et al., Eliminating springback error in U-shaped part forming by variable blankholder force, *Journal of Materials Engineering and Performance*, 11, 64–70, February 2002.

55. S. Yossifon and J. Tirosh, Deep drawing with a fluid pressure assisted blankholder, *Proceedings of the Institution of Mechanical Engineers Part B-Journal of Engineering Manufacture*, 206, 247–252, 1992.

56. T. W. Ku et al., Implementation of backward tracing scheme of the FEM to blank design in sheet metal forming, *Journal of Materials Processing Technology*, 111, 2001, pp. 90–97.

57. J. Cao et al., Next generation stamping dies—Controllability and flexibility, *Robotics and Computer Integrated Manufacturing*, 17, 49–56, 2001.

58. S. Kok and N. Stander, Optimization of a sheet metal forming process using successive multipoint approximations, *Structural Optimization*, 18, 277–295, December 1999.

59. B. Budiansky and N. M. Wang, On the swift cup test, *Journal of the Mechanics And Physics of the Solids*, 14, 357–374, 1966.

60. W. F. Hosford and R. M. Caddell, *Metal Forming: Mechanics and Metallurgy*. Prentice Hall, Englewood Cliffs, NJ, 1993.

61. L. B. Shulkin et al., Blank holder force (BHF) control in viscous pressure forming (VPF) of sheet metal, *Journal of Materials Processing Technology*, 98, 7–16, January 2000.

62. J. Liu et al., Evaluation of sheet metal formability, viscous pressure forming (VPF) dome test, *Journal of Materials Processing Technology*, 98, 1–6, January 2000.

63. J. Shang, Electromagnetically assisted sheet metal stamping 4 99 mb. Available: http://rave.ohiolink.edu/etdc/view?acc%5Fnum=osu1158682908, 2006.

64. Y. Hashida and R. H. Wagoner, Experimental analysis of blank holding force control in sheet forming, in *Sheet Metal Stamping Symposium*, September 15–18, 1997, Warrendale, PA, 1993, pp. 93–100.

65. I. Eguia et al., Improved Crimp Joining onto Mandrels with Undulating Surfaces, in *Proceedings from 2nd Annual International Conference on High Speed Forming (ICHSF) 2004*, March 31–April 1, 2004, Dortmund, Germany, 2004.

66. W. S. Hwang et al., Joining of copper tube to polyurethane tube by electromagnetic pulse forming, *Journal of Materials Processing Technology*, 37, 83–93, February 1993.

67. Y. B. Park et al., Design of axial/torque joint made by electromagnetic forming, *Thin-Walled Structures*, 43, 826–844, May 2005.

68. Y. B. Park et al., Joining of thin-walled aluminum tube by electromagnetic forming (EMF), *International Journal of Automotive Technology*, 6, 519–527, October 2005.

69. P. Barreiro et al., Development and effects of residual stresses in joints produced by electromagnetic compression and its implication on the mechanical properties, *Materials Science Forum*, 524–525, 485–490, 2006.

70. S. Golovashchenko, Methodology of design of pulsed electromagnetic joining of tubes, in *TMS: Innovations in Processing and Manufacturing of Sheet Materials*, 2001.

71. B. Reed, Torque Tube Assembly at Boeing Georgia, in *TMS Fall Meeting 1997*, Cleveland, OH, 1997.

72. S. F. Golovashchenko, Numerical and experimental results on pulsed tubes calibration, in *Sheet Metal Forming Technology*, San Diego, CA, 1999, pp. 117–127.

73. S. Golowin et al., Application of a uniform pressure actuator for electromagnetic processing of sheet metal, *Journal of Materials Engineering and Performance*, 16, 455–460, August 2007.

74. M. Kamal, *A uniform pressure electromagnetic actuator for forming flat sheets*. Available: http://rave.ohiolink.edu/etdc/view?acc%5Fnum=osu1127230699, 2005.

75. M. Kamal et al., Agile manufacturing of a micro-embossed case by a two-step electromagnetic forming process, *Journal of Materials Processing Technology*, 190, 41–50, July 2007.

76. M. Kamal et al., Design, construction, and applications of the uniform pressure electromagnetic actuator, in *Proceedings from 2nd Annual International Conference on High Speed Forming (ICHSF) 2006*, Dortmund, Germany, March 20–21, 2006, p. 217.

77. M. Kamal et al., Replication of microfeatures by electromagnetic launch and impact, in *Proceedings of the First International Conference on Micro-Manufacturing*, 2006, p. 197.

78. J. R. Johnson et al., Coupling experiment and simulation in electromagnetic forming using photon doppler velocimetry, *Steel Research International*, 80, 359–365, May 2009.

79. H. Wang, Title, unpublished.

80. I. V. Volobuev and A. V. Legeza, Phase-transformations in joints produced by magnetic-pulse welding, *Welding Production*, 19, 12–15, 1972.

81. E. S. Karakozov et al., Interaction of metals in magnetic impulse welding, *Welding Production*, 24, 5–7, 1977.

82. Z. A. Chankvetadze, Technological features of magnetic pulse welding, and conditions under which the joint is formed, *Automatic Welding USSR*, 32, 26–28, 1979.

83. V. A. Glouschenkov et al., The European Institute for the Joining of Materials, in *Proceedings of the 6th International Conference on "Joining of Materials" (JOM-6)*, April 5–7, 1993, Helsingor, Denmark, 1993, pp. 473–484.

84. V. F. Karpouchin et al., The European Institute for the Joining of Materials, in *Proceedings of the 5th International Conference on "Joining of Materials" (JOM-5)*, May 1991, Helsingor, Denmark, 1991, pp. 241–245.

85. M. Marya et al., Microstructures at aluminum-copper magnetic pulse weld interfaces, *Material Science Forums*, 426–432, 4001–4006, 2003.

86. M. Marya and S. Marya, Interfacial microstructures and temperatures in aluminium-copper electromagnetic pulse welds, *Science and Technology of Welding and Joining*, 9, 541–547, 2004.

87. T. Aizawa et al., Application of magnetic pulse welding for aluminum alloys and SPCC steel sheet joints, *International Conference on New Frontiers of Process Science and Engineering in Advanced Material*, November 24–26, 2004, Kyoto, Japan, 2004.

88. T. Aizawa et al., Application of magnetic pulse welding for aluminum alloys and SPCC steel sheet joints, *Welding Journal*, 86 (5), 119–124, 2007.

89. V. Shribman et al., Magnetic pulse welding produces high-strength aluminum welds, *Welding Journal*, 81, 33–37, April 2002.

90. D. Winter, *Auto World*, 35 (1), 57, 1999.

91. M. Pezzutti, Innovative welding technologies for the automotive industry, *Welding Journal*, 79, 43–46, June 2000.

92. Y. Livshitz et al., in *Magnesium-2000: 2nd International Conference on Magnesium Science and Technology*, Magnesium Research Institute/Ben-Gurion University of the Negev, 2000.

93. A. Durgutlu et al., Examination of copper/stainless steel joints formed by explosive welding, *Materials & Design*, 26, 497–507, 2005.

94. F. Grignon et al., Explosive welding of aluminum to aluminum: Analysis, computations and experiments, *International Journal of Impact Engineering*, 30, 1333–1351, November 2004.

95. Y. Li et al., Morphology and structure of various phases at the bonding interface of Al/steel formed by explosive welding, *Journal of Electron Microscopy*, 49 (1), 2000, 5–16.

96. K. Hokamoto et al., New explosive welding technique to weld. aluminum alloy and stainless steel plates. Using a stainless steel intermediate plate, *Metallurgical Transactions A-Physical Metallurgy and Materials Science*, 24, 2289–2297, October 1993.

97. K. Hokamoto et al., A new method for explosive welding of Al/ZrO_2 joint using regulated underwater shock wave, *Journal of Materials Processing Technology*, 85, 175–179, January 1999.

98. A. Abe, Numerical study of the mechanism of wavy interface generation in explosive welding, *JSME International Journal Series B-Fluids and Thermal Engineering*, 40, 395–401, August 1997.

99. A. Ben-Artzy et al., Interface phenomena in aluminium-magnesium magnetic pulse welding, *Science and Technology of Welding and Joining*, 13, 402–408, July 2008.

100. A. Stern and M. Aizenshtein, Bonding zone formation in magnetic pulse welds, *Science and Technology of Welding and Joining*, 7, 339–342, October 2002.

101. Y. Zhang et al., Numerical simulation and experimental study for magnetic pulse welding process on AA6061-T6 and Cu101 sheet, in *10th International LS-DYNA Users Conference*, Detroit, MI, 2008.

102. S. D. Kore et al., Effect of process parameters on electromagnetic impact welding of aluminum sheets, *International Journal of Impact Engineering*, 34, 1327–1341, August 2007.

103. M. Watanabe and S. Kumai, Interfacial morphology of magnetic pulse welded aluminum/aluminum and copper/copper lap joints, *Materials Transactions*, 50, 286–292, February 2009.

104. Y. Zhang et al., Microstructure characterisation of magnetic pulse welded AA6061-T6 by electron back-scattered diffraction, *Science and Technology of Welding and Joining*, 13, 467–471, July 2008.

105. K. Okagawa and T. Aizawa, Impact seam welding with magnetic pressure for aluminum sheets, 2004, pp. 231–236.

106. C. Prothe, Dynamic Materials, 2008.

19 Electrically Assisted Manufacturing

Wesley A. Salandro and John T. Roth

CONTENTS

19.1 INTRODUCTION

The goal of any manufacturing engineer is to produce quality parts as efficiently and cost-effectively as possible. For many of the common engineering materials, this can be accomplished rather easily; however, it has proven challenging with many of the stronger, more lightweight materials that are being implemented into today's manufacturing processes. These materials, such as high-strength aluminum-, magnesium-, copper-, and titanium-based alloys, all possess great strength-to-weight ratios, but their limited formability makes them impractical for use in many real-world applications that require complex part geometries. Currently, the main downfall in using these materials to make complicated shapes is the fact that one entire blank cannot be used to form the shape, but numerous simpler parts must first be formed and then attached using screws, rivets, or welds.

High production costs and poor part quality issues can result from attaching smaller, simpler parts together, making the disadvantages of using these materials outweigh their great strength-to-weight characteristics. To this end, formability-enhancing techniques are used to increase the overall efficiency of the manufacturing process, thus increasing the applicability of these materials and allowing more complex part geometries to be formed from single blanks rather than attaching many smaller components together. Three of the most prominent manufacturing methods are hot working, incremental forming, and superplastic forming. There are benefits and consequences to using each of these manufacturing processes. In the following sections, these processes will be compared to a novel manufacturing process known as electrically assisted manufacturing (EAM), which has been proven to increase process efficacy.

19.1.1 Hot Working

Hot working is defined as the deformation of a material at an elevated temperature. As part of this process, the metal is heated above its recrystallization temperature, thus increasing the formability of the material. Advantages to this process include decreased flow stress and increased ductility. This is one of the simplest manufacturing methods because all that is required is a heat source, such as a heater or furnace. In many cases, however, these benefits come at the expense of part quality. One key disadvantage includes less dimensional accuracy, due to uneven thermal expansion resulting from temperature gradients within the material. Moreover, a rougher surface finish (resulting from an oxide layer developing on the outside of the part) is another consequence of using this process. Also, as the size of the workpiece increases, larger furnaces will be needed, proving to be more costly and taking up a larger footprint on the shop floor. Nonetheless, regardless of the minor fluctuations in part quality and cost, this relatively simple and effective process makes it a desirable choice when holding rough tolerances where secondary finishing operations will likely follow.

Using the stress vs. strain graph in Figure 19.1, the effects of hot working can be compared to a room-temperature (i.e., cold working) compression test when forging Ti-6Al-4V. Due to hot working, the compressive flow stress was decreased and the amount of achievable compressive displacement prior to fracture was increased when compared to cold-working conditions.

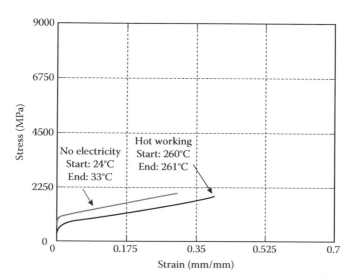

FIGURE 19.1 Hot working of Ti-6Al-4V in compression.

19.1.2 INCREMENTAL FORMING

Incremental forming is a type of manufacturing process in which a metallic part is deformed in small steps with a possible minor heat treatment (i.e., a process anneal) performed in between steps [1]. This type of manufacturing is especially beneficial when forming sheet metal and is widely used in the automotive and aircraft industries. The major advantages to this process are the large amounts of deformation and the decrease in the required deformation forces that can be obtained. These advantages are possible because of the minor heat treatments performed in between the increments of deformation. The treatments eliminate the effects of cold work or strain hardening by causing recrystallization to occur, thus resulting in a new, overall weaker material. Aside from the benefits, this process does have its disadvantages. A big downfall is low dimensional accuracy, since the part must be continuously removed and refixtured before and after the heat treatments. The decreased accuracy arises from the fact that the part will not be fixtured in the same position once removed and reinstalled. Also notable, this process can be very time consuming, depending on variables such as the number of heat treatments and their respective durations, as well as the depth of desired deformations. Notorious for having long production times, incremental forming may not be an optimum process for high production or high precision manufacturing; however, materials can be formed to great distances and complex shapes can be achieved.

19.1.3 SUPERPLASTIC FORMING

Superplastic forming involves heating a material to extremely high temperatures (roughly two-thirds of its melting temperature) when deforming. This process can produce tremendous elongations of up to 2000%, coupled with greatly reduced flow stress [2]. Other advantages include being able to form precise complex shapes in which minimal or no residual stresses are present. Also, lower strength tooling and fixtures can be used since the required forces for deformation are minimized. This process is great for forming complex shapes; however, it also has its disadvantages. First, the superplastic forming process is only applicable for very fine-grained alloys (less than 10–15 μm), such as some aluminum (5083-FG and 7475), titanium (Ti-6AL-4V), and magnesium alloys (Mg-AZ31B). These small grains allow grain boundary sliding (GBS) to occur at elevated temperatures, which is the deformation mechanism responsible for the huge elongation increases. Another consequence of this process is that extremely slow strain rates must be used (10^{-4} to $10^{-2} s^{-1}$). Similar to the incremental forming technique, the superplastic technique may not be practical for many high-production manufacturing applications and can be classified as a batch-forming process. Neglecting the limited number of applicable materials and the slow strain rate that is required, this process is capable of producing precise complex geometrical parts with little or no finishing operations needed.

19.1.4 ELECTRICALLY ASSISTED MANUFACTURING

Electrically assisted manufacturing (EAM) is the process of applying electricity to a metallic workpiece in order to increase its workability, and thus greatly improve the overall efficiency of the respective manufacturing process it is being applied toward. The beneficial workability effect that the electricity exerts on a material is known as the "electroplastic effect." Two main advantages resulting from this effect are reduced flow stress and increased elongation. It must be noted that, although the elongation increase and force decrease are normally less than those arising from superplastic forming, the EAM technique is applicable to a much wider range of materials than superplastic forming. In general, EAM effects are comparable to, if not greater than the effects of incremental forming and hot working. Another significant advantage is this method's ability to reduce/eliminate springback, a challenging obstacle to overcome in sheet metal forming. Also, electricity can be applied to the workpiece without halting the deformation process, unlike the incremental forming technique where refixturing is required. Furthermore, no heater/furnace is needed, as in hot working, nor is the EAM process limited to a slow strain rate, like in superplastic forming. Considering this, the EAM process is a highly viable alternative to any of the hot working, incremental forming, and superplastic techniques.

However, like all processing techniques, EAM has some disadvantages and challenges that exist within the process and its implementation toward industrial use. First, in order for the electricity to reach the part, there must be some type of electrical applicator in contact with the conductive workpiece at all times. For manufacturing processes where the workpiece is stationary, like forging or stamping, this can be done rather easily. Conversely, when implementing the EAM technique on manufacturing processes with spinning workpieces, like friction welding two workpieces together, an applicator system, which is in continuous contact with the workpiece while not becoming entangled in the part, must first be designed. In addition, since the workpiece is subjected to electrical flow, all personnel and machinery components should be insulated from the electricity. However, the isolation of the electricity is a manageable concern because the voltage is nearly zero (millivolt range). Thus, the main precaution that is necessary is to limit the electricity's exposure to sensitive electronic devices. It must be noted that extreme caution should always be used when working with electricity. Isolating the electricity, however, can prove to be a challenging task in some cases, since most machinery components are comprised of conductive metals and the common insulating materials (e.g., nylons, rubbers, ceramics, or plastics) cannot withstand the mechanical demands placed on these components. Outweighing these issues, in most cases, are the tremendous benefits, such as decreased flow stress, increased ductility, and reduced springback that can be accomplished using the EAM method.

19.2 HISTORICAL DEVELOPMENT OF EAM

As early as the 1950s, scientists and engineers began researching the effects of electricity on the properties of different materials. The research, however, did not begin by examining electricity's effects on metals. Instead, salts were examined, followed by crystals and metals. Beginning in 1959, Machlin et al. examined electricity's effect on group 1A salts (NaCl) by utilizing 3-point bending tests and room-temperature compression tests. This research determined that the applied electric current significantly affected the salt's ductility, flow stress, and yield strength [3]. In subsequent work, in 1967, Nabarro dedicated a chapter of his book toward the effect of an applied electric current on materials [4]. Troitskii et al., in 1969, studied how electrons influenced dislocation movement and multiplication with different alloys of zinc, tin, lead, and indium. With data from both tension and compression tests, he was able to conclude that flow stress within the materials could be lowered by using pulsed electricity [5]. Years later in 1982, Klimov et al. explained that electric current may have a specific effect on a metal's structure that is unrelated to Joule heating [6]. Then in 1988, a microstructural analysis by Xu et al. demonstrated that a continuous electric current in titanium materials caused the recrystallization rate and grain size of the material to increase. Both AC and DC were tested in this research [7]. Moving forward, Chen et al. and Conrad et al. studied phase and compound-related phenomena. In 1998, Chen et al. developed a relationship between electric flow and the formation of intermetallic compounds (Sn/Cu and Sn/Ni systems) [8]. On the other hand, in 2000, Conrad et al. determined that very high current density short duration electrical pulses [9] affect the plasticity and phase transformations of metals and ceramics.

In 2005, Heigel et al. performed more microstructural work, examining alterations in the 6061 aluminum alloy caused by direct current [10]. In 2007, Andrawes et al. were able to conclude that electrical current has the potential to significantly reduce the required deformation energies without excessively heating the workpiece in a uniaxial tensile deformation process [11]. Also in 2007, Perkins et al. examined the effects of a continuously applied current on many metals (aluminum-, copper-, iron-, and titanium-based alloys) while undergoing an upsetting process [12]. Findings from this research prove that the application of electricity during the deformation process reduced the flow stress, minimized springback, and increased forgeability, which led to greater deformations prior to cracking. As an example, Figure 19.2 displays compressive stress–strain profiles of Al 2024-T6, clearly showing how forgeability is enhanced with the increase of current density. From the

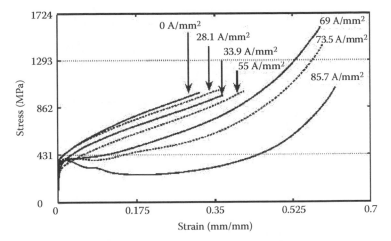

FIGURE 19.2 Increased compressive deformation using continuous EAM.

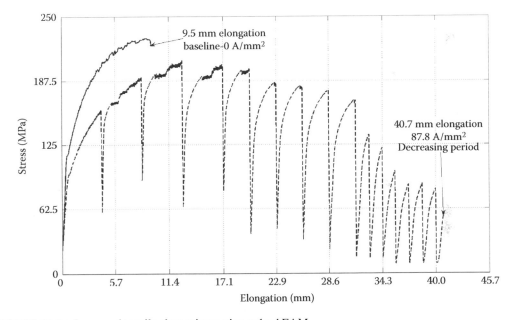

FIGURE 19.3 Increased tensile elongations using pulsed EAM.

figure, one can notice that at current densities greater than 55 A/mm², formability was significantly increased and the specimens were formed completely without fracturing.

Again, in 2007, Ross et al. studied the effect of a continuous current on the same metals as Perkins, now for uniaxial tension tests [13]. It could be concluded that the yield strength, flow stress, and elastic modulus were all reduced. Also, because of strain weakening, the overall energy of deformation was decreased. However, one important finding of this research was that, due to the continuously applied current, the achievable elongations of the specimens decreased (an opposite effect of using continuously applied current in compression). This led to experimentation with pulsed electricity, rather than leaving it applied continuously, for tensile applications.

In 2008, Roth et al. experimented with this pulsing technique on 5754 aluminum alloy while searching for optimal pulsing parameters for the respective alloy [14]. In this tensile study, flow stress was decreased, and extreme elongations of over 400% were achieved. Figure 19.3 shows an example of these elongations with respect to a baseline tensile test conducted without electricity.

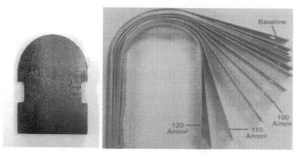

FIGURE 19.4   Springback reduction/elimination using EAM.

TABLE 19.1
Materials Tested Using EAM

Base Material	Alloys Examined
Aluminum	2024, 5052, 5083, 5754, 6061, 6111, and 7075
Copper	C11000, C36000, and C46400
Iron	304 stainless steel and A2 steel
Titanium	Ti 6AL4V and RC-130B titanium
Magnesium	Mg AZ31B-O

Furthermore, Salandro et al. examined pulsed electricity's effect on different heat treatments of two aluminum alloys (5052 and 5083) and concluded that flow stress decreases were apparent throughout all tests, while the elongation increases were proportional to the time/temperature of the corresponding heat treatment [15]. Similar tensile results were found when pulsing electricity on Mg AZ31B-O sheet metal specimens [16]. This research also noted that a linear and inverse relationship exists between current density and pulse duration. Specifically, it was found that constant elongation levels could be attained using high current density/short pulse duration or low current density/long pulse duration settings. Salandro et al. also examined the effect of electrical pulsing on 5052 aluminum alloy sheet metal channel formation [17]. From this work, it was found that adding EAM to the channel forming process increased achievable elongation by up to 25% and notably decreased the required machine forces.

Originally, Perkins et al. discovered that electricity could reduce or eliminate springback in compression specimens when applied during the deformation process. In 2009, using flattening and shape-retention tests, Green et al. also demonstrated that a single DC pulse could be used to reduce or eliminate springback in 6111 aluminum alloy sheet metal after deformation, but prior to removal from the die [18]. Figure 19.4 shows how greater amounts of current led to larger springback reductions (when compared to a non-pulsed baseline test), and eventually even led to complete springback elimination. Also, longer pulse durations were proven to have the same effect with lower current densities. The round die, of which the sheet metal strips were bent around, is shown to the left of the specimens.

As seen from the historical review, many different metallic materials have been tested using the EAM technique. Materials include aluminum-, copper-, ferrous-, titanium-, and magnesium-based alloys. A full listing of the different materials that EAM was tested on can be found in Table 19.1. By examining the table, it can be seen that EAM has been tested on a great variety of engineering materials. All of these materials experienced formability enhancements when the EAM technique was applied.

19.3 ELECTRICAL THEORY

As mentioned earlier, three main manufacturing benefits can result from the application of electricity: decreased flow stress, increased achievable elongation, and the reduction/elimination of springback. Springback in metal parts can be reduced/eliminated using two methods: by applying the electricity to the part during deformation or by applying the electricity after deformation has occurred but before the part is removed from the die. Because of this, the two similarly based theories (when electricity is applied during and after deformation) will be explained separately.

19.3.1 Effect of EAM while Deforming

It is known that metals are held together by electron clouds. The bond strength of a metal is directly related to the "pull" on the electron clouds. With this being said, it is of no surprise that the application of electricity will affect a metal's formability. From the historical research mentioned above, many theoretical conclusions have been developed, which explain how electricity improves the ability to deform a material (the "electroplastic effect"). Electrical current can be described as the flow of electrons through any conductive medium (i.e., metal blanks, sheets, plates, rods, etc.). Currently, electricity's effect on materials can be summarized using a multipart theory focusing on localized heating effects and the direct effects of the moving electrons.

The first part of this theory is related to the resistivity of the material. Each material has a set resistance to electron flow, which is affected by lattice obstacles such as voids, dislocations, grain boundaries, impurities, and stacking faults:

$$\rho = \rho_T + P_v C_v + P_I C_I + P_D C_D + P_s C_s$$

where
ρ_T is the thermal resistivity
C_v and C_I are the concentrations of vacant lattice sites and interstitial ions
C_D is the density of dislocations
C_s is the area of stacking faults
P_v, P_I, P_D, and P_s are the material-specific weightings [4]

As the number and size of these lattice obstacles are increased, the resistivity of the respective material is also increased.

When plastically deforming a material, dislocations move throughout the lattice by way of diffusion. Considering this, in order for diffusion to occur, there must be sufficient energy to (1) break current bonds, (2) distort the local lattice, and (3) allow the dislocations to move. Conversely, without the movement of dislocations, plastic deformation cannot take place in most cases (twinning is an exception). Commonly, the dislocations tend to pile up at various lattice obstacles, forming regions of high dislocation density throughout the material, thus making the part harder to deform (i.e., increasing the flow stress).

Electrical current moving through a part can help free "stuck" dislocations by indirect heating effects or direct effects caused by electron–dislocation contact. Beginning with the indirect effects, electrons moving throughout the lattice will tend to scatter off of flaws within the lattice and reduce the strength of the metallic bonds within the metal by way of localized resistive heating. In doing this, it does not take as much energy for the dislocations to continue moving since the metal's bonds are weaker because of the heat. Also, due to heating at the lattice flaws, the local lattice around the flaw will expand more. This reduces local strain fields (reducing the amount of strain cancellation that occurs when dislocations pass by, reducing residence time) and also leads to easier dislocation movement since the lattice does not have to be distorted as much. The final effect of localized resistive heating is also a result of the lattice expanding around the hotter flaws. When the lattice expands, the bonds holding it together are spaced farther apart. This makes them weaker because their distances are greater, thus easing dislocation movement. Of interest is that, since the amount of resistive heating is proportional to the amount of lattice obstacles, a material with greater resistance will require less electricity to achieve equivalent formability effects from the electricity as a material with a lower resistivity.

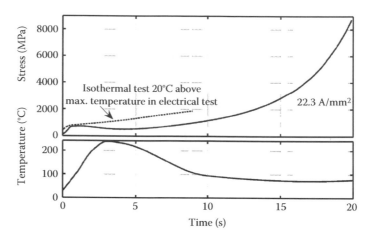

FIGURE 19.5 Isothermal testing with transient EAM temperature profile.

Since a large portion of the formability enhancement from EAM depends on localized resistive heating, it is important to distinguish localized resistive heating from globalized resistive heating. When electricity is run through a metallic part, heating will always occur because of the extra energy induced into the parts. This is known as global resistive heating, which is the compilation of many different sites of localized resistive heating within the material's lattice. It must be noted that the localized resistive heating within the microstructure is much higher than the global resistive heating witnessed at the part's surface. This supports the postulation by Klimov et al. in that EAM has a much greater effect on the behavior during deformation, in comparison to working the material at an elevated temperature. Perkins et al. have proven this concept experimentally by performing isothermal tests that were run at the maximum temperatures reached during an upsetting EAM process. From this, it was determined that the isothermal tests did not show nearly the amount of formability enhancement as the tests using the applied electricity. The stress plot in Figure 19.5 displays this phenomenon, and from the transient temperature plot below the stress plot, the maximum temperature reached during EAM was only held for a second or two, whereas the isothermal part was held at the elevated temperature during the entire test.

Now the direct effects of electrons will be discussed. Given the fact, the dislocations are also lattice flaws, electrons will scatter off of the dislocations, thus pushing them. This pushing effect provided by the electrons reduces the energies required for deformation and helps to extend elongations. Kravchenko, in his explanation of electroplasticity, stated that, if there is an electric current flowing through a material and the electrons are traveling at a faster rate than the dislocations within the lattice, the energy from the electrons would be transferred to the dislocations, thus making the material flow easier [19]. In other words, when the electrons scatter off of the dislocations, the electrons transfer extra energy to these flaws, and this energy is used to help propel the dislocations through the lattice. As an additional effect, the dislocations will also experience heating due to the scattering of the electrons (since dislocations are also defects). As the hotter dislocations are pushed through the cooler lattice, the dislocations will heat the local lattice, thus further enhancing the localized resistive heating effect and increasing overall formability, while also reducing the strain fields situated around the dislocation. Furthermore, since this effect is directly related to the electrons interacting with the dislocations, this effect moves with the dislocation; thereby continuously assisting the dislocation's motion, even when not relatively close to imperfections in the lattice.

With the application of electricity, greater amounts of free electrons are available in the lattice. Metals are composed of many metallic bonds that consist of ion cores surrounded by electron clouds. When bonds are broken, as in plastic deformation, electrons move from one region to another. As electricity is applied to a metal, excess electrons are being introduced into the material. With the extra

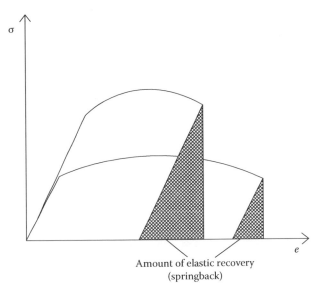

FIGURE 19.6 Tensile springback reduction using EAM while deforming.

electrons present in the metal's electron cloud, the metallic bonds can break and reform much easier, thus making plastic deformation easier by way of reduced flow stress and increased elongations.

Springback, the tendency of a metal to lose a portion of its formed shape after removal from a die, is the direct result of the stored elastic energy within a part. Comparatively, due to regions of increased dislocation density, this internal elastic energy is generated from the stretching of metallic bonds within the material's lattice structure. Both the retained elastically stretched bonds and the force from the repulsive nature of aligned dislocations cause the material to move in the direction opposite to that of the previously applied deformation force, thus altering the part's shape after removal from the die. Using EAM while deforming the material, reductions in springback are achieved mainly due to the significant reductions in flow stress that are apparent because of the EAM technique. From Figure 19.6, two tensile stress–strain curves can be seen. The larger curve represents stress seen in cold-working conditions, as compared to the smaller curve, which is representative of stress in an EAM process. It is known that elastic recovery within a part is directly related to the material's respective modulus of elasticity (note that the modulus slope is the same as the slope of elastic recovery). Considering this, if the material's flow stress can be reduced by the EAM process, then the amount of elastic recovery will also be reduced (the smaller curve has less springback).

19.3.2 Effect of EAM after Deformation

Aside from applying electricity to the workpiece during deformation, the EAM technique can also be utilized after deformation, but prior to the workpiece being removed from the die. It must be noted that this technique can be used to reduce/eliminate springback effects in sheet metal parts. Applying the electricity after deformation, however, will not reduce the deformation forces/energies or improve the material's formability. Macroscopic-level and microscopic-level examinations must be considered in order to fully explain the means by which electricity affects the reduction/elimination of springback after deformation has already occurred.

On the macroscopic level, the electricity that is applied to the workpiece instantaneously lowers the material's yield point, thus converting some of the stored internal elastic energy (that which exceeds the new lower yield due to the presence of the electricity) into permanent plastic deformation. On the microscopic-level, two phenomena are accountable for the rapid decrease in yield strength. First, localized resistive heating takes place because the electrons meet resistance at different lattice obstacles (i.e., voids, grain boundaries, and impurities) and this atomic-level heating results in an expansion of the local lattice and puts the bonds holding the lattice together into a weaker state, thus allowing for easier dislocation movement throughout the lattice. The second effect arises from the flowing electrons pushing the dislocation lines and transferring excess energy to the dislocations, hence reducing the required force for dislocation motion and minimizing the material's yield strength in the process. It must be noted that the effect arising from the "push" on the dislocation lines is not thought to be as significant as the localized heating effect due to the fact that the slip planes on which the dislocations move are not oriented consistently with the path of flowing electrons or the axis of deformation. Both of these effects allow the dislocations to move away from high stress regions where "pile-ups" have occurred during deformation, essentially nearly instantaneously stress relieving the material.

Recently, in 2009, research by Green et al. confirmed this theory from his springback study performed on 6111 aluminum alloy sheet metal strip specimens [18]. From the research, it was concluded that the amount of flattening or the ability of the part to retain its current shape depended highly on the magnitude of current density or the duration of the applied electrical pulse. Considering that each specimen was only pulsed once for a maximum of 3 s, the EAM technique for springback is much faster and effective when compared to processes such as an anneal or stress relief, which may require several hours.

19.4 MANUFACTURING APPLICATIONS FOR EAM

Although yet not widely used in industry, research on the EAM process has proven that it has the potential to be implemented on many traditional manufacturing processes. Furthermore, the EAM process can easily be adapted to most major manufacturing processes, such as bulk deformation processes (forging, rolling, and extrusion), sheet metal processes (stretch forming and deep drawing), and joining processes (friction welding and stir friction welding). All that is needed is a power supply, a method of applying the electricity, and a way of insulating personnel and machinery from the electricity. For each process, the effects and issues related to the application of the EAM technique will be discussed.

From Figure 19.7, a schematic of the research setup used in research by Roth et al. can be seen. This schematic is specifically for compression testing; however, the setup can be easily modified for use in tensile testing, simply by relocating the insulating materials and changing the fixturing method. For this setup, a Lincoln arc welder is used to supply power to the specimen. Used in conjunction with this welder is a thermally cooled resistor system, which controls the amount of current flowing through the specimen. In the figure, the + and − welder leads are shown in their respective locations within the test setup. The next component within the testing setup consists of a

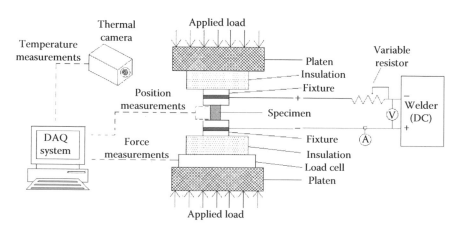

FIGURE 19.7 Experimental setup for EAM compression testing.

method of insulating the testing machine/personnel from the electric current, thus causing all of the electricity to flow through the specimen. Insulating materials are located between the top/bottom platens and fixtures. The type of insulation used is haysite-reinforced polyester, which is extremely rigid and strong when loaded in compression. The precise location of the insulation materials may vary depending on the exact testing machines used. The final component for testing is a method of gathering test data (i.e., a data acquisition system). In the figure, data collection is represented using dashed lines. Three types of data must be collected: thermal, force, and position. The thermal data, used to monitor a specimen's temperature during testing, is collected using a PC that is directly linked to an infrared thermal imaging camera. The force vs. position data is gathered using PC compatible software that is onboard the testing machine. Using initial specimen dimensions, the force/position data can easily be converted to engineering stress/elongation data.

19.4.1 BULK DEFORMATION PROCESSES

Bulk deformation processes make up a large portion of all manufacturing methods used today. Of all the bulk deformation processes, forging, rolling, and extrusion will be examined. Before beginning, it must be described how the electricity is applied to the workpiece and techniques that will maximize the electricity's effect on the workability of the material. One rule of thumb is that the electricity does not need to be applied to the entire workpiece, but rather, it only needs to be present in the areas where significant deformation will take place. In addition, it is preferred for the electrical flow lines (paths along which electrons travel) to be in the same direction as the anticipated deformation. This is because the electrons will most effectively aid in deformation by pushing the dislocation lines in the same direction required for plastic deformation. Another important aspect of the electricity is the fact that its effects are strain rate sensitive. More specifically, as faster strain rates (or die speeds) are utilized, greater amounts of current are required to ensure increased formability. This aspect is illustrated in Figure 19.8, which displays force vs. position data for EAM Mg compression tests [20]. In the figure, a constant current density of 20 A/mm^2 was coupled with four different platen speeds. From the plot, it can be seen that, as the platen speed was increased, the ductility of the metal was decreased. It is essential to understand these key concepts of electricity/electrical flow prior to implementing the EAM technique on any specific manufacturing process.

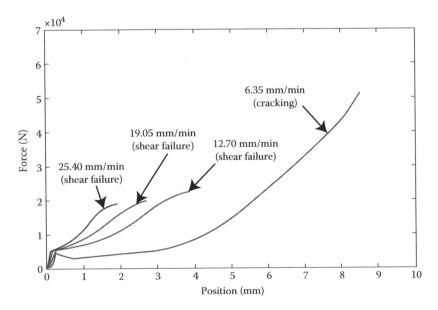

FIGURE 19.8 Strain rate dependency of EAM.

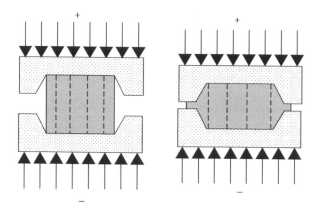

FIGURE 19.9   Forging process using EAM.

19.4.1.1 Forging

Forging is one of the oldest manufacturing processes (dating back to 5000 BC) in which a compressive force is used to plastically deform a material [2]. Large deformation forces are a characteristic of this process and engineers are constantly searching for ways that allow easier deformations and greater deformation lengths. Forging is normally conducted at extremely high temperatures (i.e., hot working temperatures), and the die and equipment costs of the process are large, since these components must be comprised from materials that possess great thermal and strength characteristics. By applying an electric current to a traditional forging process, reduced deformation forces and increased compressive displacements can be achieved. Additionally, it must be noted that the workpiece temperatures reached using the EAM technique are below each material's respective operating temperature, and are significantly less than temperatures reached during hot working. Perkins et al., using a variety of materials, determined that a continuously applied electrical current provides the best results when performing upsetting tests. From the examination of Figure 19.9, it can be seen that the electricity is applied directly to the dies with the metallic workpiece completing the closed circuit. It must be noted that the dashed lines in the figures are flow lines describing the paths along which the electricity will travel through the workpiece. In this case, electricity should travel through the whole workpiece since the entire slug is being deformed. Also, by applying electricity to the top/bottom dies and forcing the electricity to run vertically, the electrons will push the dislocation lines along the axis of primary deformation; thus creating premium workability conditions for the process.

As much as the electricity appears to be ideal for this type of manufacturing process (due to the reduced flow stress and increased deformation depths), there are prevalent issues and implementation challenges related to applying the EAM technique. First, the dies to which electricity is applied should be insulated from the rest of the forge in order to ensure that the electricity flows through the workpiece and to isolate the machine and operator. Since most insulation materials do not possess strengths as great as metallic conductive materials, the best way to insulate a forging process is to place the insulation materials away from the dies, where large surface areas are present so the load is spread out and high temperatures are not reached. The main objective of the insulation is to ensure that the workpiece experiences all of the imposed electrical flow for the best formability possible. As previously stated, since the voltage is so low when using EAM, people and machines should be unaffected by electrical contact; however, any electrically sensitive devices may become affected. Another issue is the fact that a lubricant is commonly used in forging operations as a way to reduce frictional effects. Since this lubricant is applied to the dies and workpiece, it may affect how well the electricity flows from the dies to the workpiece. Two possible solutions exist. First, the process could be run without any lubrication; however, material flow problems may exist because of excess friction. Another solution would be to devise new advanced lubricants, specific to the EAM process that will allow for enhanced conductivity.

In the following few paragraphs, a study by Jones et al., which highlights the promising effects of EAM, will be discussed [20]. As part of this research, MgAZ31B-O specimens were forged using the EAM technique. Two different types of forging were used (upsetting and impression die forging). The force vs. position plot and a figure containing the compression specimens corresponding to each plot can be seen in Figures 19.10 and 19.11. From Figure 19.10, it can be noted that the smaller current densities (10 and 20 A/mm²) had little formability effect and only slightly reduced the required deformation forces compared to a baseline test. Conversely, the higher current densities (30 and 40 A/mm²) displayed significant forgeability improvements. The required machine forces were greatly reduced, while elongation increases of over 450% were obtained (since the 40 A/mm² specimen never fractured, the maximum achievable improvement has not been fully established).

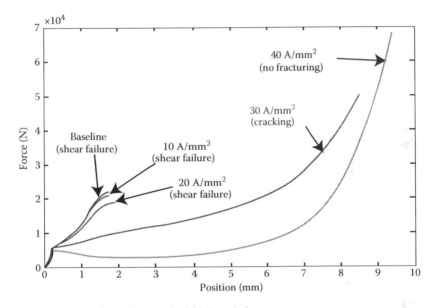

FIGURE 19.10 Upsetting using a die speed of 25.4 mm/min.

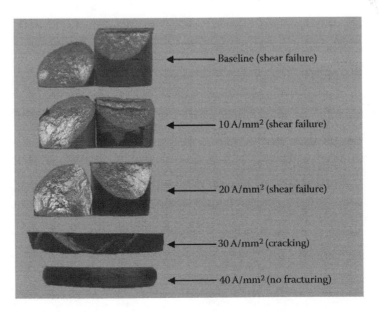

FIGURE 19.11 Upsetting specimens.

In addition, the test conducted at 40 A/mm² showed strain weakening (between 0.5 and 3.0 mm of elongation), where the required machine force decreased while the strain increased.

Figure 19.11 displays the resulting specimens (presented in Figure 19.10) after testing. From the figure, it can be concluded that the lower current densities, as well as the baseline test, resulted in shear failure. This occurred since the electrical threshold, point where electricity enhances forgeability, was not reached yet and the material could not continue deforming. Also seen in the figure, the test specimen used at 30 A/mm² failed due to cracking. However, when using a current density of 40 A/mm², a specimen could be completely formed without failure, something that cannot be easily accomplished with this magnesium alloy.

As part of the research, Jones et al. also examined electricity's effect on impression die forging, a common type of forging for many industrial applications. The force vs. position plot and specimen pictures from this part of the research can be seen in Figures 19.12 and 19.13, respectively. Examining Figure 19.12, the lower current densities (10 and 20 A/mm²) produced minor formability improvements, as was the case with upsetting. Also seen with upsetting, the higher current densities (30 and 40 A/mm²) achieved great forgeability improvements with significantly reduced deformation forces. Additionally, elongations exceeding 260% were produced when using a current density of 40 A/mm² (again, without fracture, the actual limit has not been established).

As mentioned, Figure 19.13 displays the specimens used for the impression die forging research. The baseline, 10, and 20 A/mm² specimens all failed due to shear fracture. The specimens run using current densities of 30 and 40 A/mm² could be formed without any type of failure. Cracking was not even visible in the flash of the specimens. The effects highlighted in this research prove that the EAM technique has the ability to transform a hard-to-deform brittle material, such as Mg AZ31B-O, into an easily workable metal. Without extreme amounts of energy and time, no other process could easily accomplish this feat.

Although postulated theories were covered previously, the specific mechanisms that enable electricity to increase a metal's formability have not been fully established. Also discussed previously, it is known that one effect of EAM is resistive heating. However, by comparing EAM tests to isothermal tests of similar temperatures, it can be seen that the heating effect is not the sole reason for the formability increases prevalent in EAM processes. In fact, the electricity only causes a specimen's temperature to increase for a very short period of time. From Figure 19.14, which shows the transient temperature

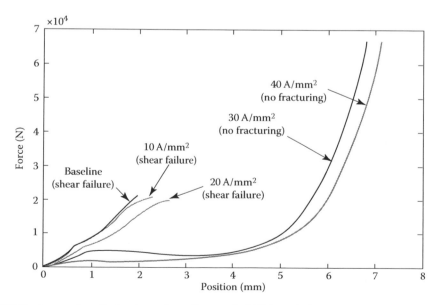

FIGURE 19.12   Impression die forging using a die speed of 25.4 mm/min.

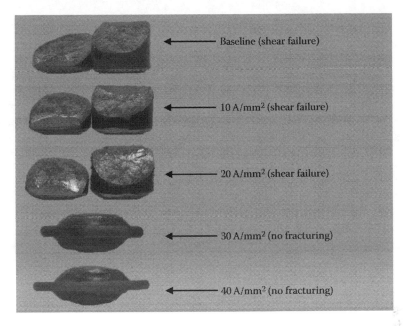

FIGURE 19.13   Impression die forging specimens.

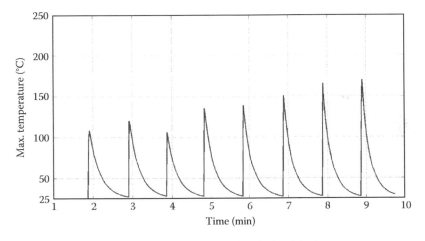

FIGURE 19.14   Transient specimen temperature profile (electrical pulsing).

of an electrically pulsed Mg tensile specimen, it can be seen that the specimen temperature increases instantaneously at the onset of each electrical pulse and decreases sharply after each pulse. The particular test in the figure actually cooled to near-room temperature between successive pulses.

In research by Ross et al., temperature profiles were developed for EAM-compressive and EAM-tensile processes (Figures 19.15 and 19.16) [21]. Figure 19.15 displays the thermal profile of an EAM compression specimen partway through the test. From the figure, it can be noted that the temperature profile is relatively narrow and there are no unexpected hot spots or abnormalities. Worth mentioning is that the temperature is significantly greater at the middle of the specimen, where the cooler fixtures have less of an effect. Figure 19.16 shows the thermal profile of an EAM tensile specimen partway through a test. After reviewing the figure, one can notice that this thermal profile is dramatically different than the compressive thermal profile previously explained. In tension, the entire specimen experienced significant heating, in which the center of the specimen was slightly hotter than the fixed ends.

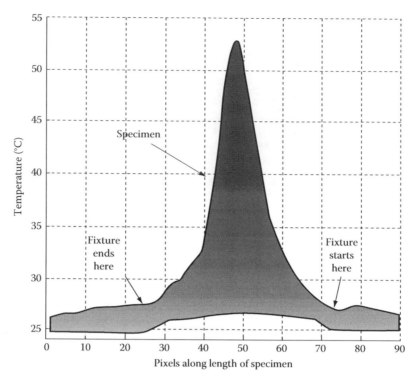

FIGURE 19.15 EAM compressive temperature profile.

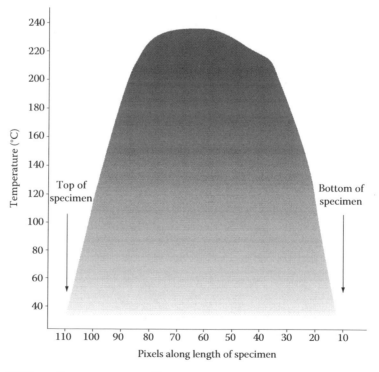

FIGURE 19.16 EAM tensile temperature profile.

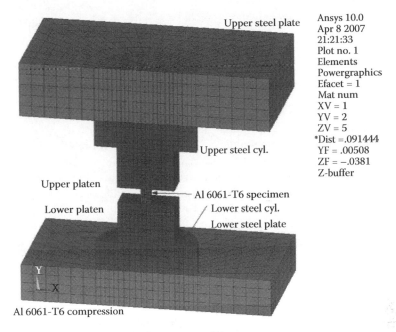

Ansys 10.0
Apr 8 2007
21:21:33
Plot no. 1
Elements
Powergraphics
Efacet = 1
Mat num
XV = 1
YV = 2
ZV = 5
*Dist =.091444
YF = .00508
ZF = −.0381
Z-buffer

Al 6061-T6 compression

FIGURE 19.17 FE model geometry and boundary conditions.

Research is ongoing in search of the actual reasons for the effects of EAM. The difference in behavior has also been established through FEA simulations through research by Kronenberger et al., where a FE model simulated the EAM compression of Al 6061-T6511 specimens [22]. From Figure 19.17, the model consisted of outer platens and inner steel fixtures, which contact the test specimen. Sandwiched between the platens and fixtures was a thermoset material (haysite reinforced polyester), which insulated the platens from any electricity. After being constructed, this model was then tested and verified thermally and mechanically.

The first test for the model was to simply test the ability of the model to predict the deformation behavior of the material when compressed at room temperature. This test verified the model's ability to represent strain hardening and the effects of the heat generated during deformation. Figure 19.18 presents the curve predicted by the simulation, along with the corresponding actual experimental results. As can be seen, the FE model was very accurate, when compared to the experimental test.

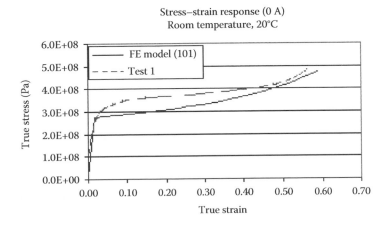

FIGURE 19.18 Stress–strain baseline comparison.

The stress vs. strain curve generated by the FE model was only slightly lower than the experimental stress vs. strain curve.

Next, the model's ability to predict the deformation behavior at elevated temperatures was examined. Figure 19.19 presents the comparison between the model's predicted behavior and the actual behavior when the metal is compressed under an elevated isothermal temperature of 142°C. As was the case with the cold working test, the FE model once again accurately predicted the stress vs. strain profile, when compared to the experimental test. Worth mentioning is that the FE model was again slightly lower at all strains than the experimental test.

Finally, the model was tested under simulated EAM conditions. More specifically, for the model, it was assumed that the AL 6061-T6511 specimens were compressed while a current density of 60 A/mm^2 was applied, as seen in Figure 19.20. From this, it can be noted that the FE model, which only accounted for resistive heating, and the experimental test did not produce the same stress vs. strain curvatures. From the figure, the modulus of elasticity and the general amount of elongation between the FE model and experimental model are similar; however, the experimental test displayed significant strain weakening effects, which were not accounted by the FE model. Based on this, it can be concluded that the resistive heating that occurs during the electrical tests does not fully account for the complete formability improvements found when deforming under an applied electrical current.

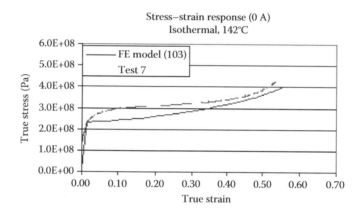

FIGURE 19.19   Isothermal test and model comparison.

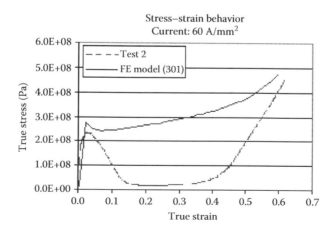

FIGURE 19.20   Stress vs. strain plot—60 A/mm^2.

19.4.1.2 Rolling

Rolling is another popular manufacturing process where a material's thickness is reduced by a set or sets of rollers that impose compressive forces on the material. As pointed out by Kalpakjian, rolling "accounts for about 90% of all metals produced by metalworking processes." Currently, the workpiece is usually heated prior to rolling the material in order to help reduce the required deformation forces. As previously stated for compression processes, the applied electricity will significantly reduce the material's flow stress, increase ductility, and reduce springback effects. Figure 19.21 shows two conceptual images of how electricity can be applied to a rolling process and the pre- and post-electrical flow lines for each method. The first method involves running the electricity through the rollers so that the electricity will travel through the material at the point of deformation. The second method incorporates the need for applicators that would transfer the electricity from a power supply to the material. The key with applying electricity to the rolling process is that the electricity must be uniformly distributed through the material prior to deformation; otherwise, nonuniform deformation may occur. A profound benefit of using the EAM technique compared to hot working is the fact that the entire roll of material does not need to be heated prior to rolling. With the EAM techniques proposed, the applicators could simply touch the material at some point before each set of rollers, or for the other method, the rollers could provide electricity just prior to deformation, thereby significantly reducing the energy costs. Additionally, the negative material effects of hot working (i.e., dimensional variations, scaling, etc.) can be avoided since the temperature of the material after the EAM application will only be a fraction of the material's temperature when the hot working method is used. Also, since easier deformation is achieved through electrical application, the effects of roller deflection and roll flattening can be dramatically reduced, thus allowing for more accurate rolled products and less wear on the rolling mill components. Additionally, due to the ductility increases from EAM, greater deformation per roller pass can be achieved. This will increase the efficiency of the rolling process and the general size of a rolling facility could be reduced.

Before employing EAM in high-speed production, a few obstacles must be overcome. One concern is to check whether common rolling lubricants are compatible with an applied electric current. This problem would be significant in the method where electricity is applied through the rollers; however, this obstacle could be remedied for the applicator method by applying the electricity prior to applying the lubricant to the rolling material. If these lubricants do not conduct electricity, the overall effect of the EAM technique (lower forces and greater elongations) could be dramatically lessened to the point where it is simply not beneficial. Considering this, conductive lubricants should

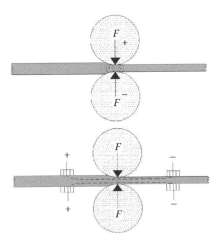

FIGURE 19.21 Rolling process using EAM.

be applied when implementing this hybrid rolling process. Another implementation obstacle is to formulate a sufficient method of insulating the rollers or applicators from the rest of the machinery. In rolling, high speeds and large forces are reached. Many insulation materials tend to wear out easily, so the best method of insulating the rolling process is to place the insulation materials away from the deformation zone so that the loading is spread out.

19.4.1.3 Extrusion

Extrusion is a very popular and practical manufacturing process for producing long continuous parts with a constant cross section. In the extrusion process, a billet or slug is placed into a chamber where a ram forces it through a die in order to produce the part. Currently, many products are extruded, such as angle iron, I-beams, and pipes. With this process, large compressive and frictional forces are generated; hence, extrusion is usually performed at elevated temperatures. Also, there are four main types of extrusion (direct, indirect, hydrostatic, and impact extrusion) that make extrusion useful in manufacturing for various applications where different part quality needs are required.

Considering the characteristics of the extrusion process, the EAM technique should have tremendous formability effects. Figure 19.22 shows three methods to apply electricity when extruding (the direct extrusion process is pictured). From the concept image, the first method shows the electricity being applied through the ram, as it is always in contact with the workpiece. The second method involves applying the electricity to the side of the blank before it is significantly deformed. The third method pertains to applying the electricity so that it travels vertically across the deformation zone (at the lead-in angles). Using insulation, the dies can be arranged for maximum formability. Examining the shape of the die, the majority of deformation takes place at the angle leading up to the die cavity. Knowing this, one way to get the most use from the applied current is to insulate the dies everywhere except at the angles leading into the die cavity. For the second EAM implementation method, a small portion of the die in the insulated region would have to be conductive so the electricity could be applied to the blank. This way, the flow lines will flow into the billet and then will be directed toward the "lead-in" angles, thus placing the electricity where it is needed most, the location where the greatest amount of deformation occurs. The third method is possibly the most efficient since it only places the electricity in the

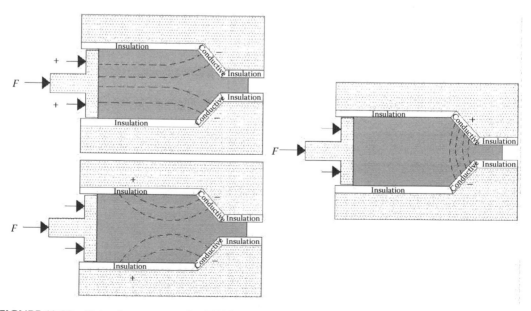

FIGURE 19.22 Extrusion process using EAM.

deformation region. In doing this, the electricity will lower the material's flow stress and increase ductility, thus allowing for lower force/energy requirements and a more efficient overall extrusion process.

Due to these effects, many aspects of the extrusion process can be enhanced. First, the need to heat the billet prior to extrusion will be reduced or possibly eliminated. This would result in less energy required to extrude, and a manufacturing company could potentially save floor space by eliminating the need for some or all of the accompanying furnaces or annealing ovens. Also, since deformation forces will be reduced, larger billets/slugs could be extruded using smaller capacity presses. This would allow companies to use their already-existing extrusion setup to extrude larger and stronger billets than they otherwise could not handle, thus expanding their market potential to larger workpieces. Finally, the overall quality and accuracy of the extruded part will increase with the application of the EAM technique. Without the extreme heat introduced to the workpiece and the EAM technique's ability to reduce springback, greater dimensional accuracy could be held, and the tendency of the extruded material to crack would be decreased due to the ductility improvements achieved by EAM.

As was true with the other bulk deformation processes, several design challenges are apparent in the creation of this hybrid process. First, the die would need to be constructed from both insulating and conductive materials in order to route the electrical flow to the region of maximum deformation (although the use of coatings may be an alternative). This type of die (including both materials) would be difficult and expensive to construct. Additionally, the application of insulating materials to the die may significantly reduce the maximum stress permitted on the die, if the chosen material does not perform well in compression, potentially making this EAM-implemented extrusion process only applicable for small parts made with weaker materials. Also, as was the case for forging and rolling, new lubricants that are compatible with electricity may need to be employed. It must be noted that the EAM technique will not be easily applicable for all types of extrusion. For example, hydrostatic extrusion, where a liquid separates the ram and billet, would not be easily adapted to an EAM process.

19.4.2 Sheet Metal Processes

Sheet metal manufacturing is a multibillion dollar industry worldwide. The most common industries that use large amounts of sheet metal are the automotive and aircraft industries, where sheet metal is used mainly for the shell and various panels of each vehicle. Both being very competitive industries (automotive and aircraft), there is a constant desire for the leading auto and aviation companies to produce lighter and more efficient cars or planes. Currently, the simplest way to do this is to use lighter and stronger sheet metal materials as compared to what is currently used. This means incorporating the use of many high-strength aluminum, magnesium, and titanium alloys into production due to their great strength-to-weight ratios. However, formability is extremely limited on many of these materials. This is where EAM plays a huge role. Unlike rolling and extrusion, significant research has been devoted toward studying electricity's effect on sheet metal processes. The two most prevalent processes that will be discussed are stretch forming and deep drawing. In both of these processes, the electricity will pass through the entire sheet metal workpiece. All the same electrical characteristics as bulk deformation processes hold true for sheet metal processes also. These include the fact that electricity is strain rate sensitive (requiring greater current densities at higher speeds), as well as optimizing workability by aligning the flow lines with the direction of deformation. One prevalent difference is that, for sheet metal applications that are mainly tensile loadings, research has proven that pulsed electricity, rather than continuously applied electricity, leads to greater formability. This allows for much more freedom when applying the electricity to the workpiece, since the current density, pulse duration (time that the pulse is applied), and pulse period (time between each successive pulse) can all be varied to achieve optimum behavior.

19.4.2.1 Stretch Forming

Stretch forming is a well-known sheet metal–forming process where the ends of the sheet metal are firmly clamped and the desired shape is created by deforming (or "stretching") the material with a die. Since the material is stretching, the workpiece is mostly in tension during this process. Pulsed electricity, rather than continuously applied, is used for tensile processes to help reduce the material's susceptibility to neck when forming. As one could imagine, it can be difficult to get thin sheets of various materials to stretch great distances. To aid in this, companies use the hot working, incremental forming, and even superplastic forming methods; however, each of these has disadvantages. Using the EAM technique for stretch forming, the corresponding material's flow stress is lowered, while the achievable elongation is increased, allowing for less required deformation energies and greater deformation depths. Additionally, if the sheet metal part is pulsed after deformation, the amount of springback within the part can be reduced, allowing the part to hold its shape better after being removed from the die. From Figure 19.23, one can see that electricity could be applied two ways. First, the electricity could be imposed on the workpiece through the blank holders, which hold the material in place during deformation. Also, the die could be designed such that the electricity could travel from the die to each blank holder, hence flowing through the workpiece along the way. The electrical flow lines depict the electricity moving throughout the entire piece of sheet metal, and this is needed because most of the sheet metal will be required to deform in order to form the final shape.

Many advantages arise from the application of electricity. First, the forming process does not need to be halted in order to apply the EAM technique. This saves time and should allow for more number of accurate parts, since the workpiece does not need to be removed and refixtured partway

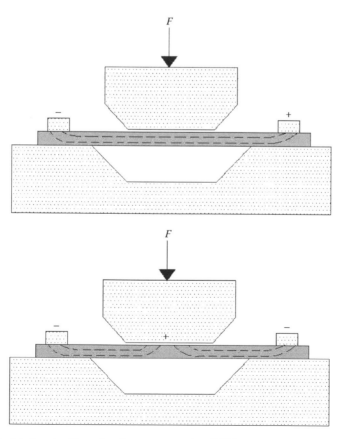

FIGURE 19.23 Stretch forming/deep drawing process using EAM.

through the process. The next benefit arising from EAM are the stress reductions, allowing for near-zero flow stress conditions to be achieved. This essentially means that, with the EAM technique, almost zero stress is required for plastic deformation to continue. The other prominent effect of EAM is the massive elongation increase that can be obtained, allowing for great deformation lengths in comparison to other manufacturing methods. The last advantage is that springback can be reduced or even eliminated in sheet metal specimens simply by applying a single electrical pulse to the material after deformation [17]. Due to the benefits of the electricity, two new avenues have opened in sheet metal processing. First, since greater formability limits have been established using EAM, more complex parts can be made from a single sheet metal blank rather than having to manufacture several smaller simpler parts and use a fastening method to attach them. This is ideal in the aircraft industry, where large single sheet metal components are sometimes required. The second avenue is opened because of the electricity's ability to allow for lightweight and stronger materials (i.c., magnesium and titanium) to be used in sheet metal forming. As previously stated, these materials possess limited formability characteristics; however, with the help of applied electricity, the formability of these metals increases significantly.

Along with the advantages, some issues are apparent in stretch forming as well. The most notable effect is the presence of a diffuse neck in tensile specimens that were pulsed with electricity while being formed. A diffuse neck is simply a geometric characteristic resulting from nonuniform deformation arising from geometrical constraints on material flow. From Figure 19.24, a picture of a diffuse neck in Al 5083 tensile specimens, constructed from strain grid measurements, can be seen. It is clearly shown that the pulsed tests display nonuniform deformation, thus resulting in diffuse necks. This could be detrimental in forming because unwanted material thinning could occur in various areas of the sheet metal due to the nonuniform deformation caused by EAM.

As was the case with the bulk metal processes, lubrication and insulation challenges may exist. In research by Salandro et al., the effects of die coating and lubrication were briefly examined [21]. Figure 19.25 shows three baseline tests (non-EAM) that were run at the beginning, middle, and the end of testing on one particular die. As testing progressed, the die condition diminished and its coating wore, thus leading to steady and noticeable decreases in formability during the testing.

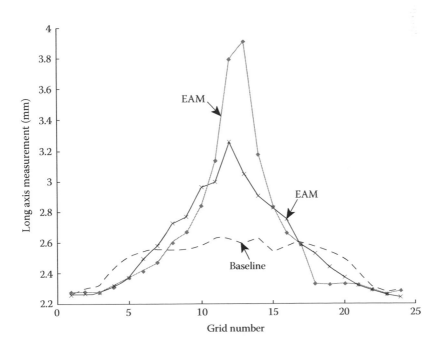

FIGURE 19.24 Displacement analysis showing diffuse neck.

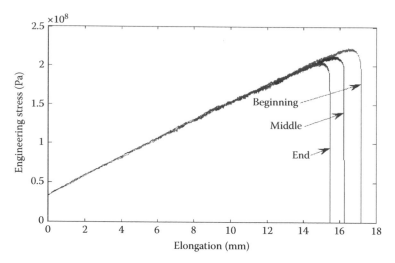

FIGURE 19.25 Die coating wear effects.

These formability decreases prove that die condition and lubrication play a significant role in sheet metal manufacturing.

19.4.2.2 Deep Drawing

Deep drawing is a process in which a flat piece of sheet metal is deformed into a cylindrical or box-like component [2]. Since its creation, during the 1700s, deep drawing has been used to make products such as beverage cans, automotive paneling, and virtually any type of metal container. Because this process is similar to stretch forming, Figure 19.23 will also be used to represent deep drawing. The only profound difference between the two processes is the fact that, in deep drawing, the workpiece is allowed to move (or draw) into the die cavity to which it is forming. With the workpiece allowed to move, parts requiring great depths can be produced from single blanks. Just as the processes are similar, electricity's effects are similar as well. The applied electricity will weaken the material and increase its ductility, thus allowing the material to stretch further prior to pulling new material through the blank holders. In doing this, deformation forces will be minimized and greater deformation depths can be achieved. Additionally, the EAM deep drawing process should take less time to form a part when compared to the conventional method; hence, process efficiency is increased.

The same disadvantages are apparent, including any type of lubrication issues, coupled with the task of devising an insulation method for the dies. The most prominent obstacle to overcome is the nonuniform elongation that can occur as a result of not applying the electricity to the entire workpiece. In stretch forming, the electricity was applied through the blank holders; however, for deep drawing, electricity may or may not be beneficial if applied here since the material will be moving through the blank holders and contact issues could result. Another way to apply the current would be through the dies, specifically the top die. The disadvantage of this is that if the surface of the top die does not contact the entire surface of the workpiece, the electricity may be concentrated in a certain area of the workpiece, thus increasing the chances of nonuniform deformation. On the other hand, the ability to control the electricity's concentration within different portions of the workpiece may be useful when dealing with a localized tearing issue.

19.4.3 Joining Processes

Joining processes, as the name describes, are processes by which two or more components are attached or assembled after being manufactured. These processes are extremely important because almost all manufacturing processes require components to be mated together (with the exception of single component manufacturing processes). Some examples of when joining methods may be utilized are such cases as the product being impossible to form as a single part, or the fact that the product must be frequently disassembled for maintenance over the course of its lifetime [2]. Some types of joining processes include welding, soldering, brazing, adhesive bonding, and a variety of mechanical fastening methods. Two welding methods, friction welding and friction stir welding, will be discussed as applicable processes for EAM implementation.

19.4.3.1 Friction Welding

Friction welding is a manufacturing process in which solid rods or hollow tubes are joined using mechanical energy. This energy is created by friction because of spinning one of the workpieces. More specifically, one workpiece is rotated at a fast rate while the other is held stationary. When the two parts touch, large frictional forces create intense heating effects, causing the materials to deform, thus forming a joint between the parts. Considering the fact that this process relies heavily upon material characteristics such as yield stress and ductility, the EAM technique should increase process efficiency when implemented. Figure 19.26 depicts a potential way of implementing the EAM technique on the friction welding process. In the figure, the electricity (+lead) is applied through a set of rollers that remain in contact with the workpiece at all times. An opposite charge is applied to the stationary workpiece, enabling the electricity to flow through the area of deformation when the two workpieces touch. Since the application of electricity to metals reduces flow stress and causes formability enhancements, many manufacturing advantages may arise. First, the material will become easier to deform with the application of electricity. This will significantly reduce the required deformation forces and may speed up the joining process. Also, since formability is increased through EAM, other materials that possess low formability characteristics, may now be applicable to this process.

Although, significant benefits are possible with EAM, many implementation obstacles exist. From the figure, one can notice the materials highlighted in black. These are insulating materials designed to fully isolate the metallic workpiece from the machine. With this in mind, one challenge lies with the fact that these insulating materials must possess great strength and durability characteristics. Another design challenge deals with the fact that the material's flow stress is decreased by EAM. With this being said, the workpieces (especially the one spinning) have the potential to buckle when exposed to the large axial forces, if they become too weak. Furthermore, the amount of axial or rotational force able to be imposed on the material may be limited by the actual strength of the material. When considered, it will be most beneficial to apply the current to each workpiece as close to their junction point as possible. This would only weaken the portions of each workpiece that are to deform, allowing the rest of each workpiece to remain rigid.

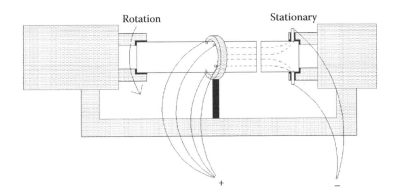

FIGURE 19.26 EAM friction welding concept image.

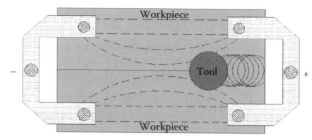

FIGURE 19.27   EAM friction stir welding concept image.

19.4.3.2 Friction Stir Welding

Friction stir welding is a process in which two materials are joined by using a rotating tool. Due to the effects of heating and mechanical deformation resulting from the spinning tool, welds can be created. It must be noted that this process can produce strong and ductile joints, and it is frequently used when conventional welding is not applicable, as on traditionally difficult-to-weld materials, such as aluminum alloys. Furthermore, this type of joining process is used on applications that are long and flat, such as sheets and plates. As with the friction welding process, the friction stir welding process also depends on the yield stress and ductility of the material. With this being said, the EAM technique should prove to be beneficial toward this process as well. Figure 19.27 displays a schematic of a friction stir welding process and describes how EAM can be implemented. In this case, electricity will flow evenly into and out of both workpieces from the C-shaped clamps (notice the dashed flow lines). This electricity will lower the flow stress of the parts, thus allowing the spinning tool to mate the two parts by mixing them together. The lowered flow stress helps to decrease the energy required to join the parts. Also, the enhanced formability created by the electricity will allow the implementation of less ductile materials into the friction stir welding process.

Prior to implementation of EAM friction stir welding in industrial applications, research is needed to explore the effects that electricity has on this process. The need for strong and durable insulating materials is necessary for this process. Also, since the tool is normally spun using mill-type machinery, an insulated collet must be devised in order to free the machine and its components from any electrical flow. The main challenge is finding insulating materials that can withstand the intense heating created by this process. Also, since friction stir welding is a rather slow process, it may be costly to obtain a power supply that can output relatively high current densities for extended durations.

19.5 MICROSTRUCTURE ANALYSIS OF EAM

In order to see the effects on the workpiece when applying electricity to a manufacturing process, microstructure examinations have been performed. Two recent studies were conducted by Roth et al. and McNeal et al. examining Al 5754 specimens and Mg AZ31B-O specimens, respectively. Both studies were conducted on sheet metal specimens that were deformed in uniaxial tension prior to the analysis. Moreover, both studies concentrate on slightly different microstructure aspects.

In the study by Roth et al., changes in the grain size and number of microvoids resulting from EAM were noted. Figure 19.28 below displays the baseline microstructure for the Al 5754 material. From the figure, there are no microvoids visible. Looking at Figure 19.29, which displays the microstructure of a specimen exposed to continuous electricity, one can see that there are a few microvoids that have formed. However, the amount of microvoids formed using continuous electricity is far less than the amount of voids formed using hot working or superplastic forming. Additionally, from Figure 19.30, which shows a test specimen deformed using pulsed electricity, one can see that

FIGURE 19.28 Al 5754 baseline microstructure.

FIGURE 19.29 Al 5754 continuous EAM microstructure.

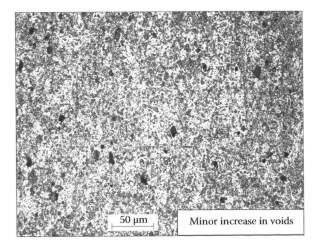

FIGURE 19.30 Al 5754 pulsed EAM microstructure.

an increased number of microvoids are visible. Although the number of voids increased, there are still less voids from the EAM process than would result from most other manufacturing methods. As a part of this study, it was also determined that the application of the electricity did not notably alter the grain size of the specimens in this particular research.

McNeal et al. conducted the second microstructure study. In this study Mg AZ31B-O sheet metal specimens were examined and grain size, grain elongation, and twinning were analyzed [23]. Figure 19.31 displays a deformed baseline micrograph. From the figure, twinning is visible. Figure 19.32 shows the microstructure of a deformed and pulsed EAM specimen. Apparent in this figure is that the presence of twinning has been eliminated due to the electric current. Also, the average grain size was increased in the EAM specimen. In Figure 19.33, another deformed and pulsed EAM specimen micrograph can be seen. As was the case with the previous micrograph (Figure 19.32), twinning was eliminated with EAM. However, in this case, the average grain size within the specimen (with different pulsing parameters) was decreased when compared to the baseline specimen (unlike the other pulsed micrograph). From this research, two conclusions can be made. First, the EAM technique eliminated twinning in all specimens to which it was applied. Second, the EAM technique has the ability to modify (increase or decrease) grain size by using different pulsing parameters (current density, pulse duration, pulse period, etc.).

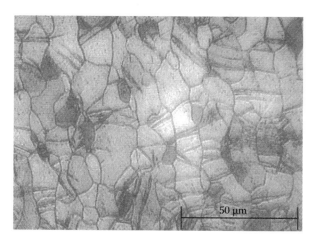

FIGURE 19.31   Mg AZ31B-O baseline microstructure.

FIGURE 19.32   Mg AZ31B-O pulsed EAM microstructure (increased grain size).

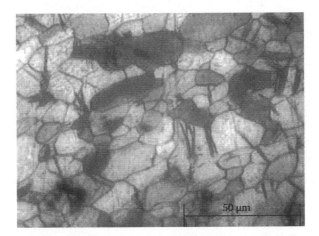

FIGURE 19.33 Mg AZ31B-O pulsed EAM microstructure (decreased grain size).

While both of the studies present a preliminary framework, there is still a lot that can be learned about EAM through microstructure evaluations. With this being said, the electroplastic effect that is occurring at the microstructural level during EAM needs to be examined to a greater depth.

19.6 CONCLUSIONS

By the completion of this chapter, the advantages and disadvantages of EAM should be apparent. Shown in Table 19.2 are the most prominent advantages and disadvantages of EAM when applied to any manufacturing process.

The advantages from the table combine to produce three proven effects of EAM. First, the required deformation forces needed to form a particular part are decreased because of EAM. Specifically, EAM lowers the flow stress of a material, thus allowing it to be deformed at a reduced force as compared to non-EAM forming. Second, the achievable elongation of a particular part can be increased using EAM. This is true for both tension and compression applications; however, different methods of application (pulsed or continuously applied) are needed. Third, it has been proven that EAM will reduce or eliminate springback effects in sheet metal specimens. All three of these effects contribute toward improvements in general part formability and allow for a larger material processing window, which is not achievable using traditional manufacturing

TABLE 19.2
Advantages and Disadvantages of EAM

Advantages	Disadvantages
Lower required machine forces	Must have an electrical applicator that contacts the workpiece at all times
Improved process efficiency	Need to devise methods to insulate machines and personnel
Increased material formability	May not be compatible with some manufacturing lubricants
Applicable to most conductive metals	
Can be applied without stopping process	EAM process is not completely understood
Ability to reduce/eliminate springback	
Can obtain different microstructures	

techniques. Additionally, EAM proves to be a versatile process, since it can be applied during deformation and the technique has the ability to alter the microstructure of the formed material.

Since EAM is a new and emerging manufacturing technique, there are still many areas to research regarding the various aspects of this process. As was previously stated, the theory behind the electroplastic effect is not entirely due to temperature effects; however, an alternative theory has yet to be fully developed. More microstructure examinations of EAM-processed workpieces will assist in further developing the current theories and help explain the electrical interactions at the microstructural level. Also, the application of EAM should be considered for all applicable manufacturing processes (not just the ones described in this work). EAM is broadly applicable to most deformation-based processes, thus making it potentially applicable toward hundreds of current and even future manufacturing processes. EAM is a relatively simple and effective way to increase manufacturing efficiency, and has the potential to have a positive impact on the manufacturing industry. However, before it can be confidently used in industry, the EAM technique will need to be further researched.

QUESTIONS

Q.19.1 Discuss the advantages and disadvantages of each technique for improving a metal's workability.

Q.19.2 Describe the difference between global and local resistive heating.

Q.19.3 What are the advantages to increasing a metal's maximum achievable elongation?

Q.19.4 What are the advantages to reducing the forces of deformation?

Q.19.5 For each manufacturing process, find five major companies for which the process is a major portion of their company's production.

Q.19.6 What are the advantages and disadvantages of using magnesium instead of titanium when producing automobile panels?

ABOUT THE CONTRIBUTING AUTHORS

Wesley A. Salandro is a student researcher, whose research is in the manufacturing and materials science disciplines. He obtained his education at Penn State Erie, The Behrend College in Erie, Pennsylvania. His specialty is his research in sheet metal forming, where he is a lead author on a few conference and journal publications. Additionally, several more of Wes' technical papers are under review for conferences and journals. As a second primary research subject, Wes has conducted research on tool wear monitoring. He is currently a coauthor on a conference article that focuses on wireless tool monitoring. In his research, Wes has worked with industry leaders, such as Ford Motor Company and Pacific Northwest National Laboratory. Prior to attending college, Wes spent several years working as a machinist and competing in motocross racing. He has used the experience and knowledge obtained from machining and rac-ing and applied it toward his research. Currently, Wes is attending Clemson University, South Carolina to obtain his PhD in automotive engineering. He has plans to use his PhD working in industry or becoming employed at a university where he can educate others on manufacturing and materials science.

John T. Roth is currently an associate professor of mechanical engineering at Penn State Erie, The Behrend College, Erie, Pennsylvania and has been with The Pennsylvania State University since 2001. He received his PhD degree in mechanical engineering at Michigan Technological University, Houghton. As a part of his investigations into EAM, Dr. Roth has 27 technical publications, 7 invited talks, 3 patents, and has collaborated with 11 university, industrial, and national laboratories. In addition, Dr. Roth's research interests span a broad range of other materials and manufacturing applications (including process modeling and prognostics; metal cutting, cryogenic heat treatments; and sensors and sensing), where he has over 50 other technical publications.

REFERENCES

1. Golovashchenko, S.F., A. Krause, and A.J. Gillard (2005). Incremental forming for aluminum automotive technology. *2005 ASME International Mechanical Engineering Congress and Exposition, IMECE2005-81069*, Orlando, FL, p. 7.
2. Kalpakjian, S. (1997) "Forging," "rolling," "superplastic forming," "deep drawing," "joining processes." *Manufacturing Processes for Engineering Materials*, 3rd edn., Addison Wesley Publishing Co., Menlo Park, CA, pp. 295–296, 324–325, 421–422, 424, 770, 801–802.
3. Machlin, E.S. (1959). Applied voltage and the plastic properties of "brittle" rock salt. *Journal of Applied Physics*, 30(7), 1109–1110.
4. Nabarro, F.R.N. (1967). *Theory of Crystal Dislocations*, Oxford University Press, chapter IX.
5. Troitskii, O.A. (1969). Electromechanical effect in metals. *Pis'ma Zhurn. Experim. Teoret. Fiz.*, 10, 18.
6. Klimov, K.M. and I.I. Novikov (1982). The "electroplastic effect." *Problemy Prochnosti*, 2, 98–103 (A.A. Baikov Institute of Metallurgy, Academy of Sciences of the USSR, Moscow, Translated).
7. Xu, Z.S., Z.H. Lai, and Y.X. Chen (1988). Effect of electric current on the recrystallization behavior of cold worked alpha-Ti. *Scripta Metallurgica*, 22, 187–190.
8. Chen, S.W., C.M. Chen, and W.C. Liu (1998). Electric current effects upon the Sn/Cu and Sn/Ni interfacial reactions. *Journal of Electronic Materials*, 27, 1193.
9. Conrad, H. (2000). Effects of electric current on solid state phase transformations in metals. *Materials Science & Engineering*, A287, 227–237.
10. Heigel, J.C., J.S. Andrawes, J.T. Roth, M.E. Hoque, and R.M. Ford (2005). Viability of electrically treating 6061 T6511 aluminum for use in manufacturing processes. *Transactions of the North American Manufacturing Research Institute of SME*, 33, 145–152.
11. Andrawes, J.S., T.J. Kronenberger, J.T. Roth, and R.L. Warley (2007). Effects of DC current on the mechanical behavior of AlMg1SiCu. *Material and Manufacturing Proceedings*, 22(1), 91–101.
12. Perkins, T.A., T.J. Kronenberger, and J.T. Roth (2007). Metallic forging using electrical flow as an alternative to warm/hot working. *Journal of Manufacturing Science and Engineering*, 129(1), 84–94.
13. Ross, C.D., D.B. Irvin, and J.T. Roth (2007). Manufacturing aspects relating to the effects of DC current on the tensile properties of metals. *Journal of Engineering Materials and Technology*, 129(2), 342–347.
14. Roth, J.T., I. Loker, D. Mauck, M. Warner, S.F. Golovashchenko, and A. Krause (2008). Enhanced formability of 5754 aluminum sheet metal using electric pulsing. *Transactions of the North American Manufacturing Research Institute of SME*, 36, 405–412.
15. Salandro, W.A., J.J. Jones, T.A. McNeal, J.T. Roth, S.T. Hong, and M.T. Smith (2008). Effect of electrical pulsing on various heat treatments of 5xxx series aluminum alloys. *International Manufacturing Science & Engineering Conference, MSEC 2008-72512*, Evanston, IL, p. 10.
16. Salandro, W.A., A. Khalifa, and J.T. Roth (2009). Tensile formability enhancement of magnesium AZ31B-O alloy using electrical pulsing. *Transaction of the North American Manufacturing Research Institute of SME*, 37, 387–394.
17. Salandro, W.A. and J.T. Roth (2009). Formation of 5052 aluminum channels using electrically-assisted manufacturing (EAM). *International Manufacturing Science & Engineering Conference, MSEC 2009-84117*, West Lafayette, IN, p. 9.

18. Green, C.R., T.A. McNeal, and J.T. Roth (2009). Springback elimination for Al-6111 alloys using electrically-assisted manufacturing (EAM). *Transactions of the North American Manufacturing Research Institute of SME*, 37, 403–410.

19. Kravchenko, V. (1966). *JETP (USSR)*, 51, 1676.

20. Jones, J.J. and J.T. Roth (2009). Effect on the forgeability of magnesium AZ31B-O when a continuous DC electrical current is applied. *International Manufacturing Science & Engineering Conference, MSEC 2009-84116*, West Lafayette, IN, p. 10.

21. Ross, C.D., T.J. Kronenberger, and J.T. Roth (2009). Effect of DC on the formability of Ti-6AL-4V. *Journal of Engineering Materials and Technology*, 131(3), 11.

22. Kronenberger, T.J., D.H. Johnson, and J.T. Roth (2009). Coupled multifield finite element analysis model of upsetting under an applied direct current. *Journal of Manufacturing Science and Engineering*, 131, 1–8.

23. McNeal, T.A., J.A. Beers, and J.T. Roth (2009). The microstructural effects on magnesium alloy AZ31B-O while undergoing an electrically-assisted manufacturing process. *International Manufacturing Science & Engineering Conference, MSEC 2009-84377*, West Lafayette, IN, p. 10.

20 Laser-Assisted Machining

Shuting Lei, Yung C. Shin, and Frank E. Pfefferkorn

CONTENTS

20.1 INTRODUCTION

Laser-assisted machining (LAM) represents an important and growing field in the area of thermally assisted machining (TAM), also known historically as hot machining. LAM is a hybrid machining process that integrates a laser source into a conventional machining setting. Lasers are used to elevate the workpiece temperature at the cutting zone to facilitate the material removal process by a cutting tool. LAM excludes those thermal machining processes in which the heating source is the same as the "cutting tool," such as the laser in laser cutting and the plasma in plasma cutting. Also excluded are such machining processes in which heat generation is intrinsic (e.g., due to plastic deformation at the shear zone). The purpose of heating in LAM is to reduce the hardness and shear strength of the workpiece material as well as to improve ductility in case of hard and brittle materials, and to assist the cutting tool in the material removal process. Ideally, only the material at the shear zone would be rapidly heated to a sufficiently high temperature and most of the heat input will be taken away by the chip. Thus, the heating source is required to satisfy the following conditions: (1) high power density to attain a high temperature quickly without thermal damage to the substrate material; (2) easy controllability of heating spot size and position; and (3) for practical reasons a heating source that is safe, inexpensive, and adaptable to production-type machine tools. The expected gains from LAM include reduced cutting force and power consumption, improved surface finish, longer tool life, and higher productivity.

TAM has evolved to the present state from more than 100 years ago, when the first patent was issued to Tilghman [1] on cutting metals by electrical resistance heating. Since then, TAM has been driven by the emergence of new materials and advanced by the development of new heating sources. In particular, with the development of laser technology in the 1960s, high-power industrial laser systems began to attract the attention of metal cutting researchers. Since the late 1970s, LAM has gradually taken the center stage in TAM research. Initially, LAM was considered for difficult-to-machine metals such as Inconel 718 and Ti6Al4V [2–7]. A large LAM research program was sponsored by the Defense Advanced Research Projects Agency (DARPA) in the late 1970s [2,3]. Both high-power continuous wave CO_2 and pulsed Nd:YAG lasers were used in machining aerospace alloys including Inconel 718 and Ti6Al4V. A significant increase in the material removal rate (MRR) with simultaneous reduction in tool wear and cutting force was reported.

Research in TAM continued through the early 1980s but waned due to the lack of evidence for economic benefits until the 1990s due to the findings that very-high-power lasers are needed for successful LAM [7]. Mostly due to the need for machining hard and brittle structural ceramics, the interest in laser- and plasma-assisted machining was resurrected in the early 1990s. Although laser beams have the advantage of high controllability and high power density, a plasma arc may be a preferred method for workpiece heating from an economical viewpoint. Hence, interest in plasma-assisted machining has continued until today as evidenced by the considerable publications in this area [8–13]. In a more recent study by Lopez de Lacalle et al. [14], plasma-assisted milling was considered for three different heat resistant superalloys (Inconel 718, Haynes 25, and Ti6Al4V). A significant reduction in the cutting force and an increase in the tool life were obtained for Inconel 718 and Haynes 25 without detrimental effects on surface integrity. However, the tool life was reduced for Ti6Al4V and the machined surface became more brittle due to structural changes caused by inadvertent heat treatment by the plasma arc. Thus, plasma-assisted milling was not recommended for this titanium alloy.

On the LAM front, new developments have continued to emerge around the world as manufacturing engineers search for ways to reduce the cost of shaping hard materials. As the development and engineering applications of structural ceramics has accelerated since the 1980s, LAM has taken on the new challenge of machining fully sintered and dense ceramics. Since the 1990s, studies on the LAM of advanced ceramics have been initiated at the Fraunhofer Institute of Production Technology in Aachen, Germany [15–18] and Purdue University [19–26]. In recent years, LAM has gained greater acceptance as an economically viable method of shaping difficult-to-machine materials and has resulted in a greater number of groups worldwide conducting research in this area. Besides ceramics, difficult-to-machine metals and composites are once again considered for LAM with new laser technology development. For example, Skvarenina and Shin [27] applied LAM to improve the machinability of compacted graphite iron; Anderson and Shin [28] used LAM for P550, a difficult-to-machine stainless steel due to its high nitrogen content; Anderson et al. [29] showed that the cost of machining Inconel 718 could be improved with LAM over conventional hard turning; and Dumitrescu et al. [30] conducted the laser-assisted turning of AISI D2 tool steel.

Table 20.1 provides an overview of the types of lasers, machining methods, cutting tools, workpiece materials, and benefits that have been investigated in some representative LAM work. The following sections will focus on the LAM of structural ceramics, difficult-to-machine metals, and metal matrix composites. Finally, some thoughts are given on the challenges and future directions in LAM.

TABLE 20.1
Laser-Assisted Machining Methods and Main System Components

Process	References	Cutting Tool	Work Material	Benefits
Turning with CO_2 laser	[17,22,23,31]	PCBN	Si_3N_4	↑ machinability, good surface finish, long tool life, compressive residual stress
Turning with CO_2 laser	[25]	PCBN	PSZ	↓ specific cutting energy, ↑ tool life
Turning with CO_2 laser	[32]	Carbide	Mullite	↑ tool life, good surface integrity
Turning with combined CO_2 and Nd:YAG laser	[28,29]	Ceramic	Inconel 718, P550,	↓ force, ↓ specific cutting energy, ↑ tool life, ↑ surface finish, ↑ material removal rate, ↓ machining cost
Turning with CO_2 or pulsed Nd:YAG laser	[3,6,33]	Carbide	Inconel 718, Ti6Al4V, hardened steel	↓ force, ↓ specific cutting energy, ↑ tool life, ↑ surface finish
Planing with Nd:YAG laser	[34]	PCBN	Al_2O_3	↓ force, ↑ surface quality
Turning and orthogonal machining with HPDL	[30]	TiN/TiAlN coated carbide	AISI D2 tool steel	↓ force, ↓ chatter, ↑ tool life
Milling with HPDL	[35,36]	TiAlN coated Carbide, PCBN	Si_3N_4	↑ machinability, good surface finish
Micromilling with Nd:YAG laser	[37]	Carbide	6061-T6 aluminum, 1018 steel	↓ force, ↑ chip load, ↓ specific cutting energy
Micromilling with Ytterbium fiber laser	[38]	TiAlN coated carbide	H-13 mold steel	↓ force
Micromilling with Nd:YAG laser	[39]	Carbide	Ti6Al4V, Stainless steel 316	↓ force, ↑ tool life, ↓ specific cutting energy, good surface finish, ↓ burr formation

20.2 LASER-ASSISTED MACHINING OF STRUCTURAL CERAMICS

Figure 20.1 shows the basic principle of LAM as well as its implementation for turning and milling operations. A focused laser beam is positioned ahead of the cutting tool to provide localized intense heating, so that the material reaches a high temperature at the time it is to be removed. A model-based process design for LAM is illustrated in Figure 20.2 to show the issues that need to be addressed to realize successful LAM. While the ultimate goal is to achieve high quality parts with high productivity, one necessary condition is the selection of optimum operating conditions, which, in turn, relies heavily on the basic understanding of the coupled laser heating and material removal process in LAM. In the following sections, progresses made in experimental investigation (material removal mechanism, machinability, surface integrity), numerical modeling, and simulation are summarized.

20.2.1 MATERIAL REMOVAL MECHANISM

The objective of LAM is to locally preheat the workpiece in order to modify its deformation behavior in such a way that is beneficial to the material removal process. For the LAM of fully-sintered structural ceramics, the goal of preheating is to increase material ductility and thus induce more plastic deformation during the machining process through alteration of the mechanical

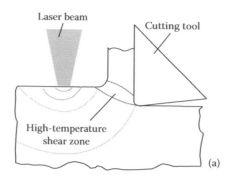

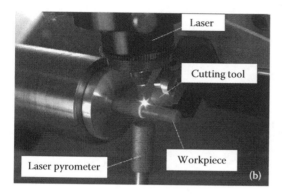

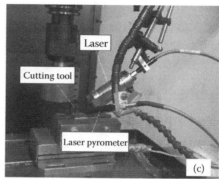

FIGURE 20.1   (a) Principle of LAM, (b) laser assisted turning system, (From Lei, S. et al., *Trans. ASME J. Manuf. Sci. Eng.*, 123, 639, 2001. With permission.) and (c) laser assisted milling system. (From Lei, S. and Pfefferkorn, F.E., A review of thermally assisted machining, *Proceedings of MSEC2007, 2007 ASME International Conference on Manufacturing Science and Engineering*, Atlanta, GA, 2007, 12 p. With permission.)

property of the bulk material or a specific phase. This is especially true for hot-pressed silicon nitride ceramics. König and Wageman [15] found that the temperature at the onset of quasi-plastic deformation and hence the minimum material removal temperature required was approximately 1100°C for silicon nitride. Subsequently, König and Zaboklicki [17] suggested that the softening of an amorphous glassy phase at the silicon nitride grain boundaries was responsible for the quasi-plastic deformation of silicon nitride in LAM. Rozzi et al. [19,22] further confirmed that the material removal mechanism of silicon nitride in LAM was dominated by plastic deformation rather than brittle fracture by experimentally obtaining semi-continuous chips and long continuous chips at material removal temperatures between 1150°C and 1300°C and above 1330°C, respectively (Figure 20.3a and b). It was suggested by Lei et al. [40] that the material removal mechanism of silicon nitride in LAM involves material oxidation, melting, and vaporization under the laser irradiated area; plastic deformation in the primary shear zone; and segmentation of chips due to the initiation, coalescence, and propagation of intergranular microcracks. The temperatures that have been achieved during the LAM of silicon nitride are above the glass transition temperature of the glassy phase (usually ~10 wt.%). Therefore, it is believed that during cutting, the ductility of the glassy phase enables pencil-like silicon nitride grains to reorient themselves to accommodate the cutting operation.

LAM was applied to other types of ceramics such as pressureless sintered mullite (65% alumina and 35% silica) ceramics by Rebro et al. [32] and partially stabilized zirconia (PSZ) by Pfefferkorn et al. [25]. Semi-continuous and continuous chips were generated for mullite (Figure 20.3c and d),

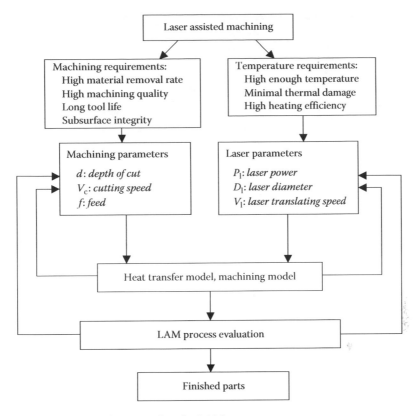

FIGURE 20.2 Model-based design procedure for LAM.

which showed that the material removal mechanism was mainly viscous flow and the plastic deformation of material at elevated temperatures. Unlike silicon nitride, PSZ has no intergranular glassy phase, and the material removal during LAM is believed to be a combination of brittle fracture ahead of the cutting tool due to the intense laser heating and plastic deformation via high-temperatures and high-pressure dislocation motions (Figure 20.3e).

20.2.2 Machinability—Cutting Force, Tool Wear, and Material Removal Rate

Most of the LAM work to date focuses on machinability studies. Comparison with conventional machining for ceramics is seldom done because machining fully sintered ceramics at room temperature is usually not a feasible process. LAM makes single point machining of ceramics feasible with attainable benefits such as long tool life and high MRR. An extensive experimental investigation was conducted by Lei et al. [23] for the LAM of silicon nitride. The results showed that both cutting force and tool wear depend on workpiece temperature as shown in Figures 20.4 and 20.5. Cutting force decreases with increasing temperature, while an optimal temperature exists for tool wear, at which the hardness difference between the tool and the workpiece is believed to be the largest and the tool is still chemically stable. For polycrystalline cubic boron nitride (PCBN) inserts, tool life corresponding to a typical tool wear criterion of VB = 0.3 mm is projected to be 42 min for $T_{mr,se} \approx 1410°C$. Tool wear occurs mainly on the flank face, and crater wear on the rake face is negligible. The results suggest that the mode of tool wear is a combination of adhesion and abrasion. Similar trends in cutting force and tool wear were reached by Rebro et al. [32] for mullite ceramics. Rebro et al. [24] compared the MRRs for three different ceramic materials as shown in Figure 20.6. It should be noted that one key measure for LAM to become a viable industrial process is that its

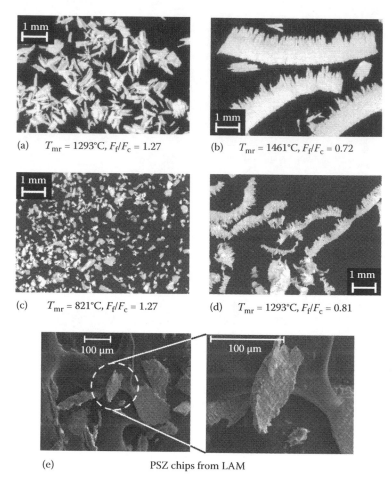

(a) $T_{mr} = 1293°C$, $F_f/F_c = 1.27$ (b) $T_{mr} = 1461°C$, $F_f/F_c = 0.72$

(c) $T_{mr} = 821°C$, $F_f/F_c = 1.27$ (d) $T_{mr} = 1293°C$, $F_f/F_c = 0.81$

(e)   PSZ chips from LAM

FIGURE 20.3   Chip forms for silicon nitride (a, b), mullite (c, d), and partially stabilized zirconia (e). (From Rebro, P.A. et al., *Trans. NAMRI/SME*, 30, 153, 2002.)

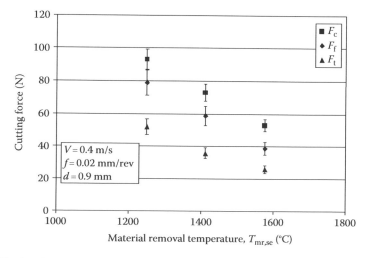

FIGURE 20.4   Cutting force vs. temperature for LAM of silicon nitride. (From Lei, S. et al., *Trans. ASME J. Manuf. Sci. Eng.*, 123, 639, 2001. With permission.)

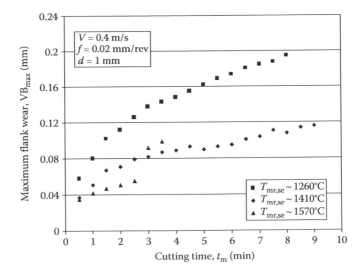

FIGURE 20.5 Maximum flank wear of PCBN insert vs. temperature for LAM of silicon nitride. (From Lei, S. et al., *Trans. ASME J. Manuf. Sci. Eng.*, 123, 639, 2001. With permission.)

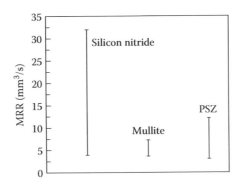

FIGURE 20.6 Ranges of MRR. (From Lei, S. et al., *Trans. ASME J. Manuf. Sci. Eng.*, 123, 639, 2001. With permission.)

MRR needs to be much larger than grinding for comparable surface roughness. Figures 20.6 and 20.7 show that for silicon nitride ceramics the MRR can reach over $30\,mm^3/s$ while maintaining a roughness value of $R_a \approx 0.5\,\mu m$. The MRR compares favorably for LAM when put into contrast with two investigations on the grinding of silicon nitride ceramics [41,42] in which the MRR was found to be about $1.3\,mm^3/s$ for a corresponding surface roughness of $R_a \approx 0.3\,\mu m$.

20.2.3 SURFACE INTEGRITY—SURFACE ROUGHNESS, MICROSTRUCTURE, SURFACE DAMAGE, AND RESIDUAL STRESS

Surface roughness from LAM was found to be comparable with but slightly higher than that of ground surfaces [17]. A comparison between LAM and grinding in terms of average surface roughness is shown in Figure 20.7 for three types of ceramics [24]. It is believed that LAM can produce a similar surface finish as conventional grinding but at a higher MRR.

One concern in LAM is the potential physical, chemical, and mechanical alterations of the machined surface. Machining induced strength reduction is a big problem for ceramic components. A ceramic's strength is strongly related to the amount of surface/subsurface damage and residual

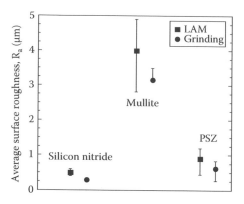

FIGURE 20.7　Surface roughness: LAM vs. grinding. (From Lei, S. et al., *Trans. ASME J. Manuf. Sci. Eng.*, 123, 639, 2001. With permission.)

FIGURE 20.8　SEM image of a silicon nitride surface machined by LAM. (From Lei, S. et al., *Trans. ASME J. Manuf. Sci. Eng.*, 123, 639, 2001. With permission.)

stress. LAM holds an advantage over grinding because in theory subsurface cracks can be eliminated in LAM with its more ductile nature in material removal. Lei et al. [23] showed that a silicon nitride ceramic surface produced by LAM consists of irregular glassy phase materials, cavities resulting from grain pullout, and dislodged β-Si_3N_4 grains, as shown in Figure 20.8. For comparison, a ground surface is shown in Figure 20.9. The surface is characterized by large regions of microfracture [43]. Similarly, plastic deformation in LAM of PSZ is obvious compared with grinding as shown in Figure 20.10 [25,66].

Residual stress affects the material strength (fatigue strength in particular) and thus is an important quantity for ceramic components. It is known that grinding creates a surface with relatively large compressive residual stress compared with machining. LAM studies showed that LAM can also impart compressive residual stress on the machined surface of silicon nitride ceramics [17,31]. Direct evidence of the effect of LAM on material strength was provided by Klocke and Bergs [44] in their LAM investigation. A comparative rotary bending test was conducted for silicon nitride (Si_3N_4) rods machined by LAM and grinding. The bending strength for parts machined by LAM was greater than for the ground parts. The residual stress measured after the LAM of silicon nitride ceramics by Tian and Shin [31] also clearly showed compressive stress in both axial and circumferential directions (Figure 20.11). All this evidence demonstrates that LAM can be used as an effective and economically viable process for shaping ceramic parts.

FIGURE 20.9 SEM image of a ground silicon nitride surface with significant microfracture area. Arrows point at four typical areas: (1) microfracture area, (2) smeared area, (3) area covered with debris, and (4) smooth area. (From Xu, H.H.K. et al., *J. Mater. Res.*, 11(7), 1717, 1996. With permission.)

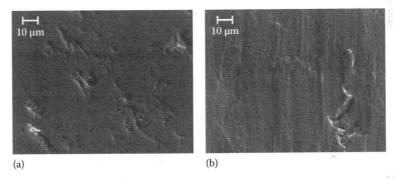

(a)   (b)

FIGURE 20.10 SEM micrograph of: (a) a ground partially stabilized zirconia surface and (b) a partially-stabilized zirconia surface machined by LAM. (From Lei, S. and Pfefferkorn, F.E., A review of thermally assisted machining, *Proceedings of MSEC2007, 2007 ASME International Conference on Manufacturing Science and Engineering*, Atlanta, GA, 2007, 12 p. With permission.)

20.2.4 THEORETICAL/NUMERICAL MODELING

20.2.4.1 Thermal Modeling

The significance of determining temperatures in the workpiece is clear for successful LAM, since the benefits of LAM can be achieved and optimized only when the workpiece is heated properly. Although analytical solutions continue to be attempted for the highly transient and three-dimensional temperature field in LAM [45], the complexity of the problem usually demands the use of numerical means for a detailed understanding. A transient, three-dimensional thermal model was first developed by Rozzi et al. [20] for a rotating cylindrical workpiece subjected to a translating laser heat source, and later this model was extended to the workpiece undergoing laser-assisted turning with material removal [46,47]. Solved by a numerical scheme based on the finite volume method, this model provided the temperature field in an opaque, homogeneous (silicon nitride) workpiece during LAM. With the numerical solution approach, this model was able to account for the following aspects: incident laser flux distribution, conduction within the workpiece, convection and radiation on all surfaces, heat and mass loss resulting from material removal, temperature dependent thermo-physical properties of the workpiece, and heat generation due to machining. Empirical formulation and experimental data were used to determine the thermal effects of machining, i.e., heat

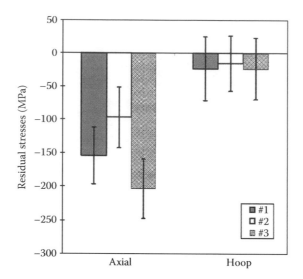

FIGURE 20.11 Measured residual stress in the surface layer of the silicon nitride workpieces produced by LAM. (From Tian, Y. and Shin, Y.C., *J. Am. Ceramic Soc.*, 89(11), 3397, 2006. With permission.)

generations due to the workpiece plastic deformation and the friction between the workpiece and the tool flank, respectively. Numerical simulations showed that the heat generation from machining had a much smaller effect on the transient temperature field during LAM compared with the laser heating, while the laser power and feed rate were the most significant parameters affecting the temperature field and material removal temperature.

Surface temperature predictions given by the model were compared with transient surface temperature measurements by a laser pyrometer [48]. The laser pyrometer measured the surface temperature and emissivity of the workpiece concurrently at the wavelength of 0.865 μm, at which silicon nitride is opaque. In general, it was shown that the temperature histories predicted by the thermal model agreed reasonably well with the experimental measurements under all operating conditions.

Later on, Pfefferkorn et al. [26] extended the model to a semitransparent, homogeneous material undergoing LAM. Two approaches, the discrete-ordinates method and the diffusion approximation, were used to account for the internal radiation inside the semitransparent workpiece. The predicted temperatures by the two approaches were shown to be close to each other and were validated by temperature measurements with a long-wavelength pyrometer [49]. The diffusion approximation, which used effective thermal conductivity to account for internal radiative transfer under the optically thick assumption, was chosen for further investigations [25] because it consumed less computer time and compared more favorably with the measurements.

Tian and Shin [50] reformulated the standard enthalpy-based heat transfer model into a transient three-dimensional temperature-based iterative method based on Lagrangian formulation to simulate the temperature distributions of the workpiece undergoing LAM. By adopting the technique of partial activation and deactivation of control volumes, the new modeling technique was capable of simulating LAM with material removal during profiling of a part with complex geometry, similar to the actual process. Tian et al. [35] also presented a transient, three-dimensional prismatic thermal model applicable to laser-assisted milling. These models have been validated using surface temperature measurements by an infrared camera, which showed good agreement (Figure 20.12).

20.2.4.2 Thermo-Mechanical Modeling

Although the workpiece temperature is a very critical parameter in LAM, the quality of the machined part eventually depends on the coupled thermo-mechanical material removal process.

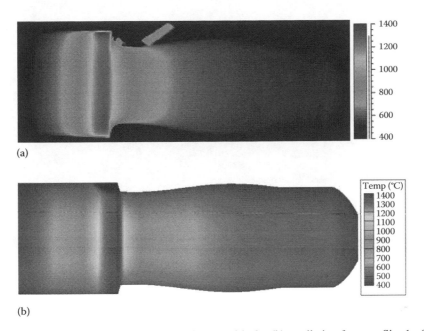

(a)

(b)

FIGURE 20.12 Comparison of the (a) infrared image with the (b) prediction for case Sin_1 after 56 s of LAM of silicon nitride. (From Tian, Y. and Shin, Y.C., *Trans. ASME J. Manuf. Sci. Eng.*, 128, 425, 2006. With permission.)

Hence, modeling and simulation of the thermo-mechanical LAM process is needed for surface damage assessment and process optimization. Since it has been qualitatively shown that the chip formation of silicon nitride (a typical structural ceramic) during LAM occurs under the combined effects of glassy phase flow, grain redistribution, and microcrack evolution, LAM simulation needs to treat the initiation and propagation of a large number of microcracks. In view of the limitations of some existing numerical approaches such as finite element analysis (FEA), a two-dimensional distinct element method (DEM) was used together with FEA to simulate the material removal process of silicon nitride ceramics in LAM [51,52]. DEM models the ceramic workpiece using bonded circular particles. Particle clusters are used to model the microstructure of silicon nitride. Bond breakage during simulation represents crack formation and propagation. FEA was used for thermal modeling to determine the temperature field in the workpiece. The temperature distribution was input to DEM indirectly by altering the mechanical properties of the workpiece. Figure 20.13 shows a simulation result. The simulation results demonstrate that DEM can reproduce many experimental observations of ceramic machining, including the initiation and propagation of cracks, chip formation process, and material removal mechanisms.

In a recent study, Tian and Shin [53] adopted a multiscale finite element modeling approach to quantitatively characterize the deformation and crack evolution in silicon nitride undergoing LAM. To consider the workpiece heterogeneous microstructure and crack evolution in silicon nitride machining, the workpiece material was modeled with continuum elements imbedded in thin interfacial cohesive elements. The continuum elements simulate the deformation of the bulk workpiece while the interfacial cohesive elements account for the initiation and propagation of intergranular cracks. The model reveals that discontinuous chips form by the propagation of cracks in the shear zone while the machined surface is generated by the plastic deformation of the workpiece material under confined high pressure (see Figure 20.14). The simulated cutting forces, chip morphology, and subsurface integrity compared favorably with the corresponding experimental observations.

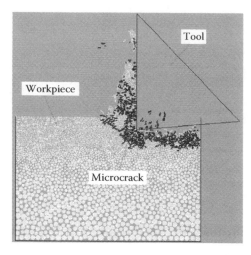

FIGURE 20.13   Distinct element simulation of LAM. (From Lei, S. and Shen, X., Finite element analysis of temperature distribution and distinct element simulation of laser assisted machining of silicon nitride ceramics, *Proceedings of the 2005 NSF Design, Service and Manufacturing Grantees and Research Conference*, Scottsdale, AZ, January 3–6, 2005.)

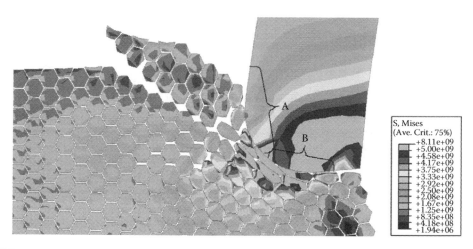

FIGURE 20.14   Simulated chip formation of using the multi-scale modeling approach. (From Tian, Y. and Shin, Y.C., *Trans. ASME J. Manuf. Sci. Eng.*, 129, 287, 2007. With permission.)

20.3   LASER-ASSISTED MACHINING OF METALS AND METAL MATRIX COMPOSITES

Unlike structural ceramics, most metals are machinable by conventional methods. However, many materials such as superalloys and hardened steels exhibit poor machinability due to their high strain hardening capabilities or high hardness. Some candidates of metals for LAM include the following:

- Nickel-based superalloys, such as Inconel 718 and MP35N
- Hardened steel
- Titanium alloys
- Compacted graphite iron
- Stainless steels
- Metal matrix composites
- Many others

Although the feasibility of LAM in metals has been demonstrated previously by various researchers, there are some fundamental issues to be addressed to achieve effective LAM. First, most metals exhibit poor absorptivity compared with structural ceramics and hence an effective means of coupling laser energy with the workpiece is necessary; second, the workpiece integrity after performing LAM must be preserved; and lastly, it is necessary to optimize the LAM process to establish economical benefits over conventional machining. In general, designing LAM for metals must consider the following issues:

- Efficiency of laser heating, i.e., absorptivity
- Tool life
- Cutting force
- Surface finish
- Microstructure of the finished surface
- Resultant residual stress

20.3.1 MATERIAL REMOVAL MECHANISM

Most metals exhibit lower yield or tensile strength at elevated temperatures. Figure 20.15 depicts the change of yield strength as a function of temperature for MP35N and Inconel 718. In the case of Inconel 718, there is a rapid reduction of yield strength beyond 700°C. If material removal is performed around 800°C, one can easily see that machining can be more easily performed. On the contrary, MP35, although stronger at room temperature, shows a decrease in strength at a lower temperature. Therefore, the key to successful LAM is finding the optimal temperature region according to the given cutting conditions.

As workpiece temperature in the shear zone increases, the specific cutting energy usually decreases as expected (Figure 20.16). With LAM, various researchers have reported a reduction of cutting force up to 40%–50% for 100Cr6 and Ti6Al4V [54], around 40% for pure titanium [55], ~20% for AISI D2 [30], and 15%–20% for Inconel 718 [29], CGI [27], and P550 [28]. Reduced cutting forces alleviate the catastrophic failure of cutting tools and also minimize or eliminate chatter problems.

This reduction of force will also contribute to a reduction of temperature rise in the primary and secondary shear zones. Since tool wear depends on the temperature and cutting velocity, a further increase in workpiece temperature will eventually accelerate tool wear as shown in Figure 20.17, although it might continually decrease cutting force. Therefore, in designing LAM conditions for

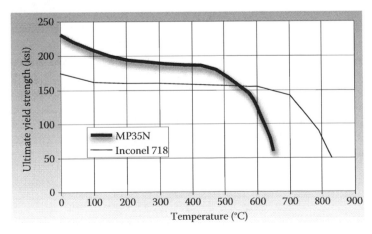

FIGURE 20.15 Yield strength vs. temperature for MP35N and Inconel 718. (From Anderson, M.C., 2005, Laser-assisted machining of difficult to machine metals, MSME thesis, Purdue University, West Lafayette, IN.)

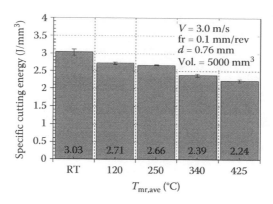

FIGURE 20.16 Specific cutting energy vs. workpiece temperature. (From Anderson, M. and Shin, Y.C., *Proc. Inst. Mech. Eng. B J. Eng. Manuf.*, 220(B12), 2055, 2006. With permission.)

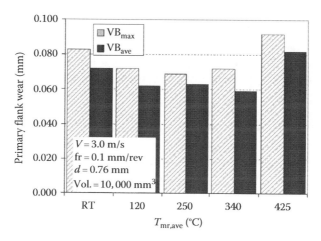

FIGURE 20.17 Flank wear vs. workpiece temperature. (From Anderson, M. and Shin, Y.C., *Proc. Inst. Mech. Eng. B J. Eng. Manuf.*, 220(B12), 2055, 2006. With permission.)

metals, it is critical to find the optimal material removal zone temperature where one can get the best combination of results in terms of cutting force, surface finish, tool wear, and surface integrity.

Chip removal plays a vital role in determining the quality of surface finish, and also in the amount of notch wear. Chips formed during LAM are usually more continuous with less curls. Also saw-tooth patterns, which are the characteristics of chips for some high-temperature alloys at higher speeds, are less prominent. The higher temperatures increase the ductility of the material and decrease the amount of notching that occurs in the cutting tool. Under a suitable operating condition of LAM, tool wear is usually greatly improved. While it varies, up to five- to sixfold tool life improvement has been reported. In the LAM of AISI D2 tool steel using carbide tools [30], catastrophic tool failure was avoided and machining chatter was also suppressed due to the thermal softening of the workpiece.

20.3.2 SURFACE INTEGRITY—SURFACE ROUGHNESS, MICROSTRUCTURE, SURFACE DAMAGE, AND RESIDUAL STRESS

LAM generally improves surface finish as the material removal temperature increases. This is due to the thermal softening and suppression of strain hardening, thereby facilitating easy chip formation.

The microstructure of finished parts can be preserved under a good operating condition of LAM. Through post inspection by optical microscope and XRD, it has been demonstrated that the microstructure of parts produced by LAM can remain unaffected [27,28]. Similarly it has been shown that surface and subsurface hardness does not have to be affected by optimal laser heating.

Residual stress after the LAM of metals has been shown to depend on the workpiece material. Germain et al. [54] reported that surface residual stress becomes less compressive or tensile with laser power in the case of Ti6Al4V, indicating that too much heating despite lower cutting force might negatively affect the residual stress. In the LAM of other materials, it has been reported that residual stress after LAM is comparable with those after conventional machining [29].

20.4 ECONOMIC VIABILITY OF LAM

20.4.1 HEATING EFFICIENCY

Since a high-power laser represents a considerable portion in the capital investment for a LAM system, heating efficiency should be maximized so to minimize the laser power requirement. Recently, Pfefferkorn and Lei [56] introduced a metric for defining the efficiency of a particular TAM process. The efficiency is defined as the fraction of energy deposited into the workpiece (i.e., absorbed by the workpiece) that is actually used for heating the material that is being removed. The analysis of a few LAM studies indicates that the heating efficiency is generally below 25% for current LAM practices, which means that most of the laser power is lost to outside the cutting zone. A 100% heating efficiency, although unattainable, means that all the laser energy is deposited to only the material at the cutting zone. Using this metric, the efficiency gap between an actual LAM process and the ideal one can be assessed to motivate improvements. Different laser systems and heating strategies can be compared against this metric for energy efficiency evaluations.

20.4.2 ECONOMIC ANALYSIS

In 1981, Tipnis et al. [7] concluded that the LAM of large metal objects (e.g., tank gun barrels) was not economically viable. This was a result of the high capital investment required for a 15 kW laser in 1981, the relatively low efficiency of CO_2 laser at the time, and the low absorptivity of the metal surfaces being studied. Since then, laser technology and the application of lasers for preheating workpieces have improved significantly. Less energy needs to be deposited in a workpiece if the laser spot is located closer to the primary shear zone. Besides the conventional (easiest) practice of locating the laser spot on the outer circumference of the unmachined workpiece during turning, investigations of heating the machining chamfer [33] and directing the beam through the tool directly onto the tool-chip interface [57] have been reported. Improvements in laser technology have resulted in a significant reduction in the equipment cost, improved efficiency, and a greater variety of wavelengths to choose from. For example, metals absorb a greater amount of incident laser energy at 0.8–0.9 μm wavelength (diode laser) than at 1.06 and 10.6 μm wavelengths emitted by Nd:YAG and CO_2 lasers, respectively. New high-power diode lasers also have efficiencies of up to 60% far surpassing the 10%–15% laser efficiency that Tipnis et al. had to assume in 1981.

Recently, Anderson et al. [29] demonstrated the economic benefits of LAM for Inconel 718 with a 1.5 kW CO_2 laser and a 500 kW Nd:YAG laser. Compared with conventional machining with carbide tools and ceramic tools, LAM could produce a significant cost reduction due to the attendant high MRR and extended tool life as shown in Figure 20.18. Similar economic benefits were demonstrated for the LAM of P550 [28]. These studies clearly indicate that LAM provides economic benefits, despite the added cost of laser, in addition to many other benefits such as improved surface finish, reduced cutting force, and improved subsurface integrity. In the case of silicon nitride, a cost reduction of over 70% was shown when compared with conventional diamond grinding [58].

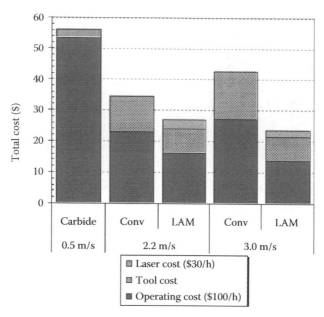

FIGURE 20.18   Total cost comparison when machining 1 m length (or 58,000 mm³) by conventional machining and LAM. (From Anderson, M. et al., *Int. J. Mach. Tools Manuf.*, 46, 1879, 2006. With permission.)

20.5   OTHER PROGRESS IN LAM AND RELATED PROCESSES

20.5.1   LASER-ASSISTED MILLING

Both experimental investigation and numerical modeling have been initiated for laser-assisted milling of ceramics. Tian et al. [35] showed the feasibility of laser-assisted milling of ceramics through experiments. The laser-assisted milling experiments designed with the assistance of their thermal model were successfully conducted on silicon nitride ceramics using TiAlN coated carbide end mills. Good surface roughness and acceptable tool wear were achieved. Yang and Lei [36] also demonstrated that laser-assisted milling can significantly increase the machinability and productivity of silicon nitride ceramics. In their end milling tests, the authors observed good surface finish and diminishing edge chipping as the workpiece temperature increased.

20.5.2   LASER-ASSISTED MICROMACHINING

Jeon and Pfefferkorn [37] applied LAM to the micro end milling of 6061-T6 aluminum and 1018 steel. The authors experimentally found that laser heating could significantly improve the productivity of micro-milling for the two investigated materials. Laser assistance enabled increased chip loads, reduced cutting forces, and reduced specific cutting energy but it could increase burr formation and surface roughness. At the same time, Singh and Melkote [38] reported a new laser-assisted mechanical machining process for micromachining applications. The process attempts to overcome the limitations of low tool stiffness and bending strength in pure mechanical micro cutting and the geometry limitations of pure laser micromachining. It is similar to laser-assisted shaping. The study revealed that low depths of cut thermal expansion of the tool and/or workpiece can offset the effect of thermal softening by effectively increasing the depth of the cut. Recently, Shelton and Shin [39] evaluated the laser-assisted micro milling capabilities for Ti6Al4V and 316 stainless steel alloys using 100 μm diameter end mills in slotting operations. Laser-assisted mechanical micromachining (LAMM) was found to improve the R$_a$ of 316 stainless steel by up to 37% at feedrates greater than 20 mm/min. LAMM also reduced the wear rates over conventional micromachining.

20.5.3 LASER-ASSISTED DRILLING

Jen et al. [59] conducted an investigation of laser-assisted drilling. Carbon steel (0.1% carbon content) was used as the workpiece material and a CW CO_2 laser was employed with the beam taking a ring shape. The finite element method was used for numerical simulation of the process. A comparison between experimental and numerical results showed good qualitative agreement between theory and experiment within the limits of the (rather large) uncertainty with which material properties are known to date. The result of using laser heating was an increase in the feed that could be used.

20.5.4 OTHER RELATED PROCESSES

Laser-assisted grinding of advanced ceramics was experimentally attempted by Westkämper [60] and Marinescu [61]. Zhang and Shin [62] demonstrated that the application of laser heating during the dressing of vitrified CBN grinding wheels could reduce the wear of the diamond dresser several times while still producing the dressed wheel topography similar to those by conventional mechanical dressing techniques. Although not a material removal process, Tian and Shin [63] adopted the LAM principle in laser-assisted burnishing of MP35 and 4140 alloy steel and demonstrated its advantages, such as reduced wear of the burnishing tool and increased subsurface hardness.

20.6 CHALLENGES AND FUTURE DIRECTIONS

Although advantages of TAM over conventional machining have been well documented over the years, limited success has been found in terms of industrial applications. Current TAM activities suggest that opportunities exist especially for LAM, as a promising cost-effective alternative method for structural ceramics. Besides the high MRR and superior surface integrity in comparison with grinding, LAM has clear advantages when it comes to parts with complex geometry, as demonstrated by Tian and Shin [31,50], as well as by parts like engine valves and bearings [44,64]. However, to be used in an industrial environment, LAM (as an integrated system) needs to be reliable, repeatable, easy to control, and economical. To achieve this goal, the following research areas need to be advanced.

Process optimization. LAM is a complex machining process because laser heating adds several more parameters to the process, which makes process optimization much more difficult. Optimization requires a good understanding of the process physics and effects of the operating parameters on part quality. Such understanding needs valid modeling and simulation of the process. While thermal modeling has been quite successful, more progress is needed for modeling the thermo-mechanical LAM process, in particular for predicting the subsurface integrity as a function of operating parameters. Based on these accurate thermal models and new advances in the thermo-mechanical modeling of LAM processes, it will be necessary to develop a good optimization strategy.

New process development. To take full advantage of LAM process capabilities, different machining processes need to be explored for targeted applications of LAM. Most of the current efforts are on laser-assisted turning. Laser-assisted milling and drilling are just beginning to be attempted.

System integration. Because of the high cost of high-power laser systems, creative LAM systems are needed to offset the added cost. For example, time-sharing of an expensive laser with a few machining centers (areas) was introduced in the early years of LAM [2]; Hügel et al. [65] demonstrated laser-integrated turning and milling centers. On the other hand, LAM at the microscale has the advantage of low system cost.

Economic justification. The economic feasibility of LAM needs to be further evaluated to include a wider range of materials, part shapes, and processes. Tipnis et al. [7] applied macroeconomic models to LAM for a few selected metallic components from industry using what is now outdated laser efficiency and cost data. More recently, Anderson et al. [28] and Anderson and Shin [29] determined that the laser-assisted turning of Inconel 718 and P550 is economically beneficial. These studies indicate that more systematic studies to justify the economic benefits for different advanced materials, particularly for parts with complex shapes and requiring multiple machining processes, are needed to make LAM more widely acceptable to various industrial applications.

20.7   TOWARD THE IDEAL LAM PROCESS—A LOOK INTO THE FUTURE

The ideal LAM process would have near 100% heating efficiency, large MRR, long tool life, high part quality, and user-friendly control interface. Despite the numerous innovations and persistent research throughout the history of TAM, we are still far from realizing the ideal process. Two issues stand out as major obstacles down the road: an efficient heating scheme and a user-friendly control system. Heating efficiency depends on the heat source and heating method. While lasers have been established as a premier heating source, efficiently heating the workpiece demands novel strategies (even more so for milling and drilling operations) as well as help from the fundamental understanding of the physical process through valid modeling and simulation. For use in the production environment, a robust, reliable, and safe control system has to be developed so that no special knowledge is needed to operate the system. Finally, it may be preferred for the heating and control system to be designed as an add-on apparatus that can be seamlessly integrated into an existing system. Even though considerable efforts are needed to turn a new generation of LAM into a viable industrial machining technology, with the recent rapid progress in the development of LAM, the future for LAM is bright.

QUESTIONS

Q.20.1   Compare laser and plasma arc as heating sources in TAM, i.e., what are the advantages and limitations of each?

Q.20.2   What types of materials are suitable candidates to be considered for LAM?

Q.20.3   What are the basic requirements for cutting tool materials in LAM?

Q.20.4   What in your opinion would be an ideal TAM process? What are the major challenges between current TAM processes and the ideal process?

Q.20.5   Compare LAM and the grinding of structural ceramics and what makes LAM a promising alternative to grinding?

Q.20.6   Suggest ways to improve heating efficiency in LAM.

Q.20.7   Compared with laser-assisted turning, what additional challenges does laser-assisted milling/drilling pose in terms of workpiece heating?

ABOUT THE CONTRIBUTING AUTHORS

Dr. Shuting Lei joined Kansas State University, Manhattan, in 1999 as an assistant professor in the Department of Industrial and Manufacturing Systems Engineering and was promoted to associate professor in 2005. His research focus is to develop LAM processes and innovative cutting tool technologies to enable high-performance machining techniques for difficult-to-machine materials such as structural ceramics and composites. His group has conducted both laser-assisted turning and milling experiments of silicon nitride ceramics and studied tool wear, surface integrity, and material removal mechanisms. He has published more than 40 papers in archival journals and conference proceedings in the area of advanced manufacturing processes. Dr. Lei earned his PhD in the School of Mechanical Engineering at Purdue University, West Lafayette, Indiana. Prior to this, Dr. Lei received his MS and

BS in mechanical engineering at Tsinghua University, Beijing, China. He then worked at Tsinghua University as a lecturer and research engineer involved in teaching and research in ceramic machining and metal cutting.

Professor Yung C. Shin currently is the Donald A. and Nancy G. Roach Professor of Advanced Manufacturing at Purdue University, West Lafayette, Indiana. He received his PhD from the University of Wisconsin in Madison in 1984 and worked as a senior project engineer at the General Motors Technical Center in Warren, Michigan, from 1984 to 1988 and as an assistant professor at Pennsylvania State University, University Park from 1988 to 1990. In 1990, he joined the School of Mechanical Engineering at Purdue University. His research areas include laser processing of materials, intelligent and adaptive control of manufacturing processes, dynamics of machine tools, high-speed machining, machining of advanced materials, process monitoring, and automation. He has published over 220 papers in archived refereed journals and refereed conference proceedings and has authored chapters in several engineering handbooks, coedited two books, and recently coauthored a book entitled *Intelligent Systems: Modeling, Optimization, and Control*. He established the Center for Laser-based Manufacturing in 2003 and currently is serving as its director. He also has served as the chair of the Systems, Measurement, and Control Area in the School of Mechanical Engineering, and as the associate editor for the *ASME Journal of Manufacturing Science and Engineering*. He received the 2007 ASME Blackall Machine Tool and Gage Award, the annual best paper award.

Frank E. Pfefferkorn is an associate professor in the Department of Mechanical Engineering at the University of Wisconsin–Madison. He joined the UW–Madison faculty in 2003 as an assistant professor and was promoted to associate professor in 2009. He received his PhD in mechanical engineering from Purdue University, West Lafayette, in 2002 on the topic of laser-assisted machining of ceramics. Prior to his appointment in Madison, Frank was a postdoctoral research associate in the Nanoscale Thermo-Fluids Laboratory at Purdue University. The goal of Pfefferkorn's research is to develop and apply a science-based understanding of manufacturing processes (heat transfer, material behavior, machinability, tribology, etc.) in order to increase performance and offer new and improved manufacturing tools to the industry. He has published more than 20 papers in archival journals and more than 30 conference papers on advanced manufacturing processes. Pfefferkorn has active research projects on laser-assisted micro end milling of metals and ceramics, laser-assisted friction stir welding, and thermal efficiency of thermally assisted manufacturing processes. He has received a Research Initiation Award and the 2007 Kuo K. Wang Outstanding Young Manufacturing Engineer Award from the Society of Manufacturing Engineers.

REFERENCES

1. Tilghman, B. C., 1889, U.S. Patent 416,873.
2. Rajagopal, S., 1982, Laser-assisted machining of tough materials, *Laser Focus*, 18(3), 49–54.
3. Rajagopal, S., Plankenhorn, D. J., and Hill, V. L., 1982, Machining aerospace alloys with the aid of a 15 kW laser, *Journal of Applied Metalworking*, 2(3), 170–184.

4. Bass, M., Beck, D., and Copley, S. M., 1978, Laser assisted machining, *Fourth European Electro-Optics Conference*, Utrecht, the Netherlands, pp. 233–240.

5. Copley, S. M., 1985, Laser applications, *Handbook of High-Speed Machining Technology*, Chapman & Hall, New York.

6. Komanduri, R., Flom, D. G., and Lee, M., 1985, Highlights of the DARPA advanced machining research program, *Journal of Engineering for Industry*, 107(11), 325–335.

7. Tipnis, V. A., Ravignani, G. L., and Mantel, S. J., Jr., 1981, Economic feasibility of laser assisted machining (LAM), *Transactions of NAMRI/SME*, 9, 547–552.

8. Kitagawa, T. and Maekawa, K., 1990, Plasma hot machining for new engineering materials, *Wear*, 139, 251–267.

9. Kitagawa, T., Maekawa, K., and Kubo, A., 1988, Plasma hot machining of high hardness metals, *Bulletin of Japanese Society of Precision Engineering*, 22(2), 145–151.

10. Kozhura, L. M. and Osarenren, J. O., 1991, The optimization of rotary cutting plasma assisted machining conditions of hard coatings, *International Journal of Production Research*, 29(1), 199–208.

11. Novak, J. W., Shin, Y. C., and Incropera, F. P., 1994, Assessment of plasma enhanced machining for improved machinability of Inconel 718, *Manufacturing Science and Engineering*, ASME PED-Vol. 68-1, pp. 443–451, Symposium on "Nontraditional Manufacturing Processes in the 1990s," Chicago, IL.

12. Shin, Y. C. and Kim, J. N., 1996, Plasma enhanced machining of Inconel 718, *Proceedings of the Manufacturing Science and Engineering, Symposium on Advanced Machining and Finishing Processes of Ceramics, Composites and High Temperature Alloys, ASME IMECE*, Atlanta, GA, pp. 243–249.

13. Madhavulu, G. and Ahmed, B., 1994, Hot machining process for improved metal removal rates in turning operations, *Journal of Materials Processing Technology*, 44(3–4), 199–206.

14. Lopez de Lacalle, L. N., Sanchez, J. A., Lamikiz, A., and Celaya, A., 2004, Plasma assisted milling of heat-resistant superalloys, *Transactions of the ASME, Journal of Manufacturing Science and Engineering*, 126(2), 274–285.

15. König, W. and Wageman, A., 1991, Fine machining of advanced ceramics, in P. Vincenzini (ed.), *Proceedings of the Seventh International Meeting on Modern Ceramics Technologies*, Montecatini, Italy, pp. 2769–2783.

16. König, W. and Wagemann, A., 1993, Machining of ceramics components: Process-technological potentials, in S. Jahanmir (ed.), *Proceedings of the International Conference on Machining of Advanced Materials*, Gaithersburg, MD, NIST Special Publication, Vol. 847, pp. 3–16.

17. König, W. and Zaboklicki, A. K., 1993, Laser-assisted hot machining of ceramics and composite materials, *International Conference on Machining of Advanced Materials*, Gaithersburg, MD, NIST Special Publication, Vol. 847, pp. 455–463.

18. Weck, M. and Kasperowski, S., 1997, Integration of lasers in machine tools for hot machining, *Production Engineering (London)*, 4(1), 35–38.

19. Rozzi, J. C., Incropera, F. P., and Shin, Y. C., November 1997, Transient, three-dimensional heat transfer model for the laser assisted machining of ceramic materials, *Proceedings of the ASME Heat Transfer Division*, HTD-Vol. 351, pp. 75–85, ASME IMECE, Dallas, TX.

20. Rozzi, J. C., Incropera, F. P., and Shin, Y. C., 1998, Transient thermal response of a rotating cylindrical silicon nitride workpiece subjected to a translating laser heat source: II—Parametric effects and assessment of a simplified model, *Transactions of the ASME, Journal of Heat Transfer*, 120(4), 907–915.

21. Rozzi, J. C., Pfefferkorn, F. E., Incropera, F. P., and Shin, Y. C., 1998, Transient thermal response of a rotating cylindrical silicon nitride workpiece subjected to a translating laser heat source: I—Comparison of surface temperature measurements with theoretical results, *Transactions of the ASME, Journal of Heat Transfer*, 120(4), 899–906.

22. Rozzi, J. C., Pfefferkorn, F. E., Shin, Y. C., and Incropera, F. P., 2000, Experimental evaluation of the laser assisted machining of silicon nitride ceramics, *Transactions of the ASME, Journal of Manufacturing Science and Engineering*, 122(4), 666–670.

23. Lei, S., Shin, Y. C., and Incropera, F. P., 2001, Experimental investigation of thermo-mechanical characteristics in laser assisted machining of silicon nitride ceramics, *Transactions of the ASME, Journal of Manufacturing Science and Engineering*, 123, 639–646.

24. Rebro, P. A., Pfefferkorn, F. E., Shin, Y. C., and Incropera, F. P., 2002, Comparative assessment of laser-assisted machining for various ceramics, *Transactions of NAMRI/SME*, 30, 153–160.

25. Pfefferkorn, F. E., Shin, Y. C., Incropera, F. P., and Tian, Y., 2004, Laser-assisted machining of magnesia-partially-stabilized zirconia, *Transactions of the ASME, Journal of Manufacturing Science and Engineering*, 126(1), 42–51.

26. Pfefferkorn, F. E., Incropera, F. P., and Shin, Y. C., 2005, Heat transfer model of semi-transparent ceramics undergoing laser-assisted machining, *International Journal of Heat and Mass Transfer*, 48, 1999–2012.

27. Skvarenina, S. and Shin, Y. C., 2006, Laser-assisted machining of compacted graphite iron with microstructural analysis, *International Journal of Machine Tools and Manufacture*, 46, 7–17.

28. Anderson, M. and Shin, Y. C., 2006, Laser-assisted machining of an austenitic stainless steel: P550, *Proceedings of the Institution of Mechanical Engineers, Part B (Journal of Engineering Manufacture)*, 220(B12), 2055–2067.

29. Anderson, M., Patwa, R., and Shin, Y.C., 2006, Laser-assisted machining of Inconel 718 with an economic analysis, *International Journal of Machine Tools and Manufacture*, 46, 1879–1891.

30. Dumitrescu, P., Koshy, P., Stenekes, J., and Elbestawi, M. A., 2006, High-power diode laser assisted turning of AISI D2 tool steel, *International Journal of Machine Tools and Manufacture*, 46(15), 2009–2016.

31. Tian, Y. and Shin, Y. C., 2006, Laser-assisted machining of damage-free silicon nitride parts with complex geometric features via in-process control of laser power, *Journal of American Ceramic Society*, 89(11), 3397–3405.

32. Rebro, P. A., Shin, Y. C., and Incropera, F. P., 2002, Laser-assisted machining of reaction sintered mullite ceramics, *Transactions of the ASME, Journal of Manufacturing Science and Engineering*, 124(4), 875–885.

33. Copley, S. M., 1983, Laser shaping of materials, lasers in materials processing, *Proceedings of ASM Conference*, Los Angles, CA, Vol. 8301, pp. 82–86.

34. Cheng, C. W. and Kuo, C. P., 2006, An investigation of laser-assisted machining of Al_2O_3 ceramics planing, *International Journal of Machine Tools and Manufacture*, 47, 452–461.

35. Tian, Y., Wu, B. X. and Shin, Y. C., 2008, Laser-assisted milling of silicon nitride and Inconel 718, *Transactions of the ASME, Journal of Manufacturing Science and Engineering*, 130, 031013.

36. Yang, B. and Lei, S., 2008, Laser assisted milling of silicon nitride ceramic: A machinability study, *International Journal of Mechatronics and Manufacturing Systems*, 1(1), 116–130.

37. Jeon, Y. and Pfefferkorn, F. E., 2008, Effect of laser preheating the workpiece on micro-end milling of metals, *Journal of Manufacturing Science and Engineering*, 130, 1–9.

38. Singh, R. and Melkote, S. N., 2005, Experimental characterization of laser-assisted mechanical micromachining (LAMM) process, *Proceedings of IMECE2005, 2005 ASME International Mechanical Engineering Congress and Exposition*, Orlando, FL, 8 p.

39. Shelton, J. A. and Shin, Y. C., 2008, An experimental evaluation of laser-assisted micromilling of two difficult to machine alloys, *ASME International Conference on Manufacturing Science and Engineering, MSEC_ICMP2008-72246*, October 8–10, 2008. Evanston, IL.

40. Lei, S., Shin, Y., and Incropera, F., 2000, Deformation mechanisms and constitutive modeling for silicon nitride undergoing laser-assisted machining, *International Journal of Machine Tools and Manufacture*, 40(15), 2213–2233.

41. Xu, L. M., Shen, B., and Shih, A. J., 2005, Vitreous bond silicon carbide wheel for grinding of silicon nitride, *International Journal of Machine Tools and Manufacture*, 46(6), 631–639.

42. Mcspadden, Jr., S. B., Ott, R. D., Watkins, T. R., Wereszczak, A. A., Brinkman, C. R., Andrews, M., Fix, B., and Thiele, J., 2001, Addressing the Manufacturing Issues Associated with the Use of Ceramic Materials for Diesel Engine Components, Oak Ridge National Laboratory.

43. Xu, H. H. K., Jahanmir, S., and Ives, K., 1996, Material removal and damage formation mechanisms in grinding silicon nitride, *Journal of Materials Research*, 11(7), 1717–1724.

44. Klocke, F. and Bergs, T., 1997, Laser-assisted turning of advanced ceramics, *Proceedings of SPIE: Rapid Prototyping and Flexible Manufacturing*, 3102, 120–130.

45. Gutierrez, G. and Araya, J. G., 2004, Analytical solution for a transient three-dimensional temperature distribution in laser assisted machining processes, *Proceedings of the ASME Heat Transfer/Fluids Engineering Summer Conference 2004, HT/FED 2004*, Charlotte, NC, Vol. 3, pp. 1055–1063.

46. Rozzi, J. C., Pfefferkorn, F. E., Incropera, F. P., and Shin, Y. C., 2000, Transient, three-dimensional heat transfer model for the laser assisted machining of silicon nitride: I—Comparison of predictions with measured surface temperature histories, *International Journal of Heat and Mass Transfer*, 43, 1409–1424.

47. Rozzi, J. C., Incropera, F. P., and Shin, Y. C., 2000, Transient, three-dimensional heat transfer model for the laser assisted machining of silicon nitride: II—Assessment of parametric effects, *International Journal of Heat and Mass Transfer*, 43, 1425–1437.

48. Pfefferkorn, F. E., Rozzi, J. C., Incropera, F. P., and Shin, Y. C., 1997, Surface temperature measurement in laser-assisted machining processes, *Experimental Heat Transfer*, 10(4), 291–313.

49. Pfefferkorn, F. E., Incropera, F. P., and Shin, Y. C., 2003, Surface temperature measurement of semi-transparent ceramics by long-wavelength pyrometry, *Transactions of the ASME, Journal of Heat Transfer*, 125(1), 48–56.

50. Tian, Y. and Shin, Y. C., 2006, Thermal modeling for laser-assisted machining of silicon nitride ceramics with complex features, *Transactions of the ASME, Journal of Manufacturing Science and Engineering*, 128, 425–434.

51. Lei, S. and Shen, X., 2005, Finite element analysis of temperature distribution and distinct element simulation of laser assisted machining of silicon nitride ceramics, *Proceedings of the 2005 NSF Design, Service and Manufacturing Grantees and Research Conference*, Jan. 3–6, 2005, Scottsdale, AZ.

52. Lei, S. and Yang, B., 2005, Distinct element simulation of ceramic machining: Material removal mechanism, *Transactions of NAMRI*, 33, 485–492.

53. Tian, Y. and Shin, Y. C., 2007, Multiscale finite element modeling of silicon nitride ceramics undergoing laser-assisted machining, *Transactions of the ASME, Journal of Manufacturing Science and Engineering*, 129, 287–295.

54. Germain, G., Morel, F., Lebrun, J.-L., Morel, A., and Huneau, B., 2006, Effect of laser assistance machining on residual stress and fatigue strength for a bearing steel (100Cr6) and a titanium alloy (Ti 6Al 4V), *Materials Science Forum*, 524–525, 559–574.

55. Sun, S. J., Harris, J., and Brandt, M., Jun. 2008, Parametric investigation of laser-assisted machining of commercially pure titanium, *Advanced Engineering Materials*, 10(6): 565–572.

56. Pfefferkorn, F. E., Lei, S., Jeon, Y., and Haddad, G., 2009, A metric for defining the energy efficiency of thermally-assisted machining, *International Journal of Machine Tools and Manufacture*, 49, 357–365.

57. Patten, J. and Dong, L., 2007, Laser heating and thermally enhanced deformation of silicon, *Proceedings of the International Conference on Manufacturing Science and Engineering*, Atlanta, GA, Oct. 15–18, 2007.

58. Tian, Y. 2006, Laser-assisted machining of silicon nitride with complex features, PhD dissertation, Purdue University, West Lafayette, IN.

59. Jen, T. C., Chen, Y. M., and Tuchowski, F., 2004, Experimental and numerical studies of laser-assisted drilling processes, *Proceedings of the ASME Heat Transfer/Fluids Engineering Summer Conference 2004, HT/FED 2004*, Charlotte, NC, Vol. 3, pp. 1015–1023.

60. Westkämper, E., 1995, Grinding assisted by Nd:YAG lasers, *CIRP Annals*, 44(1), 317–320.

61. Marinescu, I. D., 1998, Laser-assisted grinding of ceramics, *Interceram*, 47(5), 314–316.

62. Zhang, C. and Shin, Y. C., 2002, A novel laser-assisted turing and dressing technique for vitrified CBN wheels, *International Journal of Machine Tools and Manufacture*, 42, 835.

63. Tian, Y. and Shin, Y. C., 2007, Laser-assisted burnishing, *International Journal of Machine Tools and Manufacture*, 47(1), 14–22.

64. Klocke, F., Markworth, L., and Borsdorf, R., 2001, High performance machining of aerospace materials, *Proceedings of International Conference on Competitive Manufacturing*, Stellenbosch, South Africa, pp. 281–288.

65. Hügel, H., Wiedermaier, M., and Rudlaff, T., 1995, Laser processing integrated into machine tools—design, application, economy, *Optical and Quantum Electronics*, 27(12), 1149–1164.

66. Lei, S. and Pfefferkorn, F. E., 2007, A review of thermally assisted machining, *Proceedings of MSEC2007, 2007 ASME International Conference on Manufacturing Science and Engineering*, Atlanta, GA, 12p.

67. Anderson, M. C., 2005, Laser-assisted machining of difficult to machine metals, MSME thesis, Purdue University, West Lafayette, IN.

21 Polishing and Magnetic Field–Assisted Finishing

Takashi Sato and Hitomi Yamaguchi

CONTENTS

21.1 INTRODUCTION

Components of precision devices, such as those found in optical, communication, and medical applications, require high form accuracy and fine surface finish. One strategy for manufacturing highly precise components is to employ the precision cutting or grinding processes to form the desired shape and the loose-abrasive finishing processes with extremely low material removal rates to smooth the surface. Lapping and polishing are loose-abrasive finishing processes that are among the commonly used methods for implementing this strategy.

Lapping uses a tool, called a lap, which has a reverse shape of the desired form and is made of a hard material such as cast iron. The lap rubs the abrasive, generally suspended in fluid, against the workpiece surface and copies the tool form to the workpiece while smoothing the surface. In the case of polishing, relatively soft tools, such as pads made of soft cloth or resin, are used to rub the abrasive slurry against the workpiece surface to achieve extremely smooth surfaces. These processes are widely used in fabricating extremely smooth surfaces that are required for optical lenses, bearings, dies and molds, silicon wafers, and other precision components. However, lapping and polishing are not practical for the treatment of complex surfaces such as sharp corners, deep recesses, sharp projections, free-form surfaces, and the interiors of complicated components. This is because of the limitation of the flexibility and motion of the tool configuration.

Many complex-shaped workpieces found in critical applications, such as the components in aerospace, biomedical, and semiconductor industries, require highly finished surfaces to achieve their desired functions, and many of these are currently finished by hand despite increased production costs. To make things worse, some of these are virtually unreachable by conventional techniques, and the lack of finishing technology is as an obstacle to the technology innovation. It is, therefore, desired to develop new technologies to overcome these difficulties.

The combination of a magnetic field with the mechanical action of a magnetic tool against a workpiece gives rise to the magnetic field–assisted finishing (MAF) process. The magnetic tools can be introduced into areas hard to reach by conventional technologies, and, by means of magnetic manipulation, they exhibit relative motion against the workpiece surface needed for finishing. This shows potential for overcoming problems associated with more conventional finishing processes.

The idea was initially developed in the United States in the 1930s. Since then, research on the process has been undertaken in the former Soviet Union, Bulgaria, the German Democratic Republic (now Germany), Poland, and the United States. The first patent known to us was granted in the United States in 1940s, and full-scale research on MAF started to appear in the 1960s [1–3]. In the 1980s, research groups succeeded in putting the process into practical use. In the 1990s, as its mechanisms and advantages became better understood, the applicability of the process gained public recognition.

A new type of magnetic fluid, a mixture of magnetic particles (such as finely divided carbonyl iron particles) and a liquid (such as oil), was developed in the 1940s in the United States for use in hydraulic actuators, dashpots, and clutches [4]. This was the introduction of magneto-rheological fluid (MR fluid). In 1960s, colloidal suspensions containing magnetic nanoparticles were first synthesized and called magnetic fluid (MF) [5]. In addition to processes using either magnetic abrasives [3] or a mixture of ferrous particles and abrasives, processes using MF, MR fluid, and, more recently, magnetic compound fluid (MCF) [6] have attracted considerable attention in nanometer-scale finishing.

Sections 21.2 through 21.5 explain various magnetic media and outline the finishing processes using magnetic abrasives, MF, MR fluid, and MCF. Their advantages and representative examples of their applications are also introduced. Section 21.6 summarizes the distinctive features of the MAF process, and Section 21.7 concludes this review.

21.2 POLISHING AND FINISHING USING MAGNETIC ABRASIVES

Magnetic abrasives produced by various methods can be found in open literature, and they can be divided broadly into three groups: (1) mechanical mixtures of abrasives and magnetic powders, (2) composites, and (3) cast materials [7]. The mechanical mixtures of abrasives and magnetic powders (Group 1) include abrasive-coated magnetic powders. The manufacture of composite abrasives (Group 2) involves sintering a mixture of magnetic and abrasive powders, and then crushing and milling the resultant compacts. Cast abrasives (Group 3) are made by atomizing melts containing iron, carbon, certain carbide-forming elements, and other alloying additives.

The magnetic abrasives shown in Figure 21.1 consist of iron particles and Al_2O_3 (white alumina: WA) abrasive grains. The composite ingot is made by the thermite process using aluminum powder and iron oxide powder. The ingot is then mechanically crushed and sieved to make the finished magnetic abrasives. The Al_2O_3 grains are located both inside and outside the resulting magnetic abrasives. The magnetic abrasives in practical use have a mean diameter of 80 μm, and the contained Al_2O_3 abrasive grains are smaller than 10 μm [7].

In a magnetic field, ferrous particles (including magnetic abrasives) are suspended by magnetic force and linked together along the lines of magnetic flux, as shown in Figure 21.2. Because the magnetic flux flows unimpeded through the nonferrous workpiece material, it is possible to influence the motion of a ferrous particle—even if the particle is not in direct contact with a magnetic pole—by controlling the magnetic field. The ferrous particle chains connected by magnetic force offer the advantage of a flexible configuration. This unique behavior of the ferrous particles enables the application of the finishing operation not only to easily accessible surfaces but also to areas that are hard to reach by means of conventional mechanical techniques.

FIGURE 21.1   SEM photo of WA magnetic abrasive.

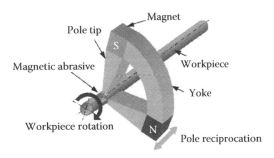

FIGURE 21.2   Schematic of magnetic tool behavior in a magnetic filed.

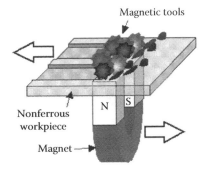

FIGURE 21.3   Schematic of processing principle. (From Yamaguchi, H. et al., *ASME J. Manuf. Sci. Eng.*, 129(5), 885, 2007. With permission.)

Figure 21.3 shows a schematic of the internal finishing process for tubes using a work-rotation system [8]. Permanent magnets generate the magnetic field needed for attracting the magnetic abrasive to the finishing area, pressing it against the inner surface of the workpiece. In a non-uniform magnetic field, the magnetic force F acts on the magnetic abrasives and is shown in Equation 21.1:

$$F = V\chi H \text{ grad } H \tag{21.1}$$

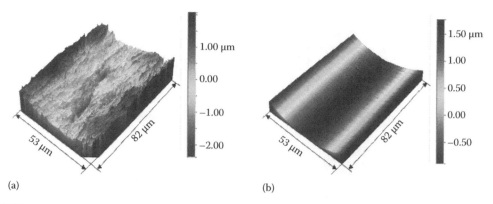

FIGURE 21.4   Three-dimensional shape of unfinished and finished surfaces of stainless steel tubes (0.5 mm outer diameter, 0.4 mm inner diameter) measured by optical profiler. (a) Unfinished surface (0.26 μm *Ra*) and (b) finished surface (0.002 μm *Ra*). (From Yamaguchi, H. et al., *ASME J. Manuf. Sci. Eng.*, 129(5), 885, 2007. With permission.)

where
   V is the volume of the magnetic abrasive
   χ is the susceptibility
   H and grad *H* are the intensity and gradient of the magnetic field, respectively

If the tangential component of the magnetic force acting on the magnetic abrasives is larger than the friction force between the magnetic abrasives and the inner surface of the workpiece, the magnetic abrasives show smooth relative motion against the inner surface of the workpiece when the workpiece is rotated at high speed. Material is removed from the surface by magnetic abrasives as a result of this relative motion, and the surface is smoothed. Moreover, manipulating the poles along the workpiece axis causes the abrasive mixture to move in the axial direction following the pole motion, thereby efficiently finishing a wide area of the inner workpiece surface.

Figure 21.4 shows an example of the surface of SUS304 stainless steel capillary tube (0.5 mm outer diameter, 0.4 mm inner diameter) finished using this method [8]. This illustrates the feasibility of the process to achieve nearly uniform internal finishing of capillary tubes. Magnetic abrasive finishing is also applicable for the removal of burrs generated by drilling or other operations [9,10].

Abrasiveless finishing processes facilitate the post-finishing rinsing operation; this makes MAF an environmentally friendly process. Based on this concept, gas-atomized magnetic abrasives and electroformed magnetic abrasives were developed, which feature micro-sharp cutting edges made of base metals on their surfaces, instead of including conventional abrasives [11,12]. In tests, both types of developed magnetic abrasives achieved smoothly finished stainless steel tube surfaces comparable to the surfaces produced by the conventional magnetic abrasives.

21.3   POLISHING AND FINISHING USING MAGNETIC FLUID

21.3.1   Magnetic Field–Assisted Fine Finishing

Magnetic fluids are stable colloidal suspensions of permanently magnetized particles, such as magnetite. Brownian motion keeps the particles, which are about 10 nm in diameter, from settling under gravity, and a surfactant covers each particle to create short-range steric (steric means relating to the spatial arrangement of atoms) repulsion between particles, which prevents particle agglomeration in the presence of nonuniform magnetic fields [13]. Magnetic fluids were developed by Papell in 1965 to magnetically control fuel flow for the Apollo project in the zero-gravity conditions of space [5]. They have been practically applied as seal components and are routinely used in voice coils, dampers, and rotary brakes.

The application of magnetic fluid to polishing and finishing processes was undertaken by Kurobe and Imanaka et al. in 1983 [14]. The resultant process, called *magnetic field–assisted fine finishing*, was developed as a new lapping technique for the controllable finishing of materials, such as semiconductors and ceramics in the fields of electronics and precision machinery. Magnetic fluid fills the groove cut in the brass disk, the polisher (a 1 mm thick rubber sheet) covers the magnetic-fluid-filled groove, and a water-based polishing compound is supplied over the polisher. Electromagnets that are placed above and below the disk create a magnetic field when DC voltage is applied. The polishing pressure is generated by the ferrous particles that are attracted by the magnetic field and push polishing compound (via the polisher) against the work surface. The polishing action is achieved when the upper pole that is connected to the work and the disk rotates in opposite directions. It has been experimentally demonstrated that flexible surface finishing is feasible using this method, and the surface roughness and stock removal rate are controllable by varying the current to the electromagnet.

21.3.2 Magnetic Float Polishing and Magnetic Fluid Grinding

Tani and Kawata in 1984 developed *magnetic float polishing* (MFP) that makes use of the magnetohydrodynamic behavior of the magnetic fluid (MF) [15]. Figure 21.5 shows the processing principle. The polishing compound is a MF suspended SiC grains, which make up 40% (by volume) of the MF. When a vessel filled with the compound is placed in a magnetic field, the nonmagnetic abrasives experience buoyant forces. The forces generate the polishing pressure, and the relative motion between the work and compound in this investigation was the result of fixing the work to the spindle and the vessel to the table of a vertical milling machine. It was found that the efficiency of this finishing process is improved by increasing the relative volume of the abrasives, which increases the viscosity of the compound.

In methods using abrasive grains suspended in magnetic fluid (MF), the polishing force is mainly dependent upon the buoyant force experienced by the abrasive grains. However, this does not generally provide an efficient polishing force and results in a low material removal rate. To increase the polishing force generated in the MF process, in 1988 Umehara and Kato proposed a *magnetic fluid grinding* process using a floating pad to push the abrasive grains against the target surface more strongly than in other processes [16].

Figure 21.6 shows the processing principle. The floating pad helps to increase the process-removal rate and enables fine control of the surface and forms accuracy, such as the surface roughness and sphericity of ground silicon nitride (Si_3N_4)-bearing balls. A batch of 46 Si_3N_4 balls was finished to

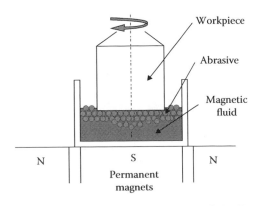

FIGURE 21.5 Concept of magnetic float polishing. (Courtesy of Prof. Tani.)

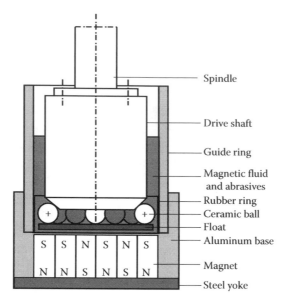

FIGURE 21.6 Schematic diagram of magnetic fluid grinding. (From Umehara, N. et al., *Int. J. Mach. Tools Manuf.*, 46(2), 151, 2006. With permission.)

a final diameter of 0.7500 in. with an average sphericity of 0.27 mm (best value: 0.15 mm) and an average surface finish, *Ra* of 8.2 nm (best value 6.7 nm) [17]. Moreover, this process has also been applied for ceramic rollers, ceramic plates, aluminum alloy cylinder ends, ceramic cylinder surfaces, and the inner surface of steel pipe.

21.3.3 POLISHING USING MAGNETIC INTELLIGENT COMPOUND

One technique for making a solid polishing tool with a controlled abrasive distribution is to freeze a mixture of abrasive grains and water-based MF under the influence of a magnetic field. The material-removal rate and surface-roughness improvement achieved using the described frozen tool are sufficient for the tool to be applied to mold surface polishing [18]. However, to maintain its configuration, the frozen tool is useful only in a cold room. In order to expand the application of this polishing tool, a room-temperature tool was necessary, and the *MAGnetic intelligent compound* (MAGIC), which is made of a mixture of magnetic particles, abrasives, and polymers, was developed in response to this need [19].

Figure 21.7 shows the procedure for making the MAGIC tool. The advantages of the tool can be summarized as follows [19]:

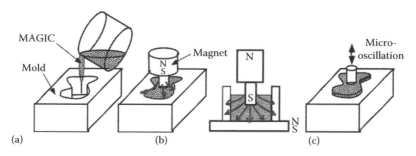

FIGURE 21.7 Manufacturing process and polishing concept. (a) Pouring hot-MAGIC to the mold, (b) cooling with magnetic field normal to the polishing surface, and (c) polishing by micro-oscillation of MAGIC. (Courtesy of Prof. Umehara.)

1. Mold surfaces can be finished quickly, because abrasives are well distributed without any agglomerations.
2. The bonding material is so soft that deep scratches cannot be generated by the finishing process.
3. The pad can be recycled when worn out; the MAGIC and abrasives can be used many times.
4. Complicated surfaces can be finished by pouring and oscillating the MAGIC.

21.4 POLISHING PROCESSES USING MAGNETO-RHEOLOGICAL FLUID

21.4.1 MAGNETO-RHEOLOGICAL FINISHING

Magneto-rheological fluids (MR fluids) typically consist of micron-sized, magnetically polarizable particles dispersed in a carrier medium—most commonly mineral oil or silicone oil. Particle chains form when a magnetic field is applied. In the presence of a shear force, the equilibrium between the forming and breaking of the particle chains corresponds to the fluid's yield strength. MR fluids flow freely when no magnetic field is present, and they exhibit shear-thinning characteristics. MR fluids should not be confused with colloidal ferrofluids that contain particles that are about 1000 times smaller than those found in typical MR fluids [20].

The initial discovery and development of MR fluids and devices is credited to Rabinow, who studied them at the U.S. National Bureau of Standards (now named the National Institute of Standards and Technology, NIST) in the 1940s [4]. Several companies currently market MR fluids, and numerous research projects have been aimed at the practical applications of MR fluids. These research efforts have resulted in the application of MR fluids to the manufacture of dampers, clutches, brakes, shock absorbers, etc. Among recent significant research efforts are the applications of MR fluids to the sub-nanometer-scale polishing and finishing of components, such as glass lenses, ceramics, small-diameter tubes, etc. Due to their complex shapes, these components are difficult to polish and finish using traditional techniques.

The development of the *magneto-rheological finishing* (MRF) process was initiated by Kordonski and Prokhorov et al. in 1992 [21,22]. The MRF process was introduced commercially in 1998 and has dramatically changed the manufacture of precision optics [23]. The developed process utilizes polishing slurry based on MR fluid that can be mixed, pumped, and conditioned in their liquid state, but an applied magnetic field causes a viscosity change to a semisolid state to create a stable and conformable polishing tool. A typical composition of an MR fluid is 36% carbonyl iron, 6% abrasive (cerium oxide), 3% stabilizer, and 55% water [24].

Figure 21.8 shows the MRF processing principle. A workpiece is fixed at some distance from a moving surface, so that the workpiece surface and the moving surface form a converging gap.

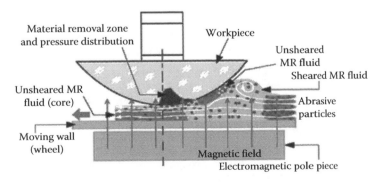

FIGURE 21.8 A schematic diagram of the MRF processing principle. (From Tricard, M. et al., *Proceedings of 2003 ASME International Mechanical Engineering Congress & Exposition* (*IMECE 03*), Washington, DC, pp. 1–10, 2003. With permission.)

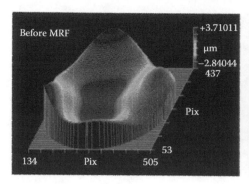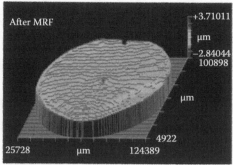

FIGURE 21.9 Prime silicon wafer flatness is improved after using MRF. Shown (left) is the surface figure of a 100 mm dia. 450 μm thick silicon substrate before MR finishing. MRF was able to improve the surface flatness from 6.55 to 0.81 μm (right) with roughness of ~0.2 nm rms. (From Tricard, M. et al., *Proceedings of 2003 ASME International Mechanical Engineering Congress & Exposition (IMECE 03)*, Washington, DC, pp. 1–10, 2003. With permission.)

An electromagnet placed below the moving surface generates a nonuniform magnetic field in the vicinity of the gap. The MR fluid is delivered to the moving surface just above the electromagnetic poles then pressed against the surface by the magnetic field gradient and the fluid becomes a Bingham plastic before it enters the gap. Thereafter, the shear flow of plastic MR fluid occurs through the gap, resulting in the development of high stresses in the interface zone and thus, material removal over a portion of the workpiece surface [25].

The process, which has been adopted by major manufacturers of precision optics, can make significant improvements to the surface roughness and flatness. In terms of both form accuracy and micro-roughness, the MRF process has demonstrated the ability to produce optical surfaces to tight tolerances. The surfaces, including aspheres, can be made with materials ranging from glass/glass ceramics (including fused silica, ULE, and Zerodur) to single-crystalline materials (including silicone and calcium fluoride) or polycrystalline materials (including SiC). An example is shown in Figure 21.9 [25].

21.4.2 Magneto-Rheological Fluid-Based Slurry for Internal Finishing

A new type of slurry—referred to as *MRF-based slurry*—was developed in 2006 by Yamaguchi et al. [26] and is especially appropriate for the internal finishing process of piping systems in micro- and nanotechnological industrial devices. The major feature of the MRF-based slurry is that the abrasives are smaller than the iron particles so that the iron particles can trap as many abrasive particles as possible.

The processing principle is illustrated in Figure 21.10. The MRF-based slurry is introduced inside the work and is attracted to the finishing area by the field generated by magnetic poles (e.g., electromagnetic coils or permanent magnets). When the work is rotated and oscillated along its axis, there is relative motion between the abrasives and the work surface, and the entire inner workpiece surface can be finished. Under typical finishing conditions, the MRF-based slurry experiences a magnetic force, a centrifugal force, a friction force against the work surface, and the force of gravity; these combine and result in the dynamic behavior shown in Figure 21.10b. This behavior continuously displaces the cutting edges of the abrasives in the MRF-based slurry that improves the finishing efficiency [27].

Figure 21.11 shows an example of the surface roughness improvement of copper tubes (Ø 19.05 × Ø 17.05 × 100 mm) finished by MRF-based slurry. The inner surface was finished to 2.9 nm *Ra*. A high exciting current enhances the magnetic field intensity; therefore, the diamond abrasives in the

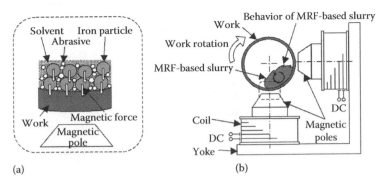

(a)   (b)

FIGURE 21.10 A schematic diagram of the processing principle of internal finishing using MRF-based slurry. (From Sato, T. et al., *Key Eng. Mater.*, 329, 249, 2007. With permission.)

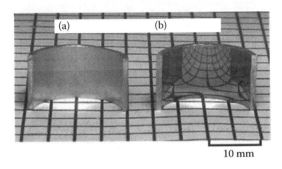

FIGURE 21.11 Example of internal finishing of copper tubes using MRF-based slurry. (a) Unfinished surface and (b) finished surface.

MRF-based slurry cut deeply into the surface and the material removal is increased. A lower exciting current creates a smoother finished surface with shallower intermittent cutting marks on the surface of the work that is the case illustrated in Figure 21.11. If the exiting current becomes too low, MRF-based slurry adheres to the surface of the work and does not generate relative motion because the magnetic force cannot overcome the friction against the target surface. Under such low magnetic field conditions, the surface was barely finished using MRF-based slurry.

21.4.3 Magneto-Rheological Abrasive Flow Finishing

Magneto-rheological abrasive flow finishing (MRAFF) was developed by Jha et al. in 2004 as a new precision finishing process using MR fluid for complicated geometries [28,29]. The process mechanism is shown in Figure 21.12. In the MRAFF process, a piston extrudes a magnetically stiffened slug of MR polishing fluid back and forth through or across the passages formed by the workpiece and fixture. Abrasion selectively occurs only where the magnetic field is applied across the workpiece surface, keeping the other areas unaffected. The rheological behavior of the polishing fluid changes from nearly Newtonian to Bingham plastic and back when entering, traversing, and exiting the finishing zone, respectively. The abrasive cutting edges, held by carbonyl iron chains, rub the workpiece and shear the peaks from its surface. The amount of material sheared from the workpiece surface peaks by the abrasive grains depends on the bonding strength of the field-induced structure of the MR polishing fluid and the extrusion pressure applied through the piston.

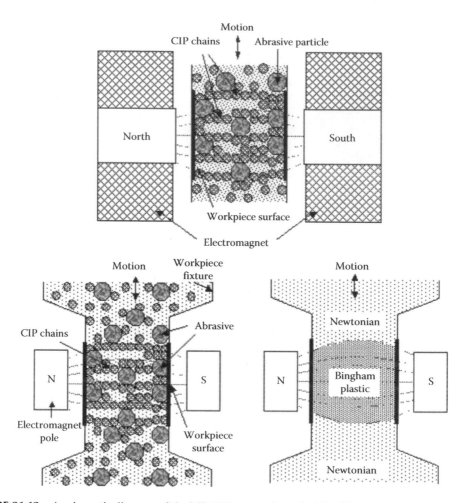

FIGURE 21.12 A schematic diagram of the MRAFF processing principle. (Courtesy of Prof. Jain.)

This process makes use of determinism and in-process controllability of the rheological behavior of an abrasive-laden medium to finish intricate shapes. The desired properties of the base MR fluid used in MRAFF are as follows:

1. The fluid should be thermally stable and should have a high boiling point.
2. It should be non-corrosive and nonreactive with the employed magnetic and abrasive particles.

The representative MR fluid slurry consists of carbonyl iron particles (20 vol%; 6 μm in mean diameter), silicon carbide abrasive (20 vol%; mesh size #800, #1000, #1200, or #1500), and organic medium (60 vol%).

21.5 POLISHING PROCESS USING MAGNETIC COMPOUND FLUID

Magnetic compound fluid (MCF) that consists of MF and MR fluid was developed by Shimada et al. in 2001 as an intelligent fluid [6]. The apparent viscosity of MF in a magnetic field is lower than that of MR fluid in the same conditions. However, the stability of the particle distribution in MF is greater than in MR fluid. Therefore, the apparent viscosity and the particle distribution stability in MCF can be targeted by changing the mixing ratio of MF and MR fluid. This point must be the greatest advantage

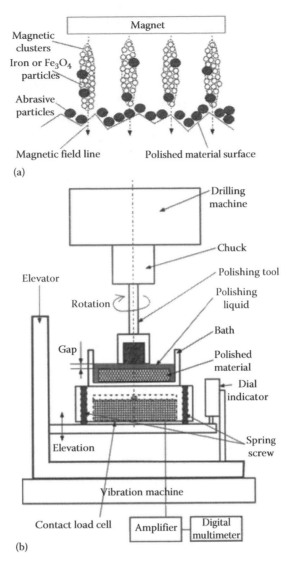

FIGURE 21.13 (a) Schematic diagram of MCF float polishing. (b) Schematic diagram of MCF float polishing. (Courtesy of Prof. Shimada.)

of MCF. The structure of MCF is made up of chain-shaped magnetic clusters consisting of magnetic particles of different sizes. Magnetic particles of MF (a few nm in diameter) surround the magnetic particles of MR fluid (a few μm in diameter), and long clusters of magnetic particles are formed. MCF has been applied to dampers and haptic sensors made using MCF conductive rubber [30,31].

Research into the application of MCF to float polishing was conducted in 2002 [32]. Figure 21.13a illustrates a schematic of chain-shaped clusters consisting of MCF mixed with abrasive, and Figure 21.13b schematically illustrates *MCF float polishing*. Chain-shaped magnetic clusters, composed of nanometer-sized magnetic particles and micrometer-sized magnetic particles, are created along the lines of magnetic flux as the MCF is subjected to a magnetic field. The dimensions of the clusters depend on the composition of the MCF, the strength of the magnetic field, and the method by which the field is applied. The nonmagnetic abrasive particles are trapped by the clusters or distributed between clusters, and á-cellulose fiber has been used interspersed with the clusters to increase the viscosity. When the magnet is moved, the clusters and abrasive exhibit relative motion against the work surface, and work material is removed by the micro-cutting action of the abrasive particles. This process can be diversely applied to free-form metal surfaces, ceramic, glass, resin surfaces, the inner surfaces of tubes, etc.

Magnetic polishing liquid (MPL) is made by mixing abrasive and MCF [33]. MPL, a mixture of aqueous MF (6.2 wt%), 1.2 μm carbonyl iron powder (9.47 wt%), 3 μm pure alumina (Al_2O_3) abrasive particles (6.19 wt%), and á-cellulose fiber (6.02 wt%) with distilled water (72.12 wt%), was applied to the polishing of SUS304 stainless steel disks. The surface roughness decreased from an initial value of 0.09 μm *Ra* to a final value of 0.067 μm *Ra*. These experiments resulted in the following conclusions [33]:

1. The MPL needs to be replenished with water at an appropriate rate. Otherwise, the MPL will dry and its polishing ability will decrease. Under the applied experimental conditions, the appropriate adding rate was about 8 mL/5 min.
2. Stronger and larger magnetic fields enhance the surface-roughness improvement rate.
3. The gap between the working surface and the MPL dish bottom should be set up so that the working surface is as close to the MPL swell top as possible.

21.6 REFLECTION ON THE PROCESS CHARACTERISTICS

The following characteristics distinguish the MAF process from traditional loose-abrasive processes:

1. *Mechanical material removal by magnetic tools (with or without abrasive) under the influence of a magnetic field*: The material removal action is due to the relative motion between the magnetic tools (with or without abrasive) and the workpiece surface. As with traditional loose-abrasive processes, the abrasive material, size, and cutting-edge condition are chosen in MAF to match the workpiece material and produce the desired surface quality. This process can machine both ductile and brittle materials, e.g., alloys (steels, stainless steels, copper, brass, magnesium, nickel, cemented carbide, etc.), ceramics (aluminum oxide and silicon nitride), silicon, quartz, and glasses. It can achieve surface roughness values ranging from angstroms to micrometers.
2. *Flexibility of the magnetic tool chain configuration*: The magnetic tools are magnetized following the lines of magnetic force, generating magnetic tool chains suspended by magnetic force. At the contact area between a magnetic tool chain and the workpiece surface, the magnetic tool chain configuration is modified to follow the workpiece form. The flexibility of the configuration, including the chain stiffness, can be controlled by the magnetic field distribution at the machining area. This enables the surface finishing of free-form surfaces and other complex-shaped workpieces. Fine control of the finishing pressure is obtained by manipulating the magnetic field, and tools used for the traditional loose-abrasive process, such as laps or polishing pads, are not required. Moreover, the flexibility of the magnetic tool configuration permits relocation of the magnetic tools. This encourages the introduction of sharp cutting edges and removing chips and thereby improves the finishing efficiency. However, this process, which can be categorized as a pressure-copying process, cannot significantly modify the form accuracy of a workpiece.
3. *Remote controllability of the magnetic force acting on the tool and its motion in a magnetic field*: The magnetic flux flows unimpeded through a nonferrous workpiece. The magnetic tools introduced on one side of a nonferrous workpiece can be suspended by magnetic force generated by magnets placed on the other side of the workpiece. The motion of the magnetic tools can be manipulated from outside the workpiece by moving the magnets; this capability is called remote controllability. This process enables surface finishing inside the nonferrous workpiece or that in the areas hard to reach by

conventional tools, e.g., bent tubes, capillary tubes, internal passage of components, and microgrooves.

4. *Dependency of the process feasibility on the magnetic field and the magnetic properties of the workpiece*: The magnetic field at the finishing area is determined by geometry and magnetic properties of the magnets and workpiece; this affects the force acting on the magnetic tools and, thus, their behavior. Since these parameters play prominent roles in determining finishing performance, the process feasibility must depend on these properties. For example, these processes may have difficulties in finishing thick nonferrous workpieces or inside ferrous workpieces, as both conditions diminish the magnetic field in the machining area.

21.7 CONCLUSIONS

Since the MAF process was introduced in the 1930s, numerous process principles and magnetic media have been developed to materialize the process in different ways. These developments have been made in response to the changing needs of the commercial market to improve the quality of our lives. This chapter has provided a brief history of the MAF process, and a selection of recent process developments has been presented showing diverse magnetic media, including magnetic abrasives and slurries using smart fluids, such as MF, MR, MCF, and MPL.

Finishing processes using alternating magnetic field have also been developed and used for surface and edge finishing in practical use. Moreover, new hybrid finishing methods created by the combination of the magnetic field–assisted mechanical action and chemical reactions are being studied intently. Although this chapter cannot cover all the research and development related to MAF, the authors hope that this chapter will be of assistance in understanding MAF and in supporting future technology innovation.

QUESTIONS

Q.21.1 List and briefly explain the various magnetic media used for MAF.

Q.21.2 Explain the processing principle of MAF using magnetic abrasives.

Q.21.3 Describe the advantages of MAF that distinguish it from the traditional loose-abrasive finishing process.

Q.21.4 Describe the limitations of the MAF process.

Q.21.5 Describe several industrial applications of MAF.

Q.21.6 Why can this process solve some of the trickiest polishing needs? Can you generalize it in the language of energy field manufacturing? Start with the list of involved energy fields, materials, and intelligence.

Q.21.7 To bring a surface to extremely high finish, multiple steps are necessary. Suppose you are given a raw circular copper plate, and you are asked to bring the surface finish to <1 nm, list the necessary processing steps.

ABOUT THE CONTRIBUTING AUTHORS

Takashi Sato is an assistant professor in the Department of Systems Science and Technology at Akita Prefectural University, Akita, Japan. He received his PhD in precision engineering from Utsunomiya University, Tochigi, Japan. His current research interest includes magnetic field–assisted finishing using functional fluid.

Hitomi Yamaguchi has served as associate professor in the University of Florida's (Gainesville, Florida) Department of Mechanical and Aerospace Engineering since October 2007. Previously, she served as associate professor at Utsunomiya University, Tochigi and Research Associate at the University of Tokyo (both in Japan) and as research engineer at Extrude Hone Corporation (in Irwin, Pennsylvania). Her research topics include precision machining processes, especially magnetic field–assisted finishing (MAF) processes.

REFERENCES

1. Baron, Y. M., *Magnetic-Abrasive and Magnetic Finishing of Parts and Cutting Tools*, Mashinostroenie, Leningrad, Russia, 1986 [translated into Japanese by Endo K., Nisso Tsushinsha, 1988].
2. Coats, H. P., Method of and Apparatus for Polishing Containers, U.S. Patent 2,196,058, 1940.
3. Ruben, H. J., *Advances in Surface Treatments*, Vol. 5, Niku-Lari, A., Ed., Pergamon Press, Oxford, U.K., p. 239, 1987.
4. Rabinow, J., The magnetic fluid clutch, *AIEE Transactions*, 67, 1308–1315, 1948.
5. Papell, S. S., Low viscosity magnetic fluid obtained by the colloidal suspension of magnetic particles, U.S. Patent 3,215,572, 1965.
6. Shimada, K., Fujita, T., Oka, H., Akagami, Y., and Kamiyama, S., Hydrodynamic and magnetized characteristics of MCF (magnetic compound fluid), *Journal of the Japan Society of Mechanical Engineers*, 67 (664), 3034–3040, 2001 [in Japanese].
7. Yamaguchi, H. and Hanada, K., Development of spherical magnetic abrasive made by plasma spray, *ASME Journal of Manufacturing Science and Engineering*, 130 (3), 031107-1–031107-9, 2008.
8. Yamaguchi, H., Shinmura, T., and Ikeda, R., Study of internal finishing of austenitic stainless steel capillary tubes by magnetic abrasive finishing, *ASME Journal of Manufacturing Science and Engineering*, 129 (5), 885–893, 2007.
9. Ko, S., Baron, Y. M., Chae, J. W., and Polishuk, V. S., Development of deburring technology for drilling burrs using magnetic abrasive finishing method, in *Proceedings of International Conference on Leading Edge Manufacturing in 21st Century*, Japan Society of Mechanical Engineers, Niigata, Japan, pp. 367–372, 2003.
10. Yin, S. and Shinmura, T., Vertical vibration-assisted magnetic abrasive finishing and deburring for magnesium alloy, *International Journal of Machine Tools and Manufacture*, 44 (11–12), 1297–1332, 2004.
11. Saito, T., Koike, K., Yamato, H., Kuwana, A., Suzuki, A., Yamaguchi, H., and Shinmura, T., Development of gas-atomized magnetic tools, *Key Engineering Materials*, 291–292, 287–290, 2005.
12. Suzuki, A., Kuwana, A., Saito, T., Koike K., Yamato H., Yamaguchi, H., and Shinmura, T., Development of magnetic polishing tools using electroforming, *Journal of the Japan Society of Mechanical Engineers C*, 71 (711), 3326–3331, 2005 [in Japanese].
13. Zahn, M., Magnetic fluid and nanoparticle applications to nanotechnology, *Journal of Nanoparticle Research*, 3, 73–78, 2001.
14. Kurobe, T., Imanaka, O., and Tachibana, S., Magnetic field-assisted fine finishing, *Bulletin of the Japan Society for Precision Engineering*, 17 (1), 49–50, 1983.
15. Tani, Y. and Kawata, K., Development of high-efficient fine finishing process using magnetic fluid, *Annals of the CIRP*, 33 (1), 217–220, 1984.
16. Umehara, N. and Kato, K., A study on magnetic fluid grinding (1st report, The Effect of the Floating Pad on Removal Rate of Si_3N_4 Balls), *Transactions of the Japan Society of Mechanical Engineer C*, 54 (503), 1599–1604, 1988 [in Japanese].
17. Umehara, N., Kirtane, K., Gerlick, R., Jain, V. K., and Komanduri, R., A new apparatus for finishing large size/large batch silicon nitride (Si_3N_4) balls for hybrid bearing applications by magnetic float polishing (MFP), *International Journal of Machine Tools & Manufacture*, 46 (2), 151–169, 2006.
18. Kato, K., Umehara, N., and Suzuki, M., A study of hardness of the frozen magnetic fluid grinding wheel, *Journal of Magnetism and Magnetic Materials*, 201, 376–379, 1999.
19. Umehara, N., MAGIC polishing, *Journal of Magnetism and Magnetic Materials*, 252, 341–343, 2002.

20. Carlson, J. D., Catanzarite, D. M., and St. Clair, K. A., Commercial magneto-rheological fluid devices, *International Journal of Modern Physics B*, 10 (23–24), 2857–2865, 1996.
21. Kordonski, W. I., Adaptive structures based on magnetorheological fluids, in *Proceedings of the Third International Conference Adaptive Structure*, Wada, B. K., Natori, M., and Breibtbach, E., Eds., San Diego, CA, pp. 13–27, 1992.
22. Prokhorov, I. V., Kordonski, W. I., Gleb, L. K., Gorodkin, G. R., and Levin, M. L., New high-precision magnetorheological instrument-based method of polishing optics, *OSA of Workshop Digest*, 24, 134–136, 1992.
23. Kume, T., Tricard, M., Dumas, P., and Murphy, P., Sub-aperture approaches for precision polishing and metrology, in *Proceedings of International Conference on Leading Edge Manufacturing in 21st Century*, Japan Society of Mechanical Engineers, Nagoya, Japan, 3, pp. 1217–1222, 2005.
24. Kordonski, W. I. and Goloni, D., Fundamentals of magnetorheologial fluid utilization in high precision finishing, *Journal of Intelligent Material Systems and Structures*, 10 (9), 683–689, 1999.
25. Tricard, M., Dumas, P. R., Golini, D., and Mooney, T., Prime silicon and silicon-on-insulator (SOI) wafer polishing with magnetorheological finishing (MRF), in *Proceedings of 2003 ASME International Mechanical Engineering Congress & Exposition (IMECE 03)*, Washington, DC, pp. 1–10, 2003.
26. Yamaguchi, H., Shinmura, T., Sato, T., Taniguchi, A., and Tomura, T., Study of surface finishing process using magneto-rheological fluid (MRF)-development of MRF-based slurry and its performance, *Journal of the Japan Society for Precision Engineering*, 72 (1), 100–105, 2006 [in Japanese].
27. Sato, T., Yamaguchi, H., Shinmura, T., and Okazaki, T., Study of internal magnetic field assisted finishing for copper tubes with MRF (magneto-rheological fluid)-based slurry, *Key Engineering Materials*, 329, 249–254, 2007.
28. Jha, S., Jain, V. K., and Komanduri, R., Effect of extrusion pressure and number of finishing cycles on surface roughness in magnetorheological abrasive flow finishing (MRAFF) process, *The International Journal of Advanced Manufacturing Technology*, 33, 725–729, 2007.
29. Jha, S. and Jain, V. K., Rheological characterization of magnetorheological polishing fluid for MRAFF, *The International Journal of Advanced Manufacturing Technology*, DOI:10.1007/s00170-008-1637-8, 2008.
30. Shimada, K., Shuchi, S., Kanno, H., Wu., Y., and Kamiyama, S., Magnetic cluster and its applications, *Journal of Magnetism and Magnetic Materials*, 289, 9–12, 2005.
31. Zheng, Y. and Shimada, K., Research on a haptic sensor made using MCF conductive rubber, *Journal of Physics: Condensed Matter*, 20, 204148–204152, 2008.
32. Shimada, K., Matsuo, Y., Yamamoto, K., and Zheng, Y., A new float-polishing technique with large clearance utilizing magnetic compound fluid, *International Journal of Abrasive Technology*, 1 (3–4), 302–315, 2008.
33. Wu, Y., Tetsuka K., Katoa M., Shimada K., and Wong Y. C., A new magnetic polishing liquid (MPL) for contact force-free surface finishing II: Detailed performance in metal surface finishing, *International Journal of Applied Electromagnetics and Mechanics*, 25, 89–94, 2007.

22 Tribology and Surface Engineering: Scientific and Technological Bases for Energy Efficiency

Q. Jane Wang, Dong Zhu, and Jian Cao

CONTENTS

22.1 INTRODUCTION

Tribology, named by the British Lubrication Engineering Working Group in 1966, is the science and technology of interacting surfaces in relative motion and related subjects and practices [1]. It is a multidisciplinary science closely related to materials science and engineering, surface science and engineering, physics, chemistry, and mechanical science and engineering. Tribological technologies that include friction control, wear reduction, lubrication, and tribological interface fabrication have played irreplaceable and indispensable functions in the development of modern machinery.

Surfaces are the key in the study and practice of tribology. Motion and power transmissions in a machine system are accomplished through surface contact in relative motion. The surfaces of parts under the relative motion form a contact pair, in which they are subject to loading and interaction. A contact may be between two solid surfaces, which defines a dry contact, or in the presence of a lubrication, which defines a lubricated contact. The contact between two convex surfaces is counterformal, such as the external gear-tooth surface contact, and the contact between a convex and a concave surface is conformal, such as the journal-bearing surface contact. The contact happens between two nominally flat surface areas, such as those of a die and a workpiece. Surface interaction in relative motion results in rubbing, in which power loss and surface wear occur. Lubrication by a third medium has been an effective means to reduce friction and wear.

A surface pair under contact and relative motion with or without lubrication forms a tribological interface. Friction and wear are inevitable because of rubbing. Relative motion is everywhere and in all mechanical systems, and so are tribological interfaces, surface contact, rubbing, friction, and wear. Wear induced material loss may alter designed surfaces and change their performance. Excessive wear can increase friction and induce surface failure, such as scuffing, galling, seizure, and contact fatigue, which is detrimental to machine operation. Replacing damaged parts further consumes materials and energy. Therefore, tribological design is closely related to industrial energy efficiency.

Energy and environmental problems require minimizing energy consumption via more efficient machines that run with less energy consumption and more durable materials. Accomplishing this goal requires tribological technologies, and engineering surfaces are critical to lubrication enhancement, friction reduction, and wear life improvement. This chapter introduces the concepts of surfaces, friction, wear, and lubrication. It also reviews the research results of the effects of surfaces on friction and surface technologies for friction reduction.

22.2 ENGINEERING SURFACES

An engineering surface may be referred to as the surface of a machine component obtained through an engineering or fabrication process. Although no clear theory has been developed for the depth to which a surface should be defined, top layers whose material structures are different from the bulk are of general concern.

Engineering surfaces are complex with complicated material, chemical, physical, and topographic properties. First, surfaces are complicated in topography. They are rough in nature, no matter which machining method is used for surface finish and how smooth they are polished. Irregularities representing the deviation from an ideally smooth surface compose a micro or nanoscopic view of "mountainous" peaks and valleys. The peaks, or summits, are usually named as asperities; the overall surface appearance may be referred to as topography, and the structure of surface geometry as texture.

The cross section of a rough surface is a two dimensional profile, which can be conveniently used to demonstrate the statistical description of surface roughness. The deviation of surface irregularities from the centerline is considered to be the measure of roughness. If the probability density function of a surface height is defined as the frequency count $Fc_i(z)$ divided by the height datum interval, Δz:

$$\phi(z_i) = \frac{Fc_i(z)}{\Delta z} \tag{22.1}$$

The mean, μ, or the mathematical expectation, is

$$\mu = \int_{-\infty}^{\infty} z\phi(z)dz = \frac{1}{N}\sum_{1}^{N} z_i \tag{22.2}$$

The centerline average, R_a, is one of the indications of surface roughness and is defined as

$$R_a = \int_{-\infty}^{\infty} |z - \mu| \phi(z)dz = \frac{1}{N}\sum_{1}^{N} |z_i - \mu| \tag{22.3}$$

The standard deviation, or the root-mean square (RMS) roughness, R_q, is also a commonly used measurement of surface roughness, which is defined as

$$R_q = \left(\int_{-\infty}^{\infty} (z - \mu)^2 \phi(z)dz\right)^{1/2} = \left(\frac{1}{N}\sum_{1}^{N} (z_i - \mu)^2\right)^{1/2} \tag{22.4}$$

Surface topography is very important to machine component lubrication because the two mating surfaces are the boundaries of the lubricant flow in between, and the real contact of surfaces occurs at asperities, not the nominal area of contact. Asperities may support a substantial portion of the applied load. Understanding surface topography is the first step to surface design.

Surface observations can be conducted with a scanning electron microscope (SEM), a scanning tunneling microscope (STM), and an atomic force microscope (AFM), as well as a variety of optical microscopes. Surface details can be digitized with the assistance of an AFM, an STM, a white-light interferometer, a stylus profilometer, or a laser profilometer. Shown in Figure 22.1 are several samples of surfaces from typical finish processes taken with a white-light interferometer [2].

Surfaces are complicated in the point of view of materials. A surface may have several layers, and the properties of these layers may be different from that of the bulk material. Bhushan [3] considers that, from the top to the bulk, there exist a physisorbed layer, a chemisorbed layer, a chemically reacted layer, a Beilby layer, and then a heavily deformed layer and a lightly deformed layer. He also indicates that the Beilby layer is produced by melting and surface flow during machining of molecular layers that are subsequently hardened by quenching as they are deposited on the cool

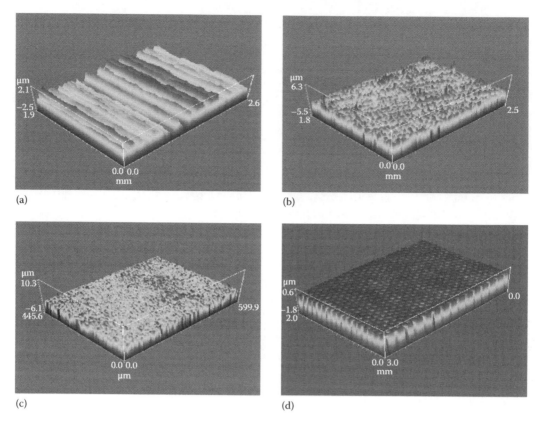

FIGURE 22.1 Sample surfaces. (a) Ground surface ($R_q = 1.14 \mu m$). (b) Honed surface ($R_q = 0.70 \mu m$). (c) Polished surface ($R_q = 0.30 \mu m$). (d) Dimpled surface ($R_q = 0.30 \mu m$). (From Wang, Q. and Zhu, D., *ASME J. Tribol.*, 127, 722, 2005. With permission.)

underlying material. Buckley [4] described a metal surface with three layers on a bulk metal: a worked layer next to an oxide layer that is underneath an adsorbed layer of gases and water moisture. Although these considerations are not identically the same, they all indicate the complex nature of materials of engineering surfaces. A cross sectional view of a surface may reveal the structure of layers and different zones of the subsurface with the information of layer formation. The properties of one layer may be different from those of another, and layer responses to loading and rubbing should be different as well. Understanding and characterization of surface layers are important steps in contact and lubrication analyses [5].

Layer structure may be observed with an SEM, a transmission electron microscope (TEM), and an x-ray diffraction (XRD). Material and chemical properties of layers may be characterized by means of an energy dispersive x-ray spectroscope (EDX), x-ray photoelectron spectroscope (XPS), Auger electron spectroscope (AES), ion-scattering spectrometer (ISS), electron probe microanalyzer (EPMA), x-ray fluorescence (XRF), as well as a secondary ion mass spectrometer (SIMS) [5].

Surfaces vary in a tribological process. A surface may go through a running-in process, through which mating asperities may become conformal at the microscale. Wear may occur if solid surfaces rub. Contact fatigue and surface tribochemistry may all contribute to material loss. Wear causes surfaces to evolve and deviate away from the initially made topography. The surface variation may be favorable or unfavorable to a designed tribological process. The micrographs shown in Figure 22.2 are for the variation of diamond-like carbon (DLC) coated surfaces with different surface textures in a tribological process. They clearly reveal the effect of surface topography on surface wear, and the potential for surface design for wear reduction.

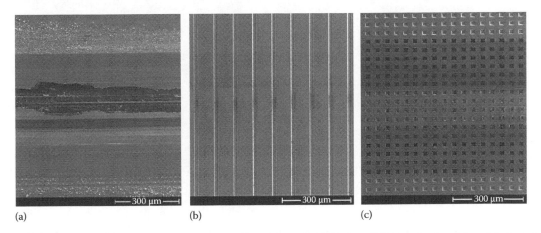

(a)   (b)   (c)

FIGURE 22.2 Variation of DLC coated surfaces after 20,000 cycles sliding under boundary-lubricated conditions. In each SEM figure, both unworn (top edge area) and worn surfaces (central, appearing as a wear track) are shown. (a) Flat DLC coated surface, (b) 20 mm wide groove texture, and (c) 20 mm wide square depression texture. (From Pettersson, U. and Jacobson, S., *Tribol. Int.*, 36, 857, 2003. With permission.)

Surface engineering indeed requires two aspects, creation of surface topography through surface finishing, fabrication, and preparation of preferred material properties through surface material processing. The topography and the surface material together contribute to the performance, such as friction, wear, and lubrication, of a tribological interface.

22.3 FRICTION

Friction occurs when an object is moving, sliding or rolling, or both, on the surface of another. Friction also exists when an object intends to move, but the surfaces in contact are macroscopically stationary. The former is called the kinetic friction, while the latter the static friction. An idealized model of friction may be described by a body of mass m under the combined action of gravity and a pushing force, F. The static friction equals the applied force before a motion is established. It reaches the maximum, F_{fs}, when the body is about to move, which is usually called the motion impending point. The static friction coefficient, $\mu_s = F_{fs}/N$, is defined at this point, and only for this point. This model also indicates that friction will become the kinetic friction as soon as the body moves, and that, in general, the kinetic friction, $\mu_k N$, is usually smaller than the maximum static friction, $\mu_s N$. However, friction is much more complicated than what is mentioned here.

22.3.1 Classic Laws of Friction

Amontons' law of friction. Amontons established the first law of friction [7–9], which states that friction is directly proportional to the applied load (normal force) and is independent of the apparent area of contact between the interacting bodies. The same friction should occur whenever the narrow or the wide edge of a brick is used as the contact surface. This is true because it has been known that surfaces are rough and that the real area of contact counts for load support, not the apparent area of contact. The real area of contact is related to the mechanical properties and surface topography of the materials in contact. Thus, friction is related to the interface of two materials, not just one of them. It should be indicated that the friction measurement made by Amontons was for dry friction, and described as the "force required to overcome friction and initiate sliding" [9].

Amonton's contribution defines the coefficient of friction. When a body starts to move, the friction at this moment is considered to be proportional to the normal force on the surface of the body, as shown below, which is the maximum static friction. Constant μ_s is the static friction coefficient with subscript s for "static":

$$F_{fs} = \mu_s N \tag{22.5}$$

Coulomb's law of friction. The Coulomb friction coefficient, or the kinetic friction coefficient, is defined for sliding friction. In many cases, kinetic friction is nearly independent of the speed of sliding. The Coulomb sliding friction is expressed by an equation similar to Equation 22.5 with the use of the kinetic friction coefficient, $\mu = \mu_k$:

$$F_f = \mu N \tag{22.6}$$

This equation indicates a constant friction for an interface under a given load, regardless of any other conditions. However, many reports indicate that sliding friction may vary with speed and other factors. Friction dependence on speed is especially visible when a lubricant is used at a tribological interface.

Euler first drew the clear distinction between the static and kinetic friction coefficient in 1750 and mathematically indicated that the kinetic friction coefficient is generally lower than the static friction coefficient [9].

22.3.2 Mechanisms of Friction

Adhesion, ploughing, and deformation are the commonly accepted major mechanisms of friction of engineering materials [10]. For micro- and nanoscale components, where the surface/body ratio is large, the electrostatic, chemical, Van der Waals, and capillary effects may strongly affect friction. Rabinowicz [11] reviewed and explained the micro/macroscopic origin of friction, and Krim [12] discussed the origin of friction at the atomic level. The mechanics of friction have been investigated by means of tribological testing, numerical simulations with the assistance of the continuum and molecular dynamics theories, and theoretical studies with the help of statistical or fractal roughness modeling [13–21]. However, friction is complicated due to the complex nature of a tribological interface and the interactive mechanical, physical and tribo-chemical processes [22,23].

22.3.2.1 Nanoscale Friction

The model by Muser et al. [16] considers that surface energy penalty increases exponentially as the local separation between two surfaces decreases, and the normal and lateral forces on a static contact equals the derivatives of this energy penalty. The static friction coefficient corresponds to the maximum of the force, or derivative, ratio, which is analogous to the classic expression of the macroscopic static friction coefficient, $\tan \alpha$. The work by Gao et al. [15] on non-adhering interfaces reveals that both computer molecular dynamics simulations and nanoscale experiments verified the accuracy of the proportionality of friction to normal load. Muser et al. [16] suggested that adsorbed molecules and other "third bodies" might provide the key to understanding macroscopic friction measurements. The Frenkel–Kontorova–Tomlinson type of models [23–27] are often used to explain nanoscale friction to include the effect of material elasticity.

22.3.2.2 Adhesion Theory of Friction

This theory [28] works relatively well for intimate metal–metal contact. Contact starts at the tips of some asperities. Once the pressure reaches the hardness of the metal, plastic flow occurs and the contact area expands (called junction growth) until it is sufficient to support the load. Strong adhesion may take place in the junction that is sometimes called "cold welding." Shear will happen when the applied tangential force is sufficiently large. The load supported by the junction, based on the ideal elastic–plastic stress–stain model without considering work hardening, equals the product of the area of the junction by the yield pressure, p_0, or the average contact pressure for fully developed plastic flow of the material:

$$W = Ap_0 \tag{22.7}$$

The force required to shear the junction is the same as the friction force that is the product of the area of the junction by the shear strength of the weaker material of the junction. If τ_y is the shear strength of the weaker material of the junction, the adhesive friction coefficient is

$$\mu = \frac{F}{W} = \frac{\tau_y}{p_0} = \frac{\tau_y}{\text{yield pressure}} \tag{22.8}$$

However, frictional shear also contributes to yield. Therefore, either the maximum shear strength or the von Mises stress should be used to identify the plastic flow. One needs to be cautious in choosing the yield pressure. The rule of thumb is to use the shear strength of the surface layer for τ_s and the yield pressure of the bulk material for p_0.

Classic mechanics indicates that for a typical metallic material, surface plastic flow would occur when the average pressure, $p_m = p_0$, is about $6\tau_y$ or $3\sigma_y$. Actually, τ_y/p_0 is in the range of 0.17–0.2 ($\tau_y \sim 0.5 - 0.577\sigma_y$). Friction coefficients of 0.2 may be true for some clean metals in air [10]. However, many metallic interfaces show higher friction in air and much higher friction in a vacuum.

The exact solution for a three-dimensional (3D) case relies on numerical calculations. However, simplified models can be assumed with the introduction of a junction coefficient, α, and another constant, k. Assuming the shear force is F, the normal load is W, and the real area of contact is A, then

$$k^2 = \left(\frac{W}{A}\right)^2 + \alpha\left(\frac{F}{A}\right)^2 \tag{22.9}$$

When the shear force is zero, $k = W/A = p_0$, can be obtained. Then the equation for the modified adhesion theory is

$$p_0^2 = p^2 + \alpha\tau^2 \tag{22.10}$$

Here, p is the average pressure that causes the junction to yield.

22.3.2.3 Ploughing Friction

The ploughing friction theory [28] works relatively well for a harder material abrading a softer one, such as scratching a hard diamond tip on a metal surface. The friction is the yield pressure multiplied by the projected area of the tip, or the area of the material to be displaced, and the normal load acts upon the vertically projected contact and load-supporting area of the tip, which is the area of the effective tip–material interaction.

22.3.3 EFFECTS OF SURFACE ROUGHNESS AND TOPOGRAPHY ON NON-LUBRICATED FRICTION

Although the classic theories of friction did not show the direct link between friction and surface topography, friction is related to the real area of contact. Tabor [29] summarized three main elements involved in a typical friction process of non-lubricated contacts, (1) the true contact area of rough surfaces, (2) the nature and strength of junctions formed at a contact interface, and (3) the way in which the materials of the contact asperities are sheared and ruptured. Apparently, (1) indicates the connection between friction and surface roughness, and asperity distribution as well, and, (2) and (3) suggest a correlation between friction and topography, especially the asperity shape and orientation, if the material conditions are the same.

Many models of friction of rough surfaces are available either based on a given asperity friction or certain physical mechanisms, such as yielding [19–21] or asperity interlocking [18]. The modeling work by Chen and Wang [20] on static friction of commonly used engineering materials is based

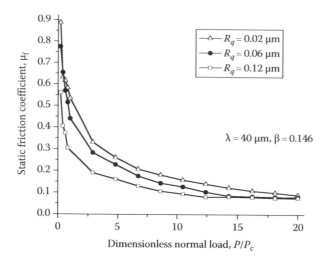

FIGURE 22.3 Effects of surface RMS roughness, R_q, on static friction coefficient for a copper ball on a sapphire half-space. (From Chen, W.W. and Wang, Q., *ASME J. Tribol.*, 131, 021402-1, 2009. With permission.)

on the Boussinesq–Cerruti integral equations and relates surface tractions to elastic-perfectly-plastic displacements, localized hardness and shear strength without asperity interlocking [20], and the Taber constant for junction growth [19,29]. This work investigated the behavior of static friction involving rough surfaces and revealed the dependence of the static friction coefficient on surface roughness. The results suggested that the interface of a sphere contact involving a surface of larger RMS roughness may render a smaller static friction coefficient and that the surface roughness effect becomes less obvious under a heavier normal load, as shown in Figure 22.3.

Figures 22.4 and 22.5 try to explain the phenomena at the sliding inception moment identified using the curves of the sticking area versus the tangential force. Full sliding occurs when the sticking area vanishes, and thus the static friction forces can be determined. It turns out that the static friction between smooth surfaces is larger than that involving a rough surface. Variations of the contact area as a function of the tangential force are plotted in Figure 22.5. The initial contact

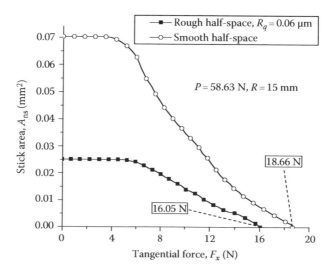

FIGURE 22.4 Identification of the sliding inception moment for the contacts of a copper sphere and a steel half-space under the normal load of $P = 58.63$ N. (From Chen, W.W. and Wang, Q., *ASME J. Tribol.*, 131, 021402-1, 2009. With permission.)

area of the rough-surface contact case is smaller than that of the smooth surface case. However, the contact involving a rough surface has a larger growth percentage of the contact area at the sliding inception.

The effect of surface textures on the non-lubricated friction of a poly(dimethylsiloxane) (PDMS) elastomer made with different surface textures (Figure 22.6) has been investigated [30] at

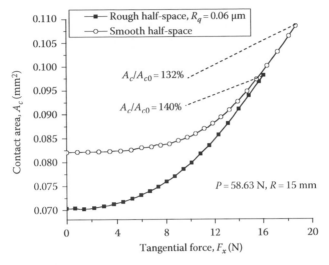

FIGURE 22.5 Variations of contact area with the increasing tangential force for the contacts of a copper sphere and a steel half-space under the normal load of $P = 58.63$ N. (A_{c0} is the initial contact area under the normal load alone.) (From Chen, W.W. and Wang, Q., *ASME J. Tribol.*, 131, 021402-1, 2009. With permission.)

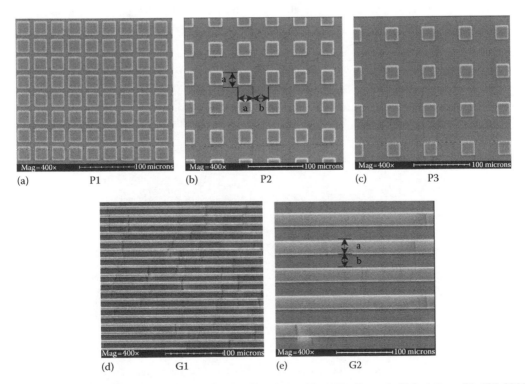

FIGURE 22.6 SEM images of the textured PDMS surfaces. (From He, B. et al., *Tribol. Lett.*, 31, 187, 2008. With permission.)

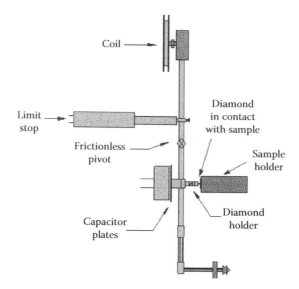

FIGURE 22.7 The NanoTest pendulum system. (From Smith, J. F. and Zhang, S., *Surf. Eng.*, 16(2), 143, 2000. With permission.)

both the macro- and microscales using a micro-material nanoindentation–scratching system [31] (Figure 22.7). Friction tests were conducted with a stainless-steel bearing ball with a diameter of 1.6 mm (macroscale tests) and a Rockwell diamond tip with a radius of curvature of 25 μm (microscale tests) under normal loads of 5, 10, and 25 mN and with a sliding speed of 1 μm/s. The coefficient of friction (COF) on the pillar-textured surface was found to be much lower than that on the smooth surface of the same material, and it was reduced by about 59% at the macroscale tests and 38% at the microscale tests. The reduction of COF can be attributed to the reduced contact areas due to the existence of textures.

It is interesting to view the effect of surface orientation on friction of the interface involving this type of materials. Two extreme cases are the surfaces with parallel and perpendicular groove orientations, along (the former) and perpendicular to (the latter) the direction of the relative sliding motion. The frictions in other directions are expected to be in between these two extreme cases [32–34].

Figure 22.8a shows the friction obtained from the macroscale tests with the bearing ball sliding in the parallel and perpendicular directions on two different groove-textured PDMS surfaces

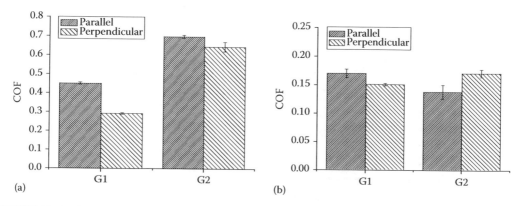

FIGURE 22.8 COFs for the groove-textured surfaces in the parallel and perpendicular directions obtained from the macroscale tests. (a) Macroscale tests. (b) Microscale tests. (From He, B. et al., *Tribol. Lett.*, 31, 187, 2008. With permission.)

(hereafter referred to as G1 and G2 surfaces) under 10 mN normal load and 1 μm/s scanning speed. The friction corresponding to the parallel sliding direction is higher than that to the perpendicular direction for both G1 and G2 surfaces. The maximum sticking length measured in the sticking area along the direction of the motion tendency might have played a role. The static friction force can be defined as the tangential force at the moment when the maximum sticking length drops below the critical value. Because the sliding speed used in the experiments was very low, the sliding testing could be treated as a quasi-static process, and a static friction analysis, such as the one by Chen and Wang [20] can be applied. The relationships between the maximum sticking length and the tangential force suggest that the static friction force along the groove is larger than that across the groove geometry.

The friction coefficients for the G1 surface in both of the parallel and perpendicular directions are smaller than those for the G2 surface are, and this is likely due to the difference in the groove dimensions between the G1 and G2 surfaces that leads to different contact area when mating with the bearing ball. The groove size, a, and spacing, b, of the G1 surface are about 5 and 10 μm, while those of the G2 surface are 25 and 25 μm, respectively. Therefore, the surface texture density, defined as the ratio of the pattern size over the period, of the G2 surface is about 50%, while that of the G1 surface is only 33%. As a result, the G2 surface offers more surface contact within the same contact radius, and this should be one of the reasons for its higher friction.

Figure 22.8b presents the friction obtained from the microscale tests with the Rockwell tip sliding in the parallel and perpendicular directions on the G1 and G2 surfaces under the 10 mN normal load and 1 μm/s scanning speed. Similar friction anisotropy effect can be observed on the G1 surface. However, the behavior of the G2 surface is different. The friction coefficient in the perpendicular direction is higher than that in the parallel direction. Here the increased corrugation of the surface profile along the direction perpendicular to the groove texture plays a role. Apparently, friction anisotropy is scale dependent.

The effects of rough surface orientation on friction have also been observed by Zhang and Komvopoulos [35], Gangopadhyay and McWat [36], and Li et al. [37], and the friction anisotropy reported by the first group is mainly for the material and topography effects while that of the latter two groups involve lubrication. Furthermore, the series of friction measurements on the surfaces of different hardness and topography reported in the work of Li et al. [37] reveal that roughness amplitude and surface hardness are two important factors that affect the frictional performance of textured surfaces, mainly through asperity deformation and wear.

22.4 WEAR

22.4.1 CLASSIC THEORIES ON THE MECHANISMS OF WEAR

Wear may be defined as the progressive loss or removal of material at surfaces due to rubbing. Mechanisms of wear, such as adhesive wear, abrasive wear, corrosive wear, oxidative wear, polishing, plastic deformation, and fatigue wear, etc. have been recognized and some mathematically described to a certain degree.

22.4.1.1 Non-Lubricated Adhesive Wear [38]

Assuming that a rough surface of spherical asperities interacts with a smooth and rigid surface, the junction formed has an area of πr^2 and the material lost due to wear is likely to be the volume of the crown, $2\pi r^3/3$. If the wear rate is defined as the volume loss of material per unit sliding distance, then, for a total of n asperities [38],

$$Q = \frac{n\pi r^2}{3} \tag{22.11}$$

However, the normal load may be $W=Ap_0=np_0\pi r^2$. Not all contact produces wear, and not all asperities are worn at the same rate. It is reasonable to introduce a coefficient for wear, which is sometimes called the probability of wear, k. Therefore, the above equation becomes

$$Q = \frac{kW}{p_0} = \frac{kW}{H} \tag{22.12}$$

For many metals, the yield pressure is about the hardness value, H.

This equation states that wear is proportional to load but inversely proportional to the yield stress or hardness of the softer material of the interface. The volume of materials loss, which is the wear rate by distance, is proportional to the distance of travel. Similar to friction coefficient, this wear coefficient is not a material property, rather a property of an interface under certain operating conditions.

22.4.1.2 Lubricated Adhesive Wear [39]

Rowe modified Archard's law of adhesive wear and pointed out that in a lubricated interface, only the metal–metal contact area, A_m, contributes to wear [39]. A parameter, β, is introduced to characterize the effectiveness of lubricant coverage, and $\beta=A_m/A$. However, for a 3D stress case, the pressure is p_0, or the yield pressure at the interface, usually smaller than p_0, is the yield pressure of the material.

Using the maximum shear stress theory, the yield pressure is $p=p_0/(1+\alpha\mu^2)^{1/2}$, where α is a constant mentioned in the previous section, the modified adhesive wear rate becomes

$$Q = k_m\beta(1+\alpha\mu^2)^{1/2}\frac{W}{p_0} \tag{22.13}$$

22.4.1.3 Abrasive Wear [1]

Abrasive wear usually refers to the material removal by hard asperities (or particles) in a ploughing process. The material volume displaced by the hard conical tip in traveling unit distance is $rh=rr\cot\theta$. Because the total normal load is $W=n\pi r^2 p_0/2$, the wear rate is then

$$Q = nr^2\cot\theta = \frac{2W\cot\theta}{\pi p_0} = \frac{k_a W}{H} \tag{22.14}$$

In this equation, k_a is a coefficient for abrasive wear to be determined by the geometry of the hard asperities (particles) and the probability of wear.

22.4.2 Empirical Wear Laws

Many empirical wear laws have been developed. Wear maps have been produced for design references [40,41]. Among the empirical wear laws, Harris et al. [42–46] conducted extensive studies of the abrasive wear of steel by hard coatings using a pin-on-disk tribotester, that is, the steel ball was worn due to the abrasion of hard asperities of the rotating disk coated by a hard coating. An average abrasion rate during n cycles $\overline{A(n)}$ is defined as the total volume of the steel removed divided by the total distance traveled, which is the same as the wear rate in principle:

$$\overline{A(n)} = \frac{V}{d} = \frac{V}{2\pi rn} \tag{22.15}$$

In this equation

V is the volume of steel removed

d is the distance traveled

n is the number of cycles

Dry sliding wear tests were run at room temperature with a humidity of $35\% \pm 15\%$. Using a track radius of approximately 9 mm, the B_4C coated disk was rotated at approximately 10 cm/s with a load of $100 \times g$ for a given number of cycles, ranging from 1 to 30,000. The ball was then removed from the testing apparatus and its surface was wiped clean with a tissue soaked in ethanol to remove loose debris. The surface area of the ball wear scar, totaling typically 0.25 mm² (500×500 µm), was digitized using an optical profilometer to show the newly worn area using a spatial sampling resolution of 1.0 µm. The worn coating track was inspected with an atomic force microscope (AFM) with a spatial resolution of 0.167 µm per pixel in the horizontal dimension. Fresh coupons and fresh balls were used for each test.

Following a power law scaling relationship, the average abrasion rate of the steel balls decreases with increasing sliding distance. Because of the linear relationship shown in this figure, it can be shown that the instantaneous abrasion rate on the nth cycle A_n is given by [43]

$$A_n = A_1 \left(n^{1+\beta} - (n-1)^{1+\beta} \right) \tag{22.16}$$

where

A_1 is the abrasion rate on the first cycle

β is the slope of the line in this figure, approximately -0.78

The rate of wear of the steel is initially large, but the above equation shows that after 30 cycles (1.7 m sliding distance) of wear, the instantaneous abrasion rate has dropped by a factor of more than 60.

A further study of the same type of material pair during the first cycle of wear reveals that in the range $0 < n \leq 1$, $A_p(n)$ is independent of the sliding distance. Its variation somehow obeys the Archard's wear law. The wear coefficient provides a means of comparing the severity of wear processes for different systems, while the quantity (W/H) is the real area of contact for fully plastic asperities. Although the overall size of the contact area increases with sliding distance, the size of the contact area experiencing plastic deformation remains nearly constant. The measured value of the plastic deformation area for $0 < n \leq 1$ was 4.6×10^{-5} mm². Using the measured value of the plastic deformation area, the value of k_a is found as 4.3×10^{-2}. The majority of the contact area was indeed found to be elastic. Figure 22.9 plots the entire wear process: initiation (the first cycle) and further wear ($n > 1$). A nearly flat initiation stage is shown (Table 22.1).

The wear simulation model is suggested in [46] based on work reported in [43]

$$V = A_1 dn^{\beta}. \tag{22.17}$$

The wear volume is easily obtained for any given cycle simply by knowing the abrasiveness of the first cycle and the rate at which the abrasiveness decreases with increasing cycles on a log–log scale. Although the values of A_1 and β vary with the specific set of testing conditions, this equation itself is directly independent of these conditions. In other words, this equation is valid for a multitude of experimental conditions and it is only necessary to directly investigate the properties of the abrasiveness in order to determine the wear volume of the abraded material. Although a wide range of A_1 and β is observed from the review of 50 wear publications, similarities do exist within each general material category, as illustrated in Figure 22.10, which compares the approximate range of the wear model results for each general category of the tested materials [46].

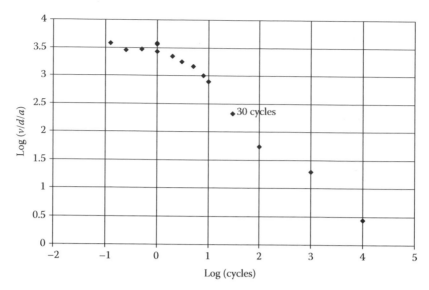

FIGURE 22.9 Average abrasive rate per unit of plastic deformation versus number of ball-on-disk cycles for the 52100 steel balls against a B4C coating. (From Siniawski, M. et al., *Tribol. Lett.*, 15, 29, 2003. With permission.)

TABLE 22.1
Summary of Several Empirical and Theoretical Wear Laws[a]

	α	β	γ	Additional Terms	Primary Application
Empirical					
Lewis (1968)	1	1	0		Adhesion of filled PTFE and piston rings
Khrushchov and Babichek (1970)	1	1	1		Micro-cutting of metals
Rhee (1970)	A	β	0	Exponential function of *t*	Adhesion with thermal effects
Lancaster (1973)	1	1	0	Includes wear rate correction factors	Filled thermoplastics and filled PTFE
Larsen–Basse (1973)	1	1	0	Defined in terms of impact frequency	Thermal fatigue and carbide polishing
Moor, Walker, and Appl (1978)	1	1.8	0	$p=p$ (rock volume removed/distance)	Wear of diamond inserts and rotary drag bits
Theoretical					
Holm (1946)	1	1	1		Adhesion
Archard (1953)	1	1	1		Adhesion
Kragelsky (1965)	>1	1	0		Fatigue
Rabinowicz (1971)	1	1	1		Abrasion/fretting
Harricks (1976)	1	1	0		Fretting

Sources: Zhu, D. et al., *ASME J. Tribol.*, 129, 544, 2007; Goryachev, I. G., *Contact Mechanics in Tribology*, Kluwer, Dordrecht, the Netherlands, 191–201, 1998. With permission.

[a] With corresponding exponential constants for pressure (load in the case of empirical expressions), α, sliding velocity, β, and hardness, γ.

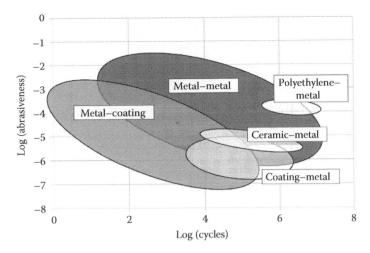

FIGURE 22.10 Approximate range of wear model results for all of the examined cases. (From Siniawski, M. et al., *Wear*, 262, 883, 2007.)

Consider the wear rate as the volume loss of material per unit sliding distance. Equations 22.12 through 22.14, as well as many others, actually have the same, or similar, format. Converting the sliding distance into time, the wear rate per unit time can be defined as $q = Q/u$, where, u is the sliding speed. Upon reviewing a large number of previously developed wear models, researchers believed that the most common of these expressions could be summarized in the form of a single equation, in which the wear rate, q, is a function of contact pressure, p, relative sliding velocity, u, and material hardness, H [41,47–49]. Note that the conversion between load and pressure is by area, which is covered by constant k:

$$Q = k \frac{p^{\alpha} u^{\beta}}{H^{\gamma}} \tag{22.18}$$

22.4.3 SIMULATION OF WEAR OF ROUGH SURFACES

A wear model can be integrated into a contact simulation process or a lubrication modeling system at the asperity level. For the latter, wear can be treated as an additional term in the expression for the instantaneous film thickness, and the film thickness is the sum of the geometry of the contacting bodies, roughness of the surfaces, surface deformation, and surface modifications due to wear. Wear of the two surfaces should be introduced separately because their material properties may differ resulting in different wear behavior. Zhu et al. [49] developed a numerical wear simulation scheme based on Equation 22.18 for the prediction of wear in mixed lubrication. Such an integrated wear simulation approach allows numerical representation of surface evolution at different wear phases and correlation of the wear performance of different surfaces with material hardness, sliding speed, applied load, and wear coefficient.

22.5 INTRODUCTION TO LUBRICATION

22.5.1 REYNOLDS EQUATION AND HYDRODYNAMIC LUBRICATION

The Reynolds equation below is the essential equation for the basic relationship between the film thickness and the hydrodynamic pressure. It describes the characteristics of fluid flow through a small clearance bounded by two solid surfaces. In deriving the Reynolds equation the following are assumed: the fluid is Newtonian and the flows are laminar; there is no interfacial slippage; the body force is negligible; the thickness of the fluid film (lubricant film) is very small and the derivative with respect to this direction is more important than others; the curvature effect is insignificant, the pressure across the fluid film can be treated as constant, and the solid surfaces are rigid and ideally smooth:

$$\frac{\partial}{\partial x}\left(\frac{h^3}{\eta}\frac{\partial p}{\partial x}\right)+\frac{\partial}{\partial y}\left(\frac{h^3}{\eta}\frac{\partial p}{\partial y}\right)=6U\frac{\partial h}{\partial x}+12\frac{\partial h}{\partial t} \tag{22.19}$$

where

h is the lubricant film thickness between the two surfaces

p the hydrodynamic pressure

U the sum of the speeds of the two surfaces

η the viscosity of the lubricant

The coordinates, x and y, are along and perpendicular directions of the motion. This equation describes three types of flows involved in a hydrodynamic lubrication: the pressure driven flows, shown by the terms at the left side of the equation, and the shear and squeezing flows, shown by the first and second term at the right side of the equation.

The above equation defines three conditions to be satisfied for hydrodynamic lubrication: the overall geometry should have a convergent wedge, the lubricant must have a certain viscosity, and the two surfaces must be in a relative motion (sliding, rolling, or squeezing). Once all these conditions are satisfied and properly designed, a pressurized lubrication film can be established between the two solid surfaces to separate them from contact and direct rubbing.

22.5.2 Elastohydrodynamic Lubrication (EHL)

Surfaces under high pressure should experience deformation. The lubrication with the contribution from surface elasticity is in the category of elastohydrodynamic lubrication, or EHL in short. EHL is especially the case for components under a lubricated counterformal contact. It is well known from elasticity that when two elements are in a counterformal contact, the contact pressure is highly localized, causing the surfaces to deform. Such deformation opens the gap between the surfaces, or increases the film thickness. Therefore, the Reynolds equation should be solved with an elasticity component. Shown in the following equations are the deformations for plane-strain line-contact EHL problems and three dimensional point-contact EHL problems, respectively [50]. The Flamant solution for the line contact EHL problems is

$$u_z(x)=-\frac{4}{\pi E'}\int_\Omega \ln\left|\frac{\xi-x}{\xi-x_r}\right|p(\xi)d\xi \tag{22.20a}$$

and the Boussinesq solution for elliptical contact EHL problems is

$$u_z(x,y)=-\frac{4}{\pi E'}\int_\Omega p(\xi,\zeta)\frac{d\xi d\zeta}{\sqrt{(x-\xi)^2+(y-\zeta)^2}} \tag{22.20b}$$

with

$$\frac{2}{E'}=\frac{1-v_1^2}{E_1}+\frac{1-v_2^2}{E_2}$$

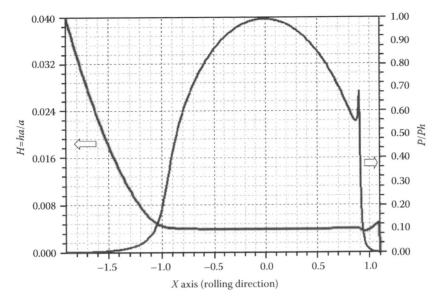

FIGURE 22.11   Typical EHL film thickness and pressure distributions in a line contact problem, where ha is the central film thickness and a the Hertz contact semi-width.

Both pressure and temperature affect lubricant viscosity. A commonly used viscosity–pressure relationship, the Barus equation, is given below:

$$\eta = \eta_0 e^{\alpha p} \tag{22.21}$$

Bair et al. [51] suggested that a piezo–viscous parameter that together with the ambient viscosity would quantify the Newtonian rheology so that the film thickness may be calculated accurately. Several studies have been conducted using viscosity measured in viscometers. The isothermal EHL film thickness and friction agree well with those measured with a thin-film tribometer [51,52].

The density of a lubricant may also vary with pressure. Following is a density–pressure relationship for mineral oils developed by Dowson–Higgison [53]:

$$\frac{\rho}{\rho_0} = 1 + \frac{0.6p}{1+1.7p} \tag{22.22}$$

Numerically solving Equations 22.18 through 22.22 should yield the pressure and film thickness at an EHL interface. Figure 22.11 illustrates a typical solution set of the pressure and film thickness distributions for a line EHL contact along the centerline in the motion direction. The central part of the film is largely flat. A film neck may appear close to the existing portion of the EHL interface, and a pressure spike should accompany the film necking phenomenon.

22.5.3   EFFECT OF ROUGHNESS ON LUBRICATION

The rough nature of engineering surfaces has to be considered in solving lubrication problems. A great deal of efforts have been made toward the understanding and mathematical description of the effect of surface roughness on lubrication. Among them are the statistical flow models, such as those by Christensen [54], Patir and Cheng [55], Harp and Salant [56], and Sahlin [57]. Briefly shown here is Patir and Cheng's average Reynolds equation [55]:

$$\frac{\partial}{\partial x}\left(\phi_x \frac{h^3}{\eta}\frac{\partial \bar{p}}{\partial x}\right)+\frac{\partial}{\partial y}\left(\phi_y \frac{h^3}{\eta}\frac{\partial \bar{p}}{\partial y}\right)=6(U_1+U_2)\frac{\partial \bar{h}_\text{T}}{\partial x}+6\left(U_1-U_2\right)R_q\frac{\partial \phi_s}{\partial x}+12\frac{\partial \bar{h}_\text{T}}{\partial t} \qquad (22.23)$$

Here, U_1 and U_2 are surface speeds. Compared with Equation 22.18, the Patir–Cheng average equation shown above involves three flow factors to modify the lubricant flows, with ϕ_x and ϕ_y, as the pressure-flow factors, and ϕ_x as the shear-flow factor. The pressure-flow factors may be unity when the film thickness is much larger than the height of surface asperities, or

$$\phi_x, \phi_y \to 1 \quad \text{as} \quad \frac{h}{R_q} \to \infty \; (h \to \infty \text{ or } R_q \to 0)$$

Because x and y are orthogonal directions, one only needs to define one of these two flow factors. The other can be determined with the same method.

An additional flow is introduced by the shear-flow factor, ϕ_s, which is found to be a function of the non-dimensional film thickness, h/R_q, and the roughness orientation, λ, as well as a shear-flow factor for a single surface, Φ. Obviously, ϕ_s is for the combined effect of the two mating rough surfaces:

$$\phi_s = \left(\frac{R_{q1}}{R_q}\right)^2 \Phi_s\left(\frac{h}{R_q}, \lambda_1\right) - \left(\frac{\sigma_2}{R_q}\right)^2 \Phi_s\left(\frac{h}{R_q}, \lambda_2\right) \qquad (22.24)$$

The additional flow is

$$q = \frac{U_1 - U_2}{2}R_q\phi_s \qquad (22.25)$$

Physically, if one of the surfaces is smooth, for example, $R_{q1}=0$, the equation for the shear-flow factor becomes

$$\phi_s = -\left(\frac{R_{q2}}{R_q}\right)^2 \Phi_s\left(\frac{h}{R_q}, \lambda_2\right) = -\Phi_s \qquad (22.26)$$

If $U_2 > U_1$, the rough surface is moving faster and there is an increase in the total flow. On the other hand if $U_2 < U_1$, the smooth surface is moving faster and there is a decrease in the total flow.

The detailed approach for flow factor determination can be found in the paper by Patir and Cheng [55].

22.5.4 REGIME AND FRICTION OF LUBRICATION

Different status of lubrication may involve different friction. The Stribeck curve is widely used for the presentation of friction transitions in lubrication. The concept of systematically describing friction variation in lubrication was due to Professor Richard Stribeck, who in 1902 confirmed the existence of a minimum friction through his extensive journal-bearing friction experiments. Ludwig Gumbel, later in 1914, summarized the Stribeck results into a single curve by means of dimensionless parameters. Mayo Hersey [58] showed that friction due to viscous shear was a unique function of the product of viscosity (η) by rotational speed (N) divided by the average load (P), which is called the Hersey number. Such a diagram with friction coefficient plotted against the Hersey number is now commonly known as the Stribeck curves. Professor Dowson [9] summarized

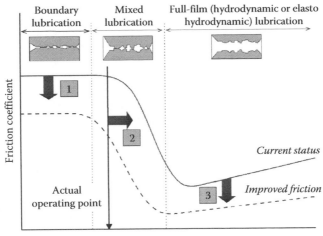

FIGURE 22.12 Regimes of lubrication and friction variation with a typical Stribeck curve (solid line) and a potential Stribeck curve subject to possible friction reduction (dashed line). (From Martini, A. et al., *Tribol. Lett.*, 28, 171, 2007. With permission.)

the history of the development of the Stribeck curve concept. Recently, Wang et al. [59] reviewed the recent work on Stribeck curve, presented the results of a research on the friction of a journal-bearing conformal contact system and suggested a form of 3D Stribeck surface with considerations of the surface roughness, material elasticity, and thermoelasticity.

The full-film lubrication, or the hydrodynamic lubrication, is for the lubrication where the lubricant film successfully separates the solid surfaces completely. The partial-film lubrication, or the mixed lubrication, means that the lubricant film is insufficient to completely cover the entire surfaces. There exists some solid asperity contact in a mixed lubrication. When the lubricant film is further reduced, the asperity contact may dominate and the lubricant film is no longer continuous in most of the interface region. The surfaces may be covered only by some absorbed, either physically or chemically, lubricant film. This regime of lubrication is called the boundary-layer lubrication, or simply the boundary lubrication. Figure 22.12 below presents these regimes of lubrication in terms of the well-known Stribeck curve.

The Stribeck curve can be a guide for the development of technologies for friction reduction that is also shown in Figure 22.12. Friction reduction may be done for a particular operation in a certain regime of lubrication shown in the Stribeck curve. However, a typical machine system should go through start and stop and therefore experiences at least two lubrication regimes. In order to lower the overall level of friction, all the major factors reflected by the curve should be considered. Martini et al. [60] suggested three primary means of minimizing friction in a lubricated interface, which are to (1) reduce contact sliding friction by improving boundary film performance in the interface contact regions, (2) reduce the percentage of surface contact by optimization of the operating conditions and surface topography, and (3) reduce hydrodynamic friction by improving lubricant rheological properties and surface design. Here, surface design, both topography and materials, plays a central role.

22.6 FRICTION AND WEAR IN MANUFACTURING PROCESSES

Friction and wear problems in many manufacturing processes, including metal forming, metal cutting, often involve the generation of new surfaces due to large deformation. Therefore, many test apparatuses in an abstract form have been developed over the years to specifically simulate the deformation mechanism occurring in various manufacturing processes, for example, the combined forward conical cup and backward straight can extrusion test, tube expansion test, ring compression test, sheet pull out test, etc. A summary of the lubricants and tribo-systems used for metal forming can be found in a keynote paper at the 2010 60th CIRP General Assembly [61].

Prediction of friction and wear behavior has been conducted by many researchers as indicated in the rest of this chapter. Another way to control friction in a manufacturing process is through a real-time feedback control. The objective is to provide a more consistent restraining force to the workpiece so that a consistent part can be obtained. In this setup, sensors are placed to measure the reaction force and/or temperature. Based on the resulting measurements, the corresponding force input is altered to achieve a prescribed value or trajectory. The control system can be a simple proportional-integral controller or a more advanced neural network controller. Briefly, a neural network mimics the structure of the brain. Through a set of carefully conducted experiments, the network is trained to identify small or large variations in the process and then acts accordingly to achieve a preset goal. Examples of metal forming processes that have benefited from neural network springback control include a rebar [62] and an air bending process [63]. A control system using artificial neural networks and a stepped binder force trajectory to minimize and control springback in a channel forming process was discussed in [64,65]. However, the neural network is known to be good for interpretation, but not for extrapolation. When variations occurred in the normal expected zone, the use of a neural network can be very effective.

22.7   EMERGING SURFACE ENGINEERING TECHNOLOGIES FOR ENERGY EFFICIENCY

Friction reduction requires the antifriction design of tribological interfaces. Material technologies that can result in a low friction surface layer, lubricant technologies that permit low viscous dissipation and *in situ* formation of a lubricious boundary film, surface treatment technologies that may favorably alter the mechanical and chemical properties of the surface material, and texturing/finishing technologies that can render preferred surface topography for lubrication enhancement and stress reduction are all helpful toward efficiency improvement. This section reviews the tribological applications of several of these technologies with an emphasis on surface texturing.

22.7.1   COATING TECHNOLOGIES FOR LOW FRICTION

The combination of high yield pressure and low surface shear strength tends to produce low friction. A hard substrate coated by a soft layer may be a good choice for low-friction design. With a soft coating, the surface layer yields anyway, leaving the substrate to support the load. However, shear occurs in the soft layer, where the shear strength is low.

Low friction may be achievable by means of coatings under certain conditions. There have been extensive investigations of different lubricious coatings that can substantially reduce friction. Molybdenum disulfide (MoS_2) and diamond like carbon (DLC) are two of these coatings. MoS_2 has a layer structure similar to graphite. When properly oriented and in an inert environment, MoS_2 exhibits a friction coefficient below 0.01 [66]. However, in the presence of oxygen, this low friction property disappears. The relative softness of MoS_2 also makes it unsuitable for demanding tribological applications.

Carbon's versatility is due to the different ways that carbon atoms can bond to each other and to other elements. Carbon may appear in the form of graphite, where carbon atoms bond strongly to each other within a plane but weakly between planes. Graphite is soft, electrically conducting, and opaque. In diamond, the bonding is strong in all directions. Diamond is the hardest known material, electrically insulating, and transparent from the far ultraviolet to the far infrared. Diamond films with excellent protective properties can be produced by vacuum deposition but the optimum substrate temperature for coating is about 900°C that severely limits the range of the substrates to which diamond can be applied. At near room temperature, an amorphous carbon-containing coating can be produced, in which a proportion of the carbon atoms are bonded as in diamond, resembling diamond in some ways, and thus DLC coatings are formed. The mechanical and tribological properties of DLC films are very close to those of diamond. Moreover, these films are chemically inert in most aggressive environments, and may be deposited with densities approaching that of diamond. Importantly, DLC films can be produced at room temperature that makes them particularly attractive for applications where the substrate cannot experience elevated temperatures [67,68].

Erdemir et al. at the Argonne National Laboratory, United States, showed that the DLC coatings deposited using a patented plasma-enhanced chemical vapor deposition technique can attain friction coefficients as low as 0.001 in inert environments [69,70]. They indicated that the coefficient of friction between two sliding DLC films decreases with increasing hydrogen incorporation. In general, the higher the H/C ratio is, the lower the friction coefficient. When all surface carbon atoms are fully hydrogenated, the interaction between two DLC film surfaces is analogous to that of two methane molecules, which is weak, thus resulting in low friction. In air, the friction coefficient of DLC becomes one order of magnitude higher than that in an inert gas [71]. Freyman ct al. [68] developed a DLC that is insensitive to the environment; with the addition of an optimum amount of sulfur. More advancement of the DLC and other new coatings for low friction has also been reported [72].

22.7.2 Advanced Surface Treatments

Altering the surface properties for low friction, low wear, and high corrosion resistance is the goal of surface treatments. Typical surface treatment methods include the plating processes, such as electrodeposition and electroless plating, and the diffusion processes, such as carburing and nitriding. Budinski [73] gave an in-depth review of many surface treatment methods, techniques, and their applications. Recently, several approaches, the friction-stir process, laser glazing, and laser shock peening have drawn a lot of attention from researchers and engineers.

22.7.2.1 Friction Stir Process

Friction welding is conducted by moving one body with respect to the other in a contact under a compressive force. Friction heating at the interface softens both surfaces. Extrusion of the junction will leave the clean material from each component along the original interface. With this process, no molten material is generated and the weld is formed in the solid state. The friction stir process uses the same principle for surface treatment, leaving surfaces with finer grains and improved fracture resistance [74].

22.7.2.2 Laser Glazing and Laser Shock Peening

Laser shock peening can produce higher magnitude of compressive residual stresses in a deeper depth than the traditional shot peening does [75]. It is accomplished by using a high-energy pulsed laser to create a high amplitude stress wave on the surface to be treated and make the surface layer to plastically deform, thus developing a residual compressive stress. Laser glazing uses a high-power laser to melt the top layer of a surface. Rapid cooling and re-solidification can result in a much harder surface with modified microstructures. Research has shown that both laser glazing and laser shock peening can result in wear resistant surfaces [76].

22.7.3 Surface Texturing

Interfacial conditions may become more severe as the compact design prevails. Well-designed features of textured surfaces may contain micro-reservoirs to retain the lubricant always in a full-film state and act as micro-bearings to enhance lubrication. Creating surfaces with controlled micro-geometry can be an effective approach to improve the performance of tribological interfaces. Specially designed textures may be found on the surfaces of many mechanical components. Early efforts on textured surfaces in metal forming were reported by Hector and Sheu [77]. More recent studies on surfaces produced by laser-beam processes for mechanical seals and a few auto parts can be found in [78–84]. Experimental studies have been conducted to investigate the tribological performance of textured surfaces. The demands for optimal surface design have stimulated the development of texturing methods and the research on performance understanding.

22.7.3.1 Surface Texturing Methods

Several surface texturing techniques are available for research and engineering applications. The mechanical processing techniques such as machining and forming, electric discharge machining (EDM), vibration turning, electrochemical machining (ECM), electron beam texturing, and lithography etching, as well as shot peening, have all been used for texturing surfaces; however, each has significant disadvantages in efficiency or accuracy.

Micromachining is able to generate pre-described 3D surface dimples in almost any shape on any surface with ultrahigh precision (as high as 1 nm resolution [85]. However, it is a much slower process compared with laser texturing. Microforming is capable of efficiently producing high precision 3D dimples on large surfaces, but suffers from tool wear problems, especially when processing hard workpiece materials.

Microstamping creates surface textures through micro-presses of desired shape. One of the notable features is that compressive stresses can be built into the workpiece surface.

EDM is independent of workpiece hardness, but requires the workpiece to be conductive. It is difficult to control the dimple shape, as tool wear in the EDM is inevitable. In addition, forces due to pressure within the discharge bubble, as well as electrostatic and electromagnetic forces, can cause workpiece distortion when machining very small features [86].

ECM creates a stress-free, undamaged surface and has the major advantage of zero tool wear. On the other hand, the ability to confine the erosion to a narrow zone appears to be the major stumbling block to its wider use for high precision applications [86].

Electron beam texturing has the advantage of higher degree of control; however, it requires vacuum and is more expensive. Micro molding [87,88] through lithography etching is a commonly used technique in microfabrication of textured surfaces. For example, textured surfaces with square pillars and grooves of different dimensions mentioned before were fabricated by the micro-molding technique. Apparently, this technique works mainly for small flat samples and requires a clean room.

A lathe vibration turning approach has been developed for rotational surfaces and for the convenience in linking with the current manufacturing technologies that can be directly applied to machine dimples on cylindrical surfaces [85,89]. This vibro-mechanical texturing (VMT) technique is based on a standard, single point turning process where a work piece is mounted in a lathe and rotated while a cutting tool engages the surface causing the shearing removal of the surface material. The amount of material removed depends on the depth of cut and the nose radius of the engaging cutting tool. The VMT technique makes use of an advanced tool positioning system that allows for controlled micropositioning of the cutting tool along the axis of engagement, allowing for active control of the depth of cut. By appropriately controlling the tertiary motion of the tool relative to the work piece, depressions of various sizes and shapes are machined onto the surface.

Since this NMT can be done on a standard lathe, all turning operations required for the manufacture of the part should be performed using the same setup with the positioning stage in static engagement for the initial turning, and vibro-texturing would be performed in the final tool pass, saving operation time and cost. Also, since the VMT equipment is retrofitted on a standard lathe, the equipment cost is considerably lower than other standalone texturing platforms.

The VMT process can make use of a standard CNC lathe to control the workpiece rotation and tool carriage position. The tool mount can be replaced with an advanced micropositioning stage

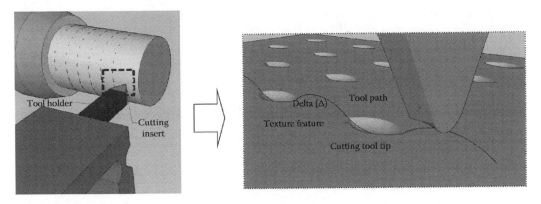

FIGURE 22.13   VMT system and process. (a) Cutter and workpiece, (b) enlarged view.

(MPS) to control the tool engagement. The MPS may consist of three main components: (1) parallel box spring, (2) piezo element, and (3) cutting tool. The box spring, machined from spring steel using a wire cut electric discharge process, gives the main structure of the positioning stage and houses the piezo-actuating element (Figure 22.13).

However, this method can only be used for texturing surfaces of revolution. Since it uses a piezo-actuator controlled sharp tool, there are also limitations imposed by the geometric and kinematic characteristics of the system.

Laser-based texturing. Laser surface texturing can, in principle, be accomplished by two methods: (1) fixed focus system (similar to a XY-plotter), or (2) 3D scan head. The first method provides a much smaller spot size. It is more accurate, but slower. With a stationary laser, the workpiece will have to be moved in relation to the focal point of the system while maintaining the beam orthogonal to the surface and the focal point on the surface. The second method uses galvanometers and focusing optics to direct and focus the beam. This system is considerably faster, but the spot size is much larger, hence its ability to precisely control the topography of the surface is diminished. From a production standpoint, the second method would be preferable [90].

As an example, the laser texturing system (Figure 22.14) at Northwestern University consists of a diode pumped Nd: YVO4 picosecond laser with 1064 and 532 nm wavelengths, as shown below.

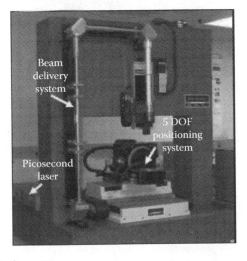

FIGURE 22.14   A laser texturing system.

The laser has a variable repetition rate from 10 to 500 kHz. Average power with the 1064 wavelength is 2 W and the peak power can be as high as 20 MW. The beam is delivered to the machining area via fully enclosed tubing and optics. The minimum beam diameter is approximately 10 μm. It has a positioning stage with five degrees of freedom (X, Y, Z, B, and C) with an accuracy of 10 nm. The system also includes a CCD camera capable of inline monitoring of the machined surface. The entire stage is mounted on a large granite base, providing vibration isolation.

The positioning stage movement is G-code programmable and the position synchronized output allows external triggering of the laser beam in coordination with the stage movement. This allows laser micromachining of complex shapes and patterns.

22.7.3.2 Texture Effect on Friction in Lubrication

Many researchers have reported the benefit of surface texture on friction reduction. However, the level of the reduction defers from case to case due to the complication of details of surfaces, materials, contact type and size, as well as operation conditions.

Case 1: Friction in a boundary lubrication condition, Davis et al. [90]. The flat-on-flat friction testing experimental setup consists of a thin strip of sheet metal that is pulled across the flat bottom end of a steel cylinder with a support shaft underneath. A thin sheet of polytetrafluoroethylene was used between the strip and the support shaft to reduce the friction force between these components. A normal force was applied to the cylinder through an adjustable counterweight, pressing the cylinder bottom surface and strip together. A small amount of mill oil lubricant was applied to each strip before testing. The sliding speed of the strip was actuated by a computer-controlled rack with adjustable speed. A digital force gauge was used to monitor the force required to pull the strip at a constant rate. Figure 22.15 shows a picture of the actual setup. A variable normal force was applied between the bottom of a steel cylinder and a thin strip of sheet steel. The strip was pulled at a constant rate, and the pulling force was recorded. The support shaft is free to rotate.

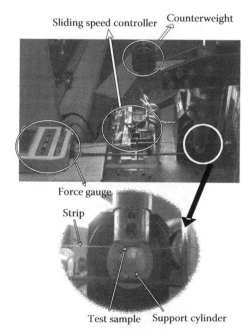

FIGURE 22.15 Flat-on-flat friction testing setup. Normal force and sliding speed are controlled via adjustable counterweight and computer controller. A digital force gauge is used to monitor the friction force. The test sample is loaded into the upper fixture. (From Davis, T. et al., Experimental friction study of micro-scale laser-textured surfaces, in *International Workshop on Microfactories*, 2008.)

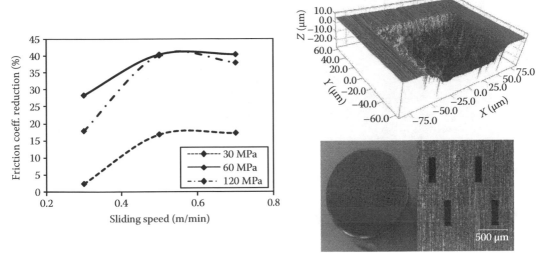

FIGURE 22.16 Laser textured surface and results of the friction test. (From Davis, T. et al., Experimental friction study of micro-scale laser-textured surfaces, in *International Workshop on Microfactories*, 2008.)

The laser system shown in Figure 22.14 was used to create rectangular dimples measuring $400\,\mu m$ in length, $100\,\mu m$ in width, and $25\,\mu m$ in depth, on a flat surface of a cylindrical steel sample. A 2 mm wide strip of DP600 sheet steel was pressed onto the textured sample with a known contact pressure. The strip was then pulled across the surface of the sample with a constant sliding speed. The force required to pull the strip was used to calculate the coefficient of friction between the strip and the sample. In comparison with a baseline non-textured sample, it was found that the textured sample resulted in a friction reduction across all experimental conditions studied. The largest reductions of up to 40% were observed for conditions with larger contact pressure and faster sliding speed (Figure 22.16).

Case 2: Lubrication and friction transitions, Kovalchenko et al. [91]. The work reported by Kovalchenko et al. [9] is a group of experimental studies on the friction behavior of circular dimpled surfaces of a few depths, densities, and diameters (Table 22.2). The tests were conducted with a pin-on-disk apparatus with flattened balls. Figures 3 and 4 of their paper reported the friction coefficients measured at different disk rotation speeds for two lubricants, and the friction coefficients with the low-viscosity lubricant are reproduced in Figure 22.17. Disks 3, 4, and 8 seem to have given the lowest friction overall when the lubricant with the lower viscosity was used, while disk 9 always

TABLE 22.2
Disks Tested by Kovalchenko et al.

Disk	1	2	3	4	5	6	7	8	9
Depth of dimples (μm)	—	—	5.5	5.5	4.5	5	6.5	4.0	6.5
Surface roughness between dimples R_a (μm)	0.01	0.12	0.03	0.02	0.05	0.06	0.07	0.9	0.09
Diameter of dimples (μm)	—	—	78	76	80	58	120[a]	58	140[a]
Distance between dimples (μm)	—	—	200	200	200	80–100	200	200	200
Dimple density (%)	—	—	12	12	13	14	30[a]	7	40[a]

Source: Kovalchenko, A. et al., *Tribol. Trans.*, 47, 299, 2004. With permission.

[a] Measured optical data; area includes bulges around dimples (unlapped).

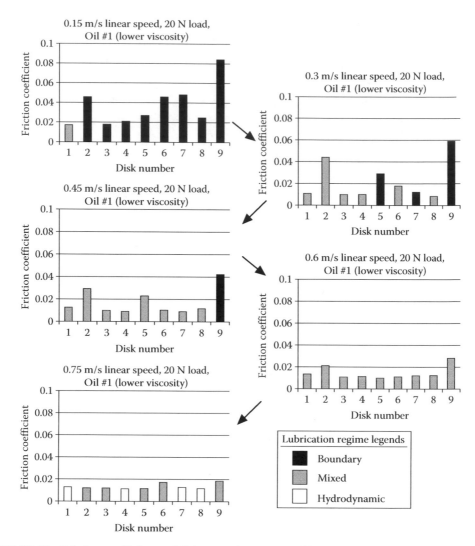

FIGURE 22.17 Friction coefficients of disks at various speeds, 20 N load, and a low-viscosity lubricant. (From Kovalchenko, A. et al., *Tribol. Trans.*, 47, 299, 2004. With permission.)

produced the highest friction. The dimple data show that disk 9 has the deepest dents (6.5 μm), largest size (140 μm) and the highest density (40%). Disk 8, on the other hand, has the smallest size (58 μm), most shallow dents (4 μm), and the lowest density (7%). Disks 3 and 4 are in between. The density and depth suggested the preference for shallower dimples and lower densities (3%–12%). It should be mentioned that the dimple size is not an independent factor, but it is rather related to density.

Figure 22.17 shows two important facts. (1) Friction is very sensitive to surface topography, including both textures and machined asperities. (2) Although the friction produced by using disk 9 is the highest, it is rapidly reduced as the speed is increased, which changes the lubrication status from the boundary to the mixed lubrication regime.

Case 3. Dimple depth estimation, Nanbu et al. [92]. The dimple depth, d, may be determined based on existing reports [2,92–95] that suggested the use of a reasonably shallow depth. It can also be analytically estimated based on the theory of hydrodynamic lubrication [96] and the projected operating film thickness. The maximum load capacity of a slider bearing can be expressed in the following form:

$$W_{\max} = k \frac{\eta V L B^2}{h_2^2} \tag{22.27}$$

where

h_2 the film thickness
V the sliding speed
L the width of the slider
B the length of the slider

Parameter k varies for different slider bearings determined by the optimized ratio of the inlet film thickness over the outlet film thickness, h_1/h_2. In this study, h_2 can also be the film thickness. The film ratio, h_1/h_2, may be expressed by $1 + d/Hc$ using Hc, the central film thickness, to approximate the film thickness in most of the EHL region. Thus, the depth of a texture may be approximated by $d \sim Hc(h_1/h_2 - 1)$. If Hc is about 1.6 μm, and the optimal values of the film ratio, h_1/h_2, for different sliders should be in the range of 1.87–2.2; thus, the texture depth should be in the range of 1.4–1.76 μm.

Another way to estimate the dimple depth is by using the inlet suction mechanism. Olver et al. [97] and Fowell et al. [98] recently reported this micro-hydrodynamic lubrication mechanism in the study of micro-lubrication and loading lifting in textured bearings. Based on the inlet suction mechanism, the volume flow for a parallel bearing with flat bottom dimples can be shown [98] as

$$Q = \frac{U h_0}{2} + \frac{U h_0^3}{2} \frac{h_1/h_0 - 1}{(h_1/h_0)^3 (L_1 + L_2)/L + 1} \tag{22.28}$$

where

U is the sliding speed
h_0 and h_1 are the film thickness in flat and pocket section, respectively
L_1 and L_2 are length of the leading and trailing flat section, respectively
L is the length of the pocket section

The optimal pocket depth can be found by maximizing the volume flow, Q:

$$\frac{dQ}{d(h_1/h_0)} = 0 \tag{22.29}$$

With $(L_1 + L_2)/L$ between 1 and 10, the ratio of (h_1/h_0) should be about 1.7 in order to have the largest volume flow rate for the case mentioned before. This result is consistent with that described in the earlier part of this section.

However, a practical depth cannot be too shallow because the fabrication convenience and wear must be considered.

22.8 CONCERNS ON SURFACE TEXTURE APPLICATION AND POSSIBLE FUTURE DEVELOPMENTS

22.8.1 CONDITION DEPENDENT TEXTURE PERFORMANCE AND THE NEED FOR MODEL-BASED SIMULATION

Compared with smooth surface contact, friction, and lubrication, where materials and operating conditions are the major concern, and compared with machined surfaces of random asperities, where the topographic effect is generally averaged, texturing adds several new dimensions of complexity for consideration: (1) texture shape, size, depth, density, and patterns of distribution [2,99], (2) materials of the surface layer that contains the texture, (3) interfacial contact type and orientation of the host surfaces relative to the texture orientation, (4) one surface textured or both surfaces textured, and (5) operating conditions. The effects of load and speed are especially important for lubricated cases as shown by the Stribeck curve mentioned before. Terms mentioned in (1) are independent design parameters; however, all together they determine the mechanism of surface micro-hydrodynamic lifting if lubricated, and surface traction distribution if non-lubricated. Most importantly, the first three decide the feature of a surface micro bearing at the micro-, or even the nanoscale. Weaker materials or damaged surface materials will result in a quick loss of the designed features due to wear or micro-fatigue. So far, single-surface texturing cases have been widely studied, but very few consider paired textured surfaces [100]. Many wait to be explored.

A model-based simulation system may be the tool for the surface and precise mixed-lubrication design. This tool can be used to conduct most of the investigations through numerical simulations. Only a limited amount of experimental work is required for verification. The concept of the virtual texturing technology in reference [2] is repeated here, which is illustrated in Figure 22.18. The following parts are included in virtual texturing: (1) Virtual surface texture generation based on application requirements; (2) Contact and lubrication analyses for performance evaluation; (3) Efficiency, life, and surface evolution prediction; (4) Surface texture modification and optimization; and (5) Necessary experimental verification. In this procedure, the results from virtual surface texture generation (1) and contact and lubrication analyses (2) can be used in texture modification and optimization (3). The kernel of virtual texturing is the model-based simulation tool that should effectively link the surface micro-geometry to material and lubricant properties on the one hand and to the lubrication performance of the interface on the other hand. Fluid dynamics and contact mechanics are parts of the effective means to model the interfaces of the textured surfaces. The model system should be deterministic and multi-scaled, at least micro-to-macro, because statistic models cannot catch the geometric factors while single-asperity models fail to grasp the effect of a texture matrix.

CFD modeling. Computational Fluid Dynamics (CFD) is applied to model the hydrodynamics of surfaces with fluid pockets [101,102]. Sahlin et al. [101] conducted a two-dimensional steady-state CFD analysis of micro-patterned surfaces in the lubrication of two parallel walls, where the stationary wall is patterned and the moving wall is smooth. The comparison of the results from the CFD

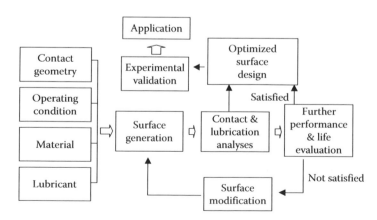

FIGURE 22.18 Logic of the virtual surface texturing technology. (From Wang, Q. and Zhu, D., *ASME J. Tribol.*, 127, 722, 2005. With permission.)

modeling [101,102] with that from the solution of the Stokes equation indicates that clear pressure differences appear when the Reynolds number is larger than 10. The results show that the introduction of a micro-pattern on one of the walls gains a pressure rise in the fluid domain and that such pressure rise increases with the Reynolds number (for the Reynolds number larger than 10). The load capacity increases with the width and depth of the pattern; however, at a certain depth, vertex appears in the micro-pattern.

Mixed-EHL modeling. Surface features should help form micro-bearings. Textures favoring lubrication may not be designed too deep for regular applications because significant pressure variations due to dents should be avoided. In addition, shallow features introduce less interruption to the surface and subsurface material. Most machined surfaces contain shallow topographic features. Dimpled surfaces, as an example of patterned surfaces, may be considered as micro-slider bearings of different shapes with revolved geometry. Typical dimples, for example, have large size-to-depth ratios. The edges of dimples may be compared with step bearings that are proven to be the desirable geometry of slider bearings [96]. According to the hydrodynamic lubrication theory and the engineering practice with step bearings, grooved bearings, and seals, turbulence caused by the edges of surface features may be insignificant, and, in many cases, may be negligible. Reynolds equation is still applicable to the lubrication of many textured surfaces designed for lubrication significance.

One of the mixed lubrication model is Hu and Zhu's mixed-elastohydrodynamic (EHL) model [103,104] that is capable of describing the mixed lubrication of rough surfaces. This model links the micro-geometry of surfaces with the macro-lubricated contact interface and integrates the micro-EHL effect into global load-supporting capacity. The film thickness is computed from the deformed average gap; and flows and asperity contacts are treated in one unified model.

Cavitation deserves attention. The effect of modeling accuracy may be affected whether a proper cavitation model is included or not [97,101,105] although contradictory results may be found in literature.

22.8.2 Wear and Texture Durability

Wear of the surface material alters the originally made surface and situates the surface in a continuously changing process. It is difficult to predict the exact topography of a worn surface due to complicated deformation and material removal, as well as wear debris spreading. Such complication challenges researchers for durable low-friction surface design and reliable prediction of surface evolution in a tribological environment.

22.8.3 Fatigue Life

Not many publications have mentioned the effect of textures on the fatigue life of textured surfaces in a tribological contact process. However, the fatigue life of a textured surface should be carefully evaluated, especially for the case of a counterformal contact where high contact pressure and cyclic loading are encountered. The textures should be made free of any stress raisers, such as cracks and sharp profile discontinuities. Compressive surface stresses should be preferred, and any means to enhance the fatigue life is welcome.

22.8.4 Cost

Texture adds additional fabrication cost. Even the most convenient vibro-turning requires additional processes after a regular surface finish. Therefore, it is important to pursue benefit gaining larger than the corresponding investment in manufacturing.

22.8.5 COMBINED TECHNOLOGIES

If a surface material enhancement technology, or a surface mechanical or chemical treatment process, can be combined with surface texturing, much greater friction reduction may be accomplished, and much stronger textured surfaces may be created. However, such a combination will further increase the cost, and the design and evaluation should be carefully done.

22.8.6 MULTIPURPOSE SURFACE TEXTURES

Surfaces may be created with several mechanical, chemical, and physical functions, such as low friction, enhanced lubrication, selective light transmission, water- or oil-phobicity, and strong corrosion resistance. Research on the feasibility and functionality evaluation should be conducted.

22.9 CONCLUSIONS

This chapter reviews the concepts and theories of surfaces, friction, wear, and lubrication, as well as some recent research results on surface technologies for friction reduction. Because motion and power transmissions in a machine system are accomplished through surface contact in relative motion, surfaces hold the key for efficient mechanical system design. Tribology and surface engineering are the scientific and technological bases for energy efficiency.

QUESTIONS

Q.22.1 What is a tribological interface? Think about contact and relative motion, can you think about a machine that does NOT have a tribological interface?

Q.22.2 Friction consumes energy. How does wear consume energy?

Q.22.3 How is friction variation reflected by the Stribeck curve? Is friction a material property?

Q.22.4 How does wear affect the performance of a surface? How is wear related to the property of the material of a surface in contact and rubbing?

Q.22.5 What are the essential concerns on friction and wear? How should a tribological interface be designed for low friction and wear?

ABOUT THE CONTRIBUTING AUTHORS

Q. Jane Wang received her PhD in mechanical engineering at Northwestern University, Evanston, Illinois, United States, in 1993. She is now a professor in the same department. She is a fellow of the American Society of Mechanical Engineers (ASME) and the Society of Tribologists and Lubrication Engineers (STLE), and Society of Automotive Engineers (SAE). She received a Ralph R. Teeter Educational Award from the Society of Automotive Engineers in 2000, a CAREER Award from US National Science Foundation in 1997, and the Captain Alfred E. Hunt Award for Best Papers from STLE in 1997. She is an associate editor of *Tribology Transactions*. Her research includes analyses of extreme-condition tribology problems, theories of and methods for contact and interfacial mechanics, numerical simulations of frictional heat transfer and mixed-thermo-elasto-hydrodynamic lubrication, thermal-tribological designs of machine elements and their surfaces, multidisciplinary modeling of asperity contact, and industrial applications of models and surface designs for friction reduction, lubrication enhancement, and failure prevention. She and her group are working on theories of contact, lubrication, and friction of engineering surfaces of various materials, model-based simulations of several lubrication problems, and surface designs.

Dong Zhu received his PhD in mechanical engineering from Tsinghua University in Beijing, China in 1984. He was a research fellow of Northwestern University, Evanston, Illinois, United States during 1986–1989. He was employed by the Aluminum Company of America in 1990 as a staff research engineer. Later he joined Eaton Corporation, Corporate R&D Center in Southfield, Michigan, in 1994, conducting engineering research and product development as a principal engineer and a program manager. He is now a guest professor at State Key Laboratory of Tribology, Tsinghua University, China, and is also providing consulting and software services as a consultant and an owner of Tri-Tech Solutions, Inc., Mount Prospect, Illinois. He is a fellow of the American Society of Mechanical Engineers (ASME) and the

Society of Tribologists and Lubrication Engineers (STLE). He received numerous awards, including the Edmond E. Bisson Award from the STLE in 2003 for the best technical contribution in 2001–2002, and the First Class Award of National Science and Technology Advancement, National Education Commission of China, 1989. He is an associate editor of *Tribology Transactions*. His research interests include contact and lubrication analyses for engineering applications, surface modification technologies such as PVD and thermal spray coatings, precise machining processes, and surface texturing. He has extensive knowledge and experience in engine and drivetrain tribology, design optimization, and failure analyses, as well as problem solving for heavy-duty mechanical components.

Professor Jian Cao has been with the Northwestern University, Evanston, Illinois, since she received her PhD degree in solid mechanics from M.I.T., Cambridge, Massachusetts in 1995. She received her M.S. degree from M.I.T. in 1992 and the dual B.S. degrees in materials engineering and automatic control from Shanghai Jiao Tong University, China in 1989. She is now a professor and the director of Graduate Studies at the Department of Mechanical Engineering. During her tenure at the Northwestern, she took a 1 year leave at General Motors in FY1996 and a 2 year leave at the National Science Foundation in FY2004 and 2005 as a program director. During that short 2 year period at NSF, under her leadership, the *World Technology Evaluation Study on Micromanufacturing* was conducted and co-sponsored by four government agencies and more than ten NSF programs.

Prof. Cao's primary interests are in the mechanics analysis and design of macro/micro metal forming and composite sheet forming processes. Her research contributions in the understanding of material instability in the deformation process have advanced the predictability of numerical simulations, and they were implemented in companies' engineering practices. Her research also includes intelligent process design algorithms considering process variations and utilizing real-time sensors, efficient simulation tools, and more recently on incremental forming process, micro-texturing engineering surfaces, and on the size effect in material deformation and friction behavior in microforming processes. She has more than 200 technical publications and she has given more than 70 invited talks.

Professor Cao has approached her research in manufacturing based on solid mechanics combined with intelligent control. The overarching link has earned her many honors and awards.

Prof. Cao is a recipient of *the General Electric Foundation Professorship, ALCOA Foundation Award, the National Science Foundation CAREER award, Society of Automotive Engineers (SAE) Ralph R. Teetor Educational Award, Society of Manufacturing Engineers (SME) Outstanding Young Manufacturing Engineer Award, the Young Investigator Award from the Japan-US Flexible Automation*, and most recently *the Young Investigator Award from the ASME Applied Mechanics Division*. She is named a Chang Jiang Chair Professor in 2009 by the State Council of the Chinese government to acknowledge her special contributions in the field of materials processing.

Prof. Cao is a fellow of the American Society of Mechanical Engineers and an associate member of the International Academy for Production Engineering (CIRP). She was an associate editor for the ASME Journal of Manufacturing Science and Engineering from 2003 to 2009. Prof. Cao is an associate editor of the ASME Journal of Applied Mechanics. She is an active member of ASME and has held more than 10 positions within ASME, including currently as the chair of Committee on Administration and Finance and the vice-chair of Manufacturing Technology Group, and previously as chair of the Executive Committee of ASME Manufacturing Engineering Division. In addition to her involvement with ASME, Prof. Cao is the secretary of the Executive Board of SME North America Manufacturing Research Institute, a member of the Scientific Board of European Scientific Association for Material Forming, and a co-director of the NSF Summer Institute on Nanomechanics, Nanomaterials, and Micro/Nanomanufacturing.

REFERENCES

1. Arnell, R. D., Davies, P. B., Halling, J., and Whomes, T. L., *Tribology, Principles and Design Applications*, Springer-Verlag, New York, 1991.
2. Wang, Q. and Zhu, D., Virtual texturing: Modeling the performance of lubricated contacts of engineered surfaces, *ASME Journal of Tribology*, 127, 722–728, 2005.
3. Bhushan, B, *Principles and Applications of Tribology*, Wiley Interscience, New York, 1999.
4. Buckley, D. H., Properties of surfaces, in *Handbook of Lubrication*, Vol. II, Editor, Booser, E. R., CRC Press, Boca Raton, FL, 2000.
5. Wang, Q. and Cheng, H. S., Chapter 2, Fundamental principles of lubrication, in *ASTM Handbook of Automotive Lubricants and Testing*, Editors, G. Totten and S. Tung, accepted.
6. Pettersson, U. and Jacobson, S., Influence of surface texture on boundary lubricated sliding contacts, *Tribology International* 36, 857–864, 2003.
7. Amontons, G., *De La Resistance Causee dans les Machines*, Memoires de l'Academic Royale, A, (*Paris*), 257–262, 1699.
8. Landman, U., Frictional forces and Amontons' law: From the molecular to the macroscopic scale, *The Journal of Physical Chemistry B*, 108, 3410–3425, 2004.
9. Dowson, D, *History of Tribology*, Professional Engineering Publishing, London, U.K., 1998.
10. Halling, J., *Principles of Tribology*, McMillian Press, London, U.K., 1978.
11. Rabinowicz, E., *Friction and Wear of Materials*, John Wiley & Sons, New York, 1995.
12. Krim, J., Surface science and the atomic-scale origins of friction: What once was old is new again, *Surface Science*, 500, 741–758, 2002.
13. Greenwood, J. A. and Williamson, J. B. P., Contact of nominally flat surfaces, *Proceedings Of the Royal Society, London A*, 295, 300, 1966.
14. Greenwood, J. A., *Fundamentals of Friction: Macroscopic and Microscopic Processes*, Editors, Singer, I. L. and Pollock, H. M., Kluwer, Dordrecht, the Netherlands, pp. 37–56, 1992.
15. Gao, J., Luedtke, W. D., Gourdon, D., Ruths, M., Israelachvili, J. N., and Uzi, L., Frictional forces and Amontons' law: From the molecular to the macroscopic scale, *Journal of Physical Chemistry B*, 108 (11), 3410–3425, 2004.
16. Muser, M., Urbakh, M., and Robbins, M., Statistical mechanics of STATIC and low-velocity kinetic friction, *Advances in Chemical Physics*, 126, 2003, edited by Prigogine, I. and Rice, S.A.
17. Persson, B. N. J., Rubber friction: Role of the flash temperature, *Journal of Physics-Condensed Matter*, 18 (32), 7789–7823, 2006.
18. Farhang, K. and Lim, A., A kinetic friction model for viscoelastic contact of nominally flat rough surfaces, *ASME Journal of Tribology*, 129 (3), 684–688, 2007.

19. Chang, L., and Zhang, H., A mathematical model for frictional elastic-plastic sphere-on-flat contacts at sliding incipient, *ASME Journal of Applied Mechanics*, 74, 100–106, 2007.
20. Chen, W. W. and Wang, Q., A numerical static friction model for spherical contacts of rough surfaces, influence of load, material, and roughness, *ASME Journal of Tribology*, 131, 021402-1–021402-8, 2009.
21. Etsion, I., Discussion: A mathematical model for frictional elastic-plastic sphere-on-flat contacts at sliding incipient (Chang, L., and Zhang, H., *ASME J. Appl. Mech.*, 74, 100–106), *Journal of Applied Mechanics-Transactions of the ASME*, 74 (5), 1057–1057, 2007.
22. Ludema, K., *Friction, Wear, and Lubrication. A Textbook in Tribology*, CRC Press, Boca Raton, FL, 1996.
23. Blau, P., *Friction Science and Technology*, Marcel Dekker, New York, 1996.
24. Weiss, M. and Elmer, F. J., Dry friction in the Frenkel–Kontorova–Tomlinson model: Static properties, *Physical Review B, Condensation Matter*. 1996 Mar 15;53(11):7539–7549, 1996.
25. Kim, W. K. and Falk, M. L., Atomic-scale simulations on the sliding of incommensurate surfaces: The breakdown of superlubricity, *Physical Review B*, vol. 80, Issue 23, id. 235428, 2009.
26. Gyalog, T. and Thomas, H., Friction between atomically flat surfaces, *Europhysics Letters*, 37, 195–200; doi:10.1209/epl/i1997-00132-1, 1997.
27. Meyer, E., Overney, R. M., Dransfeld, K., and Gyalog, T., *Nanoscience, Friction and Rheology on the Nanometer Scale*, World Scientific, River Edge, NJ, 2002.
28. Bowden, F. P. and Tabor, D., *The Friction and Lubrication of Solids*, Oxford University Press, Oxford, U.K., 1950.
29. Tabor, D., Friction—The present state of our understanding, *ASME Journal of Lubrication Technology*, 103, 169–179, 1981.
30. He, B., Chen, W., and Wang, Q., Surface texture effect on friction of a micro-textured poly(dimethylsiloxane) (PDMS), *Tribology Letters*, 31, 187–197, 2008.
31. Smith, J. and Zhang, S., High temperature nanoscale mechanical property measurements, *Surface Engineering*, 16 (2), 143–146, 2000.
32. Singh, R., Melkote, S. N., and Hashimoto, F., Frictional response of precision finished surfaces in pure sliding, *Wear*, 258 (10), 1500–1509, 2005.
33. Menezes, P. L., Kishore, and Kailas, S. V., Effect of roughness parameter and grinding angle on coefficient of friction when sliding of Al-Mg alloy over EN8 steel, *ASME Journal of Tribology*, 128 (4), 697–704, 2006.
34. Smith, J. F. and Zhang, S., High temperature nanoscale mechanical property measurements, *Surface Engineering*, 16 (2), 143–146, 2000.
35. Zhang, H. and Komvopoulos, K., Anisotropic behavior of nanotextured surfaces, *Proceedings of the 2008 STLE/ASME International Joint Tribology Conference*, Miami, FL, IJTC2008-71146, 2008.
36. Gangopadhyay, A. and McWat, D., The effect of novel surface textures on tappet shims on valvetrain friction and wear, *Tribology Transactions*, 51 (2), 221–230, 2008.
37. Li, Y. R., Shakhvorostov, D., Lennard, W. N., and Norton, P. R., A novel method for quantitative determination of ultra-law wear rates of materials, Part II: Effects of surface roughness and roughness orientation on wear, *Tribology Letters*, 33, 63–72, 2009.
38. Archard, J. F., Contact and rubbing of flat surfaces, *Journal of Applied Physics*, 24, 981–988, 1953.
39. Rowe, C. N., Some aspects of the heat of absorption in the function of bounary lubricant, *Transactions of American Society of Lubrication Engineers*, 9, 100–111, 1966.
40. Lim, S. C., Lee, S. H., Liu, Y. B., and Seah, K. H. W., Wear maps for some uncoated cutting tools, *TriboTest*, 3 (1), 67–88, 1996.
41. Hsu, S. M., Shen, M. C., and Ruff, A. W., Wear prediction for metals, *Tribology International*, 30, 377–383, 1997.
42. Harris, S. J., Weiner, A., and Meng, W., Tribology of metal-containing diamond-like carbon coatings, *Wear*, 211, 208–217, 1997.
43. Harris, S. J. and Weiner, A. M., Scaling relationships for the abrasion of steel by diamondlike carbon coatings, *Wear*, 223, 31–36, 1998.
44. Harris, S. J., Krauss, G., Siniawski, M., Wang, Q., Liu, S., and Ao, Y., Surface feature variations observed in 52100 steel sliding against a thin boron carbide coating, *Wear*, 249, 1004–1013, 2001.
45. Siniawski, M., Harris, S., and Krauss, G., Wang, Q., and Liu, S., Wear initiation in 52100 steel sliding against a thin boron carbide coating, *Tribology Letters*, 15, 29–41, 2003.
46. Siniawski, M., Harris, S., and Wang, Q., A universal wear law for abrasion, *Wear*, 262, 883–888, 2007.
47. Goryachev, I. G., *Contact Mechanics in Tribology*, Kluwer, Dordrecht, the Netherlands, pp. 191–201, 1998.

48. Meng, H. C. and Ludema, K. C., Wear models and predictive equations: Their form and content, *Wear*, 181–183, 443–45, 1995.
49. Zhu, D., Martini, A., Wang, W., Hu, Y., Lisowsky, B., and Wang, Q., Simulation of sliding wear in mixed lubrication, *ASME Journal of Tribology*, 129, 544–552, 2007.
50. Hamrock, B, *Fundamentals of Fluid Film Lubrication*, McGraw Hills, New York, 1994.
51. Bair, S., Liu, Y., and Wang, Q., The pressure-viscosity coefficient for Newtonian EHL film thickness with general piezoviscous response, *ASME Journal of Tribology*, 128, 624–631, 2006.
52. Liu, Y, Wang, Q., Krupka, I., Hartl, M., and Bair, S., The shear-thinning elastohydrodynamic film thickness of a two-component mixture, *Journal of Tribology*, 130, 021502-1–021502-7, 2008.
53. Dowson, D. and Higginson, G. R., *Elastohydro-Dynamic Lubrication*, Pergamon Press, Oxford, U.K., 1966.
54. Christensen, H., Stochastic models for hydrodynamic lubrication of rough surfaces, *Proceedings of the Institution of Mechanical Engineers*, 184 (Part 1, 55), 1013–1026, 1970.
55. Patir, N. and Cheng, H. S., An average flow model for determine effects of three dimensional roughness on partial hydrodynamic lubrication, *ASME Journal of Lubrication Technology*, 100, 12–17, 1978.
56. Harp, S. and Salant, R., An average flow model of rough surface lubrication with inter-asperity cavitation, *ASME Journal of Tribology*, 123, 134–143, 2001.
57. Sahlin, F., Lubrication, Contact Mechanics and Leakage between Rough Surfaces, PhD Thesis, Luleå University of Technology, Luleå, Sweden, 2008.
58. Hersey, M. D., The laws of lubrication of horizontal journal bearings, *Journal of Washington Academic Science*, 4, 542–552, 1914.
59. Wang, Y., Lin, C., Wang, Q., and Shi, F., Development of a set of stribeck curves for conformal contacts of rough surfaces, *Tribology Transactions*, 49, 526–535, Reprinted in *Tribology and Lubrication Technology*, 63 (4), 32–40, 2006 (as Editor's Choice).
60. Martini, A., Zhu, D., and Wang, Q., Friction reduction in mixed lubrication, *Tribology Letters*, 28, 171–181, 2007.
61. Bay, N., New, Environmentally friendly tribo-systems for metal forming, Keynote, CIRP General Assembly, *Annuals of the CIRP*, 59 (2), 2010.
62. Dunston, S., Ranjithan, S., and Bernold, E., Neural network model for the automated control of springback in rebars, *IEEE Expert/Intelligent Systems & Their Applications*, 11 (4), 45–49, 1996.
63. Inamdar, M., Date, P. P., Narasimhan, K., Maiti, S. K., and Singh, U. P., Development of an artificial neural network to predict springback in air vee bending, *International Journal of Advanced Manufacturing Technology*, 16 (5), 376–381, 2000.
64. Kinsey, B., Cao, J., and Solla, S., Consistent and minimal springback using a stepped binder force trajectory and neural network control, *ASME Journal of Engineering Materials and Technology*, 122, 113–118, 2000.
65. Viswanathan, V., Kinsey, B. L., and Cao, J., Experimental implementation of neural network springback control for sheet metal forming, *ASME Journal of Engineering Materials and Technology*, 125, April 2003.
66. Donnet, C., le Mogne, T., and Martin, J. M., Superlow friction of oxygen-free MoS_2 coatings in ultrahigh vacuum. *Surface and Coatings Technology*, 62, 406–411, 1993.
67. Donnet, C., and Erdemir, A., Diamond-like carbon films, in *Tribology of Diamond-like Carbon Films, Fundamentals and Applications*, Editors, Donnet, C. and Erdemir, A., Springer, New York, pp. 1–12, 2007.
68. Freyman, C., Chen, Y., and Chung, Y. W., Synthesis of carbon films with ultralow friction in dry and humid air, *Surface and Coatings Technology*, 201, 164–167, 2006.
69. Erdemir, A., Eryilmaz, O. L., and Fenske, G., Synthesis of diamond-like carbon films with superlow friction and wear properties, *Journal of Vacuum Science and Technology A*, 18, 1987, 2000.
70. Erdemir, A., Eryilmaz O. L., Nilufer, I. B., and Fenske, G., Synthesis of superlow-friction carbon films from highly hydrogenated methane plasmas, *Surface and Coatings Technology*, 133, 448, 2000.
71. Heimberg, J. A., Wahl, K. J., Singer, I. L., and Erdemir, A., Superlow friction behavior of diamond-like carbon coatings: time and speed effects, *Applied Physics Letter*, 78, 2449, 2001.
72. Erdemir, A. and Martin, J.-M., New materials and coatings for superlubricity and near-wearless sliding, *Proceedings of the ASME/STLE International Joint Tribology Conference, IJTC 2007*, San Diego, CA, pp. 1069–1071, 2008.
73. Budinski, K., *Surface Engineering for Wear Resistance*, Prentice Hall, Englewood Cliffs, NJ, 1998.
74. Kwon, Y., Shigematsu, I., and Saito, N., Production of ultra-fine grained aluminum alloy using friction stir process, *Materials Transactions*, 44, 1343–1350, 2003.

75. Ding, K. and Ye, L., *Laser Shock Peening: Performance and Process Simulation*, Woodhead Publishing Ltd., Cambridge, U.K., 2006.
76. Aldajah, S. H., Ajayi, O. O., Fenske, G. R., and Xu, Z., Effect of laser surface modifications on tribological performance of 1080 carbon steel, *ASME Journal of Tribology*, 596–604, 2005.
77. Hector, L. G. and Sheu, S., Forced energy beam work roll surface texturing science and technology, *Journal of Materials Processing and Manufacturing Technology*, 2, 63–117, 1993.
78. Etsion, I., Kligerman, Y., and Halperin, G., Analytical and experimental investigation of laser-textured mechanical seal faces, *Tribology Transaction*, 42, 511–516, 1999.
79. Etsion, I., Improving tribological performance of mechanical seals by laser surface texturing, in *Proceedings of the 17th International Pump Users Symposium*, Houston, TX, pp. 17–22, 2000.
80. Wang, X., Kato, K., Adachi, K., and Aizawa, K., The effect of laser texturing of SiC surface on the critical load for the transition of water lubrication mode from hydrodynamic to mixed, *Tribology International*, 34, 703–711, 2001.
81. Kligerman, Y. and Etsion, I., Analysis of the hydrodynamic effects in a surface textured circumferential gas seal, *Tribology Transaction*, 44, 472–478, 2001.
82. Guichelaar, K., Folkert, I., Etsion, and Pride, S., Effect of micro-surface texturing on breakaway torque and blister formation on carbon-graphite faces in a mechanical seal, *Lubrication Engineering*, 58, 18–23, 2002.
83. Wang, Y. and Tung, S. C., Scuffing and wear behavior of aluminum piston skirt coatings against aluminum cylinder bore, *Wear*, 225–229, 1100–1108, 1999.
84. Ye, Z., Zhang, C., Wang, Q., Cheng, H. S., Wang, Y., Tung, S., and He, X., An experimental investigation on piston skirt scuffing: piston scuffing apparatus, scuffing experiments, and scuffing mechanism exploration, *Wear*, 257, 8–31, 2004.
85. Wang, C. J. H., Min, S., Ehmann, Kornel, F., Surface topography control in single-point cutting, in *S.M. Wu Symposium*, Beijing, China, 1994.
86. Altan, T., Lilly, B., and Yen, Y., Manufacturing of dies and molds, *CIRP Annals - Manufacturing Technology*, 50 (2), 405–423, 2001.
87. Jackman, R., Wilbur, J., and Whitesize, G. M., Fabrication of submicron features on curved substrates by microcontact printing, *Science*, 269, 664–666, 1995.
88. Madou, M. J., *Fundamentals of Microfabrication*, CRC Press, Boca Raton, FL, 1997.
89. Greco, A., Raphaelson, S., Ehmann, K., and Wang, Q., Surface texturing of tribological interfaces using the VMT method, *Journal of Manufacturing Science and Engineering*, 131 (6), 061005, 2009.
90. Davis, T., Zhou, R., Pallav, K., Beltran, M., Ehmann, K., Cao, J., and Wang, Q., Experimental friction study of micro-scale laser-textured surfaces, in *International Workshop on Microfactories*, Evanston, IL, 2008.
91. Kovalchenko, A., Ajayi, O., Erdemir, A., Fenske, G., and Etsion, I., The effect of laser texturing of steel surfaces and speed-load parameters on the transition of lubrication regime from boundary to hydrodynamic, *Tribology Transaction*, 47, 299–307, 2004.
92 Nanbu, T., Ren, N., Yasuda, Y., Zhu, D., and Wang, Q., Micro textures in concentrated conformal-contact lubrication: Effects of texture bottom shape and surface relative motion, *Tribology Letters*, 29, 241–252, 2008.
93. Wang, X., Kato, K., and Adachi, K., The lubrication effect of micro-pits on parallel sliding faces of sic in water, *Lubrication Engineering*, 58 (8), 27–34, 2002.
94. Mourier, L., Mazuyer, D., Lubrecht, A. A., and Donnet, C., Transient increase of film thickness in micro-textured EHL contacts, *Tribology International*, 39, 1745–1756, 2006.
95. Křupka, I. and Hartl, M., The effect of surface texturing on thin EHL lubrication films, *Tribology International*, 40, 1100–1110, 2007.
96. Pinkus, O. and Sternlicht, B., *Theory of Hydrodynamic Lubrication*, McGraw-Hill, Inc., New York, 1961.
97. Olver, A. V., Fowell, M., T., Spikes, H. A., and Pegg, I. G., Inlet suction, a load support mechanism in non-convergent, pocketed, hydrodynamic bearings, *Proceedings of IMechE Part J: Journal of Engineering Tribology*, 220 (2), 105–108, 2006.
98. Fowell, M., Olver, A. V., Gosman, A. D., Spikes, H. A., and Pegg, I. G., Entrainment and inlet suction: Two mechanisms of hydrodynamic lubrication in textured bearings, *ASME Journal of Tribology*, 129 (2), 336–347, 2007.
99. Etsion, I., State of the art in laser surface texturing, *ASME Journal of Tribology*, 127 (1), 248–253, 2005.
100. Wang, Q., Zhu, D., Zhou, R., and Hashimoto, F., Investigating the effect of surface finish and texture on mixed EHL of rolling and rolling-sliding contacts, *ASME Tribology Transactions*, 51, 748–761, 2008.

101. Sahlin, F., Glavatskin, S., Almqvist, T., and Larson, R., Two-dimensional CFD-analysis of micro-patterned surfaces in hydrodynamic lubrication, *ASME Journal of Tribology*, 127, 96–102, 2005.
102. Brajdic-Mitidieri, P., Gosman, A. D., Ioannides, E., and Spikes, H. A., CFD analysis of a low friction pocketed pad bearing, *ASME Journal of Tribology*, 127, 803–812, 2005.
103. Hu, Y. and Zhu, D., A full numerical solution to the mixed lubrication in point contacts, *ASME Journal of Tribology*, 122, 1–10, 2000.
104. Zhu, D. and Hu, Y. Z., A computer program package for the prediction of EHL and mixed lubrication characteristics, friction, subsurface stresses and flash temperatures based on measured 3-D surface roughness, *Tribology Transactions*, 44, 383–390, 2001.
105. Meng, F. M., Zhang, Y. Y., Hu, Y. Z., and Wang, H., On elastic deformation, inter-asperity cavitation and lubricant thermal effects on flow factors, *Tribology International*, 40 (7), 1089–1099, 2007.

Part V

Toward Intelligent EFM

23 Metrology and Quality Control

Kevin Harding

CONTENTS

23.1 INTRODUCTION

Modern tools of manufacturing add new flexibility to how parts can be made. Multiple axes of motion, multi-pass operations, fine control in some areas, and fast sweeps in others are all means to improve the speed, quality, and flexibility of manufacturing. A key set of tools that is needed to work within this new multidimensional environment is metrology, and this metrology tool set must be up to the task of providing the type of information needed to control the energy field based manufacturing systems.

In old times of metal cutting manufacturing, metrology was often left as a last step in the manufacturing process. The part was designed based upon two-dimensional views and a fixed set of primitive features such as holes, flat surfaces, or edges. As each feature was made, there might be a go, no-go check such as using a plug gage to verify if a drilled hole was of the right diameter, but few other in-process measurements were done. When the part was complete, a limited set of key parameters might be checked using micrometers or surface plate tools such as mechanical gages (see Figure 23.1). But ultimately, the check of the correctness of the part was purely functional. Did the part fit where it was supposed to fit, and if not, could we do minor adjustments (without measurement) to make it fit?

For many years, many automotive parts would be sorted into large, medium, and small bins. As a system like an engine was assembled, parts would be tried out. If a cylinder was a little large for the bored hole that was made, one would try the smaller size. This fitting process was commonplace, and accommodated the many manual operations and variability such as tool wear that would lead to small part variations.

This type of metrology began to change with the introduction of more automated processes such as CNC (computer numerical controlled) machining and robotic assembly. With CNC machining, it was possible to make parts in a much more repeatable manner. To accomplish this repeatability, touch probes, which determine a part location by touching the part, were added to many CNC machines to check for such things as part setup position and tool offsets due to either the mounting of the cutting tool or wear on the tool.

The touch probe works by using the actual machine tool's electronic scales to position cutting tools (see Figure 23.2). The probe is loaded into the spindle or tool holder of the machine tool just like any cutting tool. However, in this case the machine slowly moves the probe toward the part surface until the probe just touches the surface. The probe acts like a switch. As soon as the probe tip is displaced slightly by the touch on the part, it sends a signal telling the machine tool to stop. The machine tool then reads out the position of the touch probe using the built in position scales used by the machine to do automated machining.

FIGURE 23.1   Typical mechanical gages used for manual measurements.

FIGURE 23.2 A touch probe (left) acts as a trigger to tell the machine when to read out the position on the motion scales. Optical probes (right) can also be put on motion stages to provide a scanning capability.

This type of touch probe check allows the CNC machine to verify the position of a feature on the part, and to use any changes from the ideal location to correct or offset the path the actual cutting tool will need to take to do the desired machining. This process can be slow. On a high-value part such as a critical part in an aircraft engine, where a small mistake may mean that the part cannot be used, costing the manufacturer thousands of dollars, it is not unusual for the CNC machine to spend ten to twenty percent of the machining time checking features or positions with touch probes.

With energy field manufacturing systems, the same issues of part positioning and part processing pathways are still important, and in some cases may be critical to the manufacturing process, but there may be no inherent way of determining whether the process is even taking place. On a CNC drill, manufacturers have learned to use power monitoring to verify that the drill is actually cutting something, and it may even look for a characteristic signature of how the power to a motor should change during a processing operation. In an energy field manufacturing process such as a laser material processing, this type of monitoring based upon force or vibration feedback, resulting from the physical interaction of the tool and the part may not be possible as there may not be any such physical interaction. Different interactions, not involving contact, may be needed to monitor the process.

As with contact based manufacturing, drilling a hole in a wrong place can still be a problem with energy field manufacturing. In fact, the extra flexibility and speed offered by such systems makes monitoring the process much more important. To really gain the full value of energy field manufacturing, the metrology tools need to have flexibility, speed, and often noncontact capability. With the right information, the flexibility of energy field manufacturing can offer many advantages to correct small problems with a part during processing, providing a high-quality part every time. In many cases, even issues of tool wear become irrelevant with energy field manufacturing. Making sure the process is done right makes possible the opportunity of highly repeatable manufacturing results.

Fortunately, there is a wide range of metrology tools capable of measuring points, lines, or surfaces at speeds thousands of times faster than that of a touch probe or manual operation that can be integrated into these new energy field manufacturing systems. The rest of this chapter will review these metrology tools, and their pros and cons. Finally, we will look at how new capabilities being developed today may provide even more options for energy field manufacturing that may provide the means to completely rethink manufacturing methods and strategies.

23.1.1 Sensor Technology Justification

The advent of automated manufacturing processes has placed new demands on the controls to those processes. In the past, the human machine operator was expected to monitor the manufacturing process, and insure that the finished product was of high quality. High-quality products have long been associated with the skilled craftsman. Now, after a period of growth in automation that often compromised quality for volume, there is a new emphasis in industry on the production of "quality" product. To be competitive in today's marketplace requires not only that you make your product cheaper, but you must also make it "better" than ever before. The drive toward quality has forced a rethinking of the role of sensors in manufacturing, and how the results of those sensors are used.[1-6] The days of the skilled craftsman with the caliper in his back pocket are giving way to untended machines that must perform all of the tasks formerly done by the craftsman that were taken for granted.

Machines may be getting smarter, but they are still a long way from the sophistication of the skilled craftsman. When a person looks out the window and sees a tree, they recognize it as being a tree no matter whether it is a pine or an apple tree, in full bloom or dead. That person has used a variety of sensors and knowledge to recognize the tree. He may have used stereo perception to estimate its size and distinguish it from a painting, he may have heard leaves rustling in the breeze, or caught a whiff of apple blossom. The actual interpretation of this data about the tree has drawn upon many years of experience of seeing other trees, smelling flowers, or listening to noises in the woods. What actually distinguishes the sound that leaves make in the breeze from that of a babbling brook or a slow moving freight train? These may seem obvious questions to you or me, but a computer has no such experience base to draw upon. The sensory data received by a machine must be of a very succinct nature. The data must be unambiguous in what it means and there needs to be a clear understanding of what the machine must do with that information.

23.2   UNDERSTANDING THE PROBLEM

For a sensor to be effective as a tool for controlling quality, the implementation of the sensor must be right. At first glance, we may say we want to measure the wear of the cutting tool, but is that really what we are interested in measuring, or is it the part surface finish or shape we want to measure? A dimensional measure of a diameter may seem an obvious application for a micrometer, but what of the environment and materials handling in the system? Should the micrometer become broken or bent, we will receive incorrect data. The error may be obvious to the operator, but will not be obvious to a deaf, dumb, and blind machine. The right technology must be matched to the task. There are many ways to make a measurement, but only one of them will likely be the best, and even then may not be optimum. Beyond the technology, the implementation of sensors requires

- An organizational *strategy*, incorporating such points as management acceptance and cost justification
- *Training* of and understanding by the operators who must maintain the equipment
- *Interfacing* to the environment of the physical plant, users, equipment, etc.
- Finally some *means of using* the information provided by the sensor

A sensor without a useful "receptacle" for the sensed data is like a leaky faucet, at best an annoyance and at worst a waste of money.

The purpose of sensing and metrology is to measure some parameters that will help the manufacturing process, either by keeping the machines at their peak through machine monitoring, or by verifying the quality of the finished product at each step of operation to minimize the cost of a mistake. It has been said that "any good inspection system should be self-obsoleting." Throwing away the bad parts is at best a stopgap measure in most cases. To ensure quality, we would like to improve the process so that it does not make bad parts in the first place! Once we no longer make bad parts, we should no longer need to sort the parts.

23.2.1 BASIC TERMS FOR SENSOR TECHNOLOGY

The first step in applying sensors is to understand the language. There are many good references that describe these terms in more detail, so only a general review will be given here.[1,2,7]

23.2.1.1 Repeatability

Of primary interest in an automated process is the issue of repeatability. A sensor can have a high precision, that is, put out very small numbers or many decimal places, but if the same physical quantity gives rise to a different number each time, the output cannot be used to control the process or insure quality. Repeatability is effectively *a measure of how reliable the results are over the long haul. To repeat a number does not ensure that it is correct* in the eyes of the technical community at large, *but at least the number is consistent.*

Example 1—Photoelectric proximity

A typical photoelectric proximity sensor has a repeatability of 0.001 in. This means that if a particular part is brought in proximity of the sensor in a consistent manner over and over, the sensor will produce a particular signal, typically a simple switch closure. The switch will close at the same part position each time. However, if the part is brought to the sensor from a different direction, the sensor's switch will likely close at an entirely different part location. The sensor was repeatable, but it does not alone tell whether the part is in the correct place.

Example 2—Electrical scales

Electrical scales are used on many systems to measure linear distances. Such a scale may have a repeatability of 1 μin., but an accuracy of 50–80 μin. (2 μm). At a particular location on the scale, the sensor will produce the same reading very consistently, but the relation between that point and some other point on the scale is only correct, by conventional standards, to the 0.00008 in. In this case, the repeatability alone is not sufficient to provide the information we want.

23.2.1.2 Resolution

A more often quoted number as related to measurement is resolution. In the terms of metrology, the resolution is the ability of the system to distinguish two closely spaced measurement points. In simple terms, *resolution is the smallest change you can reliably measure.* What prevents this measure from being reliable is typically noise. If the signal associated with a small change in the measurement is overshadowed by noise, then sometimes we will measure the change, sometimes we will not measure the change, and sometimes we will measure the noise as being a change in the measurement and therefore it will not be repeatable.

Example 1—Photoelectric proximity

A rating of resolution for a photoelectric proximity sensor might be 0.01 in., but still have a repeatability of 0.001 in. In the case of a proximity sensor, the resolution indicates to how small of a change of part position the "switch closure" of the sensor can be adjusted. This does not mean the sensor will actually measure the change, just as the human eye cannot measure stars in the sky, but the sensor will detect it.

Example 2—Electrical scales

The resolution of an electrical scale is generally set by the counting mechanism used to read the scale. In this case, one count may be on the order of 20 millionths of an inch (20 μin.), but four counts would be needed to make a reliable reading. Therefore, the resolution is necessarily better than the actual usable measurement obtained from the sensor.

23.2.1.3 Accuracy

The issue of accuracy is an even more difficult one to address. To metrology, accuracy requires that the number be traceable to some primary standard, accepted by the industry and justifiable by the laws of physics. Accuracy is the means to ensure that two different sensors provide numbers which relate to each other in a "known" manner. When the supplier makes a part to some dimension and tolerance, the OEM builder wants to be able to measure that part and get the same results; otherwise, the part may not fit mating parts made by other suppliers.

Example—Electrical scale

The accuracy of the scale was given before as around 0.00008 in. If we have two scales with this accuracy and we measure a common displacement, they should both provide the same reading. In fact, if we compare the reading of the scale for any displacement it can measure, we should be able to compare that number against any other sensor of the same or better accuracy, such as a laser interferometer, and get the same reading. Accuracy provides the only common ground for comparing the measurements across many sensors and from company to company. In comparison, the optical proximity, for this reason, is not accurate at all, but rather just self-consistent.

The need for common numbers is the reason for industry wide "standards" of measurement. When the woodworker is making that cabinet you ordered for your dining room, he can make the door fit just right and not need to know exactly what size the door is to make it fit. The woodworker is using the same measurements for the door and the opening, even if it is just a piece of cut wood, so it does not matter if his measurements do not match anyone else's. He needs resolution, but not accuracy. He is effectively inspecting the part to fit, not to tolerance. When a similar situation occurs between a supplier of car doors and the auto manufacturer such that the doors are made to one measurement and the door openings to another, it requires the time and expense of a worker with a "big hammer" to bring the two measures into agreement. A common standard of measurement is not being used in this example, so the measurements are not accurately related.

Obviously, if a number is not repeatable or resolvable, it cannot really be proven to be accurate. A popular rule of thumb is to use the "rule of 10," or what I call the "wooden ruler rule." That is, if you need to know that a dimension is good to a certain number, you need to measure it to 10 times better than the resolution of that number to ensure that it is accurate. I call this the wooden ruler rule because the number 10 seems to relate more to the resolution of divisions on a wooden ruler, or the number of fingers of the metrologist than any statistical significance.

An apt statistical rule of measurement is the 40% rule (ala Nyquist sampling), which says that you must sample the number to within 40% to know which way to round it for the final answer. The 40% rule seems to inherently imply a rule of 10 in any case, but it can slow down the "runaway specification." When these rules of 10 start getting piled on top of each other, a measure can easily become overspecified to the extent that you may be measuring to a factor of a hundred times more accuracy than the process can manufacture to in any case, leading to the leaky faucet of information.

As an example of how rules of 10 pile up, consider a part that must be correct to 0.01 in. A tolerance of 0.001 in. would be placed on the part's dimensions to ensure 0.01 accuracy. To ensure meeting the 0.001 in., that part is measured to an accuracy of 0.0001 in. In order to assure the 0.0001 in. accuracy, the sensor is required to have a resolution of 0.00001 in. or 10 millionths of an inch, to measure a part whose dimension is important to 0.01 in., a factor of a thousand times coarser. The actual measurement tolerance, and not the rule of thumb, is what we must keep in mind when specifying the sensor needed.

An interesting question arises when a surface dimension is specified to be measured to an accuracy that is much finer than the surface finish of the surface. Since we want the number to be repeatable by anyone, perhaps we must ask whether we measure the top of the surface finish "hills," the bottom of the surface finish "valleys," or perhaps "which" hill or valley we should measure. It is for this reason that a location relative to a common datum (e.g., 1 in. from the leading edge) should be specified for the measurement for the tolerance to be meaningful.

23.2.1.4 Dynamic Range

The *dynamic range relates to the "range of measurements" that can be made by the sensor.* There is often also a standoff (the physical dimension from the sensor to the part), which is not part of dynamic range, and a working range which is the high and the low value of the measurements. *The working range of a sensor divided by the resolution, the smallest change that can be measured, gives an indication of the dynamic range.* If the dynamic range is 4000 to 1, this implies there are 4000 resolvable elements that can be distinguished by the sensor. If this is now read out as 8 bits of information, which only describes 256 numbers, the significance of the 4000 elements is moot unless only a limited part of the entire range is used at a time.

Example 1—Optical proximity

The working range of a typical proximity sensor might be 3 in. This means the sensor will detect a part as far away as 3 in., as well as closer. However, once set to a particular detection level, the proximity sensor tells nothing about where the part is within that range. Proximity sensors are inherently on–off devices and as such do not have a range of measurements or dynamic range to speak of, but rather only a static standoff range.

Example 2—Electrical scale

Scales typically produce a measurement along their entire distance of use. So, if we have a measurement resolution of 0.00004 in., and a working length of 4 in., the scale would have a dynamic range of 100,000 to one. Since with a scale we are concerned with actual accuracy, it may be more relevant to consider the dynamic range with respect to the usable number produced. So, if we have an accuracy of 0.00008 in., and a range of 4 in., the dynamic range is just 50,000 to 1 (about 16 bits of information). The dynamic range of measurement sensors is typically very important in considering how good of a measure you can obtain over some range. With many modern sensors, this range is in fact limited to the digital data that can be used, so a 16 bit sensor can only describe 64,000 measurements over whatever range you chose to measure. Beyond the basic range, sensors such as scales are often cascaded together to obtain a larger dynamic range.

23.2.1.5 Speed of Measurement vs. Bandwidth

A similar question of dynamic range arises with respect to speed of measurement. When we speak of the *speed of measurement* or the rate at which we can make a measurement, we are referring to the *rate at which actual data points can be completely obtained* to the extent that they are usable as a measurement of the part. The *bandwidth of the sensor* is not necessarily the speed at which data can be obtained, but *relates to the electrical or other operating frequencies of the detector.*

For example, a 2000 element array of detectors can have a bandwidth of 5 MHz, but to get the measurement requires that the 2000 elements be read out, each in 200 ns, sequentially, and before that happens the detectors may require some integration time to obtain the energy or force they are sensing. The result would be a detector that can be read out every 1 or 2 ms, with a bandwidth of 5 MHz. The bandwidth will generally limit the signal to noise ratio that can be expected. For optical detectors, the response is actually specified for a specific bandwidth, and changes as the square root of the bandwidth. If you want to know the speed of the data output, ask for the data output rate and not the bandwidth. Speed can be overspecified by looking at the wrong number. Many controllers can only respond on multi-millisecond or multi-second time frames, so using a detector that tells you of impending disaster a millisecond before it happens, becomes hindsight in reality.

23.3   PROCESS CONTROL SENSORS—BACKGROUND

Modern production lines are making and moving parts at speeds much faster than any other time in history. The standards of Six Sigma quality have demanded much better control than ever over even small, cosmetic defects. Industries such as primary metals, automotive, textiles, and even plastic extruders have found that having about the right dimensions and being "functional" is just not enough. Manufacturers are finding that any appearance of defects, be they pits, scratches, or a bad overall appearance can mean rejections of full lots of product, costing millions of dollars to a company, and affecting their bottom line. At modern speeds of production and tight defect tolerances, human inspectors have trouble keeping up with production. Studies have shown that even after 2 h of such work, the human inspector becomes distracted. The same mind that provides for high defect discrimination can "fill in" missing pieces, even when they are not present. After seeing 1000 parts with a hole in the center, part 1001 will appear as though it has a hole, whether it does or not.

Computers and the Internet have provided the tools to deal with large amounts of information very quickly. The same limitation that requires that a task be completely spelled out for a computer ensures that it will find that missing hole in part 1001 as consistently as in part 1,000,001. In addition, the simple act of reporting a variation, inherent to the philosophy of statistical process control, becomes a quick transfer of data over Internet lines, in the digital form needed for SPC software. So, computer based inspection and monitoring not only affords the programmable flexibility demanded by flexible energy field manufacturing, but also provides the quick data collecting and tracking abilities needed for high-speed repetitive operations.

Simple sensors such as touch probes have been used in traditional metal cutting machines for some years. There are many instances where sparse occasional data is all that is needed, and as such, touch probes are a reasonable tool to use. A touch probe, however, does not provide any measurement itself, but it is merely a switch that says, "I have touched something." The measurement actually comes from a machine axis, such as on a traditional milling machine. With the advent of energy field manufacturing, the machines often do not have the traditional tool holder, and may have a much different type of axis system than is needed to slowly approach and touch a part with a touch probe. So, although touch probes can still be a viable tool, the flexibility and speed of noncontact probes will generally be a better fit with the demands of energy field manufacturing methods.

Noncontact sensors made for large standoffs include optical systems such as machine vision, laser based probes, and 3-D mapping systems.[3,4,8] We will explore the details of these optical based systems within the context of fast, flexible manufacturing methods such as energy field manufacturing, then contrast some of the application challenges and errors with the contact based systems.

23.3.1   MACHINE VISION SENSORS OVERVIEW

Manufacturing has employed contact probes and gages in regular use since the turn of the twentieth century. Coordinate measurement machines (CMMs) have gone from slow, laboratory systems to automated factory floor systems. But even with those improvements, 100% inspection is rarely feasible with CMMs alone. Many fixed gages have now become computerized as well, providing a dedicated part gage, with computer output at the speeds needed. For loading these gages, robotic systems are able to load and unload parts in a highly repeatable manner, so good that they have revolutionized the electronics fabrication industry. But this option means a dedicated set of gages for each part, demanding rooms full of gages, and billions of dollars in expenses each year.

FIGURE 23.3 The electronics industry has made extensive use of machine vision for part inspection to allow automated processing and assembly. Verifying all the leads are in the right place on a chip and the chip number allows a robot to automatically place it on a PC board.

At the billion dollar costs of fixed electronic gages, the small batch run envisioned as the main tool of flexible manufacturing systems just is not economically feasible. Even with these computerized advances, the high speed and high tolerances of new parts have pushed past the limits of these more traditional sensors. The flexibility of machine vision to check hundreds of points on one part, then a different set of points on the next part, all in a matter of seconds has provided a capability not before available with traditional fixed gages.

The progression of machine vision as a tool in process control has not been an overnight occurrence.[9] Early applications of machine vision as a sorting tool and part ID aid were little more than hundred thousand dollar bar code scanners. High-speed, low-cost, and flexible changeover in the fast moving computer and semiconductor industries has acted as a catalyst to increase the speed of these machine vision systems. Early machine vision systems using simple processor chips progressed to dedicated ICs, gate arrays, digital signal processing (DSP) chips, and now integrated Internet devices. The dynamic nature of the electronics and semiconductor market segment has kept these areas as the largest current application of machine vision, still accounting for over 50% of sales in a multibillion dollar worldwide machine vision market today (see Figure 23.3). The competition for tighter quality control will push vision technology into even the most conservative metal cutting and forming operations, and when applied to energy field manufacturing may offer a natural marriage of fast, new manufacturing technologies.

23.3.2 MACHINE VISION OPERATION TECHNOLOGY

Machine vision can actually be any system where visual information is sensed and analyzed by a computer system for determining something about the manufacturing process. Such a system typically consists of a light source to highlight the features, some optics to image, a video camera to sense the scene, a digitizer to move the video into the computer format, and a computer system with analysis software (see Figure 23.4). By industry consensus, machine vision strictly relates only to the application of this technology to the manufacturing environment that has been the mainstay of vision technology for the past 25 plus years. However, it is very telling of the maturity of the technology that applications in other areas such as medical image analysis, transportation, and security are being seriously pursued today.

In operation, the lighting and optics system creates an image on the video camera of some region of interest (ROI) or field-of-view (FOV). The image is recorded by the video camera and digitizer to create an array of picture elements or "pixels," each of which denotes one point in the image. The image might then be smoothed by means of averaging or filtering, and then segmented to isolate key "features" of interest.

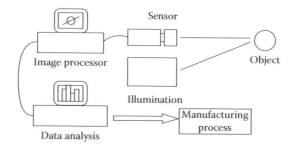

FIGURE 23.4   A typical machine vision system.

The value of each pixel may be a gray level of light, or may be binarized to be either dark or light. The analysis then typically uses simple tools that relate the pixels to each other by identifying edges, counting the number of dark or light pixels in an area, or looking for certain patterns in the arrangement and values of the pixels.

There have been many advances made in dedicated computer systems and user-friendly software over the years. Easy-to-program icon-based systems for computers have also helped to provide much easier to use software for machine vision. Some of the new vision algorithms have just been refinements of old ideas on new, faster computer platforms, where others are new ways of doing things. Camera systems have also improved over recent years, fueled by the use of home digital cameras. The combination of camera technology and computers has facilitated a whole new area of machine vision referred to as smart cameras. These smart cameras have onboard processors that can perform a powerful, if limited set of operations for such defined operations as part ID, location, and simple measurements, then communicate the results over an Internet connection. Let us consider where the technology of machine vision is today, and where that technology may apply to energy field manufacturing in the future.

23.3.3   Making Machine Vision Work

The area of machine vision has seen quite a few advances in the past 25 years.[9] The first step, and perhaps the least developed today, is getting a good image to analyze. The problem is not that the technology for producing high-quality imaging does not exist; it is more that the machine vision market has not yet reached the type of commodity volumes, such as home cameras, that make investing in making many products for the machine vision market profitable. However, recent years has seen a few notable exceptions.

The lighting and optics have long been widely recognized as an essential first step to a successful vision application. A clean image can make the difference between an easy-to-analyze scene, and an unreliable failure (see Figure 23.5). Parts are not designed like optical elements, with well-defined optical characteristics. They are designed as bearings, pumps, fasteners, and any number of other items, the appearance of which is at best secondary. However, the optical performance of the part has a major impact on the ultimate performance of the machine vision system.

FIGURE 23.5   The lighting can make features stand out, or not be seen at all.

23.3.4 MACHINE VISION LIGHTING

There are many lighting techniques that have been developed both by accident and by design that can be used in a vision system.[10] The objective of the lighting is to make the features of interest stand out in the image. Typically, this means the features of interest must be evident in a black and white image. Black and white cameras still provide the best resolution, cost, and flexibility for machine vision. Even for colored parts, separating colors with specific color filters typically provides better control than what can be realized with a color video camera.

Determining what is required in the image outlines the task and defines the limitation of performance that can be expected from the viewing system. A simple shape identification task may not need the high resolution needed to accurately gage a small feature to small tolerances. As the resolution performance of the lens system is approached, the lens degrades the contrast of the image until the small dimensional changes are washed out of the image. An initially low-contrast image produced by the lighting further degrades the limiting resolution of the viewing system.

There are many other considerations in a machine vision application such as mechanical vibrations, fixturing, and space limitations. However, getting the lighting and optics right goes a long way toward a successful application. To facilitate getting the right image for machine vision, there have been a wide range of tools that have appeared on the market in the past few years. On the lighting side, some of these tools have included

- Diffuse lighting modules, both on axis and surrounding like a tent that help to decrease the sensitivity to local shiny spots or surface irregularities
- Directional lighting modules, including line illuminators, dark field illuminators, and collimated illuminators that highlight surface textures, some point defects such as scratches, and surface irregularities such as flatness issues
- High-frequency lights to permit asynchronous image capturing
- Highly stabilized lights and high light uniformity, both of which provide for a more repeatable image

Perhaps the most useful advance as can be applied to on-machine monitoring and sensor for energy based manufacturing is the use of LEDs. The improved brightness and longer operation life of LEDs achieved in the past years have made them a good alternative to halogen lamps that put out heat and have limited lifetimes or fluorescent lamps that remained too bulky to fit in a machine environment and expensive to customize. Although not yet powerful enough to be used in many larger field-of-view applications, and still a more expensive option than incandescent lights, LEDs offer a degree of flexibility of design, and a potential ruggedness of manufacture that is needed for on-machine types of applications. Making an odd shaped line, a spotlight, or a surrounding light that can be adjusted by quadrants are all tools now in use in machine vision using LEDs as the light source (see Figure 23.6). To illuminate a large part with a thousand LEDs may not be desirable. However, as the application to energy based manufacturing often implies looking on a more detailed scale, the size issue has not been a big limitation. As LEDs make their way into more consumer products such as cars and appliances, this market is likely to both consolidate to stronger companies, and likewise grow and become more cost competitive.

23.3.5 MACHINE VISION OPTICS

In the optics area, manufacturers have also found it financially beneficial to address the machine vision needs, primarily on two fronts. The first area is just the availability of new high-resolution lenses with better-corrected fields as needed for high-accuracy inspections in a machine environment. In the earlier days of machine vision, it was often argued that the "software" could "correct"

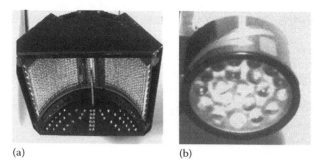

(a)   (b)

FIGURE 23.6   LED light module made for machine vision provide custom lighting from diffuse surrounding light to highly directional light: (a) surrounding LED light and (b) LED spot light.

the image. However, these corrections are something that takes up processing time, computing power, and in many cases are only partially successful. Now, several companies have introduced lines of small, highly corrected, c-mount lenses that reduce aberrations such as field curvature and distortion. Such optical errors are often up to 10% of the field in security camera lenses. New machine vision lenses make measurements at thousandths of an inch level both meaningful and cost effective with a vision system.

The other advance in the imaging area has been the availability of telecentric optics. Telecentric lenses provide a means first to present a uniform intensity image onto the camera if the subject is uniformly lit. It provides this uniform collection by using a telecentric stop to provide a uniform cone angle of collection of light as shown in Figure 23.7. This may not sound like much, but in fact, most common lenses vignette the light that is not in the middle of the image field, causing the light at the edges to be reduced. Just looking at an image, most people might not even see this effect, as people are very tolerant to light level changes. However, given a vision system that is thresholding the light at some level in order to do the inspection, this light level variation is at least a nuisance. Even with adaptive thresholding, the limited dynamic range of cameras and the processing time taken for such filtering are both things that would be nice to not use up on something like light uniformity, given that the light sources themselves have become much better at presenting a uniformly lit field.

The second benefit telecentric optics provides (the ones telecentric in object space) is to produce an image that does not change in magnification for small shifts in the object distance (shown in Figure 23.8). This means the system can be much more tolerant to shifts in the position of the part under inspection, both within the field and in distance, without the character of the image

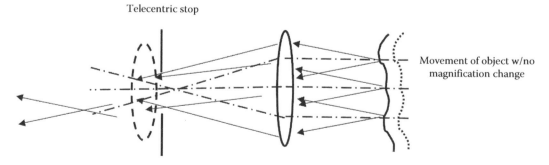

Telecentric stop

Movement of object w/no
magnification change

FIGURE 23.7   A telecentric optical system collects a constant perspective, magnification and cone of light across the part.

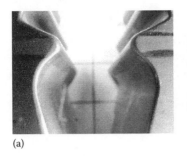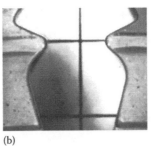

(a) (b)

FIGURE 23.8 A perspective view of a feature with depth such as a straight hole of slot is difficult to analyze because the walls are visible, whereas in a telecentric image the view looks directly down the walls with the same angle of view across the field, providing a clean 2-D projection of the feature: (a) perspective view of a straight slot and (b) telecentric view of a slot.

changing, in terms of either perspective view or magnification. This means a measurement made in a machine tool can be tolerant to changes in distance to the part, without losing accuracy. Telecentric optics do have their limitations. To be completely effective, there is some loss of light due to larger f-numbers, and the lens system does need to be larger than the part under inspection. But once again, with the great majority of energy based manufacturing applications being the inspection of small part areas or features, telecentric optics has become a commonly used tool today.

23.3.6 CAMERAS

The next piece in a machine vision system is the camera. In the early days of machine vision, the application engineer was stuck with cameras made for closed circuit video, or needed to pay very high prices for higher quality cameras. The typical analog camera provided 50–80 usable gray levels, with a varying degree of noise that was often different from one camera to the next. Today, the consumer market has helped to push the digital revolution to the old camera industries. Many vision systems today use, or at least have the capability to use, digital cameras that provide better stability, higher dynamic range, and more pixels.

Cameras with over 1000 by 1000 pixels of resolution are commonplace today. A camera offering 10 or 12 bits of pixel depth, that is, a light dynamic range of over 1000 to 1 (even allowing for a few counts of noise) are within the price range of many machine vision applications. This has made it easier, and more reliable to cover larger fields or obtain better resolution without complicated multi-camera systems. The advent of new communication options for cameras, including firewire, high-speed Internet, USB and the camera oriented camera link (which offers very high image speeds) has also made it easier to install cameras in the manufacturing environment, network them together, and ultimately collect more data.

But even with all these great advances in the camera options of digital interfaces, more pixels, and lower noise, the big impact of growing digital capabilities today is in the form of "smart" cameras. A smart camera combines a video camera with an on-board processor and memory to create a vision system that can fit in the palm of your hand. Smart cameras in one form or another have been around for some time, going back to early systems that were very simple in rather clunky boxes. These early systems could typically do only one of a handful of operations at one time. Simple edge detectors that could find the distance between two edges were useful in measuring alignments or gaps in assembled parts. Basic blob recognition provided simple optical character verification, or verification of any shape pattern. But these systems were so limited, they were often little more than high-end bar code scanners, but at a much higher price.

23.3.7 SMART CAMERAS—MACHINE VISION IN THE PALM OF YOUR HAND

Today, smart cameras have a much larger range of operations (see Figure 23.9), using more memory and processing power (typically Pentium class) than older desktop computers had even 10 years ago. The types of operations available on a typical smart camera system can include

- Identifying part position and rotation (allows for part position variation)
- Analysis of multiple edge locations including counts, separations, and angles between edges
- Blob analysis to match complicated patterns, including doing full optical character reading (not just verification)
- Providing a wide range of outputs ranging from simple logic outputs to detailed numerical reports of fits to tolerances, amount of errors, and statistical information

In many cases, these cameras are networked together through Internet connections. Since most of the processing is local, only the results or daily reports need to go over the network, removing the need for separate dedicated computers.

The degree of sophistication and cost of these smart cameras is still fairly wide, ranging from modern versions of the simple pattern matchers, now in the $1,000 range rather than $10,000, to full systems costing a few thousands. In general, the software is made user friendly, using pull down menus and icons to set up applications rather than C-code and low-level communication protocols. However, these systems do not do all the operations that are needed.

Complicated operations like image preprocessing, morphological operations, Fourier analysis or similar mathematical analysis, and correlations are typically beyond what is reasonable to do with smart cameras. However, the current capabilities of smart cameras allow them to perform a vast number of "good" vision applications well. Applications such as basic part measurements, simple inspections, hole finding or counting, are very reasonable to do both in effort and cost using smart cameras. What does this mean to the industry? Many of those applications that just could not be justified because the vision system itself was going to cost $30,000–$60,000, can now be done with a smart camera at an equipment cost of a tenth of these amounts. It is still important to do a good application engineering job, but the hardware investment is less costly now.

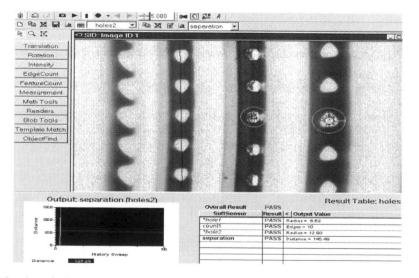

FIGURE 23.9 A typical smart camera interface showing the counting of holes, sizing of holes, and finding the separation of two features is typical of the types of operations of a smart camera.

23.3.8 MACHINE VISION SOFTWARE

This brings us to the third basic building block of machine vision, the software. Not too many years ago, to set up any vision application meant either being a good programmer using C or a similar language, or worse yet, learning a dedicated vision language, built around mnemonics, command strings, and controller codes. Basic C-code libraries are still available today, and in fact have become fairly inclusive, providing a wide range of capabilities for the programmer. However, for many applications, vision systems offer a user interface with access to all of the functions, filters, communications, and the like without writing a single line of code. In some systems, the user interface will generate an executable file (often in C or C++) for use in production, just to provide better speed of operation. Simple additions of setups are often available through small scripts, usually based upon visual basic or visual C++, with clear examples and instructions for the end user.

Few vision systems use any form or proprietary hardware today. They may use multiple processors, high-speed memory and some fast graphics, but it is not often that the end user needs to know how to do low-level programming of any type to get the system to work properly. Even the camera setups, which at one time often required the writing of special configuration files, is often plug and play using the standards mentioned above. The high speed of modern computers has replaced the high-end workstations costing $100,000, with a simple system selling for less than $500. The exception to this rule of non-dedicated hardware and software includes some of the high-speed production systems in continuous process applications such as roll-to-roll or primary metals operations where the standard PC computer may not yet be up to the speeds needed.

The user-friendly nature of machine vision software has brought the setup, programming or changes into the capabilities of plant engineers without extensive programming experience, as well as many shop floor maintenance personnel who can now easily maintain systems that at one time could only be serviced by a high-level programmer not typically available in a plant. New parts programming rarely needs to be done by the vendor anymore (though most are willing to do so). This puts the control and schedule of new part introduction, tolerance changes, or the ability to add new checks to diagnose manufacturing issues within the hands of the production manager. This ease of use of modern machine vision along with the lower prices has made vision systems more attractive for many industrial inspection, alignment, and simple gaging applications that in the past may have been too expensive to purchase, to maintain, and to setup for many applications.

23.4 3-D SENSORS—OVERVIEW

Just as we can now easily scan a 2-D image into computer memory, the tools are commercially available to do the same with a 3-D part. There are many tools available for digitizing 3-D shapes. Some applications may require a very high density of data over very small areas to capture a complex shape. A quarter or dime would be an example of such a part. For other applications, the sizes in question may be quite large, with only minimal variations from one area to another (see examples in Figure 23.10). There are systems available on both ends of this spectrum and at many points in between. Choosing the best tool for a particular job is the challenge to be met by the designer. Many of these systems have been made to address a range of applications from robot guidance to surface structure analysis.[11,12] No single system is likely to ever address all the possible applications for 3-D contouring in the near future.

For example, a system capable of describing the work area of a robot doing welding may be looking at an area of a few square meters to a resolution of a few millimeters, while a surface-laser treatment system may be looking at a few millimeters to submicron levels. The density of data is not the same for all applications either. If the concern is the presence of high spots on a part that may lead to cracking, then the sensor cannot skip points. In the case of many robotic manufacturing applications, only the distance to the part, and where one or two edges are located are important. In these latter cases, perhaps just a line or a small array of a dozen points will be sufficient.

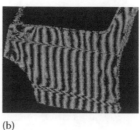

(a)   (b)

FIGURE 23.10   For small features, very high density of 3-D data may be needed, while on larger parts high resolution may be needed, but not as much of area resolution: (a) 3-D of Lincoln memorial on a penny and (b) 3-D of shape errors on a car door.

One of the early applications of using 3-D information was the Consight system developed by General Motors.[13] The purpose of this system was not to measure the 3-D shape of the part, but rather to take advantage of the known difference in the three dimensional shape to sort the parts. The parts in this application were gray metal castings on a gray conveyor belt. These parts were hard to distinguish using only 2-D images. The 2-D silhouette was not necessarily different, and the features of the gray, cast parts were too low in contrast to pick out of a typical 2-D view. In this case, the density of data needed was small. A single white light projected from an angle provided a changing cross section silhouette of the part shape. This information was sufficient for the task.

In some cases, a sensor made for low data density can be used to build up the data. Scanning a sensor that measures one point or a line of points can be used to build up a full, 3-D shape. In the case of a complicated shape like an airfoil surface or plastic molding, building up the shape may be a long process one point at a time, suggesting the need for more of a full-field data collection sensor if real time data is needed to control the shaping process. This does not mean it is necessarily desirable to work with the maximum number of points at all times. A typical video frame has a quarter of a million data points. If there is a depth associated with each data point in such an image, there is, indeed, a large amount of data, more than may be practical to handle in the time available in a production operation.

Because of the variety of applications for three-dimensional sensing, there is a variety of systems available.[11] These sensors can perhaps be broken into a few basic types:

- Point scanning sensors measure only the specific points of interest, typically in a serial fashion,
- Line sensors provide a single line of points in the form of a cross section of the contour of interest,
- Full-field sensors provide an *X*, *Y*, and *Z* map of all the points in the scene that must then be analyzed down to the information of interest.

Each of these types of sensors has been developed using technology that is suited to the application. In some cases, the technology is capable of multiple modes of operation (finding a point on a surface, or defining the full surface) as well, but this often stretches the technology into a field overlapping other technologies. There has not to date been any single sensor in industrial applications that does everything. The result has been an assortment of sensors finding their best fit to specific applications.

23.4.1   Discussion of 3-D Technologies

Before we address the performance of specific sensors, it is useful to establish the basic technologies in use. There are methods that can be used to find the distance to an object.[13–27] A simple version is to focus a beam of light on the object at a given distance. As the object surface moves closer or more distant, the spot on the object surface will enlarge, with the size of the spot being directly proportional to the change in surface height. This method has not seen much industrial use, so will not be further explored at this time. Some of the other methods, such as the scanning and full-field methods, have seen commercial success and have the potential to be used in process control as well as detailed gaging functions.

23.4.2 POINT TRIANGULATION

The most popular commercial versions of range finding use the triangulation method where a beam of light is projected onto the object's surface at some angle and the image of this spot or line of light is viewed at some other angle (see Figure 23.11). As the object distance changes, a spot of light on the surface will move along the surface by

$$\left(\text{change in spot position}\right) = \left(\text{change in distance}\right) \times \left(\tan\left(\text{incident angle}\right) + \tan\left(\text{viewing angle}\right)\right)$$

A wide range of commercial gages exist that can provide a single point of measurement based upon this triangulation principle. To make a discrete point measurement as a process control tool, such a sensor can be directed at the location of interest, with a wide range of possible standoff distances and send data at thousands of points per second in most cases. In order to obtain a contour map, these systems are typically scanned across the part.[27–29] The scanning has been addressed both by scanning the entire sensor head in a mechanical manner, and by using scanning mirrors. Some of the mirror-based systems can collect a full field of data at nearly the rate of data of a video camera. Resolution of a few microns to tens of microns has been realized with point-based triangulation sensors.

Most triangulation gages today use laser light. When a laser beam is incident on an opaque, rough surface, the microstructure of the surface can act as though it is made of a range of small mirrors, pointing in numerous directions. These micromirrors may reflect the light off in a particular direction, or may direct the light along the surface of the part. Depending on how random or directional the pointing of these micromirrors may be, the apparent spot seen on the surface will not be a direct representation of the light beam incident on the part surface. The effects of a laser beam reflected off a rough surface include[28]

- Directional reflection due to surface ridges
- Skewing of the apparent light distribution due to highlights
- Expansion of the incident laser spot due to micro surface piping

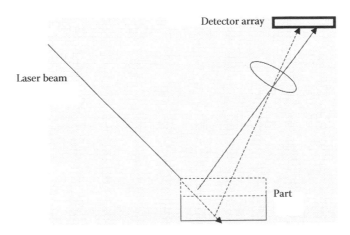

FIGURE 23.11   A triangulation based system using a point of light to obtain distance.

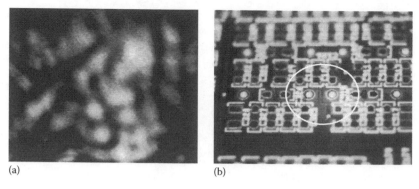

(a)   (b)

FIGURE 23.12   Laser light does not always provide a clean spot to use for measurement. Scattering surfaces or translucent surfaces can provide an uncertain spot location. (a) Speckle on a scattering surface and (b) laser light diffusing into a circuit board.

The result of this type of laser reflection or "speckle" is a noisy signal from some surfaces such as shown in Figure 23.12. Trying to determine the centroid of such a signal, will likely lead to some errors in the measurement. In like manner, there can be a problem with translucent surfaces such as plastics or electronics circuit boards. For translucent surfaces, the laser light will scatter through the medium and produce a false return signal. For a laser-based sensor, a smooth, non-mirror like, opaque surface produces the best results. Just as a contact probe has problems measuring a soft or delicate part (such as a gasket of a metal foil part), laser probes must be adapted to measure optically unfriendly parts. There have been a number of methods developed for dealing with such parts with laser gages that are typically based upon restricting the view of the surface to only those areas where the laser beam should be seen and using smart data processing. Restriction of the view is perfectly reasonable since the laser probe is only measuring a specific point on the part.

An active variation of restricting the view uses synchronized scanning.[29] In the synchronized scanning approach (see Figure 23.13), both the laser beam and the viewing point are scanned across the field. In this manner, the detector only looks at where the laser is going. This method does require an active scan, but can be made more selective to what view the detector sees. The view cannot be completely restricted with synchronized scanning if an array or a lateral effect photodiode is used.

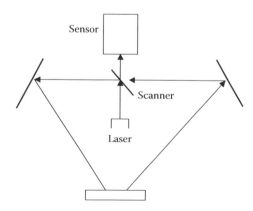

FIGURE 23.13   A synchronized scanning system, limiting the range of view.

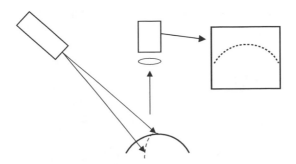

FIGURE 23.14   A line of light-based sensor showing the surface profile.

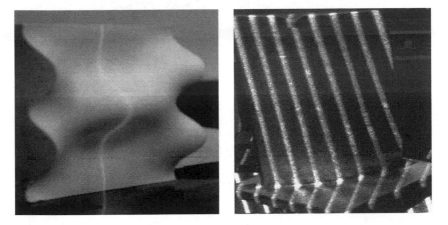

FIGURE 23.15   One or more lines of light on a part provide cross sections of the shape.

23.4.3   LINE TRIANGULATION

In contrast to a single spot of light, if a line is projected onto the surface by imaging or by scanning a beam, as shown in Figure 23.14, the line will deform as it moves across a contoured surface as each point of the line moves as described above.[18–23] The effect is to provide an image of a single profile of the part (see Figure 23.15). In applications requiring only a profile measurement, these techniques offer good speed of measurement. If the full contour is of interest, then the line is scanned over the part, requiring a video frame of data for each profile line of interest.

23.4.4   AREA TRIANGULATION AND MOIRE CONTOURING

A popular extension in industry for the individual line of light system has been the use of multiple lines or patterns such as reticles, to cover more area at a time.[19,20] These patterns can take the form of encoded dot patterns, distorted grid lines, or simple gratings. One special case of structured lighting using simple gratings is moire contouring.[11,30,31] In the case of moire contouring, it is not the grating lines that are analyzed directly, but rather the result when the initial grating as seen on the part is beat against a secondary or submaster grating. The resulting beat pattern or moire pattern creates lines of constant height that will delineate the surface the same way that a topographic map delineates the land (see Figure 23.16). This beat effect provides an extra leverage, since the grating line changes need not be directly detected, and data is available at every point in the field to be captured

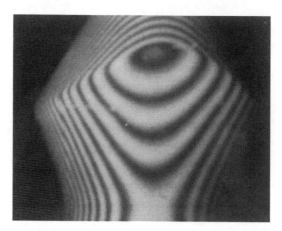

FIGURE 23.16 The moire beat lines delineate the shape of a plastic soap bottle.

within a single video image. This leverage can be useful in special applications such as flatness monitoring, as the depth resolution can be made much finer than in the X–Y plane.

The optical system for a moire system is more complicated than that of simple structured light. So, only in some specific applications where very high-depth resolution is needed, such as sheet metal flatness, has this technology been used. The other drawbacks of a moire contour include the difficulty in distinguishing a peak from a valley, ambiguity over steps, and the large amount of data generated. With the current commercial systems and computing technology, most of these issues regarding moire and structured light in general have been addressed. The methods of analyzing such patterns have been well established.[32–42] In fact, many commercial structured light systems, which directly analyze a projected grid pattern, use the same type of analysis as found in interferometry in the optics industry. Interferometry provides nanometer level resolutions, which are typically beyond most applications in manufacturing, and so will not be further explored here.

23.4.5 CURRENT APPLICATIONS OF LASER PROBES

Triangulation based distance sensors have been around since the time of ancient Egypt. Modern sensors have resolutions approaching a few microns. The most common industrial uses are in semi-fixed sensing operations where a fixed set or a few fixed sets of points are measured in a fixture. Entire car bodies, engine blocks, or other machined parts are measured by this means. For the purpose of reverse engineering, the flexibility of the "scanning" triangulation sensor offers some attractive capabilities.

The individual laser probes have seen nearly 10-fold improvement in resolution in the past few years. The application of such probes in energy-based manufacturing has been a great benefit as a feedback control in systems where monitoring the force of a mechanical contact may not be possible. Scanning and fixed triangulation systems have been used to contour large structures such as airplanes, airfoil shapes, and flatness of rolled metal. The large area systems primarily have used one or multiple lines of light to obtain a cross-sectional profile at a time. In many cases, these line sensors are connected with a machine tool axis of motion to extend the working range of the sensor. The resolutions of such systems need typically be less than a millimeter, and more typically is around 2.5 µm (0.0001 in.).

Because of the long time the triangulation based systems have been around, and the well ordered nature of the line profile, there has been very good progress in adapting this technology to the needs of energy field manufacturing such as welding. A number of systems are available with direct CAD interface capabilities, and would be capable of generating CAM type data as well. There is extensive second source software available that permits the large "clouds" of data to be reduced to CAD type of information for direct comparison to the computer data of the part. Most such comparisons have been largely specialized in nature, but as computer power increases, the user friendliness of such software is increasing.

Scanning triangulation sensors have been used in the manufacturing of small parts, such as precision parts made by laser machining.[43,44] The resolutions for these smaller sensors have been in the micron range, over distances of a few millimeters at a time, with data rates approaching a megahertz. Dedicated inspection systems that work like the full-field coordinate measurement systems for small parts are commercially available, gaining wide use particularly in the electronics industry (see Figure 23.17). The use of these sensors in energy-based manufacturing has been a significant tool in the electronic data transfer of dimensional information.

Full-field structured light systems, based upon projected grids by direct sensing of the grid or related to moire are also commercially available. The primary application of this type of sensor has been the contouring of continuously curved non-prismatic parts such as turbine airfoils, sheet metal, clay models, and similar shaped parts, as shown in Figure 23.18.

Special compact sensors for use on machine tools are also available with this technology. Most of the applications of this technology has been on applications requiring dense data, but have also been engineered to enhance video data for the purpose of 3-D "comparator" type measurements on objects ranging from rolled sheet metal (for flatness) to car bodies. Coverage of $2\,m^2$ areas at a time has permitted very high-speed relative measurements of large structures to several micron resolutions. Typical resolutions of commercial full field structured light systems are in the range

FIGURE 23.17 Point scanning system with small X–Y table.

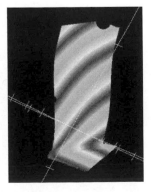

FIGURE 23.18 Structured light 3-D systems are available, used for mapping continuous shapes such as a compressor blade.

of submillimeter to several microns (down to 0.0001 in.), with data collection taking from 1 to 5 frames of video in a few seconds or less. This technology has benefited greatly from the advances in computing power due to the large amounts of data involved (up to a quarter of a million data points in a few seconds). These systems have also been interfaced to provide direct CAD data inputs. Area based structured light systems offer better speed in applications requiring dense data over complex shapes as opposed to selected regions of a part.

23.5 APPLICATION ERROR CONSIDERATIONS

As discussed previously, noncontact probes may be better suited for energy field manufacturing than touch probes. Touch probes are used in traditional machining for process control. The tool holders and mechanical scales are not adaptable enough to use touch probes in energy field systems. However, to best understand the errors that may be encountered in any process control, we will examine both contact and noncontact probes.

Both touch based probes and optical probes have certain errors associated with their operation. The errors tend to be inherent in the nature of the sensor. Each sensor technology has operations it is good at measuring, but it has problems with other operations. In the case of touch probes, measuring any feature on a sharply curved surface, be it the diameter of a hole or going around a corner edge, requires more points to compensate for how the touch probe makes measurements. In the case of optical probes, the biggest errors tend to come from the edges of parts, either because the probe cannot see past the edge or because the measurement point is larger than the edge. Understanding what these basic errors are with the probes used for the control of a process is an important step in correctly applying the technology and getting useful data to control the process.

23.5.1 CONTACT PROBE ERRORS

In the case of touch probes, the operations are geared toward the types of potential errors specifically found in these sensors. Touch probes used on machine tools have errors associated with the direction of touch.[45–47] These errors fall into two categories. First, since the end of the touch probe is a ball of finite size, the measurement that the machine tool axis provides must be combined with the offset of the radius of the ball and added to the measurement to offset the measure in the direction of the normal of touch to the ball. Of course, knowing precisely what the angle of the normal of touch can be a difficult question. As a sphere, the touch can be in any direction over nearly 360°. Therefore, in operation additional points are taken around the first touch point to try to establish the local plane of the object. The orientation of the plane of the object is used to determine the direction of offset of the measured values.

Much work has been done to minimize these touch offset errors, both in determining the minimum number of points needed to establish the direction of touch, as well as the means to devise durable small point touch probes to reduce this potential of error on high-precision machines. However, as can be seen in Figure 23.19, there remain many error conditions that may still provide

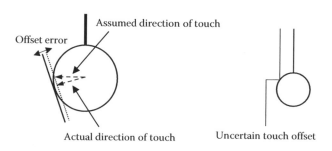

FIGURE 23.19 Errors from touch direction on contact probes.

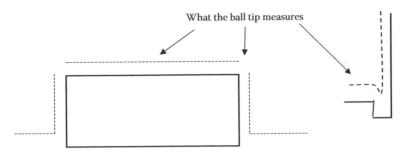

FIGURE 23.20   Actual shapes versus what is measured by a touch probe.

an erroneous reading. Features such as corners and small holes necessarily remain a problem for touch probes. A sharp corner location is typically inferred from the intersecting surfaces forming the corner (see Figure 23.20.)

The second type of error associated with direction of touch is the so-called lobing errors present in many touch probes (see Figure 23.21.). The lobing error is the result of the design and operation of the probe. The probe responds faster to the touch in some directions than in other directions. The result is an additional error that is systematic and consistent with respect to the orientation of the touch probe. Any calibration test must map the response of the probe at a full range of angles and approach speeds. Touch probes are typically tested using a sphere of known size. By finding the center of the sphere, the errors associated with lobing and ball touch angle can be corrected.

For most point probes used on a machine tool, the measurement is actually being made using the scales on the machine. A touch probe itself does not really provide any measurement directly; it only acts as a switch to indicate when to take a measurement. There are available analog touch probes that provide some small measurement range directly. If part of the measurement comes from the movement of a machine tool and some from a sensor, the alignment and calibration of one source of measurement to the other is very important to the overall performance of the measurement. In either case, the machine scales are playing a significant role in the measurement of the part. The machine axis themselves are what is often used to do material processing, and as such any measurement made with them will be self-consistent, whether they are right or wrong.

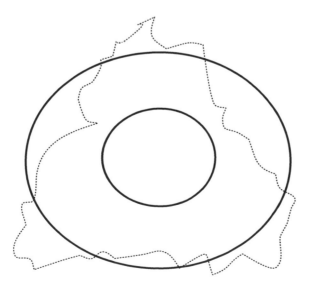

FIGURE 23.21   Typical lobing errors from a touch probe.

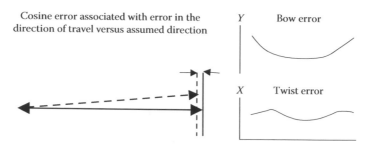

FIGURE 23.22   Cosine errors due to axis alignment errors.

23.5.2   MACHINE AXIS ERRORS

The errors associated with the linear axes of the machine include errors in the read out of the stages, as linear errors in X, Y, and Z, as well as the squareness of these three axes. The specific nature of these errors is unique to the machine tool operation. Scale errors tend to be linear, often as a result of the axis not being in line with the assumed direction, but rather at a small angle. The result of a small angle in the axis is referred to as the cosine error effect (see Figure 23.22).

The straightness of the Cartesian motion axis of the machine tool can also contribute to the cosine error. However, the motion axis alignment is more a design parameter than something that can be fixed by some user alignment. That is, the axis may actually be slightly bowed or twisted, due to mechanical sagging of the beam carrying the cutting head or tool holder. In addition, the initial straightness of the ways used to build the machine may not be perfect. Because the composite performance is important, touch probe positions on machine tools are usually calibrated using a ball bar. The balls at the end of the bars permit the errors of the touch probe to be taken into account; while the length and various angle positions of the bar, tests out the machine scale accuracies and the squareness of the movements of the machine axis.

23.5.3   NONCONTACT PROBE ERRORS

In the case of noncontact gaging systems, the potential causes of errors are different, requiring different types of tests to isolate. Unlike the touch probe on a machine tool, whose variations tend to be not in what movements or errors it may posses, but rather the particulars of how it makes these movements, optical measurement systems are much more varied in the basics of what they do. The variety of 3-D optical systems might be classified into three basic areas[48]:

1. Radial scanners measure the distance along a line of sight from some central location, such as laser, radar, or conoscopic systems. Errors in these systems are related to errors in scan angles.[49] The base coordinate system is typically R theta in this case (Figure 23.23a).

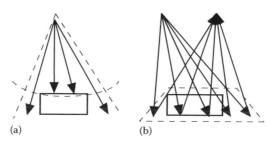

(a)   (b)

FIGURE 23.23   Coordinate systems formed by different 3-D optical scanning methods: (a) radial scan coordinate system and (b) triangulation coordinate system.

A special case of such a scanner would be when the scan center is at infinity. In this case, the scan is telecentric or parallel. There is no angle effect if the scan is parallel, but the linear translation may have a small error.

2. Triangulation based systems, such as point laser probes and structured light probes rely on obtaining information from two angles of view. Both views can be passive as with a stereo view, or one view can be active in the form of a projected dot or pattern of some sort. The coordinate system with triangulation systems is typically taken as Cartesian, but in fact is at best trapezoidal (see Figure 23.23b). The errors associated with the triangulation-based systems tend to produce a field that is curved or saddle shaped. The errors in a curved field include magnification effects and the change in the triangulation angle. Both the magnification and angle can change with position across the field, and with changes in distance.[50,51] The interaction of the two or more optical systems must be taken into account when addressing the actual calibration.

3. Interferometric based systems, such as classic interferometry or so-called phase shift structured light systems make measurement based upon the distance light travels relative to some reference surfaces (real or virtual). In this case, the calibration is tied to the real or effective wavelength being used to measure this difference. Moire contouring is an example that can be analyzed using interferometric analysis based on an effective period of light (typically much longer than the optical wavelength). Moire is also a triangulation method subject to the variations and constraints of magnification changes and angle of view.

Clearly when applying a noncontact 3-D system to an application currently done by machine tools, the very basic question of what coordinate system is being used must be answered. Machine tools typically are built around three axes all perpendicular to each other. An optical 3-D system may have a curved measurement area, one that is trapezoidal, or even spherical. Much of this variation in coordinate system is accommodated for in the calibration routines of the sensor. It is not necessarily the case that a spherical coordinate system is incorrect, but typically, parts are specified in square Cartesian coordinates.

In order to apply optical methods, the coordinates are translated from the inherent system coordinates of the optical sensor into the equivalent Cartesian space native to the machining operation. Such transformations always have their errors and approximations. In the case of a trapezoidal or spherical measurement, this may mean reducing the accuracy of the measurement to that obtained in the worst area of the measured volume. This worst area is typically the points furthest or most off-center from the sensor. If machines were initially made with spherical coordinate geometries, then the transition to some types of optical based measurement tools might be a simpler task. For some energy field manufacturing systems, this might be an option. One type of coordinate system is not necessarily superior to the other; it is just a matter of what is being used.

In applying optical based measurement systems to on-machine operations, the other primary issue is how optical based measurements handle edges. We have already described the potential errors that occur when a touch probe goes over an edge, and the uncertainty in offsets that can arise depending on the angle of attack to the surface. Optical probes that are based upon finding the center of a laser spot typically have just the opposite problem from a touch probe. As the laser spot goes over the edge (see Figure 23.24), part of the spot is no longer seen by the sensor. The center of the spot actually seen is not in the same location as it would be if the whole spot were visible. The result is a measurement suggesting that there is a raised lip on the edge that is not really there.

Typically, a laser spot is less than 50 μm, and perhaps only a few microns in size. Even so, this finite spot size produces an offset error that increases as the spot goes across the edge. The actual centroid calculation may depend on the intensity of the spot, the surface finish, the shape of the spot, and the algorithm used to estimate the center. For many optical based systems (other than

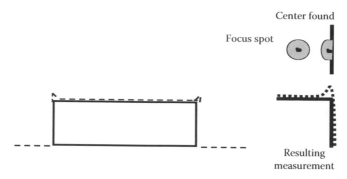

FIGURE 23.24 Edge error associated with many laser probes.

interferometric or laser radar), this edge lift off exists whether there is a real edge, or just a transition from a bright to a dark area.

Clearly, 3-D systems that rely more on area based averaging will have more of a problem with how close they can measure an edge before errors start to come into play. Some methods, such as the interferometric or phase step based systems that calculate a range at each pixel, can typically measure closer to the edge than a system that uses a spot or a line that may be many pixels wide. Such differences in offset errors and how systems see edges often mean one type of system is seen as superior over the other. A sensor that can measure closer to a physical edge may be judged better than one that can only measure to within a millimeter of the edge. That fact is, just as touch probes can be used around an edge and the offset compensated for in the analysis, the same can be said about the optical probes. The correction for the edge offset is different for optical versus touch probes, but not less predictable for either method of measurement.

For example, in the case shown in Figure 23.24, the laser spot is shown as being a round spot, with the measurement based upon finding the centroid of that spot. Therefore, we can predict that the spot centroid error will change as a quadratic function of the form

$$\text{delta}(Z) = Z + P \times \frac{X^2}{R^2},$$

where
P is the triangulation factor of $Z(X)$
R is the spot radius
X is the displacement past the edge

As the laser spot hits the bottom of an edge, some of the spot will highlight the sidewall. Depending on the steepness of the wall, this may then lead to the complimentary effect of a rounded bottom corner that follows the form:

$$\text{delta}(Z) = Z - P \times \frac{R^2}{X^2},$$

which is just the inverse as seen on the top of the edge (see Figure 23.24). This basic form agrees fairly well with what is typically seen from experimental data of this type. Once the laser spot center has moved one-half of the spot diameter from the wall, the spot is completely on the bottom, and a correct measurement is available directly. This correction to triangulation sensors assumes that the triangulation angle has not been occluded by the wall that would block the beam. Occlusions going

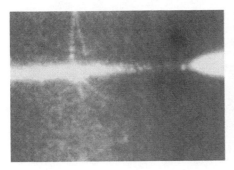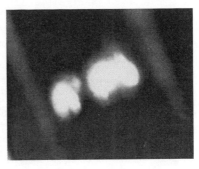

FIGURE 23.25   Two laser spots showing part of the spot bouncing off a sidewall (in the left image), and one being broken into two bright, but irregular spots (right image) as the laser spot seen in a groove.

past an edge are really more of a problem than the liftoff, since there is no data to correct. For this reason, many triangulation sensors that are used for this type of application will view the laser spot from two or more perpendicular directions to avoid occlusion issues.

The point of this discussion is to show that the errors from optical sensors near the edge are both understandable and predictable and can be corrected in the same manner as touch probes accommodating the ball radius. As an additional complication, if the edge causes a bright glint of light, the error in a standard centroid based triangulation system can be compounded. Some manufacturers monitor the change in light level to recognize such glint conditions, either to reject the data or to attempt to correct that spot.

In like manner, if the side of a step is not steep, then light may be seen from the detector on the side wall as shown in the left image in Figure 23.25, creating a very elongated spot, and again increasing the error. A groove may appear to be two spots, as shown on the right side of Figure 23.25, confusing the interpretation of the centroid.

These reflection problems are a function of the surface finish and the geometry of the edge, so are more difficult to predict. For phase and frequency based sensors, a bright glint is typically not a problem assuming the sensor has the dynamic light range to accommodate the extra light. For that matter, if the triangulation sensor can be used with the plane of triangulation along the edge, then relatively little offset would be seen. However, depending on how the spot is sensed, an area-based system can still misinterpret the Y displacement as a change in Z.

23.5.4   MEASURING TO DATUM

In a similar manner to optical glints, a touch probe can easily be thrown off by a small local nick or burr on the edge, or an uneven cut at the bottom of a groove. The same basic problems exist for contact or noncontact sensors, only the interpretation changes. The different way of interpreting the errors of contact versus noncontact sensors leads to difficulties in replacing one type of sensor with the other. The challenge is to understand how to verify that the optical, contact, or any other methods are in fact measuring the same point.

When any feature on a part is measured, it is measured at a location, and relative to some predefined references or datum points. The datum point may be on the part, on the fixture holding the part, or somehow defined by the shape or fit of the part. Machine based measurements commonly start a measurement by locating a few key datum features such as planes or holes in the part. These are features a simple point sensor can define with a minimal number of points. For production applications, the part is typically referenced off a fixed position fixture, pre-located on the processing machine tool. To the extent that the fixture is repeatable and kinematic, the machine axis or other measurement device such as a hard gage, can expect to make measurements in the correct, prescribed positions. If the fixture is off, or the part does not sit right in the fixture, say due to a

TABLE 23.1
Overview Summary of Primary Sensor Types of On-Machine Monitoring

Sensor	Data Rate (pts/s)	Resolution (Typical)	Issues
Touch probe	1/s	na	Uses scales to measure
Point optical probe	500–20,000 Hz	1–3 μm	Reflections, edges
Laser line probe	20,000 Hz	10–50 μm	Laser speckle
Machine vision	To 50,000 (to 100 images per second)	5–100 μm	Resolution depends on field-of-view
3-D structured light	>1 million (large data sets)	10–50 μm	Resolution depends on field-of-view

slight error in the edge geometry of the part, the system will make a good measurement, just not in the right place. Clearly, from the discussion above, the way in which a measurement can be miss-positioned can be very different when using a contact sensor versus a 3-D optical gage. In both cases, understanding how to correct these errors is the key to making good measurements.

23.6　SUMMARY AND THE FUTURE

We have discussed a wide range of possible sensors for use with energy field manufacturing operations. The intent of the use of any of these sensors would be to control the process in a situation where there is no traditional contact with the part being machined or formed. The possible sensors range from point contact probes, currently in wide use on traditional metal cutting machines, non-contact point laser probes, 2-D machine vision camera based measurement systems, and full-three dimensional mapping systems. A summary of these methods and typical capabilities is shown in Table 23.1.

The right sensor for an application is very dependent on the nature and amount of data needed to provide feedback to the process. To monitor a few key points, a touch or point laser probe can provide sufficient feedback, and is in wide use in many industries today as a process control tool.

Laser line probes are typically used in continuous process applications such as extrusions where only a contour section really matters to the process control. A common use of these sensors is to monitor welds as they are being formed.

Machine vision is widely used as a feature inspection tool, including applications such as aligning and verifying holes made by electro-discharge machining (EDM) and laser drilling. The full-field, structured light 3-D systems are still new on the market and are primarily being used to verify only the first parts made in production. However, the speed of 3-D systems is such that monitoring a fast manufacturing operation is practical.

The processing capabilities of computers will continue to make any of these sensors faster, easier to interface to manufacturing systems, and easier to interpret. The combination of fast 3-D sensors with energy field manufacturing has the potential to enable completely automated processes that go from drawing to finished product. The capability exists today to make a 3-D copier machine that would work as easily as a 2-D document copier. Such a device could completely change the way we do manufacturing in the future.

QUESTIONS

Q.23.1　Discuss the operation and the limitations of current on-machine probing means using contact probes. Why might such methods be not viable with energy field manufacturing methods?

Q.23.2　Describe the general implementation issues that need to be considered for any sensor application to be successful. How might this apply to areas in your own work?

Q.23.3 Review the basic terms of repeatability, resolution, accuracy, dynamic range, and speed, and describe how you might use each feature to decide if a particular sensor is appropriate to a given application.

Q.23.4 Describe the basic building blocks of a machine vision system. What are some applications that machine vision may be good for and what applications may not be right for machine vision?

Q.23.5 What is a smart camera and when might you want to use one?

Q.23.6 Discuss the pros and cons of the various ways of collecting noncontact 3-D data (point scanning, line sensors, and area sensors) and the types of applications that would be appropriate for each.

Q.23.7 What are the applications of laser probes and what do you need to consider in these applications to be successful? What types of errors might occur with one of your applications?

ABOUT THE CONTRIBUTING AUTHOR

Kevin Harding is a principal scientist at GE Research. At the R&D center, he leads the activity in Optical Metrology and provides guidance to a wide range of optical technology projects, doing work for many of the GE businesses. Before joining GE, as director of the Electro-Optics Lab at the Industrial Technology Institute for 14 years, Kevin built the electro-optics business doing over 200 projects and spinning-off a number (6) of commercial products. Within the technical community for over 25 years, he developed and chaired new technical conferences and workshops including the industry recognized workshop of Optics and Lighting for Machine Vision that he taught for over 15 years. Kevin Harding is internationally recognized for his expertise in 3-D measurement technology, and has been recognized for this work by many organizations including

Society of Manufacturing Engineers (SME), Outstanding Young Engineer Award 1989
Engineering Society of Detroit, Leadership Award 1990
Automated Imaging Association, Leadership Award, 1994
SME (Society of Manufacturing Engineers), Eli Whitney Productivity Award 1997

Kevin Harding has published over 100 technical papers, taught over 60 short courses and tutorials to industrial and academic audiences, as well as an NTU video course on Optical Metrology, contributed sections to 6 books, and received over 35 patents. He has served as association chair, society committee chair and conference chair for over 20 years working with the SPIE, LIA, ESD, SME, and OSA. Kevin was the 2008 President of SPIE-the International Society for Optics and Photonics as well as a fellow of the SPIE.

REFERENCES

1. C.W. Kennedy and D.E. Andrews, *Inspection and Gaging*, Industrial Press, New York (1977).
2. E.O. Doebelin, *Measurement Systems Application and Design*, McGraw-Hill Book Company, New York (1983).
3. P. Cielo, *Optical Techniques for Industrial Inspection*, Academic Press, Boston, MA (1988).
4. A.R. Luxmore, Ed., *Optical Transducers and Techniques in Engineering Measurement*, Applied Science Publishers, London, U.K. (1983).
5. R.G. Seippel, *Transducers, Sensors, and Detectors*, Prentice-Hall, Reston, VA (1983).

6. R.P. Hunter, *Automated Process Control Systems*, Prentice-Hall, Englewood Cliffs, NJ (1978).
7. L. Walsh, R. Wurster, and R.J. Kimber, Eds., *Quality Management Handbook,* Marcel Dekker Inc., New York (1986).
8. N. Zuech, *Applying Machine Vision*, John Wiley & Sons, New York (1988).
9. K.G. Harding, The promise and payoff of 2D and 3D machine vision: Where are we today?, *Proceedings of SPIE, Two- and Three-Dimensional Vision Systems for Inspection, Control, and Metrology*, Vol. 5265, B.G. Batchelor and H. Hugli, Eds., pp. 1–15 (2004).
10. K. Harding, Machine vision lighting, *The Encyclopedia of Optical Engineering*, Marcel Dekker, New York (2000).

SENSOR REFERENCES

11. K. Harding, Overview of non-contact 3D sensing methods, *The Encyclopedia of Optical Engineering*, Marcel Dekker, New York (2000).
12. E.L. Hall and C.A. McPherson, Three dimensional perception for robot vision, *SPIE Proceedings*, 442, 117 (1983).
13. M.R. Ward, D.P. Rheaume, and S.W. Holland, Production plant CONSIGHT installations, *SPIE Proceedings*, 360, 297 (1982).
14. G.J. Agin and P.T. Highnam, Movable light-stripe sensor for obtaining three-dimensional coordinate measurements, *SPIE Proceedings*, 360, 326 (1983).
15. K. Melchior, U. Ahrens, and M. Rueff, Sensors and flexible production, *SPIE Proceedings*, 449, 127 (1983).
16. C.G. Morgan, J.S.E. Bromley, P.G. Davey, and A.R. Vidler, Visual guidance techniques for robot arc-welding, *SPIE Proceedings*, 449, 390 (1983).
17. G.L. Oomen and W.J.P.A. Verbeck, A real-time optical profile sensor for robot arc welding, *SPIE Proceedings*, 449, 62 (1983).
18. K. Harding and D. Markham, Improved optical design for light stripe gages, *SME Sensor '86*, Detroit, MI (1986).
19. B.F. Alexander and K.C. Ng, 3-D shape measurement by active triangulation using an array of coded light stripes, *SPIE Proceedings*, 850, 199 (1987).
20. M.C. Chiang, J.B.K. Tio, and E.L. Hall, Robot vision using a projection method, *SPIE Proceedings*, 449, 74 (1983).
21. J.Y.S. Luh and J.A. Klaasen, A real-time 3-D multi-camera vision system, *SPIE Proceedings*, 449, 400 (1983).
22. G. Hobrough and T. Hobrough, Stereopsis for robots by iterative stereo image matching, *SPIE Proceedings*, 449, 94 (1983).
23. N. Kerkeni, M. Leroi, and M. Bourton, Image analysis and three-dimensional object recognition, *SPIE Proceedings*, 449, 426 (1983).
24. C.A. McPhenson, Three-dimensional robot vision, *SPIE Proceedings*, 449, 116 (1983).
25. J.A. Beraldin, F. Blais, M. Rioux, and J. Domey, Signal processing requirements for a video rate laser range finder based upon the synchronized scanner approach, *SPIE Proceedings*, 850, 189, (1987).
26. F. Blais, M. Rioux, J.R. Domey, and J.A. Baraldin, Very compact real time 3-D range sensor for mobile robot applications, *SPIE Proceedings*, 1007, 330 (1988).
27. D.J. Svetkoff, D.K. Rohrer, B.L. Doss, R.W. Kelley, and A.A. Jakincius, A high-speed 3-D imager for inspection and measurement of miniature industrial parts, *SME Vision '89 Proceedings*, Chicago, IL (1989).
28. K. Harding and D. Svetkoff, 3D laser measurements on scattering and translucent surfaces, *SPIE Proceedings*, K. Harding, Ed., Philadelphia, PA, Vol. 2599, p. 2599 (1995).
29. M. Rioux, Laser range finder based on synchronized scanners, *Applied Optics*, 23(21), 3827–3836 (1984).
30. K. Harding, Moire interferometry for industrial inspection, *Lasers and Applications*, 1–73 (Nov. 1983).
31. K. Harding, Moire techniques applied to automated inspection of machined parts, *SME Vision '86*, Detroit, MI (June 1986).
32. A.J. Boehnlein and K.G. Harding, Adaption of a parallel architecture computer to phase shifted moire interferometry, *SPIE Proceedings*, 728, 183 (1986).
33. H.E. Cline, A.S. Holik, and W.E. Lorensen, Computer-aided surface reconstruction of interference contours, *Applied Optics*, 21(4), 4481 (1982).

34. W.W. Macy Jr., Two-dimensional fringe-pattern analysis, *Applied Optics*, 22(22), 3898 (1983).

35. L. Mertz, Real-time fringe pattern analysis, *Applied Optics*, 22(10), 1535 (1983).

36. L. Bieman, K. Harding, and A. Boehnlein, Absolute measurement using field shifted moire, *SPIE Proceedings, Optics, Illumination and Image Sensing for Machine Vision*, D. Svetkoff, Ed., Boston, MA, Vol. 1614, p. 259 (1991).

37. M. Idesawa, T. Yatagai, and T. Soma, Scanning moire method and automatic measurement of 3-D shapes, *Applied Optics*, 16(8), 2152 (1977).

38. G. Indebetouw, Profile measurement using projection of running fringes, *Applied Optics*, 17(18), 2930 (1978).

39. D.T. Moore and B.E. Truax, Phase-locked moire fringe analysis for automated contouring of diffuse surfaces, *Applied Optics*, 18(1), 91 (1979).

40. R.N. Shagam, Heterodyne interferometric method for profiling recorded moire interferograms, *Optical Engineering*, 19(6), 806 (1980).

41. M. Halioua, R.S. Krishnamurthy, H. Liu, and F.P. Chiang, Projection moire with moving gratings for automated 3-D topography, *Applied Optics*, 22(6), 850 (1983).

42. K. Harding and L. Bieman, Moire interferometry gives machine vision a third dimensiona, *Sensors*, p. 24 (October 1989).

43. K.G. Harding, and S.-G.G. Tang, Machine vision method for small feature measurements, *Proceedings of SPIE, Two- and Three-Dimensional Vision Systems for Inspection, Control, and Metrology II*, K.G. Harding, Ed., Vol. 5606, p. 153–160 (2004).

44. K.G. Harding and K. Goodson, Hybrid, high accuracy structured light profiler, *SPIE Proceedings*, 728, 132 (1986).

ERROR REFERENCES

45. S.D. Phillips, Performance evaluations, in *CMMs & Systems*, J.A. Bosch, Ed., Marcel Dekker, Inc., New York (1995), pp. 137–226, Chapter 7.

46. S.D. Phillips, B.R. Borchardt, G.W. Caskey, D. Ward, B.S. Faust, and D. Sawyer, A novel CMM interim testing artifact, *Cal Lab*, 1(5), 7 (1994); also in *Proceedings of Measurement Science Conference*, Pasadena, CA (1994).

47. S.D. Phillips and W.T. Estler, Improving kinematic touch trigger probe performance, *Quality* (April 1999).

48. K.G. Harding, Current state-of-the-art of contouring techniques in manufacturing, *Journal of Laser Applications*, 2(2–3), 41–48 (Fall 1990).

49. I. Moring, H. Ailisto, T. Heikkinen, A. Kilpela, R. Myllya, and M. Pietikainen, Acquisition and processing of range data using a laser scanner-based 3-D vision system, *Proceedings of SPIE, Optics, Illumination, and Image Sensing for Machine Vision II*, D. Svetkoff, Ed., Vol. 850, pp. 174–184 (1987).

50. K.G. Harding, Ed., Calibration methods for 3D measurement systems, *Proceedings of SPIE, Machine Vision and Three-Dimensional Imaging Systems for Inspection and Metrology*, Vol. 4189, p. 239 (2000).

51. K.G. Harding, Ed., Sine wave artifact as a means of calibrating structured light systems, *Proceedings of SPIE, Machine Vision and Three-Dimensional Imaging Systems for Inspection and Metrology*, Vol. 3835, pp. 192–202 (1999).

24 Microsensors for Manufacturing Processes

Xiaochun Li

CONTENTS

24.1 INTRODUCTION

24.1.1 OVERVIEW

Effective monitoring and diagnosis of manufacturing processes are of critical importance for manufacturing processes. If critical conditions are continuously monitored and controlled, problems can be detected and solved during the processing cycle, resulting in less damage to tools, less downtime for machine repair and maintenance, higher productivity, and less energy consumption. Driven by a widespread need for better process monitoring and diagnosis techniques, significant efforts have been taken to advance sensors and data analysis technology for manufacturing processes. However, current technology still evidently lags behind practical needs. The conventional sensors used are normally large in size and are either attached to the surface that might be far away from critical locations to avoid interference with the operation of the machine or destructively inserted into critical locations through appropriate channels in the components. As a result, it is difficult to provide measurement with a high-spatial and temporal resolution at distributed critical locations.

Significant development of micro thin film sensor technology has occurred in recent years. Due to their small sizes, distributed micro thin film sensors could be incorporated into manufacturing systems, particularly dies, molds, inserts for die-casting, stamping, forging, and injection molding without interfering with normal operations. Moreover, microsensors respond to changes much quicker than ordinary sensors. If microsensors can be embedded at critical locations not accessible to ordinary sensors, tremendous benefits can be achieved since both the spatial and temporal resolution of in-process sensing systems can be improved significantly. Due to their small size, microsensors can potentially be embedded without impairing the integrity of structures. Embedded sensors are also protected from damages caused by extraneous harsh environments.

Challenges of sensor embedding arise from the fact that most structures used in hostile industrial environments (such as those of manufacturing, energy utilization, automotive, and oil exploration and extraction) are metallic, and are usually made through "hostile" material processing (high temperature and stress).

24.1.2 MICROSENSORS FOR MANUFACTURING

Effective monitoring and diagnosis of manufacturing processes is of critical importance in reducing operation costs, improving product quality, and shortening response time. Tremendous efforts have been put forth by various researchers in recent years to advance microsensors technology to improve manufacturing processes. Such efforts not only focus on developing new designs of microsensors, but also on how to package and/or integrate microsensors into in-process manufacturing environment. The rapid development of microelectromechanical systems (MEMS) technologies in recent years has provided a high degree of spatial miniaturization and integration of electromechanical components that enable integrated sensing and control in manufacturing.

MEMS-based sensors have been developed for monitoring the operation status of manufacturing equipment and determine the process characteristics. They offer increased miniaturization and better performance over conventional devices and most commonly are used for pressure and acceleration measurement [1,2]. Their micro-scaled dimensions allow for more effective integration into manufacturing equipment or workpiece. Eaton and Smith [1] summarized the history of micromachined pressure sensors, and Gao and Li [2] reviewed the state of art of recent researches on micromachined pressure and acceleration sensors. Extensive research and development work have been done for these two types of microsensors. Micromachined pressure sensors [3–9] see application in automotive systems, manufacturing process control and medical diagnostics, while micromachined accelerometers [10–18] are widely used to measure vibration for monitoring the condition of the operating machines.

MEMS-based micro-acoustic emission (AE) sensor, although not as popular as pressure and acceleration microsensors, was studied for ultraprecision manufacturing [19,20]. Conventional AE sensors are widely used in manufacturing for structural flaws detection, monitoring damage progression, and real-time process monitoring. They can be effectively used for ultraprecision machining processes in diamond turning, grinding and lapping, and chemical mechanical polishing (CMP) [21–23]. Micro-AE sensors that may be embedded in the structures and tools have been developed to improve performance when used for detection of weak stress waves emitted by small defects.

24.1.3 SENSOR EMBEDDING

Embedding sensors during layer-wise fabrication of composite materials have emerged and been intensively studied in recent years. Both thin film [24–27] and fiber optic sensors [28–33] have been embedded in such materials. Du and Klamecki [24,25] applied embedded small piezoelectric sensing elements in a surface that is below the surface of interest to measure the tooling–workpiece interface loads for process monitoring. Minimum disturbance of the surface and the process was required. According to Friswell and Inman [26], structures with multiple embedded sensors will provide greater spatial resolution than a single sensor.

MEMS sensors have been implemented to measure the strain in composite structures. Hautamaki et al. [34] fabricated and embedded micro-electromechanical strain sensors in a fiber-reinforced laminated composite plate. All devices survived and yielded repeatable responses to uniaxial tension loads applied over 10,000 cycles.

A recent trend is to investigate the feasibility of embedding wireless sensors into functional composite structures; and wireless sensor embedding [34–37] and fiber Bragg grating sensors embedding [38–40] have become popular. Neural control has been applied to control the distributed sensors and actuators. Krantz et al. [35,36] studied remotely queried wireless embedded microsensors in composites. Hautamaki et al. [34] embedded MEMS sensors for measuring strain in composites, and these small-scale sensors were designed to function as part of a wireless sensing network. Pereira et al. [37] investigated methodology for intelligent sensing and wireless communication in harsh environments, focusing on recent efforts to test and characterize the performance of MEMS inertia sensors and the characterization of battery-free embedded sensors in munitions.

Comparing with sensor embedding in composite, sensor embedding in metal is much more challenging due to the lack of an effective method to integrate sensors into metal structures without damaging the sensors, and the high demands on mechanical and chemical stability of sensor structures imposed by harsh manufacturing environment, that is, high temperature and high strain/stress. The research on sensor embedding in metals has been rather limited; however, recent studies [13,41,42] conducted at Stanford University shows a promising advance in this area. They fabricated thin-film sensors on the internal surfaces of metallic parts and their subsequent embedding as the parts are completed through solid freeform fabrication (SFF). The sequence of their process is illustrated in Figure 24.1.

The thin film sensors were built on the internal surface of the part. Sensors were insulated to the surrounding metal by alumina films, and covered with the electroplated protective layers of copper and nickel. The part was then completed with high-temperature deposition (high-power laser cladding or welding) with modest success. Several severe problems were discovered during the sensor embedding process. First, pinholes were often present in the insulation layers and caused short circuit in the sensor layer. They are mainly due to the problem of compatibility between microfabrication and dirty manufacturing environment. Particularly, it was a matter of cleanness when a specimen is transferred from the "dirty" manufacturing environment to the clean environment of thin film processing. Second, severe defects were found including cracks in the thin-film sensor and insulation layers, and delamination between the layers. It is believed that such defects were caused by the high temperature gradient that produces severe stresses during material deposition. In addition, fabricating thin film sensors on a half-completed part was found very difficult due to the limitation in specimen geometry that can be made by the majority of microfabrication equipment.

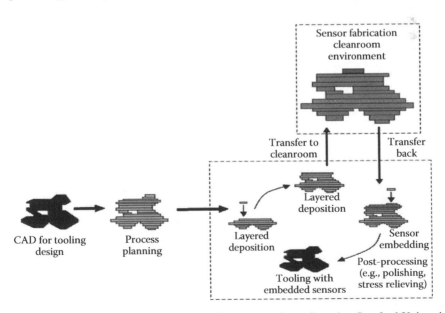

FIGURE 24.1 Sensor embedding during rapid tooling—research conducted at Stanford University. (From Golnas, A., Thin-film thermo-mechanical sensors embedded in metallic structures, PhD thesis, Stanford University, Palo Alto, CA, December 1999; Li, X.C. et al., *Proc. SPIE Int. Soc. Opt. Eng.*, 3986, 160, 2000; Li, X.C., Embedded sensors in layered manufacturing, PhD thesis, Stanford University, Palo Alto, CA, 2001.)

Electroplating-based sensor-embedding techniques can also embed sensors into electroplatable materials, such as nickel, some nickel alloys, gold, platinum, etc., that have various applications in harsh engineering environments. However, it is much desired to extend the sensor embedding approach to other important engineering materials such as stainless steels, titanium alloys, super alloys, etc., and in some cases, even to ceramic materials. To accomplish this goal, alternative viable embedding methods must be explored. Diffusion bonding is a solid-state joining process capable of joining a wide range of metal and ceramic combinations via atomic migration at the interfaces to produce both small and large components.

In the following, we will elaborate on a new sensor-embedding method, namely, the diffusion-bonding method, and demonstrate its applications in the research of mechanical milling. Embedded micro/nanosensors yield great potential to numerous engineering processes.

24.2 SENSOR EMBEDDING USING DIFFUSION BONDING METHOD

Diffusion-bonding method is capable of joining a wide range of metal and ceramic combinations to produce both small and large components. The process is dependent on a number of parameters, in particular, time, applied pressure, bonding temperature, and method of heat application. In its simplest form, diffusion bonding involves holding pre-machined components under load at elevated temperatures usually in a protective atmosphere or vacuum. The loads used are usually much below those that would cause macrodeformation of the parent material(s) and temperatures of $0.5-0.8T_m$ (where T_m = melting point in K) are employed. Bonding time can range from 1 to 60+ min, depending on the materials being bonded, the joint properties required, and the remaining bonding parameters [43].

Diffusion bonding was investigated as a viable embedding method. Using the developed thin-film system, micro thin-film strain gages were designed and fabricated on AISI 304 stainless steel (304SS) substrates and further embedded into 304SS structures through diffusion bonding. The sensors were then examined by the functionality test and metallurgical characterization.

To demonstrate the effectiveness of the metal-embedded sensors for manufacturing process study, high-temperature PdCr thin-film strain gage array was designed and fabricated using the developed method, and then tested in a vertical end-milling process.

24.2.1 SENSOR FABRICATION PROCEDURE

The developed dielectric multilayer ($Al_2O_3/Si_3N_4/Al_2O_3$) was used for sensor fabrication. The complete thin film system is shown in Figure 24.2. Palladium-13 wt.% chromium (commonly referred to PdCr) was selected as the sensing film and as resistive strain gage for its stable properties at high temperatures. PdCr resistive strain gage was developed by NASA [44–46]. It is capable of $\pm2000\,\mu\varepsilon$ at 1250 K and $\pm3000\,\mu\varepsilon$ at 1100 K with less than 10% error in the full-scale reading. Among the properties that make it ideal for high-temperature applications, PdCr forms a self-protecting chromium oxide (Cr_2O_3) layer to reduce film oxidation (thus enhancing its chemical stability), and it remains in solid solution up to the tested temperature range of 1250 K (metallurgical stability). One important parameter of the resistive strain gages for high-temperature application is the temperature coefficient of resistance (TCR). TCR describes the effects temperature has on strain for a given sensor material. PdCr film has a nearly linear response to the temperature change, that is, near-constant temperature coefficient of resistance (TCR or α), up to 1100°C.

Because this embedding method targets the materials that are not electroplatable, the previous batch production method is no longer applicable. The fabrication process has to directly start from the target substrate instead of a silicon wafer as the film transfer medium. However, the deposition techniques for the individual layer of the films are the same as used in the batch production method.

Since microelectronic grade stainless steel substrates are not available commercially, conventional grinding and polishing techniques were employed to obtain a smooth surface with an RMS

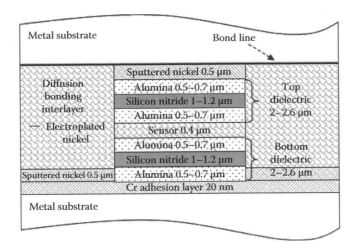

FIGURE 24.2 The structures of embedded micro thin-film sensors.

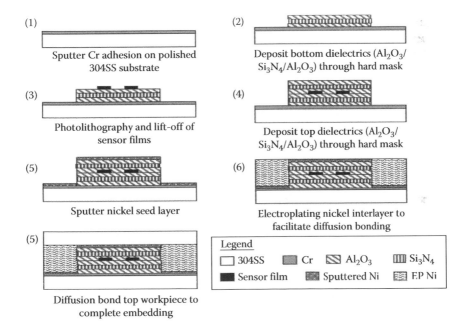

FIGURE 24.3 Sensor fabrication procedures.

(root-mean-square) roughness less than 100 nm that would be suitable for microfabrication of the microsensors in this study. The substrate surface roughness was measured using a Zygo New View white-light interferometer.

The fabrication process is illustrated in Figure 24.3. A polished 304SS substrate was first thoroughly degreased and cleaned with acetone, isopropanol, and deionized water, sequentially. A 20 nm titanium layer was sputtered onto the sample surface to promote adhesion for the later dielectric layers.

Next, through a shadow mask alumina (Al_2O_3) was deposited in an e-beam evaporator at a deposition rate of about 12 nm/min to a thickness of 0.6 μm. A 1.2 μm thick silicon nitride (Si_3N_4) layer was then deposited by plasma-enhanced chemical vapor deposition (PECVD) using SiH_4, NH_3 (both diluted in N_2), and N_2O at a deposition rate of 35 nm/min. Another 0.6 μm thick layer of alumina was evaporated on the silicon nitride before the sensor film sputtering.

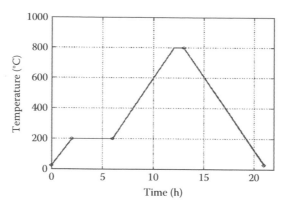

FIGURE 24.4   Diffusion bonding thermal cycle.

Photo resists (Shipley LOR and 1813) were patterned with standard photolithography techniques, followed by a PdCr film sputtering and lift-off procedure to obtain the strain gage.

On top of the sensor film, a dielectric multilayer of $Al_2O_3/Si_3N_4/Al_2O_3$ was again deposited through a patterned hard mask. In the next step, a 20 nm thick chromium adhesion-promoting layer and 500 nm thick nickel layer were sputtered on top of the dielectric multilayer. Next, the sample was coated with photo-resist AZ4620 that was then patterned to serve as the nickel electroplating mold. The area outside the dielectric layer was electroplated with nickel in a nickel sulfate bath to level the sample surface. The nickel layer serves as an interlayer for diffusion bonding.

At this point, the sensor fabrication in a cleanroom environment is completed and sensors are ready for diffusion bonding for embedding. The diffusion bonding process was carried out in a hot press machine. The bonding temperature schedule is shown in Figure 24.4. The schedule included a 4 h dwelling at 200°C, in order to bake out the moisture in the bonding chamber to achieve better vacuum level. The temperature ramp rate was at 100°C/h; the vacuum level of 2×10^{-3} Pa (1.5×10^{-5} Torr) was reached at 800°C; and the bonding pressure was at 4 MPa.

To examine the interdiffusion between the films, cross-sectional scanning electron microscopy (SEM) with energy dispersive spectroscopy (EDS) were conducted on the bonded samples using a LEO GEMINI 1530 SEM with a Schottky-type field emission and EDS x-ray analyzer. The SEM operating voltage was set at 15 kV. The results are shown in Figure 24.5. EDS mapping does not show signs of interdiffusion.

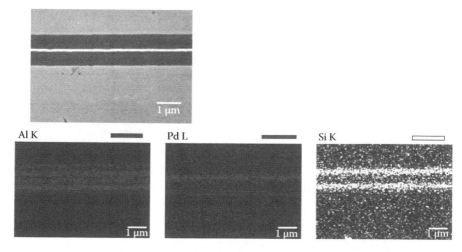

FIGURE 24.5   Cross-sectional SEM and EDS of thin film system after diffusion bonding.

24.2.2 Sensor Design for Milling Test

A layer-by-layer vertical milling testing scheme shown in Figure 24.6 was designed to evaluate the sensor performance. In this scheme, an end mill (carbide tool, $\phi = 6.35\,mm$) cut the workpiece with a feed rate of $100\,\mu m/s$. The spindle speed was set at 300 rpm.

The microsensors were embedded $735\,\mu m$ beneath the top surface of the workpiece. The $730\,\mu m$ thick material on the top of the sensors is cut through in seven steps, with $100\,\mu m$ each for the first six cut and $130\,\mu m$ for the last cut, after which only $5\,\mu m$ thick 304SS remained above the sensor dielectric films.

A sensor array consisting of 4 PdCr strain gages were designed for this study. Figure 24.7 illustrates the sensor design. Of the four gages, two (Gage #1 and #2) were aligned in the radial direction of the cutting circle and the other two (Gage #3 and #4) were aligned in tangential direction. They were designed in such a way to study if there is a significant difference of strain magnitude in the

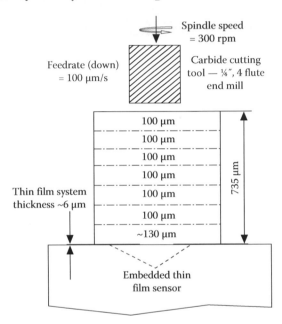

FIGURE 24.6 The seven-step layer-by-layer vertical milling test on workpiece with embedded TFSG array.

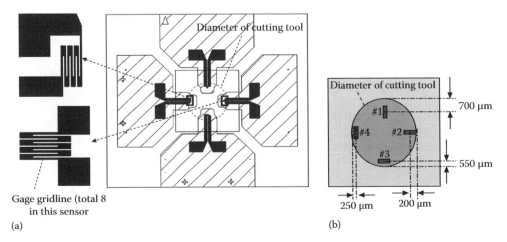

FIGURE 24.7 Sensor design: (a) detailed drawing of sensor design and (b) schematic of sensor placement in relation to the diameter of the cutting tool.

TABLE 24.1
Sensor Location and Orientation

Sensor No.	Orientation	Distance to Edge (mm)	Distance to Center (mm)
1	Radial	0.700	2.475
2	Radial	0.200	2.975
3	Tangential	0.550	2.625
4	Tangential	0.250	2.925

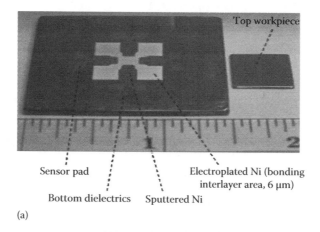

(a)

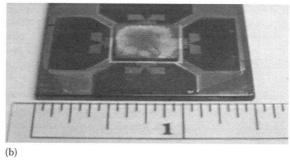

(b)

FIGURE 24.8   Pictures of a sample prepared for milling test: (a) before embedding and (b) after embedding.

different directions relative to the material removal direction during the cutting. In addition, the distances of the sensors to the center of the cutting circle are slightly different, as shown in Table 24.1, to study the spatial distribution of temperature and strain.

Gage #1 and #2 have a 225 μm by 400 μm overall gage area that consists of eight gridlines, while Gage #3 and #4 have a 195 μm by 400 μm overall gage area that consists of seven gridlines. However, all gages have the same gridline width at 15 μm and total gridline length in 1600 μm. The slight difference of their design was to accommodate their orientations. Figure 24.8 shows pictures of a sample before and after the embedding process.

24.3 CHARACTERIZATION OF DIFFUSION BONDING EMBEDDED SENSORS

The static response and thermal response of the embedded PdCr gages were characterized, respectively. For static response, its gage factor (GF) was calibrated in a cantilever beam-bending test.

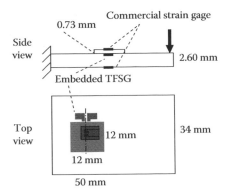

FIGURE 24.9 Schematics of static response calibration of TFSG.

Two commercial constant strain gages from Micro Measurement (Model: EA-06-090DH-350 with a room temperature gage factor of 2.105) were used to obtain the true strain at the embedded PdCr thin film strain gages.

The schematic of the experimental setup is shown in Figure 24.9. One commercial gage was attached on the surface of the top substrate (12 mm by 12 mm, 0.73 mm in thickness) and another one was attached to the bottom surface of the bottom substrate (50 mm by 34 mm, 2.6 mm in thickness). Two substrates were diffusion bonded together to form an integral part with the thin-film strain gages (TFSG) embedded between the bonding surface. Both gages have the same X–Y location and orientation as the embedded gage. All three, strain gages were connected to the Wheatstone bridge circuit in a quarter bridge configuration.

The sensor leads were attached by using solder, and were connected to a NI 6070E data acquisition system. The output of the commercial gages was used to calculate the applied strains on the TFSG. The substrate was fixed with clamps at one end as a cantilever beam and mechanical force was applied to the end of the substrate.

At each given loading, the output of the two commercial gages were used to calculate the applied strain at TFSG, which in turn, were correlated to the voltage output to obtain the TFSG gage factor. The strain outputs from the two commercial gages at five different loadings are plotted against substrate thicknesses in Figure 24.10, where the bottom surface is treated as the reference plane. From the curves, the neutral axis of this bonded structure can easily be identified at 1.37 mm from the reference plane. Since the TFSG were embedded at 2.6 mm (bottom substrate thickness) from the reference plane, the strains at 2.6 mm were interpolated from those curves to calibrate the voltage readings registered by TFSG. The average gage factor is found to be 1.601 with a standard deviation of 0.012.

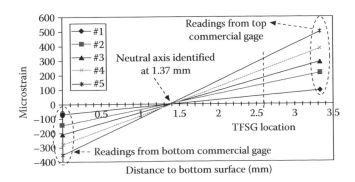

FIGURE 24.10 TFSG calibration—static response.

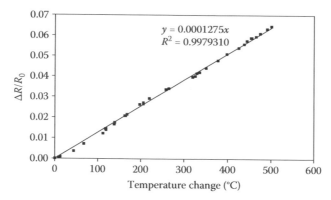

FIGURE 24.11 TFSG calibration—thermal response.

The thermal response of the PdCr TFSG was calibrated in a temperature controlled tube furnace. K-type thermocouples were directly attached to the substrate for more accurate temperature readings at the gage. Figure 24.11 plots the resistance change of TFSG against temperature change using room temperature (23°C) as the reference, that is, 0°C temperature change. Considering the definition of GF = $(\Delta R/R)/\varepsilon$, the thermal sensitivity of the gage is approximately 80 $\mu\varepsilon$/°C. Both the gage factor and TCR obtained in this study are close to the values previously reported from NASA's research [44–46].

24.4 MILLING TEST FOR EMBEDDED SENSORS

Milling experiments were performed using a fresh carbide tool to evaluate the effectiveness of the embedded sensor array for *in situ* monitoring. The milling conditions are shown in Table 24.2. Changes in the resistance of each TFSG are detected as changes in the output voltage of a quarter-bridge circuit. The bridge excitation voltage is 5 V. A 50 Hz low-pass filter was applied before the A/D converter. The sampling frequency was 1 kHz. The registered voltages are converted into strains using the calibrated gage factor at 1.601. Figure 24.12 shows pictures of a sample after the milling test. TFSGs still remained embedded and functional, but the remaining 304SS on top of the sensors was very thin. Figure 24.13 shows pictures of a sample after the milling test. The milling was carried out in seven steps. Immediately after each step, the end mill tool was retracted to let the workpiece completely cool down before the next cut.

24.4.1 Typical Signal Output Patterns from Embedded Sensors

Two types of signal patterns were observed from the cutting tests. Each sample contains only one pattern throughout the seven steps of cutting. Figure 24.13 shows one typical strain output pattern of TFSG in vertical milling. The sensor outputs change quickly at the beginning and the ending of the cutting. There are dynamic changes in strain during the cutting. The fact that the strain decreases over a period to zero after cutting suggests that the static part of the signal is purely the thermal strains that are caused by cutting.

This made it possible to separate the thermal strain and the mechanical strain during cutting by decomposing the signal into static and dynamic parts, where the static strain can be converted to temperature using the thermal response curve shown in

TABLE 24.2
Tool Parameter and Cutting Conditions

Carbide end mill
Square end
Diameter, 6.35 mm
Vertical feed, 20 μm/rev
Spindle speed, 300 rpm
Feed rate, 100 μm/s
Initial temperature, 23°C

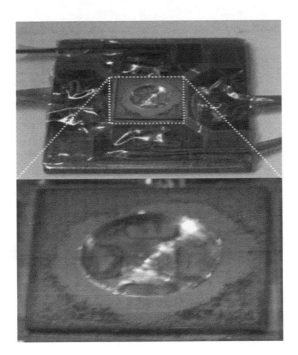

FIGURE 24.12 Pictures of a sample after milling test.

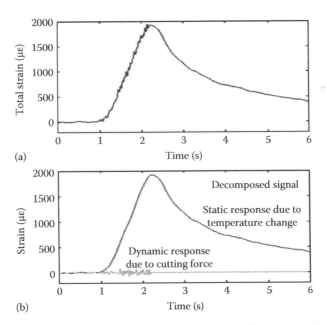

FIGURE 24.13 Typical signal pattern during vertical milling: (a) milling from 600 to 500 μm above the sensor and (b) decomposed signal: top—below 1 Hz, bottom—1 to 50 Hz.

Figure 24.11. By neglecting the gage factor change due to temperature, the thermal strain can be approximately converted to temperature change at the rate of 80 μɛ/°C.

The static part of the signal was obtained by applying a 1 Hz low-pass FFT (fast Fourier transformation) filter to the original signal that is shown as the top curve in Figure 24.13b. The dynamic part, shown as the green curve in Figure 24.13b, was obtained by using a 1 Hz high-pass FFT filter. A close examination of the dynamic signal found that it shows about 20 peaks/s, that is, a 20 Hz

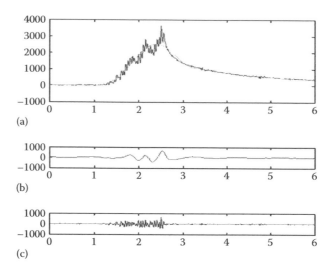

FIGURE 24.14 Typical signal pattern during vertical milling with strong 1–6 Hz component. (a) Signal from 430 to 330 mm above the sensor, blue: raw data, green: static component (<1 Hz); (b) decomposed signal: 1–6 Hz; and (c) decomposed signal: >6 Hz.

component. It is not surprising to find that the peak number matches the number of times a cutting edge passed over the top of a sensor per second, recalling that the spindle speed was 300 rpm, which is equivalent to 5 Hz, and a 4-flute end-mill cutter was used for the test. It is a clear indication that each dynamic strain peak was a direct result of the shearing force generated from the material removal by a cutting edge.

The other signal pattern was found in the majority of the experiments. It is essentially the same as the first pattern except that signals of this pattern contain a strong frequency component between 1 and 6 Hz. This component became stronger and more prominent than the 20 Hz component when the cutting was vertically approaching the embedded sensors. Figure 24.14 shows signal decomposition of such pattern from the cutting step from 430 to 330 μm. The decomposed signal consists of three components, static (1 Hz low pass), 1–6 Hz, and 6 Hz and higher. The peak strain of 1–6 Hz component nearly doubles the magnitude of the 20 Hz component. While the 1–6 Hz component consistently presents itself in each cutting step, it is not clear how it relates to the material removal of the cutting. However, one logical explanation is that the source of this component was from tool eccentricity (runout). For a milling tool, the runout would produce a frequency response having the dominant component a 1-per-rev frequency. In this case, the 300 rpm spindle speed would producing 5 Hz, which appeared to match the observed 1–6 Hz component. An evidence of the runout is that the hole diameter is 200–300 μm larger than the nominal diameter of the cutter (0.25″, that is, 6.35 mm).

24.4.2 Temperature and Strain Distribution

After separating the static and the dynamic components, the time-resolved temperatures and mechanical strains of each cutting step can be obtained. The peak temperatures from each cutting step registered by each sensor were plotted against the distance to the sensor when the corresponding cutting step concluded, as shown in Figure 24.15a, and their locations are indicated in Table 24.1.

For example, the data point at distance "635 μm" was obtained in the first step cutting, where the cutting started at 735 μm (initial material thickness above the sensors) vertically away from the sensors and concluded at 635 μm, when the peak temperature from this step was reached. The cutting depth for each of the first six steps was 100 μm. Therefore the peak temperatures were seen at 635,

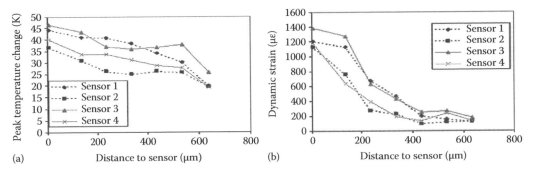

FIGURE 24.15 Distribution of peak temperature and dynamic strain from milling test. (a) Peak temperature changes vs. distance to sensor (μm). (b) Dynamic strain vs. distance to sensor (μm).

535, and 135 μm that correspond to steps 1, 2, …, and 6. The cutting depth of the seventh step was 130 μm, which started from 135 μm vertically above the sensors and concluded at 5 μm. At the end of the cutting, there was about 5 μm thick 304SS and 3 μm thick multilayer dielectric film, remaining over the sensor films.

From the distribution curve of "peak temperature change," higher temperature changes were reached when the cutting reaches closer to the sensor. However, the magnitude of the temperature changes was quite low even when cutting concluded at very close to the sensor, that is, 37°C–47°C at 5 μm. It is most likely due to the low cutting speed and small cutting depth. It also appears that the closer the sensor to the cutting center, the higher the peak temperature obtained, while the orientation of the sensor has little effect.

Similarly, Figure 24.15b shows a typical peak dynamic strain distribution of each cutting step. The dynamic strain used here only contains the frequency component above 6 Hz that is apparently only due to the material removal instead of the thermal effect (<1 Hz) and the possible cutting tool eccentricity effect (1–6 Hz). As expected, the peak dynamic strains are very low when cutting is to take place far away from the sensor. The magnitude increased significantly faster within 400 μm to the cutting surface, reaching above 1000 με within 10 μm. When comparing sensors of the same orientation, sensor #1 versus #2, and sensor #3 versus #4, it is clear that the closer the sensor to the cutting center, the higher the peak strains. If only considering the sensor locations while ignoring the orientation factor, this trend is still generally true, but less distinguishable. For instance, sensor #1 was 150 μm closer to the cutting center than sensor #3, but both saw very similar peak strain magnitudes.

When comparing sensor #1 versus #3 (in opposite orientation, 150 μm location difference to cutting center), and sensor #2 versus #4 (in opposite orientation, 50 μm location difference to cutting center), it suggests that the sensor orientation has an insignificant effect: dynamic strain induced from material removal during cutting presents similar magnitude in both radial and tangential directions.

24.5 SUMMARY

Diffusion bonding was proved a viable method for sensor embedding. The thin film system sustained an 800°C and 4 MPa bonding process to embed sensors into 304SS. The success of embedding was validated by the sensor functionality test and metallurgical characterization. The method can be used to produce both "sensors in workpiece" and "sensors in tool" for the studies of various harsh-environment manufacturing processes.

As a demonstration of the potential applications for the metal-embedded sensors, PdCr strain gage array was designed, fabricated, and then embedded into 304SS. The samples were then tested in a vertical milling process. Time- and spatial-resolved data were obtained from the test. Because temperature changes monotonically during or after cutting, while mechanical strain changes dynamically during cutting, thermal strains and mechanical strains were able to be separated by decomposing the raw data into appropriate frequency components. In turn, temperature changes were obtained by converting from thermal strain using the calibrated sensor thermal response curve.

Higher temperature changes were reached when the vertical cutting reaches closer to the sensor. However, the magnitude of the temperature changes was quite low even when cutting concluded at very close to the sensor (within 10 μm), most likely due to the low cutting speed and small cutting depth.

All of the tests presented dynamic strains at the frequency matching the frequency of flute passing over the sensors, which is a clear improvement over existing sensing methods. In some tests, a low frequency (1–6 Hz) strain component was identified. This signal component was not clearly related to the removal of material by cutting; however, it was most likely due to the tool runout, as suggested by the larger diameter of the hole than the nominal tool diameter.

A general trend found in the experiments is that vertical milling produces higher peak temperatures and dynamic strains closer to the cutting center than horizontal milling. The dynamic strains induced from material removal during cutting presents similar magnitude in both radial and tangential directions.

The milling tests demonstrated the ability of using this method to correlate the machining parameters to temperatures and strains within the workpiece that will be very valuable for the fundamental understanding of various manufacturing processes. It can also be used for benchmarking numerical modeling and simulations. In numerical modeling, many assumptions have to be made. The credibility of the model depends on these assumptions. However, many of these assumptions are difficult to validate. The work reported in this chapter provides direct experimental results that can be compared with numerical results, thus serve as a tool of model calibration. A combined, concurrent theoretical and experimental approach can be adopted with tentative theories being used to design critical experiments and experimental results (with embedded sensors) being used to suggest improvements to the theory. It would be a breakthrough in fundamental studies of the mechanics of cutting since it would be possible, for the first time, to obtain realistic measurements of stresses and temperatures in the cutting zone of the tool.

QUESTIONS

Q.24.1 How will micro/nanotechnology impact conventional manufacturing processes?

Q.24.2 How will micro/nanosensors be applied for energy efficiency in various industrial processes?

Q.24.3 How can engineers accelerate the use of micro/nanotechnology in practical engineering practice?

Q.24.4 How can educators implement more emerging technologies into curriculum for next generation engineers?

Q.24.5 How to mass-produce human accessible devices with desirable micro/nano-functionalities?

ABOUT THE CONTRIBUTING AUTHOR

Professor Xiaochun Li is a leading expert in the fabrication and embedding/packaging of thin film micro/nano-electrical and photonic sensors into metal and ceramic structures for transfer into mechanical subsystems and manufacturing applications. His knowledge provides solid expertise to study micro/nanosensors embedding into ceramic materials for cutting inserts applications. Professor Li currently serves as the founding director of the University Center for Structurally Integrated Micro/Nano Systems (SIMNS). Dr. Li received CAREER award from National Science Foundation (NSF) in 2002, and the Jiri Tlusty Outstanding Young

Manufacturing Engineer Award from the Society of Manufacturing Engineers in 2003. He currently is a professor and the Villas Associates Scholar at University of Wisconsin–Madison. He was awarded the prominent 2008 Howard F. Taylor Award from the American Foundry Society (AFS). He has published more than 100 technical papers. He received his PhD from Stanford University, California, in 2001.

REFERENCES

1. Eaton, W.P. and Smith, J.H. (1997), Micromachined pressure sensors: Review and recent developments, *Proceedings of SPIE—The International Society for Optical Engineering*, 3046, 30–41.
2. Gao, R. and Li, Z. (2004), Micromachined microsensors for manufacturing, *IEEE Instrumentation & Measurement Magazine*, 7(2), 20–26.
3. Chatzandroulis, S., Tsoukalas, D., and Neukomm, P.A. (2000), A miniature pressure system with a capacitive sensor and a passive telemetry link for use in implantable applications, *Journal of Microelectromechanical Systems*, 9(1), 18–23.
4. Chavan, A.V. and Wise, K.D. (2001), Batch-processed vacuum-sealed capacitive pressure sensors, *Journal of Microelectromechanical Systems*, 10(4), 580–588.
5. Fuller, L.F. and Sudirgo, S. (2003), Bulk micromachined pressure sensor, *Proceedings of Biennial University/Government/Industry Microelectronics Symposium*, Boise, ID, pp. 317–320.
6. Guckel, H. (1991), Surface micromachined pressure transducers, *Sensors and Actuators A: Physical*, 28(2), 133–146.
7. Lin, L., Chu, H.C., and Lu, Y.W. (1999), A simulation program for the sensitivity and linearity of piezoresistive pressure sensors, *Journal of Microelectromechanical Systems*, 8(4), 514–522.
8. Wang, C.C., Gogoi, B.P., Monk, D.J., and Mastrangelo, C.H. (2000), Contamination-insensitive differential capacitive pressure sensors, *Journal of Microelectromechanical Systems*, 9(4), 538–543.
9. Zhou, J., Dasgupta, S., Kobayashi, H., Wolff, J.M., Jackson, H.E., and Boyd, J.T. (2001), Optically interrogated MEMS pressure sensors for propulsion applications, *Optical Engineering*, 40(4), 598–604.
10. Biefeld, V., Buhrdorf, A., and Binder, J. (2000), Laterally driven accelerometer fabricated in single crystalline silicon, *Sensors and Actuators A: Physical*, 82(1), 149–154.
11. Burrer, C., Esteve, J., and Lora-Tamayo, E. (1996), Resonant silicon accelerometers in bulk micromachining technology—An approach, *Journal of Microelectromechanical Systems*, 5(2), 122–130.
12. Li, L., Xu, Y., Zhao, Y., Liang, C., Wei, T., and Yang, Y. (1999), Micromachined accelerometer based on electron tunnelling, *Proceedings of SPIE—The International Society of Optical Engineering*, 3891, 121–125.
13. Li, X.C., Golnas, A., and Prinz, F. (2000), Shape deposition manufacturing of smart metallic structures with embedded sensors, *Proceedings of SPIE—The International Society for Optical Engineering*, 3986, 160–171.
14. Liu, C.H. and Kenny, T.W. (2001), A high-precision, wide-bandwidth micromachined tunneling accelerometer, *Journal of Microelectromechanical Systems*, 10(3), 425–433.
15. Partridge, A., Reynolds, J., Chui, B., Chow, E., Fitzgerald, A., Zhang, L., Maluf, N., and Kenny, T. (2000), A high-performance planar piezoresistive accelerometer, *Journal of Microelectromechanical Systems*, 9(1), 58–66.
16. Plaza, J., Collado, A., Cabruja, E., and Esteve, J. (2002), Piezoresistive accelerometers for MCM package, *Journal of Microelectromechanical Systems*, 11(6), 794–801.
17. Seshia, A.A., Palaniapan, M., Roessig, T.A., Howe, R.T., Gooch, R.W., Schimert, T.R., and Montague, S. (2002), A vacuum packaged surface micromachined resonant accelerometer, *Journal of Microelectromechanical Systems*, 11(6), 784–793.
18. Xie, H. and Fedder, G. (2000), CMOS z-axis capacitive accelerometer with comb-finger sensing, *Proceedings of IEEE MEMS*, Miyazaki, Japan, pp. 496–501.
19. Eccardt, P., Niederer, K., and Fischer, B. (1997), Micromachined transducers for ultrasound applications, *Proceedings of Ultrasonics Symposium*, Toronto, Canada, vol. 2, pp. 1609–1618.
20. Pedersen, M., Olthuis, W., and Bergveld, P. (1998), High-performance condenser microphone with fully integrated CMOS amplifier and DC–DC voltage converter, *Journal of Microelectromechanical Systems*, 7(4), 387–394.
21. Dornfeld, D.A., Lee, Y., and Chang, A. (2003), Monitoring of ultraprecision machining processes, *International Journal of Advanced Manufacturing Technology*, 21, 571–578.

22. Dornfeld, D.A. and Cai, H.G. (1984), An investigation of grinding and wheel loading using acoustic emission, *Transactions of ASME, Journal of Engineering for Industry*, 106(1), 28–33.

23. Tonshoff, H.K., Wulfsberg, J.P., Kals, H.J., Konig, W., and Van Luttervelt, C.A. (1998), Development and trends in monitoring and control of machining processes, *Annals of the CIRP*, 37(2), 611–622.

24. Du, H. and Klamecki, B.E. (1993), Characterization of force sensors embedded in surfaces for manufacturing process monitoring, *ASME Manufacturing Science and Engineering*, 64, 207–216.

25. Du, H. and Klamecki, B.E. (1999), Force sensors embedded in surfaces for manufacturing and other tribological process monitoring, *Journal of Manufacturing Science and Engineering*, 121(4), 739–748.

26. Friswell, M.I. and Inman, D.J. (2000). Sensor validation for smart structure, *Journal of Intelligent Material Systems and Structure*, 10(12), 973–982.

27. Ogisu, T., Nomura, M., Andou, N., Takaki, J., Song, D., and Takeda, N. (2000), Development of damage suppression system using embedded SMA foil sensor and actuator, *Proceedings of SPIE—The International Society for Optical Engineering*, 3991, 62–73.

28. Chang, C. and James, J.S. (1993), Metal-coated optical fiber damage sensors, *Proceedings of SPIE—The International Society for Optical Engineering*, 1918, 138–144.

29. Foedinger, M.I., Rea, D., Sirkis, J., Wagreich, R., Troll, J., Grande, R., Davis, C., and Vandiver, T.L. (1998), Structural health monitoring of filament wound composite pressure vessels with embedded optical fiber sensors, *International SAMPE Symposium and Exhibition-Proceeding*, 43(1), 444–457.

30. Kim, K., Breslauer, M., and Springer, G.S. (August 1992), Effect of embedded sensors on the strength of composite laminates, *Journal of Reinforced Plastics and Composites*, 11(8), 949–958.

31. Lawrence, C.M. (1997), Embedded fiber optical strain sensors for process monitoring of composites, PhD thesis, Stanford University, Palo Alto, CA.

32. Sirkis, J.S., Chang, C., and Smith, B.T. (1994), Low velocity impact of optical fiber embedded laminated graphite/epoxy panels, *Journal of Composite Materials*, 28(14), 1347–1370.

33. Spillman, W.B. (1994), Fiber optic embedded sensors, *Proceedings of IEEE LEOS Annual Meetings*, Boston, MA, vol. 2, pp. 230–231.

34. Hautamaki, C., Zurn, S., Mantell, S.C., and Polla, D.L. (2000), Embedded microelectromechanical systems (MEMS) for measuring strain in composites, *Journal of Reinforced Plastics and Composites*, 19(4), 268–277.

35. Krantz, D. and Belk, J.H (1997), Remotely-queried wireless embedded microsensors in composites, *Proceedings of SPIE—The International Society for Optical Engineering*, 3044, 219–226.

36. Krantz, D., Belk, J.H., Biermann, P.J., Dubow, J., Gause, L.W., Harjani, R., Mantell, S., Polla, D., and Troyk, P. (1999), Project update: Applied research on remotely-queried embedded microsensors, *Proceedings of SPIE—The International Society for Optical Engineering*, 3673, 157–164.

37. Pereira, C.M., Mattice, M.S., and Testa, R. (2000), Intelligent sensing and wireless communications in harsh environment, *Proceedings of SPIE—The International Society for Optical Engineering*, 3990, 194–203.

38. Jin, X.D., Sirkis, J.S., Chung, J.K., and Venkat, V.S. (March 1998), Embedded in-line fiber etalon/Bragg Grating hybrid sensors to measure strain and temperature in a composite beam, *Journal of Intelligent Material Systems and Structures*, 9(3), 171–181.

39. Murukeshan, V.M., Chan, P.Y., and Ong, L.S. (January 2001), Intracore fiber Bragg grating for strain measurement in embedded composite structures, *Applied Optics*, 40(1), 145–149.

40. Murukeshan, V.M., Chan, P.Y., Ong, L.S., and Seah, L.K. (2000), Cure monitoring of smart composites using fiber Bragg grating based embedded sensors, *Sensor and Actuators, A: Physical*, 79(2), 153–161.

41. Golnas, A. (December 1999), Thin-film thermo-mechanical sensors embedded in metallic structures, PhD thesis, Stanford University, Palo Alto, CA.

42. Li, X.C. (2001), Embedded sensors in layered manufacturing, PhD thesis, Stanford University, Palo Alto, CA.

43. O'Brien, R.L. (ed.) (1991), *Welding Handbook*, 8th edn., *Welding Processes*, vol. 2, American Welding Society, Miami, FL.

44. Hulse, C.O., Bailey, R.S., Grant, H.P., and Przybyszewski, J.S. (1987), High Temperature Static Strain Gage Development Contract, NASA CR-180811.

45. Hulse, C.O., Bailey, R.S., Grant, H.P., and Przybyszewski, J.S. (1991), High Temperature Static Strain Gage Development, NASA CR-189044.

46. Wrbenek, J.D. and Fralick, G.C. (2006), Developing Multilayer Thin Film Strain Sensors with High Thermal Stability, NASA TM-2006-214389.

25 Engineering Design and Design for Manufacturing

Zuozhi Zhao

CONTENTS

This chapter is about design and the information flow from design to manufacturing. Design covers a broad spectrum ranging from architecture design, industrial design, software design, fashion design to engineering design, civil design, and aerospace design. Although they all share some common elements such as the design philosophies, the design methods, the design techniques, and the design tools, the nature of the end output differs a lot. We will concentrate on engineering design only in this chapter.

In this chapter, a reference frame for engineering design is established first as the base for further discussion. The big picture of the engineering design research in academia and the design practice in industrial sectors is briefly covered. The intention is not to introduce the concepts and methods of the engineering design but to provide a reference frame that serves as a guide to further details about each method or technique, if the audience is interested. Then the use of computers in engineering design and manufacturing is briefly introduced since nowadays the computer is an indispensable part of engineering design. It is followed by the discussion of the importance of considering manufacturability in the early design stage. Next, knowledge management in design and manufacturing is discussed. Finally, the challenges that face the current manufacturing enterprises are discussed and the advancement of technology is predicted.

25.1 ESTABLISH THE ENGINEERING DESIGN REFERENCE FRAME

25.1.1 WHAT METHODS AND TECHNIQUES DID WE LEARN AT SCHOOL?

My major in undergraduate study was mechanical engineering. Here are some of the courses that were assigned to me (and others):

- Design principles, which covers the design of gears (how to calculate the pitch, etc.), cams, and other mechanical mechanisms
- Mechanics, which introduces the concepts of stress and strain, how to calculate displacement, etc.
- Fatigue analysis, which covers creep, buckling, low-cycle fatigue, high-cycle fatigue, fretting, and so on
- Drafting, which teaches the students to draw the part and assembly drawings using the "T" ruler on A0, A1, and A2 paper
- Manufacturing process introduction and internship. I went through traditional processes such as turning, milling, broaching, grinding, forging, welding, casting, and CNC
- Optimization, statistics, and probabilities
- Computer-aided design, which teaches the students to use AutoCAD for 2D drawing and 3D modeling

As I can recall today, there were no courses on market research, which identifies the customer requirements; no courses on idea generation that systematically introduce the methods to generate concepts to satisfy the customer requirements, such as brainstorming, TRIZ, morphological charts, gallery methods, etc.; no courses on design techniques such as quality function deployment (QFD) that help translate the customer requirements to engineering design specifications; and more importantly, there were no courses on the systemic design approach to consider the product life cycle issues during engineering design.

After spending years of doing my PhD in the engineering design field, and working in GE Global Research Center on design-related projects for different products in aerospace, energy, and healthcare sectors, one idea gradually came into my mind: there should be an engineering design course to introduce the reference frame of engineering design. This will help students to visualize the big picture at the beginning of their career. And, down the road, they would know where, when, how, and why to apply certain methods or techniques they learned at school. Examples include optimization, probabilistic design, statistical analysis, numerical simulation, finite element analysis, computational fluid mechanics and solid mechanics, and so on.

25.1.2 What Kinds of Artifacts Are We Designing?

Even after we limit the scope of this chapter to engineering design, it still covers a very broad spectrum. The end products could be consumer or industrial products. Examples of consumer products include as simple as a kitchen knife, or as complex as a sports car. In this case, the ergonomics and esthetics play roles as important as functionality and reliability; thus, the end customers' preferences need to be modeled and the criteria of a "good" design are more subjective. For industrial products, examples include a simple artifact such as a cantilever beam, and complex engineering systems such as aircraft engines, airplanes, and gas turbines. The evaluation criteria for good design are relatively more objective.

25.1.3 What Methods Do We Use to Drive the Design?

Although the design requirements might be different for different products, the procedure of realizing the design is the same, from most researchers and industry practitioners' perspectives of design. They all start with the customer requirements, sometimes explicit and quantitative, and sometimes implicit and qualitative, and then the requirements are translated into engineering design specifications. From the design specifications, the design concepts are generated. This is the most exciting step in engineering design since it requires a lot of creativity and experience/knowledge to come up with one or more good concepts. Then the concept drives the working principles and system behaviors. There might be alternative ways to realize the same concept, just as there might be multiple concepts that can satisfy the same set of design specifications and customer requirements. Once the principles and behaviors are determined, the system configuration, the materials, and the geometry are designed based on theories, analysis, simulation, and testing if necessary. In the later stage of design, more details are added such as dimensions, tolerances, and surface finish. Then the design is passed to the manufacturing department or outside vendors for production and assembly.

Figure 25.1 depicts the above procedures with emphasis on the iteration nature of engineering design.

The above-described design process is a systematic and traditional way of viewing the engineering design. The bible of this perspective is *Engineering Design* by Pahl and Beitz [1]. In this view, design activities can be grouped into three states: the conceptual design in which the ideas and concepts are conceived; the preliminary design in which the principles, the behaviors, and the configurations are generated; and the detail design stage in which the detailed geometry, dimensions, and tolerances are decided.

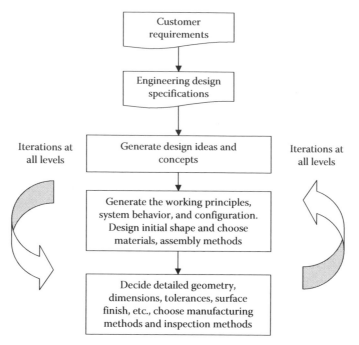

FIGURE 25.1 Engineering design phase.

Historically, engineering design is usually viewed as a problem solving procedure. According to Hazelrigg [2], "In the physical sciences or engineering, a problem is a question posed in the following form: given a state of nature at one time, determine or predict the state of nature at another time, or given elements of the state of nature, determine others." It is assumed that the current state is known in all relevant details and the laws of nature that apply are known. The designer has the necessary resources to apply the laws of nature and reach a solution.

In late 1990s, more and more design researchers started viewing engineering design as a decision-making process that requires rigorous evaluation of design alternatives [3,4]. Many publications in decision-based design (DBD) have emerged in recent years and a workshop is held online [5]. Such a viewpoint is consistent with the definition of decision making as (1) there is a range of possible actions, (2) each action is characterized by a set of consequences, some of which are beneficial and others less so, and (3) the decision maker is required to weigh up the pros and cons before arriving at a preferred action, and to do this he/she might use a range of decision criteria and rules [6].

As mentioned above, at each stage, there are decisions to be made: the choice of different concepts, different working principles, different configurations, different materials, different geometries, different dimensions and tolerances, and different manufacturing processes and assembly methods. It certainly requires experience, domain knowledge, as well as the design techniques to make a good engineering decision. At the same time, it requires a rational decision-making method to make an informed decision, based on the pros and cons. The choice of different judging criteria and the decision-making method would give totally different outcomes, given the same engineering knowledge and experiences.

25.1.4 DESIGN TYPES: NEW, ADAPTIVE, OR ROUTINE

From the previous discussion, we know that design activities can be divided into different stages. To make things more complicated, we now introduce another dimension into the reference frame of engineering design. In general, the design tasks can be classified into the novel design, the adaptive design, and the routine design. The novel design is to design something new. For example, consider the birth of the iPod, the first car, the first gas turbine, and the first airplane. When designing something new, there are no similar designs that could be referred to, and the designer(s) has to start from scratch. It requires a sharp sense of the new requirements, significant accumulation of domain knowledge and experience, and sometimes a fallen apple. The adaptive type is to improve the current design with the use of another principle, another material, another configuration, or another manufacturing process. It can deliver more functions, enhance the performance, improve the quality, or bring down the cost. The routine design is to scale up or down a similar design, and deliver the product based on similar requirements, similar specifications (only the value changed), similar principles, and similar materials. It requires the least effort of new analysis, simulation, and testing but can dramatically shorten the time to market.

25.1.5 DISTRIBUTED DECISION-MAKING NATURE OF ENGINEERING DESIGN

There exist two types of engineering decision making: independent decision making and linked decision making. In the former case, the decision maker is given only one problem each time. Whatever the decision he/she makes, the outcome will not influence other decisions he/she may make in the future. However, this is seldom the case in engineering design. At every design stage, the designer has to make a choice from a set of alternatives. Every decision made will significantly influence the way in which the design will develop from that point on.

Engineering decision making has a distributed nature (linked decision making) and involves all design stages including the selection of the concept, the configuration, the material, the geometry, and the process plan. At each stage, alternatives are generated, analyzed, and selected. The distributed nature of engineering design decision making is illustrated by the tree structure in Figure 25.2. In the context of engineering, such a decision tree is never fully explored due to the limitation of developing time and cost. Designers are reluctant to return to the previous stage to search more design alternatives when proceeding along the design process. At the early design stage, very little information is available to the designer. As he/she proceeds with the design process, more and more information will be known to help the designer make better decisions. However, decisions have to be made in the conceptual and embodiment design stages. Thus, the ability to handle different types of uncertainty in decision making becomes extremely important.

25.1.6 DETERMINISTIC, UNDER RISK, OR UNDER UNCERTAINTY?

One of the most important factors in engineering design decision making is the degree of uncertainty. Whenever the designer makes a decision, he or she is performing a prediction of the effect

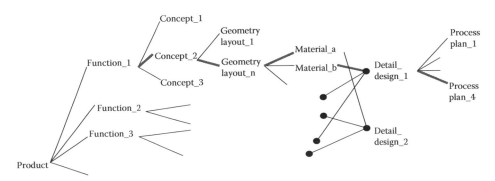

FIGURE 25.2 Distributed nature of engineering design decision making.

of future events in technical feasibility, economic viability, and the trade-off between them. To make a successful prediction, good information (previous experience, outcomes from the similar circumstance, design knowledge, expertise, etc.), proper methods, and, sometimes, good intuition are needed.

The simplest scenario is a deterministic engineering design decision making. In such an ideal case, there exist mass data on previous circumstances and the technology involved is static. All geometry information, material properties, manufacturing process parameters, market change, customers' preference, development and manufacturing cost, etc. can be estimated exactly and the future events are perfectly predictable.

Engineers and economists will be very happy if they can make decisions under certainty. However, such situations are very few, if any. The engineering and economic real-life is fraught with uncertainties. Often errors introduced by ignoring uncertainties are large, especially in the early design stage. In engineering practice, techniques such as risk analysis and uncertainty handling are widely used. When there is limited amount of data on previous circumstances and we do know something about the technology involved, we may ascribe probabilities to the occurrence of future events and get the expected result to make a decision. The most difficult situation is that there are no previous data available at the current stage and the technology is highly volatile. No probabilities can be ascribed to the outcomes. We call such a situation the design decision making under uncertainty. The difference between risk and uncertainty is defined as following:

Risk: Probabilities of the occurrence of future events are assumed known.
Uncertainty: Probabilities cannot be ascribed to the occurrence of future events because of limited information.

25.1.7 ESTABLISH THE REFERENCE FRAME FOR ENGINEERING DESIGN

Through the above discussion, we gradually established the engineering design with the following dimensions:

- Design stages: market research → customer requirements → design specifications → conceptual design → preliminary design → detail design → manufacturing, assembly, inspection → maintenance → field support → disposal, and iterations in between
- Design types: novel design, adaptive design, and routine design
- Design methods: systematic design and decision-based design
- Artifact type: consumer products and industry products
- Design decision making: deterministic, under risk, and under uncertainty

Another branch of engineering design studies how designers perform the design tasks. It is called design research methods and covers the design protocols study, case studies, etc. It is beyond the scope of this chapter.

The status of engineering design research in academia is introduced through an international conference. In Appendix 25.A, one of the most popular international conferences on engineering design is introduced. With the current threads of research fields, the audience can gain a sense of what is going on in academic research.

25.1.8 HOW DO BIG MANUFACTURING ENTERPRISES TRAIN THEIR NEW DESIGNERS?

As we mentioned above, the engineering students learn the discrete pieces of design knowledge elements and techniques at school. Once they enter a big manufacturing company, it takes years for new designers to acquire the capability to take on real design tasks independently. Often the companies

offer lots of training, in a formal or an informal way. The training usually includes the familiariza-
tion of the complex engineering product, the learning of design procedures which includes several
stages with several design reviews, and the study of company design philosophies, design manuals
or standards, and design tools including commercial CAD/CAM/CAE tools or in-house developed
CFD or optimization packages.

After gaining the necessary design knowledge and skills, he or she now can take on a real design
task. Usually it requires the designer to combine all he/she learned during the past to reach a solution. Often, creativity and innovation always play a big role to distinguish a great designer from
a group. A little different from the toy-problems at school, now the designer has to consider how
to collaborate with other designers. If he/she is designing a component, he/she has to consider the
system level requirement and negotiate with other component owners on the interface or interaction
issues. After the design specifications are generated, the designer needs to decide the steps to come
up with the final design (configuration, geometry, materials, dimensions, tolerances, etc.).

The questions he/she needs to answer include, What analysis do I need to perform? Aerodynamic?
Thermal? Structural integrity? Fatigue and failure? What is the impact of each on the final geometry
and material selection? What is the sequence of performing these analysis steps? How much iteration do I need? How to define a "good" design so that I can stop? What is the cost incurred based
on my design? Is the design manufacturable and can it be assembled? Below are two examples to
illustrate the real design scenarios: optimization and uncertainty.

25.1.8.1 Optimization

Now he/she has to consider multiple conflicting objectives when applying an optimization, since
in most cases the design has to be a comprised solution. Designers cannot always improve the performance without incurring more cost or development time. At school, we learned that the typical
optimization problem in engineering design is a process of choosing the value of design variable(s)
to maximize or minimize a desired objective function while satisfying the constraint conditions.
Although this is the simplest case in the whole spectrum, it is by no means a simple problem and
has very important meaning in engineering design and manufacturing. For example, for a high-
volume part, small savings in weight or cost will result in substantial savings for the company.
There is a rich literature on applying mathematical optimization techniques to solve the various
engineering problems (parametric design, size/shape/topology optimization, etc.) In the early days,
gradient-based optimization methods (linear programming, quadratic programming, generalized
reduced gradient method, sequential quadratic programming, augmented Lagrangian method, etc.)
were explored broadly and deeply. However, engineering design procedure often involves not only
analytical calculations but also numerical simulations and catalog selections; sometimes, it is very
difficult or not feasible to get the gradient of the objective function. Therefore, non-gradient methods are developed for these problems. With the progress of computer techniques, the development of
different types of non-gradient methods (genetic algorithms, simulated annealing, etc.) is growing
very quickly. With the help of analytical techniques as well as numerical optimization, the designers can improve their designs to a great deal. Moreover, optimization does not explicitly represent
preferences. It assumes a relation between the preference and the objective functions and a relation
between the preference and the constraint functions.

Real engineering design problems often require satisfying many conflicting and incommensurable
objectives simultaneously. In most cases, these objectives are opposing (e.g. maximize product quality and minimize cost) and it is unlikely to optimize these different objectives by the same alternative
parameter choices. Hence, the trade-offs among the conflicting objectives have to be considered. In
multi-objective engineering design, as preferences are specific to the designers or the company, the
"optimal" value becomes personalized. This is quite different from classical optimization where
there is only a single objective function and several constraints. An interesting question arises when
designers try to handle multiple objectives simultaneously: shall the designer aggregate all objectives

into an overall function based on his/her preferences before searching for the optimal solution or shall he/she search for some optimal solutions first and then articulate the preferences? Such a question leads to the classification of the methods based on the timing of eliciting the preference information.

25.1.8.2 Making Decisions under Risk

Randomness is everywhere in engineering design and manufacturing. Most physical variables used in engineering design are in fact random variables inside a range. In mass-production, every part manufactured with the same design will be different. The sources of these variations are material properties (density, yield strength, modulus of elasticity, etc.), variation caused by the manufacturing process (drift in machine settings, dimensions, residual stresses, etc.), and variation due to environment (temperature, humidity, etc.). Moreover, there exist design analysis risks and marketing risks.

Fortunately, this random nature of engineering design and manufacturing were recognized a long time ago, and different philosophies, methods, and techniques have been developed to help engineers handle these random variations.

25.1.8.3 Probabilistic Design

Risk caused by random variables has traditionally been modeled by probability theory. It is developed on an axiomatic basis and allows for the clear representation and processing of variations. The basic assumption of traditional probability theory is that there exists a known probability value for an event to happen. Such probabilities may be elicited from experimental evidence, expert opinion, or a combination of these. Statistics has become a useful tool in many engineering practices. Engineers apply statistics to analyze the data developed during simulation and modeling in the design process. Statistic techniques include hypothesis testing, statistical interval estimation (confidence interval, tolerance interval, prediction interval, etc.), analysis of variance (ANOVA), factorial and fractional factorial design of experiments (DOE), and regression analysis [7]. The sources of uncertainties are multiple and common distributions for design variables are the normal, lognormal, Poisson, uniform, triangular, exponential, and Weibull distributions. Normal distribution and mean values of variables are widely used due to their simplicity; however, probabilistic design also studies how to make calculations with other probability distributions.

A standard probabilistic approach is illustrated as follows:

- Explicitly identify uncertainties
- Use statistics, knowledge, and judgment to best estimate the range of individual uncertainty and its distribution
- Perform calculations and simulations
- Perform sensitivity analysis

25.1.8.4 Utility Analysis and Risk Profile

With the introduction of probability, the decision making is not solely "objective" any more. Now engineering decision making involves the decision-maker's risk attitude. With the same expected value, different decision makers may make quiet different decisions. Some people avoid risk at all and never take a risk in their choice. Others prefer risk to certainty and they believe the luck is always on their side. However, most people are somewhere in between.

Von Neumann and Morgenstern [8] proposed a set of axioms that imply the existence of utilities. Utility theory assumes that each individual has a measurable preference among alternatives. Such a preference is represented by utility. Utility is used to describe the concept of individual preferences and treated as a numerically measurable quantity. Utility curve or function reflects an individual's attitude toward risk and it can be constructed through lottery questions about the decision maker's preferences for the mutually exclusive consequences of an alternative. Risks are handled by using objective probabilities, based on statistic assessments or on analysis of previous decisions made in the similar circumstances. The fundamental assumption in utility theory is that the decision maker will choose the alternative with the highest expected utility if an appropriate utility is assigned to each possible consequence and the expected utility of each alternative is calculated. Von Neumann and Morgenstern's axioms provide a normative theory of decision making under risk.

25.1.8.5   Robust Design

As we point out, the criterion used in the traditional optimization is the objective function that is defined by the decision maker himself. In utility analysis, the final criterion is the utility function of some attribute (consequence and criterion). Hazelrigg proposed some favorable properties of a design method in the ninth DBD workshop [9] and one of these properties is that the design method should not impose preferences on the designer. However, the method we discuss next seems to violate this rule but works well and is accepted by many industry sectors.

The robust design method, also called the Taguchi method [10], developed by Dr. Genichi Taguchi, has placed great emphasis on the importance of minimizing variation from nominal value. The main concept of the Taguchi method is the definition of "robustness." Robustness is the degree to which a product's quality is insensitive to the variations in manufacture and operation conditions. The quality issue has been moved upstream to the design stage and quality is designed into the product and not inspected into it. The signal to noise (S/N) ratio is proposed as the criterion to evaluate a design. The Taguchi method has many advantages. The product designed by the Taguchi method is more likely to continue to satisfy the customer even when it is subject to unusual conditions of the real world or factory floor. The reliability of the product can be predicted and optimized and the cost may be reduced by using looser tolerances and cheaper materials without compromising too much on performance. However, there is no theoretical basis for the Taguchi method. The basic assumption is that the ideal product is the one insensitive to variations and the only criterion to select a design is the signal to noise ratio. It is a preference imposed to the decision maker and may not always be the case. For example, some objective (attribute) must be optimized to attract the customer and cannot be compromised.

25.2   USE OF COMPUTER IN ENGINEERING DESIGN AND MANUFACTURING

The use of the computer has dramatically changed the way of how designers conceive and realize their designs.

25.2.1   3D AND 2D MODELING

Before the development of computer-aided design (CAD) systems, the designers imagined their design in 3D in their minds and then used pencils and paper to scratch the shape, and then drew a formal 2D engineering drawing to communicate the design intent. It is very difficult for somebody who has no formal training to conduct such a task since it requires three-dimensional imagination to design a complex artifact. With the aid of CAD systems, designers now can create and visually see the 3D shape and further morph it to the final desired shape.

CAD modeling systems have matured through the years from 2D and 3D wireframe representation, to CSG (constructive solid geometry) and boundary representation (BRep), and to a combination of both (Figure 25.3). In the mid 1980s and early 1990s, the concept of "feature" has gained the attention of researchers since it is natural to create a geometry through the Boolean operations of different features, from as simple as a block, a cylinder, a cone, a hole, to as complex as a freeform surface and other user defined features. Nowadays almost all CAD systems provide the functionality of feature modeling, besides the traditional 3D modeling operations such as extrude, revolve, loft, and sweep along a path. The parametric design has been used widely to control the geometry and dimensions, so that it becomes much easier for designers to change the part shape in a short time. However, sometimes when the geometry becomes very complex, it might take hundreds of modeling steps to achieve the final shape. Then the robustness and consistency of the CSG history-based

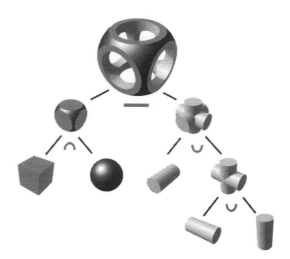

FIGURE 25.3   Constructive solid modeling. (Adapted from Wikimedia Commons.)

modeling becomes a headache. In recent years, CAD systems such as SolidEdge and Unigrahpics NX6 introduced the technique of Synchronized Modeling, which provides the freedom for designers to change the shape on the fly, without consulting the geometry and construction history, thus enhancing the flexibility to morph a complex geometry. However, such a technique is still in the early stage and its robustness needs to be improved along the way. The leading CAD modeling tool vendors include CATIA (used by Boeing), ProE (used by Siemens AG), and Unigraphics (used by GM and GE).

25.2.2 COMPUTER-AIDED ENGINEERING (CAE)

With the rapid progress in CPU speed and hard disk storage space, the use of computer-based analysis and simulation has become increasingly popular. In the past, very limited analysis could be done since it required a supercomputer to run the simulation. In the 1970s, the finite element analysis (FEA) was used mainly in big companies or national labs in aerospace, automotives, defense, and nuclear industries since only the mainframe computers can handle the computation cost and time. With the rapid development of personal computer (PC) and commercial CAE software tools, now designers can run a lot of analysis on their desktops, laptops, and workstations.

CAE has become a very important tool for designers to validate their design, to help choose the design parameters on geometry features and materials, and to simulate the manufacturing processes and assembly methods. Here are just a few examples on the application areas of CAE: tolerance analysis, structural integrity, thermal analysis, vibration analysis, fluid flows, acoustics analysis, transient analysis, failure mode prediction (low-cycle fatigue, high-cycle fatigue, contact analysis, creep analysis, etc.), and the interactions of the two or more.

Examples of commercial CAE tools include ANSYS, FLUENT, Abaqus, and FEMLAB, to name a few.

25.2.3 COMPUTER-AIDED MANUFACTURING (CAM)

The invention of computer numerical control (CNC) machines has brought the manufacturing into a new era. Now, the 3D shape that is defined exactly in the CAD system can be realized in the manufacturing system to certain accuracy. Today, high-speed and high-accuracy CNC machines are used widely around the globe.

In general, for traditional machining and some of the non-traditional machining, the CAM system consists of two types: tool-path generation system from CAD, and tool path/machining simulation system to detect the collision among the stock, tools, fixtures, and other machine elements.

The 3D shape of the artifact can now be exactly defined in the CAD system from a mathematical perspective, thanks to the development of the non-uniform rational b-spline (NURBS) surface representation. Today every geometric entity in CAD, for example, a body, a face, an edge, or a vertex, has a mathematical description, no matter the shape is prismatic or freeform.

To achieve a predefined shape from a stock on a machine, a tool path is needed to finish the job. To translate the shape from CAD to the real machine movement, a continuous curve or surface has to be discretized. Thus, a point on the surface in the CAD coordinate system (CS) has to be translated into a machine move that includes the translation along the machine axes X, Y, and Z and the rotation around the machine axes A or B, depending on the machine configuration. A point on a freeform surface, with 6 degrees of freedom ($xyzijk$), can be realized by a five-axis CNC machine (XYZAB, XYZBC, or XYZAC) through the development of a post processor. The post processor takes ($xyzijk$) from the CAD output (usually a neutral format, cutter location file, CLS) and machine configuration file and generates the G-code (the code to control the machine movement).

At present, all major CAD/CAM systems such as Unigraphics and PorE can automatically generate the gouge-free tool path for prismatic shapes and simple freeform shapes. However, to machine a complex free form shape, the user still needs to manually or by using in-house codes adjust the tool path to avoid gouging. Currently major CAD/CAM systems can handle the regular shape cutters. If a customized shape cutter is used, more development work is needed to generate a gouge free tool path. Besides Unigraphics, the representative systems on tool path generation include MasterCAM, DELCAM, and NRep.

Validating the generated tool path on a real machine is very costly and time consuming. What if there is a gouge or collision? So, machine simulation software tools were developed and are used widely to help CNC operators and process planners to validate their jobs. The most widely used simulation tools include VERICUT and UG machine simulation module. In these tools, the machine is virtually built, and then a kinematics model is established to simulate the machine moves. Then the CAD models of the fixtures, the tools, and the stock are brought on to the virtual machine. After loading the tool path (G-code), the software tool can detect the collisions and gouges based on predefined algorithms that offer huge help for the user.

25.3 DESIGN FOR MANUFACTURING

With increased global competition, the pressure to get quality products to the market in time and at competitive cost is ever increasing. To achieve these objectives, design and manufacturing must work together but conflicts often arise. Design for manufacturing (DFM) involves trade-offs between design objectives and manufacturing cost/efficiency. However, contemporary software packages consider DFM in a unilateral (manufacturing centric) way. Most research in DFM has focused on manufacturability analysis using *ad hoc* methods and measures of manufacturability that can lead to bad decisions. Moreover, uncertainties in estimating design attributes and manufacturing costs are not properly accounted for.

25.3.1 WHY IS DFM IMPORTANT?

Traditionally, design and manufacturing engineers work in different departments. The designer would finish the entire design from the concept to the very details and then pass the design drawings or the CAD model to the manufacturing department. Ideally, the designer is supposed to be familiar with manufacturing processes. He/she should have some knowledge about potential manufacturing problems and be aware of typical design features that are difficult to manufacture and try to avoid them. However, this is not the case. In the 1960s, manufacturing workshop courses disappeared in design students' curricula in the United States [11]. As a result, manufacturability analysis of the design has been neglected over the years. Substantial consideration has been given to the design of products for performance (functionality, quality, aesthetics, ergonomics, etc.). The designers believe that once their high-performance design is manufactured and launched into the market, the customers will buy it and the company will make profits from their "successful" design. However, since the designers ignore the manufacturability of the part, sometimes it is not possible to manufacture the part or the design incurs very high manufacturing cost and long delivery time. In the former case the design fails because of the manufacturing problems; and in the latter case, customers may not buy the product since the price is too high. They probably will turn to other products whose performance may not be that good but satisfy the minimum requirements and whose price is much lower.

As Figure 25.4 shows, such a situation is named as "over-the-wall approach" where the designer and the manufacturer sit on different sides of the wall. In the designer's side, the attitude is "we design it, you build it" and in the manufacturer's side the response usually turns out to be "it is difficult or even impossible to make it, you have to redesign it." They lack mutual communications and before the final product can be manufactured, many time-consuming iterations have to be done between design and manufacturing. Commonly the designer is very reluctant to change the original design since all the design details have been decided and it is difficult to make some changes. Thus, the manufacturing side always complains "we never can get designers to make changes, we'll just wait until we get it to make it manufacturable."

25.3.2 How Is It Done Today?

DFM gradually emerged as an important way to address the "over the wall" design. DFM is usually carried out in the following ways:

1. *Cross-functional teams*

   One way to address the manufacturability problem in the early design stage is to use the integrated design team approach that tries to involve all disciplines as early and often as possible. The design team includes both designers and manufacturers and the manufacturability of parts is considered throughout the design process.

   According to a survey conducted by the Society of Manufacturing Engineers [12] on mid-sized manufacturing companies ($20–200 million annual sales), the make-up of the design teams is illustrated in Figure 25.5. It can be seen from the figure that many

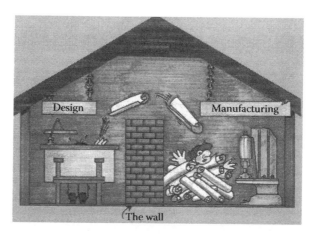

FIGURE 25.4 "Over the wall" design. (From Boothroyd G. et al., *Product Design for Manufacture and Assembly*, Marcel Dekker, New York, 1994. With permission.)

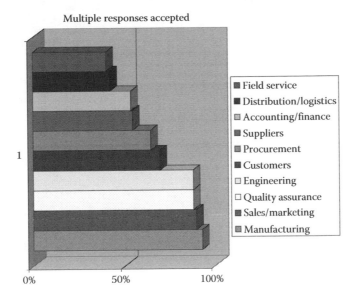

FIGURE 25.5 Composition of the design team. (Data from Owen, J.V., *Manuf. Eng., SME,* 122(4), 60, 63, 1999.)

areas get involved in the design team and the folks from manufacturing are almost always on board.

Using a cross-functional approach, the product and the process can be simultaneously designed and thus the over-the-wall design is prevented. Cross-functional teams rely on experts from several domains who can discuss trade-offs and deal with unusual situations. They may be ideal for large projects. However, there exists management difficulties for such an approach and the project size, duration, or geographical separation may make such teams infeasible.

On the one hand, the cross team is good; on the other hand, interdisciplinary education is needed to break the barriers of cooperation. The ability to cross the borders is especially important for project leaders.

2. *Design/DFM manuals*

In order to help their designer be more aware of manufacturing considerations, many companies such as General Electrics and Boeing [13–15] have developed Design/DFM manuals for the designers to follow when they perform design tasks. The content of these manuals is summarized from experience related to their own products and manufacturing resources. These manuals typically contain data on preferred feature shapes, sizes, tolerances, etc.

Leading professional societies and other experts have also published many guidelines and textbooks [16–18] that mainly catalog some typical design configurations that may cause manufacturability problems and some other general rules.

Company DFM manuals can be extremely valuable to designers, particularly because they pertain to major cost drivers for the company's specific products. Other general guidelines in the textbooks are also helpful for designers when they want to check the manufacturability of the design. However, manuals are "passive"—they could sit on shelves and not used. Such manuals contain guidelines for only the most common features. Also, these manuals are manufacturing oriented—the recommendations ignore design requirements. Finally, it is hard to get quantitative measure of cost and uncertainty in manufacturing.

3. *Software tools*

Some commercial DFM software tools are available in the market. A well-known one is DFMA by Boothroyd and Dewhurst, Inc. [19]. A suite of tools is provided: DFM software for machining, DFM software for sheet metalworking, DFE software to assess environmental impacts, DFS software for service improvement, concurrent cost for cost estimation and their best known DFA module for evaluating assembly. A sample interface of DFMA is shown in Figure 25.6.

Vendors create software tools, so the ratings are not specific to any one company's products or processes. The knowledge and logic used in these systems are transparent to the user. It is difficult to determine the reliability of the answers. The interpretation of the results is also not clear. For example, if a design scores 150 and another scores 300, is the first design twice as expensive to manufacture? The validity of the ratings (DFM metrics) has not been established by the vendors. Of course, the software also has no knowledge of design requirements that may justify higher costs. On the other hand, software can be cost-effective for manufacturability analysis in small teams or for individual designers since there might not be manufacturing expert on the design team. If DFM software is integrated with CAD, one can create an "active" mechanism for evaluating a design.

25.4 KNOWLEDGE IN ENGINEERING DESIGN AND MANUFACTURING

25.4.1 DATA, INFORMATION, AND KNOWLEDGE

What are data, information, and knowledge? Different people from various backgrounds have different viewpoint such as decision-making point of view and problem-solving point of view. Even different dictionaries define them in quite different ways:

Free On-Line Dictionary of Computing [20]

Data: Numbers, characters, images, or other method of recording, in a form that can be assessed by a human or (especially) input into a computer, stored, and processed there, or transmitted on some digital channel. Computers nearly always represent data in binary.

Information: Data on its own has no meaning, only when interpreted by some kind of data processing system does it take on meaning and become information.

FIGURE 25.6 DFMA software interface. (From Boothroyd Dewhurst, Inc., Design for manufacture, http://www.dfma.com/software/dfm.htm. With permission.)

Knowledge: Knowledge differs from data or information in that new knowledge may be created from existing knowledge using logical inference. If information is data plus meaning then knowledge is information plus processing. People or computers can find patterns in data to perceive information, and information can be used to enhance knowledge. Since knowledge is prerequisite to wisdom, we always want more data and information. But, as modern societies verge on information overload, we especially need better ways to find patterns.

The American Heritage Dictionary of the English Language, Fourth Edition

Data: Factual information, especially information organized for analysis or used to reason or make decisions.

Information: Knowledge derived from study, experience, or instruction; knowledge of specific events or situations that has been gathered or received by communication; intelligence or news; A collection of facts or data: statistical information.

Knowledge: The state or fact of knowing. Familiarity, awareness, or understanding gained through experience or study. The sum or range of what has been perceived, discovered, or learned. Specific information about something.

Considering the context of engineering design in this chapter, I will take the way of defining data and information in the *Free On-line Dictionary of Computing* and give my own definition of knowledge as following:

Knowledge is know-what, know-how, and know-why.
Know-what: facts, concepts, behavior, entity-relationship, etc.
Know-how: methods, procedures, rules, ideas, etc.
Know-why: beliefs, values, judgment, decisions that were made, etc.

However, defining data, information and knowledge in absolute terms implies data is always data, information is always information, and knowledge is always knowledge. Defining in relative terms implies that, for example, information can be data for some users and knowledge for others. Data, information, and knowledge are relative concepts rather than absolute terms. The use of knowledge is what distinguishes the experienced and novice designers.

25.4.2   IMPORTANCE OF KNOWLEDGE ENGINEERING FOR MANUFACTURING ENTERPRISES

Since most companies have access to similar technology, processes, cost estimation system, material management system infrastructure, etc., the major thing that separates them is the intellectual capital that each company holds and the important data and information accumulated over the years through experiments and real product performance. On top of the company databases, technical reports, design standards, and documents, to record and store the valuable experience and knowledge of the experts in the company is very important since there will be a loss of knowledge along with the loss of the experienced person. As to "experience," this knowledge needs to be identified, captured, stored, and made available to novice designers.

The traditional way to store engineering design and manufacturing knowledge has taken many forms including books, manuals/guidelines, drawings, reports, and information embedded in software tools and databases. However, advanced automation technologies and computational tools need to be developed to assist collaboration between designers, manufacturing engineers, managers, business employees, etc. The key objective that design automation technologies have to achieve is the capability to retain intellectual capital of the manufacturing enterprise. The task of modeling engineering design and manufacturing knowledge requires the identification of the knowledge together with a method of representing that knowledge. Some issues arise as follows:

What kind of knowledge needs to be captured?

In what form is the knowledge stored?

How is the captured knowledge (index, retrieval, etc.) organized?

How efficiently can designers use the captured knowledge?

How does one perform reasoning to generate new knowledge?

How does one update the knowledge base in a real-time manner?

How does one lower the maintenance cost?

How will knowledge flow from design to manufacturing?

Some difficulties exist in today's enterprise knowledge management. Knowledge management has traditionally been on an individual basis rather than on an enterprise basis. The knowledge has in general been managed in an *ad hoc* manner that leads to knowledge loss throughout the product life cycle (e.g., after the expert of a design team left, other designers in the enterprise do not know the original design rationale and intent). Besides the technology difficulties (design rationale capture, etc.), the competing nature of employee relationships hinders the enterprise from retaining its intellectual capital. The employees tend to hide away rather than share their knowledge if the management team does not foster a good system for ensuring and rewarding the ones who share the knowledge.

25.4.3 KNOWLEDGE REPRESENTATION

Knowledge representation involves not only representation but also storage and retrieval of the knowledge. Moreover, knowledge representation is coupled with reasoning. Two issues need to be solved: How to express what we know? How to reason with what we express?

Different knowledge representation schemes are summarized in Table 25.1.

TABLE 25.1
Knowledge Representation Schemes

Knowledge Representation Scheme	Characteristics	Scope
Semantic nets, associative networks	A representation based on a structure of linked nodes (concepts) and arcs (relations) connecting the nodes; Unstructured node-link graphs; no axioms to support reasoning; no semantics to support interpretation;	Know-what
Frames	Structured semantic nets; object-oriented descriptions; class-subclass taxonomies	Know-what
Production rule	If-then inference rules; situation-action rules; hybrid procedural-declarative representation; basis for expert systems; rules are independent of each other, thus, conflict may occur; easy for domain experts to formulate their experience	Know-what know-how
Object-oriented (OO)	Useful in modeling many real-world situations in which real objects can be put in a one-to-one correspondence with program classes and objects	Know-what
Symbolic logic, fuzzy logic, Bayesian probabilistic reasoning, qualitative reasoning	Knowledge representation coupled with reasoning	Know-what know-how Know-why
Case-based reasoning (CBR)	History-based method, refer to previous knowledge	Know-what know-how
Model-based reasoning	Descriptive models, Predictive models	Know-what know-how know-why

The author believes there is no single representation scheme that is suitable for all the situations; thus, we need to choose carefully under a given circumstance.

25.4.4 KNOWLEDGE ENGINEERING IN ENGINEERING DESIGN AND MANUFACTURING

25.4.4.1 Engineering Design Knowledge

Knowledge in product design involves many aspects such as: the design process, artifact, designers, specific field knowledge, resources, requirements, specifications, functions, forms (geometry, feature, material), behaviors, design philosophy, constraints, relationships, design rule, design rationale, etc.

Among all, capturing the design rationale is one of the most important issues in supporting an engineering design process. Representing and storing rationale or intent in a shared knowledge base allows knowledge to flow in a geographic or temporal distributed environment, in which person-to-person communication of information is inhibited. Design rationale is important because it makes explicit the fact that decisions were made, and why they were made.

25.4.4.2 Manufacturing Knowledge

Manufacturing knowledge today spans a vast spectrum, from manufacturing process capability/constraint, precedence, algorithms/heuristics of performing feature recognition, process planning and manufacturing time/cost estimation, to design for manufacturing (DFM) tactics and strategies. There are numerous production rules and process planning methods associated with each manufacturing process. Meanwhile, new manufacturing technology, processes and equipments can emerge in response to new customer requirements and competitive challenges. Some difficulties exist in today's enterprise manufacturing knowledge management. The valuable expertise and experience are deployed in different personnel and nobody can be the expert in every manufacturing process and workshop, not to mention the diverse process plans and cost estimation techniques. Hence, it is important for the enterprise to systematically identify and store the valuable knowledge of manufacturing experts and make it available to the designers.

For manufacturing-process planning, a knowledge-based system (KBS) is desired. Usually a human planner manually designs the manufacturing process plans. It is very time-consuming and the quality is highly dependent on the capability of the planner. How does one retrieve this knowledge from the domain experts and put it into an automation system? The computer-aided process planning (CAPP) system has been a research topic for several decades but there is still a long way to go.

Manufacturing process knowledge involves the shape producing capability and technological constraints for each of the available manufacturing processes. Production rules and frames are the most popular knowledge representation schemes.

25.4.4.3 Research Challenges in Knowledge Engineering

Capturing, storing, retrieving, and applying the knowledge of the experts to help other designers and manufacturing engineers remain a research challenge. Current knowledge-based engineering (KBE) software systems employ a combination of the "production rules" and the "object-oriented knowledge representation" as the representation paradigm. The representation scheme of "know-why knowledge" needs to be developed.

In developing KBE applications, the knowledge engineer and domain expert are usually not the same person. The knowledge in the applications is only accessible for the knowledge engineer who understands the KBE programming language. Training in KBE coding is expensive and the expertise is only available in a small number of people. When knowledge is to be used in another application, there is no easy way to reuse it. Thus, a comprehensive and open architecture knowledge system needs to be developed to make it easy for the domain experts to input their knowledge.

Another challenge is the integration of traditional engineering software tools with knowledge-based applications. Mechanisms for indexing, searching, and retrieving design cases, design rationale, and other types of knowledge are needed.

25.4.5 PRODUCT, PROCESS, AND RESOURCE INFORMATION MODEL

To achieve truly integrated product design and manufacturing, information model of product design, manufacturing process, and enterprise resources need to support multiple levels of abstraction for multi-directional communication.

25.4.5.1 Information Modeling Methods

Information modeling includes data and activity models [21]. An activity model describes a process activity and its sub-activities, as well as the data associated with the activity. A data model defines data elements and the relationships among them. A data element describes its attributes and relationships (e.g., inheritance, aggregation, and classification) to other data elements. Information models specify the context, the application of data, and data definitions.

Activity modeling methods include IDEF0 and process charts. Data modeling methods include entity-relationship method (ER), function modeling method, and object-oriented (O-O) method. The ER approach emphasizes on identifying the entities, their attributes, and the relationships among the entities. The function modeling approach focuses on decomposing system functionality and the information flow between different objects. The O-O approach defines the object as the basic element that contains both data and functions, and thus it is easy to model complex objects and this provides good extensibility. The modeling language based on the above methods adopts either graphical form or textual form. Examples include the integrated computer-aided manufacturing (ICAM) definition language 1 extended (IDEF1X), the EXPRESS language, EXPRESS-G (graphical), and the unified modeling language (UML), and extensible markup language (XML).

25.4.5.2 Product Data Model

In the past, the product data model contained geometry information only (STL, IGES). Ideally, a uniform product information model exists in the whole product life cycle. It is the logical accumulation of all relevant information concerning a given product during the product life cycle. Such a product data model should include the following information [22,23]: design requirements, market data, specifications, artifact, sub-artifact, functions, form (geometry, feature, and material), behaviors, design rationale, constraints, relationships, and design change request. On top of it, companies prefer to have the information security levels in place to hide the design intent information from outside vendors and only supply the necessary data to finish the manufacturing job.

As the international standard for the exchange of product model data, STEP is able to model geometry, manufacturing features in machining domain (AP224), and some non-geometry information such as specifications (tolerance, surface finish, etc.), administration data, assembly relations, etc. However, functions, behaviors, and design rationale have not been considered.

25.4.5.3 Manufacturing Process Model

A process data model describes a process activity and its sub-activities, and the associated data. The model needs to include the following data: manufacturing features, process change requests, manufacturing process knowledge, process models, operations sheets, scheduling package, etc. [24].

Manufacturing process models concern the capabilities of a manufacturing process, including shape producing capabilities, dimensions, tolerance and surface quality capabilities, geometric and technological constraints (tool interference, tool slipping, and cutting force), and manufacturing cost [25].

Traditionally, there are two methods for process shape producing capability modeling: process-based methods and part-based methods. In a process-based method, machine tools, fixture devices, cutting tools, and kinematics motions as well as operation precedence in the manufacturing process are utilized to capture the capability of the process. In a part-based method, feature types, attributes, and numbers in a machining process are adopted to define its capability. The capability of a manufacturing process can be also expressed in the form of constraints. Constraints can be classified into three levels: universal level constraints, shop level constraints, and machine level constraints [26].

25.4.5.4 Enterprise Resource Model

Manufacturing resources are defined as the equipments that enable industry to turn raw materials into marketable products [27]. Different kinds of representations of manufacturing resources have been employed by a variety of software tools that perform various tasks: CAPP, process simulation, tool selection, cost estimation, manufacturability analysis, etc. The resource model should include: tooling/materials, mach inability data, resource descriptions, equipment/labor, materials knowledge, material stock descriptions, equipment orders, tooling/materials orders, bill of materials, equipment availability, resources available, resource requirements, tooling designs, etc. [24]. Several manufacturing resource models were developed in the MO system [28], IMPPACT project [29], and NIST rapid response manufacturing (RRM) project [30].

25.4.5.5 National Standard for Product, Manufacturing Process, and Resource Model

The recent effort on AP240 [31] reflects the trend that the product model, manufacturing process, and resource model will become standardized in the near future. Thus, the manufacturing enterprise in the future could build its information model based on the standard to make collaboration easy at different levels.

25.5 CHALLENGES AND FUTURE TRENDS

25.5.1 CHANGES ARE COMING

The next 20–30 years will bring the following changes to the manufacturing enterprise:

Changes in Technology

- Broadband and wireless network is available and large amount of data can be transferred in a short time.
- There will be widespread availability and distribution of information.
- It will require huge data management: tera to peta to exabytes (10^{12} to 10^{15} to 10^{18} bytes).
- Effective communication and information visualization will become available.
- Human–machine interface will be extended from keyboard, mouse, and the screen to virtual reality environment.
- New materials, manufacturing processes, equipments, and tools emerge.

Changes in Marketplace and Products

- It becomes routine to embrace the globalization of markets and competition.
- There will be increasing customer expectations: customers become more picky and demanding.
- The demand to decrease product realization time increases.
- Government and society will put pressure on life cycle issues such as environmental and liability issues.
- Products, processes, and services become more complex.
- Smart product is required.
- There will be increasing knowledge intensity in products, technology, and workforce.

Other Changes

- The enterprise recognizes the limited resources.
- There will be environmental replenishing and resource limitation.
- More efficient global transportation systems are developed.
- The enterprise has its departments deployed around the world.

All of these changes will bring big opportunities and challenges to the manufacturing enterprise. Customized products need to be delivered with a decent price and high quality within a short time. Thus, high requirements are set on better quality, lower cost, and shorter time to market. In other words, the enterprise needs to achieve more with limited resource and time.

With the improved understanding of product design process, design methods, and material and manufacturing processes, the scientific foundation for integrated product realization will form. The product realization process will be supported by automation tools, seamless information exchange, and methods of global optimization.

25.5.2 Future Trends

The research challenges are represented by specifying where we are and where to go in this section, as Tables 25.2 through 25.7 show. "*Where we are*" tries to summarize the current-status of different activities during engineering design. "*Where to go*" indicates the scenarios we would like to see for

TABLE 25.2
Product Definition and Cost Estimation

Element	Where We Are	Where to Go
Market research	Paper survey; online survey; interview; Focus group; benchmarking	Continuously and real time extract customer preference form all kinds of interacting activities between enterprise and customer on global basis
Mapping	Use of quality function deployment (QFD)	Use of formal method which defines product holistically
Cost estimation	Long delays between data collection and utilization; activity-based costing, scaling method, cost-driver estimation	Continuously and real time collecting data and estimating cost at all levels; advanced prediction model–based estimation

TABLE 25.3
Product Design

Element	Where We Are	Where to Go
Conceptual design	Brainstorming is the dominating method to generate ideas; no CAD tools support conceptual design	More formal and valid methods support for idea-generation; conceptual CAD is embedded in commercial CAD systems
Design optimization	Optimization is done at one stage and locally	Model-based virtual prototyping helps designer optimize designs considering all product life cycle issues
Design for X	Cross-functional group, book, manual, guidelines, few computer tools with limited capability	DfX automation tools available at all levels
Geometry modeling	2D and 3D modeling of physical geometry using keyboard and mouse	Virtual reality–based modeling the desired geometry by hand or voice
Design process	*Ad hoc* processes based on individuals	Design process is recorded and validated using enterprise-wide and shared knowledge base
Design support tools	Analysis and simulation tools are used in detail design stage; some parametric design tools based on constraint solvers	Product design and analytical simulation totally automated and tools are available at each design stage
Design methods	Axiomatic; systematic; TRIZ...	Scientific design method

TABLE 25.4
Manufacturing Process: Planning, Inspection, and Quality Control

Element	Where We Are	Where to Go
Process planning	Experienced expert manually designs process plans; some CAPP systems with limited capability are used. Optimal process is not guaranteed; process simulations are used in some companies, but mainly for troubleshooting; MRP/ERP	Optimal process plan is generated automatically; virtual prototyping and validation of processes and equipment
Integrated product and process design	The importance is realized by industry but there are very few methods and tools	Formal method of integrated product and process design
Monitoring and control	Done automatically at equipment level	Done at integrated system level and optimize automatically

TABLE 25.5
Information Infrastructure

Element	Where We Are	Where to Go
Ontology	In some domains, design ontology manufacturing ontology	Standard ontology for all enterprise activities across product life cycle on global basis
Networking secure	Secured firewall protects the enterprise intranet from internet	Secure data and knowledge communications at different level
Product model	Current product data model mainly include geometry data; data in form of text, spreadsheets, and graphics; limited product data (geometry, tolerance, etc.) exchange in same-brand system; use of national standards such as STEP to exchange product data	Product data model include product needs, market data, design process knowledge, design knowledge, design change requests, and physical data; multi-media data model; product data can be seamlessly exchanged among different tools along all life cycle on global basis; product data is real-time transferable
Process model	*Ad hoc*	National standard
Resource model	*Ad hoc*	National standard
Data mining	Limited capability	Mature technology
Interoperability	Partially	Full interoperability of systems

TABLE 25.6
Knowledge Engineering

Element	Where We Are	Where to Go
Design rationale and intent capture	Poor	Automatic capture without distracting the designer
Knowledge-based systems (KBS)	CAD system with macros and scripts; rule-based expert system KBS with limited reasoning capability	No stand-alone KBS any longer. The functionality of KBS will be embedded into CAD/CAM/CAE tools, in other words, PLM tools.
Shared data and knowledge base	STEP; ambiguous terminology; limited security control	Universal product data standards; common ontology; adaptive multi-level security

TABLE 25.7
Decision-Making Strategies

Element	Where We Are	Where to Go
Decision making	*Ad hoc* decision making methods; no theoretical basis; *ad hoc* measures; spatially and temporally distributed decisions are very poorly handled; infeasible decision-making process; big gap between proposed methods and decision-making practice in industry; decisions are made separately in different departments without considering other aspects; group preference	Holistic decision making; sound theoretical basis; proper metric is used; every decision is appropriately executed in a spatially and temporally distributed decision-making environment; feasible procedure; enterprise-wide decision strategy is established. The problem of group decision making is solved.
Decision support system (DSS)	Very few tools based on different methods, for example, ACCORD based on Bayesian method, ExpertChoice based on analytical hierarchy process (AHP)	Decision support tools are available at each level and can help decision maker make smart choices

TABLE 25.8
Decision-Making Strategies

	One Criterion (Objective)	Multi Criteria (Objectives)
Decision making under certainty	Mathematical optimization methods (gradient and non-gradient based)	Matrix method; analytic hierarchy process (AHP); quality function deployment (QFD); multi-objective optimization
Decision making under risk and uncertainty	Probabilistic design; probabilistic optimization; Taguchi method Utility analysis	Fuzzy set method; analytic hierarchy process (AHP); multi-attribute utility method (MAU); decision tree; Bayesian method
Distributed and linked decision making; group decision making	?	?

engineering design in the future. The scenarios will remain as the research topics for universities and other research institutes.

The importance of decision making needs to be emphasized to make a company a high-tech, global manufacturing enterprise. A deep exploration of this issue helps build a scientific formalism for design. Many factors will finally affect the success of a company, but sometimes many of these factors are beyond the company's control except for the decisions it makes. Currently there are many decision-making methods developed in academia, however, the application of these in the industry is very poor. A summary of the major decision-making methods and tools is given in Table 25.8 based on [32].

25.6 SUMMARY

In this chapter, the focus was on the reference frame for engineering design, the relationship between design and manufacturing, the importance of knowledge and decision making, and the current status in academia and industry. Finally, the possible future of engineering was discussed.

It would fulfill the author's purpose, if after reading the chapter, the reader has a rough idea on the context of his/her design task; what type of design he/she is performing; what activities to conduct; what potential risks or uncertainties to handle; what tools, techniques, or methods to use; how to use the data, information and knowledge; and how to make decisions.

APPENDIX 25.A: ASME IDETC AND CIE CONFERENCE

The (ASME) IDETC is one of the most popular international conferences on engineering design. It attracts researchers from academia as well as industry sectors to present the latest ideas and progress on design philosophies, design methods, design techniques, and the use of computers in engineering design. It is worthwhile to introduce the full spectrum of the conference to show the aspects that the current engineering design covers.

The conference consists of four sub-conferences:

- Design Automation Conference (DAC)
- Computers and Information in Engineering Conference (CIE)
- International Conference on Design Theory and Methodology (DTM)
- Design for Manufacturing and the Lifecycle Conference (DFMLC)

DAC invites papers in all areas of design automation related to mechanical and engineered systems, including design representation, design optimization, design evaluation, and design integration. CIE provides a forum for enhancing the practice of engineering by understanding the application of emerging technologies that impact critical engineering issues of representation, product design and product development, exchange, management, and integration of information throughout the entire engineering product and process life cycle. DTM promotes research and dissemination of knowledge in such topics as the scientific theories of design, foundations for design environments, models of design processes, design education methods, design management, and other areas extending the understanding of and application of the design process. DFMLC brings together researchers, practitioners, and educators from universities, government organizations, and industry in the design for manufacturing and the life cycle areas together to share latest results in the field and define new challenges.

In each conference, there are many different technical threads and the audience might get lost. However, if we recall the reference frame we just established for engineering design, we can easily put each thread into the proper category.

Here are the topic areas for each conference in the year 2008 [33]:

CIE-1: Integrated Systems Engineering
CIE-2: CTESA: Computational Technologies for Engineering Sciences Applications
CIE-3: CAPD: Computer-Aided Product Development
CIE-4: CIEd: Computers in Education
CIE-5: CES: Computers in Energy Systems
CIE-6: EIM: Enterprise Information Management
CIE-7: VES: Virtual Environments and Systems
CIE-8: NTC: Non-Traditional Computing
CIE-9: Agent-Based Modeling
CIE-11: Inverse Problems in Science and Engineering
CIE-12: Emerging Computational Methods for Product Design
CIE-13: Product Lifecycle Management
CIE-14: Emotional Engineering
CIE-15: Self-Optimizing Mechatronic Systems
CIE-16: Volumetric, Airborne and Tangible Visualization
CIE-17: Prognostics and Health Management—PHM
DAC-1: Artificial Intelligence in Design
DAC-2: Computer-Aided Nano-design
DAC-3: Conceptual Design Methods
DAC-4: Concurrent/Collaborative Design
DAC-5: Data-Driven Risk Management for System Design

DAC-6: Decision Making in Engineering Design
DAC-7: Decomposition Methods in Design
DAC-8: Design and Development of Reconfigurable Systems
DAC-9: Design for Market Systems
DAC-10: Designing for Human Variability
DAC-11: Design Languages and Grammars
DAC-12: Design Optimization Algorithms
DAC-13: Direct Digital Manufacturing
DAC-14: Geometric Modeling and Algorithms for Design and Manufacturing
DAC-15: Innovative Industrial Applications, Developments, and Perspectives
DAC-16: Knowledge-Based Systems in Design
DAC-17: Managing Design and Analysis Processes
DAC-18: Metamodel-Based Design Optimization
DAC-19: Multidisciplinary Design Optimization
DAC-20: Multi-Scale Computational Design of Product and Materials
DAC-21: Optimal Design in/of Energy Systems
DAC-22: Product and System Optimization
DAC-23: Product Family and Product Platform Design Optimization
DAC-24: Robust Design
DAC-25: Simulation-Based Design under Uncertainty
DAC-26: Structural/Topology Optimization and Its Applications
DAC-27: System of Systems Design and Optimization
DAC-28: Validating Predictive Models in Engineering Design
DAC-29: Virtual Product Creation and Innovations
DAC-30: Visualization and Virtual Reality in Design
DAC-31: Web-Based Design and Optimization
DAC-32: DAC Industry Panel
DAC-33: DAC Plenary Session
DAC-34: Collaborative and Automated Assembly Design
DAC-35: Kinematics and Mechanism Design Automation
DTM-1: Analogies in Design
DTM-2: Artificial Intelligence in Design
DTM-3: Application of Modular Design
DTM-4: Metrics and Modular Design
DTM-5: Representing Qualitative Aspects of Design
DTM-6: DTM's 20 Years: Reminisce, Reflect, Move Ahead
DTM-7: Product Architecture Design Methods
DTM-8: Risk, Complexity, and Decision Making
DTM-9: Functional Representations in Design
DTM-10: Flexibility and Reuse in Design
DTM-11: Widely Applicable Techniques for Improving Design Practice in Industry
DTM-12: Creativity and Biological Models in Design
DFMLC-1: Theoretical Foundations for Design and Manufacturing Integration
DFMLC-2: Integrated Product and Process Development Processes
DFMLC-3: Virtual, Collaborative, and Distributed Environment for Design and Manufacturing
        Integration
DFMLC-4: Part Design and Process Planning Integration
DFMLC-5: Manufacturability Analysis
DFMLC-6: Manufacturing and Lifecycle Cost Analysis
DFMLC-7: Emerging Life Cycle Design for X Methods

DFMLC-8: Sustainable Engineering
DFMLC-9: Design for: Mass Customization, Service, Layered Manufacturing, and Quality
DFMLC-10: Assembly and Product Family Design
DFMLC-11: Robust Design and Management
DFMLC-12: Nano-manufacturing

APPENDIX 25.B: ENGINEERING DESIGN IN YEAR 2030

In March 2004, a group of about 50 researchers from U.S. universities, leading research institutes, industry sectors, and government agencies gathered in Gold Canyon in Arizona to review the engineering design research in the past 25 years, and to envision the trends of engineering design for the next 25 years [34]. There were scholars from top tier universities such as Carnegie Mellon, Northwestern, Virginia Tech, Georgia Tech, MIT, University of Maryland, University of Utah, University of Michigan, University of Southern California, University of Texas, University of Pennsylvania, Iowa State University, Arizona State University, University of California at Berkeley, Stanford University, and Michigan University; experts from big companies including General Motors, Ford Motor Co., Boeing Co., John Deere Co., Dassault Systems, and Hewlett-Packard; and experts from U.S. government agencies such as National Science Foundation (NSF), NASA Jet Propulsion Laboratory (JPL), and National Institute of Standards and Technology (NIST).

In this workshop, engineering design is defined as a socially mediated, technical activity that creates and realizes products, systems, and services that respond to human needs and social responsibilities. The committee of Engineering Design in 2030 (ED2030) particularly addressed three aspects of the engineering design: the innovation, the social-technical aspect and the computing and IT infrastructure. They summarized the research done in the field and made the following recommendations from the above three perspectives [24]:

25.B.1 INNOVATION

ED2030 recommends funding projects that study the Innovation process and create tools and methods to support it. By doing so, new product families might be created: for example, medical products, communication products, travel related products, smart house, and entertainment products, and transportation systems that are and feel safe.

25.B.2 SOCIAL ASPECTS

It is recommended that the NSF should establish a new multi-disciplinary research initiative on the social aspects of engineering design. This initiative should seek major breakthroughs in basic knowledge regarding how human and social dynamics influence technical design decisions that involve multiple stakeholders and have wide societal implications.

25.B.3 COMPUTING AND IT INFRASTRUCTURE

As discussed above, the knowledge used during design (rationale) is currently not captured in any consistent form so that it can be reused other than by the person who captured it. ED2030 recommends to NSF that that the engineering design program, on its own, and in collaboration with synergistic programs in CISE, invest in three broad areas: design informatics, collaborative and integrated design environments and smart CAD/CAE tools.

Figure 25.7 shows the perspective of design automation from a NSF director on engineering design. There are two directions to take, enhancing our state of knowledge and increasing the level of automation.

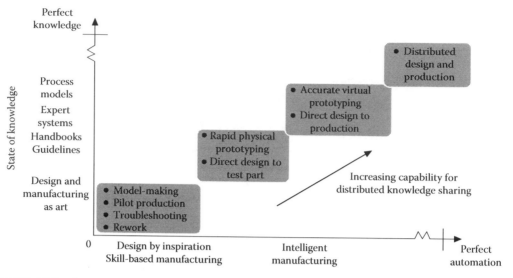

FIGURE 25.7   State of design automation. (Adapted from NSF, ED2030: Strategic plan for engineering design, Final report, *Workshop on Engineering Design in Year 2030*, Gold Canyon, AZ, March 26–29, 2004.)

QUESTIONS

Q.25.1   What is the scope of engineering design?

Q.25.2   Define data, information, and knowledge.

Q.25.3   What are the basic procedures in engineering design according to a systematic approach?

Q.25.4   Illustrate the decision-based design.

Q.25.5   How many dimensions does an engineering design reference frame have?

Q.25.6   Why is design for manufacturing important?

Q.25.7   How do engineers perform design for manufacturing?

Q.25.8   List the types of design knowledge and manufacturing knowledge and find the common set.

Q.25.9   What information or knowledge flows from design to manufacturing?

Q.25.10 How many types of CAM software tools are there? Illustrate the functions of each.

Q.25.11 What are the application areas of CAE?

Q.25.12 How does one choose a good knowledge representation scheme?

Q.25.13 How does one choose a good decision-making strategy?

Q.25.14 How are decisions in engineering design linked to each other?

ABOUT THE CONTRIBUTING AUTHOR

Dr. Zuozhi Zhao gained his PhD in the field of engineering design and design for manufacturing from the Arizona State University, Tempe. He has won the Best Paper Award by Design Engineering Division Design for Manufacturability Committee of American Society of Mechanical Engineers at the 7th Design for Manufacturing Conference. After graduation, he joined GE Global Research Center, Niskayuna, New York, as a research engineer performing design and manufacturing related research projects with GE Aircraft Engines and he had worked on the development

of Boeing 787 Dreamliner. After 4 years with GE Global Research, he is now a program manager at SLC Corporate Technology leading the effort to establish a new engineering design center in China.

REFERENCES

1. Pahl, G. and Beitz, W., 1996, *Engineering Design: A Systematic Approach*, London, U.K./New York: Springer.
2. Hazelrigg, G. A., 1996, *Systems Engineering: An Approach to Information-Based Design*, Upper Saddle River, NJ: Prentice Hall.
3. Hazelrigg, G. A., 1998, A framework for decision-based engineering design, *ASME Journal of Mechanical Design*, 120, 653–658.
4. Thurston, D. L., 1991, A formal method for subjective design evaluation with multiple attributes, *Research in Engineering Design*, 3 (2), 105–122.
5. http://dbd.eng.buffalo.edu/; *Decision Based Design Open Workshop*, sponsored by National Science Foundation.
6. Sen, P. and Yang, J. B., 1998, *Multiple Criteria Decision Support in Engineering Design*, London, U.K.: Springer.
7. Ullman, D., 1997, *The Mechanical Design Process*, 2nd edition, New York: McGraw-Hill Companies, Inc.
8. Von Neumann, J. and Morgenstern, O., 1953, *The Theory of Games and Economic Behavior*, 3rd edition, Princeton, NJ: Princeton University Press.
9. Hazelrigg, G. A., 2000, Design based design lecture, *9th DBD Workshop Presentation*.
10. Taguchi, G., 1987, *System of Experimental Design: Engineering Methods to Optimize Quality and Minimize Cost*, Dearborn, MI: UNIPUB/Kraus International Publications.
11. Boothroyd, G., Dewhurst, P., and Knight, W., 1994, *Product Design for Manufacture and Assembly*, New York: Marcel Dekker.
12. Owen, J. V., 1999, Best practices, best plants, *Manufacturing Engineering*, 122 (4) 60, 63–64.
13. General Electric Co., 1960, *Manufacturing Producibility Handbook*, Schenectady, NY: GE Manufacturing Services.
14. Boeing Co., December 1990, *Design for Producibility*, BDM-1009.
15. Boeing Co., 1995, *Machined Part Design, Tolerances, Roughness and Producibility*, BDM-1326.
16. R. Bakerjian, editor, 1992, *Design for Manufacturability*, Vol. 6, *Tool and Manufacturing Engineers Handbook*, Dearborn, MI: Society of Manufacturing Engineers.
17. Bralla, J. G., editor, 1999, *Design for Manufacturability Handbook*, New York: McGraw-Hill.
18. Dixon, J. and Poli, C., 1995, *Engineering Design and Design for Manufacturing*, Conway, MA: Field Stone Publisher.
19. Boothroyd Dewhurst, Inc., Design for manufacture, http://www.dfma.com/software/dfm.htm
20. Howe, D., editor, Free on-line dictionary of computing, http://foldoc.org/
21. Algeo, M. B., Feng, S., and Ray, S, 1994, A state-of-the-art survey on product design and process planning integration mechanisms, NISTIR 5548, National Institute of Standards and Technology, Gaithersburg, MD.
22. Ahmed, S., Blessing, L., and Wallace, K., The relationships between data, information and knowledge based on a preliminary study of engineering designers, ASME Design Engineering Technical Conferences and Computers in Engineering Conference (DETC1999), Sept 12–15, 1999, Paper Number DETC1999/DTM-8754, Las Vegas, NV.
23. Wigg, K. M., 1999, Successful knowledge management is an integrated whole—Not an assembly of individual pieces, in *Knowledge '99*, vol. 1, London, U.K., pp. 179–202.
24. Lee, Y. T., 1999, An overview of information modeling for manufacturing systems integration, NISTIR 6382. National Institute of Standards and Technology, Gaithersburg, MD.
25. Gao, J. X., and Huang, X. X., 1996, Product and manufacturing capability modeling in an integrated CAD/process planning environment, *International Journal of Advanced Manufacturing Technology*, 11, 43–51.
26. Feng, C. X and Kusiak, A., 1995, Constraint-based design of parts, *CAD*, 27 (5), 343–352.
27. Jurrens, K., Fowler, J., and Algeo, M. B., 1995, Modeling of manufacturing resource information, Requirements Specification, NISTIR 5707, National Institute of Standards and Technology, Gaithersburg, MD.

28. Lapointe, L., Laliberty, T., and Bryant, R., 1993, System description document for the manufacturing optimization (MO) systems, CDRL No. 0002AV-5, Defense Advanced Research Projects Agency, Washington, DC.

29. Bjorke, O. and Myklebust, O., editors, 1992, *IMPPACT—Integrated Modelling of Products and Processes using Advanced Computer Technologies*, Trondheim, Norway: Tapir Publishing.

30. Manufacturing Engineering Laboratory, National Institute of Standards and Technology, http://www.mel.nist.gov/rrm/fy97/jul97mrmodel.exp

31. ISO 10303-240:2005 Industrial automation systems and integration—Product data representation and exchange—Part 240: Application protocol: Process plans for machined products.

32. Zhao, Z., 2002, Design decision making in global manufacturing enterprise in 2020, Design Contest Paper, NSF Travel Grant.

33. American Society of Mechanical Engineers, 2008, *ASME International Design Engineering Technical Conferences*, http://www.asmeconferences.org/IDETC08/

34. NSF, 2004, ED2030: Strategic plan for engineering design, Final Report, NSF. *Workshop on Engineering Design in Year 2030*, March 26–29, 2004, Gold Canyon, AZ.

26 Epilogue: The Implementation and the Future of Intelligent Energy Field Manufacturing

Wenwu Zhang

CONTENTS

We have discussed the methodology of intelligent energy field manufacturing (EFM), and covered a wide range of interdisciplinary processes as well as classical manufacturing processes. All the chapters tried to convey to the readers the methodological thinking in addition to the technical details.

Now you should have understood that EFM is not something new, it is how nature functions all the time. Human engineering injects intelligence into the EFM processes. Although we can talk about manufacturing processes case by case, we really should treat them as processes of energy and material interactions.

We want to point out that intelligent EFM is quickly evolving. You can and you should be part of this evolution. In the epilogue of this book, we introduce a framework for the implementation of this important engineering methodology, and give a wild guess of the future of engineering.

26.1 A FRAMEWORK OF IMPLEMENTATION

26.1.1 THE BLACK BOX QUESTION OF ENGINEERING

Let us restate the black box question of engineering mentioned in Chapter 2. Figure 26.1 illustrates the task to be solved. Y_0 and Y_N are the current and final status of products/processes, respectively, which are functions of time t, space r, and other factors, X. The black box is the route we need to develop to link Y_0 and Y_N in the (X, r, t) space. The task of intelligent EFM is to optimize and execute the engineering route linking Y_0 and Y_N based on the objectives, resources, and constraints from customer, manufacturing engineering, society, and environment, that is, using the new criteria of engineering optimization. This optimization is realized through EFM methods. Note that (X, r, t) is the optimization space. The more freedom of optimization we have, that is, the wider range we take into account in X, the higher chance of finding the best solution.

A framework of intelligent EFM implementation trying to solve the black box question is illustrated in Figure 26.2.

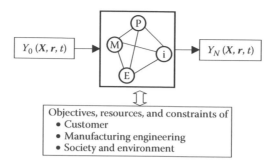

FIGURE 26.1 The black box challenge of engineering.

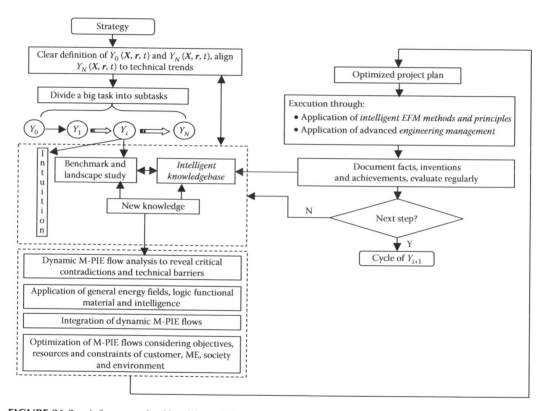

FIGURE 26.2 A framework of intelligent EFM implementation.

Step 1: From strategy to subtasks

When an organization decides its strategy of development, the desired future status of product or process is only vaguely known. The engineering team should study the technical trends and the competitive landscape to understand and define the optimization space (X, r, t), then give a clear definition of Y_0 and Y_N. A practical management plan is worked out by dividing a big task into reasonable subtasks or milestones $Y_0, Y_1, ..., Y_i, ..., Y_N$.

Step 2: Making use of both our heritage and our intuition

It is not rare to see frequent reinvention of the wheel in R&D. Normally, R&D is carried out by subset knowledge–based intuition, such as the brainstorming results from a group of experts. A historically solved problem may come up again. Due to the dynamics within organizations, mismatched talents, improper documentation, and historical precedence, waste of resources during problem solving is pervasive.

The first thing that differentiates intelligent EFM from other methodologies is that intelligent EFM focuses on the intelligent and self-growing application of the knowledge base. An intelligent knowledge base is set up and updated. It allows efficient and thorough study of benchmarking and landscape. Thus, when an engineering question is asked, historical and state-of-the-art knowledge is checked systematically and thoroughly. Such a knowledge base ideally covers physical and chemical effects, publications, patents, documentation of tests, World Wide Web, internal web, and the current mindset. Software tools should be developed to facilitate this process.

Unlike TRIZ, which emphasizes a systematic approach of creative problem solving, intelligent EFM references the complete set of a human being's intelligence heritage, and emphasizes the equal importance of individual inspiration and intuition.

This step may require modifications to subtask planning. This step outputs the improved definition of the optimization space (X, r, t) and a better understanding of the competition landscape. Note that the intelligent knowledge base is updated to include the new knowledge from both internal and external sources.

Step 3: The dynamic M-PIE flow optimization
Knowing the landscape and the optimization space, the current product or process goes through the dynamic M-PIE flow analysis to reveal the critical contradictions and technical barriers. Solutions are found through the applications of general energy field, general logic functional materials, and general intelligence. The M-PIE flows are then integrated and optimized based on the objectives, resources, and constraints of customers, manufacturing engineering, society, and environment.

The integration of energy fields offers more freedom to solve the problem and a greater chance for finding the best solution. Close loop control is an integration effort trying to integrate information with the process. In intelligent EFM, we push this tacit trend to a higher level. Step 3 outputs the optimized engineering plan.

Step 4: Execution
The optimized plan is then executed through the implementation of EFM methods and principles, and through advanced management. These methods and principles are covered in Part I of this book, and are further illustrated with specific examples in the following chapters. Energy field generators, database, and software tools are produced in this process.

Successful execution also requires a new style of leadership and management, referred to as advanced engineering management. The concepts of M-PIE flow integration and optimization are relatively new. The fusion of intelligent EFM with many other modern management methodologies requires systematic training. Current and earlier generations of engineers and, accordingly, engineering leaders were trained to be used to the conventional ways of R&D. Project leaders should be trained to facilitate rather than impede the execution of intelligent EFM.

Step 5: Evaluate, improve, and move forward
The execution status is regularly evaluated and documented, and these documents become the internal part of the intelligent knowledge base. The documentation covers successes such as patents and technical achievements, as well as failures and other facts of the project.

Good documentation and an updated knowledge base will lower the impact of talent dynamics. This will also increase the efficiency of resource usage. For example, the database contains information of various energy-field generating and measuring devices. In any large research organization, it is common to find many idle pieces of equipment after projects are terminated. The intelligent knowledge base naturally provides information for sharing the equipment, material, and many other things. This principle may reduce waste and increase revenue immediately. Based on the evaluation results, the team decides to go to the next step or go back and improve the current step.

This cycle goes on and moves the project to the final step.

Clearly, intelligent EFM is an open system, in which the power and knowledge base grows with time. Intelligence is an invaluable resource of an organization. The ability to continuously integrate with the progress in all areas of human intelligence is needed in future engineering.

26.2   THE FUTURE OF INTELLIGENT EFM: TOWARD LIFE-LIKE MULTIDIMENSIONAL MANUFACTURING

In this book, we purposely put more weight on non-mechanical processes. This is not to say that mechanical processes are less important in the future. The major reason for this arrangement is that there are already many good books covering this part. I learned those mechanical processes in the undergraduate program, and almost forgot how glorious these processes are. While non-mechanical processes are "cool" niche processes, the classic mechanical processes are still the major work horses of industry.

We argued that we should treat all energy fields equally in engineering optimization. Mechanical energy, explored first and most to date by the energy field people, has been used in many ways. These include mechanical turning, milling, grinding, broaching, sawing, tapping, shearing, forging, stamping, rolling, embossing, bending, forming, shot peening, joining, polishing, etc. Other energy fields, especially the less developed energy fields, should learn from the more established processes. As shown in Table 26.1, it would be a good practice to fill in the table to check whether the matrix can generate meaningful processes. Note that we emphasize the integration of energy fields to achieve the optimal results. A single energy field may do the work, but it is always better to use the assistance of other energy fields. Mechanical milling or grinding uses lubricants, but can you say it is a pure mechanical process? We have seen examples of ECM/EDM or waterjet cutting, drilling, milling, grinding, etc., but how about using laser or ultrasonic energy for grinding, rolling, or bending? Instead, how about using hybrid processes for joining, finishing, rolling, embossing, turning, etc.? We discussed how to manipulate energy fields in Chapter 4.

With the above thinking, we can more easily borrow ideas among various disciplines. Chemical engineering uses the inherent property of materials, converting chemicals from one state to another. The same thing can be said for materials science and engineering. Is mechanical engineering any different? To me, they are all about laying out the proper boundary conditions to channel materials

TABLE 26.1
Matrix of Process Development

Energy Fields		Manipulation of Energy Fields					
		Milling	Grinding	Rolling	Forming	Joining	Finishing etc.
Mechanical	Integrated	?	?	?	?	?	?
EM		?	?	?	?	?	?
Thermal		?	?	?	?	?	?
Photonic		?	?	?	?	?	?
Acoustic		?	?	?	?	?	?
ECM/EDM		?	?	?	?	?	?
Waterjet		?	?	?	?	?	?
E-beam		?	?	?	?	?	?
X-ray		?	?	?	?	?	?
Plasma		?	?	?	?	?	?
Ion-flow		?	?	?	?	?	?
Pressure/vacuum		?	?	?	?	?	?
Gravity		?	?	?	?	?	?
Medium etc.		?	?	?	?	?	?

into the desired states. It is like leading sheep to feed, to grow, and to go home. We do not need to teach the sheep how to walk or eat; we give orders and guide them. That is why we say all energy fields and material systems have certain levels of intelligence; engineering is about using human intelligence to guide them from one state to another.

Have you pondered how powerful life is? It was Labor Day when I wrote this epilogue. I took my kids to climb Prospect Mountain around Lake George the previous day. We saw pine trees growing out of the rock. My kids were excited when they were in the wild. In engineering, we say we are in a 4D world (3D space plus 1D time). Is life 4D? A small seed can grow into apples, cucumbers, or a baby. All it needs are suitable energy fields and material supply. There is intelligence in these processes. I will not say my baby is only 4D. A tree is more than 4D. So is a rock. All these extended dimensions are opportunities of engineering optimization and innovation. When we systematically optimize engineering solutions with the above philosophy, we are practicing intelligent EFM.

We can observe two trends in engineering. One is that engineering is increasingly divided into finer branches, the other is that there is an increasing need to cross the borders and integrate. New engineering frontiers, such as nanotechnology, bioengineering, space exploration, renewable energy, etc., require the high-level integration and cooperation of multiple disciplines. They also require new engineering methodologies. I hope the discussion in this book can help answer this need.

Manufacturing, be it in any engineering discipline, is becoming increasingly intelligent. We are in the era of migrating from 2.5D manufacturing toward truly 3D manufacturing. When we integrate more sensing and control into our products, we gradually go beyond 3D manufacturing. Computers, cell phones, and cars are not 3D products. Eventually, we will learn more from the life process, marching toward life like multidimensional manufacturing does. There are already some processes and products mimicking life structures, such as the drag reduction swimsuits, superhydrophobic surfaces, planes, etc. Understanding the big trend can be helpful for engineering studies.

26.3   WENWU'S HYPOTHESIS

If you are a good thinker, you can try to validate my hypothesis.

Although we are built upon them or we are immersed in them all the time, people were not aware of the existence of atoms, EM wave, or x-rays until science revealed them. So, for many things, we will not really see them until we know them. For energy fields, are there still things that we rely on all the time yet we do not recognize them? I will try to propose one.

We define energy as the ability to do work, and work is done whenever the state(s) of a system is changed, and work is always done by certain energy field(s). Now imagine there are two groups of students of equal number, same gender composition, and same age. The two groups are sent to the same room sequentially to build a great five-floor building using the same type of Lego pieces. The two groups are not allowed to communicate.

So here, the inputs are the same (an order), the initial states of our test are the same or very similar. However, the results of the two groups could be very different. Therefore, between these two groups, there is a difference in work. The questions are

How is this work difference created?
How do we measure this work?
Is this factor a kind of force or field?
Does this force or field interact with other forces or fields that we are more familiar with? If so, how?

My answer is that the difference in intelligence created the different results, and thus created work. Since anything that can do work is a kind of force or field, we can call it intelligence energy field. Yes, I believe this intelligence energy field could interact with the other fields, but we do not really know how yet. We learn to do things. We use our knowledge to guide the energy–material–intelligence

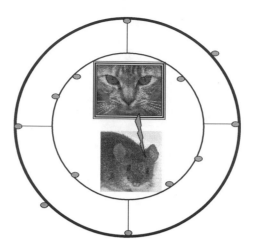

FIGURE 26.3   Experiment to test the hypothesis of intelligence energy field.

flows toward desired directions. This is something we rely on and are immersed in all the time, yet this is something engineering and science have not systematically studied yet, and without knowing it well, we could not see it well.

So, Wenwu's hypothesis is as follows:

Intelligence is a kind of energy field that can interact with other energy fields, and this can be experimentally validated.

I do not know when we can validate this hypothesis yet. Suppose the intelligence field can interact with the gravity field. We can devise the following experiment to validate it, as shown in Figure 26.3.

Consider an isolated sphere, such as a two-layered hollow metallic sphere immersed in water and pulled by springs from multiple directions. Inside the sphere, we put a mouse. We give the mouse food and a comfortable place to live. Furthermore, we play a video over and over again. The majority of the video shows happy scenes for the mouse, then, within a short period, the mouse will hear scary sounds and see horrible images. Remember the sphere is totally isolated to the rest of the world. The thing that changes with time is the video—a set of intelligence. We treat the mouse, the detection system (the dots in the picture), and the sphere system as our "hardware," the video as a controlled input variable, and the detected states of the various energy fields as the output.

If the mouse is a piece of rock, the detected states might change very little. But as it is a mouse, a living thing with a high level of intelligence, we can expect a cyclic change of the energy fields. We can also expect the adaption of the mouse to the same excitation over time. I do not know whether a scared mouse weighs heavier than a happy mouse, whether it will change the EM field, etc. I have not had the chance to do this experiment yet. This test should be very interesting.

You may have a better design to validate the hypothesis. How about directly testing human beings in a flying airplane with a piece of fake hijacking information? Well, it is too complicated, too dangerous, and may run out of control.

Thanks for your interest in this book.

Wenwu Zhang
Schenectady, New York

Index